Shape and Dimensions	Area	Centroid Location	Area Moment of Inertia	Radius of Gyration

RECTANGULAR AREA

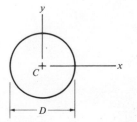

$A = bh$

$\bar{x} = 0$

$\bar{y} = 0$

$I_x = \frac{1}{12} bh^3$

$I_y = \frac{1}{12} hb^3$

$r_x = \dfrac{h}{\sqrt{12}}$

$r_y = \dfrac{b}{\sqrt{12}}$

TRIANGULAR AREA

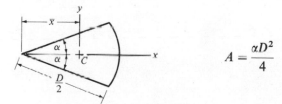

$A = \frac{1}{2} bh$

$\bar{y} = \frac{1}{3} h$

$I_x = \frac{1}{36} bh^3$

$r_x = \dfrac{h}{\sqrt{18}}$

CIRCULAR SECTOR AREA

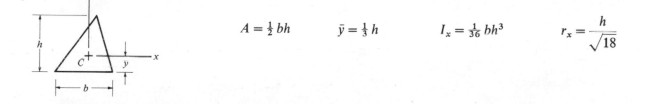

$A = \dfrac{\alpha D^2}{4}$

$\bar{x} = \dfrac{D}{3}\dfrac{\sin \alpha}{\alpha}$

$\bar{y} = 0$

$I_x = \dfrac{D^4}{64}\left(\alpha - \dfrac{1}{2}\sin 2\alpha\right)$

α = π LEADS TO CIRCULAR AREA

$A = \dfrac{\pi D^2}{4}$

$\bar{x} = 0$

$\bar{y} = 0$

$I_x = I_y = \dfrac{\pi D^4}{64}$

$r_x = r_y = \dfrac{D}{4}$

Engineering Mechanics of Materials

The colorful photoelastic data displayed on the cover represent an isochromatic fringe pattern in a uniaxially stressed tube sheet. The isochromatic fringes represent lines along which the principal stress difference is a constant, and the order of the fringes gives the magnitude of the difference. The whole field isochromatic fringe pattern thus provides a complete stress map which permits accurate stress analysis of components with complex geometry.

—Courtesy of DEAN JAMES W. DALLY, College of Engineering, University of Rhode Island

Engineering Mechanics of Materials

B. B. Muvdi Bradley University

J. W. McNabb Bradley University

Macmillan Publishing Co., Inc.
New York

Collier Macmillan Publishers
London

Macmillan Publishing Co., Inc.
866 Third Avenue, New York, New York 10022

Collier Macmillan Canada, Ltd.

Library of Congress Cataloging in Publication Data

Muvdi, B. B.
   Engineering mechanics of materials.

   Bibliography: p.
   Includes index.
   1. Strength of materials. 2. Mechanics, Applied.
I. McNabb, J. W., joint author. II. Title.
TA405.M95   620.1'123   79-4664
ISBN 0-02-385750-1 (Hardbound edition)

Printing: 2 3 4 5 6 7 8   Year: 0 1 2 3 4 5 6

*To our families
for their patience
and encouragement*

PREFACE

Galileo's "Dialogue Concerning Two New Sciences," published in 1638, marked the beginning of the study of two important subjects, dynamics and mechanics of materials, which have been of primary importance in the scientific and technological revolution, which is now only about 350 years old. A representative set of the fundamental concepts, principles, and methods of mechanics of materials that have been developed during the three and a half centuries since Galileo's time are presented in this textbook, written for sophomore and junior engineering students. The authors make no claim that the coverage is exhaustive or encyclopedic. However, their combined teaching experience of approximately 50 years has enabled them to produce a thoroughly planned introduction to the subject intended for those with a background in rigid body mechanics.

Throughout this book, a "quantum of information" could be defined as "presentation of theory, followed by examples illustrating the theory, followed by homework problems." Theoretical discussions have been written concisely and with the beginner in mind rather than being directed to faculty members or research and development specialists. Faculty members may wish to supplement the theory provided here with their own background and experience. This approach will be easier with this text than with one of encyclopedic coverage. Examples have been carefully chosen and discussed such that students will be able to follow each step of the solutions with a minimum of difficulties and then proceed to test their own knowledge by applying basic concepts, principles, and methods in solving homework problems assigned by the instructor.

Examples are included in the text to enable the beginner to develop a thorough knowledge of elementary mechanics of materials. Instructors will find over 800 homework problems from which to make selections, which means that the text can readily be used for a number of semesters.

A working knowledge of differential and integral calculus together with rigid body mechanics (primarily statics) are the main requisites for understanding the ideas presented in this text. In contrast to idealized rigid bodies considered in statics and dynamics, in mechanics of materials the bodies are considered to be deformable, and these deformations are studied when known forces are applied to the bodies. The molecular structure of matter, although well established by experiment, will not be of primary importance in this course because the concepts of mechanics of materials are based upon "continuous media"—a convenient continuous idealization of the true nature of matter which has proved satisfactory for engineering analysis and design.

This text differs from more traditional mechanics of materials texts in several major ways. Primarily, the organization of the material differs from other texts in that the first nine chapters are concerned with analysis. Design considerations are omitted until Chapter 10. This will enable the instructor to focus on the basic concepts without being hampered by difficulties and details arising in the design process. Also significant is the fact that all the material pertaining to statically indeterminate members is treated in a single chapter, Chapter 9. This method of organization has the distinct advantage of providing a valuable review and further reinforcement of the basic concepts of mechanics of materials during the development of methods for the solution of statically indeterminate problems in Chapter 9 and during the development of the basic design concepts in Chapter 10.

For those instructors desiring a more traditional coverage of topics, this text permits a convenient selection of sections from Chapters 9 and 10 to augment coverage of earlier chapters. After study of the first three chapters is completed, Section 9.2, Statically Indeterminate Members Under Axial Loads, and Section 10.3, Design of Axially Loaded Members, could be covered. Then, following Chapter 4 dealing with torsion, Section 9.3, Statically Indeterminate Members Under Torsional Loads, and Section 10.4, Design of Torsional Members, could be presented. Similar selections of sections from Chapters 9 and 10 are feasible following coverage of Chapter 6. Discussion of Section 10.6, Design of Columns, could conveniently follow coverage of topics from Chapter 8, Column Theory and Analyses.

Two other novel features are the extensive use of force diagrams for many members to be analyzed or designed and the accompanying use of Mohr's circle for stress and strain. Force diagrams for axially and torsionally loaded members as well as the traditional shear and moment diagrams for beams are developed in Chapter 1 using equilibrium equations. General developments of stress and strain concepts and equations are given in Chapter 2, together with the Mohr's circle semigraphical approach for principal stresses and strains. Extensive use is made of the Mohr's circle construction, particularly for stress analysis, in the remaining chapters. Student understanding will be greatly enhanced by extensive use of force diagrams and Mohr's circle construction in solving homework problems in subsequent chapters.

In view of the fact that the international system of units, referred to as SI (Système International), is now beginning to gain acceptance in this country, it was decided to use it in this book. However, it is realized that a complete transition from the British gravitational to SI units will be a slow and costly process which may last as long as 20 years and possibly longer. Consider, for example, the recent 1977 estimates of the Federal Highway Administration, an agency of the federal government, which show that the cost of transforming all federally financed roadway signs from British gravitational to SI units would amount to $100 million. Because of this high cost, coupled with much public protest, and in order to study the effects of similar Canadian programs, the process of transforming highway signs has been delayed. Other factors that will play a significant role in slowing down the transformation process is the existing literature of engineering research and development, plans, and calculations, as well as structures and production machinery, that have been conceived and built using largely the British gravitational system of units. Thus the decision was made to use both systems of units in this book, and approximately one-half of the examples as well as one-half of the homework problems are stated in terms of the traditional British gravitational system while the remainder are given in terms of the emerging SI system of units.

It should be further noted that the authors have made use of dimensionless equations and dimensionless plotting of variables where feasible. This approach, long used in scientific and engineering reports, will have increasing utility during the transition period, since the information from such equations and plots can be utilized with any system of units. An example of nondimensionalized equations and plots is given in Section 6.9, Deflections Associated with Shearing Deformations. Students will be well advised to study this and other such examples and urged to employ these techniques whenever possible in presenting information, since such results can be readily used with any system of units.

First courses from this text would be of either three or four semester hours' duration or two sequential courses each of three quarter hours' duration. A suggested three-semester-hour course would cover with few omissions the first nine chapters, with added selections from the remaining five chapters. A suggested four-semester-hour course would cover with few omissions the first 10 chapters, with added selections from the remaining four chapters. Two possible sequential courses, each for three quarter hours' credit, would cover with few omissions the first 12 chapters. Also, two possible sequential courses, each for three semester hours' credit, would cover with few omissions the entire 14 chapters. Thus a typical first course of three semester hours' duration would cover the first 10 chapters of the book with the following omissions: Sections 2.8, 2.9, 4.5, 4.6, 4.7, 5.5, 6.8, 6.9, 6.10, 7.6, 8.5, 8.7, 9.7, and 10.7. All of these sections, as well as those contained in Chapters 11, 12, 13, and 14 that are judged to be not essential for an understanding of the basic concepts of mechanics of materials, are starred.

The authors wish to acknowledge with deep appreciation the typing of this manuscript by Mrs. Fred Kahrs and Mrs. Sharon Ciota and the assistance provided by A. H. Al-Zahrani, Gary Zika, and Glenn Zika during the preparation of the Solutions Manual. The authors also appreciate the valuable comments and suggestions made by Milton E. Raville of Georgia Institute of Technology, Dale R. Carver of Louisiana State University, Clarence Maday of North Carolina State University, and Rodney Schaefer of the University of Missouri at Rolla.

Peoria, Illinois

B. B. M.
J. W. M.

CONTENTS

11 Introduction to System Design 536

12 Analysis and Design for Inelastic Behavior 573

13 Analysis and Design for Impact and Fatigue Loadings 610

14 Selected Topics 649

Appendixes

Engineering Mechanics of Materials

Chapter 1 Internal Forces in Members

1.1 Introduction

The analysis or design of structural and machine components is based upon a knowledge of the variation of internal forces throughout such members. Fundamental concepts and equations of statics provide the necessary background for determination of these internal forces. The equations of equilibrium together with the free-body-diagram concept are used to determine internal forces in members subjected to applied force systems. Emphasis is placed upon the construction of diagrams that reveal the variation of internal forces as functions of coordinates measured along member lengths. Equations and procedures are developed for the construction of axial force, torque, shear, and moment diagrams.

Sign conventions are used for two purposes in analyzing external and internal forces acting on members. Equilibrium equations require arbitrary sign conventions when they are written and internal axial force, torque, shear and moment require sign conventions when they are plotted. The reader will note that if the sign convention for a given equilibrium equation (which is arbitrary) is reversed from an initial choice, this reversal is algebraically equivalent to multiplication of the equation by −1. Of course, multiplication of an equation by −1 does not change its solution. Sign conventions for internal axial force, torque, shear, and moment will be introduced in this chapter and employed for plotting these quantities throughout this book. The reader should not confuse these latter sign conventions with those arbitrarily chosen each time an equilibrium equation is written.

Examples are solved in detail to illustrate the methods developed for construction of these internal force diagrams. Appendix B provides information on connections and reactions at supports that will be useful in the proper construction of free-body diagrams.

1.2 Axially Loaded Members in Equilibrium

Consider a system of forces P_1, P_2, P_3, P_4 applied to the member of varying cross section shown in Figure 1.1(a). The action line of each force is directed along the

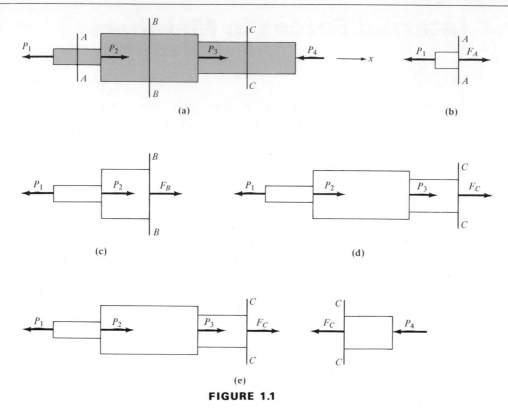

FIGURE 1.1

longitudinal axis of the member and passes through the centroid of each cross section of the member. The complete member is in equilibrium, and if we direct an x axis along the member, the available equation of equilibrium may be written $\sum F_x = 0$. Choosing the positive sense to the right, we obtain

$$-P_1 + P_2 + P_3 - P_4 = 0$$

In order for the problem to be a statically determinate one, only one of these forces may be regarded as unknown (e.g., P_4). Solving for P_4 yields

$$P_4 = -P_1 + P_2 + P_3$$

In the study of mechanics of materials, it is necessary to first determine the forces acting inside a member such as the one shown in Figure 1.1(a). In order to investigate these internal forces, imagine cutting planes passed perpendicular to the member axis (x) such as the planes denoted by $A–A$, $B–B$, and $C–C$. Since the entire member is in equilibrium, any part of it must also be in equilibrium. The free-body diagrams shown in Figure 1.1(b), (c), (d), and (e) are each in equilibrium. Forces F_A, F_B, and F_C which are internal with respect to the entire body are external when shown on these free-body diagrams. Each of these forces is assumed to be tensile at the appropriate section and a negative sign will indicate a compressive force. In order to determine F_A, F_B, and F_C, apply the equation $\sum F_x = 0$ to each of the free-body diagrams shown.

Figure 1.1(b):

$$F_A - P_1 = 0$$
$$F_A = P_1$$

Figure 1.1(c):

$$F_B + P_2 - P_1 = 0$$
$$F_B = P_1 - P_2$$

Figure 1.1(d):

$$F_C + P_3 + P_2 - P_1 = 0$$
$$F_C = P_1 - P_2 - P_3$$

An alternative approach for determining F_C, for example, is to consider the right free-body diagram shown in Figure 1.1(e):

$$-F_C - P_4 = 0$$
$$F_C = -P_4$$

In general, either left or right free-body diagrams may be chosen for determination of the internal forces. The choice is one of convenience. It is readily seen that the right free-body diagram of Figure 1.1(e) is simpler to work with than the one in Figure 1.1(d). These internal forces are distributed over the entire cross-sectional area of the member at the cut sections, but consideration of this distribution is deferred to Section 3.2.

Example 1.1
Determine the internal forces acting inside the member in Figure 1.2 and plot their variation along the member length.

Solution. Construct a free-body diagram of the entire member and orient a vertical axis positive upward as shown in Figure 1.2(a) and write $\sum F_y = 0$ in order to find P, the reaction at the base of the member.

$$10 - 2(6) + 8 - 2(2) - P = 0$$

$$\boxed{P = 2 \ kN}$$

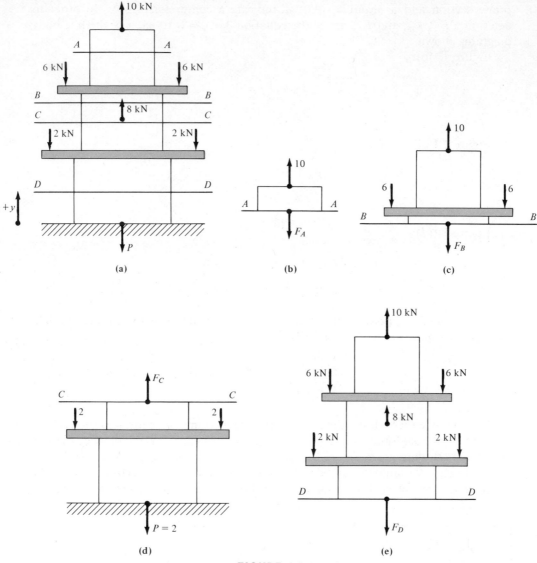

FIGURE 1.2

This value of P will be used for a check on other calculations or for the force acting on the free-body diagram of the bottom part of the member. It is preferable to use this compared to the free-body diagram of the top part of the member. Consider free-body diagrams of Figure 1.2(b), (c), (d), and (e) in order to compute F_A, F_B, F_C, and F_D, respectively. In each case $\uparrow \ \sum F_y = 0$.

Figure 1.2(b):

$$10 - F_A = 0$$

$$\boxed{F_A = 10 \; kN(T)}$$

Figure 1.2(c):

$$10 - 2(6) - F_B = 0$$

$$\boxed{F_B = -2 \; kN(C)}$$

Figure 1.2(d): Note the choice of the bottom free body in this case. Of course, the top free body would yield the same result.

$$F_C - 2(2) - 2 - 0$$

$$\boxed{F_C = 6 \; kN \; (T)}$$

Figure 1.2(e):

$$10 - 2(6) + 8 - 2(2) - F_D = 0$$

$$\boxed{F_D = 2 \; kN \; (T)}$$

These values are plotted in Figure 1.3. This plot is referred to as the axial force diagram for the member shown and will be utilized throughout this text.

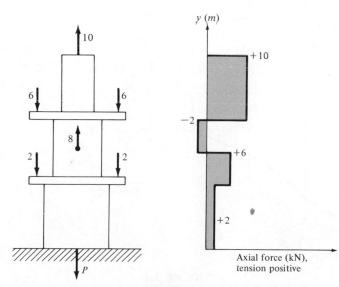

FIGURE 1.3

1.3 Variable Axial Loading—Internal Force Relationships

Axial loadings are not always applied externally at given cross sections, as depicted in Figure 1.1, but may be applied in some variable fashion along the longitudinal member axis as indicated in Figure 1.4. The axial load intensity expressed in force per unit length and denoted by f is a function of the longitudinal coordinate x. In addition to the distributed loading f, one or more concentrated loadings such as P may also be applied to the member. These externally applied axial loadings are balanced by R at the left end of the member, A.

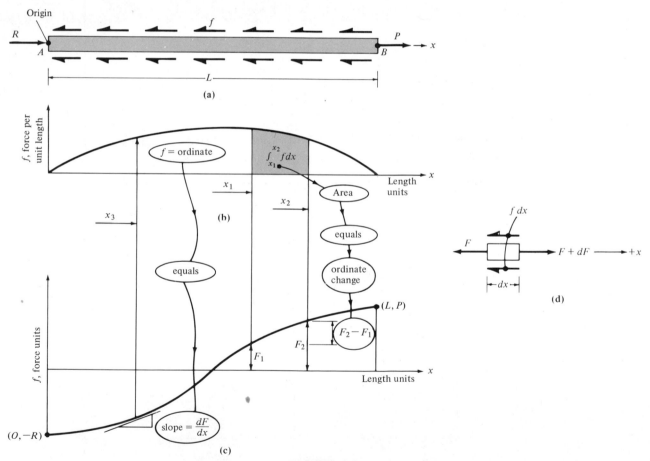

FIGURE 1.4

Consider a short segment of the member in Figure 1.4(d) and write the force equilibrium equation for an axis oriented along the member.

$$\rightarrow \sum F_x = 0$$

$$F + dF - F - f\,dx = 0$$

$$f = \frac{dF}{dx} \tag{1.1}$$

The applied axial load intensity f at any section of the loaded member equals the rate of change of the internal axial force F with respect to the longitudinal coordinate x. Graphically, this is as depicted in Figure 1.4(b) and (c). The ordinate f to the f–x curve at any point such as $x - x_3$ is equal to the slope of the F–x curve at the same point.

Multiply Eq. 1.1 by dx and write $dF = f\,dx$. Then, integrate both sides of the equation:

$$\int_{F_1}^{F_2} dF = \int_{x_1}^{x_2} f\,dx$$

The left side of this equation may be integrated to give

$$F_2 - F_1 = \int_{x_1}^{x_2} f\,dx \tag{1.2}$$

Equation 1.2 states that the change in internal axial force between any two sections of an axially loaded member equals the area under the f–x diagram between the same two sections. Graphically, this is depicted in Figure 1.4(b) and (c), where the two sections have been chosen at distances x_1 and x_2 from the left end of the member. Internal tensile forces are positive and compressive ones are negative. Positive applied axial forces of intensity f act as shown in Figure 1.4(a).

Example 1.2
A rod made of two different materials (A and B) is suspended from an overhead frame as depicted in Figure 1.5(a). Rod A weighs $w_A = 6$ lb/ft and rod B weighs $w_B = 4$ lb/ft. Construct the axial force diagram for this rod and note the maximum force in the rod. Discuss the results in terms of Eqs. 1.1 and 1.2.

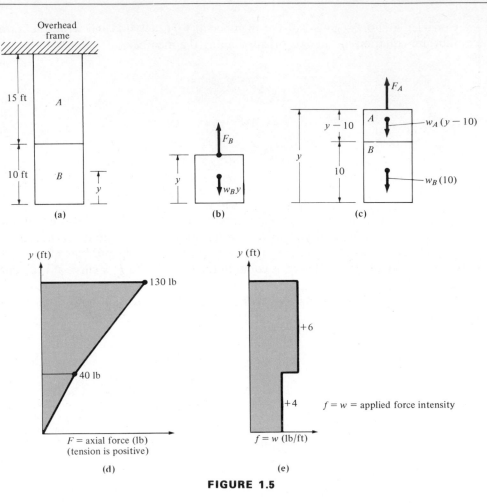

FIGURE 1.5

Solution. In this problem $f = w_A$ or $f = w_B$, the applied gravity loadings. Consider the free-body diagram shown in Figure 1.5(b):

$$\uparrow \ \sum F_y = 0$$

$$F_B - w_B y = 0$$

with $w_B = 4$ lb/ft. This becomes

$$F_B = 4y \qquad (0 \le y \le 10)$$

F_B is a linear function of y which varies from 0 lb when $y = 0$ ft to 40 lb when $y = 10$ ft. Now, consider the free-body diagram shown in Figure 1.5(c):

$$\uparrow \ \sum F_y = 0$$

$$F_A - w_B(10) - w_A(y - 10) = 0$$

With $w_A = 6$ lb/ft and $w_B = 4$ lb/ft this becomes

$$F_A = 6y - 20 \qquad (10 \le y \le 25)$$

F_A is also a linear function of y which varies from 40 lb when $y = 10$ ft to 130 lb when $y = 25$ ft.

These linear functions for F_A and F_B are plotted versus y as shown in Figure 1.5(d). This plot is the axial force diagram for the member. We observe that the axial force in the rod reaches a maximum value of 130 lb at its top, where it is connected to the overhead frame.

Discussion of the results in terms of Eqs. 1.1 and 1.2 requires reference to Figure 1.5(d) and (e). The ordinate to the f–y curve in Figure 1.5(e) for rod B equals 4 lb/ft, which equals the slope of the F–y curve in Figure 1.5(d) given by

$$\frac{\Delta F}{\Delta y} = \frac{40 - 0}{10} = 4 \text{ lb/ft}$$

To better understand Eq. 1.2, consider rod A. The difference in ordinates to the F–y curve between the ends of this rod is given by $130 - 40 = 90$ lb, which equals the area under the f–y curve for this rod: (6 lb/ft) (15 ft) = 90 lb.

The reader is urged to complete this discussion by considering rod A, Eq. 1.1, and rod B, Eq. 1.2.

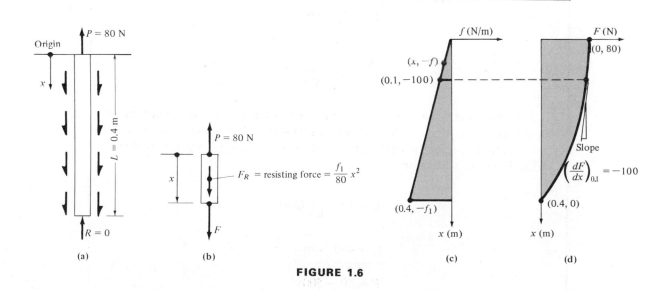

FIGURE 1.6

Example 1.3

A soil sampling device for possible use in planetary exploration is depicted in Figure 1.6(a). As withdrawal impends, the force $P = 80$ N and the soil resistance is assumed to vary linearly as shown in Figure 1.6(c). Neglect the effects of gravity and assume a zero reacting force R at the bottom of the device. Determine the maximum value of $f = f_1$ at the bottom of the device and construct the F–x curve, where F is the internal force in the rod. Illustrate the meaning of Eq. 1.1 by selecting an x value of 0.1 m.

Solution. Consider the free-body diagram shown in Figure 1.6(b). The origin is chosen at the planet's surface and x is measured positive downward. The resulting resisting force F_R may be determined by first referring to Figure 1.6(c) and expressing f as a function of x by similar triangles:

$$\frac{f}{x} = \frac{f_1}{0.4} \quad \text{or} \quad f = \frac{f_1}{0.4} x$$

Then

$$dF_R = f \, dx$$

Integrating yields

$$F_R = \int f \, dx = \frac{f_1}{0.4} \int_0^x x \, dx = \frac{f_1}{0.8} x^2$$

$\downarrow \ \sum F_x = 0$

$$F + F_R - 80 = 0$$

$$F = 80 - F_R$$

$$F = 80 - \frac{f_1}{0.8} x^2$$

To determine f_1, we make use of the boundary condition at the bottom of the device: $x = 0.4$ m, $F = 0$. Since the reacting force $R = 0$, the device will not be subjected to an internal axial force at its lower end. Applying this condition to the F versus x function:

$$0 = 80 - \frac{f_1}{0.8} (0.4)^2$$

$$\boxed{f_1 = 400 \text{ N/m}}$$

$$F = 80 - 500x^2$$

This F–x curve, which is parabolic, is plotted in Figure 1.6(d). The straight line shown in Figure 1.6(c) has a negative slope and its equation is given by

$$f = \frac{-f_1}{0.4} x = \frac{-400}{0.4} x = -1000x$$

Equation 1.1 will be written for $x = 0.1$ m,

$$f = \frac{dF}{dx}$$

$$f_{10} = -1000(0.1) = -100 \text{ N/m}$$

$$F = 80 - 500x^2$$

$$\frac{dF}{dx} = -1000x$$

$$\left(\frac{dF}{dx}\right)_{0.1} = -1000(0.1) = -100 \text{ N/m}$$

At $x = 0.1$ m, the ordinate to the f–x curve is -1.00 N/m, which equals the slope of the F–x curve as depicted in Figure 1.6(c) and (d).

The reader should apply Eq. 1.2 to Figure 1.6(c) and (d) between limits $x = 0$ to $x = 0.4$ m.

Homework Problems

1.1 Determine the force P required for equilibrium of the body shown in Figure H1.1 and plot the axial force diagram for the member. Choose an origin for the lengthwise coordinate and clearly show this on a sketch of the member.

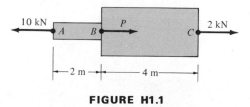

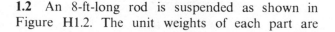

FIGURE H1.1

1.2 An 8-ft-long rod is suspended as shown in Figure H1.2. The unit weights of each part are $w_A = 2$ lb/ft, $w_B = 3$ lb/ft, and $w_C = 4$ lb/ft. Draw appropriate free-body diagrams to determine the internal axial forces in each part of the rod. Use an origin at O and an upward-directed length coordinate. Plot these axial force functions versus length along the rod. What is the maximum axial force in the rod?

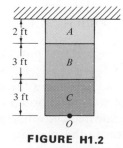

FIGURE H1.2

1.3 The bar shown in Figure H1.3 is attached to a column at its right end and loaded as shown. Find the reaction R required to maintain equilibrium of the bar. Draw the free-body diagrams required to determine the axial forces in this bar, then plot the axial force diagram for it.

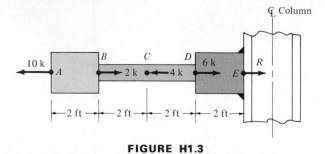

FIGURE H1.3

1.4 The bar shown in Figure H1.4 is suspended as shown. It is subjected to two applied forces and each segment has unit weights as follows: $w_A = 6$ lb/ft and $w_B = 4$ lb/ft. Draw the appropriate free-body diagrams required to express the axial forces in this bar as functions of a length coordinate measured from the origin O at the bottom of the bar. Draw the axial force diagram for this bar alongside a sketch of it. What is the maximum axial force in the bar?

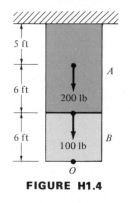

FIGURE H1.4

1.5 The member shown in Figure H1.5 is subjected to the system of forces shown and is fastened to a support at point A. Determine the reaction at the support consistent with equilibrium of the member. Draw the appropriate free-body diagrams required to find the axial forces in this member. Then construct

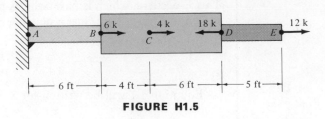

FIGURE H1.5

the axial force diagram for the member. Use a length coordinate measured from an origin at point A.

1.6 Construct an axial force diagram for the member shown in Figure H1.6. Choose an origin at point A and measure a length coordinate along the axis of the member. Sketch all free-body diagrams required to evaluate the axial forces in the various parts of the body.

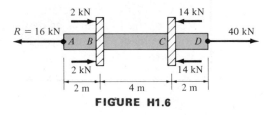

FIGURE H1.6

1.7 The body depicted in Figure H1.7 is composed of two parts welded together. Part A weighs 100 N/m, part B weighs 80 N/m, and $P = 620$ N. Choose an origin at O and measure x positive downward, then write equations for the internal axial force as a function of x for the entire body. Plot the axial force diagram for this body and note the largest axial force

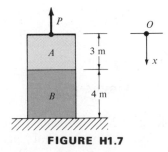

FIGURE H1.7

in the body and state whether it is a tensile or compressive force.

1.8 A steel beam 30 ft long is suspended from its upper end as shown in Figure H1.8. Choose an origin at the top or bottom of the member and write an equation that expresses the internal force in the member as a function of a coordinate measured along the beam length. Plot the axial force diagram for the beam. Steel weighs 490 lb/ft³ and the cross-sectional area of the beam is 30 in.².

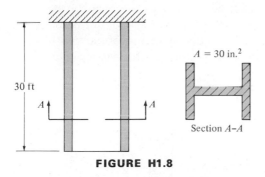

FIGURE H1.8

1.9 A 20-ft-long vertical member *AB* of a metallic truss is shown in Figure H1.9. The metal weighs 163 lb/ft³ and the cross-sectional area of the member is 10 in.². Assume that one-half the total weight of the member is supported by each welded end *A* and *B*. Choose an origin at *A* and express the axial force in the member due to its own weight as a function of *x*

measured along the member. Plot the axial force diagram for the member, note the largest values of this force, and state whether they are tensile or compressive.

1.10 A rod of a drilling rig weighs 15 lb/ft and a 30-ft length is suspended from its top and held vertically as preparations for drilling are made. Choose an origin at the top or bottom of the rod and express the axial force as a function of a lengthwise coordinate. Plot the axial force diagram for the suspended rod.

1.11 A 42-ft-long steel piling weighing 200 lb/ft was damaged during driving and must be pulled from the earth as shown in Figure H1.11. The pile is about to move upward when the pulling force is 20 tons. The pulling force is resisted by the pile weight and the skin friction forces developed on the surface of contact of the pile with the soil. Assuming that the skin friction forces to be uniform over the pile surface, determine these forces denoted by *f* which are expressible in pounds per foot of pile length. Choose an origin at the top of the pile and express the internal force in the pile as a function of a lengthwise coordinate *x* measured from the origin. Plot the axial force diagram for the pile. All unknown values are to be determined for the situation depicted when the pulling force is 20 tons and the pile is in equilibrium prior to being pulled from the earth.

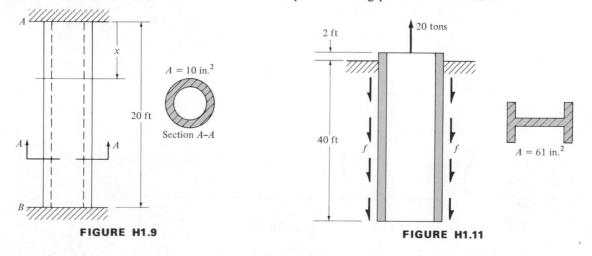

FIGURE H1.9

FIGURE H1.11

1.4 Torsionally Loaded Members in Equilibrium

The member depicted in Figure 1.7 is in equilibrium when subjected to concentrated external torques T_1, T_2, T_3, and T_4. To determine the resisting internal torques T_A, T_B, and T_C, we need to draw the free-body diagrams shown in Figure 1.7(b), (c), and (d). A single equilibrium equation applies to each free-body diagram depicted in Figure 1.7. This equation states that the sum of the torques (or moments) about the longitudinal axis of the member equals zero. Symbolically, we write $\sum T_x = 0$, where the subscript x refers to the longitudinal axis of the member as depicted in Figure 1.7. The sign convention for positive torques is shown in Figure 1.7(b). The double-headed arrows represent positive torques by means of the right-hand rule. If the thumb is pointed in the direction of the vector, an observer looking

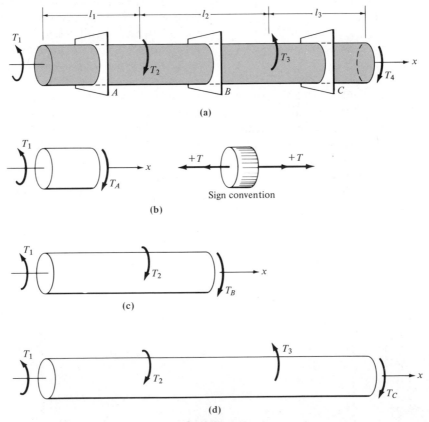

FIGURE 1.7

toward the head of the vector would observe fingers curled counterclockwise, which defines the sense of the torque. A positive resisting torque vector will always point away from the body considered and never toward the body.

Figure 1.7: $\sum T_x = 0$

$$-T_1 + T_2 - T_3 + T_4 = 0$$

Regarding the torque T_4 as unknown, we may determine it from the preceding equation:

$$T_4 = T_1 - T_2 + T_3$$

Figure 1.7(b): $\sum T_x = 0$

$$T_A - T_1 = 0$$
$$T_A = T_1$$

Figure 1.7(c): $\sum T_x = 0$

$$T_B + T_2 - T_1 = 0$$
$$T_B = T_1 - T_2$$

Figure 1.7(d): $\sum T_x = 0$

$$T_C - T_3 + T_2 - T_1 = 0$$
$$T_C = T_1 - T_2 + T_3$$

A plot of T_A, T_B, T_C versus the longitudinal coordinate x is referred to as the torque diagram for the member. In general, either a left or right free-body diagram may be chosen for determination of the internal torques. These internal torques are distributed over the entire cross-sectional area of the member at the cut sections, but consideration of this distribution is deferred to Section 4.2.

Example 1.4
Refer to Figure 1.7 with $T_1 = 20$ k-in., $T_2 = 30$ k-in., $T_3 = 25$ k-in., $l_1 = 20$ in., $l_2 = 30$ in., and $l_3 = 30$ in. Determine T_4, T_A, T_B, and T_C, and plot the torque diagram for this member.

Solution. Use the equations written above and substitute the numerical values to solve for the unknown torques.

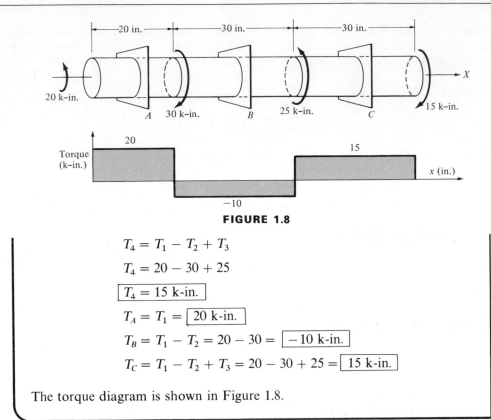

FIGURE 1.8

$$T_4 = T_1 - T_2 + T_3$$

$$T_4 = 20 - 30 + 25$$

$$\boxed{T_4 = 15 \text{ k-in.}}$$

$$T_A = T_1 = \boxed{20 \text{ k-in.}}$$

$$T_B = T_1 - T_2 = 20 - 30 = \boxed{-10 \text{ k-in.}}$$

$$T_C = T_1 - T_2 + T_3 = 20 - 30 + 25 = \boxed{15 \text{ k-in.}}$$

The torque diagram is shown in Figure 1.8.

1.5 Variable Torsional Loading – Torque Relationships

Torques are not always applied externally at given cross sections as depicted in Figure 1.7 but may be applied in some variable fashion along the longitudinal member axis as indicated in Figure 1.9. The torque intensity denoted by q is a function of x. This externally applied torsional loading is balanced at the member ends by T_A and T_B as shown in Figure 1.9(b).

Consider a short segment of the member depicted in Figure 1.9(d) and write the torque (or moment) equilibrium equation for this segment of length dx:

$$\sum T_x = 0$$

$$T + dT - T - q\,dx = 0$$

$$q = \frac{dT}{dx} \tag{1.3}$$

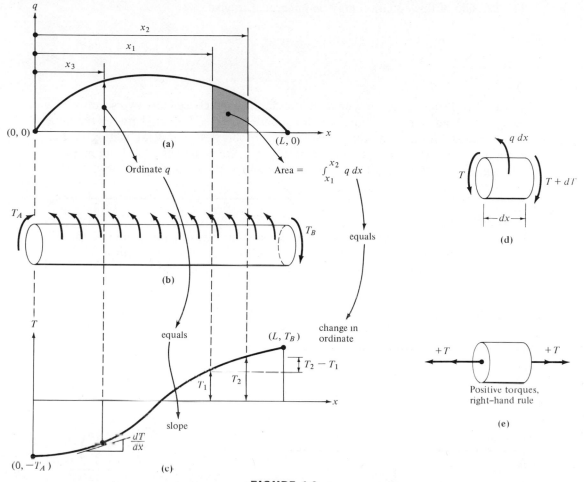

FIGURE 1.9

The applied torque intensity q at any section of the loaded member equals the rate of change of the internal torque T with respect to the longitudinal coordinate x. Graphically, this is depicted in Figure 1.9(a) and (c).

The ordinate q to the q–x curve at any point such as $x = x_3$ is equal to the slope of the T–x curve at the same point.

Multiply Eq. 1.3 by dx and write

$$dT = q\, dx$$

Then, integrate both sides of the equation:

$$\int_{T_1}^{T_2} dT = \int_{x_1}^{x_2} q\, dx$$

The left side of this equation may be integrated to give

$$T_2 - T_1 = \int_{x_1}^{x_2} q\,dx \tag{1.4}$$

Equation 1.4 states that the change in internal torque between any two sections of a torsionally loaded member equals the area under the q–x diagram between the same two sections. Graphically, this is depicted in Figure 1.9(a) and (c), where the two sections have been chosen at distances x_1 and x_2 from the left end of the member. The sign convention for positive internal torques is shown in Figure 1.9(e).

Example 1.5
A shaft 4 m long is fixed at both ends as shown in Figure 1.10(a) and is loaded by a torque of uniform intensity of $q = 100$ N-m/m. Apply and illustrate the significance of Eqs. 1.3 and 1.4. Let $x_3 = 1$ m, $x_1 = 2$ m, and $x_2 = 3$ m.

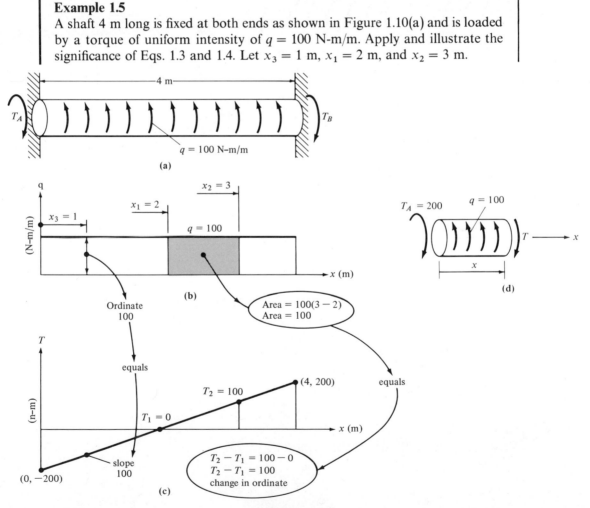

FIGURE 1.10

Solution. By symmetry, the end resisting torques are equal to each other and in turn equal to half the total applied torque:

$$T_A = T_B = \frac{ql}{2} = \frac{100(4)}{2} = 200 \text{ N-m}$$

Note by the sign convention of Figure 1.9(e) that T_A is a negative torque. To express the internal torque T as a function of x, refer to Figure 1.10(d) and write the torque equilibrium equation with respect to the x axis:

$$\sum T_x = 0$$

$$T - qx + 200 = 0$$

$$T = -200 + qx$$

This function is plotted in Figure 1.10(c) with $q = 100$ N-m/m. If this function is differentiated with respect to x, the resulting equation is given by

$$\frac{dT}{dx} = q \qquad \text{where } q = 100 \text{ N-m/m}$$

In this special case $q = $ constant and thus the slope of the T–x curve is a constant that equals 100 N-m/m regardless of the value of x in the interval $(0 \le x \le 4)$.

Apply Eq. 1.4:

$$T_2 - T_1 = \int_{x_1}^{x_2} q \, dx = 100 \int_2^3 dx = 100(3 - 2) = 100 \text{ N-m}$$

Example 1.6

Shaft AB is 10 ft long and is free at the left end and fixed torsionally at the right end. It is subjected to a variable torque as shown in Figure 1.11(a) and (b). The intensity of torsional loading is given by $q = 2x$. Determine the internal torque T as a function of x and plot the resulting function.

Solution. From Eq. 1.3,

$$q = \frac{dT}{dx}$$

where

$$q = 2x$$

$$dT = 2x \, dx$$

Integrate both sides of this equation using the condition at the free end A where $x = 0$, $T = 0$ for lower limits on the integrals.

$$\int_0^T dT = 2 \int_0^x x \, dx$$

$$\boxed{T = x^2}$$

This function is plotted in Figure 1.11(d).

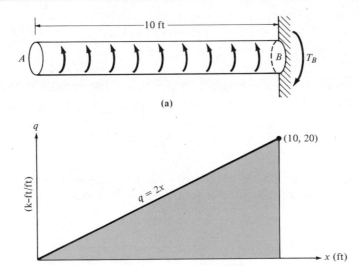

(a)

(b)

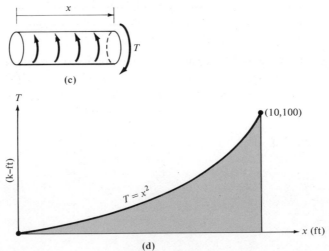

(c)

(d)

FIGURE 1.11

Homework Problems

1.12 Draw the torque diagram for the shaft depicted in Figure H1.12.

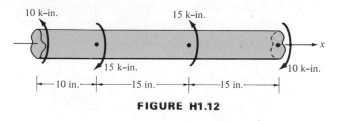

FIGURE H1.12

1.13 A cantilever shaft 3 m long is free at end A and fixed at end B. It is subject to distributed external torques which vary as shown in Figure H1.13. Determine the internal torque T as a function of x and plot the function.

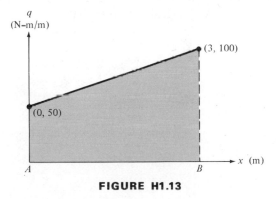

FIGURE H1.13

1.14 A shaft is subjected to an externally applied torque q between B and C as shown in Figure H1.14.

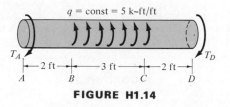

FIGURE H1.14

Determine the reacting torques at A and D. Find T versus x and plot the torque diagram for this shaft.

1.15 An airplane wing is 60 ft from wing tip to fuselage connection, as shown in Figure H1.15. In addition to bending and shear, the wing is subjected to external torques due to air pressures given by $q = 50 + 1.2x^2$, where q is measured in lb-ft/ft and x is measured in ft. Determine the internal torque T as a function of x and plot the function.

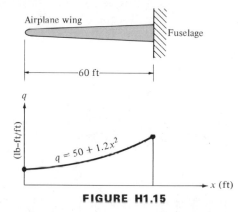

FIGURE H1.15

1.16 The shaft depicted in Figure H1.16 is torsionally free at A and fixed at B. The q versus x plot is shown in the figure. Determine the reacting torque at B and the internal torque T as a function of x. Plot T versus x.

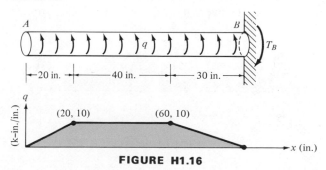

FIGURE H1.16

1.17 Draw the internal torque diagram for the shaft depicted in Figure H1.17.

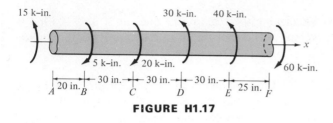

FIGURE H1.17

1.18 The shaft depicted in Figure H1.18 is subjected to an external torque that varies sinusoidally as shown. It is torsionally fixed at both ends and by symmetry the reacting torques T_A and T_B are equal. Determine the internal torque T as a function of x and sketch this function. Determine q and T for $x = 1$ m.

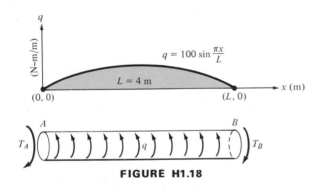

FIGURE H1.18

1.19 A cantilever shaft AB which is 40 in. long is shown in Figure H1.19. It is subjected to a torque of 20 k-in. at A and a uniformly distributed torque of 2 k-in./in. over its full length. These applied

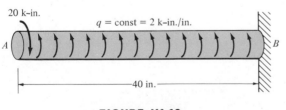

FIGURE H1.19

torques are resisted by a torsional reaction at B. Determine the internal torque T as a function of x measured from A along the shaft. Plot the q–x and T–x functions.

1.20 A hollow shaft is subjected to the torques shown in Figure H1.20. Draw the torque diagram for this shaft.

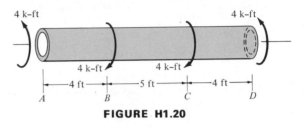

FIGURE H1.20

1.21 A stepped shaft is subjected to the torques shown in Figure H1.21. These applied torques are resisted at D. Compute the reacting torque at D and draw the torque diagram for this shaft.

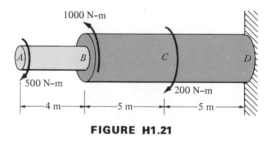

FIGURE H1.21

1.22 A hollow shaft is subjected to the torques shown in Figure H1.22. Draw the torque diagram for this shaft.

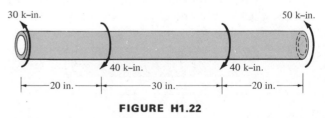

FIGURE H1.22

1.23 Two shafts have torques applied to them as shown in Figure H1.23(a) and (b). Draw the torque diagrams for each of these shafts. Compare these diagrams and comment on the effect of reducing the length BC over which the constant torque is applied to the shaft of Figure H1.23(b) while q is increased proportionately such that the total applied torque over this shaft segment remains at 20 k-ft.

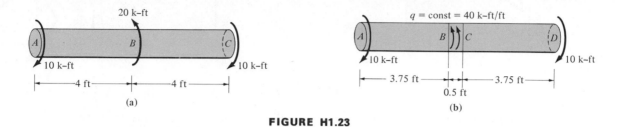

(a)

(b)

FIGURE H1.23

1.6 Shear and Bending Moment in Beams

Consider a member loaded by a system of coplanar forces as shown in Figure 1.12. All forces are applied transversely, that is, in the y direction with respect to the longitudinal or x axis of the member. Loadings may be concentrated, such as P_1; uniform, such as w_1; or may vary in some other fashion, as indicated by $w_3 = f(x)$. Members loaded as described and which are long compared to their cross-sectional dimensions are termed *beams*. The beam is in equilibrium under the action of the applied system of forces and the reactions R_1 and R_2. Given complete information about the applied system of forces, the equations of equilibrium may be applied in order to determine the reactions R_1 and R_2 acting on this statically determinate beam.

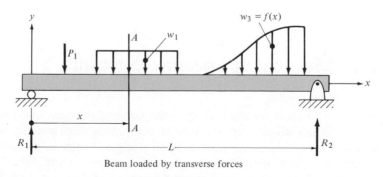

Beam loaded by transverse forces

FIGURE 1.12

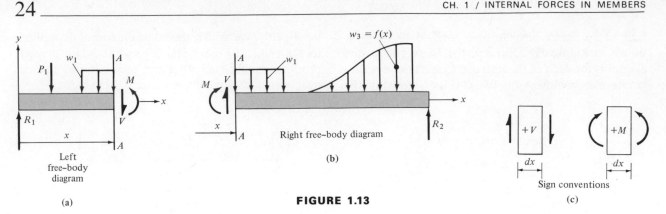

FIGURE 1.13

To determine the internal effects of the applied forces, imagine a cutting plane, A–A, passed perpendicular to the x axis of the beam and isolate either a left part or a right part of the beam as shown in Figure 1.13(a) and (b), respectively.

To maintain these bodies in equilibrium, a force (V) and a couple (M) are required, as shown in Figure 1.13(a) and (b). The force V is termed the *shear force* at section A–A and the couple M is termed the *bending moment* at section A–A. The V and M in Figure 1.13(a) are the force and couple applied to the left part of the beam by the right part of the beam to maintain it in equilibrium. Alternatively, V and M in Figure 1.13(b) are the shear force and moment applied to the right part of the beam by the left part of the beam to maintain it in equilibrium. Equilibrium of either part is required, since the entire beam is in equilibrium, and thus any part of the beam we wish to consider must also be in equilibrium under the action of external and internal forces (V and M).

Section A–A is passed at any distance x from the origin at the left end of the beam. The variable x may take on any value from 0 to L, where L is the length of the beam, and the shear V and the bending moment M are functions of this variable x. The positive senses for V and M are depicted in Figure 1.13(a) and (b). Shear and moment positive senses may also be represented on a beam segment of differential length as shown in Figure 1.13(c).

1.7 Shear and Moment at Specified Sections

Development of a clear understanding of the concept of shear and moment at each section of a beam is best accomplished by computing these quantities at arbitrarily specified sections of a given beam. Careful study of Examples 1.7 and 1.8 will enable the student to draw appropriate free-body diagrams and apply the equations of equilibrium in order to compute shear and moment at designated sections of a given beam. This section is followed by a more general discussion dealing with shear and moment relationships and functions.

Example 1.7

For the beam loaded as depicted in Figure 1.14, determine the shear and moment at sections *A–A*, *B–B*, and *C–C*, which are located 3, 8, and 18 ft, respectively, from the left end of the beam.

Solution. First, determine the reactions R_1 and R_2 by applying the equations of equilibrium to the free body of the entire beam shown in Figure 1.14. Sum the moments of the forces with respect to an axis through the right end of the beam (point *E*) in order to eliminate the reaction R_2 from the equation. A single unknown, R_1, appears in this equation.

$$\circlearrowright \quad \sum M_E = 0$$

$$R_1(20) - 5(15) - 2(10)(5) = 0$$

$$R_1 = \tfrac{1}{20}(75 + 100) = 8.75 \text{ k } (\uparrow)$$

$$\uparrow \quad \sum F_y = 0$$

$$R_1 - 5 - 2(10) + R_2 = 0$$

$$R_1 = 8.75$$

$$R_2 = 16.25 \text{ k } (\uparrow)$$

Sum the moments of the forces with respect to an axis through the left end of the beam (point *O*) in order to check the solution for the reaction R_2.

$$\circlearrowleft \quad \sum M_0 = 0$$

$$R_2(20) - 2(10)(15) - 5(5) = 0$$

$$R_2 = 16.25 \text{ k } (\uparrow) \qquad \text{(checks value above)}$$

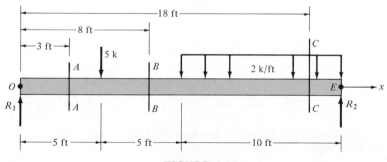

FIGURE 1.14

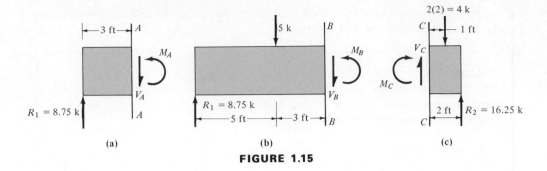

FIGURE 1.15

To determine the shear and moment at section A–A, construct the free-body diagram shown in Figure 1.15(a). Positive, unknown shear and moment at section A–A are denoted V_A and M_A. This part of the beam to the left of section A–A is in equilibrium, since the entire beam is in equilibrium.*

$\uparrow \ \ \Sigma F_y = 0$

$$8.75 - V_A = 0$$

$\boxed{V_A = 8.75 \text{ k} \qquad \text{downward on left fbd}}$

$\circlearrowleft \ \ \Sigma M_A = 0$

$$8.75(3) - M_A = 0$$

$\boxed{M_A = 26.25 \text{ k-ft} \qquad \text{ccw on left fbd}}$

Shear and moment at section A–A are both positive.

To determine the shear and moment at section B–B, construct the free-body diagram shown in Figure 1.15(b). Positive, unknown shear and moment at section B–B are denoted V_B and M_B.

$\uparrow \ \ \Sigma F_y = 0$

$$8.75 - 5 - V_B = 0$$

$\boxed{V_B = 3.75 \text{ k} \qquad \text{downward on left fbd}}$

*Abbreviations used in the following equations are: fbd, free-body diagram; cw, clockwise; and ccw, counterclockwise.

$\circlearrowleft \quad \sum M_B = 0$

$$8.75(8) - 5(3) - M_B = 0$$

$\boxed{M_B = 55.0 \text{ k-ft} \qquad \text{ccw on left fbd}}$

Shear and moment at section B–B are both positive.

To determine the shear and moment at section C–C, construct the free-body diagram shown in Figure 1.15(c). The right part of the beam is preferred over the left part of the beam in this case, since fewer forces act on the right part and the equations of equilibrium will be simpler to write.

$\uparrow \quad \sum F_y = 0$

$$+V_C - 2(2) + 16.25 = 0$$

$\boxed{V_C = -12.25 \text{ k} \qquad \text{downward on right fbd}}$

$\circlearrowleft \quad \sum M_C = 0$

$$M_C + 4(1) - 16.25(2) = 0$$

$\boxed{M_C = 28.5 \text{ k-ft} \qquad \text{cw on right fbd}}$

Review of this solution clearly reveals that the numerical values for shear and moment depend upon the location of the beam section, which is located by the variable x. In general, shear and moment are functions of x and these functional relationships will be presented in Section 1.8.

Example 1.8

For the beam loaded as depicted in Figure 1.16, determine the shear and moment at section A–A located 8 ft from the left end of the beam.

Solution. First, determine the reactions of R_1 and R_2 by applying the equations of equilibrium to the free body of the entire beam, which is shown in Figure 1.16(b). The resultant applied force equals the area under the linear force distribution diagram:

$$\tfrac{1}{2}(12)(200) = 1200 \text{ lb} \downarrow$$

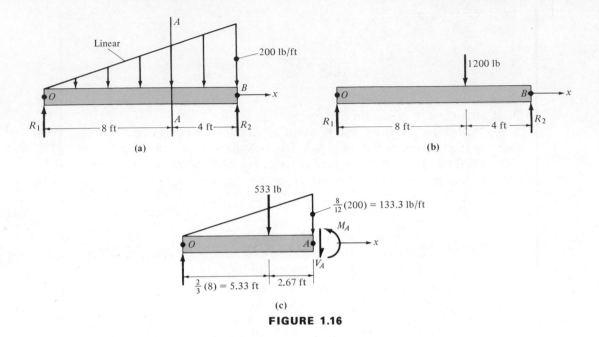

FIGURE 1.16

This resultant is applied to the beam at the centroid of the linear (or triangular) force distribution diagram which lies $\frac{2}{3}(12) = 8$ ft from the left end of the beam. Sum the moments of the forces with respect to an axis through the right end of the beam (point B) in order to eliminate the reaction R_2 from the equation. A single unknown R_1 appears in this equation.

$$\circlearrowright \quad \sum M_B = 0$$

$$R_1(12) - 1200(4) = 0$$

$$R_1 = 400 \text{ lb}$$

Sum the vertical forces acting on the beam.

$$\uparrow \quad \sum F_y = 0$$

$$R_1 - 1200 + R_2 = 0$$

$$R_1 = 400$$

$$R_2 = 800 \text{ lb}$$

A distributed coplanar force system acting on a rigid body may be replaced by a single resultant force without altering the external effect of the forces acting on the body. The reactions R_1 and R_2 are external forces acting on the complete beam and thus the technique employed above is valid. However, V_A and M_A are internal forces with respect to the entire beam, and a replacement of the distributed force system is only valid when these forces are applied externally as shown in Figure 1.16(c) on the left part of the beam. The resultant applied force equals the area under the linear force distribution diagram:

$$\tfrac{1}{2}(8)(133.3) = 533 \text{ lb} \downarrow$$

Sum vertical forces acting on the left part of the beam in Figure 1.16(c).

$$\uparrow \quad \sum F_y = 0$$

$$-V_A + 400 - 533, = 0$$

$$\boxed{V_A = -133 \text{ lb} \qquad \text{up on the left fbd}}$$

Sum the moments of the forces acting on the left part of the beam of Figure 1.16(c) with respect to an axis through point A.

$$\circlearrowleft \quad \sum M_A = 0$$

$$M_A + 533(2.67) - 400(8) = 0$$

$$\boxed{M_A = 1778 \text{ lb-ft} \qquad \text{ccw on left fbd}}$$

The shear is negative and the moment is positive at section A–A. Note that the shears and moments for the beams depicted in Figure 1.16(a) and (b) would, in general, be different as the section location variable, x, varies over its range of values.

Homework Problems

1.24 Refer to Figure 1.14 and compute the shear and moment at distances of 4, 9, and 15 ft from the left end of the beam. Include free-body diagrams for each solution.

1.25 Refer to Figure 1.16(a) and compute the shear and moment at distances of 2 and 10 ft from the left end of the beam. Include free-body diagrams for each solution.

1.26 Refer to Figure H1.26 and determine the reactions and the shear and moment at section *A–A*.

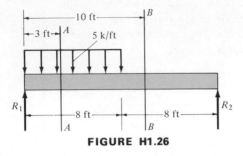

FIGURE H1.26

1.27 Refer to Figure H1.26 and determine the reactions and the shear and moment at section *B–B*.

1.28 Refer to Figure H1.28 and determine the reactions and the shear and moment at sections *A–A* and *B–B*. Make comments regarding symmetry or antisymmetry of loads, shears, and moments for this beam.

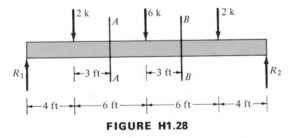

FIGURE H1.28

1.29 Find the shear and moment at section *A–A* of the cantilever beam shown in Figure H1.29.

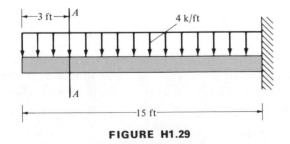

FIGURE H1.29

1.30 For the beam depicted in Figure H1.30 determine the reactions at *B* and *C* and the shear and moment at section *A–A* by using:
(a) The left part of the beam.
(b) The right part of the beam.

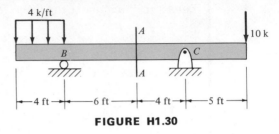

FIGURE H1.30

1.31 Determine the reactions at *B* and *C* acting on the beam shown in Figure H1.31. The couple of 1200 lb-ft is externally applied to the beam at point *D*. Find the shear and moment at section *A–A* by using:
(a) The left part of the beam.
(b) The right part of the beam.

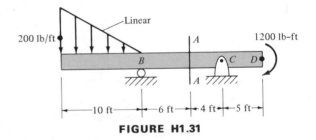

FIGURE H1.31

1.32 For the beam depicted in Figure H1.32, determine the reactions at *B* and *C* and the shear and moment at section *A–A* by using:
(a) The left part of the beam.
(b) The right part of the beam.

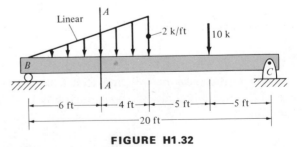

FIGURE H1.32

1.33 Determine the reactions at D and E acting on the beam depicted in Figure H1.33. Then find the shear and moment at sections A–A and B–B. Make comments regarding symmetry or antisymmetry of loads, shears, and moments for this beam.

1.34 The cantilever beam shown in Figure H1.34 is free at the left end and fixed at the right end. Determine the shear and moment at sections A–A and B–B by using the left part of the beam in both cases. Why is this easier than using the right part of the beam?

1.35 Determine the reactions at C and E acting on the beam shown in Figure H1.35. The hinge at D provides another equation of equilibrium required to determine the reactions. This equation is $\sum M_D = 0$. Then, find the shear and moment at section A–A by using:
(a) The left part of the beam.
(b) The right part of the beam.

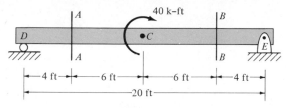

FIGURE H1.33

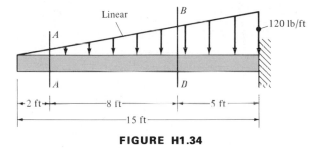

FIGURE H1.34

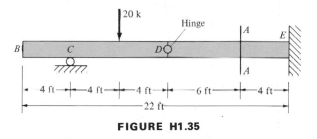

FIGURE H1.35

1.8 Shear and Moment Relationships

In order to write shear and moment equations and understand their plots, relationships will be developed among loading, shear, and moment. Consider a differential length of beam shown in Figure 1.17(d). This beam segment is in equilibrium under the action of the force system shown. The downward force applied to the beam segment is stated as $w\,dx$ since any change in w over the length dx would contribute to a second-order term, and this term would vanish in the limit if a rigorous viewpoint were adopted.

$$\uparrow \quad \sum F_y = 0$$

$$+V - (V + dV) - w\,dx = 0$$

$$w = -\frac{dV}{dx} \tag{1.5}$$

Graphically, Eq. 1.5 may be interpreted by referring to Figure 1.17(a) and (b). The ordinate to the loading diagram (w) equals the negative of the slope of the shear diagram $-dV/dx$ at corresponding points.

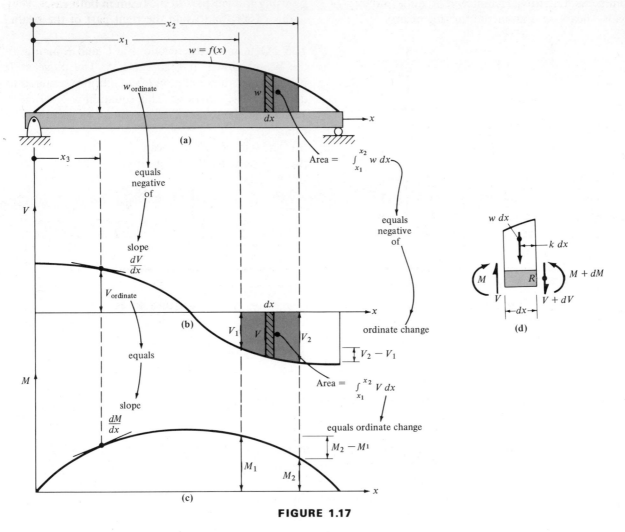

FIGURE 1.17

Now apply the moment equilibrium equation to the differential beam element shown in Figure 1.17(d).

$$\circlearrowright \quad \sum M_R = 0$$

$$V\,dx + M - (M + dM) - w\,dx(k\,dx) = 0$$

The term $w\,dx(k\,dx)$ is a second-order differential term and may be dropped.* The moment arm of the applied force $(w\,dx)$ with respect to the right side of the beam element is $k\,dx$, where k is a constant less than unity. The value of k is not required, since the higher-order term is dropped to yield

$$V = \frac{dM}{dx} \qquad \textbf{(1.6)}$$

Graphically, Eq. 1.6 may be interpreted by referring to Figure 1.17(b) and (c). The ordinate to the shear diagram (V) equals the slope of the moment diagram (dM/dx) at corresponding points.

Differential Eqs. 1.5 and 1.6 may be expressed in integral form and interpreted graphically by reference to Figure 1.17. Equation 1.5 may be written as $dV = -w\,dx$ and then integrated between limits x_1, x_2 and V_1, V_2. Thus

$$\int_{V_1}^{V_2} dV = -\int_{x_1}^{x_2} w\,dx$$

$$V_2 - V_1 = -\int_{x_1}^{x_2} w\,dx \qquad \textbf{(1.7)}$$

Graphically, Eq. 1.7 may be interpreted by referring to Figure 1.17(a) and (b). The change in ordinate to the shear diagram $(V_2 - V_1)$, taken between two x values x_1 and x_2, equals the negative of the area under the loading diagram between the same two x values. Equation 1.6 may be written as $dM = V\,dx$ and then integrated between limits x_1, x_2 and M_1, M_2.

$$\int_{M_1}^{M_2} dM = \int_{x_1}^{x_2} V\,dx$$

$$M_2 - M_1 = \int_{x_1}^{x_2} V\,dx \qquad \textbf{(1.8)}$$

* This is not an approximation. Replace the differentials by deltas as follows:

$$V\,\Delta x + M - (M + \Delta M) - w\,\Delta x(k\,\Delta x) = 0$$

Divide through by Δx: $V - (\Delta M/\Delta x) - w(k\,\Delta x) = 0$.

$$V = \lim_{\Delta x \to 0} \frac{\Delta M}{\Delta x} = \frac{dM}{dx}$$

Graphically, Eq. 1.8 may be interpreted by referring to Figure 1.17(b) and (c). The change in ordinate to the moment diagram $(M_2 - M_1)$, taken between two x values x_1 and x_2, equals the area under the shear diagram between the same two x values.

Example 1.9

A simply supported beam 10 m long shown in Figure 1.18 is loaded by a uniform load of 200 N/m. Apply and illustrate the significance of Eqs. 1.5 to 1.8. Let $x_3 = 2$ m, $x_1 = 6$ m, and $x_2 = 8$ m.

Solution. The reactions are equal by symmetry. Each reaction equals half the total downward applied force.

$$R_1 = R_2 = \tfrac{1}{2} \int_0^L w \, dx = \tfrac{1}{2} \int_0^{10} 200 dx = 1000 \text{ N} \uparrow$$

Refer to the free-body diagram shown in Figure 1.18(d) and write the equations of equilibrium in order to express the shear and moment as functions of x.

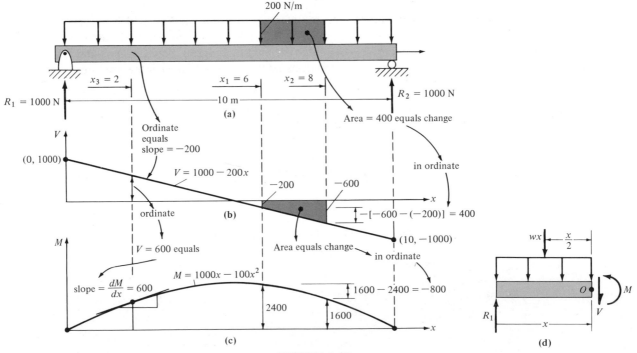

FIGURE 1.18

$$\uparrow \quad \sum F_y = 0$$

$$R_1 - wx - V = 0$$

Since $R_1 = 1000$,

$$V = 1000 - 200x \qquad (0 < x < 10)$$

This equation is plotted in Figure 1.18(b) and is referred to as the *shear diagram*.

$$\circlearrowright \quad \sum M_0 = 0$$

$$R_1 x - wx\left(\frac{x}{2}\right) - M = 0$$

$$M = 1000x - 100x^2 \qquad (0 \le x \le 10)$$

This equation is plotted in Figure 1.18(c) and is referred to as the *moment diagram*.

Differentiate the shear equation with respect to x to obtain $dV/dx = -200$. $w = 200$ N/m $(0 < x < 10)$ and Eq. 1.5, $dV/dx = -w$ is satisfied. The downward load intensity is constant over the beam length and the shear diagram has a constant slope equal to the negative of this load intensity.

Differentiate M with respect to x to obtain $dM/dx = 1000 - 200x$. By Eq. 1.6, $V = dM/dx$. By comparison with the equation obtained previously for V, note that $V = dM/dx$ for $(0 \le x \le 10)$. In particular, for $x_3 = 2$ m, $V_2 = 1000 - 200(2) = 600$ N.

Now consider Eqs. 1.7 and 1.8, which are the integrated forms of Eqs. 1.5 and 1.6. Use limits of $x_1 = 6$ m and $x_2 = 8$ m.

$$V_2 - V_1 = -\int_6^8 200dx = -200x \Big|_6^8 = -400 \text{ N}$$

which is the negative of the area under the load diagram between $x_1 = 6$ and $x_2 = 8$.

$$M_2 - M_1 = \int_6^8 (1000 - 200x)\,dx = 1000x - 100x^2 \Big|_6^8 = -800 \text{ N} - \text{m}$$

which is the area under the shear diagram between $x_1 = 6$ and $x_2 = 8$ m.

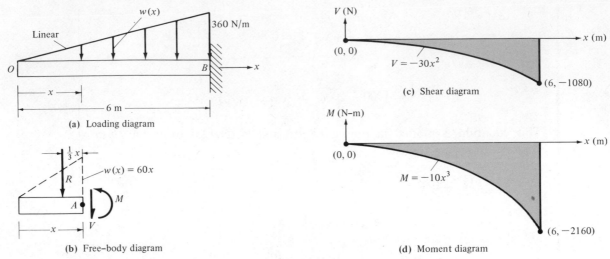

(a) Loading diagram

(b) Free–body diagram

(c) Shear diagram

(d) Moment diagram

FIGURE 1.19

Example 1.10
A cantilever beam 6 m long shown in Figure 1.19 is subjected to a linearly varying loading which has a maximum ordinate of 360 N/m at the fixed end on the right. Write the shear and moment as functions of x for this beam and draw the shear and moment diagrams.

Solution. Choose an origin O at the free left end of the cantilever and express the loading as a function of x by using the similarity of triangles shown in Figure 1.19(a).

$$\frac{w(x)}{x} = \frac{360}{6}$$

$$w(x) = 60x$$

Refer to the free-body diagram shown in Figure 1.19(b) and compute the resultant force R due to the linearly distributed force applied to this portion of the beam of any length x. Recall that the resultant force equals the area under the force distribution diagram, which is triangular in this case. Thus

$$R = \tfrac{1}{2}(60x)(x) = 30x^2$$

This resultant acts through the centroid of the force distribution diagram, which is at a distance of $x/3$ from point A, located at the cut section of the beam.

$$\uparrow \quad \textstyle\sum F_y = 0$$

$$-V - R = 0$$

$$V = -R = -30x^2$$

The shear function is parabolic and the shear diagram is shown in Figure 1.19(c).

$$\circlearrowleft \quad \textstyle\sum M_A = 0$$

$$M + R(\tfrac{1}{3}x) = 0$$

$$M = -\tfrac{1}{3}Rx = -\tfrac{1}{3}(30x^2)x = -10x^3$$

The moment function is cubic and the moment diagram is shown in Figure 1.19(d).

The reader should refer to Figure 1.19 and the equations noted as he studies the following statements. At O the ordinate to the w–x curve is zero, which equals the slope of the V–x curve at O, Eq. 1.5. The area under the w–x curve from O to B is $\tfrac{1}{2}(6)(360) = 1080$ N, which equals the negative of the ordinate to the V–x curve at point B, Eq. 1.7. At B the ordinate to the w–x curve is 360 N/m, which equals the negative of the slope of the V–x curve at B: $V = -30x^2, dV/dx = -60x, dV/dx|_{x=6} = -360$ N/m, Eq. 1.5. At O the ordinate to the V–x curve is zero, which equals the slope of the M–x curve at O, Eq. 1.6. The area under the V–x curve from O to B, $\int_0^6 -30x^2\, dx = -2160$ Nm, equals the ordinate of the M–x curve at B, Eq. 1.8. At B the ordinate to the V–x curve is -1080 N, which equals the slope of the M–x curve at B: $M = -10x^3$, $dM/dx = -30x^2, dM/dx|_{x=6} = -1080$ N, Eq. 1.6.

Example 1.11

Determine the reactions at A and D for the beam depicted in Figure 1.20, and then construct the shear and moment diagrams for this beam.

Solution. Refer to Figure 1.20(a) and compute the reactions:

$$\circlearrowright \quad \textstyle\sum M_D = 0$$

$$R_A(45) - 20(35) - 40(15) + 2(30)(15) = 0$$

$$R_A = 8.89 \text{ k} \uparrow$$

$$\uparrow \quad \textstyle\sum F_y = 0$$

$$R_A - 20 - 40 - 2(30) + R_D = 0$$

$$R_D = 120 - 8.89 = 111.11 \text{ k} \uparrow$$

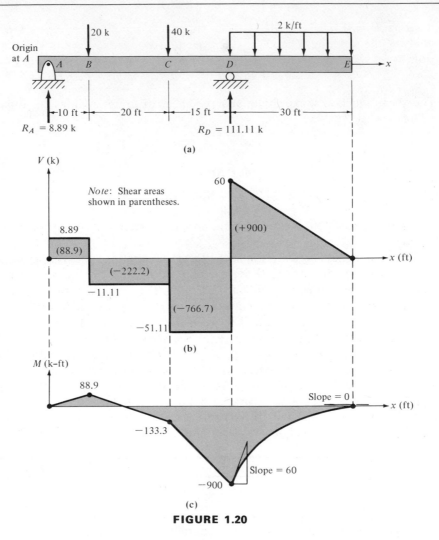

FIGURE 1.20

Write the moment equilibrium equation with respect to point A in order to check this value of R_D.

$$\circlearrowleft \quad \sum M_A = 0$$

$$R_D(45) - 2(30)(60) - 40(30) - 20(10) = 0$$

$$R_D = 111.11 \text{ k} \uparrow \quad \text{(checks value above)}$$

To determine the ordinates to the shear diagram of Figure 1.20(b), consider parts of the beam to the left of a given section and construct their free-body diagrams in order to sum vertical forces acting on them. These free-body diagrams are not shown here, but the student is urged to draw them during study of this solution. The left upward reaction must be balanced by an equal downward shear force acting on a left segment of the beam obtained by passing a vertical cutting plane anywhere between A and B ($0 < x < 10$). The shear force between A and B is $+8.89$ k and remains constant because there are no external forces between A and B and the weight of the beam is ignored in this problem. Now consider sections between B and C (i.e., $10 < x < 30$):

$$\uparrow \quad \sum F_y = 0$$

$$8.89 - 20 - V = 0$$

$$V = -11.11 \text{ k}$$

Segment CD ($30 < x < 45$):

$$\uparrow \quad \sum F_y = 0$$

$$8.89 - 20 - 40 - V = 0$$

$$V = -51.11 \text{ k}$$

Cut a section just to the right of the reaction at D:

$$\uparrow \quad \sum F_y = 0$$

$$8.89 - 20 - 40 + 111.11 - V = 0$$

$$V = 60 \text{ k}$$

At the free end E the shear must be zero. What curve should connect the ordinate of $+60$ at D and 0 at E? Recall Eq. 1.5: $w = -dV/dx$. Between D and E, $w = \text{const} = 2$ k/ft. Thus $dV/dx = \text{const} = -2$ k/ft. The slope of the curve joining these two ordinates to the shear diagram is a constant and, therefore, a straight line is the proper curve.

The moment diagram will be constructed by referring to the shear diagram and use of Eqs. 1.6 and 1.8. Areas under each part of the shear diagram are shown in Figure 1.20(b) and appear in parentheses. By Eq. 1.8, the area under the shear diagram between any two x values equals the change in ordinate to the moment diagram between the same two points. For example, the area under the shear diagram between A and B equals 88.9 k-ft, which equals the moment at B, since the moment at A equals zero because a frictionless hinge supports the beam at A. A straight line connects the moment diagram ordinate at A to the ordinate at B because $V = dM/dx$ and the shear has the constant value of 8.89 k between A and B. Continuing in a similar fashion, we obtain the shear and moment values shown in the following tabulation:

Range of x Values	Moment at	+	Area Under Shear Diagram	=	Moment at
BC	B 88.9		−222.2		= −133.3 C
CD	C −133.3		−766.7		= −900.0 D
DE	D −900.0		900.0		= 0.0 E

The moment at E must equal zero, since the beam is free of applied forces on the cross section at its right end, E. As for AB, the moment curves for BC and CD are sloping straight lines, since the shear is constant over each of these segments also. Finally, consider the segment DE. At D, the ordinate to the shear diagram is $+60$ and from $V = dM/dx$ the slope of the moment diagram at D is $+60$, while at E the ordinate to the shear diagram is zero and from $V = dM/dx$ the slope of the moment diagram at E is zero. The curve connecting these two end slopes consistently is the parabola shown between D and E on the moment diagram.

Example 1.12

Refer to Figure 1.21 and determine the reactions acting on the beam shown. Choose an origin at A and write the shear and moment equations for this beam. Plot the shear and moment diagrams.

Solution. Draw the free-body diagrams shown in Figure 1.21(b) and note the four unknowns as follows: M_A and V_A at the fixed end A, V_C at the hinge C, and R_D the reaction at D. Two equations of equilibrium are available for each of the two free bodies, making a total of four equations available for solving for the four unknowns.

FREE-BODY DIAGRAM CDE:

$$\circlearrowleft \quad \sum M_C = 0$$

$$R_D(5b) - 6wb(8b) = 0$$

$$R_D = \tfrac{48}{5}wb \uparrow$$

$$\uparrow \quad \sum F_y = 0$$

$$R_D - V_C - 6wb = 0$$

$$V_C = \tfrac{48}{5}wb - 6wb$$

$$V_C = \tfrac{18}{5}wb \downarrow \text{ on } CDE$$

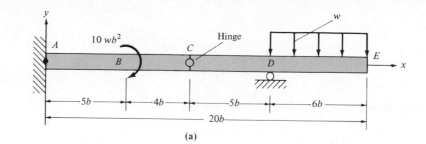

(a)

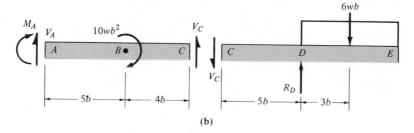

(b)

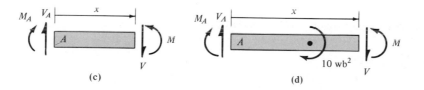

(c) (d)

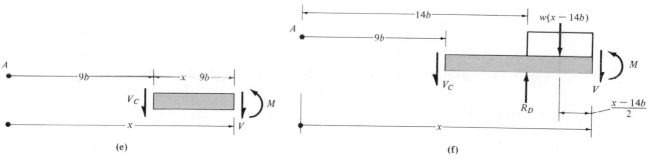

(e) (f)

FIGURE 1.21

FREE-BODY DIAGRAM ABC:

$\circlearrowleft \quad \sum M_A = 0$

$$-M_A + V_c(9b) - 10wb^2 = 0$$

$$M_A = +\tfrac{112}{5}wb^2$$

$\uparrow \ \sum F_y = 0$

$$+ V_A + V_C = 0$$

$$V_A = -V_C = -\tfrac{18}{5}wb \ (\downarrow)$$

FREE-BODY DIAGRAM OF FIGURE 1.21(c):

$\uparrow \ \sum F_y = 0$

$$-V + V_A = 0$$

$$V = V_A = -\tfrac{18}{5}wb \qquad (0 < x < 5b)$$

$\circlearrowleft \ \sum M = 0$

$$M - M_A - V_A x = 0$$

$$M = M_A + V_A x = \tfrac{112}{5}wb^2 - \tfrac{18}{5}wbx \qquad (0 \le x \le 5b)$$

FREE-BODY DIAGRAM OF FIGURE 1.21(d):

$\uparrow \ \sum F_y = 0$

$$-V + V_A = 0$$

$$V = V_A = -\tfrac{18}{5}wb \qquad (5b < x < 9b)$$

$\circlearrowleft \ \sum M = 0$

$$M - M_A - V_A x - 10wb^2 = 0$$

$$M = M_A + V_A x + 10wb^2 = \tfrac{162}{5}wb^2 - \tfrac{18}{5}wbx \qquad (5b \le x \le 9b)$$

FREE-BODY DIAGRAM OF FIGURE 1.21(e):

$\uparrow \ \sum F_y = 0$

$$-V - V_C = 0$$

$$V = -V_C = -\tfrac{18}{5}wb \qquad (9b < x < 14b)$$

$\circlearrowleft \ \sum M = 0$

$$M + V_C(x - 9b) = 0$$

$$M = -V_C(x - 9b) = -\tfrac{18}{5}wb(x - 9b) \qquad (9b \le x \le 14b)$$

FREE-BODY DIAGRAM OF FIGURE 1.21(f):

$$\uparrow \quad \sum F_y = 0$$

$$-V - V_C + R_D - w(x - 14b) = 0$$
$$V = -V_C + R_D - w(x - 14b)$$
$$V = -\tfrac{18}{5}wb + \tfrac{48}{5}wb - w(x - 14b)$$
$$V = 6wb - w(x - 14b) \qquad (14b < x \le 20b)$$

$$\circlearrowleft \quad \sum M = 0$$

$$M + V_C(x - 9b) - R_D(x - 14b) + \frac{w}{2}(x - 14b)^2 = 0$$

$$M = -\frac{18}{5}wb(x - 9b) + \frac{48}{5}wb(x - 14b) - \frac{w}{2}(x - 14b)^2$$

$$M = 6wbx - 102wb^2 - \frac{w}{2}(x - 14b)^2 \qquad (14b \le x \le 20b)$$

The shear and moment diagrams which are plots of the foregoing derived functions are shown in Figure 1.22. In conclusion, several checks of these

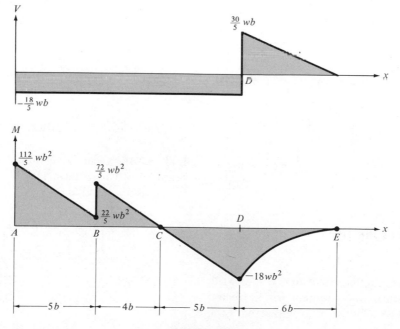

FIGURE 1.22

diagrams will be noted that will prove useful in solving similar problems. Proceeding left to right in constructing the shear diagram, note that the shear at the right end equals zero. This must be true, since the right end of the beam is free of any applied force. Construction of the moment diagram yields a moment of zero at *C*, which must be true at the hinge located there, and the moment at the free right end *E* must also vanish. The student should relate these equations for shear and bending moment to the diagrams shown in Figure 1.22.

Homework Problems

1.36 Determine the reactions at *A* and *C* and draw the shear and moment diagrams for the beam depicted in Figure H1.36.

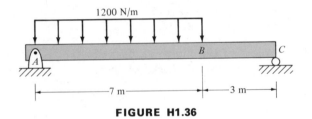

FIGURE H1.36

1.37 Determine reactions at *A* and *D* and draw the shear and moment diagrams for the beam depicted in Figure H1.37.

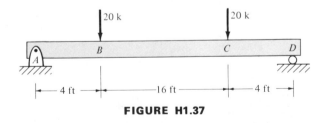

FIGURE H1.37

1.38 Determine the reactions at *A* and *C* and write the equations for shear and moment for the beam shown in Figure H1.38. Choose an origin at *A* for the *x* coordinate.

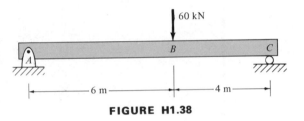

FIGURE H1.38

1.39 Find the reactions at *B* and *D* acting on the beam shown in Figure H1.39 and then compute the shear and moment at a section 5 ft to the right of *B*.

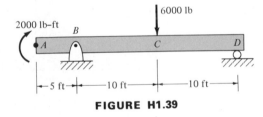

FIGURE H1.39

1.40 Draw shear and moment diagrams for the beam shown in Figure H1.40 after determining the reactions at *A* and *B*.

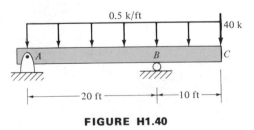

FIGURE H1.40

1.41 Determine the reactions at the right end of the cantilever beam shown in Figure H1.41. Construct shear and moment diagrams for this beam and check the reaction values computed for the right end.

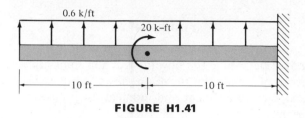

FIGURE H1.41

1.42 Determine the reactions at B and D acting on the beam shown in Figure H1.42. Construct shear and moment diagrams for this beam.

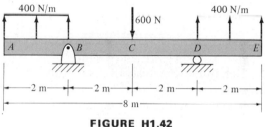

FIGURE H1.42

1.43 Find the reactions at B and F acting on the beam of Figure H1.43. Construct shear and moment diagrams for this beam.

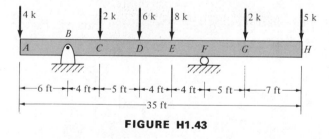

FIGURE H1.43

1.44 Find the reactions at B and C acting on the beam of Figure H1.44. Then, determine the shear and

moment at a section 1 m to the left of B and at a section 3 m to the right of B.

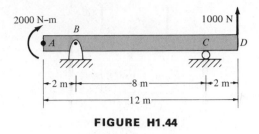

FIGURE H1.44

1.45 Determine the reaction at B, the shear at the hinge C, and the force and couple reactions at E acting on the beam of Figure H1.45. Draw shear and moment diagrams for the beam.

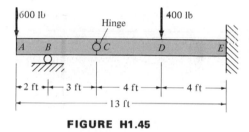

FIGURE H1.45

1.46 Find the reactions at B and E and then draw shear and moment diagrams for the beam shown in Figure H1.46.

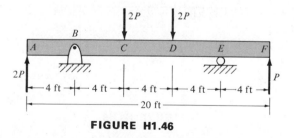

FIGURE H1.46

1.47 Determine the reactions at A and E in terms of P for the beam shown in Figure H1.47. Then choose an origin at A for the x coordinate and write the shear and moment equations for each segment of the beam.

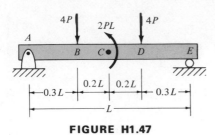

FIGURE H1.47

1.48 Choose an origin at A as shown in Figure H1.48 and write shear and moment equations for this cantilever beam.

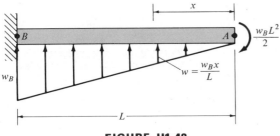

FIGURE H1.48

1.49 Determine the reactions acting on the beam shown in Figure H1.49 at points B and C. Draw shear and moment diagrams for this beam.

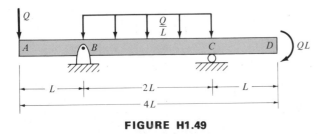

FIGURE H1.49

1.50 For the beam shown in Figure H1.50, draw shear and moment diagrams after finding the reactions at points A and D.

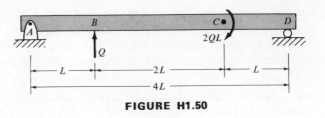

FIGURE H1.50

1.51 Determine the upward reaction intensity w in terms of P and L for the beam shown in Figure H1.51, and then draw shear and moment diagrams for the beam.

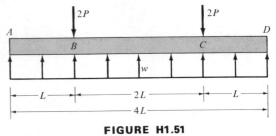

FIGURE H1.51

1.52 Determine the reactions at A and B for the beam shown in Figure H1.52. Choose an origin at A and write shear and moment equations for this beam.

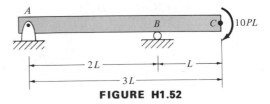

FIGURE H1.52

1.53 Choose an origin at point A as shown in Figure H1.53 after determining the reactions at A and B.

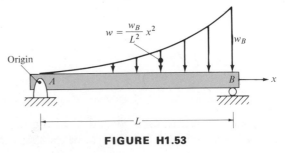

FIGURE H1.53

Then write the shear and moment equations for this beam. Check the right reaction by evaluating the shear equation for $x = L$.

1.54 Find the reactions at B and E acting on the beam shown in Figure H1.54. Then, draw shear and moment diagrams for the beam.

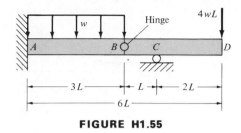

FIGURE H1.55

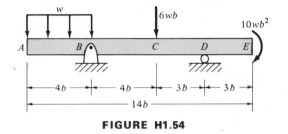

FIGURE H1.54

1.55 Determine the reaction at A (a vertical force and a couple), the shear at hinge B and the reaction at C acting on the beam shown in Figure H1.55. Choose an origin at A and write the shear and moment equations for the beam.

1.9 Combined Loadings and Associated Diagrams

Axial, torsional, and transverse loadings associated with shear and moment have been discussed separately in previous sections of this chapter, but frequently the engineer encounters combinations of these loadings. In such cases, the individual

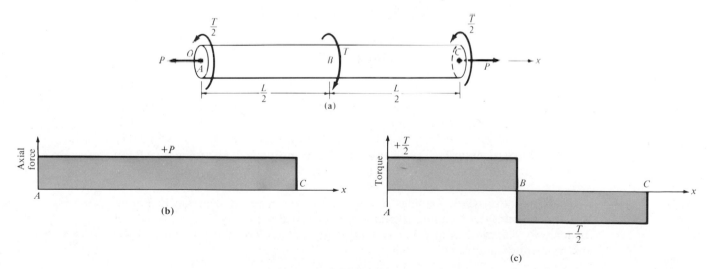

FIGURE 1.23

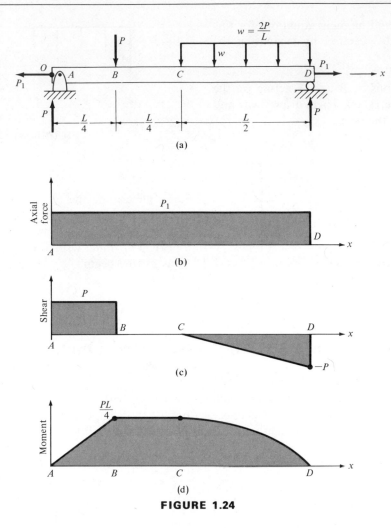

FIGURE 1.24

effects are each considered to act separately and the appropriate diagram constructed for axial force, torque, shear, or moment.

A member subjected to both axial forces and torques is depicted in Figure 1.23(a). The axial force diagram for this member is shown in Figure 1.23(b). Figure 1.23(c) shows the torque diagram that depicts the variation of internal torque along the member.

Axial and transverse loadings are applied to the member depicted in Figure 1.24(a). The transverse loadings P and w are associated with shear and moment and the force P_1 is applied axially. Figure 1.24(b), (c), and (d) show the axial force, shear and moment diagrams, respectively.

In Figure 1.25(a), the member $ABCDE$ is subjected to the axial force Q, torque T, and transverse loads P and w associated with shear and moment. The axial force diagram is shown in Figure 1.25(b) and the torque, shear, and moment diagrams are shown in Figure 1.25(c), (d), and (e), respectively.

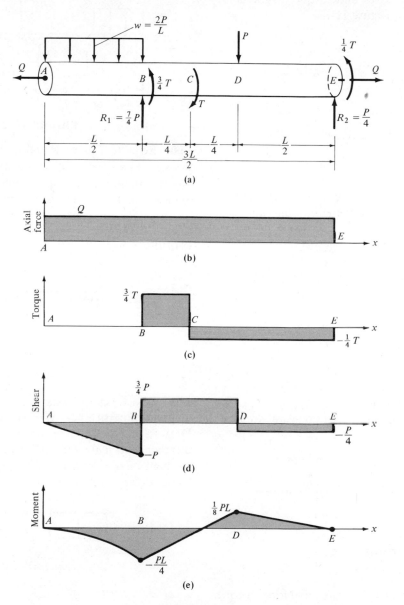

FIGURE 1.25

Homework Problems

1.56 Draw the axial force and torque diagrams for the member shown in Figure H1.56.

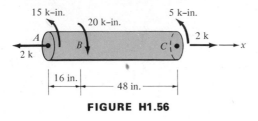

FIGURE H1.56

1.57 Compute the reactions at A and C acting on the loaded member shown in Figure H1.57. Construct the axial force, shear, and moment diagrams for ABC.

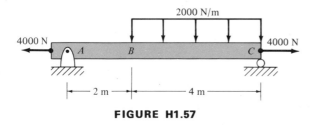

FIGURE H1.57

1.58 Compute the reactions at A and C acting on the loaded member shown in Figure H1.58. Construct the axial force, shear, and moment diagrams for ABC.

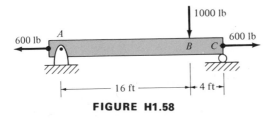

FIGURE H1.58

1.59 Compute R_1 and R_2. Draw axial force, torque, shear, and moment diagrams for the member shown in Figure H1.59.

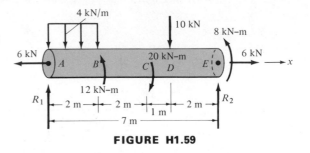

FIGURE H1.59

1.60 Refer to Figure H1.60 and compute the reactions at A and D. Construct the axial force, shear, and moment diagrams for this member. Draw a free-body diagram of the left half of this member and calculate the axial force, shear, and moment at the center section. Compare your results to those shown on your diagrams.

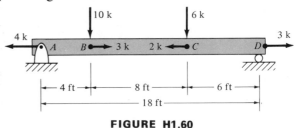

FIGURE H1.60

1.61 Construct the axial force, shear, and moment diagrams for member ABC shown in Figure H1.61. Draw a free-body diagram of segment AB of this member and calculate the axial force, shear, and moment at the cut section. Compare your results to those shown on your diagrams.

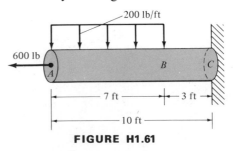

FIGURE H1.61

1.62 Find the reactions at A and C acting on the beam depicted in Figure H1.62. Draw axial force, shear, and moment diagrams for this beam. Cut a vertical section 5 ft to the right of A and compute the axial force, shear, and moment acting at this section by considering both left and right free-body diagrams. Compare these values to those shown on your diagrams.

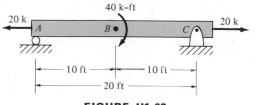

FIGURE H1.62

1.63 Draw the torque, shear, and moment diagrams for the member AB depicted in Figure H1.63. Sketch the member and clearly show the reactions acting at B.

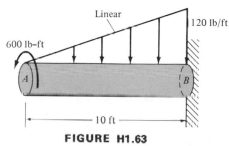

FIGURE H1.63

1.64 Determine the reactions acting at A and C on the member depicted in Figure H1.64. Construct axial

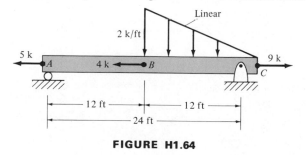

FIGURE H1.64

force, shear, and moment diagrams for ABC. Draw a free-body diagram consisting of segment AB by passing a vertical section through the member just to the left of point B and compute the axial force, shear, and moment at this section. Compare these values to those shown on your diagrams.

1.65 The bar shown in Figure H1.65 consists of two segments, A and B, welded together. Segment A weighs 10 lb/ft and B weighs 20 lb/ft. Choose an origin at the bottom of the bar and direct a positive x axis upward. Plot axial force, shear, and moment functions versus the longitudinal coordinate x for this bar.

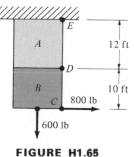

FIGURE H1.65

1.66 For the member depicted in Figure H1.66, find the reactions at B and C. Construct axial force, shear, and moment diagrams. Draw a free-body diagram consisting of a 3-ft-long segment to the right of A and compute the axial force, shear, and moment acting on this vertical section.

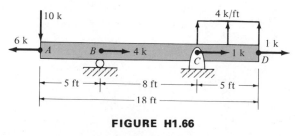

FIGURE H1.66

1.67 Construct axial force, torque, shear, and moment diagrams for the beam depicted in Figure H1.67. Use a left free-body diagram 1 m long to

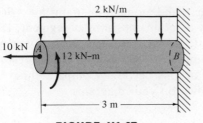

FIGURE H1.67

determine the axial force, shear, and moment acting on a vertical section. Relate these values to your diagrams.

1.68 Is it possible to determine the vertical reaction at A and the cable tension in order to maintain the equilibrium of the member shown in Figure H1.68? If not, revise the magnitude of the horizontal force of 4 k in order to maintain the body in equilibrium. Draw axial force, shear, and moment diagrams for this member ABC. Suppose that the sense of the force applied at A were revised to act to the right; would equilibrium be maintained?

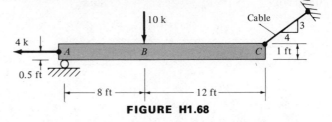

FIGURE H1.68

Chapter 2 Stress, Strain, and Their Relationships

2.1 Introduction

The concepts of stress and strain are two of the most important concepts within the subject of mechanics of materials or mechanics of deformable bodies. They are discussed in detail in this chapter, particularly as they relate to two-dimensional situations.

In the case of stress as well as in the case of strain, emphasis is placed on the use of the semigraphical procedure known as the Mohr's circle solution. The underlying mathematical concepts leading to Mohr's circle are developed and discussed. Examples are solved to illustrate the use of this powerful semigraphical method of solution, which will be used throughout this book whenever problems are encountered dealing with stress and strain analysis.

The treatment given to the concepts of stress and strain in this book differs from that in other books in several important respects, the most significant of which is the fact that the sign convention adopted for strain is compatible with the sign convention for stress as they relate to the construction of the corresponding Mohr's circles. This approach is advantageous in that it makes the construction of the stress and strain Mohr's circles identical.

The discussion relating stress to strain in this chapter is limited to the range of material behavior within which the strain varies linearly with stress. This procedure frees the students from information which, although very important, is extraneous for the time being. A more complete discussion of material behavior is provided in Chapters 3 and 4.

2.2 Concept of Stress at a Point

If a body is subjected to external forces, a system of internal forces is developed. These internal forces tend to separate or bring closer together the material particles that make up the body. Consider, for example, the body shown in Figure 2.1(a), which is subjected to the external forces $F_1, F_2, \ldots, F_i$. Consider an imaginary plane

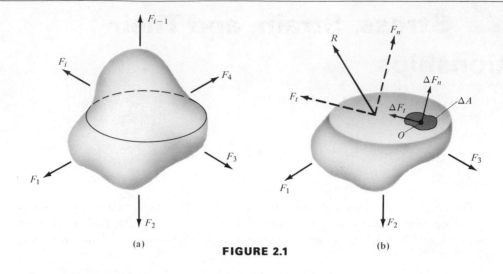

FIGURE 2.1

that cuts the body into two parts, as shown. Internal forces are transmitted from one part of the body to the other through this imaginary plane. Let the free-body diagram of the lower part of the body be constructed as shown in Figure 2.1(b). The forces F_1, F_2, and F_3 are held in equilibrium by the action of an internal system of forces distributed in some manner through the surface area of the imaginary plane. This system of internal forces may be represented by a single resultant force R and/or by a couple. For the sake of simplicity in introducing the concept of stress, only the force R is assumed to exist. In general, the force R may be decomposed into a component F_n, perpendicular to the plane and known as the *normal force*, and a component F_t, parallel to the plane and known as the *shear force*.

If the area of the imaginary plane is to be A, then F_n/A and F_t/A represent, respectively, average values of normal and shear forces per unit area called *stresses*. These stresses, however, are not, in general, uniformly distributed throughout the area under consideration, and it is therefore desirable to be able to determine the magnitude of both the normal and shear stresses at any point within the area. If the normal and shear forces acting over a differential element of area ΔA in the neighborhood of point O are ΔF_n and ΔF_t, respectively, as shown in Figure 2.1(b), then the normal stress σ and the shearing stress τ are given by the following expressions:

$$\sigma = \lim_{\Delta A \to 0} \frac{\Delta F_n}{\Delta A}$$

$$\tau = \lim_{\Delta A \to 0} \frac{\Delta F_t}{\Delta A}$$

$$(2.1)$$

In the special case where the components F_n and F_t are uniformly distributed over the entire area A, then $\sigma = F_n/A$ and $\tau = F_t/A$.

Note that a normal stress acts in a direction perpendicular to the plane on which it acts and it can be either tensile or compressive. A *tensile* normal stress is one that tends to pull the material particles away from each other, while a *compressive* normal stress is one that tends to push them closer together. A shear stress, on the other hand, acts parallel to the plane on which it acts and tends to slide (shear) adjacent planes with respect to each other. Also note that the units of stress (σ or τ) consist of units of force divided by units of area. Thus, in the British gravitational system of measure, such units as pounds per square inch (psi) and kilopounds per square inch (ksi) are common. In the metric (SI) system of measure, the unit that has been proposed for stress is the Newton per square meter (N/m^2), which is called the pascal and denoted by the symbol Pa. Because the pascal is a very small quantity, another SI unit that is widely used is the megapascal (10^6 pascals) and is denoted by the symbol MPa. This unit may also be written as MN/m^2.

2.3 Components of Stress

In the most general case, normal and shear stresses at a point in a body may be considered to act on three mutually perpendicular planes. This most general state of stress is usually referred to as *triaxial*. It is convenient to select planes that are normal to the three coordinates axes x, y, and z and designate them as the X, Y, and Z planes, respectively. Consider these planes as enclosing a differential volume of material in the neighborhood of a given point in a stressed body. Such a volume of material is depicted in Figure 2.2 and is referred to as a *three-dimensional stress element*. On each of the three mutually perpendicular planes of the stress element, there acts a normal stress, and a shear stress which is represented by its two perpendicular components.

The notation for stresses used in this text consists of affixing one subscript to a normal stress, indicating the plane on which it is acting, and two subscripts to a shear stress, the first of which designates the plane on which it is acting and the second its direction. For example, σ_x is a normal stress acting on the X plane, τ_{xy} is a shear stress acting on the X plane and pointed in the positive y direction, and τ_{xz} is a shear stress acting in the X plane and pointed in the positive z direction.

It is observed from Figure 2.2 that three stress components exist on each of the three mutually perpendicular planes that define the stress element. Thus there exists a total of nine stress components that must be specified in order to define completely the state of stress at any point in the body. By considerations of the equilibrium of the

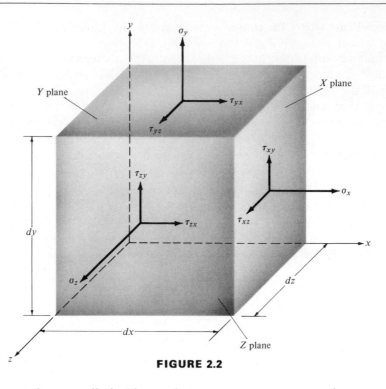

FIGURE 2.2

stress element, it can easily be shown that $\tau_{xy} = \tau_{yx}$, $\tau_{xz} = \tau_{zx}$, and $\tau_{yz} = \tau_{zy}$ so that the number of stress components required to completely define the state of stress at a point is reduced to six.

By convention, a normal stress is positive if it points in the direction of the outward normal to the plane. Thus a positive normal stress produces tension and a negative normal stress produces compression. A component of shear stress is positive if it is pointed along the positive direction of the coordinate axis and if the outward normal to its plane is also in the positive direction of the corresponding axis. If, however, the outward normal is in the negative direction of the coordinate axis, a positive shear stress will also be in the negative direction of the corresponding axis. The stress components shown in Figure 2.2 are all positive. It should be noted, however, that such a sign convention for shear stress is rather cumbersome. It is only used in the analysis of triaxial stress problems that are usually dealt with in advanced courses such as the theory of elasticity.

A complete study of the triaxial or three-dimensional state of stress is beyond the scope of this chapter, and the analysis that follows is limited to the special case in which the stress components in one direction are all zero. For example, if all the stress components in the z direction are zero (i.e., $\tau_{xz} = \tau_{yz} = \sigma_z = 0$), the stress condition reduces to a biaxial or two-dimensional state of stress in the xy plane. This state of stress is referred to as *plane stress*. Fortunately, many of the problems encountered in practice are such that they can be considered plane stress problems.

2.4 Analysis of Plane Stress

As mentioned previously, the state of stress known as plane stress is one in which all the stress components in one direction vanish. Thus, if it is assumed that all the components in the z direction shown in Figure 2.2 are zero (i.e., $\tau_{xz} = \tau_{yz} = \sigma_z = 0$), the stress element shown in Figure 2.3(a) is obtained and it is the most general plane stress condition that can exist. It should be observed that a stress element is in reality a schematic representation of two sets of perpendicular planes passing through a point and that the element degenerates into a point in the limit when both dx and dy approach zero.

From considerations of the equilibrium of forces on the stress element in Figure 2.3(a), it can be shown that $\tau_{xy} = \tau_{yx}$. Thus assume that the depth of the stress element into the paper is a constant equal to h. Since, by definition, force is the product of stress and the area over which it acts, then a summation of moments of all forces about a z axis through point O leads to the following equation:

$$\tau_{xy}(h \, dy) \, dx - \tau_{yx}(h \, dx) \, dy = 0$$

from which

$$\tau_{xy} = \tau_{yx} \qquad\qquad (2.2)$$

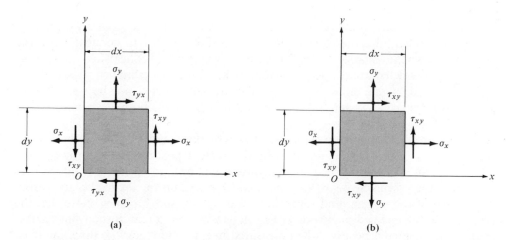

(a)　　　　　　　　　　(b)

FIGURE 2.3

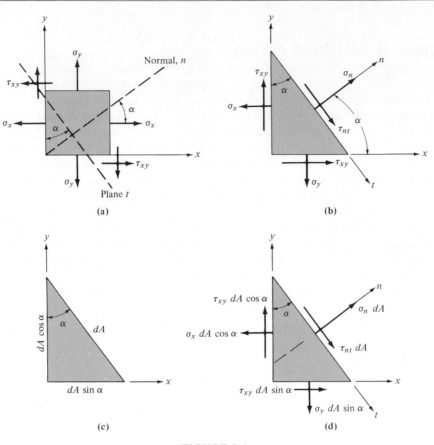

FIGURE 2.4

Note that the forces produced by the normal stresses in both the x and y directions cancel each other, and therefore their moments did not have to be included in the preceding moment equation. Also, Eq. 2.2 states a very useful conclusion—that shear stresses on any two mutually perpendicular planes through a point in a stressed body must be equal in magnitude. With this fact in mind, it will not be necessary to distinguish between τ_{xy} and τ_{yx}, and henceforth they will both be labeled τ_{xy} as shown in Figure 2.3(b).

It is desirable to be able to relate the stresses on the X and Y planes to the stresses acting on any inclined plane t defined by the angle α, positive ccw as shown in the stress element of Figure 2.4(a). The resulting relations would make it possible to determine the normal stress σ_n and the shear stress τ_{nt} on the inclined plane t from knowledge of the normal and shear stresses on the X and Y planes. Note that the normal n to the inclined plane t makes an angle α with the x axis. By convention, this angle α is considered positive when measured in a counterclockwise direction from the positive end of the x axis.

In analyzing plane stress problems, it is most convenient to use a sign convention for shear stresses different from the one discussed earlier for a triaxial stress condition. According to this sign convention, which is the one used throughout this text, a shear stress is positive if it produces clockwise rotation of the element on which it is acting and negative if it produces counterclockwise rotation. Thus τ_{xy} on the X plane in Figure 2.4(a) is positive while τ_{xy} on the Y plane (i.e., τ_{yx}) is negative. However, as was stated earlier, normal stresses are positive if tensile and negative if compressive.

Isolate the small wedge to the left of the inclined plane and construct its free-body diagram as shown in Figure 2.4(b). If one assumes the area of the inclined plane to be dA square units, then the area of the X plane would be $dA \cos \alpha$ square units and the area of the Y plane would be $dA \sin \alpha$ square units, as shown in Figure 2.4(c). Thus the forces on the three faces of the wedge produced by the normal and shear stresses acting on them can be determined in terms of dA and the trigonometric functions of the angle α, as shown in Figure 2.4(d).

Summation of forces in Figure 2.4(d) along the direction of σ_n leads to the equation

$$\sigma_n \, dA - \sigma_x(dA \cos \alpha) \cos \alpha + \tau_{xy}(dA \cos \alpha) \sin \alpha$$
$$+ \tau_{xy}(dA \sin \alpha) \cos \alpha - \sigma_y(dA \sin \alpha) \sin \alpha = 0$$

from which

$$\sigma_n = \sigma_x \cos^2 \alpha + \sigma_y \sin^2 \alpha - 2\tau_{xy} \sin \alpha \cos \alpha \qquad (2.3)$$

Using the trigonometric identities $\sin^2 \alpha = (1 - \cos 2\alpha)/2$, $2 \sin \alpha \cos \alpha = \sin 2\alpha$, and $\cos^2 \alpha = (1 + \cos 2\alpha)/2$, we can reduce Eq. 2.3 to the form

$$\sigma_n = \tfrac{1}{2}(\sigma_x + \sigma_y) + \tfrac{1}{2}(\sigma_x - \sigma_y) \cos 2\alpha - \tau_{xy} \sin 2\alpha \qquad (2.4)$$

Similarly, summing forces along the direction of τ_{nt} leads to the equation

$$\tau_{nt} = (\sigma_x - \sigma_y) \sin \alpha \cos \alpha + \tau_{xy}(\cos^2 \alpha - \sin^2 \alpha) \qquad (2.5)$$

which, after using the trigonometric identities above, reduces to

$$\tau_{nt} = \tfrac{1}{2}(\sigma_x - \sigma_y) \sin 2\alpha + \tau_{xy} \cos 2\alpha \qquad (2.6)$$

Thus, for a plane stress condition, the normal and shear stresses may be found from Eqs. 2.4 and 2.6, respectively, for any plane defined by the angle α if the stresses σ_x, σ_y, and τ_{xy} are known.

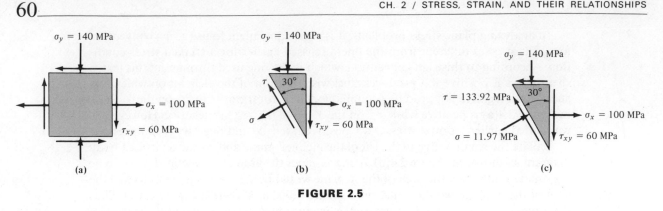

FIGURE 2.5

Example 2.1

At a point in a body, the stress condition is known to be as shown in Figure 2.5(a) (i.e., plane stress). Determine the normal and shear stresses on a plane inclined to the X plane through an angle 30° ccw (a) using equilibrium conditions, and (b) using Eqs. 2.4 and 2.6.

Solution

(a) Figure 2.5(b) shows an isolated stress element containing the plane of interest (i.e., 30° ccw from the X plane). It is assumed that both a normal and shear stress exist on this plane and the sense of these stresses is assumed as shown in Figure 2.5(b). Let the area of the inclined plane be dA. Then the area of the X plane is $dA \cos 30 = 0.866dA$ and that of the Y plane is $dA \sin 30 = 0.500dA$. Summing forces in the direction of σ, we obtain the following equation:

$$\sigma dA + \sigma_y(0.500dA) \sin 30 + \tau_{xy}(0.500dA) \cos 30$$
$$+ \tau_{xy}(0.866dA) \sin 30 - \sigma_x(0.866dA) \cos 30 = 0$$

Substituting the values of sin 30° and cos 30° and those for σ_x, σ_y, and τ_{xy} into the equation above, eliminating dA from every term, and simplifying, we obtain $\sigma = -11.97$ MPa. The negative sign indicates that the normal stress is opposite to the direction assumed in the solution, and therefore it is a compressive stress.

Also, summing forces in the direction of τ, we obtain the equilibrium equation

$$\tau dA - \sigma_y(0.500dA) \cos 30 + \tau_{xy}(0.500dA) \sin 30$$
$$- \tau_{xy}(0.866dA) \cos 30 - \sigma_x(0.866dA) \sin 30 = 0$$

which leads to $\tau = 133.92$ MPa. The positive sign indicates that the assumed sense for τ is correct.

Thus, on the inclined plane of interest, the following stresses are acting:

$$\sigma = -11.97 \text{ MPa}$$

$$\tau = 133.92 \text{ MPa}$$

as shown in Figure 2.5(c).

(b) In solving stress problems using Eqs. 2.4 and 2.6, care must be exercised to ensure that all quantities are entered with their correct signs. Thus, according to the sign convention previously established, σ_x is positive, σ_y is negative, τ_{xy} is positive, and α is positive. Therefore, from Eq. 2.4,

$$\sigma = \frac{100 - 140}{2} + \frac{1}{2}(100 + 140)\cos 60 - (60 \sin 60)$$

$$= \boxed{-11.97 \text{ MPa}}$$

and from Eq. 2.6,

$$\tau = \tfrac{1}{2}(100 + 140)\sin 60 + 60 \cos 60$$

$$= \boxed{133.92 \text{ MPa}}$$

The negative sign on the normal stress signifies compression and the positive sign on the shear stress indicates that it has to produce clockwise rotation of the stress element. These conclusions are, of course, identical to those obtained in part (a) of the solution and the reader is once again referred to Figure 2.5(c) for a complete sketch of the stress element containing the plane of interest.

2.5 Principal Stresses and Maximum Shear Stresses

It is observed from Eqs. 2.4 and 2.6 that both the normal and shear stresses depend upon the orientation of the plane defined by the angle α. The normal stress assumes a maximum value for a certain value of α and a minimum value for $\alpha + 90$.

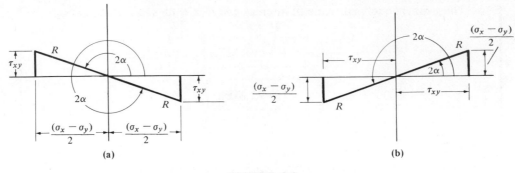

FIGURE 2.6

Thus differentiating σ_n in Eq. 2.4 with respect to α, setting the result equal to zero, and solving for 2α, one obtains the orientation of the plane for which σ_n is either a maximum or a minimum. Thus

$$2\alpha = \tan^{-1} \frac{-2\tau_{xy}}{\sigma_x - \sigma_y} \tag{2.7}$$

Equation 2.7 defines two values of 2α differing by 180°, or two values of α differing by 90°. On one of these two planes, the normal stress is a maximum, and on the second, the normal stress is a minimum. These planes are known as *principal planes* and the normal stresses acting on them as *principal stresses*. Thus principal stresses are normal stresses acting on principal planes. In the following development, it will be shown that principal planes are free of any shear stresses, and therefore another way of defining principal streses is to say that they are normal stresses acting on planes on which the shear stresses are zero. It is very important to note that principal planes are perpendicular to the xy plane (i.e., parallel to the z axis).

Having the values of 2α that define the principal planes, we can determine the values of the principal stresses from Eq. 2.4. These principal stresses will be denoted by σ_1 and σ_2.* Without loss of generality, the magnitude of σ_x may be assumed to be larger than that of σ_y, in which case Eq. 2.7 defines angles whose tangents are negative. Such angles occur in the second and fourth quadrants as shown in Figure 2.6(a), where the symbol R represents the quantity

$$\left[\left(\frac{\sigma_x - \sigma_y}{2} \right)^2 + (\tau_{xy})^2 \right]^{1/2}$$

*As will be shown in a later section, the two principal stresses in a plane may be designated as σ_1 and σ_2, σ_1 and σ_3, or σ_2 and σ_3, depending upon conditions existing in the plane.

From the geometry in Figure 2.6(a), it can be shown that

$$\sin 2\alpha = \frac{\pm \tau_{xy}}{R} \quad \text{and} \quad \cos 2\alpha = \frac{\mp (\sigma_x - \sigma_y)}{2R}$$

where the upper signs correspond to the angle in the second quadrant and the lower signs to the angle in the fourth quadrant. Substitution of these quantities into Eq. 2.4 leads to the two principal stresses as follows:

$$\sigma_1 = \frac{(\sigma_x + \sigma_y)}{2} + R = \sigma_A + R$$

$$\sigma_2 = \frac{(\sigma_x + \sigma_y)}{2} - R = \sigma_A - R$$

(2.8)

where the principal stress σ_1, by convention, is defined as the algebraic maximum and the principal stress σ_2 is defined as the algebraic minimum and the symbol σ_A is used to denote the average of the two normal stresses σ_x and σ_y.

The vanishing of the shear stress on principal planes can be shown easily by substituting the values obtained above for $\sin 2\alpha$ and $\cos 2\alpha$ into Eq. 2.6. For example, if the lower signs are used (those corresponding to the angle in the fourth quadrant), Eq. 2.6 becomes

$$\tau = -\frac{1}{2}\frac{(\sigma_x - \sigma_y)\tau_{xy}}{R} + \frac{\tau_{xy}(\sigma_x - \sigma_y)}{2R} = 0$$

Differentiating τ_{nt} in Eq. 2.6 with respect to α and setting the result equal to zero leads to the orientation of planes for which τ_{nt} is either a maximum or minimum. Thus

$$2\alpha = \tan^{-1} \frac{(\sigma_x - \sigma_y)}{2\tau_{xy}}$$

(2.9)

Equation 2.9 defines two values of 2α differing by 180°, or two values of α differing by 90°. Corresponding to one of these two angles, τ_{nt} is an algebraic maximum, and corresponding to the second, τ_{nt} is an algebraic minimum.

Equation 2.9 defines two angles whose tangents are positive. Such angles occur in the first and third quadrants as shown in Figure 2.6(b), where the symbol R represents the same quantity given previously. From the geometry in Figure 2.6(b), it can be seen that

$$\sin 2\alpha = \frac{\pm (\sigma_x - \sigma_y)}{2R} \quad \text{and} \quad \cos 2\alpha = \pm \frac{\tau_{xy}}{R}$$

where the upper signs correspond to the angle 2α in the first quadrant and the lower signs to the angle 2α in the third quadrant. Substitution of these values into Eq. 2.6 yields the algebraic maximum and algebraic minimum shear stresses, τ_1 and τ_2, respectively. Thus

$$\tau_1 = +R$$
$$\tau_2 = -R$$

(2.10)

It should be noted that the planes of maximum and minimum shear stresses (i.e., the planes of τ_1 and τ_2) contain a normal stress whose magnitude can be determined from Eq. 2.4 by substituting the values of $\sin 2\alpha$ and $\cos 2\alpha$ as obtained above. If this is done, we obtain the following relation for the normal stress σ on the planes of τ_1 and τ_2. Thus

$$\sigma = \frac{\sigma_x + \sigma_y}{2} = \sigma_A$$

(2.11)

It should also be emphasized that, like the principal planes, the planes of τ_1 and τ_2 are perpendicular to the xy plane (i.e., parallel to the z axis).

Comparison of Eqs. 2.7 and 2.9 shows that the angles 2α defined by these two equations have tangents that are negative reciprocal of each other, and therefore the angle 2α defined by Eq. 2.7 is 90° away from the angle 2α defined by Eq. 2.9. In other words, the principal planes are inclined to the planes of maximum or minimum shear by a 45° angle. Alternatively, the planes of τ_1 and τ_2 bisect the 90° angles between the two principal stresses in the xy plane. This important geometric relation is further described and illustrated by means of Mohr's circle, which is discussed in Section 2.6.

Another useful relation is obtained by adding the values of σ_1 and σ_2 as given by Eq. 2.8. The addition of these two values yields

$$\sigma_1 + \sigma_2 = \sigma_x + \sigma_y$$

(2.12)

Equation 2.12 states that the sum of the normal stresses on any two orthogonal planes through a point in a stressed body is a constant.

Example 2.2
At a point in a stressed body, there exists a plane stress condition as depicted in Figure 2.7(a). Determine (a) the principal stresses and the maximum and minimum shear stresses (i.e., τ_1 and τ_2), and (b) the planes on which these stresses act and show the stresses on properly oriented stress elements.

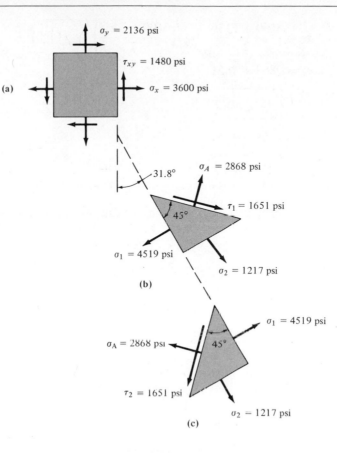

FIGURE 2.7

Solution

(a) According to the previously established sign convention, both σ_x and σ_y are positive. The shear stresses τ_{xy}, however, are negative because they are pointed in directions opposite to those used [see Figure 2.4(a)] for the derivation of the plane stress equations.* By Eq. 2.8,†

---

* Extreme care must be exercised in entering the shear stress quantities with their proper signs into the stress equations. The difficulty with the signs of the shear stresses is partially minimized by the use of Mohr's circle solution, a semigraphical approach named after Otto Mohr (1835–1918), the German engineer who first introduced it.

† Throughout, we shall use a T following a quantity to indicate tension, and a C to indicate compression.

$$\sigma_1 = \frac{3600 + 2136}{2} + \sqrt{\left(\frac{3600 - 2136}{2}\right)^2 + (1480)^2}$$

$$\sigma_1 = 2868 + 1651$$

$$= \boxed{4519 \text{ psi}(T)}$$

$$\sigma_2 = 2868 - 1651$$

$$= \boxed{1217 \text{ psi}(T)}$$

As a check on this part of the solution, we can use Eq. 2.12 to verify that

$$\sigma_x + \sigma_y = 3600 + 2136 = 5736 \text{ psi}$$

is the same as

$$\sigma_1 + \sigma_2 = 4519 + 1217 = 5736 \text{ psi}$$

Also, by Eq. 2.10,

$$\tau_1 = \sqrt{\left(\frac{3600 - 2136}{2}\right)^2 + (1480)^2}$$

$$= \boxed{1651 \text{ psi}}$$

$$\tau_2 = -\sqrt{\left(\frac{3600 - 2136}{2}\right)^2 + (1480)^2}$$

$$= \boxed{-1651 \text{ psi}}$$

It will be shown later that the values obtained for τ_1 and τ_2 in this problem are not, in fact, the absolute maximum values for the shear stress at the point.
(b) To determine the orientation of the principal planes, Eq. 2.7 will be used. Thus

$$2\alpha = \tan^{-1} - \frac{2(-1480)}{3600 - 2136} = \tan^{-1} 2.022$$

$$= 63.7°$$

and

$$\alpha = 31.8°$$

Therefore, the first principal plane is inclined to the X plane at an angle of 31.8° ccw and the second principal plane is perpendicular to the first. Also, the planes on which τ_1 and τ_2 act are inclined at 45° with the principal planes. By Eq. 2.11, the normal stress on the planes of τ_1 and τ_2 becomes

$$\sigma_A = \frac{3600 + 2136}{2} = 2868 \text{ psi}$$

These stresses as well as the geometric relations that exist between principal planes and planes of τ_1 and τ_2 are shown in Figure 2.7(b) and (c), in which the plane of τ_1 is perpendicular to the plane of τ_2.

A very useful relation is obtained when we substract σ_2 from σ_1 in Eq. 2.8. Thus

$$\sigma_1 - \sigma_2 = 2R$$

and

$$R = |\tau_{1,2}| = \tau_{\max} = \frac{\sigma_1 - \sigma_2}{2} \qquad \textbf{(2.13a)}$$

Equation 2.13a indicates that the maximum value of the shear stress τ_1 or τ_2 at a point is equal to one-half the difference between the two principal stresses at the point and acts on two mutually perpendicular planes which are parallel to the z axis and which bisect the 90° angles between these two principal stresses. It should be emphasized, however, that Eq. 2.13a may not yield the absolute value of the maximum shear stress in the element (point) considered as a particle in a three-dimensional body, even though the stresses acting at the point constitute a plane stress condition. Consider, for example, the stressed three-dimensional element shown in Figure 2.8(a) subjected to the three principal stresses σ_1, σ_2, and σ_3, where $\sigma_3 = 0$. Therefore, the element is, in reality, in a two dimensional state of stress or in a plane stress condition. By convention, it is assumed that $\sigma_1 > \sigma_2 > \sigma_3$, algebraically. Thus, since $\sigma_3 = 0$, it follows that, in this case, both σ_1 and σ_2 must be positive.

The state of stress depicted in Figure 2.8(a) can be decomposed into three two-dimensional stress situations. Considering first the plane defined by σ_1 and σ_2 and using Eq. 2.13a, we obtain $\tau_{\max} = (\sigma_1 - \sigma_2)/2$ acting on planes $EBDH$ and $AFGC$ as depicted in Figure 2.8(b) and (c). Similarly, consideration of the plane defined by σ_2 and σ_3 yields $\tau_{\max} = (\sigma_2 - \sigma_3)/2 = \sigma_2/2$ acting on planes $BGHA$ and $FDCE$ as shown in Figure 2.8(d) and (e). Finally, in the plane defined by σ_1 and σ_3, one obtains $\tau_{\max} = (\sigma_1 - \sigma_3)/2 = \sigma_1/2$ acting on planes $BCHF$ and $ADGE$ as illustrated in Figure 2.8(f) and (g). Thus there are three possible maximum values of the shear stress when the element is considered as a three-dimensional body, and the absolute maximum value of the shear stress at the point, $|\tau_{\max}|$, would obviously be the largest absolute

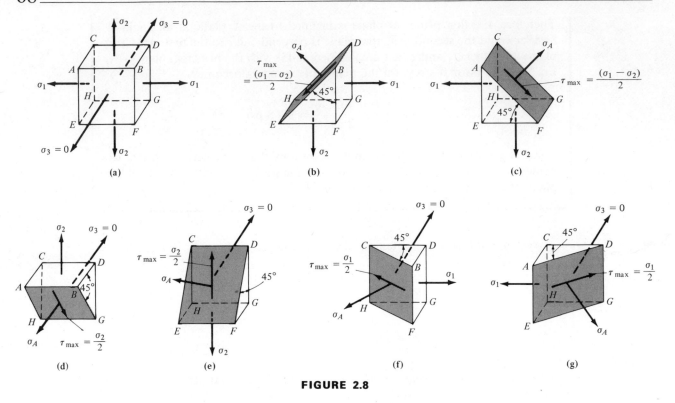

FIGURE 2.8

value among the three values discussed above. In summary, for the plane stress case where both principal stresses are positive, $|\tau_{max}| = (\sigma_1 - \sigma_3)/2$, which reduces to $\sigma_1/2$, since $\sigma_3 = 0$. If, however, the plane stress condition is such that one principal stress is positive while the second is negative, then the labeling of the three principal stresses should be such that σ_1 is positive, $\sigma_2 = 0$, and σ_3 negative in order to satisfy the requirement that, algebraically, $\sigma_1 > \sigma_2 > \sigma_3$. Examining this case for the absolute maximum shear stress as was done previously, we conclude that the same equation, namely $|\tau_{max}| = (\sigma_1 - \sigma_3)/2$, is valid. Finally, if the plane stress condition is such that both principal stresses are negative, it follows that $\sigma_1 = 0$ in order to satisfy the algebraic requirement that $\sigma_1 > \sigma_2 > \sigma_3$. In such a case, once again, $|\tau_{max}| = (\sigma_1 - \sigma_3)/2$. Thus, for the general case of plane stress, the absolute maximum value of the shear stress at a point is given by

$$|\tau_{max}| = \frac{\sigma_1 - \sigma_3}{2} \qquad\qquad \textbf{(2.13b)}$$

if σ_1 and σ_3 are interpreted as the algebraic maximum and algebraic minimum principal stresses, respectively. Furthermore, $|\tau_{max}|$ acts on two mutually perpendicular planes which are parallel to σ_2 and bisect the 90° angles between σ_1 and σ_3. It should be emphasized that the correct use of Eq. 2.13b requires that the three principal stresses be properly labeled according to the algebraic requirements that $\sigma_1 > \sigma_2 > \sigma_3$ even if we are dealing with a plane stress condition for which, by definition, one of the three principal stresses is zero. Thus the three principal stresses must be known in magnitude and in sense before they can be labeled correctly.

Example 2.3

For each of the following cases, values of the three principal stresses are given. Label these principal stresses according to the algebraic requirement that $\sigma_1 > \sigma_2 > \sigma_3$. Use Eq. 2.13b to find $|\tau_{max}|$ in each case and define the plane(s) on which $|\tau_{max}|$ acts: (a) 50 MN/m^2, 100 MN/m^2, 0; (b) -40 MPa, 0, 70 MPa; and (c) -5000 psi, 0, -9000 psi.

Solution

(a) $\sigma_1 = 100 \text{ MN/m}^2$, $\sigma_2 = 50 \text{ MN/m}^2$, and $\sigma_3 = 0$. Thus, by Eq. 2.13b,

$$|\tau_{max}| = \frac{\sigma_1 - \sigma_3}{2} = \frac{100 - 0}{2}$$

$$= \boxed{50 \text{ MN/m}^2}$$

The planes on which $|\tau_{max}|$ acts are parallel to the σ_2 axis and make 45° angles with the planes of σ_1 and σ_3 as shown in Figure 2.9(a) and (b). Also, on the plane of $|\tau_{max}|$, there is a normal stress $\sigma_A = (\sigma_1 + \sigma_3)/2 = (100 + 0)/2 = 50$ MN/m^2.

(b) $\sigma_1 = 70 \text{ MPa}$, $\sigma_2 = 0$, $\sigma_3 = -40 \text{ MPa}$. By Eq. 2.13b,

$$|\tau_{max}| = \frac{\sigma_1 - \sigma_3}{2} = \frac{70 + 40}{2}$$

$$= \boxed{55 \text{ MPa}}$$

The planes on which $|\tau_{max}|$ acts are parallel to the σ_2 axis and make 45° angles with the planes of σ_1 and σ_3, as shown in Figure 2.9(c) and (d). As in the previous case,

$$\sigma_A = \frac{\sigma_1 + \sigma_3}{2} = \frac{70 - 40}{2} = 15 \text{ MPa}$$

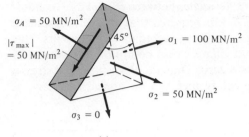

$\sigma_A = 50 \text{ MN/m}^2$

$|\tau_{\max}| = 50 \text{ MN/m}^2$

$45°$

$\sigma_1 = 100 \text{ MN/m}^2$

$\sigma_2 = 50 \text{ MN/m}^2$

$\sigma_3 = 0$

(a)

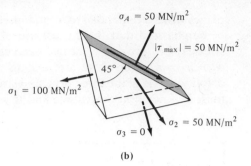

$\sigma_A = 50 \text{ MN/m}^2$

$|\tau_{\max}| = 50 \text{ MN/m}^2$

$45°$

$\sigma_1 = 100 \text{ MN/m}^2$

$\sigma_2 = 50 \text{ MN/m}^2$

$\sigma_3 = 0$

(b)

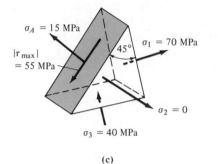

$\sigma_A = 15 \text{ MPa}$

$|\tau_{\max}| = 55 \text{ MPa}$

$45°$

$\sigma_1 = 70 \text{ MPa}$

$\sigma_2 = 0$

$\sigma_3 = 40 \text{ MPa}$

(c)

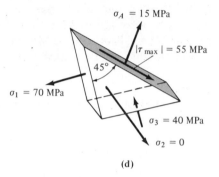

$\sigma_A = 15 \text{ MPa}$

$|\tau_{\max}| = 55 \text{ MPa}$

$45°$

$\sigma_1 = 70 \text{ MPa}$

$\sigma_3 = 40 \text{ MPa}$

$\sigma_2 = 0$

(d)

$\sigma_A = 4500 \text{ psi}$

$|\tau_{\max}| = 4500 \text{ psi}$

$45°$

$\sigma_1 = 0$

$\sigma_2 = 5000 \text{ psi}$

$\sigma_3 = 9000 \text{ psi}$

(e)

FIGURE 2.9

$\sigma_A = 4500 \text{ psi}$

$|\tau_{\max}| = 4500 \text{ psi}$

$45°$

$\sigma_1 = 0$

$\sigma_2 = 5000 \text{ psi}$

$\sigma_3 = 9000 \text{ psi}$

(f)

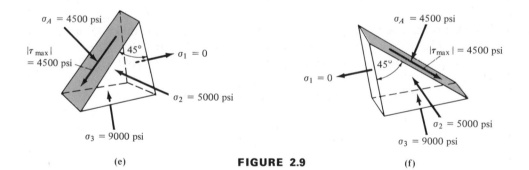

(c) $\sigma_1 = 0$, $\sigma_2 = -5000$ psi, $\sigma_3 = -9000$ psi. Therefore, by Eq. 2.13b,

$$|\tau_{\max}| = \frac{\sigma_1 - \sigma_3}{2} = \frac{0 + 9000}{2}$$

$$= \boxed{4500 \text{ psi}}$$

As in the first two cases, the planes on which $|\tau_{max}|$ acts are parallel to the σ_2 axis and make $45°$ angles with the planes of σ_1 and σ_3 as shown in Figure 2.9(e) and (f). Also on these planes,

$$\sigma_A = \frac{\sigma_1 + \sigma_3}{2} = \frac{0 - 9000}{2}$$

$$= -4500 \text{ psi}$$

2.6 Mohr's Circle for Plane Stress

As was pointed out in a footnote previously, a semigraphical procedure known as Mohr's circle solution is available for the analysis of plane stress problems. Not only does Mohr's circle minimize the difficulty with the sign of the shear stresses, but it also simplifies the solution of Eqs. 2.4 and 2.6.

Thus, if the first term in the right-hand side of Eq. 2.4 is transported and the square of the resulting equation is added to the square of Eq. 2.6, we obtain

$$\left(\sigma_n - \frac{\sigma_x + \sigma_y}{2}\right)^2 + (\tau_{nt})^2 = \left(\frac{\sigma_x - \sigma_y}{2}\right)^2 + (\tau_{xy})^2 \qquad \textbf{(2.14)}$$

which is the equation of a circle in the σ_n–τ_{nt} plane of radius

$$R = \left[\left(\frac{\sigma_x - \sigma_y}{2}\right)^2 + (\tau_{xy})^2\right]^{1/2}$$

and having its center along the σ_n axis at a distance of $(\sigma_x + \sigma_y)/2$ from the origin.

Consequently, Eq. 2.14 shows that the values of the normal and shear stresses on any plane passing through a point in a body may be represented by the coordinates of a point on the circle defined by this equation. Thus, if one considers a plane stress situation as shown in Figure 2.10(a), the stress field at this point is represented by the coordinates of points on the circumference of a circle defined by Eq. 2.14 and depicted in Figure 2.10(b). As mentioned previously, such a circle is known as *Mohr's circle* after the famous German engineer Otto Mohr. If the values of σ_x, σ_y, and τ_{xy} are known, this circle may be constructed in one of two ways:

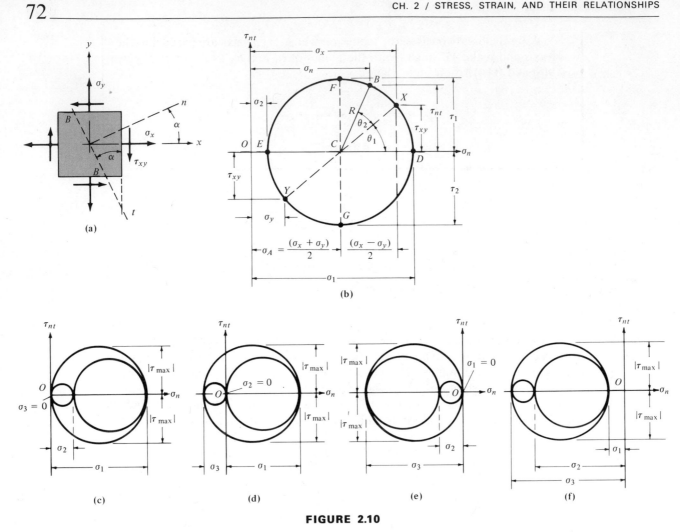

FIGURE 2.10

1. On the σ_n axis, locate the center C at a distance of $(\sigma_x + \sigma_y)/2$, and construct a circle with a radius

$$R = \left[\left(\frac{\sigma_x - \sigma_y}{2}\right)^2 + (\tau_{xy})^2\right]^{1/2}$$

as shown in Figure 2.10(b). This method, however, is not as useful for problem solving as the second method, which is developed and described in the following paragraphs.

2. As shown in Figure 2.10(a), when $\alpha = 0$, the n axis coincides with the x axis and the t axis with the y axis. Therefore, the stresses on the plane for which the x axis is

the normal (i.e., the X plane) are $\sigma_n = \sigma_x$ and $\tau_{nt} = \tau_{xy}$. These two values define point X on the circumference of Mohr's circle as shown in Figure 2.10(b). For any arbitrary value of α defining plane $B\text{–}B$, the stress condition on such a plane is represented by point B in Figure 2.10(b), whose coordinates are σ_n and τ_{nt}. From the geometry in Figure 2.10(b), we conclude that

$$\sin \theta_1 = \frac{\tau_{xy}}{R} \tag{2.15a}$$

$$\cos \theta_1 = \frac{\sigma_x - \sigma_y}{2R} \tag{2.15b}$$

$$\sin (\theta_1 + \theta_2) = \frac{\tau_{nt}}{R} \tag{2.15c}$$

Substituting for τ_{nt} in Eq. 2.15c its value from Eq. 2.6, we obtain

$$\sin (\theta_1 + \theta_2) = \frac{\frac{1}{2}(\sigma_x - \sigma_y) \sin 2\alpha + \tau_{xy} \cos 2\alpha}{R} \tag{2.15d}$$

From trigonometry,

$$\sin (\theta_1 + \theta_2) = \cos \theta_1 \sin \theta_2 + \sin \theta_1 \cos \theta_2 \tag{2.15e}$$

Substitution of Eqs. 2.15a and 2.15b into Eq. 2.15e yields

$$\sin (\theta_1 + \theta_2) = \frac{\frac{1}{2}(\sigma_x - \sigma_y) \sin \theta_2 + \tau_{xy} \cos \theta_2}{R} \tag{2.15f}$$

Therefore, by comparing Eqs. 2.15d and 2.15f, we conclude that

$$\sin \theta_2 = \sin 2\alpha \quad \text{and} \quad \cos \theta_2 = \cos 2\alpha \tag{2.15g}$$

which indicates that $\theta_2 = 2\alpha$. Thus any angle in the actual stress element is doubled when represented by Mohr's circle. Furthermore, θ_2 is the same direction (cw or ccw) as α.

When $\alpha = 90°$ in Figure 2.10(a), the n axis coincides with the y axis and the t axis with the x axis. Therefore, the stresses on the plane for which the y axis is normal (i.e., the Y plane) are $\sigma_n = \sigma_y$ and $\tau_{nt} = -\tau_{xy}$. The shear stress τ_{xy} is negative because it produces counterclockwise rotation of the element. These stresses are represented in Figure 2.10(b) by point Y, which, because of the conclusion that $\theta_2 = 2\alpha$, must be diametrically opposite to point X, since the X plane is perpendicular to the Y plane.

The foregoing conclusions may be utilized to advantage in the construction of Mohr's circle from the stress element without having to compute the radius R and the location of the center. Thus, after establishing a convenient $\sigma_n - \tau_{nt}$ coordinate system, Mohr's circle representing the stress condition depicted in Figure 2.10(a), for example, can be constructed quickly by locating two diametrically opposite points on its circumference. The straight line connecting these two points (a diameter of the circle) intersects the σ_n axis at a point that is the center of the circle. One of the two points needed is obtained from the stress condition on the X plane (i.e., $+\sigma_x$ and $+\tau_{xy}$), which locates point X in Figure 2.10(b). The second point is obtained from the stress condition on the Y plane (i.e., $+\sigma_y$ and $-\tau_{xy}$), which locates point Y in Figure 2.10(b). The line X–Y intersects the σ_n axis at point C, which is the center of Mohr's circle. With this center and radius $CX = CY$, we can complete the required construction. It should be pointed out that the construction of Mohr's circle can be executed in a free-hand manner, since the solution of a given problem can be accomplished, as will be shown in Example 2.4, by a mathematical analysis of the geometry involved.

Once the circle has been constructed, the stresses (both normal and shear) on any plane defined by the angle α may be determined. For example, the stress condition on plane B–B in Figure 2.10(a) is depicted by point B in Figure 2.10(b), where, since α is counterclockwise from the X plane, $\theta_2 = 2\alpha$ was measured counterclockwise from point X. The abscissa of point B gives the value of σ_n and the ordinate gives the value of τ_{nt}. Also, points D and E, where the circle intersects the σ_n axis, are very significant. Point D, which represents the maximum value σ_n may have, has the coordinates $(\sigma_1, 0)$, where σ_1 is the algebraic maximum principal stress. Similarly, point E, which represents the minimum value of σ_n, has the coordinates $(\sigma_2, 0)$, where σ_2 is the algebraic minimum principal stress. Also, the location of points D and E with respect to either of the two known X and Y planes (points X and Y, respectively) enables one to locate precisely the principal plane on which σ_1 is acting (represented by point D) and the principal plane on which σ_2 is acting (represented by point E). Thus, for example, since point D is located through the angle θ_1 measured clockwise from point X, it follows that the principal plane on which σ_1 acts is $\frac{1}{2}\theta_1$ clockwise from the X plane. Finally, the coordinates of point F [i.e., $\sigma_A = (\sigma_x + \sigma_y)/2$ and τ_1] give the values of the normal and shear stress on the plane where the shear stress τ_1 is the algebraic maximum and equal to the radius of the circle; and the coordinates of point G [i.e., $\sigma_A = (\sigma_x + \sigma_y)/2$ and τ_2] give the values of the normal and shear stress on the plane where the shear stress τ_2 is the algebraic minimum and also equal in magnitude to the radius R of the circle. As can be seen from Figure 2.10(b), point F represents a plane that is 45° ccw from that represented by point D. Note also that the plane represented by point G is perpendicular to that represented by point F.

Figure 2.10(b) shows a Mohr's circle for a general stress system in which both σ_1 and σ_2 are indicated as positive. Thus in this case σ_3 would be equal to zero because the problem under consideration is plane stress. Note also that this labeling of the three principal stresses satisfies the algebraic requirement that $\sigma_1 > \sigma_2 > \sigma_3$. Thus, by Eq. 2.13b,

$$|\tau_{\max}| = \frac{\sigma_1 - \sigma_3}{2} = \frac{\sigma_1}{2}$$

This same conclusion may be reached by plotting three two-dimensional Mohr's circles as shown in Figure 2.10(c). One of these three circles represents the stress condition in the σ_1–σ_2 plane, the second in the σ_1–σ_3 plane, and the third in the σ_2–σ_3 plane. Note that the circle with the greatest radius, and hence the one providing the absolute maximum shear stress, is the one in the σ_1–σ_3 plane for which the radius is $(\sigma_1 - \sigma_3)/2 = \sigma_1/2$. Thus, once the three principal stresses for a given stress condition are determined, they may be used, if desired, to construct three two-dimensional circles which conveniently represent the three-dimensional stress situation and yield the absolute maximum shear stress at the point. Figure 2.10(d), (e), and (f) shows similar constructions for other combinations of σ_1, σ_2, and σ_3. Note, however, that it is not always necessary to perform the type of construction shown in Figure 2.10(c), (d), (e), and (f) as the absolute maximum shear stress is given directly by Eq. 2.13b.

Example 2.4

The plane stress condition at a point in a structural member is shown in Figure 2.11(a). Determine by using Mohr's circle and show on appropriate sketches (a) the normal and shear stresses on a plane 30° ccw from the X plane, (b) the principal stresses and the maximum and minimum shear stresses τ_1 and τ_2, and (c) the absolute maximum shear stress $|\tau_{max}|$ in the structural member at the point in question.

Solution

(a) Construct a σ_n–τ_{nt} coordinate system with origin at point O as shown in Figure 2.11(e). Locate point X representing the X plane on which the normal stress is $+90$ MPa and the shear stress is $+40$ MPa. Thus point X, whose coordinates are (90, 40), lies in the first quadrant. Similarly, locate point Y representing the Y plane. This point lies in the fourth quadrant, since its coordinates are (30, −40). Connect points X and Y to locate the center C and construct Mohr's circle.

The plane of interest is 30° ccw from the X plane. Therefore, from the radius CX, measure an angle $2(30) = 60°$ ccw to locate point H, which represents the plane of interest and whose coordinates give the values and signs of both the normal and shear stresses on this plane. Thus, from the geometry in Figure 2.11(e),

$$OC = \sigma_A = \frac{\sigma_x + \sigma_y}{2} = \frac{90 + 30}{2} = 60 \text{ MPa}$$

$$R = [(CB)^2 + (BX)^2]^{1/2} = [(30)^2 + (40)^2]^{1/2} = 50 \text{ MPa}$$

$$2\alpha = \tan^{-1}\frac{BX}{CB} = \tan^{-1}\frac{4}{3} = 53.1°$$

$$\beta = 180 - (60 + 2\alpha) = 180 - (60 + 53.1) = 66.9°$$

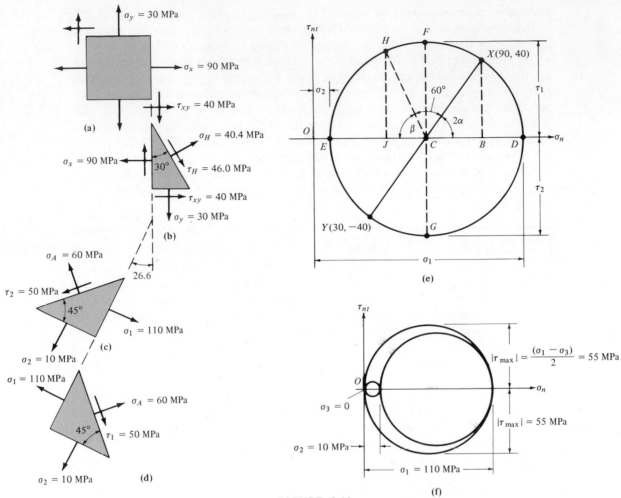

FIGURE 2.11

Thus the normal stress on the plane of interest, σ_H, and the shear stress on the same plane, τ_H, are

$$\sigma_H = OC - CJ = 60 - R \cos \beta$$

$$= 60 - 50(\cos 66.9)$$

$$= \boxed{40.4 \text{ MPa}}$$

$$\tau_H = JH = R \sin \beta = 50 \sin 66.9$$

$$= \boxed{46.0 \text{ MPa}}$$

These stresses are shown on a properly oriented element in Figure 2.11(b).
(b) Again, from the geometry in Figure 2.11(e),

$$\sigma_1 = OD = OC + R = 60 + 50$$

$$= \boxed{110 \text{ MPa}}$$

$$\sigma_2 = OE = OC - R = 60 - 50$$

$$= \boxed{10 \text{ MPa}}$$

$$\tau_1 = CF = +R$$

$$= \boxed{50 \text{ MPa}}$$

$$\tau_2 = CG = -R$$

$$= \boxed{-50 \text{ MPa}}$$

Since the angle 2α was found to be 53.1° in part (a) of the solution, the principal plane on which σ_1 acts [represented by point D in Figure 2.11(e)] is $1/2(53.1°) \doteq 26.6°$ clockwise from the vertical plane. Furthermore, the normal stresses, σ_A, on the planes of τ_1 and τ_2 are given by Eq. 2.11, or, from Figure 2.11(e), by the distance $OC = 60$ MPa. All the stresses above, as well as the geometric relations just mentioned, are shown on properly oriented elements in Figure 2.11(c) and (d).
(c) As discussed in Section 2.4, since σ_1 and σ_2 are both positive, then by Eq. 2.13a,

$$|\tau_{max}| = \frac{\sigma_1}{2} = \frac{110}{2}$$

$$= \boxed{55 \text{ MPa}}$$

Obviously, $|\tau_{max}|$ is larger in magnitude than either τ_1 or τ_2 and acts on planes that are parallel to the σ_2 axis and bisect the 90° angles between σ_1 and σ_3, which in this case is zero because the stress condition at hand is plane stress. These statements are also illustrated in Figure 2.11(f).

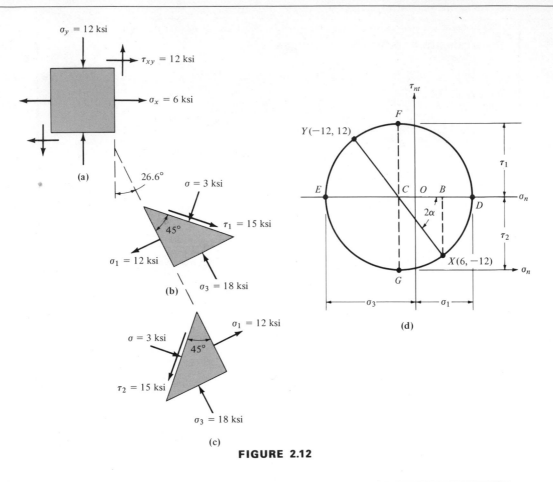

FIGURE 2.12

Example 2.5

At a point in a stressed body, the stress condition is known to be plane stress as shown in Figure 2.12(a). Use Mohr's circle solution to determine (a) the principal stresses and the maximum shear stresses τ_1 and τ_2, and (b) the absolute maximum shear stress $|\tau_{max}|$ in the structural member at the point under consideration.

Solution

(a) Mohr's circle is constructed in exactly the same manner as was done in Example 2.4, by plotting points $X(6, -12)$ and $Y(-12, 12)$ representing the stress conditions on the X and Y planes, respectively. This circle is shown in Figure 2.12(d). From the geometry contained in the circle,

$$OC = \frac{6 - 12}{2} = -3 \text{ ksi}$$

$$R = [(CB)^2 + (BX)^2]^{1/2} = (9^2 + 12^2)^{1/2} = 15 \text{ ksi}$$

$$2\alpha = \tan^{-1} \frac{BX}{CB} = \tan^{-1} \frac{12}{9} = 53.1°$$

$$\alpha = \tfrac{1}{2}(53.1) \doteq 26.6°$$

$$\sigma_1^* = OD = OC + R = -3 + 15$$

$$= \boxed{12 \text{ ksi}}$$

$$\sigma_3^* = OE = OC - R = -3 - 15$$

$$= \boxed{-18 \text{ ksi}}$$

$$\tau_1 = CF = +R$$

$$= \boxed{15 \text{ ksi}}$$

$$\tau_2 = CG = -R$$

$$= \boxed{-15 \text{ ksi}}$$

The principal plane on which σ_1 acts [represented by point D in Figure 2.12(d)] is 26.6° ccw from the X plane. Also, the normal stresses, σ_A, on the planes of τ_1 and τ_2 are equal to -3 ksi. These stresses are shown on properly oriented elements in Figure 2.12(b) and (c).

(b) In part (a) of the solution, it was determined that the two principal stresses are opposite in sign. Then, as was discussed in Section 2.4, $|\tau_{\max}|$ is given by Eq. 2.13b, which leads to the same value as obtained from the radius of Mohr's circle. Thus

$$|\tau_{\max}| = \frac{\sigma_1 - \sigma_3}{2} = \frac{12 + 18}{2}$$

$$= \boxed{15 \text{ ksi}}$$

and acts on two mutually perpendicular planes that are parallel to σ_2 and that bisect the 90° angles between σ_1 and σ_3 as shown in Figure 2.12(b) and (c).

---

* Note that because the two principal stresses are opposite in sign, the designation of the three principal stresses is such that $\sigma_2 = 0$ in order to satisfy the algebraic requirement that $\sigma_1 > \sigma_2 > \sigma_3$.

Homework Problems

Use Mohr's circle for stress in the solution of the following problems.

2.1–2.3 An element is under plane stress conditions. The normal and shear stresses on the Y plane are -50 MN/m^2 and -30 MN/m^2, respectively. The normal stress on the X plane is 10 MN/m^2. Determine the normal and shear stresses on each of the following planes and show them on properly oriented stress elements:

2.1 $30°$ cw from the Y plane.
2.2 $45°$ ccw from the X plane.
2.3 $60°$ ccw from the Y plane.

2.4–2.6 The stress condition at a point in a structural member is known to be plane stress, as shown in Figure H2.4. Determine and show on properly oriented sketches, the normal and shear stresses on each of the following planes:

2.4 $50°$ ccw from the X plane.
2.5 $20°$ ccw from the Y plane.
2.6 $70°$ cw from the Y plane.

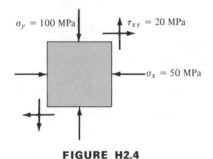

FIGURE H2.4

2.7 The plane stress condition shown in Figure H2.7 is referred to as *pure shear*. Construct Mohr's circle for this stress condition and compute the principal stresses and the absolute maximum shear stress developed. Show these stresses on properly oriented planes.

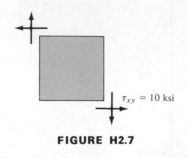

FIGURE H2.7

2.8 A plane stress condition is shown in Figure H2.8. Construct Mohr's circle for this stress condition and determine the principal stresses and the absolute maximum shear stress developed. Define the plane on which this absolute maximum shear stress acts.

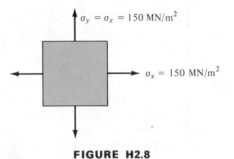

FIGURE H2.8

2.9 The plane stress condition shown in Figure H2.9(a) is known as *uniaxial tension* and that in Figure H2.9(b) as *uniaxial compression*. Construct

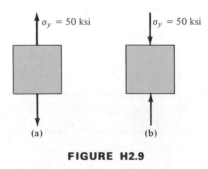

FIGURE H2.9

Mohr's circles for these stress conditions and determine for each case the principal stresses and the absolute maximum shear stress developed. Define in each case the plane on which this absolute maximum shear stress acts.

2.10 The plane stress condition at a point in a structural member is shown in Figure H2.10. Determine and show on properly oriented sketches the principal stresses and the maximum and minimum shear stresses τ_1 and τ_2. Do the stresses τ_1 and τ_2 represent the absolute maximum shear stress developed at the point? If not, determine the absolute maximum shear stress.

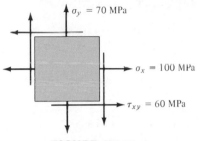

FIGURE H2.10

2.11 The plane stress condition at a point in a structural member is shown in Figure H2.11. Determine and show on properly oriented sketches the principal stresses and the maximum and minimum shear stresses τ_1 and τ_2. Do the shear stresses τ_1 and τ_2 represent the absolute maximum shear stress developed at the point? If not, determine the absolute maximum shear stress.

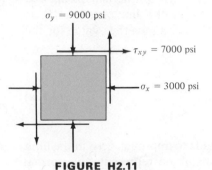

FIGURE H2.11

2.12 The plane stress condition at a point in a structural member is shown in Figure H2.12. Determine and show on properly oriented sketches the principal stresses and the maximum and minimum shear stresses τ_1 and τ_2. Do the shear stresses τ_1 and τ_2 represent the absolute maximum shear stress developed at the point? If not, determine the absolute maximum shear stress.

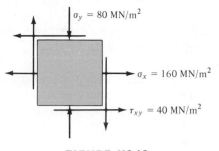

FIGURE H2.12

2.13 The principal stresses and the direction of the maximum or minimum shear stress τ_1 or τ_2 are indicated in Figure H2.13. Determine this maximum or minimum shear stress, the normal stress on its plane, and the stresses on the X and Y planes. Show all these stresses on properly oriented elements.

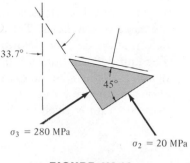

FIGURE H2.13

2.14 In the plane stress condition depicted in Figure H2.14, τ_{xy} is unknown. However, the magnitude of the maximum or minimum shear stress $\tau_{1,2}$ is

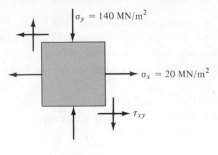

FIGURE H2.14

100 MN/m². Find the principal stresses and show them on properly oriented sketches.

2.15 In the plane stress condition shown in Figure H2.15, τ_{xy} and σ_x are unknown. However, one of the two principal stresses on planes parallel to the z axis is equal to -30 MPa, and the magnitude of the maximum or minimum shear stress τ_1 or τ_2 is 85 MPa. Find the second principal stress on planes parallel to the z axis as well as τ_{xy}, σ_x, and σ_A on the planes of τ_1 and τ_2. Show these stresses on properly oriented elements.

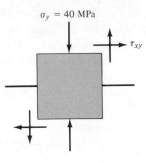

FIGURE H2.15

2.7 Concept of Strain at a Point

When a nonrigid (deformable) body is subjected to stresses, it undergoes deformations and distortions. The term *deformation* refers to the geometric changes that take place in the dimensions of the body (extensions or contractions), while the term *distortion* represents geometric changes in its shape.

Thus any line element in a body is said to undergo deformation if its length increases or decreases. The term *strain* of a line element is used to signify its unit deformation (i.e., the deformation of the line element divided by its initial length). The unit deformation of a line element is referred to as *linear* (or *normal*) *strain* and denoted by the symbol ε_n. Therefore, if the initial length of a line element AB, as shown in Figure 2.13(a), is L and its deformation BC is δ, an average value for the normal strain is given by the equation

$$\varepsilon_n = \frac{\delta}{L} \tag{2.16a}$$

If, however, the initial length of the line element is made to approach zero in the limit (i.e., point B approaches point A), the foregoing definition represents the normal strain at a point. Thus

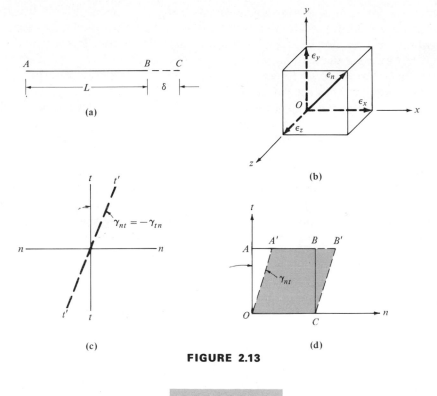

FIGURE 2.13

$$\varepsilon_n = \lim_{L \to 0} \frac{\delta}{L} = \frac{d\delta}{dL} \qquad\qquad (2.16b)$$

Note that, by definition, the normal strain ε_n is a dimensionless quantity. However, it is sometimes expressed in terms of units of length divided by units of length in order to give it a meaningful physical significance. Thus, for example, in the British gravitational system of measure, the derived unit (inches per inch) is widely used for normal strain. The sign convention used in this text is such that a positive normal strain represents an extension, while a negative normal strain represents a contraction. At any point in a body, such as point O in Figure 2.13(b), and in the absence of any distortion, the normal strain, ε_n, can be completely defined by specifying its three rectangular components, ε_x, ε_y, and ε_z, along the three rectangular axes, x, y, and z, respectively.

As was mentioned earlier, the term *distortion* signifies a change in the original shape of a body. A more precise definition of distortion, however, is needed, and for this purpose, the change in angle between two initially perpendicular line elements is considered. The change in the 90° angle between two line elements nn and tt, as shown in Figure 2.13(c), represents a distortion and is given the name *shear strain* and denoted by the symbol γ_{nt}. Note that two subscripts are assigned to the shear strain, γ, one for each of the two line elements being considered. If after distortion, line tt in Figure 2.13(c) rotates to a new position $t't'$ while line nn remains fixed, the shear strain between line nn and line tt (proceeding from nn to tt in a ccw manner) is γ_{nt} and represents a decrease in the 90° angle. On the other hand, the shear strain between line tt and line nn (proceeding from tt to nn in a ccw manner) is γ_{tn} and represents an increase in the 90° angle. This concept gives rise to a sign convention for shear strains. For the sake of consistency with the equations developed earlier for plane stress, the sign convention adopted in this text is such that a positive shear strain represents an increase in the 90° angle, while a negative shear strain represents a decrease. In Figure 2.13(c), for example, γ_{nt} is negative and γ_{tn} is positive. Note that as in the case of the normal strain, ε_n, the shear strain γ_{nt} is also a dimensionless quantity since it represents a radian measure.

In summary, a normal strain is positive if it represents an extension and negative if it represents a contraction. Also, a shear strain is positive if it represents an increase and negative if it represents a decrease in the 90° angle.

Another way to visualize the shear strain and to arrive at the same conclusion as above is to consider a small rectangular element $OABC$ in the neighborhood of point O, as shown in Figure 2.13(d). After distortion, the rectangular element assumes the position $OA'B'C$. The shear strain between the lines On and Ot is the angle γ_{nt}. For very small distortions, the angle γ_{nt} may be assumed equal to $\tan \gamma_{nt}$. Thus

$$\gamma_{nt} \doteq \tan \gamma_{nt} = \frac{AA'}{OA}. \tag{2.16c}$$

Therefore, since the shear strain is represented by the ratio of two lengths, it is a dimensionless quantity, as previously stated.

At any point in a body, such as point O in Figure 2.13(b) at the origin of a rectangular x, y, z coordinate system, six shear strain components may be specified to represent the changes in the 90° angles between the various axes. However, because of the fact that γ_{nt} is equal in magnitude to γ_{tn}, the six strain components can be reduced to three. Thus

$$\gamma_{xy} = -\gamma_{yx} \qquad \text{represents the shear strain between the } x \text{ and } y \text{ (or } y \text{ and } x) \text{ axes}$$

$$\gamma_{yz} = -\gamma_{zy} \qquad \text{represents the shear strain between the } y \text{ and } z \text{ (or } z \text{ and } y) \text{ axes}$$

$$\gamma_{zx} = -\gamma_{xz} \qquad \text{represents the shear strain between the } z \text{ and } x \text{ (or } x \text{ and } z) \text{ axes}$$

Analysis of strain in three dimensions is beyond the scope of this chapter, and only two-dimensional strain conditions will be examined and discussed in the following section.

★2.8 Analysis of Plane Strain

The relations that exist between stresses and strains will be developed and discussed in detail in a later section. However, for present purposes, it is only necessary to know that when a member is subjected to a normal stress, it experiences not only a normal strain in the direction of the applied stress (normal longitudinal strain), but also a smaller normal strain in a direction perpendicular to the applied stress (normal transverse strain). If the applied normal stress is tensile, the resulting normal longitudinal strain will be an extension, and the normal transverse strain will be a contraction. Inversely, if the applied normal stress is compressive, the resulting normal longitudinal strain will be a contraction and the normal transverse strain will be an extension. These relations are depicted schematically in a very exaggerated fashion in Figure 2.14(a) and (b), where the initial configuration is shown by the solid lines and the final configuration by the dashed lines.

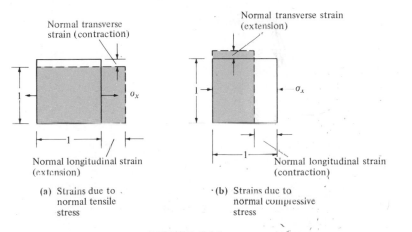

(a) Strains due to normal tensile stress

(b) Strains due to normal compressive stress

FIGURE 2.14

The term *plane strain* is used to denote a condition in which all strains in a body are parallel to a given plane. It should be noted, however, that when a member is subjected to stress, it experiences three-dimensional deformation unless its deformation is prevented in one or more directions. Thus, in order to achieve a condition of plane strain, say in the xy plane, the strain in the z direction must be prevented by physical restraint (i.e., by the application of a stress in the z direction). In such a case, $\varepsilon_z = \gamma_{yz} = \gamma_{xz} = 0$ and the only nonvanishing components of strain are ε_x, ε_y, and $\gamma_{xy} = -\gamma_{yx}$.

Consider the element subjected to a plane stress condition as shown in Figure 2.15(a). If the assumption is made that strains in the z direction are zero (i.e., a plane

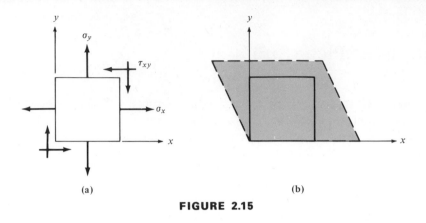

FIGURE 2.15

strain condition exists), the element undergoes normal strains in the x and y directions as well as shear strains in the xy plane. A sketch of how the element would look after experiencing these strains is shown schematically and in a very exaggerated manner by the dashed lines in Figure 2.15(b). The strains represented in Figure 2.15(b), however, can be visualized easier if we decomposed them into their separate components ε_x, ε_y, and γ_{xy}, as shown in Figure 2.16.

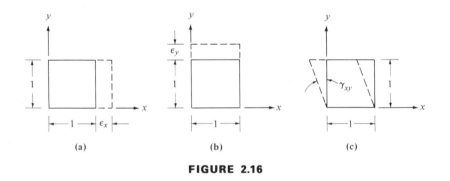

FIGURE 2.16

It is desirable to be able to relate the strains ε_x, ε_y, and γ_{xy} to the strains ε_n, ε_t, and γ_{nt}, where n and t are any pair of orthogonal axes inclined to the xy axes as shown in Figure 2.17(a). As in the case of plane stress analysis, the angle α is considered positive if counterclockwise and negative if clockwise from the positive end of the x axis. The development of the desired relations is enhanced if one considered line elements n and t as the diagonals of two adjacent rectangular elements $OABC$ and $OCDG$ whose dimensions have been chosen such that n and t are perpendicular to each other as shown in Figure 2.17(b).

If the assumption is made that the composite element $GABD$ is subjected to the plane stress condition shown in Figure 2.15(a), it would deform and distort into the configuration shown by the dashed lines. This latter configuration consists of a

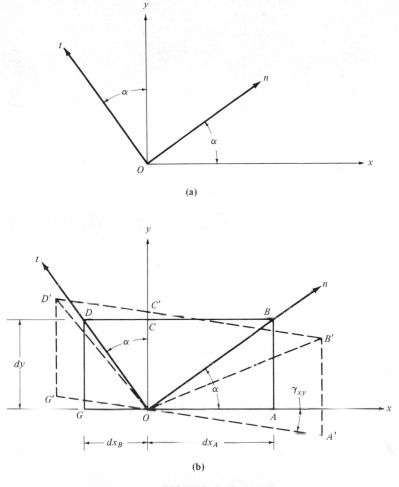

FIGURE 2.17

deformation in the x direction equal to $\varepsilon_x(dx_A + dx_B)$, a deformation in the y direction equal to $\varepsilon_y \, dy$ and a distortion due to the shear strain γ_{xy} which rotates line elements such as DB into position $D'B'$. Note that no generality is sacrificed by assuming that line element OC remains vertical.

Consider now the deformation of line element OB along the n direction. This deformation will be denoted by δ_n and is obviously the composite effect of three different deformations: one in the x direction produced by the strain ε_x, a second in the y direction produced by the strain ε_y, and the third is the result of the shear strain γ_{xy}. These three separate effects are shown in Figure 2.18. Figure 2.18(a) depicts the change in length of line element OB due to the deformation $\varepsilon_x \, dx$. For very small deformations, OB' will have essentially the same direction as OB. Therefore, the angle $OB'C$ may be assumed equal to α. Thus, for very small deformations, the increase in

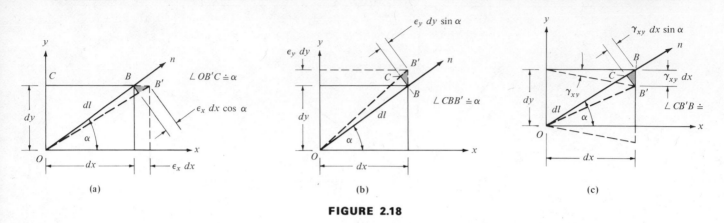

FIGURE 2.18

length of the line element OB is given by $\varepsilon_x \, dx \cos \alpha$, as shown in Figure 2.18(a). Similarly, as shown in Figure 2.18(b), the increase in length of the line element OB due to the deformation $\varepsilon_y \, dy$ is given by $\varepsilon_y \, dy \sin \alpha$. Finally, as shown in Figure 2.18(c), the rotation of the line elements due to the shear strain γ_{xy} decreases the length of OB by $\gamma_{xy} \, dx \sin \alpha$. Thus

$$\delta_n = \varepsilon_x \, dx \cos \alpha + \varepsilon_y \, dy \sin \alpha - \gamma_{xy} \, dx \sin \alpha \qquad \textbf{(2.17a)}$$

If the initial length of the line element OB is dl, then the strain of OB is given by the ratio δ_n/dl. Thus

$$\varepsilon_n = \varepsilon_x \left(\frac{dx}{dl}\right) \cos \alpha + \varepsilon_y \left(\frac{dy}{dl}\right) \sin \alpha - \gamma_{xy} \left(\frac{dx}{dl}\right) \sin \alpha \qquad \textbf{(2.17b)}$$

Since $dx/dl = \cos \alpha$ and $dy/dl = \sin \alpha$,

$$\varepsilon_n = \varepsilon_x \cos^2 \alpha + \varepsilon_y \sin^2 \alpha - \gamma_{xy} \sin \alpha \cos \alpha \qquad \textbf{(2.17c)}$$

By using appropriate trigonometric identities, Eq. 2.17c can be rewritten in the following form:

$$\varepsilon_n = \frac{\varepsilon_x + \varepsilon_y}{2} + \frac{1}{2}(\varepsilon_x - \varepsilon_y) \cos 2\alpha - \frac{1}{2} \gamma_{xy} \sin 2\alpha \qquad \textbf{(2.17d)}$$

Equation 2.17d is mathematically similar to Eq. 2.4 previously derived for plane stress conditions. This property will be utilized in a later part of this chapter. Note also that the normal strain ε_t may be obtained directly from Eq. 2.17d by letting α become $\alpha + 90°$.

The development of an equation for the shear strain γ_{nt} can be simplified if we decompose the resultant deformations and distortions shown in Figure 2.17(b) into three separate effects, as shown in Figure 2.19(a), (b), and (c). In Figure 2.19(a), the deformation in the x direction $\varepsilon_x(dx_A + dx_B)$ is seen to change the 90° angle between OB and OD by the small angles BOB' and DOD'. Since both of these angles represent an increase in the 90° angle BOD, they are positive according to the sign convention that has been established. Thus, if the initial length of OB is dl_A and that of OD is dl_B, then assuming small deformations, the increase in the angle BOD due to the x deformation becomes

$$
\Delta \angle BOD = \frac{BC}{OB} + \frac{DE}{OD}
$$

$$
= \varepsilon_x \left(\frac{dx_A}{dl_A} \sin \alpha + \frac{dx_B}{dl_B} \cos \alpha \right)
$$

$$
= 2\varepsilon_x \sin \alpha \cos \alpha = \varepsilon_x \sin 2\alpha \qquad \textbf{(2.18a)}
$$

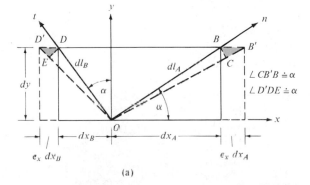

(a)

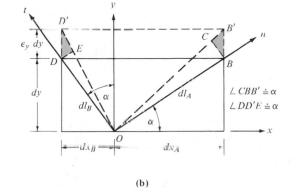

(b)

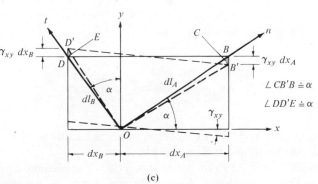

(c)

FIGURE 2.19

Similarly, Figure 2.19(b) shows that the deformation $\varepsilon_y \, dy$ changes the 90° angle between OB and OD by the small angles BOB' and DOD'. Since both of these angles represent a decrease in the 90° angle between OB and OD, they are negative quantities according to the sign convention. Therefore, the change in the angle BOD is

$$\Delta \measuredangle BOD = -\frac{BC}{OB} - \frac{DE}{OD}$$

$$= -\varepsilon_y \left(\frac{dy}{dl_A} \cos \alpha + \frac{dy}{dl_B} \sin \alpha \right)$$

$$= -2\varepsilon_y \sin \alpha \cos \alpha = -\varepsilon_y \sin 2\alpha \qquad \textbf{(2.18b)}$$

Finally, the change in the 90° angle between OB and OD due to the shear strain γ_{xy} is shown in Figure 2.19(c). This change consists of the positive small angle BOB' and the negative small angle DOD'. Therefore, the change in the angle BOD is given by

$$\Delta \measuredangle BOD = \frac{CB'}{dl_A} - \frac{DE}{dl_B}$$

$$= \gamma_{xy} \left(\frac{dx_A}{dl_A} \cos \alpha - \frac{dx_B}{dl_B} \sin \alpha \right)$$

$$= \gamma_{xy}(\cos^2 \alpha - \sin^2 \alpha) = \gamma_{xy} \cos 2\alpha \qquad \textbf{(2.18c)}$$

The resultant change in the 90° angle between OB (line n) and OD (line t) is the shear strain γ_{nt} and is the sum of the three separate changes given by Eqs. 2.18a, 2.18b, and 2.18c. Thus

$$\gamma_{nt} = \varepsilon_x \sin 2\alpha - \varepsilon_y \sin 2\alpha + \gamma_{xy} \cos 2\alpha$$

$$= (\varepsilon_x - \varepsilon_y) \sin 2\alpha + \gamma_{xy} \cos 2\alpha \qquad \textbf{(2.19a)}$$

If both sides of Eq. 2.19a are divided by 2, we obtain

$$\frac{\gamma_{nt}}{2} = \frac{1}{2}(\varepsilon_x - \varepsilon_y) \sin 2\alpha + \frac{1}{2}\gamma_{xy} \cos 2\alpha \qquad \textbf{(2.19b)}$$

Once again, it is observed that Eq. 2.19b is mathematically similar to Eq. 2.6, which was obtained previously for the case of plane stress.

Equations 2.17d and 2.19b for plane strain can be made mathematically identical with the plane stress equations 2.4 and 2.6, respectively, by satisfying the following equalities:

$$\sigma_x \equiv \varepsilon_x; \qquad \sigma_y \equiv \varepsilon_y; \qquad \tau_{xy} \equiv \frac{\gamma_{xy}}{2}$$

$$\sigma_n \equiv \varepsilon_n; \qquad \tau_{nt} \equiv \frac{\gamma_{nt}}{2}$$

(2.20)

Therefore, using the equalities in Eq. 2.20, one can develop, for example, expressions for the two principal strains ε_1 and ε_2* directly from Eq. 2.8. Thus

$$\varepsilon_1 = \frac{\varepsilon_x + \varepsilon_y}{2} + R' = \varepsilon_A + R'$$

$$\varepsilon_2 = \frac{\varepsilon_x + \varepsilon_y}{2} - R' = \varepsilon_A - R'$$

(2.21)

where $R' = \{[(\varepsilon_x - \varepsilon_y)/2]^2 + (\gamma_{xy}/2)^2\}^{1/2}$ and the symbol ε_A is used to denote the average of the two normal strains. Also, from Eq. 2.10 we can obtain the equations for the maximum and minimum shear strains, γ_1 and γ_2, respectively, as follows:

$$\gamma_1 = +2R'$$

$$\gamma_2 = -2R'$$

(2.22)

Furthermore, from Eq. 2.13a we conclude that

$$2R' = |\gamma_{1,2}| = \gamma_{max} = \varepsilon_1 - \varepsilon_2$$

(2.23a)

which indicates that the maximum value of the shear strain in the plane of ε_1 and ε_2 is equal to the difference between the two principal strains in this plane. However, for the same reasons explained earlier in the case of plane stress, Eq. 2.23a may not yield the value of the absolute maximum shear strain at a point considered as a particle in a three-dimensional body. Since, algebraically, $\varepsilon_1 > \varepsilon_2 > \varepsilon_3$, one concludes, on the basis of Eq. 2.13b developed for plane stress conditions, that the absolute value of the maximum shear strain at a point is given by

$$|\gamma_{max}| = \varepsilon_1 - \varepsilon_3$$

(2.23b)

---

* As in the case of plane stress, the two principal strains in a plane may be designated as ε_1 and ε_2, ε_1 and ε_3, or ε_2 and ε_3, depending upon conditions existing in the plane.

where ε_1 and ε_3 are interpreted as the algebraic maximum and algebraic minimum values, respectively, of the principal strains at the point. It is, therefore, important that one labels the three principal strains properly and in accordance with the algebraic requirement that $\varepsilon_1 > \varepsilon_2 > \varepsilon_3$ if Eq. 2.23b is to be used correctly. Consider, for example, the case where both principal strains in a plane are positive. Since the problem under consideration is one of plane strain (i.e., one of the three principal strains is zero), it follows that the labeling of the three principal strains at the point must be such that $\varepsilon_3{}^* = 0$. In such a case $|\gamma_{max}| = \varepsilon_1 - \varepsilon_3 = \varepsilon_1$. On the other hand, if one of the two principal strains in the plane is positive and the second negative, the labeling of the three principal strains must be such that $\varepsilon_2 = 0$. In such a case, $|\gamma_{max}| = \varepsilon_1 - \varepsilon_3$, where ε_3 is a negative quantity. Finally, if both principal strains in the plane are negative, the labeling of the three principal strains must be such that $\varepsilon_1 = 0$. In such a case $|\gamma_{max}| = \varepsilon_1 - \varepsilon_3 = -\varepsilon_3$, where ε_3 is a negative quantity. In all three cases, the absolute value of the maximum shear strain occurs between two orthogonal axes bisecting the 90° angles between ε_1 and ε_3.

It is important to note that because of the similarities between the plane stress equations and those for plane strain as has been pointed out, Mohr's circle construction which was developed for the solution of the plane stress problem is equally applicable to the plane strain problem. Only minor modifications, as implied in Eq. 2.20, become necessary, as will be shown in the following example.

Example 2.6

The plane strain condition at a point in a stressed body is given by the following values:

$$\varepsilon_x = -800 \times 10^{-6}; \qquad \varepsilon_y = -200 \times 10^{-6}; \qquad \gamma_{xy} = 450 \times 10^{-6}$$

Using Mohr's circle for strain, determine (a) the principal strains and the principal directions; (b) the maximum shear strain in the xy plane (i.e., γ_1 and γ_2) and the pair of orthogonal axes through the point corresponding to this maximum shear strain; (c) the absolute maximum shear strain at the point considered as a particle in a three-dimensional body; (d) the normal strain along a line making an angle 30° ccw from the x axis; and (e) the shear strain between two orthogonal lines in the xy plane, one of which is defined in part (d).

Solution

(a) Construct an $\varepsilon_n - \frac{1}{2}\gamma_{nt}$ coordinate system with origin at point O as shown in Figure 2.20(a). Locate point X representing the strain condition along the x axis by plotting the coordinates $\varepsilon_x = -800 \times 10^{-6}$ and $\frac{1}{2}\gamma_{xy} = \frac{1}{2}(450 \times 10^{-6})$

---

* In general, a plane stress condition leads to a strain situation in which none of the three principal strains is zero. This situation will be discussed briefly in a later part of this chapter.

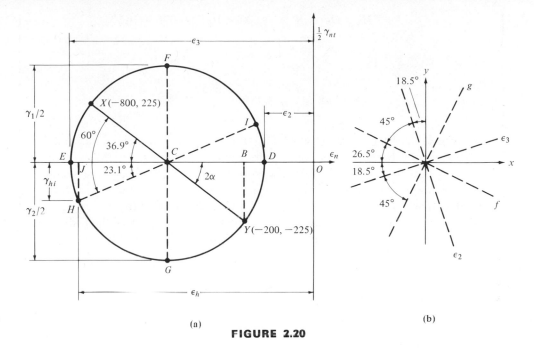

FIGURE 2.20

as shown in Figure 2.20(a). Note that the common factor 10^{-6} has been omitted from the diagram for convenience. Locate point Y representing the strain condition along the y axis by plotting the coordinates $\varepsilon_y = -200 \times 10^{-6}$ and $\frac{1}{2}\gamma_{yx} = \frac{1}{2}(-450 \times 10^{-6})$ as shown in Figure 2.20(a). Connect points X and Y to locate the center of the circle, point C, which lies along the ε_n axis, and construct Mohr's circle.

From the geometry of the circle, $OC =$ the average of the two normal strains.

$$\varepsilon_A = \frac{\varepsilon_x + \varepsilon_y}{2} = \frac{(-800 - 200) \times 10^{-6}}{2} = -500 \times 10^{-6}$$

$$R' = CX = CY = [(CB)^2 + (BY)^2]^{1/2}$$

$$= [(300)^2 + (225)^2]^{1/2} \times 10^{-6} = 375 \times 10^{-6}$$

$$\varepsilon_2{}^* = OD = OC + R' = -500 \times 10^{-6} + 375 \times 10^{-6}$$

$$= \boxed{-125 \times 10^{-6}}$$

* Note that because the two principal strains in the xy plane are both negative, and therefore algebraically less than the third principal strain, which in this case is zero, they have been labeled as ε_2 and ε_3 in order to fulfill the algebraic requirement that $\varepsilon_1 > \varepsilon_2 > \varepsilon_3$. Thus, in this problem, the strain that is zero is ε_1.

$$\varepsilon_3{}^* = OE = OC - R' = -500 \times 10^{-6} - 375 \times 10^{-6}$$

$$= \boxed{-875 \times 10^{-6}}$$

$$2\alpha = \tan^{-1} \frac{BY}{CB} = \tan^{-1} \frac{225}{300} = 36.9°$$

Therefore, since point D (representing the axis along which ε_2 occurs) is 36.9° ccw from point Y (representing the y axis), the principal direction along which ε_2 occurs is $\frac{1}{2}(36.9) \doteq 18.5°$ ccw from the y axis. Obviously, the second principal direction (i.e., the direction along which ε_3 occurs) is perpendicular to the first. These geometric relations are shown in Figure 2.20(b).

(b) The maximum shear strain in the xy plane is given by twice the radius of the circle as expressed in Eq. 2.22 and is represented by points F and G in Figure 2.20(a). Thus

$$\gamma_1 = \gamma_{fg} = 2R'$$

$$= \boxed{750 \times 10^{-6}}$$

$$\gamma_2 = \gamma_{gf} = -2R'$$

$$= \boxed{-750 \times 10^{-6}}$$

where the axes f and g bisect the 90° angles between the two principal directions, as shown in Figure 2.20(b).

(c) Since the principal strains in the xy plane have the same sign (both are negative), the absolute value of the maximum shear strain is given by Eq. 2.23b. Thus

$$|\gamma_{max}| = \varepsilon_1 - \varepsilon_3$$

$$= [0 - (-875)] \times 10^{-6}$$

$$= \boxed{875 \times 10^{-6}}$$

and occurs between two perpendicular axes bisecting the 90° angle between ε_1 and ε_3.

(d) To determine the normal strain along a line making an angle 30° ccw from the x axis, locate point H at an angle $2(30°) = 60°$ ccw from point X as shown in Figure 2.20(a). The abscissa of point H gives the value of the desired normal strain. Thus

$$\varepsilon_h = OC - R' \cos 23.1 = -500 \times 10^{-6} - 375 \times 10^{-6} \cos 23.1$$

$$= \boxed{-855 \times 10^{-6}}$$

(e) The shear strain between two orthogonal lines in the xy plane, one of which is represented by point H in Figure 2.20(a), is given by the ordinate of this point. It should be pointed out that the point on the circle representing the second orthogonal line is point I, which is diametrically opposite to point H. Thus

$$\gamma_{hi} = 2(JH) = -2R' \sin 23.1 = -2(375) \times 10^{-6} \sin 23.1$$

$$= \boxed{-294 \times 10^{-6}}$$

and

$$\gamma_{ih} = \boxed{+294 \times 10^{-6}}$$

*2.9 Strain Rosettes

In order to construct Mohr's circle for strain and determine the principal strains and their directions at a given point, it is necessary to have the values of ε_x, ε_y, and γ_{xy} at the point. Measurement of the normal strains can be readily made by means of electric resistance strain gages,* but there is no convenient way of measuring the shear strain. Consequently, shear strains are determined, indirectly, from measurements of normal strains at the point. This procedure is accomplished by the use of Eq. 2.17c, which contains the three unknown quantities ε_x, ε_y, and γ_{xy} required for the construction of Mohr's circle. Thus, if we measured the normal strain ε_n in three different directions through the point and applied Eq. 2.17c separately to each one of the three directions, we should obtain a system of three simultaneous equations that would lead to the three desired unknown quantities ε_x, ε_y, and γ_{xy}. Any configuration of electric strain gages measuring strains in at least three directions passing through a point is known as a *strain rosette*. For convenience, the two most

* Briefly, the electric resistance strain gage consists of a grid of fine wire sandwiched between two sheets of paper. This sandwich is cemented to the member for which the strain is to be measured. As the member is deformed, the wire in the strain gage changes in length, giving rise to a change in its electrical resistance. This change in resistance is a measure of the strain to which the member is subjected and may be measured by means of a Wheatstone bridge.

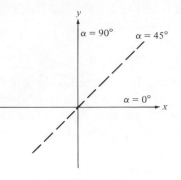

FIGURE 2.21

commonly used strain rosettes are the rectangular, or 45° strain rosette, and the equiangular, or 60° strain rosette. The rectangular strain rosette is described briefly in this section and the equiangular rosette is dealt with in one of the Problems. However, it should be pointed out that other strain gage configurations may be used.

In the *rectangular rosette*, the axes of the three strain gages are placed at an angle of 45° with respect to each other as shown in Figure 2.21. The direction of one of the three gages is taken as a reference coincident with the x axis and is labeled $\alpha = 0°$. The other two directions may then be labeled as $\alpha = 45°$ and $\alpha = 90°$ with respect to the reference axis. These relations are shown in Figure 2.21.

Applying Eq. 2.17c to the normal strains in the three directions, we obtain the following relations:

$$\varepsilon_0 = \varepsilon_x$$
$$\varepsilon_{45} = \tfrac{1}{2}\varepsilon_x + \tfrac{1}{2}\varepsilon_y - \tfrac{1}{2}\gamma_{xy} \qquad\qquad \textbf{(2.24a)}$$
$$\varepsilon_{90} = \varepsilon_y$$

Solving the relations expressed in Eq. 2.24a for ε_x, ε_y, and γ_{xy}, we obtain the following equations for the rectangular rosette:

$$\varepsilon_x = \varepsilon_0$$
$$\varepsilon_y = \varepsilon_{90} \qquad\qquad \textbf{(2.24b)}$$
$$\gamma_{xy} = \varepsilon_0 + \varepsilon_{90} - 2\varepsilon_{45}$$

Thus, having the three normal strain readings from the rectangular rosette (i.e., ε_0, ε_{45}, and ε_{90}), we may use Eq. 2.24b to determine ε_x, ε_y, and γ_{xy}, which can then be used to construct Mohr's circle and determine the principal strains as well as the maximum shear strain in the xy plane. It should be pointed out that graphical constructions exist for obtaining Mohr's circle directly from the three measured normal strains. These graphical constructions will not be discussed in this text, but the interested reader is referred to appropriate references in **Appendix A**.

Homework Problems

Use Mohr's circle for strain for the solution of the following problems.

2.16–2.20 A thin rectangular plate lies in the xy plane as shown in Figure H2.16. The following normal strains were measured parallel to the x and y axes. Assume that after deformation, the corners of the plate are still perfectly square and that $\varepsilon_z = \gamma_{xz} = \gamma_{yz} = 0$. For each of the following cases, determine the shear strain γ_{nt}.

2.16 $\varepsilon_x = 0$, $\varepsilon_y = 200 \times 10^{-6}$

2.17 $\varepsilon_x = 400 \times 10^{-6}$, $\varepsilon_y = 0$

2.18 $\varepsilon_x = 300 \times 10^{-6}$, $\varepsilon_y = 300 \times 10^{-6}$

2.19 $\varepsilon_x = -500 \times 10^{-6}$, $\varepsilon_y = 400 \times 10^{-6}$

2.20 $\varepsilon_x = -200 \times 10^{-6}$, $\varepsilon_y = 300 \times 10^{-6}$

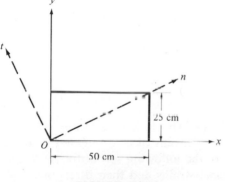

FIGURE H2.16

2.21–2.23 A rectangular xy coordinate system is constructed at point O on the surface of a body that is to be subjected to loads. Refer to Figure H2.21, assume that $\varepsilon_z = \gamma_{xz} = \gamma_{yz} = 0$, and determine for each of the following cases ε_n, ε_t, and γ_{nt}:

2.21 $\varepsilon_x = 100 \times 10^{-6}$, $\varepsilon_y = 200 \times 10^{-6}$, $\gamma_{xy} = 0$, $\alpha = 60°$

2.22 $\varepsilon_x = -200 \times 10^{-6}$, $\varepsilon_y = 400 \times 10^{-6}$, $\gamma_{xy} = 500 \times 10^{-6}$, $\alpha = 45°$

2.23 $\varepsilon_x = 0$, $\varepsilon_y = 0$, $\gamma_{xy} = 600 \times 10^{-6}$, $\alpha = 30°$

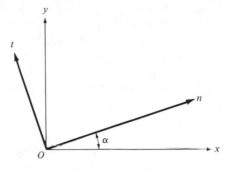

FIGURE H2.21

2.24–2.26 Refer to Figure H2.21, assume that $\varepsilon_z = \gamma_{xz} = \gamma_{yz} = 0$, and determine ε_x, ε_y, and γ_{xy} for each of the following cases:

2.24 $\varepsilon_n = 800 \times 10^{-6}$, $\varepsilon_t = -600 \times 10^{-6}$, $\gamma_{nt} = 200 \times 10^{-6}$, $\alpha = 30°$

2.25 $\varepsilon_n = -400 \times 10^{-6}$, $\varepsilon_t = -1000 \times 10^{-6}$, $\gamma_{nt} = 400 \times 10^{-6}$, $\alpha = 50°$

2.26 $\varepsilon_n = 500 \times 10^{-6}$, $\varepsilon_t = 200 \times 10^{-6}$, $\gamma_{nt} = -800 \times 10^{-6}$, $\alpha = -45°$

2.27 2.29 Refer to Figure H2.21, assume that $\varepsilon_z = \gamma_{xz} = \gamma_{yz} = 0$, and determine the angle α for each of the following cases:

2.27 $\varepsilon_n = -500 \times 10^{-6}$, $\varepsilon_x = 300 \times 10^{-6}$, $\varepsilon_y = -400 \times 10^{-6}$, $\gamma_{xy} = 800 \times 10^{-6}$

2.28 $\gamma_{nt} = 800 \times 10^{-6}$, $\varepsilon_x = 900 \times 10^{-6}$, $\varepsilon_y = 200 \times 10^{-6}$, $\gamma_{xy} = 1000 \times 10^{-6}$

2.29 $\varepsilon_n = 700 \times 10^{-6}$, $\varepsilon_t = -500 \times 10^{-6}$, $\gamma_{nt} = 300 \times 10^{-6}$, $\varepsilon_x = 300 \times 10^{-6}$

2.30–2.34 For each of the following plane strain cases, determine the principal strains as well as the maximum shear strains in the xy plane (i.e., γ_1 and γ_2). Show on a sketch the principal axes as well as the axes corresponding to the shear strains γ_1 and γ_2. Also, determine the absolute maximum value of the shear strain at the point for each of the cases.

2.30 $\varepsilon_x = 700 \times 10^{-6}$, $\varepsilon_y = 0$, $\gamma_{xy} = 500 \times 10^{-6}$
2.31 $\varepsilon_x = 200 \times 10^{-6}$, $\varepsilon_y = -400 \times 10^{-6}$,
$\gamma_{xy} = 300 \times 10^{-6}$
2.32 $\varepsilon_x = -200 \times 10^{-6}$, $\varepsilon_y = -400 \times 10^{-6}$,
$\gamma_{xy} = -300 \times 10^{-6}$
2.33 $\varepsilon_x = 0$, $\varepsilon_y = 0$, $\gamma_{xy} = 800 \times 10^{-6}$
2.34 $\varepsilon_x = 500 \times 10^{-6}$, $\varepsilon_y = -800 \times 10^{-6}$, $\gamma_{xy} = 0$

2.35 The following information is known about the plane strain condition at a point in a structural member.

$$\varepsilon_2 = -200 \times 10^{-6}, \qquad \varepsilon_3 = -1200 \times 10^{-6}$$

Refer to Figure H2.35 and determine:
(a) The maximum shear strain in the xy plane (i.e., γ_1 and γ_2) and the corresponding orthogonal axes. Show these axes on a neat sketch.
(b) The absolute maximum shear strain at the point and define the orthogonal axes corresponding to this strain.
(c) The values of ε_x, ε_y, and γ_{xy}.

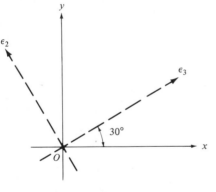

FIGURE H2.35

2.36 A plane strain condition exists at a point in a body and the following information is known: $\varepsilon_x = 200 \times 10^{-6}$, $\varepsilon_y = -1400 \times 10^{-6}$, and the maximum value of the shear strain in the xy plane is 2000×10^{-6}. Determine:
(a) γ_{xy}.
(b) The principal strains and their directions.

(c) The absolute maximum shear strain at the point, and define the orthogonal axes corresponding to this strain.

2.37 The *equiangular* (60°) *rosette* is one in which the three strain axes make 60° with respect to each other. If one of the three axes is made to coincide with the x axis as shown in Figure H2.37, prove that ε_x, ε_y, and γ_{xy} are given by the following equations:

$$\varepsilon_x = \varepsilon_0, \qquad \varepsilon_y = \tfrac{1}{3}(2\varepsilon_{60} + 2\varepsilon_{120} - \varepsilon_0),$$

$$\gamma_{xy} = -\frac{2}{\sqrt{3}}(\varepsilon_{60} - \varepsilon_{120})$$

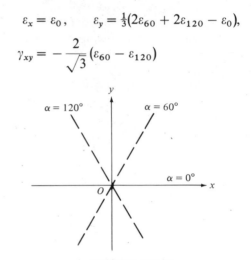

FIGURE H2.37

2.38–2.40 The following strain readings were obtained by means of a rectangular strain rosette. For each of the following sets of readings, determine the principal strains and their directions, and the maximum value of the shear strains in the xy plane (i.e., γ_1 and γ_2).
2.38 $\varepsilon_0 = 800 \times 10^{-6}$, $\varepsilon_{45} = 200 \times 10^{-6}$,
$\varepsilon_{90} = -400 \times 10^{-6}$
2.39 $\varepsilon_0 = -200 \times 10^{-6}$, $\varepsilon_{45} = 300 \times 10^{-6}$,
$\varepsilon_{90} = -500 \times 10^{-6}$
2.40 $\varepsilon_0 = 0$, $\varepsilon_{45} = 500 \times 10^{-6}$, $\varepsilon_{90} = 800 \times 10^{-6}$

2.41–2.43 The following strain readings were obtained by means of an equiangular strain rosette. For each of the following sets of readings, determine the principal strains and their directions, and the maximum value of the shear strains in the xy plane (i.e., γ_1 and γ_2).

2.41 $\varepsilon_0 = 600 \times 10^{-6}$, $\varepsilon_{60} = 500 \times 10^{-6}$, $\varepsilon_{120} = -200 \times 10^{-6}$

2.42 $\varepsilon_0 = -300 \times 10^{-6}$, $\varepsilon_{60} = -100 \times 10^{-6}$, $\varepsilon_{120} = 200 \times 10^{-6}$

2.43 $\varepsilon_0 = 0$, $\varepsilon_{60} = 200 \times 10^{-6}$, $\varepsilon_{120} = 500 \times 10^{-6}$

2.44   Measurements of strain at a point in a body subjected to plane stress were made along the directions marked a, b, and c in Figure H2.44. Develop expressions for ε_x, ε_y, and γ_{xy} in terms of ε_a, ε_b, and ε_c.

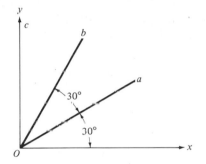

FIGURE H2.44

2.45   Three strain gages arranged as shown in Figure H2.45 gave the following readings:

$$\varepsilon_0 = 400 \times 10^{-6}, \qquad \varepsilon_{60} = 300 \times 10^{-6},$$
$$\varepsilon_{150} = -120 \times 10^{-6}$$

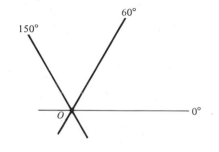

FIGURE H2.45

Determine the principal strains and their directions. Also, determine the maximum shear strains in the plane of measurements and define the corresponding axes.

2.10 Linear Stress–Strain Relations

As stated in Section 2.7, a nonrigid body experiences normal and shear strains when subjected to a general state of stress. In order to be able to analyze the behavior of a member accurately, it is necessary that the relations that exist between stresses and strains be properly examined. A complete study of such relationships will be postponed until Chapters 3 and 4. For the present, the discussion will be limited to a certain range of material behavior within which stresses and strains are linearly related.

Three cases of linear stress–strain relations will be considered in this section.

Uniaxial Stress–Strain Relations

The term *uniaxial* signifies a normal stress (tension or compression) in one direction only. Experiments show that when a prismatic bar of length L_x is subjected to a

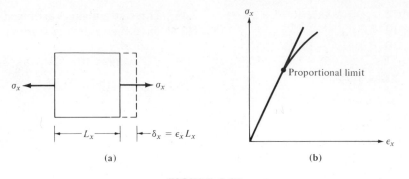

FIGURE 2.22

tensile stress σ_x, as shown in Figure 2.22(a), acting along the axis of the bar, it experiences an extensional deformation $\delta_x = \varepsilon_x L_x$ such that ε_x increases in direct relation to σ_x, as shown in Figure 2.22(b), as long as σ_x is kept within certain limits. The upper limit for σ_x for which the relationship between stress and strain is linear is known as the *proportional limit*. For values of stress above the proportional limit, the relation between stress and strain is a topic that will be discussed in Chapters 3 and 4.

The relation between stress and strain within the proportional limit can be expressed mathematically by the simple relation

$$\sigma_x = E_x \varepsilon_x \tag{2.25a}$$

in which E_x is a factor of proportionality between stress and strain and represents a unique property for a given material that has been given the name *Young's modulus of elasticity* after Thomas Young, who, in 1807, was the first to define it. It can be seen from Eq. 2.25a, since ε_x is a dimensionless quantity, that E_x has the same units as stress such as psi or ksi in the British gravitational system and N/m^2 or Pa in the SI system. Another SI unit used for this quantity is the giga pascal (10^9 pascals), which is denoted by the symbol GPa. This unit may also be expressed as GN/m^2. The relation expressed in Eq. 2.25a is referred to as *Hooke's law* in honor of Robert Hooke, who, in 1678, was the first to formulate a statement relating force to deformation.

Similarly, if the uniaxial tensile stress is in the y direction as shown in Figure 2.23(a), the resulting extensional deformation $\delta_y = \varepsilon_y L_y$ would be such that the stress σ_y is related linearly to the strain ε_y, as shown in Figure 2.23(b), provided that σ_y remains within the proportional limit for the material. Such a relation is expressed by the equation

$$\sigma_y = E_y \varepsilon_y \tag{2.25b}$$

where E_y has the same physical meaning in the y direction as E_x has in the x direction. The assumption is usually made, however, that materials exhibit identical

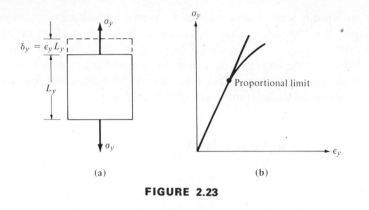

FIGURE 2.23

properties* in all directions, and therefore no distinction is necessary between the modulus of elasticity in the x direction and that in the y direction. Thus $E_x = E_y = E$, and in subsequent work, the modulus of elasticity for a given material will be denoted simply by the symbol E without a subscript. Also, Hooke's law will be simply stated as

$$\sigma = E\varepsilon \qquad (2.25c)$$

It has been pointed out earlier that when a member is subjected to a tensile normal stress, its longitudinal extensional strain is accompanied by a smaller contractional strain in a direction perpendicular (transverse) to the applied stress. Experiments have shown that the ratio of the transverse to the longitudinal strain is a constant for a given material within the range of proportionality between stress and strain. This constant is given the symbol μ and has come to be known as *Poisson's ratio*, in honor of Simeon D. Poisson, who was first to define it, in 1811. As in the case of the constant E, the assumption is made that the material exhibits identical behavior in all directions and the symbol μ without subscripts is used for Poisson's ratio regardless of the direction of the applied uniaxial stress. For example, for the uniaxial tensile stress shown in Figure 2.22(a), since the longitudinal strain is ε_x, the perpendicular (transverse) strain would be $-\mu\varepsilon_x$, which is obviously the value of the strains in both the y and z directions. Similarly, for the uniaxial tensile stress condition shown in Figure 2.23(a), since the longitudinal strain is ε_y, the perpendicular strains, or ε_x and ε_z, would be $-\mu\varepsilon_y$. Note that if the uniaxial stresses in both Figures 2.22(a) and 2.23(a) are compressive instead of tensile, the relations that have been established

---

*A material that has identical properties in all directions is known as an *isotropic* material. The term *anisotropic* is used to signify material that exhibits different properties in different directions. It should be emphasized, however, that because of manufacturing processes, all structural materials are anisotropic. Nevertheless, for most structural materials the isotropic idealization is satisfactory and leads to a great simplification in the mathematical analysis.

would apply equally well, except that the signs of the resulting strains would change. For example, if σ_x is compressive in Figure 2.22(a), ε_x would be negative and the transverse strain $\varepsilon_y = \varepsilon_z = -\mu\varepsilon_x$ would be positive. Since Poisson's ratio represents the ratio between two strains, it is a dimensionless quantity which, experimentally, has been found to vary between about 0.25 and 0.33 for materials that are commonly used in the construction of frames, structures, and machines.

Shear Stress–Shear Strain Relations

When a plane stress condition such as shown in Figure 2.24(a) is applied to an element, the resulting distortion is represented by the shear strain $\gamma_{xy} = -\gamma_{yx}$, as was discussed in Section 2.7. Because of the absence of normal stresses, this type of plane stress condition is referred to as *pure shear* and it should be emphasized that under pure shear conditions, a line element such as OA or AB does not undergo any change in length. Thus for small distortions (small values of γ_{xy}), AB' can be assumed equal in length to AB.

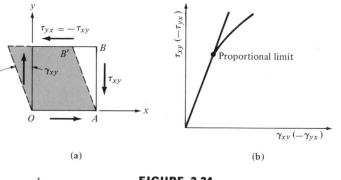

(a)   (b)

FIGURE 2.24

As in the case of normal stress–normal strain relations, experiments show that a linear relation exists, as shown in Figure 2.24(b), between the shear stress τ_{xy} (or τ_{yx}) and the shear strain γ_{xy} (or γ_{yx}) as long as the shear stress is kept within the proportional limit for the material. Thus

$$\tau_{xy} = G\gamma_{xy}$$

or

$$\tau_{yx} = G\gamma_{yx}$$

(2.26)

where G is a factor of proportionality between shear stress and shear strain and represents a unique property for a given material that has been called the *modulus of rigidity* or the *modulus of elasticity in shear*.

The three material constants, E, μ, and G, that have thus far been defined are not independent from one another. From fundamental concepts, it will be shown in Section 4.4 that these three constants are related by the equation

$$G = \frac{E}{2(1 + \mu)} \tag{2.27}$$

Relations Between General Plane Stress and Resulting Strains

The plane stress condition depicted in Figure 2.25(a) leads to the three normal strains ε_x, ε_y, and ε_z, as well as the shear strain $\gamma_{xy}(-\gamma_{yx})$, in addition to shear strains between orthogonal axes in planes perpendicular to the xy plane. The development

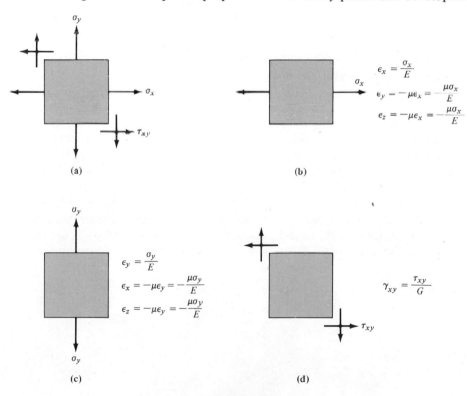

FIGURE 2.25

of the stress–strain relations for such a case can be greatly simplified by using the principle of *superposition*. In general, the principle of superposition states that if a linear relation exists between stress and strain, the resultant effect may be obtained by superposing (i.e., adding algebraically) all partial effects. Thus the plane stress condition shown in Figure 2.25(a) may be decomposed into the three partial stress conditions, which have already been discussed and which are shown in Figures 2.22(a), 2.23(a), and 2.24(a). For convenience, these three partial stress conditions are reproduced in Figure 2.25(b), (c), and (d), which also show the strains resulting from each of the three partial stress conditions. Thus, superposing the partial strains, we obtain the following relations for the normal strains*:

$$\varepsilon_x = \frac{\sigma_x}{E} - \frac{\mu\sigma_y}{E}$$

$$\varepsilon_y = \frac{\sigma_y}{E} - \frac{\mu\sigma_x}{E} \qquad\qquad (2.28)$$

$$\varepsilon_z = -\frac{\mu\sigma_x}{E} - \frac{\mu\sigma_y}{E}$$

The relation that exists between the shear stress and the shear strain in the xy plane has already been discussed and is given by Eq. 2.26.

It can be seen from Eq. 2.28 that, in general, for a plane stress condition, the resulting state of strain is not plane strain (i.e., $\varepsilon_z \neq 0$) except for the special case in which $\sigma_x = -\sigma_y$. And, as was mentioned earlier, achieving a condition of plane strain in the xy plane, for example, requires the application of a stress in the z direction in order to make $\varepsilon_z = 0$.

It is usually most convenient to invert the first two of the relations in Eq. 2.28.

---

* By a method analogous to that used in deriving Eq. 2.28, expressions for the strain ε_x, ε_y, and ε_z may be obtained for the three-dimensional state of stress in which σ_x, σ_y, and σ_z are all nonzero. These expressions are given below without proof. Thus

$$\varepsilon_x = \frac{\sigma_x}{E} - \frac{\mu}{E}(\sigma_y + \sigma_z)$$

$$\varepsilon_y = \frac{\sigma_y}{E} - \frac{\mu}{E}(\sigma_x + \sigma_z) \qquad\qquad (2.28')$$

$$\varepsilon_z = \frac{\sigma_z}{E} - \frac{\mu}{E}(\sigma_x + \sigma_y)$$

Thus Eq. 2.28 is a special case of Eq. 2.28' when σ_z is set equal to zero.

Such an inversion would result in expressions giving the normal stresses in terms of the normal strains. Thus*

$$\sigma_x = \frac{E}{1 - \mu^2}(\varepsilon_x + \mu\varepsilon_y)$$

$$\sigma_y = \frac{E}{1 - \mu^2}(\varepsilon_y + \mu\varepsilon_x) \qquad (2.29)$$

Obviously, $\sigma_z = 0$ for a plane stress condition in the xy plane.

In view of the fact that a shear stress does not contribute to normal strains, the relations expressed in Eqs. 2.28 and 2.29 apply equally well for the relations between principal stresses and principal strains. Thus†

$$\varepsilon_1 = \frac{\sigma_1}{E} - \frac{\mu\sigma_2}{E}$$

$$\varepsilon_2 = \frac{\sigma_2}{E} - \frac{\mu\sigma_1}{E} \qquad (2.30)$$

$$\varepsilon_3 = -\frac{\mu\sigma_1}{E} - \frac{\mu\sigma_2}{E}$$

* The three-dimensional counterpart of Eq. 2.29 may be obtained by solving Eq. 2.28′ for stresses in terms of strains. Thus

$$\sigma_x = \frac{E}{(1 + \mu)(1 - 2\mu)}[(1 - \mu)\varepsilon_x + \mu(\varepsilon_y + \varepsilon_z)]$$

$$\sigma_y = \frac{E}{(1 + \mu)(1 - 2\mu)}[(1 - \mu)\varepsilon_y + \mu(\varepsilon_x + \varepsilon_z)]$$

$$\sigma_z = \frac{E}{(1 + \mu)(1 - 2\mu)}[(1 - \mu)\varepsilon_z + \mu(\varepsilon_x + \varepsilon_y)]$$

(2.29′)

† The three-dimensional counterpart of Eq. 2.30 may be obtained from Eq. 2.28′ by letting the x, y, and z axes coincide with the one, two, and three principal axes, respectively. Thus

$$\varepsilon_1 = \frac{\sigma_1}{E} - \frac{\mu}{E}(\sigma_2 + \sigma_3)$$

$$\varepsilon_2 = \frac{\sigma_2}{E} - \frac{\mu}{E}(\sigma_1 + \sigma_3) \qquad (2.30′)$$

$$\varepsilon_3 = \frac{\sigma_3}{E} - \frac{\mu}{E}(\sigma_1 + \sigma_2)$$

and*

$$\sigma_1 = \frac{E}{1 - \mu^2} (\varepsilon_1 + \mu\varepsilon_2)$$

$$\sigma_2 = \frac{E}{1 - \mu^2} (\varepsilon_2 + \mu\varepsilon_1)$$

(2.31)

and for a plane stress condition $\sigma_3 = 0$.

Example 2.7

Measurements of strain at a point on the surface of a structural member resulted in the following values:

$$\varepsilon_x = 815 \times 10^{-6} \qquad \varepsilon_y = 165 \times 10^{-6} \qquad \gamma_{xy} = 1124 \times 10^{-6}$$

Assume the material to be aluminum with the properties $E = 10.5 \times 10^6$ psi and $\mu = 0.33$. Determine: (a) the principal strains in the xy plane and their directions, (b) the maximum shear strains in the xy plane (i.e., γ_1 and γ_2), (c) the stresses corresponding to the strains found in parts (a) and (b) using the appropriate stress–strain relations, and (d) the absolute maximum shear strain at the point and the corresponding absolute maximum shear stress.

Solution

(a) Mohr's circle for the given strain condition is constructed in the usual manner as shown in Figure 2.26(a) and from its geometry, the following values are determined:

* The three-dimensional counterpart of Eq. 2.31 may be obtained from Eq. 2.29′ by letting the x, y, and z axes coincide with the one, two, and three principal axes, respectively. Thus

$$\sigma_1 = \frac{E}{(1 + \mu)(1 - 2\mu)} [(1 - \mu)\varepsilon_1 + \mu(\varepsilon_2 + \varepsilon_3)]$$

$$\sigma_2 = \frac{E}{(1 + \mu)(1 - 2\mu)} [(1 - \mu)\varepsilon_2 + \mu(\varepsilon_1 + \varepsilon_3)]$$

$$\sigma_3 = \frac{E}{(1 + \mu)(1 - 2\mu)} [(1 - \mu)\varepsilon_3 + \mu(\varepsilon_1 + \varepsilon_2)]$$

(2.31′)

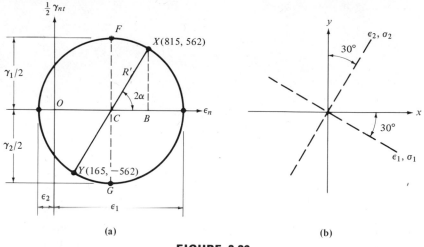

FIGURE 2.26

$OC = \frac{1}{2}(815 + 165) \times 10^{-6} = 490 \times 10^{-6}$

$R' = [(CB)^2 + (BX)^2]^{1/2} = [(325)^2 + (562)^2]^{1/2} \times 10^{-6} = 649.2 \times 10^{-6}$

$\varepsilon_1 = OC + R'$

$$= \boxed{1139.2 \times 10^{-6}}$$

$\varepsilon_2 = OC - R'$

$$= \boxed{-159.2 \times 10^{-6}}$$

$2\alpha = \tan^{-1} \dfrac{BX}{CB} = \tan^{-1} \dfrac{562}{325} = 60°$

As shown in Figure 2.26(b), the direction of ε_1 is 30° cw from the positive x axis and that of ε_2 is, of course, perpendicular to ε_1.

(b) The maximum shear strain in the xy plane is given by Eq. 2.22. Thus

$$\gamma_1 = \gamma_{fg} = 2R' = 2(649.2 \times 10^{-6})$$

$$= \boxed{1298.4 \times 10^{-6}}$$

$$\gamma_2 = \gamma_{gf} = -2R'$$

$$= \boxed{-1298.4 \times 10^{-6}}$$

(c) The principal stresses are found by using Eq. 2.31. Thus

$$\sigma_1 = \frac{E}{1 - \mu^2}(\varepsilon_1 + \mu\varepsilon_2) = \frac{10.5}{1 - 0.109}(1139.2 - 52.5)$$

$$= \boxed{12{,}806 \text{ psi}}$$

$$\sigma_2 = \frac{E}{1 - \mu^2}(\varepsilon_2 + \mu\varepsilon_1) = \frac{10.5}{1 - 0.109}(-159.2 + 375.9)$$

$$= \boxed{2554 \text{ psi}}$$

The maximum and minimum values of the shear stress associated with the two principal stresses found above are given by an equation similar to those expressed in Eq. 2.26. Thus

$$\tau_1 = G\gamma_1 = \frac{E}{2(1 + \mu)}\gamma_1 = \frac{10.5}{2(1 + 0.33)}(1298.4)$$

$$= \boxed{5125 \text{ psi}}$$

and

$$\tau_2 = G\gamma_2$$

$$= \boxed{-5125 \text{ psi}}$$

Note that the direction of the principal stresses coincides with the direction of the principal strains. This conclusion is valid in all cases where the material is assumed to be isotropic. Thus σ_1 is along ε_1 and σ_2 along ε_2 as shown in Figure 2.26(b). Also, τ_1 and τ_2 act on planes parallel to the z axis (or the σ_3 axis) that bisect the 90° angles between σ_1 and σ_2.

(d) As stated, this problem is *not* a plane strain problem (it is, in fact, a plane stress problem), and therefore a strain will be experienced in a direction perpendicular to the surfaces of measurements (i.e., the z direction). This strain will, obviously, be a principal strain and will be labeled as ε_3 momentarily until its value is determined and compared to those labeled tentatively as ε_1 and ε_2. The value of ε_3, of necessity, is equal to ε_z and is given by either Eq. 2.28 or Eq. 2.30. If Eq. 2.30 is used, we obtain

$$\varepsilon_3{}^* = -\frac{\mu\sigma_1}{E} - \frac{\mu\sigma_2}{E} = -\frac{0.33}{10.5 \times 10^6}(12{,}806 + 2554)$$

$$= \boxed{-482.7 \times 10^{-6}}$$

Thus the labeling of the principal strains has been done correctly in accordance with the algebraic requirement that $\varepsilon_1 > \varepsilon_2 > \varepsilon_3$. If this were not the case, a relabeling would become necessary in order to be able to use Eq. 2.23b correctly in determining the absolute maximum shear strain at the point. Thus, by Eq. 2.23b,

$$|\gamma_{max}| = \varepsilon_1 - \varepsilon_3 = (1139.2 + 482.7) \times 10^{-6}$$

$$= \boxed{1621.9 \times 10^{-6}}$$

The corresponding absolute maximum shear stress is obtained from a relation similar to that expressed in Eq. 2.26. Therefore,

$$|\tau_{max}| = G|\gamma_{max}|$$

$$= \frac{E}{2(1+\mu)}(|\gamma_{max}|) = \frac{10.5}{2(1+0.33)}(1621.9)$$

$$= \boxed{6402 \text{ psi}}$$

This shear stress acts on planes parallel to the ε_2 (or σ_2) axis that bisect the 90° angles between ε_1 and ε_3.

Example 2.8
Using the same data provided in Example 2.7, determine (a) the stresses σ_x, σ_y, and τ_{xy} and show them on a stress element; (b) the principal stresses and define the planes on which they act; (c) the maximum and minimum shear stresses τ_1 and τ_2, and show these stresses and the two principal stresses on a properly oriented element; and (d) the absolute maximum shear stress at the point.

* Note that the existence of a normal strain in a given direction does not necessarily mean that there is a normal stress in this direction. As a matter of fact, since this is a plane stress problem, $\sigma_3 = \sigma_z = 0$.

Solution

(a) From Eq. 2.29

$$\sigma_x = \frac{E}{1 - \mu^2}(\varepsilon_x + \mu\varepsilon_y)$$

$$= \frac{10.5}{1 - 0.109}(815 + 54.5)$$

$$= \boxed{10{,}247 \text{ psi}}$$

$$\sigma_y = \frac{E}{1 - \mu^2}(\varepsilon_y + \mu\varepsilon_x)$$

$$= \frac{10.5}{1 - 0.109}(165 + 269)$$

$$= \boxed{5115 \text{ psi}}$$

Also, by Eq. 2.26,

$$\tau_{xy} = G\gamma_{xy} = \frac{E}{2(1 + \mu)}\gamma_{xy} = \frac{10.5}{2(1 + 0.33)}(1124)$$

$$= \boxed{4437 \text{ psi}}$$

Thus a stress element can now be constructed as shown in Figure 2.27(a).
(b) Mohr's circle for stress is constructed in the usual fashion as shown in Figure 2.27(b), from which the following values are computed:

$$OC = \frac{10{,}247 + 5115}{2} = 7681 \text{ psi}$$

$$R = [(2566)^2 + (4437)^2]^{1/2} = 5126 \text{ psi}$$

$$\sigma_1 = OC + R$$

$$= \boxed{12{,}807 \text{ psi}}$$

$$\sigma_2 = OC - R$$

$$= \boxed{2555 \text{ psi}}$$

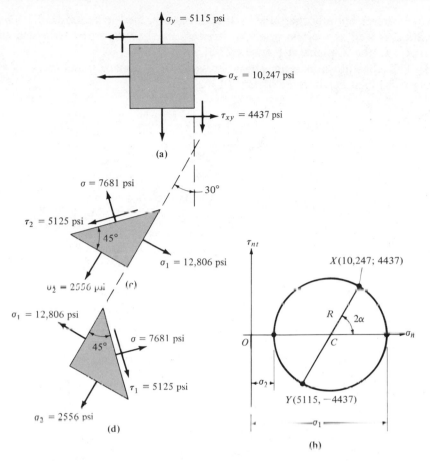

FIGURE 2.27

These values are essentially the same as those obtained in the solution of Example 2.7. Any slight difference that may exist between the two sets of answers is due to roundoff errors.

(c) From the geometry of Mohr's circle, it can be seen that

$$\tau_1 = R$$

$$= \boxed{5126 \text{ psi}}$$

$$\tau_2 = R$$

$$= \boxed{-5126 \text{ psi}}$$

which are essentially the same as the values obtained in Example 2.7. These values, as well as the two principal stresses, are shown properly oriented with respect to the X plane in Figure 2.27(c) and (d).

(d) The absolute maximum shear stress at the point on the surface of the structural member is given by Eq. 2.13b. Thus

$$|\tau_{max}| = \frac{\sigma_1 - \sigma_3}{2}$$

Since $\sigma_3 = 0$,

$$|\tau_{max}| = \frac{12,806 - 0}{2}$$

$$= \boxed{6403 \text{ psi}}$$

which is essentially the same as that determined in Example 2.7 and acts on planes parallel to the σ_2 axis that bisect the 90° angle between σ_1 and σ_3.

Homework Problems

2.46 A prismatic bar that has a square cross section 0.04 m on each side and a length of 0.25 m elongates by 0.0002 m, while its 0.04 m cross-sectional dimension decreases by 0.00001 m when subjected to a uniaxial tensile stress of 150 MPa. Determine:

(a) The modulus of elasticity for the material.
(b) The value for Poisson's ratio.
(c) The absolute maximum shear stress developed in the bar.
(d) The absolute maximum shear strain developed in the bar.

2.47 At a point in a body, the plane stress condition is as shown in Figure H2.47. If the material is such that $E = 200 \times 10^9$ N/m^2 and $\mu = 0.25$, determine:

(a) The absolute maximum shear strain at the point.
(b) The principal strains and their directions.
(c) The principal stresses and their directions.

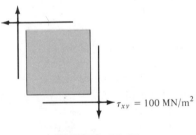

FIGURE H2.47

2.48–2.50 At a point in a body known to be in plane stress, the following values for ε_x, ε_y, and γ_{xy} are known to exist. For each of the following cases, determine:

(a) The principal stresses and their planes.
(b) The absolute maximum shear stress at the point and the plane on which it acts.

Assume that $E = 72$ GN/m^2 and $\mu = 0.3$.

2.48 $\varepsilon_x = 300 \times 10^{-6}$, $\varepsilon_y = -400 \times 10^{-6}$, $\gamma_{xy} = 800 \times 10^{-6}$

2.49 $\varepsilon_x = 900 \times 10^{-6}$, $\varepsilon_y = 200 \times 10^{-6}$, $\gamma_{xy} = 1000 \times 10^{-6}$

2.50 $\varepsilon_x = 200 \times 10^{-6}$, $\varepsilon_y = -500 \times 10^{-6}$, $\gamma_{xy} = -400 \times 10^{-6}$

2.51–2.53 At a point in a body known to be in plane stress, the following values for ε_x, ε_y, and γ_{xy} are known to exist. For each of the following cases, determine:

(a) The principal stresses and their planes.

(b) The absolute maximum shear stress at the point and the plane on which it acts.

Assume that $E = 30 \times 10^6$ psi and $\mu = 0.28$.

2.51 $\varepsilon_x = 100 \times 10^{-6}$, $\varepsilon_y = -500 \times 10^{-6}$, $\gamma_{xy} = -400 \times 10^{-6}$

2.52 $\varepsilon_x = 600 \times 10^{-6}$, $\varepsilon_y = 200 \times 10^{-6}$, $\gamma_{xy} = 300 \times 10^{-6}$

2.53 $\varepsilon_x = -100 \times 10^{-6}$, $\varepsilon_y = -300 \times 10^{-6}$, $\gamma_{xy} = 700 \times 10^{-6}$

2.54–2.56 Assume that $E = 200 \times 10^9$ N/m² and $\mu = 0.25$ and for each of the following plane stress conditions determine:

(a) The principal strains and their directions.

(b) The maximum shear strains in the xy plane (i.e., γ_1 and γ_2).

(c) The absolute maximum shear strain.

2.54 Stress condition shown in Figure H2.54.

2.55 Stress condition shown in Figure H2.55.

2.56 Stress condition shown in Figure H2.56.

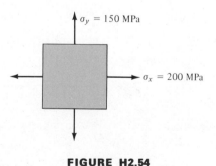

FIGURE H2.54

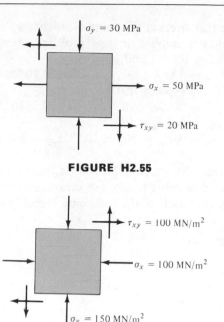

FIGURE H2.55

FIGURE H2.56

2.57–2.59 Assume plane stress conditions, and for each of the following cases, determine:

(a) The stresses σ_x, σ_y, and τ_{xy} and show them on a stress element.

(b) The principal stresses and the planes on which they act.

(c) The maximum and minimum shear stresses τ_1 and τ_2 and the corresponding planes.

(d) The absolute maximum shear stress at the point and the plane on which it acts.

Assume that $E = 200 \times 10^9$ N/m² and $\mu = 0.33$.

2.57 $\varepsilon_x = 600 \times 10^{-6}$, $\varepsilon_y = 100 \times 10^{-6}$, $\gamma_{xy} = -800 \times 10^{-6}$

2.58 $\varepsilon_x = 150 \times 10^{-6}$, $\varepsilon_y = -300 \times 10^{-6}$, $\gamma_{xy} = -500 \times 10^{-6}$

2.59 $\varepsilon_x = -200 \times 10^{-6}$, $\varepsilon_y = -400 \times 10^{-6}$, $\gamma_{xy} = -600 \times 10^{-6}$

2.60–2.62 A rectangular rosette on the surface of a structural member resulted in measurements as given below. For each of the following three cases, determine the principal stresses and the absolute maximum shear stress at the point. Note that a *plane stress* condition is implied in this problem. Assume that $E = 10.5 \times 10^6$ psi and $\mu = 0.3$.

2.60 $\varepsilon_0 = 800 \times 10^{-6}$, $\varepsilon_{45} = 200 \times 10^{-6}$, $\varepsilon_{90} = -400 \times 10^{-6}$

2.61 $\varepsilon_0 = -200 \times 10^{-6}$, $\varepsilon_{45} = 300 \times 10^{-6}$, $\varepsilon_{90} = -500 \times 10^{-6}$

2.62 $\varepsilon_0 = 0$, $\varepsilon_{45} = 500 \times 10^{-6}$, $\varepsilon_{90} = 800 \times 10^{-6}$

2.63–2.65   An equiangular rosette on the surface of a structural member resulted in measurements as given below. For each of the following three cases, deter-

mine the principal stresses and the absolute maximum shear stress at the point. Note that as in Problems 2.60 to 2.62, a *plane stress* condition is implied.

Assume that $E = 200$ GN/m^2 and $\mu = 0.25$.

2.63 $\varepsilon_0 = 600 \times 10^{-6}$, $\varepsilon_{60} = 500 \times 10^{-6}$, $\varepsilon_{120} = -200 \times 10^{-6}$

2.64 $\varepsilon_0 = -300 \times 10^{-6}$, $\varepsilon_{60} = -100 \times 10^{-6}$, $\varepsilon_{120} = 200 \times 10^{-6}$

2.65 $\varepsilon_0 = 0$, $\varepsilon_{60} = 200 \times 10^{-6}$, $\varepsilon_{120} = 500 \times 10^{-6}$

Chapter 3 Stresses and Strains in Axially Loaded Members

3.1 Introduction

In Chapter 1, several cases of members subjected to axial loads were introduced and discussed. Methods were developed for determining internal forces in axially loaded members for cases in which the applied axial load was either a constant, or varied in some fashion along the length of the member. The concept of the axial force diagram was also developed to show the way the internal axial force in a member varies along the length of the member.

In this chapter, the concepts of stress and strain developed in Chapter 2 will be discussed in relation to members that are subjected to axial loads. It should be pointed out that, in many cases, an axially loaded member is just one part of a larger frame or machine to which this axially loaded member is attached by means of a pin joint that may be assumed frictionless. Thus the force in the axially loaded member is transmitted to other parts of the frame or machine through the frictionless pin by means of *cross-shear stresses*, whose concept will be introduced and discussed by means of illustrative examples. Also, in this chapter, a detailed discussion is provided of the relations that exist between stress and strain under uniaxial conditions. Furthermore, the most significant mechanical properties of materials in use today, as obtained from the tension and the compression tests, are defined and discussed.

3.2 Stresses due to Axial Forces

Stress was defined in Section 2.2 as force per unit of area. Consider now the simple case of a prismatic member subjected to a tensile axial load P as shown in Figure 3.1(a). The internal force at any cross section within the length of the member is determined by passing a cutting plane such as $B–B$ perpendicular to the longitudinal axis of the member according to the methods developed in Section 1.2. The two portions of the member created by the cutting plane $B–B$ are isolated and their free-body diagrams constructed as shown in Figure 3.1(b) and (c). Equilibrium dictates that the internal force F_B at the cross section of interest be equal in magnitude to the applied load P.

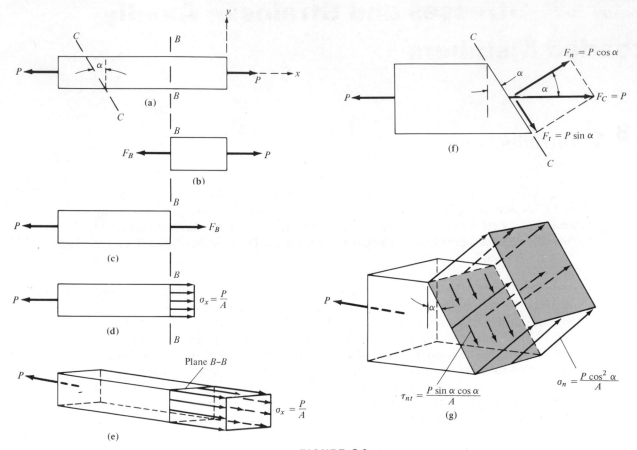

FIGURE 3.1

The force $F_B \doteq P$ acting at any cross section is perpendicular to that section and represents the resultant of an infinite number of minute forces perpendicular to the cross section and distributed in some manner throughout the cross-sectional area of the member. If the load P is applied axially and centrally, and if the material is homogeneous,* this system of minute forces can be assumed to be uniformly distributed throughout the cross-sectional area of the member. Physically, these minute forces represent the intensity of the force (i.e., force per unit of area, or stress) at any location within the cross-sectional area of the member. Since, in the case under consideration, the force per unit of area, or stress, is perpendicular to the plane on which it is acting, it is by definition a normal stress σ. Also, since the x axis is along the longitudinal axis of the member as shown in Figure 3.1(a), this stress is labeled as σ_x. If the cross-sectional area of the member is denoted by the symbol A, and since

---

* The term *homogeneous* is used to signify a material that exhibits the same properties at every point in its domain.

the force normal to this area, F_n, is, in this case, equal to $F_B = P$, the normal stress σ_x can be expressed by the equation

$$\sigma_x = \frac{F_n}{A} = \frac{F_B}{A} = \frac{P}{A} \tag{3.1}$$

Thus, in Figure 3.1(b), for example, the resultant force F_B can be replaced by a stress system σ_x as shown in Figure 3.1(d), which, when multiplied by the area A, yields the resultant force $F_B = P$. A three-dimensional schematic sketch of Figure 3.1(b) is depicted in Figure 3.1(e), in which the resultant force $F_B = P$ has been replaced by its equivalent stress system σ_x, and in which, for the sake of clarity, only a few of the components of the stress system are shown.

Cross section B–B in Figure 3.1(a), which is perpendicular to the longitudinal axis of the member, is found to be free of any shear stresses. Any other plane such as plane C–C, inclined to plane B–B through any angle α, as shown in Figure 3.1(a), is, however, subjected to a system of shear stresses in addition to normal stresses. The free-body diagram of that part of the member to the left of section C–C is shown in Figure 3.1(f). As before, equilibrium requires that $F_C = P$ act on plane C–C. In this case, however, the internal force F_C is not perpendicular to the plane of the member cut by plane C–C. Therefore, F_C may be decomposed into components F_n, normal to the plane, and F_t, parallel to the plane. The force F_n leads to a system of normal stresses given by Eq. 3.1, in which $F_n = F_C \cos \alpha = P \cos \alpha$ and the area of the section on which it acts, A_{C-C}, is equal to $A/\cos \alpha$. The force $F_t = F_C \sin \alpha = P \sin \alpha$ is the resultant of an infinite number of minute forces acting parallel to (i.e., in the plane of) the area A_{C-C} and represent the intensity of the force (i.e., force per unit of area, or stress in the plane of the area). As in the case of the normal stress, the assumption is made that this stress in the plane of the area (i.e., shear stress τ_{nt}) is uniformly distributed over the area A_{C-C} and an average value for τ_{nt} on plane C–C is obtained from the basic definition of stress given in Section 2.2. Thus

$$\tau_{nt} = \frac{F_t}{A_{C-C}} = \frac{F_C \sin \alpha \cos \alpha}{A} \tag{3.2}$$

A three-dimensional sketch showing the normal stress σ_n and shear stresses τ_{nt} on plane C–C is shown in Figure 3.1(g), in which, for the sake of clarity, only a few components of the two stress systems (i.e., normal and shear) are shown.

Consider now a prismatic bar subjected to a tensile load P, and establish an xy coordinate system such that the x axis is parallel to the longitudinal axis of the bar as shown in Figure 3.2(a). The origin of the coordinate system was located at point O, at which point the stress condition is to be investigated. Two cutting planes B–B and C–C perpendicular to the x axis and an infinitesimal distance dx apart are constructed at the point as shown in Figure 3.2(a). That part of the prismatic bar

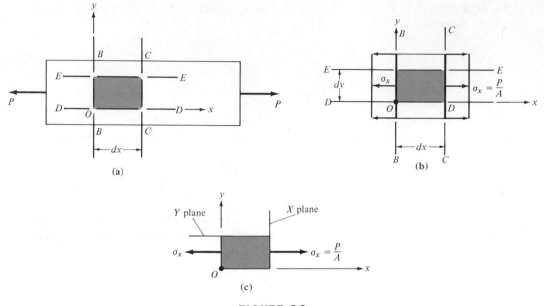

FIGURE 3.2

contained between the two parallel X planes is isolated and its free-body diagram
constructed as shown in Figure 3.2(b). The resultant forces $F_B = P$ and $F_C = P$ on
planes $B–B$ and $C–C$, respectively, have been replaced by the stresses they produce
on these two planes. These stresses are $\sigma_x = P/A$, where A is the cross-sectional area
of the prismatic bar. Two more cutting planes $D–D$ and $E–E$ perpendicular to the y
axis and an infinitesimal distance dy apart are constructed at point O as shown in
Figure 3.2(a) and (b).

The infinitesimal element contained between these two Y planes and the two X
planes previously described is now isolated as a free-body diagram and shown in
Figure 3.2(c). This free-body diagram is the stress element at point O on the surface of
the prismatic bar of Figure 3.2(a). Note that the X planes have positive normal (i.e.,
tensile) stresses σ_x acting on them but are free of any shear stresses. Thus, by
definition, the X planes are principal planes. Note also that the Y planes, being free
from shear stresses are principal planes on which the normal stress σ_y is zero. The fact
that $\sigma_y = 0$ in this particular case can be clearly established by isolating that part of
the element above plane $E–E$ in Figure 3.2(b) and noting that no forces exist to cause
normal stresses in the y direction. The condition depicted in Figure 3.2(c) is referred
to as a *uniaxial tensile stress condition*, which is a special case of the plane stress
condition. If the sense of the load P in Figure 3.2(a) is reversed (i.e., compressive
instead of tensile), the resulting stress condition at point O would be uniaxial compression (i.e., σ_x is negative).

Clearly, since the X and Y planes are principal planes, the associated normal
stresses are principal stresses. Thus in this case $\sigma_1 = \sigma_x = P/A$ and $\sigma_2 = \sigma_3 = 0$.

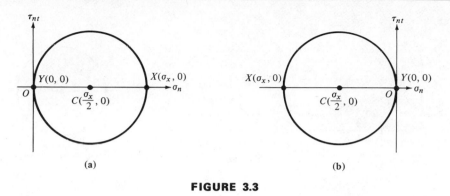

FIGURE 3.3

Mohr's circle describing the stress condition shown in Figure 3.2(c) is depicted in Figure 3.3(a) when σ_x is positive (tension) and in Figure 3.3(b) when σ_x is negative (compression).

Example 3.1

A member with a rectangular cross-sectional area $A = 2$ in. $\times$ 4 in. is subjected to a compressive axial load as shown in Figure 3.4(a). (a) Determine the principal stresses and show them on a stress element properly oriented with respect to the given xy coordinate system. (b) Find the normal and shear stresses on plane B–B, inclined to the cross section of the member through a 30° ccw angle. Show a part of the member containing this plane as a free-body diagram. (c) Compute the absolute value of the maximum shear stress developed in the member and define the plane on which it acts. Show a part of the member containing this plane as a free-body diagram.

Solution

(a) Isolate a stress element, as was explained earlier in this section, contained between two X planes and two Y planes at any point on the surface of the member. The exact location of the point to be investigated is irrelevant, since the cross-sectional area and the internal axial force are both constant. Such a stress element is shown in Figure 3.4(b), in which

$$\sigma_1 = \sigma_y = 0, \qquad \sigma_2 = \sigma_z = 0$$

and

$$\sigma_3 = \sigma_x = -\frac{80}{(2)(4)}$$

$$= -10 \text{ ksi}$$

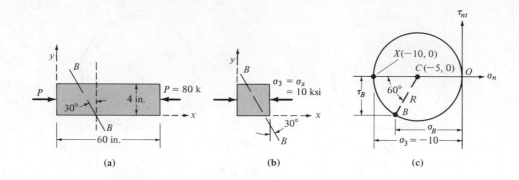

(a)   (b)   (c)

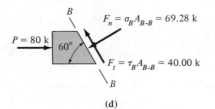

(d)

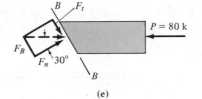

(e)

(f)   **FIGURE 3.4**   (g)

(b) The solution to this part of the problem can be obtained in one of two ways:

(i) Construct Mohr's circle for the stress condition depicted in Figure 3.4(b). This circle is shown in Figure 3.4(c). The plane of interest, plane B–B, is inclined to the X plane through a 30° ccw angle as shown in both Figure 3.4(a) and (b). Point B representing the stress condition on this plane is located on the circle in the usual manner, as shown in Figure 3.4(c). The coordinates of this point give the desired stresses. Thus

$$OC = -5 \text{ ksi} \quad \text{and} \quad R = 5 \text{ ksi}$$

$$\sigma_B = OC - R \cos 60 = -5 - 2.5$$

$$= \boxed{-7.50 \text{ ksi}}$$

$$\tau_B = -R \sin 60$$

$$= \boxed{-4.33 \text{ ksi}}$$

These stresses are transformed to forces and shown on the free-body diagram of that part of the member to the left of plane B–B, as shown in Figure 3.4(d).

The transformation of a stress to a force is accomplished by means of Eqs. 3.1 and 3.2. Thus, by Eq. 3.1,

$$F_n = \sigma_B A_{B-B}$$

$$= -\frac{7.50 \ A}{\cos 30} = -\frac{7.50(2 \times 4)}{\cos 30}$$

$$= \boxed{-69.28 \text{ k}}$$

and, by Eq. 3.2,

$$F_t = \tau_B A_{B-B}$$

$$= -\frac{4.33 \ A}{\cos 30} = -\frac{4.33(2 \times 4)}{\cos 30}$$

$$= \boxed{-40.00 \text{ k}}$$

(ii) The same conclusions for σ_B and τ_B may be reached if one isolates one part of the member containing plane B–B as a free body. That part of the member to the right of plane B–B is isolated as shown in Figure 3.4(e). Equilibrium dictates that the horizontal force F_B on plane B–B be equal in magnitude to the applied load $P = 80$ k. This force can then be decomposed into two components, F_n normal to plane B–B and F_t parallel to plane B–B as follows:

$$F_n = -F_B \cos 30$$

$$F_t = -F_B \sin 30$$

The area of plane B–B can be expressed in terms of the cross-sectional area A as follows:

$$A_{B-B} = \frac{A}{\cos 30}$$

Therefore, by Eq. 3.1,

$$\sigma_B = \frac{F_n}{A_{B-B}} = -\frac{F_B \cos 30}{A/\cos 30} = -\frac{F_B \cos^2 30}{A}$$

$$= \boxed{-7.50 \text{ ksi}}$$

and, by Eq. 3.2,

$$\tau_B = \frac{F_t}{A_{B-B}} = -\frac{F_B \sin 30}{A/\cos 30} = -\frac{F_B \sin 30 \cos 30}{A}$$

$$= \boxed{-4.33 \text{ ksi}}$$

These values are, of course, the same as those obtained by the previous solution. The reader is, once again, referred to Figure 3.4(d) for the free-body diagram of that part of the member to the left of plane $B-B$.

(c) The solution for part (a) showed that $\sigma_1 = \sigma_2 = 0$ and that $\sigma_3 = -10$ ksi. Thus, by Eq. 2.13b,

$$|\tau_{\text{max}}| = \frac{\sigma_1 - \sigma_3}{2}$$

$$= \frac{0 + 10}{2}$$

$$= \boxed{5 \text{ ksi}}$$

This absolute maximum shear stress acts on two perpendicular planes that are parallel to the σ_2 axis and bisect the 90° angle between σ_1 and σ_3. Also, by Eq. 2.11, the normal stress σ_A on such planes has a value of

$$\sigma_A = \frac{\sigma_1 + \sigma_3}{2} = \frac{0 - 10}{2}$$

$$= \boxed{-5 \text{ ksi}}$$

One such plane is shown in the three-dimensional stress element of Figure 3.4(f). Since $\sigma_2 = \sigma_1 = 0$, there exist two other perpendicular planes on which $|\tau_{\text{max}}| = 5$ ksi acts. These two perpendicular planes are parallel to the σ_1 direction and bisect the 90° angles between σ_2 and σ_3. One such plane, along with its stresses, is shown in the three-dimensional stress element of Figure 3.4(g).

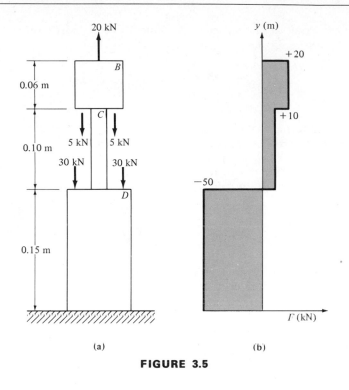

(a)

(b)

FIGURE 3.5

Example 3.2

A composite member consisting of three parts is subjected to the loads shown in Figure 3.5(a). The three members have circular cross-sectional areas such that $A_B = 4 \times 10^{-4}$ m², $A_C = 1 \times 10^{-4}$ m², and $A_D = 6 \times 10^{-4}$ m². Determine (a) the principal stress of largest magnitude in the member and specify its location and plane, and (b) the absolute maximum value of the shear stress specifying its location and plane.

Solution

(a) To determine the principal stress of largest magnitude in the composite member, one needs to compute the normal stress in each of the three parts of the member on planes perpendicular to the longitudinal axis of the member. From Example 3.1 one concludes that these normal stresses are the principal stresses in the three parts. The force diagram for the member is shown in Figure 3.5(b).

The internal axial force in part B is found to be a tensile force of 20 kN. Thus the normal stress σ_B in member B is given by

$$\sigma_B = \frac{20 \times 10^3}{4 \times 10^{-4}} = 50 \text{ MPa}$$

The internal axial force in part C is found to be a tensile force of 10 kN. Therefore, the normal stress in member C is given by

$$\sigma_C = \frac{10 \times 10^3}{1 \times 10^{-4}} = 100 \text{ MPa}$$

Similarly, the internal axial force in part D is found to be a compressive force of 50 kN, and therefore the normal stress in member D is

$$\sigma_D = \frac{-50 \times 10^3}{6 \times 10^{-4}} = -83.3 \text{ MPa}$$

Thus the absolute maximum normal stress in the composite member is

$$\boxed{\sigma_C = 100 \text{ MPa}}$$

and acts on all planes within part C of the member that are perpendicular to its longitudinal axis.

(b) Since, as was mentioned earlier, the normal stresses that have been computed are principal stresses and realizing that the other two principal stresses for each case are zero, one can compute for each of the three cases the absolute maximum shear stress from Eq. 2.13b. Thus

$$|\tau_{max}|_B = \frac{50 - 0}{2} = 25 \text{ MPa}$$

$$|\tau_{max}|_C = \frac{100 - 0}{2} = 50 \text{ MPa}$$

$$|\tau_{max}|_D = \frac{0 + 83.3}{2} = 41.7 \text{ MPa}$$

Therefore, the absolute maximum shear stress in the member is

$$\boxed{|\tau_{max}|_C = 50 \text{ MPa}}$$

and occurs anywhere within part C on planes that are inclined to the cross sections of this part through 45° angles.

Example 3.3

Refer to Example 1.3 and assume the cross-sectional area of the sampling device to be $A = 2 \times 10^{-5}$ m^2. Determine the position along the device where the normal stress is 1.5 MN/m^2. What is the absolute maximum shear stress at this location?

Solution. In Example 1.3, the internal axial force in the device was found to be

$$F = 80 - 500x^2$$

where x is measured downward from the planet's surface as shown in Figure 1.6(a). Therefore, the normal stress σ_x at any position along the device is given by

$$\sigma_x = \frac{80 - 500x^2}{2 \times 10^{-5}}$$
$$= (4 - 25x^2) \times 10^6 \text{ N/m}^2$$
$$= 4 - 25x^2 \text{ MN/m}^2$$

To obtain the value of x (i.e., the position of the cross section) for which the normal tensile stress is 1.5 MN/m^2, we equate σ_x to this value in the equation above to obtain

$$1.5 = 4 - 25x^2$$

from which

$$x^2 = 0.1 \quad \text{and} \quad x = \pm 0.316 \text{ m}$$

The negative sign is physically meaningless and the normal stress is 1.5 MN/m^2 at a distance below the surface of the planet given by

$$x = \boxed{0.316 \text{ m}}$$

The principal stresses at this location are

$$\sigma_1 = 1.5 \text{ MN/m}^2$$
$$\sigma_2 = \sigma_3 = 0$$

Therefore, by Eq. 2.13b,

$$|\tau_{max}| = \frac{\sigma_1 - \sigma_3}{2} = \frac{1.5 - 0}{2}$$

$$= \boxed{0.75 \text{ MN/m}^2}$$

and acts on planes inclined at 45° with the axis of the sampling device.

3.3 Strain and Deformation due to Axial Forces

Longitudinal or normal strain was defined in Section 2.7 as unit deformation or change in length per unit of length. For a prismatic bar such as the one shown in Figure 3.1(a), the strain ε_x along the longitudinal axis of the member can be obtained at any cross section such as B–B if the normal stress σ_x at the section is known and if the modulus of elasticity for the material is given (i.e., $\varepsilon_x = \sigma_x/E$ by Eq. 2.25). If we neglect body forces and remember that σ_x is constant throughout a given cross section and that it does not vary from cross section to cross section, and since for a homogeneous isotropic material E is constant, it follows that ε_x is constant throughout the member.

The total deformation of an axially loaded member such as shown in Figure 3.1(a) is obtained from the basic definition for strain (i.e., Eq. 2.16a). Thus the total deformation or change in length in the x direction is $\delta_x = \varepsilon_x L = \sigma_x L/E$, where L is the original length of the member. If the applied axial load is tensile, the strain ε_x would be positive and the deformation δ_x would represent an increase in the length of the member. On the other hand, if the applied axial load is compressive, the strain ε_x would be negative and the deformation δ_x would represent a decrease in the length of the member.

As was pointed out in Section 2.10, the longitudinal strain ε_x is accompanied by transverse strains in all directions perpendicular to the x axis. If the two perpendicular directions y and z are examined, we obtain the transverse strains

$$\varepsilon_y = -\mu\varepsilon_x = -\frac{\mu\sigma_x}{E}$$

and

$$\varepsilon_z = \varepsilon_y = -\mu\varepsilon_x = -\frac{\mu\sigma_x}{E}$$

where, as defined earlier, μ is Poisson's ratio for the material. Thus this uniaxial stress condition (which is a very special case of the plane stress condition) leads to a triaxial strain situation in which $\varepsilon_x = \sigma_x/E$ and $\varepsilon_y = \varepsilon_z = -\mu\sigma_x/E$. Since, as was shown in Section 3.2, σ_x is a principal stress, it follows that ε_x as well as ε_y and ε_z are principal strains. If, for example, σ_x is tension, $\varepsilon_1 = \varepsilon_x = \sigma_x/E$, $\varepsilon_2 = \varepsilon_y = -\mu\sigma_x/E$, and $\varepsilon_3 = \varepsilon_z = -\mu\sigma_x/E$. Thus, by Eq. 2.23b,

$$|\gamma_{max}| = \left(\frac{\sigma_x}{E}\right)(1 + \mu)$$

Example 3.4

The member described in Example 3.1 was found to decrease in length by 0.06 in. after the 80-k compressive load is applied. At the same time its 4-in. cross-sectional dimension was found to increase by 0.001 in. Assume a linear relation between stress and strain and determine the modulus of elasticity as well as Poisson's ratio for the material.

Solution. Since the initial length of the member is 60 in., the strain $\varepsilon_x = -0.06/60 = -1 \times 10^{-3}$. Thus, by Eq. 2.25,

$$E = \frac{\sigma_x}{\varepsilon_x} = -\frac{10}{-1 \times 10^{-3}}$$

$$= \boxed{10 \times 10^3 \text{ ksi}}$$

where the value $\sigma_x = -10$ ksi is that obtained in the solution of Example 3.1.

The transverse strain $\varepsilon_y = 0.001/4 = 2.5 \times 10^{-4}$. Thus, by the definition of Poisson's ratio given in Section 2.10,

$$\mu = \frac{\varepsilon_y}{\varepsilon_x} = \frac{2.5 \times 10^{-4}}{1 \times 10^{-3}}$$

$$= \boxed{0.25}$$

Example 3.5

Refer to the composite member described in Example 3.2 and assume that $E_B = 73 \times 10^9 \text{ N/m}^2$, $E_C = 207 \times 10^9 \text{ N/m}^2$, and $E_D = 104 \times 10^9 \text{ N/m}^2$. Determine the total deformation of the member.

Solution. Consider the deformation of member B. By Eq. 2.16a,

$$\delta_B = \varepsilon_B L_B = \frac{\sigma_B}{E_B} L_B = \frac{F_B L_B}{A_B E_B}$$

Similarly,

$$\delta_C = \frac{F_C L_C}{A_C E_C} \quad \text{and} \quad \delta_D = \frac{F_D L_D}{A_D E_D}$$

Therefore, the total deformation of the composite member is the algebraic sum of δ_B, δ_C, and δ_D. Thus

$$\delta = \sum_{i=1}^{n} \delta_i = \frac{F_B L_B}{A_B E_B} + \frac{F_C L_C}{A_C E_C} + \frac{F_D L_D}{A_D E_D}$$

where, from the solution of Example 3.2, $F_B = 20,000$ N, $F_C = 10,000$ N, and $F_D = -50,000$ N. Therefore,

$$\delta = \frac{20,000 \times 0.06}{4 \times 10^{-4} \times 73 \times 10^9} + \frac{10,000 \times 0.1}{1 \times 10^{-4} \times 207 \times 10^9}$$

$$+ \frac{-50,000 \times 0.15}{6 \times 10^{-4} \times 104 \times 10^9}$$

$$= (0.411 + 0.483 - 1.202) \times 10^{-4}$$

$$= \boxed{-3.08 \times 10^{-5} \text{ m}}$$

Example 3.6

Determine the total deformation of the sampling device described in Example 1.3 if its material is aluminum for which $E = 73$ GN/m^2 and if the relation between stress and strain is linear.

Solution. Consider a differential section of the device at a distance x below the surface of the planet, as shown in Figure 3.6(a) and (b). By definition, the differential quantity of deformation $d\delta$ is the product of the strain and the length of the differential section. Thus

$$d\delta = \varepsilon_x \, dx = \left(\frac{\sigma_x}{E}\right) dx$$

where $\sigma_x = 4 - 25x^2$ MN/m^2 from Example 3.3. Therefore,

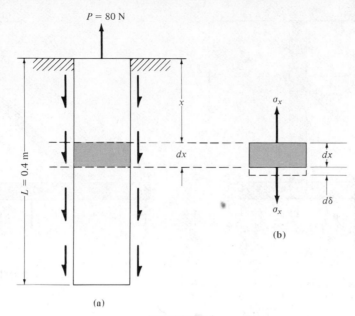

$P = 80$ N

$L = 0.4$ m

x

dx

σ_x

dx

$d\delta$

σ_x

(b)

(a)

FIGURE 3.6

$$d\delta = \frac{1}{E}\left(4 - 25x^2\right) dx$$

and

$$\delta = \int_0^L d\delta = \frac{1}{E}\int_0^L \left(4 - 25x^2\right) dx$$

$$= \frac{1}{E}\left(4L - 8.33L^3\right)$$

For $E = 73$ GN/m^2 and $L = 0.4$ m, the equation above leads to

$$\boxed{\delta = 1.461 \times 10^{-5} \text{ m}}$$

Example 3.7

The pin-connected frame shown in Figure 3.7(a) can be assumed to be weightless. Determine (a) the average value of the shear stress developed in the $\frac{1}{2}$-in.-diameter pin at A. Note the magnified details of this pin joint shown in Figure 3.7(b), (b) the cross-sectional area of rod CE if its normal stress is to have a maximum value of 15 ksi, and (c) the average value of the shear stress

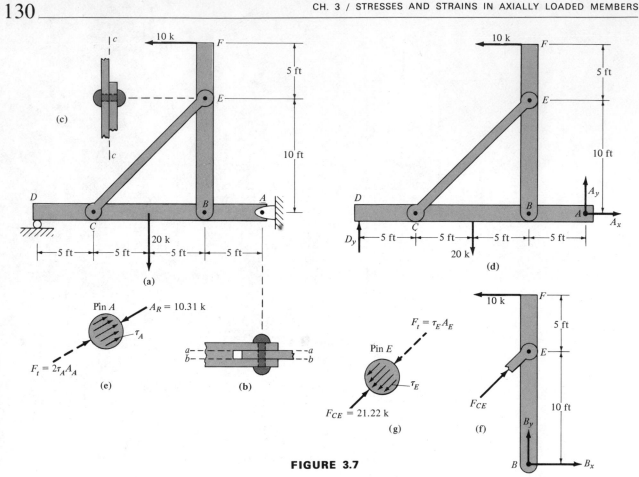

FIGURE 3.7

developed in the $\frac{3}{4}$-in.-diameter pin at E. Note the magnified details of this pin joint shown in Figure 3.7(c).

Solution

(a) Consider the free-body diagram of the entire frame as shown in Figure 3.7(d) and apply the equations of equilibrium as follows:

$$\circlearrowleft \quad \sum M_A = 0$$

$$10(15) + 20(10) - D_y(20) = 0$$

$$D_y = 17.5 \text{ k}$$

$$\rightarrow \quad \sum F_x = 0$$

$$A_x - 10 = 0$$

$$A_x = 10 \text{ k}$$

$$\uparrow \quad \sum F_Y = 0$$

$$A_y + D_y - 20 = 0$$

$$A_y = 2.5 \text{ k}$$

Therefore, the resultant force A_R at joint A is

$$A_R = \sqrt{A_x^2 + A_y^2} = 10.31 \text{ k}$$

This force is transmitted from member $DCBA$ to the support at A through a system of shear stresses (cross-shear stresses) developed at the two sections a–a and b–b in the pin at joint A shown in Figure 3.7(b). Thus two cross sections in the pin are mobilized to transmit the force $A_R = 10.31$ k from the frame to the support and the pin is said to be in *double shear*. Each of these two cross sections carries one-half of the total shear force F_t developed in the pin. A view of the free-body diagram of the pin at A is shown in Figure 3.7(e).

Therefore, the average value of the shear stress in the pin at A can be obtained by dividing the force transmitted by the pin (i.e., 10.31 k) by twice its cross-sectional area A_A. Thus

$$\tau_A = \frac{F_t}{2A_A} = \frac{A_R}{2A_A} = \frac{10.31}{2(0.1963)}$$

$$= \boxed{26.26 \text{ ksi}}$$

(b) The force in rod CE is obtained from consideration of the equilibrium of the free-body diagram of member BEF shown in Figure 3.7(f). Thus

$$\circlearrowright \quad \sum M_B = 0$$

$$10(15) - 0.707F_{CE}(10) = 0$$

$$F_{CE} = 21.22 \text{ k(C)}$$

The cross-sectional area of rod CE, A_{CE}, is given by Eq. 3.1. Thus

$$A_{CE} = \frac{F_{CE}}{\sigma_{CE}} = \frac{21.22}{15}$$

$$= \boxed{1.415 \text{ in.}^2}$$

(c) The force $F_{CE} = 21.22$ k on rod CE is transmitted to member BEF through a system of cross-shear stresses developed at section c–c in the pin at E shown in Figure 3.7(c). In this case only one cross section in the pin is mobilized to transmit the shear load F_t, and the pin is said to be in *single shear*. A view of the free-body diagram of the pin at E is shown in Figure 3.7(g). Thus the average value of the shear stress in the pin at E can be obtained by dividing the force transmitted by the pin (i.e., 21.22 k) by its cross-sectional area. Thus

$$\tau_E = \frac{F_t}{A_E} = \frac{21.22}{0.4418}$$

$$= \boxed{48.03 \text{ ksi}}$$

Example 3.8

Determine the principal stress of largest magnitude and the absolute maximum shear stress in member AB of the press shown in Figure 3.8(a). Assume AB to be weightless and to have a rectangular cross section 0.005 m × 0.01 m. If the pin at B is in double shear, determine the average value of the cross-shear stress if the pin has a diameter equal to 0.005 m. Assume frictionless conditions between the vertical column and the upper jaw of the press.

Solution. Since member AB is assumed weightless, it is a two-force member subjected to an axial force whose magnitude may be determined by isolating member BCD and constructing its free-body diagram as shown in Figure 3.8(b). Thus

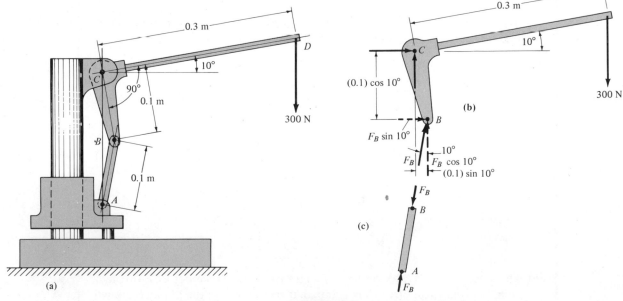

FIGURE 3.8

$$\circlearrowright \quad \sum M_C = 0$$

$$F_B \sin 10(0.1)(\cos 10) + F_B \cos 10 \, (0.1 \times \sin 10) - 300(0.3 \times \cos 10) = 0$$

$$F_B = 2591.4 \text{ N}$$

Therefore, as seen from Figure 3.8(c), member AB is subjected to a compressive axial force $F_B = 2591.4$ N. By Eq. 3.1, the normal stress on any cross section perpendicular to member AB is compression given by

$$\sigma = -\frac{F_B}{A_{AB}} = -\frac{2591.4}{5 \times 1 \times 10^{-5}}$$

$$= -51.83 \times 10^6 \text{ N/m}^2$$

$$= -51.83 \text{ MPa}$$

Since, as discussed earlier, cross sections perpendicular to the axis of an axially loaded member are free from shear stresses, they are principal planes and the stress $\sigma = -51.83$ MPa is one of the principal stresses at any point within the length AB. The other two principal stresses are obviously zero. Thus $\sigma_1 = \sigma_2 = 0$ and $\sigma_3 = 51.83$ MPa and the principal stress of largest magnitude in member AB is

$$\boxed{|\sigma_3| = 51.83 \text{ MPa}}$$

The absolute maximum shear stress in member AB is given by Eq. 2.13b. Thus

$$|\tau_{\max}| = \frac{\sigma_1 - \sigma_3}{2} = \frac{0 + 51.83}{2}$$

$$= 25.92 \text{ MPa}$$

and acts on planes making a 45° angle with the axis of member AB.

The force $F_B = 2591.4$ N is transmitted from member BCD to member AB through the pin at B, which is in double shear. Therefore, the average value of the cross-shear stress in the pin at B is

$$\tau_B = \frac{F_B}{2A_B} = \frac{2591.4}{2(1.963 \times 10^{-5})}$$

$$= \boxed{66.0 \times 10^6 \text{ N/m}^2}$$

$$= \boxed{66.0 \text{ MPa}}$$

Homework Problems

3.1 Refer to the composite member shown in Figure H1.3 and assume the cross-sectional areas of the three parts of the member to be as follows: $A_{AB} = 2$ in.^{2}, $A_{BD} = 1$ in.^{2}, and $A_{DE} = 3$ in.^{2}. Determine:

(a) The principal stress of largest magnitude in the composite member.

(b) The absolute maximum shear stress in the composite member.

(c) The total deformation of the entire member if the material is aluminum for which $E = 10 \times 10^6$ psi.

3.2 Refer to the composite member shown in Figure H1.6; assume the cross-sectional area for all parts to be the same, that is, $A_{AB} = A_{BC} = A_{CD} = 4 \times 10^{-4}$ m^2 and the material to be such that $E_{AB} = 2E_{CD} = \frac{1}{3}E_{BC} = 73 \times 10^9$ N/m^2. Determine:

(a) The principal stress of largest magnitude in the composite member.

(b) The absolute maximum shear stress in the composite member.

(c) The deformation in part BC of the member.

(d) The total deformation of the entire composite member.

3.3 Refer to the member shown in Figure H1.8 and the description given in Problem 1.8. Assume the modulus of elasticity to be $E = 30 \times 10^6$ psi. Determine:

(a) The principal stress of largest magnitude in the composite member.

(b) The absolute maximum shear stress at the upper end of the member.

(c) The total deformation of the member.

3.4 Refer to the steel piling shown in Figure H1.11 and the description given in Problem 1.11. Assume the material to have a modulus of elasticity $E = 30 \times 10^6$ psi and plot the principal stress of largest magnitude as well as the absolute maximum shear stress as functions of the lengthwise coordinate x. Determine also the extension of the piling.

3.5 A wall bracket is constructed as shown in Figure H3.5. Member BC may be assumed rigid. If the drum weighs 2000 lb, determine:

(a) The principal stress of largest magnitude in member AB.

(b) The elongation of member AB, assuming that $E_{AD} = 10 \times 10^6$ psi and $E_{DB} = 30 \times 10^6$ psi.

(c) The cross shear stress in the $\frac{1}{4}$-in.-diameter pin at A which is in double shear.

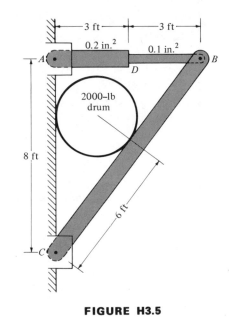

FIGURE H3.5

3.6 Assume all members in the pin-connected frame shown in Figure H3.6 to be weightless. Determine:

(a) The principal stress of largest magnitude in rod BE which has a circular cross section of radius equal to $\frac{1}{2}$ in.

(b) The absolute maximum shear stress in rod BE.

(c) The cross shear stress in the $\frac{1}{2}$-in.-diameter pin at B which is in single shear.

(d) The cross shear stress in the $\frac{1}{2}$-in.-diameter pin at D which is in double shear.

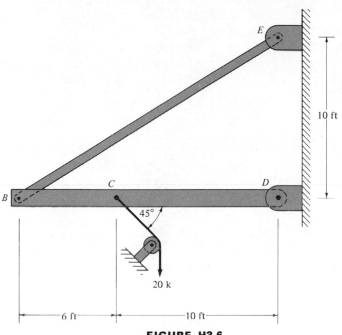

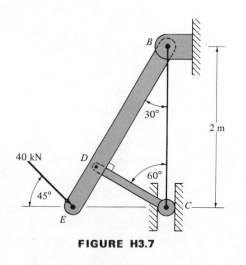

FIGURE H3.6

(b) The extension of the cable *BC* if the material is steel for which $E = 207$ GN/m².
(c) The maximum compressive stress in member *DC* if its cross section is a 0.02 m × 0.04 m rectangle.
(d) The cross-shear stress in the 0.015-m-diameter pin at *D* if it is in single shear.

3.8 The cross-sectional area of all members of the pin connected truss shown in Figure H3.8 is $A = 4 \times 10^{-4}$ m². If the material is such that $E = 207 \times 10^9$ N/m², determine:
(a) The principal stress of largest magnitude in every member of the truss.
(b) The cross-shear stress in the 0.02-m-diameter pin *D* if it is in double shear.
(c) The vertical movement of joint *F*.

(e) The deformation of rod *BE* if the material is steel for which $E = 30 \times 10^6$ psi.

3.7 Assume frictionless conditions at all supports of the frame in Figure H3.7 and determine:
(a) The diameter of the cable *BC* if the normal stress in it is to have a maximum value of 50 MPa.

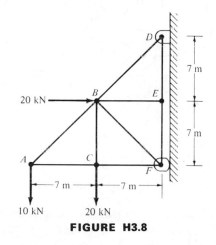

FIGURE H3.8

3.9 Determine the horizontal movement of joint *F* of the pin-connected truss shown in Figure H3.9.

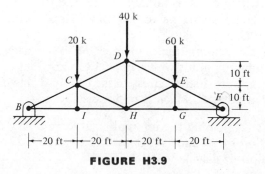

FIGURE H3.9

FIGURE H3.7

Assume the material to have a modulus of elasticity $E = 10 \times 10^3$ ksi and all members to have the same cross-sectional area $A = 1$ in.^{2}.

3.10 Determine the tensile stresses and the deformations induced in cables AB and CD in the frame shown in Figure H3.10. The applied force $F = 30$ kN is directed along the negative x axis and the cross-sectional areas of both cables are the same and equal to 5×10^{-4} m^2. Assume that $E = 207 \times 10^9$ N/m^2.

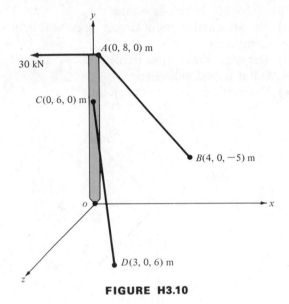

FIGURE H3.10

3.11 The wheel force on the landing gear shown in Figure H3.11 is $P = 30$ kN. Assume the system to be at rest and determine:

(a) The principal stress of largest magnitude in member BC which is assumed weightless and whose cross-sectional area is 15×10^{-4} m^2.

(b) The absolute maximum shear stress in member BC.

(c) The deformation of member BC if the material is aluminum for which $E = 73$ GN/m^2.

(d) The cross-shear stress in the 0.03-m-diameter pin at point C, which is in double shear.

3.12 Assuming the internal axial force at any cross section of the member shown in Figure H3.12 to be uniformly distributed, determine:

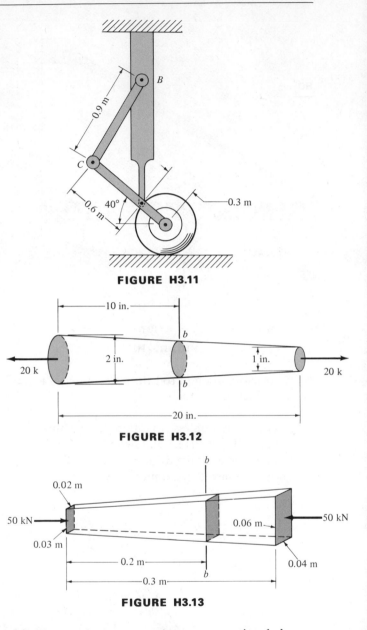

FIGURE H3.11

FIGURE H3.12

FIGURE H3.13

(a) The maximum normal stress at section b–b.

(b) The total deformation of the member if $E = 30 \times 10^6$ psi.

3.13 Solve Problem 3.12 using the information given in Figure H3.13 and assume that $E = 73 \times 10^9$ N/m^2.

3.4 Mechanical Properties of Materials Under Axial Forces

The term *mechanical properties* refers to certain material characteristics that, in general, are different for different materials. Some of these mechanical properties, including the modulus of elasticity and Poisson's ratio, have already been introduced and used in Section 2.10 during the development of linear stress–strain relations. A more complete discussion of the mechanical properties of materials is given in this section.

In general, mechanical properties of a given material are obtained by means of a tension or compression test. However, as discussed in Section 4.8, the torsion test is used to obtain shearing stress–strain characteristics of materials. Mechanical properties are, in general, obtained at room temperature under loads that are applied very gradually (i.e., static loads). It should be noted, however, that these properties are very sensitive to the effects of temperature and to the rate of application of the loads. For example, under impact loads at elevated temperatures, a material would exhibit properties entirely different from those at room temperature under static loads.

The tension or compression test is performed on a small sample of a given material known as the *test specimen*, whose dimensions have been standardized by the American Society for Testing and Materials. The design of three such standardized test specimens is shown in Figure 3.9. The load on the test specimen is applied by means of a machine known as a *universal testing machine*. A hydraulically

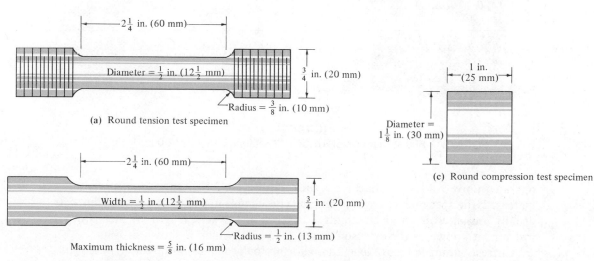

(a) Round tension test specimen

(b) Rectangular tension test specimen

(c) Round compression test specimen

FIGURE 3.9

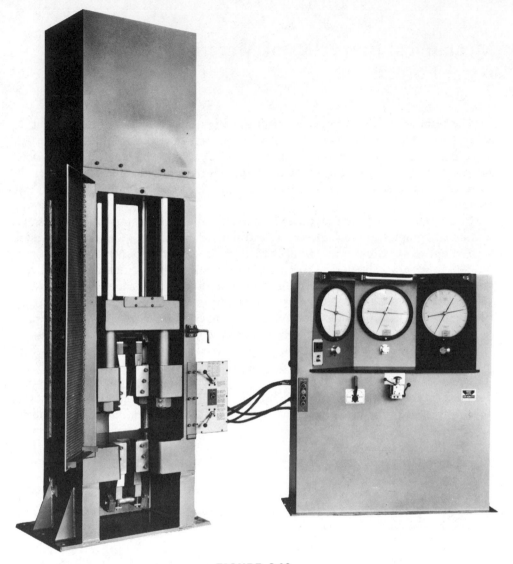

FIGURE 3.10
FORNEY LT-1000 Universal Testing Machine. Capacity: 600,000 lb.

operated universal testing machine is shown in Figure 3.10. As the load is gradually increased, the specimen changes in length. This change in length or deformation, δ, is usually measured for a given length, L, known as the gage length, by means of an instrument known as the extensometer, a picture of which is shown in Figure 3.11.

Corresponding values of load and deformation are taken while the load is gradually increased from zero until fracture in a tension test, or until fracture or some

FIGURE 3.11
The SATEC Systems, Inc., Separable Snap-On Extensomer, Model KSMD.

predetermined value of the load is reached in a compression test. Values of the load are converted into values of normal stress σ, by Eq. 3.1 and values of deformation into values of normal strain ε, by Eq. 2.16a. In determining the normal stress from Eq. 3.1, the original cross-sectional area of the test specimen is used, even though this cross-sectional area changes as the load is increased. The value of the normal stress thus obtained is known as the *engineering stress*. Also, the value of the normal strain obtained from Eq. 2.16a by using the original gage length is referred to as the *engineering strain*. Under certain conditions, it is necessary to account for the changes that occur in the cross-sectional area and in the gage length of the specimen. When such changes are taken into consideration (i.e., when the actual cross-sectional area and the actual gage length are used in Eqs. 3.1 and 2.16a), the resulting normal stress and normal strain are known as the *true stress* and *true strain*, respectively. The following discussion is limited to the engineering stress and strain, since these are the values used in the selection of materials for specific purposes, while the true stress and strain are, in general, useful in conducting research on materials.

Corresponding values of σ and ε are plotted to obtain what is known as the engineering stress–strain diagram for a given material. An engineering stress–strain diagram for a plain carbon steel specimen in tension is shown schematically in Figure

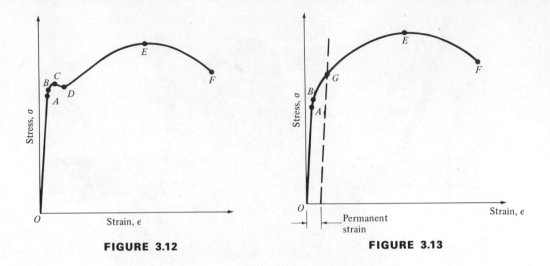

FIGURE 3.12   **FIGURE 3.13**

3.12. An engineering stress–strain diagram representing the behavior of steel alloys and aluminum alloys in tension is shown schematically in Figure 3.13. These diagrams will be used to introduce and discuss the significant mechanical properties of materials.

The diagrams shown in Figures 3.12 and 3.13 define two ranges of material behavior known as the *elastic* and the *plastic* (or *inelastic*) *ranges*. In general, the elastic range is that part of the diagram defining a linear relation between the stress and the strain (approximately segment OA in Figures 3.12 and 3.13), and it is the part of the stress–strain diagram that has already been discussed in Section 2.10 and is expressed mathematically by Hooke's law (i.e., $\varepsilon = \sigma/E$) up to the proportional limit for the material. The plastic or inelastic range is that part of the stress–strain diagram that defines a nonlinear relation between the stress and the strain and is represented by segment BF in Figures 3.12 and 3.13. Several empirical equations have been proposed to describe the inelastic relation between the stress and the strain, but the most widely used is the one known as the *Ramberg–Osgood equation*, which may be expressed as follows:

$$\varepsilon = \frac{\sigma}{E} + \left(\frac{\sigma}{B}\right)^{n} \tag{3.3}$$

where B and n are constants for a given material. The Ramberg–Osgood relations as expressed in Eq. 3.3 will be utilized in Section 8.7 to develop the use of the modified tangent modulus method for the analysis of columns.

The most significant mechanical properties will now be stated and defined. For the sake of completeness, these statements and definitions include those mechanical properties that have already been introduced in Section 2.10. Typical values of mechanical properties for some of the most commonly used engineering materials are given in Appendix C.

PROPORTIONAL LIMIT. The *proportional limit* for a given material represents the value of stress beyond which the material no longer behaves in such a way that the stress is proportional to strain. The proportional limit, σ_p, is represented by the ordinate to point *A* in Figures 3.12 and 3.13.

ELASTIC LIMIT. The *elastic limit*, σ_e, for a given material is the value of stress beyond which the material experiences a permanent deformation even after the stress is removed. Thus, if the material is loaded to any level of stress within the elastic limit and the load is then removed, it will regain its original dimensions and is said to behave *elastically*. However, if the load exceeds the elastic limit before it is removed, the material does not fully regain its initial dimensions. In such a case the material is said to experience a permanent deformation.

The elastic limit is represented by the ordinate to point *B* in Figures 3.12 and 3.13. Its determination, experimentally, is extremely difficult, and therefore its exact location on the stress–strain diagram is usually not known, even though it is generally higher than the proportional limit σ_p. For all practical purposes, however, the elastic limit σ_e and the proportional limit σ_p may be assumed to have the same value.

MODULUS OF ELASTICITY. The *modulus of elasticity*, *E*, is the constant of proportionality between stress and strain in Hooke's law as expressed in Eq. 2.25. Physically, it represents the slope of the stress–strain diagram within the proportional range of the material (i.e., the slope of the straight segment *OA* in Figures 3.12 and 3.13). The term *stiffness* is used to describe the capacity of materials to resist deformation in the elastic range and it is measured by the modulus of elasticity. For example, steels with a modulus of elasticity of about 30×10^6 psi are stiffer than aluminums, with a modulus of elasticity of about 10×10^6 psi.

YIELD POINT. The *yield point*, σ_y, is the stress at which the material continues to deform without further increase in the stress. The stress may even decrease slightly as the deformation continues past the yield point. Some materials, notably the plain carbon steels, exhibit a well-defined yield point, as shown by point *C* in Figure 3.12. If the stress decreases past this point, it is referred to as the *upper yield point*, in contrast to the *lower yield point* represented by point *D* in Figure 3.12 and beyond which the stress increases with further strain.

YIELD STRENGTH. For materials having a stress–strain diagram such as shown in Figure 3.13 (those that do not exhibit a well-defined yield point) a value of stress, known as the *yield strength* for the material, is defined as one producing a certain amount of permanent strain. Although several values of permanent strain may be used in defining the yield strength for a material, the most commonly encountered values are 0.0020 and 0.0035.

To determine the yield strength, σ_s, the assigned numerical value of permanent strain is measured along the strain axis of the stress–strain diagram to locate a point through which a line is drawn parallel to the straight portion (segment *OA*) of this diagram. The straight line is then extended until it intersects the stress–strain curve at the desired point. This construction is shown schematically in Figure 3.13, in which the ordinate to point *G* represents the value of the yield strength for the material.

ULTIMATE STRENGTH. The *ultimate strength*, σ_u, represents the ordinate to the highest point in the stress–strain diagram and is equal to the maximum load carried by the specimen divided by the original cross-sectional area. The ultimate strength is represented by the ordinate to points E in Figures 3.12 and 3.13.

FRACTURE STRENGTH. The *fracture strength*, σ_f, also known as the *rupture* or *breaking strength*, is the engineering stress at which the specimen fractures and complete separation of the specimen parts occurs. This strength is represented by the ordinate to points F in Figures 3.12 and 3.13 and is equal to the load at fracture divided by the original cross-sectional area of the specimen.

POISSON'S RATIO. *Poisson's ratio*, μ, which has already been introduced and used in Section 2.10, is a constant for a given material and is defined as the absolute value of the ratio of the transverse to the longitudinal strain. Thus in a given uniaxial situation, tension or compression, if the longitudinal strain (i.e., the strain in the direction of the applied load) is ε_{LON}, the transverse strain ε_{TR} (i.e., the strain in any direction perpendicular to the applied load) is $-\mu\varepsilon_{LON}$. Therefore, $\mu = |\varepsilon_{TR}/\varepsilon_{LON}|$.

DUCTILITY. The property of materials known as *ductility* is a measure of their capacity to deform in the plastic or inelastic range. Thus materials that exhibit large plastic deformations, represented by segment *BEF* in Figures 3.12 and 3.13, are said to be *ductile materials*. Examples of ductile materials include steel and aluminum alloys as well as the alloys of copper. Specimens of these ductile materials, when subjected to tension, undergo considerable plastic deformation. Not only do they exhibit large extensions but, after the ultimate strength is reached, they also undergo considerable reduction in the cross-sectional dimension, known as *necking*, in the region of fracture as shown in Figure 3.14(a). On the other hand, materials that fracture with little or no measurable plastic deformation are known as *brittle materials*. Examples of brittle materials include most cast irons as well as concrete. A tension specimen made of gray cast iron, for example, would exhibit a very slight extension and no appreciable necking at fracture as shown in Figure 3.14(b). Another characteristic of brittle materials is that they are generally much stronger in compression than they are in tension.

Ductility of a material is usually measured by one or both of two properties:

1. *Percent elongation*, which is defined as 100 multiplied by the change in gage length divided by the original gage length of the tension specimen.
2. *Percent reduction of area*, which is defined as 100 multiplied by the change in cross-sectional area divided by the original cross-sectional area of the tension specimen.

ENERGY ABSORPTION CAPACITY. Often materials are called upon to resist the action of dynamic and impact loads. Under such conditions, the energy absorption capacity becomes a useful tool in comparing materials for dynamic and impact applications. Two such energy quantities are defined:

1. The *modulus of resilience* represents the amount of energy per unit of volume absorbed by the material when stressed up to the proportional limit. Thus the modulus of resilience represents the amount of energy per unit of volume that the material can absorb elastically.

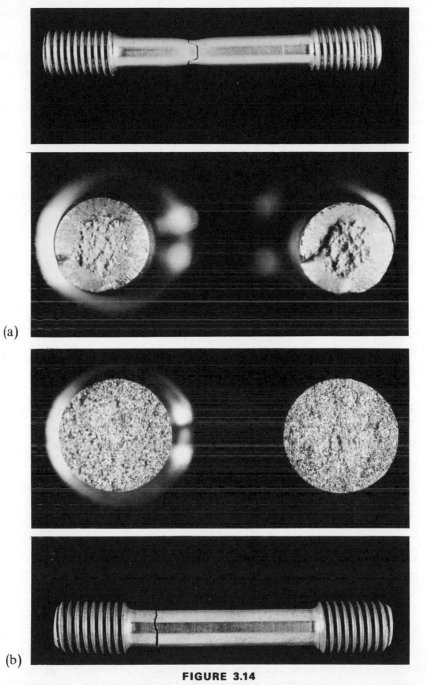

FIGURE 3.14
(a) Photograph Showing Ductile Fracture with Appreciable Plastic Deformation.
(b) Photograph Showing Brittle Fracture Lacking Measurable Plastic Deformation.

When a gradually increasing axial force F acts on a member, it produces a deformation δ. The work, dW, of a force, F, corresponding to an infinitesimal deformation $d\delta$, is, by definition, given by

$$dW = F \, d\delta \tag{3.4a}$$

Thus the work performed by the force F in deforming the member through the amount δ becomes

$$W = \int F \, d\delta \tag{3.4b}$$

If deformations are assumed to be within the proportional range of behavior, the force $F = k\delta$, where k is a constant of proportionality. Substituting $F = k\delta$ into Eq. 3.4b and integrating, one obtains

$$W = k \int \delta \, d\delta = \frac{k\delta^2}{2} = \frac{F\delta}{2} \tag{3.4c}$$

Since $F = \sigma A$ and $\delta = \varepsilon L$, where L and A are the length and cross-sectional area of the member, respectively, the work W may be rewritten as follows:

$$W = \frac{\sigma \varepsilon}{2} (AL) \tag{3.4d}$$

Note that the product AL represents the volume of the member and if both sides of Eq. 3.4d are divided by it, the resulting quantity represents the work performed per unit volume of the material. Thus

$$\frac{W}{AL} = \frac{\sigma \varepsilon}{2} \tag{3.4e}$$

Under ideal conditions, the work per unit volume W/AL, is equal to the energy per unit volume, u, stored in the member by virtue of its strained configuration. This energy u is known as the *strain energy* and, in view of the foregoing, may be expressed as follows:

$$u = \frac{\sigma \varepsilon}{2} \tag{3.5}$$

Thus the modulus of resilience for a given material is $\sigma_p \varepsilon_p / 2$, where σ_p is the proportional limit and ε_p is the strain corresponding to the proportional limit. If ε is replaced by σ/E, Eq. 3.5 may be expressed as follows:

$$u = \frac{\sigma^2}{2E} \qquad (3.6)$$

and the modulus of resilience for the material can be written $\sigma_p^2/2E$.

Examination of Eq. 3.5 reveals that the modulus of resilience is equal to the area under the stress–strain diagram up to the proportional limit, as shown in Figure 3.15(a).

2. The *modulus of toughness* is the amount of energy per unit volume that a material can absorb before fracture. Thus the modulus of toughness is a measure not only of the strength of the material, but also its ductility. By definition, then, the modulus of toughness is given by the area under the entire stress–strain diagram, as indicated in Figure 3.15(b). An approximate value for this property is obtained by estimating an average stress σ_{av} as shown in Figure 3.15(b) and multiplying it by the strain at fracture, ε_f. Thus, in terms of these two quantities, the modulus of toughness has the value $\sigma_{av}\varepsilon_f$.

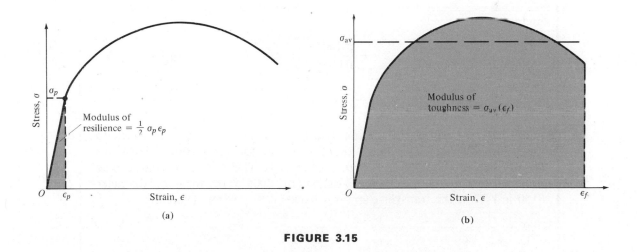

FIGURE 3.15

Example 3.9

A steel rod 0.50 in. in diameter, with a gage length of 4 in., is subjected to a gradually increasing tensile load. At the proportional limit, the value of the load was 20,000 lb, the change in the gage length was 0.014 in., and the change in diameter was 0.0005 in. Determine (a) the proportional limit, σ_p; (b) the modulus of elasticity, E; (c) Poisson's ratio, μ; and (d) the modulus of resilience.

Solution

(a) The proportional limit, σ_p, may be found from Eq. 3.1, that is, $\sigma = P/A$, where P is the load at the proportional limit, 20,000 lb, and A is the original cross-sectional area of the rod. Thus

$$\sigma_p = \frac{20,000}{(\pi/4)(0.5)^2}$$

$$= \boxed{101,859 \text{ psi}}$$

(b) The modulus of elasticity is obtained from Hooke's law as expressed by Eq. 2.25 (i.e., $E = \sigma/\varepsilon$), in which σ and ε are two corresponding values of stress and strain, respectively, within the range of proportionality for the material. The proportional limit stress has already been determined in part (a) to be 101,859 psi. The strain corresponding to this stress, ε_p, is found by dividing the deformation at the proportional level (0.014 in.) by the initial gage length (4 in.). Thus

$$\varepsilon_p = \frac{0.014}{4} = 0.0035$$

and

$$E = \frac{101,859}{0.0035}$$

$$= \boxed{29.1 \times 10^6 \text{ psi}}$$

(c) Poisson's ratio, μ, is, by definition, the ratio of the transverse strain to longitudinal strain. The longitudinal strain at the proportional limit has already been determined in part (b) to be 0.0035. The transverse strain can be obtained by dividing the change in diameter (0.0005 in.) by the initial diameter (0.50 in.). Thus the transverse strain is $0.0005/0.50 = 0.001$. Therefore,

$$\mu = \frac{0.001}{0.0035}$$

$$= \boxed{0.286}$$

(d) Modulus of resilience $= \frac{1}{2}\sigma_p \varepsilon_p$

$$= \tfrac{1}{2}(101, 859)(0.0035)$$

$$= \boxed{178.25 \text{ in.-lb/in.}^3}$$

Homework Problems

3.14 An aluminum bar with a rectangular cross section 0.01 m × 0.03 m and a gage length of 0.2 m is subjected to a tensile load of 60,000 N. The 0.03-m dimension changed to 0.02999 m and the 0.2 m gage length changed to 0.2002 m. If the behavior of the material is assumed to be within the range of proportionality between stress and strain, determine:
(a) Poisson's ratio.
(b) The final value of the 0.01 m dimension.
(c) The axial stress in the bar.
(d) The maximum shearing stress in the bar.
(e) The modulus of elasticity.

3.15 A steel bar 2 in. in diameter and 4 in. long is subjected to a compressive force. The proportional limit for this steel is 40,000 psi, its modulus of elasticity 30×10^6 psi, and its Poisson's ratio is 0.28. Determine:
(a) The maximum compressive load that may be applied without exceeding the proportional limit.
(b) The changes in diameter and in length at the proportional limit.

3.16 An aluminum rod 0.50 in. in diameter and a gage length of 2 in. is subjected to a gradually increasing tensile force. At the proportional limit, the value of the load was 5000 lb, the gage length measured 2.005 in., and the diameter measured 0.4996 in. Find:
(a) The modulus of elasticity.
(b) Poisson's ratio.
(c) The modulus of resilience.

3.17 A compressive force of 200,000 N is gradually applied to a prismatic bar whose cross section is 0.03 m × 0.04 m and whose length is 0.1 m. The 0.03-m dimension changed to 0.03003 m and the 0.1-m length changed to 0.0997 m. Determine:
(a) Poisson's ratio.
(b) The final value of the 0.04 m dimension.
(c) The modulus of elasticity for the material.

3.18 A tension test was performed on a steel specimen whose original diameter was 0.50 in. and whose gage length was 2.0 in. Corresponding values of load and deformation are given in the following tabulation. Construct the engineering stress–strain curve for this material and determine the ultimate strength and the fracture strength.

Load (lb)	Deformation (in.)	Load (lb)	Deformation (in.)
0	0	12,400	0.0170
1,600	0.0004	12,300	0.0200
3,200	0.0009	13,000	0.0270
4,800	0.0015	14,000	0.0330
6,400	0.0021	15,000	0.0400
8,000	0.0026	16,000	0.0490
9,600	0.0031	17,500	0.0670
11,200	0.0035	19,000	0.1050
12,000	0.0040	19,500	0.1500
12,400	0.0050	20,000	0.2000
12,600	0.0070	20,000	0.2500
12,800	0.0100	18,600	0.3200
12,800	0.0130	17,000	0.4000
12,600	0.0150	16,500	0.4500 (fracture)

3.19 Redraw the initial portion of the data given in Problem 3.18 using a larger scale and determine:
(a) The modulus of elasticity.
(b) The upper yield point.
(c) The lower yield.
(d) The proportional limit.

3.20 If the final diameter of the specimen of Problem 3.18 was 0.352 in., determine:
(a) The percent reduction of area.
(b) The percent elongation.

3.21 Refer to the data in Problem 3.18 and determine:
(a) The modulus of resilience.
(b) The modulus of toughness.

Chapter 4 Torsional Stresses, Strains, and Rotations

4.1 Introduction

A major part of this chapter is concerned with shearing stresses and angles of twist of circular and hollow circular cylindrical shafts subjected to torques. Mathematically, this is the simplest torsion problem and an important practical one because such sections resist torque efficiently. Principal stresses and strains in shafts are evaluated with the aid of the Mohr's circle construction. Surface elements taken from shafts are usually free of stresses and the analysis reduces to one of plane stress. Power is often transmitted through shafts of circular or hollow circular cross sections and power transmission concludes the coverage of such torsional members.

Analytical and experimental solutions for torsion of members of noncircular cross section are given next. Warping displacements and warping functions are discussed and equations for angle of twist and maximum shearing stresses are presented for elliptical, rectangular, and triangular cross sections. Experimental solutions are presented in terms of the membrane analogy, which also provides a powerful visual and qualitative approach for solution of torsional problems.

Background from Chapter 1 on construction of torque diagrams and from Chapter 2 on determination of principal stresses and strains is required for a thorough understanding of the topics presented in this chapter.

4.2 Solid Circular Shafts—Angle of Twist and Shearing Stresses

A cylindrical shaft is depicted in Figure 4.1. Assumptions are required to determine the deformation of the shaft and internal shearing stresses.

ASSUMPTION 1. Cross sections perpendicular to the longitudinal axis of the shaft are plane prior to application of the torque T and they remain plane after the torque is applied.

ASSUMPTION 2. Surface elements of the cylinder, such as BD, are assumed to remain straight lines after twisting takes place. Point B moves to B' and $B'D$ is

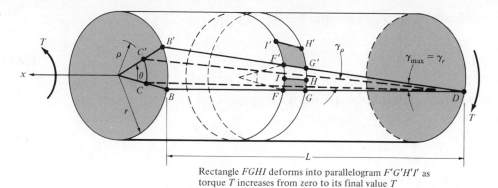

Rectangle *FGHI* deforms into parallelogram *F'G'H'I'* as
torque *T* increases from zero to its final value *T*

FIGURE 4.1

assumed to be a straight line, which is approximately correct for small angles of twist even though the true shape of *B'D* is helical. Similarly, point *C* moves to *C'* as the torque *T* is applied and interior longitudinal lines parallel to surface elements such as *BD* are assumed to remain straight during twisting of the shaft.

ASSUMPTION 3. A typical rectangle such as *FGHI* ruled on the shaft surface prior to twist will be deformed into a parallelogram *F'G'H'I'* The amount of this angular distortion of the rectangle is given by the angle *BDB'*, which represents the shear strain on the surface of the shaft between line elements *DD* and *FI*. This shear strain on the surface of the shaft is associated with the radius *r* of the shaft and is denoted by the symbol γ_r. As shown later, γ_r is, in fact, the maximum shear strain, γ_{max}, in the shaft. Since γ_{max} is a small angle, it is given by

$$\gamma_r = \gamma_{max} = \frac{BB'}{BD} \tag{4.1}$$

Similarly, inside the shaft, at a distance ρ from the *x* axis of twist,

$$\gamma_\rho = \frac{CC'}{BD} \tag{4.2}$$

Arc lengths are expressed in terms of the angle of twist

$$BB' = r\theta \tag{4.3}$$

and

$$CC' = \rho\theta \tag{4.4}$$

ASSUMPTION 4. The length of longitudinal elements of the shaft *L* is assumed to remain constant under the action of externally applied torques.

$$BD = L \tag{4.5}$$

Substitute Eqs. 4.3 and 4.5 into Eq. 4.1:

$$\gamma_r = \gamma_{max} = \frac{r\theta}{L} \tag{4.6}$$

Similarly, substitute Eqs. 4.4 and 4.5 into Eq. 4.2:

$$\gamma_\rho = \frac{\rho\theta}{L} \tag{4.7}$$

Divide Eq. 4.7 by Eq. 4.6:

$$\frac{\gamma_\rho}{\gamma_r} = \frac{\rho}{r} \tag{4.8}$$

Equation 4.8 states that the shearing strain is proportional to the distance measured radially from the axis of twist, which is the geometric axis of the cylindrical shaft. γ_ρ varies linearly from zero for $\rho = 0$ to $\gamma_r = \gamma_{max}$ for $\rho = r$.

ASSUMPTION 5. Hooke's law relates the shearing strains to the shearing stress by Eq. 2.26 for the material of which the homogeneous shaft is fabricated. This equation is repeated here without subscripts. Thus

$$\tau = G\gamma \tag{4.9}$$

In order to relate shearing stress to the radial coordinate ρ, multiply the numerator and denominator of the left side of Eq. 4.8 by the shearing modulus of elasticity, G:

$$\frac{G\gamma_\rho}{G\gamma_r} = \frac{\rho}{r} \tag{4.10}$$

By Eq. 4.9, Eq. 4.10 becomes

$$\frac{\tau_\rho}{\tau_r} = \frac{\rho}{r} \tag{4.11}$$

Equation 4.11 states that the shearing stress is proportional to the distance measured radially from the axis of twist. The stress τ_ρ varies linearly from zero for $\rho = 0$ to $\tau_r = \tau_{max}$ for $\rho = r$. Equations 4.8 and 4.11 are shown graphically in Figure 4.2.

In order to relate external torque to internal shearing stress, refer to Figure 4.3,

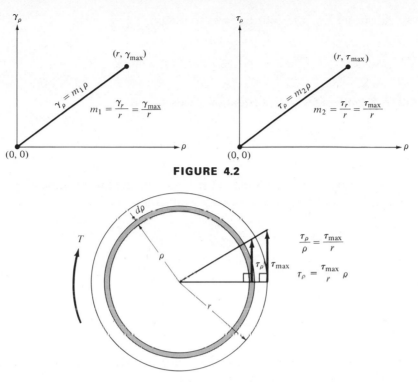

FIGURE 4.2

FIGURE 4.3

showing a typical cross section of a circular cylindrical shaft. The annular differential element of area dA is given by

$$dA = 2\pi\rho \, d\rho \tag{4.12}$$

Shearing stress τ_ρ acts over this annular area and the differential force on this differential area is given by

$$dF = \tau_\rho \, dA = \tau_\rho(2\pi\rho \, d\rho) \tag{4.13}$$

A differential part dT of the applied torque T is resisted by shearing stresses on this differential area such that

$$dT = \rho \, dF = 2\pi\rho^2\tau_\rho \, d\rho \tag{4.14}$$

Equation 4.11 solved for τ_ρ yields

$$\tau_\rho = \frac{\tau_r}{r}\rho \tag{4.15}$$

Then we substitute Eq. 4.15 into Eq. 4.14 to obtain

$$dT = \frac{\tau_r}{r} 2\pi\rho^3 \, d\rho \qquad (4.16)$$

Integration of both sides of this equation yields

$$\int dT = \frac{\tau_r}{r} \int_0^r 2\pi\rho^3 \, d\rho$$

where τ_r/r is constant for a given shaft of radius r subjected to a torque T. Figure 4.2 shows that τ_r/r is the slope of the τ versus ρ function. Therefore,

$$T = \frac{\tau_r}{r} J \qquad (4.17)$$

where

$$J = \int_{\text{area}} \rho^2 \, dA = 2\pi \int_0^r \rho^3 \, d\rho = \frac{\pi r^4}{2}$$

is the polar moment of inertia of the circular cross-sectional area. We substitute for $\tau_r/r = \tau_\rho/\rho$ from Eq. 4.11 into Eq. 4.17:

$$T = \frac{\tau_\rho}{\rho} J \qquad (4.18)$$

Solving Eq. 4.18 for τ_ρ, we obtain the shearing stress equation for circular cylindrical shafts:

$$\tau_\rho = \frac{T\rho}{J} \qquad (4.19)$$

and multiply both sides of Eq. 4.7 by G to obtain

$$G\gamma_\rho = \frac{G\rho\theta}{L} \qquad (4.20)$$

But $G\gamma_\rho = \tau_\rho$; therefore,

$$\tau_\rho = \frac{G\rho\theta}{L} \qquad (4.21)$$

Then we substitute for τ_ρ from Eq. 4.19, divide by ρ, and solve for θ:

$$\theta = \frac{TL}{JG}$$

(4.22)

Equations 4.11, 4.19, and 4.22 are the basic equations used to find the shearing stresses in solid circular shafts and the angle of twist associated with applied torques.

Example 4.1

Given: A circular shaft is shown in Figure 4.4(a). The shaft is 48 in. long and has a radius of 2.00 in. It is subjected to end torques of 8500 lb-ft and is to be fabricated of material for which $G = 12.0 \times 10^6$ psi. Find (a) the maximum

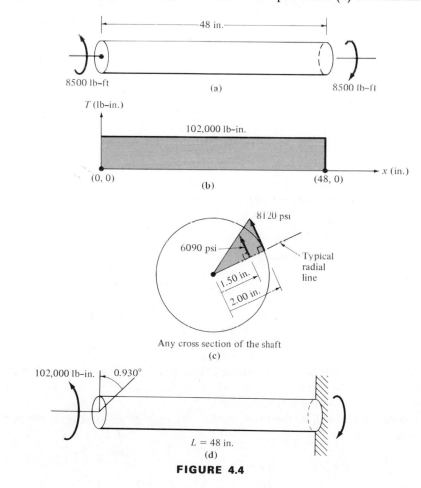

FIGURE 4.4

shearing stress in the shaft, (b) the shearing stress 1.50 in. from the center of twist of the shaft, and (c) the angle of twist of the shaft.

Solution

(a) The internal torque versus x diagram is shown in Figure 4.4(b). Over the length of the shaft the internal torque is constant and equal to 8500 lb-ft × 12 in./ft = 102,000 lb-in. Equation 4.19 is the appropriate equation for finding shearing stresses in the shaft:

$$\tau_\rho = \frac{T\rho}{J}$$

$$\tau_\rho = \tau_{max} \quad \text{for } \rho = r = 2 \text{ in.}$$

$$\tau_{max} = \tau_r = \frac{Tr}{J}$$

where

$$J = \frac{\pi r^4}{2} = \frac{\pi(2)^4}{2} = 25.13 \text{ in.}^4$$

$$\tau_{max} = \tau_r = \frac{102,000(2)}{25.13}$$

$$= \boxed{8120 \text{ psi}}$$

(b) Since shearing stress is proportional to the distance from the axis of twist, Eq. 4.11 is convenient for finding the shearing stress 1.5 in. from the center of the shaft.

$$\frac{\tau_\rho}{\tau_{max}} = \frac{\rho}{r}$$

$$\tau_\rho = \frac{\rho}{r} \tau_{max} = \frac{1.50}{2.00} (8120)$$

$$= \boxed{6090 \text{ psi}}$$

The shearing stresses are plotted along a typical radial line in Figure 4.4(c). Any cross section of the shaft is represented in the figure and this linear variation of stress would hold along any radial line.

(c) Equation 4.22 will be used to find the relative angle of rotation between the two ends of the shaft.

$$\theta = \frac{TL}{JG}$$

$$= \frac{102{,}000(48.0)}{25.13(12.0 \times 10^6)}$$

$$= \boxed{0.0162 \text{ rad} \quad \text{or} \quad 0.930°}$$

If the right end of the shaft were held fixed, the left end would rotate through an angle of 0.930° as shown in Figure 4.4(d).

Example 4.2
Refer to Figure 4.5(a) and determine the maximum shearing stress in the stepped shaft and the relative rotation between the ends A and D. Use $G = 80.0$ GPa.

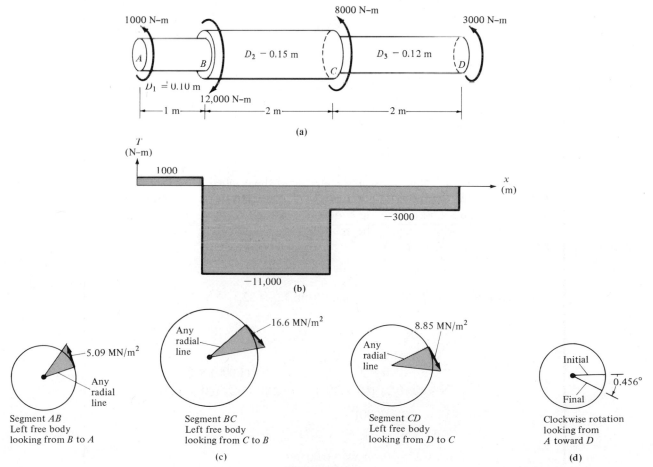

(a)

(b)

(c)

Segment AB
Left free body
looking from B to A

Segment BC
Left free body
looking from C to B

Segment CD
Left free body
looking from D to C

Clockwise rotation
looking from
A toward D

(d)

FIGURE 4.5

Solution. The torque versus x plot is shown in Figure 4.5(b), and the shearing stress will be computed for each segment of the shaft using Eq. 4.19 with $\rho = r = D/2$ in each case.

SEGMENT AB

$$\tau = \frac{T\rho}{J} = \frac{1000 \text{ N-m} \times (0.10/2) \text{ m}}{(\pi/2)(0.10/2)^4 \text{ m}^4}$$

$$= \boxed{5.09 \text{ MPa}}$$

SEGMENT BC

$$\tau = \frac{T\rho}{J} = \frac{11{,}000 \text{ N-m} \times (0.15/2) \text{ m}}{(\pi/2)(0.15/2)^4 \text{ m}^4}$$

$$= \boxed{16.60 \text{ MPa}}$$

SEGMENT CD

$$\tau = \frac{T\rho}{J} = \frac{3000 \text{ N-m} \times (0.12/2) \text{ m}}{(\pi/2)(0.12/2)^4 \text{ m}^4}$$

$$= \boxed{8.85 \text{ MPa}}$$

The maximum shearing stress, associated with the applied torques, occurs in the central region BC and equals 16.60 MPa. Shearing stress distributions are shown in Figure 4.5(c). Imagine end D of the shaft to be fixed and form the following sum, which equals the relative rotation of end A with respect to end D:

$$\theta_{AD} = \theta_{AB} + \theta_{BC} + \theta_{CD}$$

Each term of the sum will be obtained from Eq. 4.22:

$$\theta = \frac{TL}{JG}$$

$$\theta_{AD} = \frac{1000 \times 1}{(\pi/2)(0.10/2)^4(80 \times 10^9)} + \frac{(-11{,}000) \times 2}{(\pi/2)(0.15/2)^4(80 \times 10^9)}$$

$$+ \frac{(-3000) \times 2}{(\pi/2)(0.12/2)^4(80 \times 10^9)}$$

$$= 0.00127 - 0.00554 - 0.00369$$

$$= \boxed{-0.00796 \text{ rad} \quad \text{or} \quad -0.456°}$$

Signs of the torques entering the preceding equation were obtained from the torque diagram constructed as described in Section 1.4. The negative sign means that an observer at A looking toward D would observe a line scribed in the cross section at A to rotate clockwise. This angular displacement is shown in Figure 4.5(d).

4.3 Hollow Circular Shafts—Angle of Twist and Shearing Stresses

In order to apply Eqs. 4.19 and 4.22 to hollow circular shafts, we need simply to modify the equation for J, the polar moment of inertia:

$$J_1 = \frac{\pi}{2}(r_o^4 - r_i^4) = \frac{\pi}{32}(D_o^4 - D_i^4) \tag{4.23}$$

where J_1 = polar moment of inertia for annular area
  r_o = outside radius of shaft, D_o = outside diameter
  r_i = inside radius of shaft, D_i = inside diameter

Hollow shafts carry torques more efficiently than solid shafts, since material near the axis of twist, which has lower levels of stress in a solid shaft, has been removed from the hollow shaft.

Example 4.3
Refer to Figure 4.6(a) and determine the maximum shearing stress in the hollow shaft and the angle of twist of the shaft. Plot the variation of shearing stress along any radial line of the shaft at any cross section between the ends. Sketch the angle of twist of the left end relative to the right end. The following values are given: $T = 30{,}000$ lb-ft $r_o = 4.00$ in., $r_i = 2.00$ in., $L = 60$ in., and $G = 6.00 \times 10^6$ psi.

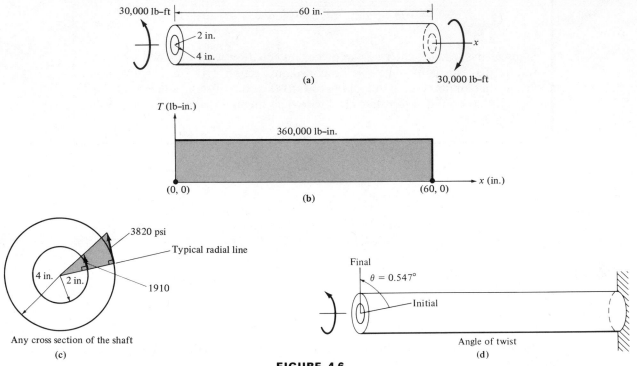

FIGURE 4.6

Solution. Apply Eq. 4.23 for a hollow shaft to obtain the polar moment of inertia of this cross-sectional area:

$$J = \frac{\pi}{2}\left(r_o^4 - r_i^4\right)$$

$$= \frac{\pi}{2}\left[(4)^4 - (2)^4\right] = 377 \text{ in.}^4$$

$$\tau_\rho = \frac{T\rho}{J}$$

with

$$\rho = r_o, \qquad \tau_\rho = \tau_{max}$$

$$\tau_{max} = \frac{360,000(4.00)}{377}$$

$$= \boxed{3820 \text{ psi}}$$

Let $\rho = r_i$ to obtain the shearing stress at the outside of the hole in the shaft.

$$\tau_i = \frac{r_i}{r_o} \tau_{max} = \frac{2}{4}(3820)$$

$$= \boxed{1910 \text{ psi}}$$

These values are plotted in Figure 4.6(c). Note the linear variation of shearing stress with the radial coordinate. The angle of twist is given by Eq. 4.22:

$$\theta = \frac{TL}{JG} = \frac{360,000(60)}{377(6 \times 10^6)}$$

$$= \boxed{0.00955 \text{ rad} \quad \text{or} \quad 0.547°}$$

This result is shown schematically in Figure 4.6(d).

Example 4.4

A hollow shaft is depicted in Figure 4.7(a), which has an inside diameter of 0.12 m over its full length and is subjected to the torques shown. Plot the variation of torque versus the x coordinate measured along the axis of the shaft. Determine the maximum shearing stresses at typical cross sections between A and B, B and C, and C and D. Compute the angle of twist of end A relative to end D of the shaft. Use $G = 100$ GPa.

Solution. The torque variation along the length of the shaft is depicted in Figure 4.7(b). Compute polar moments of inertia of the cross sections using Eq. 4.23.

SEGMENT AB

$$J = \frac{\pi}{32}(D_o^4 - D_i^4) = \frac{\pi}{32}[(0.20)^4 - (0.12)^4] = 13.672 \times 10^{-5} \text{ m}^4$$

SEGMENT BC

$$J = \frac{\pi}{32}[(0.40)^4 - (0.12)^4] = 249.292 \times 10^{-5} \text{ m}^4$$

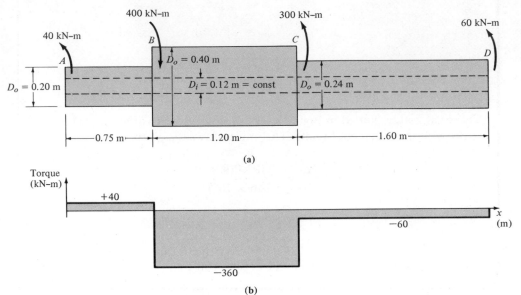

(a)

(b)

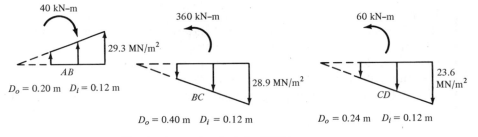

Shearing stresses along typical radial lines

(c)

FIGURE 4.7

SEGMENT *CD*

$$J = \frac{\pi}{32}[(0.24)^4 - (0.12)^4] = 30.536 \times 10^{-5} \text{ m}^4$$

Maximum shearing stresses follow from Eq. 4.19 with ρ = outside radius of each shaft segment:

$$\tau_\rho = \frac{T\rho}{J}$$

SEGMENT AB

$$\tau = \frac{40(0.10) \times 10^3}{13.672 \times 10^{-5}}$$

$$= \boxed{29.3 \text{ MPa}}$$

SEGMENT BC

$$\tau = \frac{360(0.20) \times 10^3}{249.292 \times 10^{-5}}$$

$$= \boxed{28.9 \text{ MPa}}$$

SEGMENT CD

$$\tau = \frac{60(0.12) \times 10^3}{30.536 \times 10^{-5}}$$

$$= \boxed{23.6 \text{ MPa}}$$

Plots of these shearing stresses are shown along typical radial lines in Figure 4.7(c).

Imagine end D of the shaft to be fixed and form the following sum to obtain the relative rotation of end A with respect to end D:

$$\theta_{AD} = \theta_{AB} + \theta_{BC} + \theta_{CD}$$

Each term of the sum will be obtained from Eq. 4.22:

$$\theta_{AD} = \frac{40(0.75)10^3}{13.672 \times 10^{-5}(100 \times 10^9)} + \frac{-360(1.20)10^3}{249.292 \times 10^{-5}(100 \times 10^9)}$$

$$+ \frac{-60(1.60)10^3}{30.536 \times 10^{-5}(100 \times 10^9)}$$

$$= 2.19 \times 10^{-3} - 1.73 \times 10^{-3} - 3.14 \times 10^{-3}$$

$$= \boxed{-2.68 \times 10^{-3} \text{ rad} \quad \text{or} \quad -0.154°}$$

The negative sign means that an observer at A looking toward D would observe a line scribed in the cross section of A to rotate clockwise with respect to a fixed line.

Homework Problems

4.1 Determine the maximum shearing stress in the shaft depicted in Figure H4.1. Plot the variation of shearing stress along a typical radial line of the shaft.

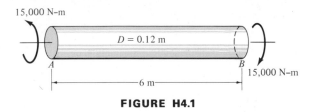

FIGURE H4.1

4.2 Plot the variation of torque versus a longitudinal coordinate meaured along the axis of the shaft depicted in Figure H4.2. Compute the maximum shearing stresses in segments AB and BC of this shaft.

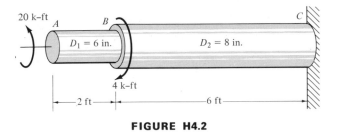

FIGURE H4.2

4.3 A hollow circular shaft is shown in Figure H4.3. Determine the maximum shearing stress in this shaft and plot the variation of shearing stress along a typical radial line of the shaft.

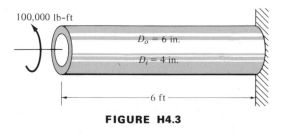

FIGURE H4.3

4.4 A stepped shaft is depicted in Figure H4.4. Plot the variation of torque versus a longitudinal coordinate measured along the axis of the shaft. Determine the maximum shearing stresses in segments AB, BC, and CD of this shaft. Use $G = 80.0$ GPa. Compute the angle of twist of end A with respect to end D of the shaft.

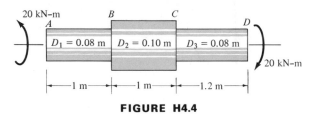

FIGURE H4.4

4.5 Plot the variation of torque versus a longitudinal coordinate measured along the axis of the shaft shown in Figure H4.5. Determine the maximum shearing stress in this shaft.

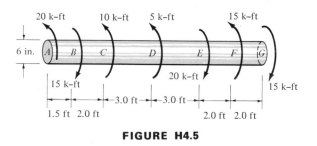

FIGURE H4.5

4.6 Determine the maximum shearing stress in the shaft depicted in Figure H4.6. Plot the variation of

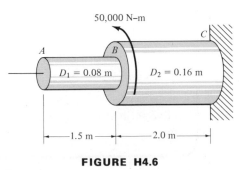

FIGURE H4.6

shearing stress along a typical radial line of the shaft. Calculate the angle of twist in degrees of:
(a) End A with respect to end C.
(b) Section B with respect to end C. Use $G = 80.0$ GPa.

4.7 A uniformly distributed torque is applied to the shaft of Figure H4.7. Plot the variation of torque versus a longitudinal coordinate measured along the axis of the shaft. Compute the maximum shearing stress in this shaft.

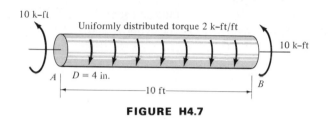

FIGURE H4.7

4.8 A hollow shaft is subjected to the torques shown in Figure H4.8. Plot the variation of torque versus a longitudinal coordinate measured along the axis of the shaft. Calculate the maximum shearing stress in this shaft. Use $G = 80.0$ GPa. Determine the angle of twist of end D with respect to end A of the shaft.

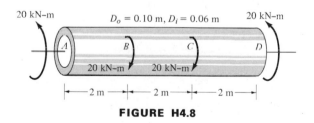

FIGURE H4.8

4.9 For the shaft depicted in Figure H4.9, plot the variation of torque versus a longitudinal coordinate

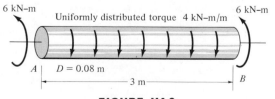

FIGURE H4.9

measured along the axis of the shaft. Calculate the maximum shearing stress in this shaft.

4.10 Determine the maximum shearing stress in each segment (i.e., AB, BC, and CD) of the shaft shown in Figure H4.10. Determine the angle of twist in degrees of end D with respect to end A of this shaft. Use $G = 12.0 \times 10^6$ psi.

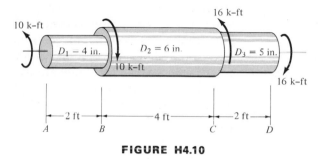

FIGURE H4.10

4.11 A uniformly distributed torque is applied to the shaft depicted in Figure H4.11. Determine the maximum shearing stresses in this shaft.

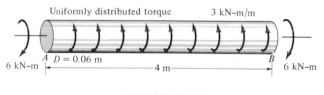

FIGURE H4.11

4.12 Plot the variation of torque versus a longitudinal coordinate measured along the axis of the shaft shown in Figure H4.12. Determine the maximum shearing stress in this shaft.

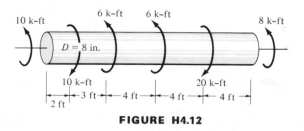

FIGURE H4.12

4.4 Principal Stresses and Strains Associated with Torsion

Although applied torques give rise to shearing stresses, the Mohr's circle construction reveals that tensile and compressive principal stresses are always associated with torsion. These principal stresses are each equal in magnitude to the shearing stress computed from the torsion formula, Eq. 4.19, with $\rho = r$.

A stressed element is shown at any point on the surface of a cylindrical shaft of Figure 4.8(a) with a longitudinal x axis oriented along an element of the cylinder and a perpendicular y axis oriented tangent to the surface. An enlarged view of this element is shown in Figure 4.8(b). On the X plane the state of stress is given by $\sigma_x = 0$, $\tau_{xy} = \tau = Tr/J$, and on the Y plane the state of stress is given by $\sigma_y = 0$, $\tau_{xy} = -Tr/J$. The surface of the shaft is free of shearing stresses and the normal stress

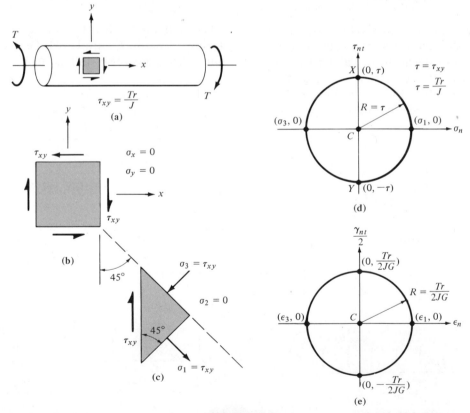

FIGURE 4.8

perpendicular to the shaft surface is also zero, which means that the term *plane stress* is appropriate, as discussed in Section 2.4. These stress values for the planes perpendicular to x and y axes are plotted in Figure 4.8(d) as points X and Y, respectively. When these points are connected by a straight line, the center of Mohr's circle, point C, lies at the origin of the $\sigma_n - \tau_{nt}$ coordinate system and the circle has a radius $R = \tau$. This circle intersects the σ_n axis at the points with coordinates $(\sigma_1, 0)$ and $(\sigma_3, 0)$ and by inspection the principal stresses are $\sigma_1 = \tau$ and $\sigma_3 = -\tau$. Stress σ_1 is tensile and stress σ_3 is compressive, and Figure 4.8(c) shows the planes on which these maximum and minimum normal stresses act. The maximum and minimum shearing stresses are given by the ordinates to the plotted points X and Y. In the special case of pure torsion, $\sigma_1, \sigma_3, \tau_{max}$, and τ_{min} all have equal magnitudes obtained from the torsion formula $\tau_{xy} = T\rho/J$ with ρ chosen equal to the shaft radius in order to obtain the largest possible value for the stress.

Provided that Mohr's circle for stress has been constructed first, it is a simple task to construct Mohr's circle for strain. Recall that $\tau_{xy} = G\gamma_{xy}$ or that $\gamma_{xy} = \tau/G$; then $\gamma = Tr/JG$. Points with coordinates $(0, \gamma/2)$, $(0, -\gamma/2)$ are plotted and the circle for strain completed as shown in Figure 4.8(e). Once again, the center of the circle lies at the origin of the coordinate system and the radius of the circle equals $Tr/2JG$. In the special case of pure torsion, the principal normal strains, ε_1 and ε_3, each have magnitudes equal to $Tr/2JG$ and the maximum and minimum shearing strains have magnitudes equal to Tr/JG. Mohr's circle for strain based upon measured normal strains on the shaft surface is discussed in Example 4.6.

An important relationship between the elastic constants (E, G, and μ) for homogeneous, isotropic materials is easily developed from information obtained from the Mohr's circles for stress and strain shown in Figure 4.8(d) and (e). Plane stress (Eq. 2.31) is

$$\sigma_1 = \frac{E}{1 - \mu^2} (\varepsilon_1 + \mu\varepsilon_3)$$

where

$$\sigma_1 = \frac{Tr}{J}, \qquad \varepsilon_1 = \frac{Tr}{2JG}, \qquad \varepsilon_3 = -\frac{Tr}{2JG}$$

Substitute these values, to obtain

$$\frac{Tr}{J} = \frac{E}{1 - \mu^2} \left[\frac{Tr}{2JG} + \mu\left(\frac{-Tr}{2JG}\right) \right]$$

Divide through by Tr/J and solve for G to obtain Eq. 2.27, which is repeated here for convenience.

$$G = \frac{E}{2(1 + \mu)} \tag{2.27}$$

The modulus of elasticity E, the shearing modulus of elasticity, or the modulus of rigidity G and Poisson's ratio μ are not independent of each other but are related by this equation. Experimental values may be checked with this equation and it may be used to write equations in alternative forms. As an example, typical values of E and μ for steel alloys are 29.0×10^6 psi and 0.30, respectively, so that

$$G = \frac{E}{2(1 + \mu)} = \frac{29.0 \times 10^6}{2(1 + 0.30)} = 11.2 \times 10^6 \text{ psi}$$

Example 4.5

Refer to Example 4.1 and construct a stress element from the shaft surface, then draw Mohr's circle for stress. Determine the principal stresses and then sketch planes on which they act. If $G = 12 \times 10^6$ psi, construct the Mohr's circle for strain and state values of the principal strains.

Solution. The shearing stress computed from the torsion formula equals 8120 psi. Figure 4.9(a) shows the appropriate plane stress element. Mohr's

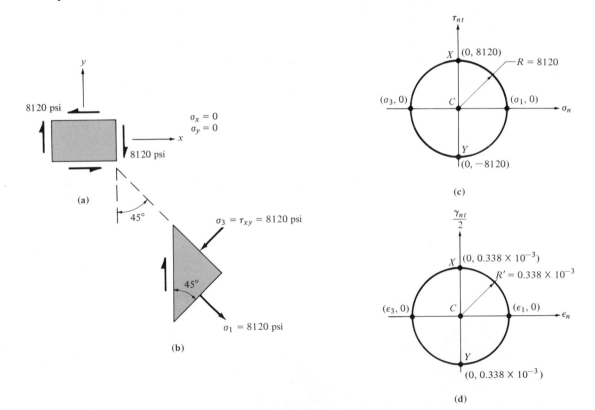

FIGURE 4.9

circle for stress is shown in Figure 4.9(c). Points X and Y with coordinates (0, 8120) and (0, −8120), respectively, were plotted for the states of stress on X and Y planes of the stress element. Points X and Y are connected, which locates the center of the circle C at the origin of the coordinate system. The radius of the circle equals 8120 psi in this case, and the principal normal stresses are $\sigma_1 = 8120$ psi and $\sigma_3 = -8120$ psi. These stresses are shown acting on their associated planes in Figure 4.9(b). In order to construct Mohr's circle for strain, compute $\gamma = \tau/G = 8120/(12 \times 10^6) = 0.676 \times 10^{-3}$ in./in. Recall that $\gamma/2$ is plotted along the vertical axis when constructing the strain circle. $\gamma/2 = 0.338 \times 10^{-3}$ in./in. Plot points X and Y in Figure 4.9(d) with coordinates as shown. Connect these points to establish the center of the circle C at the origin of the coordinate system. The radius of the circle equals 0.338×10^{-3} in./in. in this case and the principal normal strains are $\varepsilon_1 = 0.338 \times 10^{-3}$ in./in. and $\varepsilon_3 = -0.338 \times 10^{-3}$ in./in.

Example 4.6

During the course of an experiment, careful measurements of principal strains are taken on the surface of the shaft depicted in Figure 4.10(a). Directions 1 and 3 lie in a tangent plane to the shaft surface at point P and direction 2 (not shown) is perpendicular to the shaft surface. Axes 1 and 3 each make 45° angles with an x axis oriented along a longitudinal element of the circular cylindrical

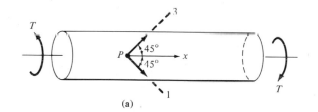

(a)

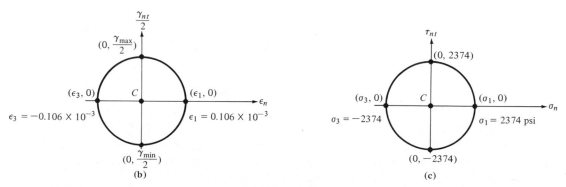

(b)

(c)

FIGURE 4.10

shaft of 2.00 in diameter. Measured results are $\varepsilon_1 = 0.106 \times 10^{-3}$ in./in. and $\varepsilon_3 = -0.106 \times 10^{-3}$ in./in. The shearing modulus of elasticity is 11.2×10^6 psi for the shaft. Draw Mohr's circles for strain and stress and state principal strains and stresses. Determine the torque T applied to the shaft when these strains were measured.

Solution. Since the stress element from the shaft surface represents a plane stress condition, $\sigma_2 = 0$ and $\sigma_1 > \sigma_2 > \sigma_3$. The strain circle shown in Figure 4.10(b) was constructed by plotting the points designated $(\varepsilon_1, 0)$ and $(\varepsilon_3, 0)$ and connecting them with a straight line. The center of the circle C lies at the origin of the coordinate system and the radius R of the circle equals 0.106×10^{-3} in./in. The maximum and minimum normal strains equal the measured values $\varepsilon_1 = 0.106 \times 10^{-3}$ and $\varepsilon_3 = -0.106 \times 10^{-3}$ in./in., respectively. From the circle $\gamma_{max}/2 = R' = 0.106 \times 10^{-3}$, or $\gamma_{max} = 0.212 \times 10^{-3}$ in./in. Similarly, $\gamma_{min} = -0.212 \times 10^{-3}$ in./in. In order to plot the stress circle, compute $\tau_{max} = G\gamma_{max} = (11.2 \times 10^6)(0.212 \times 10^{-3}) = 2374$ psi. Similarly,

$$\tau_{min} = -2374 \text{ psi}$$

The stress circle is shown in Figure 4.10(c). It was constructed by plotting the points $(2374, 0)$ and $(-2374, 0)$, which establishes the center C and the radius $R = 2374$ psi. Maximum and minimum normal stresses read from the circle are $\sigma_1 = 2374$ psi and $\sigma_3 = -2374$ psi, respectively. Their directions coincide with the measured principal strain directions 1 and 3 shown on the shaft surface at point P in Figure 4.10(a). The applied torque will be determined from the equation $\tau = T\rho/J$ with $\rho = D/2$. Since $\tau = \tau_{max} = 2374$ psi and $D = 2.00$ in., substitution yields

$$2374 = \frac{T(2.00/2)}{(\pi/32)(2.00)^4}$$

Solving for T, we obtain

$$\boxed{T = 3729 \text{ lb-in.}}$$

Homework Problems

4.13 Refer to Figure H4.1 and construct a stress element taken from a point on the surface of the shaft. Determine the maximum and minimum normal stresses at the point by constructing Mohr's circle for stress. Draw a sketch of the planes on which these stresses act.

4.14 Refer to segment *AB* of the shaft depicted in Figure H4.2 and construct Mohr's circle for stress for an element taken from the shaft surface. State maximum and minimum normal stresses for this element of segment *AB* and sketch the planes on which these stresses act.

4.15 Determine principal stresses from Mohr's circle constructions for the shaft depicted in Figure H4.3:
(a) An element on the outside surface.
(b) An element on the inside surface.

4.16 Refer to segment *BC* of the shaft depicted in Figure H4.4 and construct Mohr's circle for a surface element. State maximum and minimum normal stresses for this shaft segment and sketch the planes on which these stresses act.

4.17 Refer to Figure H4.5 and consider segment *DE* of the shaft, then construct Mohr's circle for a surface element. Determine principal stresses for this segment of the shaft.

4.18 Construct Mohr's circle for segment *BC* of the shaft shown in Figure H4.6 and find the principal stresses for a surface element of this shaft segment. Sketch the planes on which these stresses act.

4.19 Consider a stress element taken from the surface of the shaft 2 ft from the left end as shown in Figure H4.7. Assume that the shaft surface is free of applied forces at this point and construct Mohr's circle for stress. State principal stresses for this element and sketch the planes on which these stresses act.

4.20 Refer to the shaft shown in Figure H4.8 and consider segment *AB* of the shaft. Construct Mohr's circle for stress and determine principal stresses for elements taken from:
(a) The outside surface of the shaft.
(b) The inside surface of the shaft.
Sketch the planes on which these stresses act.

4.21 Refer to the shaft shown in Figure H4.9 and consider a stress element taken from the shaft surface 1 m to the left of point *B*. Assume that the surface of the shaft is free of applied forces where the element is selected. Draw Mohr's circles for stress and strain and determine principal stresses and strains for this plane stress element. Use $G = 100$ GPa.

4.22 Consider stress elements taken from the free surface of segments *AB* and *BC* of the shaft depicted in Figure H4.10. Construct Mohr's circles for stress and strain and find principal stresses and strains for these elements. Use $G = 11.2 \times 10^3$ ksi.

4.23 A uniformly distributed torque is applied to the shaft depicted in Figure H4.11. Consider a surface stress element 1 m to the right of point *A* on the shaft. Construct Mohr's circles for stress and strain for this element and determine principal stresses and strains. Use $G = 80$ GPa. Assume that the surface of the shaft is free of applied forces where the element is selected.

4.24 Select a surface element 7.00 ft from the left end of the shaft depicted in Figure H4.12. Construct Mohr's circle for stress and strain for this element and determine principal stresses and strains. Use $G = 12.0 \times 10^3$ ksi.

*4.5 Power Transmission

Torsional members are widely used to transmit power. Power *P* is defined as the time rate of doing work. Consider the shaft shown in Figure 4.11, which is subjected

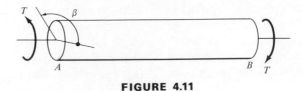

FIGURE 4.11

to end torques T while it rotates through the angle β (not to be confused with the static angle of twist θ). The time rate of change of β equals the angular velocity of the shaft, which is constant, since the end torques are balanced.

$$\omega = \frac{d\beta}{dt} = \text{const} \tag{4.24}$$

Prior to reaching the constant angular velocity ω, the shaft was accelerated angularly by applying torques which satisfied the dynamics equation for angular motion, $\sum T_0 = I_0\,\alpha$, where $\sum T_0$ equals the algebraic sum of the externally applied torques with respect to the axis of rotation, I_0 equals the mass moment of inertia with respect to the same axis, and α is the angular acceleration of the shaft. For constant ω, α equals zero, and the externally applied torques are in balance or their algebraic sum equals zero. Only shafts rotating at constant ω are considered here.

The differential work dW done by T during a differential angle change $d\beta$ is given by

$$dW = T\,d\beta \tag{4.25}$$

Power P equals the time rate of doing work:

$$P = \frac{dW}{dt} \tag{4.26}$$

Substitute Eq. 4.25 into Eq. 4.26:

$$P = T\frac{d\beta}{dt} \tag{4.27}$$

Substitute Eq. 4.24 into Eq. 4.27:

$$P = T\omega \tag{4.28}$$

A summary of notation and typical units employed in power transmission is given in Table 4.1.

TABLE 4.1

Notation and Typical Units Employed in Power Transmission

Quantity	Symbol	British Gravitational Units	SI Units
Work	W	ft-lb	N-m
Power	P	ft-lb/sec	N-m/sec
		1 hp = 550 ft-lb/sec	Watt = 1 N-m/sec
		(hp for horsepower)	kilowatt = 1000 N-m/sec
Torque	T	ft-lb	N-m
Angular velocity		rad/sec	rad/sec

Conversion factors:

To convert from	to	multiply by
hp	Watts	745.7
Watts	hp	0.001341
kW	Watts	1000.
W	kW	0.001

Static angles of twist θ of rotating shafts remain constant as the time rate of change of the rotation angle β of these shafts remains constant. Static angles are computed from Eq. 4.22. The following convention is adopted for stating the sense of static angular rotations: If the observer is positioned at the A end of a shaft and looks toward the B end of the shaft, then the shaft is assumed to be fixed at the B end. The observer views the sense in which a line scribed in the cross section at the A end of the shaft rotates with respect to a similar line, which is assumed to remain fixed at the B end of the shaft.

Example 4.7

Refer to the power transmission information given in Figure 4.12(a) and assume that the shaft is subjected to pure torsion. Construct the torque diagram for this shaft using a longitudinal coordinate with an origin at A. (Ignore gear thicknesses in the diagram.) Compute the maximum shearing stress in this hollow shaft and construct Mohr's circles for stress and strain for an outside surface element which is critical. Determine principal stresses and strains for this element and static angles of twist of the shaft segments AB and BC.

Solution. Apply Eq. 4.28 and units' data from Table 4.1 to calculate torques from the given power information.

$$P = T\omega$$

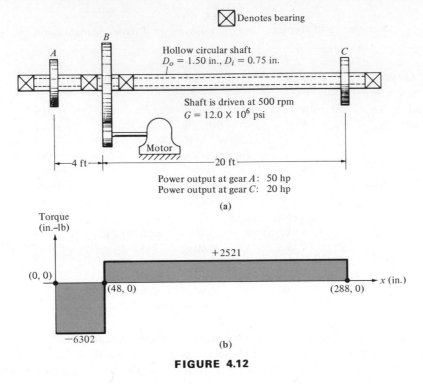

FIGURE 4.12

At gear A:

$$50 \text{ hp} \times \frac{550 \text{ ft-lb/sec}}{1 \text{ hp}} = T_A \times \frac{500 \text{ rev}}{1 \text{ min}} \times \frac{1 \text{ min}}{60 \text{ sec}} \times \frac{2\pi \text{ rad}}{1 \text{ rev}}$$

Thus $T_A = 525.2$ ft-lb.

At gear C by proportionality, since T is directly proportional to P,

$$T_C = \frac{20}{50}(525.2) = 210.1 \text{ ft-lb}$$

The power input of the motor equals the total power output if losses are neglected. At gear B, again by proportionality,

$$T_B = \frac{20 + 50}{50}(525.2) = 735.3 \text{ ft-lb}$$

As a check on the computations, consider the equilibrium condition, $\sum T = 0$, which leads to $T_B - T_A - T_C = 0$ or $T_B = T_A + T_C$. Values found for T_B, T_A, and T_C satisfy this equation, since $735.3 = 525.2 + 210.1$.

The torque versus longitudinal coordinate diagram is shown in Figure 4.12(b). The student should verify that the diagram is based upon a clockwise rotation of the motor viewed from left to right. The largest torque of 6302 in.-lb occurs in segment AB of the shaft. Equation 4.19 will be used to compute the shearing stresses acting on a stress element taken from the outside surface of the shaft.

$$\tau_\rho = \frac{T\rho}{J}$$

$$\tau_{max} = \frac{6302(1.50/2)}{(\pi/32)[(1.50)^4 - (0.75)^4]}$$

$$= \boxed{10{,}144 \text{ psi} \quad \text{or} \quad 10.1 \text{ ksi}}$$

A stress element taken from any point on the outside surface of segment AB of the shaft is shown in Figure 4.13(a). Mohr's circle for stress is shown in Figure 4.13(b), which shows all critical values.

In order to construct Mohr's circle for strain, we apply Eq.2.26. Thus $\gamma_{xy} = \tau/G = 10{,}144/(12 \times 10^6) = 8.453 \times 10^{-4}$. Recalling that $\gamma_{nt}/2$ is the appropriate ordinate for the strain circle, we refer to Figure 4.13(c) for this circle and stated critical values.

Static angles of twist (these angles are constant as the shaft rotates at a constant angular velocity ω) will be computed from Eq. 4.22.

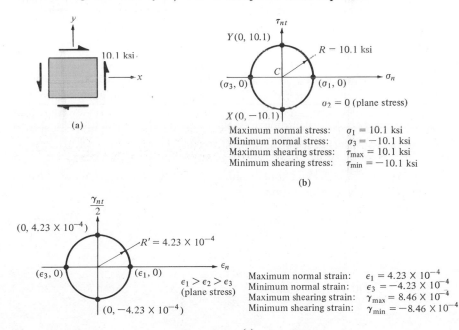

(a)

Maximum normal stress: $\sigma_1 = 10.1$ ksi
Minimum normal stress: $\sigma_3 = -10.1$ ksi
Maximum shearing stress: $\tau_{max} = 10.1$ ksi
Minimum shearing stress: $\tau_{min} = -10.1$ ksi

(b)

$\epsilon_1 > \epsilon_2 > \epsilon_3$
(plane stress)

Maximum normal strain: $\epsilon_1 = 4.23 \times 10^{-4}$
Minimum normal strain: $\epsilon_3 = -4.23 \times 10^{-4}$
Maximum shearing strain: $\gamma_{max} = 8.46 \times 10^{-4}$
Minimum shearing strain: $\gamma_{min} = -8.46 \times 10^{-4}$

(c)

FIGURE 4.13

SEGMENT *AB* (the angle of twist is clockwise as viewed by an observer looking from *A* toward *B*):

$$\theta_{AB} = \frac{TL}{JG}$$

$$= \frac{6302(48)}{(\pi/32)[(1.50)^4 - (0.75)^4](12 \times 10^6)}$$

$$= \boxed{0.0541 \text{ rad} \quad \text{or} \quad 3.10°}$$

SEGMENT *BC* (the angle of twist is counterclockwise as viewed by an observer looking from *B* toward *C*):

$$\theta_{BC} = \frac{TL}{JG}$$

$$= \frac{2521(240)}{(\pi/32)[(1.50)^4 - (0.75)^4](12 \times 10^6)}$$

$$= \boxed{0.108 \text{ rad} \quad \text{or} \quad 6.20°}$$

Example 4.8

Refer to the power transmission information given in Figure 4.14(a). Construct the torque diagram for this shaft using a longitudinal coordinate measured from the gear at *A*. Compute the maximum shearing stress in each segment of the shaft. Determine the angle of twist with respect to *B* as viewed by an observer looking from *A* toward *B* and the angle of twist of *D* with respect to *B* as viewed by an observer looking from *D* toward *B*.

Solution. Apply Eq. 4.28 and units' data from Table 4.1 to calculate torques from the given power information.

$$P = T\omega$$

AT GEAR *A*:

$$15 \text{ kW} \times \frac{1000 \text{ W}}{1 \text{ kW}} \times \frac{1 \text{ N-m/sec}}{1 \text{ W}} = T_A \times 400 \frac{\text{rev}}{1 \text{ min}} \times \frac{1 \text{ min}}{60 \text{ sec}} \times \frac{2\pi \text{ rad}}{1 \text{ rev}}$$

$$T_A = 358.1 \text{ N-m}$$

At gears *C* and *D* by proportionality, since *T* is directly proportional to *P*:

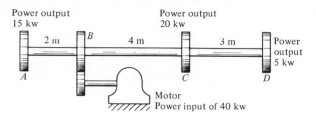

Power output
15 kw

Power output
20 kw

2 m   B   4 m   3 m   Power output 5 kw

A   C   D

Motor
Power input of 40 kw

Note: Bearings not shown. Assume pure torsion applied
to shaft.

Shaft rotates 400 rpm $G = 7.75 \times 10^{10} \text{ N/m}^2$
Segment diameter (m)
AB: 0.025
BC: 0.0375
CD: 0.020

(a)

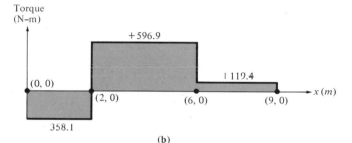

Torque
(N–m)

+596.9

$|$ 119.4

(0, 0)

(2, 0)   (6, 0)   (9, 0) x (m)

358.1

(b)

FIGURE 4.14

$$T_C = \tfrac{20}{15}(358.1) = 477.5 \text{ N-m}$$

$$T_D = \tfrac{5}{15}(358.1) = 119.4 \text{ N-m}$$

Since power input equals power output if losses are neglected, the necessary equation of equilibrium must be satisfied as follows:

$$T_B = T_A + T_C + T_D$$

$$= 358.1 + 477.5 + 119.4 = 955.0 \text{ N-m}$$

The torque versus longitudinal coordinate diagram is shown in Figure 4.14(b). The reader should verify that the diagram is based upon a clockwise rotation of the motor, viewed from left to right.

Maximum shearing stresses in each segment of the shaft will be computed with Eq. 4.19.

$$\tau_\rho = \frac{T\rho}{J}$$

SEGMENT *AB*

$$\tau = \frac{358.1(0.025/2)}{(\pi/32)(0.025)^4}$$

$$= \boxed{116.72 \text{ MPa}}$$

SEGMENT *BC*

$$\tau = \frac{596.9(0.0375/2)}{(\pi/32)(0.0375)^4}$$

$$= \boxed{57.65 \text{ MPa}}$$

SEGMENT *CD*

$$\tau = \frac{119.4(0.020/2)}{(\pi/32)(0.020)^4}$$

$$= \boxed{76.01 \text{ MPa}}$$

Static angles of twist will be computed from Eq. 4.22.

SEGMENT *AB* (the angle of twist is clockwise as viewed by an observer looking from *A* toward *B*):

$$\theta_{AB} = \frac{TL}{JG} = \frac{358.1(2)}{(\pi/32)(0.025)^4(7.75 \times 10^{10})}$$

$$= \boxed{0.241 \text{ rad} \quad \text{or} \quad 13.8°}$$

SEGMENTS *BC* AND *CD* (the angle of twist is counterclockwise as viewed by an observer looking from *D* toward *B*):

$$\theta_{DB} = \theta_{DC} + \theta_{CB}$$

$$= \frac{596.9(4)}{(\pi/32)(0.0375)^4(7.75 \times 10^{10})} + \frac{119.4(3)}{(\pi/32)(0.020)^4(7.75 \times 10^{10})}$$

$$= 0.159 + 0.294$$

$$= \boxed{0.453 \text{ rad} \quad \text{or} \quad 26.0°}$$

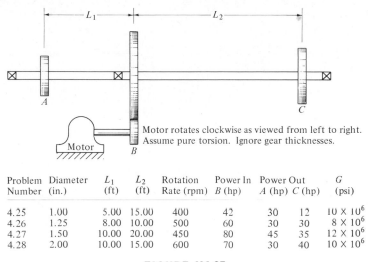

Problem Number	Diameter (in.)	L_1 (ft)	L_2 (ft)	Rotation Rate (rpm)	Power In B (hp)	Power Out A (hp)	C (hp)	G (psi)
4.25	1.00	5.00	15.00	400	42	30	12	10×10^6
4.26	1.25	8.00	10.00	500	60	30	30	8×10^6
4.27	1.50	10.00	20.00	450	80	45	35	12×10^6
4.28	2.00	10.00	15.00	600	70	30	40	10×10^6

FIGURE H4.25

Homework Problems

Given information is provided in the figures. A separate number is provided for each problem, but several problems refer to the same figure.

4.25 Refer to Figure H4.25 and construct the torque diagram using a longitudinal coordinate with an origin at A. Compute the maximum shearing stress in this shaft and construct Mohr's circles for stress and strain for an outside surface element which is critical. Determine principal stresses and strains for this element and static angles of twist of both segments of the shaft.

4.26 Refer to Figure H4.25 and construct the torque diagram using a longitudinal coordinate with an origin at A. Compute the maximum shearing stress and the static angles of twist for both segments of the shaft.

4.27 Refer to Figure H4.25 and construct the torque diagram using a longitudinal coordinate with an origin at A. Compute the maximum shearing stress in this shaft and construct Mohr's circles for stress and strain for an outside surface element which is critical. Determine principal stresses and strains for this element and static angles of twist of both segments of the shaft.

4.28 Refer to Figure H4.25 and construct the torque diagram using a longitudinal coordinate with an origin at A. Compute the maximum shearing stress in this shaft and the static angles of twist for both segments of the shaft.

4.29 Refer to Figure H4.29 and construct the torque diagram for this shaft using a longitudinal coordinate measured from the gear at A. Compute the maximum shearing stress in each segment of the shaft. Determine the angle of twist of A with respect to B as viewed by an observer looking from A toward B and the angle of twist of D with respect to B as viewed by an observer looking from D toward B.

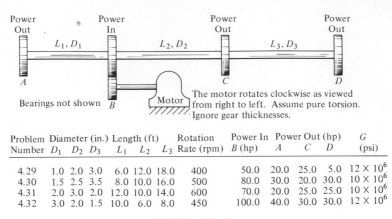

Problem Number	Diameter (in.) D_1	D_2	D_3	Length (ft) L_1	L_2	L_3	Rotation Rate (rpm)	Power In B (hp)	Power Out (hp) A	C	D	G (psi)
4.29	1.0	2.0	3.0	6.0	12.0	18.0	400	50.0	20.0	25.0	5.0	12×10^6
4.30	1.5	2.5	3.5	8.0	10.0	16.0	500	80.0	30.0	20.0	30.0	10×10^6
4.31	2.0	3.0	2.0	12.0	10.0	14.0	600	70.0	20.0	25.0	25.0	10×10^6
4.32	3.0	2.0	1.5	10.0	6.0	8.0	450	100.0	40.0	30.0	30.0	12×10^6

FIGURE H4.29

4.30 Refer to Figure H4.29 and determine the maximum shearing stress and normal stress in this shaft. State the shaft segment where these critical values occur. Determine the angle of twist of each segment of the shaft as viewed by an observer looking from A toward D.

4.31 Refer to Figure H4.29 and construct the torque diagram for this shaft using a longitudinal coordinate measured from the gear at A. Compute the maximum shearing stress in each segment of the shaft. Determine the angle of twist of A with respect to B as viewed by an observer looking from A toward B and

the angle of twist of D with respect to B as viewed by an observer looking from D toward B.

4.32 Refer to Figure H4.29 and determine the maximum shearing stress and normal stress in this shaft. State the shaft segment where these critical values occur. Determine the angle of twist of each segment of the shaft as viewed by an observer looking from A toward D.

4.33 Refer to Figure H4.33 and construct the torque diagram for this shaft using a longitudinal coordinate

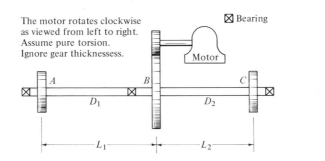

Problem Number	Diameter (m) D_1	D_2	Length (m) L_1	L_2	Rotation Rate (rpm)	Power In B (kw)	Power Out A (kw)	C (kw)	G (G Pa)
4.33	0.020	0.015	2.0	2.0	400	4.0	2.5	1.5	77.5
4.34	0.015	0.015	3.0	3.0	500	3.0	1.5	1.5	60.0
4.35	0.040	0.060	1.0	1.5	500	25.0	10.0	15.0	50.0
4.36	0.100	0.120	3.0	4.0	600	80.0	40.0	40.0	77.5

FIGURE H4.33

measured from the gear at *A*. Compute the maximum shearing stress in each segment of the shaft. Determine the angle of twist of *A* with respect to *B* as viewed by an observer looking from *A* toward *B* and the angle of twist of *C* with respect to *B* as viewed by an observer looking from *C* toward *B*.

4.34 Refer to Figure H4.33 and determine the maximum shearing stress and normal stress in this shaft. State the shaft segment where these critical values occur. Determine the angle of twist of each segment of the shaft as viewed by an observer looking from *A* toward *C*.

4.35 Refer to Figure H4.33 and determine the maximum shearing stress and normal stress in this shaft. State the shaft segment where these critical values occur. Determine the angle of twist of each segment of the shaft as viewed by an observer looking from *C* toward *A*.

4.36 Refer to Figure H4.33 and construct the torque diagram for this shaft using a longitudinal coordinate measured from gear at *A*. Compute the maximum shearing stress in each segment of the shaft. Determine the angle of twist of each segment of the shaft as viewed by an observer looking from *C* toward *A*.

*4.6 Analytical and Experimental Solutions for Torsion of Members of Noncircular Cross Sections

Treatment was limited, in preceding sections of this chapter, to solution of torsion problems for members having either circular or hollow circular cross sections. These solutions are of great practical importance because such members carry torsional loads efficiently. A primary assumption, verifiable by experiment, of these solutions is that plane cross sections of these members before twisting remain plane after twisting as shown in Figure 4.15(a).

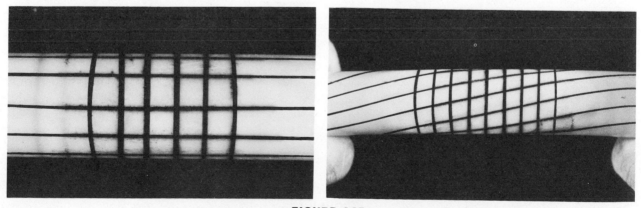

FIGURE 4.15a
A Circular, Cylindrical Shaft Before and After Twisting.

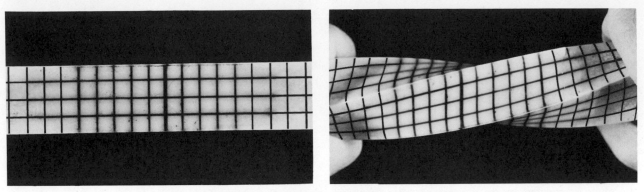

FIGURE 4.15b
A Rectangular, Prismatic Shaft Before and After Twisting.

Plane sections of members of noncircular cross section before twisting do not remain plane after twisting. Points initially in a plane are displaced in a direction parallel to the axis of twist of the member as shown in Figure 4.15(b). Such displacements, which occur in addition to in-plane displacements, are referred to as *warping*

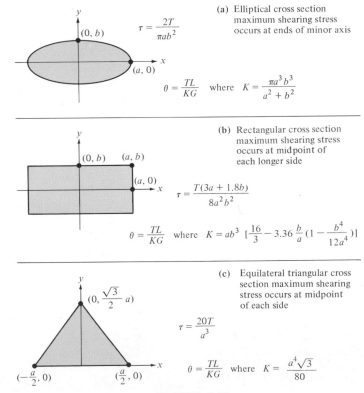

(a) Elliptical cross section maximum shearing stress occurs at ends of minor axis

$$\tau = \frac{2T}{\pi ab^2}$$

$$\theta = \frac{TL}{KG} \quad \text{where} \quad K = \frac{\pi a^3 b^3}{a^2 + b^2}$$

(b) Rectangular cross section maximum shearing stress occurs at midpoint of each longer side

$$\tau = \frac{T(3a + 1.8b)}{8a^2 b^2}$$

$$\theta = \frac{TL}{KG} \quad \text{where} \quad K = ab^3 \left[\frac{16}{3} - 3.36 \frac{b}{a} \left(1 - \frac{b^4}{12a^4} \right) \right]$$

(c) Equilateral triangular cross section maximum shearing stress occurs at midpoint of each side

$$\tau = \frac{20T}{a^3}$$

$$\theta = \frac{TL}{KG} \quad \text{where} \quad K = \frac{a^4 \sqrt{3}}{80}$$

FIGURE 4.16

displacements and the function describing them analytically is termed the *warping function*. If these displacements are prevented from taking place, normal stresses will arise that are directed parallel to the axis of twist. Warping displacements may be prevented by attaching the end of the torsional member to a relatively rigid member or structure capable of nullifying such displacements.

Mathematically, the problem of torsion of noncircular cross sections involves the solution of a partial differential equation subject to boundary conditions. Many analytical solutions have been obtained and several are shown in Figure 4.16. Equations for angle of twist and maximum shearing stresses are provided for members of these cross sections. It is assumed that warping may freely take place at the ends of the members, and thus normal stresses associated with warping do not arise. The analysis of stresses and displacements associated with warping are beyond the scope of this text.

Example 4.9

An engineer is confronted with the problem of deciding whether he should select a shaft of circular or elliptical cross section to resist a torque of 250,000 lb-in. Either shaft meets the geometric clearance requirements and both are made of the same material. The circular shaft radius is 2.00 in. and the elliptical shaft has an area equal to that of the circular shaft, with semiminor and semimajor axes of 1 in. and 4 in., respectively. This application requires a shaft that is 240 in. long. Determine maximum stresses and angles of twist of both shafts and select one shaft on this basis. Use $G = 11.2 \times 10^6$ psi.

Solution

CIRCULAR SHAFT

$$\tau_\rho = \frac{T\rho}{J}$$

$$\tau = \frac{250,000(2)}{(\pi/32)(4)^4}$$

$$= \boxed{19,900 \text{ psi}}$$

$$\theta = \frac{TL}{JG}$$

$$= \frac{250,000(240)}{(\pi/32)(4)^4(11.2 \times 10^6)}$$

$$= \boxed{0.213 \text{ rad} \quad \text{or} \quad 12.2°}$$

ELLIPTIC SHAFT (refer to Figure 4.16 for equations):

$$\tau = \frac{2T}{\pi a b^2}$$

$$= \frac{2(250,000)}{\pi(4)(1)^2}$$

$$= \boxed{39,800 \text{ psi} \quad \text{(at ends of minor axis)}}$$

$$\theta = \frac{TL}{KG} \qquad \text{where } K = \frac{\pi a^3 b^3}{a^2 + b^2}$$

$$K = \frac{\pi(4)^3(1)^3}{(4)^2 + (1)^2} = 11.827 \text{ in.}^4$$

$$\theta = \frac{250,000(240)}{11.827(11.2 \times 10^6)}$$

$$= \boxed{0.453 \text{ rad} \quad \text{or} \quad 25.9°}$$

The maximum stress in the elliptic shaft is twice that in the circular shaft and the angle of twist of the elliptic one is more than twice that of the circular one. Since both shafts have equal volumes of material and no special clearance problems arise, the circular shaft is to be preferred based upon the given information.

Example 4.10
A shaft of rectangular cross section is 0.040 m × 0.10 m with a length of 6.00 m. Determine the torque T which this shaft will safely resist if both of the following limitations are to be met:

1. Maximum shearing stress of 600 MPa.
2. Angle of twist of 15°.

Use $G = 75$ GPa.

Solution. Refer to Figure 4.16 for appropriate equations. Note that the rectangle dimensions are given as $2b \times 2a$, where $2b$ is the smaller side. In this case, $2b = 0.04$, $b = 0.02$ m, and $2a = 0.10$, $a = 0.05$ m. The maximum shearing stress occurs at the midpoint of each larger side.

$$\tau = \frac{T(3a + 1.8b)}{8a^2 b^2}$$

$$600 \times 10^6 = \frac{T[3(.05) + 1.8(.02)]}{8(.05)^2(.02)^2}$$

Solve for the torque based upon maximum shearing stress to obtain

$$\boxed{T = 25.81 \text{ kN-m}}$$

$$K = ab^3 \left[\frac{16}{3} - 3.36\frac{b}{a}\left(1 - \frac{b^4}{12a^4}\right)\right]$$

$$= 0.05(0.02)^3 \left[\frac{16}{3} - 3.36\left(\frac{0.02}{0.05}\right)\left(1 - \frac{(0.02)^4}{12(0.05)^4}\right)\right]$$

$$= 159.8 \times 10^{-8} \text{ m}^4$$

$$\theta = \frac{TL}{KG}$$

$$= \frac{15° \times \pi}{180°} = \frac{T(6)}{159.8 \times 10^{-8}(75 \times 10^9)}$$

Solve for the torque based upon angle of twist, to obtain

$$\boxed{T = 5.229 \text{ kN-m}}$$

Choose the smaller of these two torques: $T = 5.229$ kN-m. The requirement on angle of twist is critical compared to the shearing stress requirement. If the larger torque were applied to the shaft, the angle of twist would exceed 15°.

An experimental method for solving torsion problems termed the membrane analogy was introduced by L. Prandtl in 1903. It is based upon the fact that the mathematical formulation of the torsion problem and the small displacements of the homogeneous membrane give rise to the same boundary-value problems. If a membrane is supported along its edges such that the outline is the same as that of a torsional member cross section and subjected to a uniform upward pressure per unit area q, the conditions for the membrane analogy are met. Such a membrane is depicted in Figure 4.17. The tension per unit length, Q, in the membrane is assumed to be constant throughout the membrane for a given value of q.

Experiments with the membranes depend upon measurements of slopes of the membrane and ordinates to the membrane. These ordinates are used to compute the volume under the deflected membrane. The membrane analogy depends upon two concepts:

1. The slope of the membrane at any point is proportional to the shearing stress directed perpendicular to the plane in which the slope is measured.
2. Double the volume under the deflected membrane is proportional to the torque.

Membrane equations and solutions are identical to those of the torsion problem if q/Q is replaced by $2G\theta/L$. These concepts are depicted in Figure 4.18. In addition to being a powerful experimental method, the torsion analogy is also useful as a qualitative graphical aid for thinking about torsion problems.

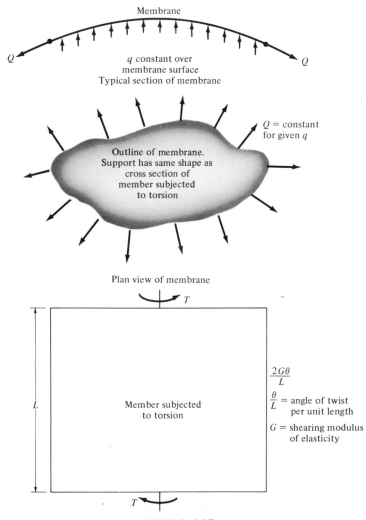

Membrane

Q

Q

q constant over
membrane surface
Typical section of membrane

$Q =$ constant
for given q

Outline of membrane.
Support has same shape as
cross section of
member subjected
to torsion

Plan view of membrane

T

L

Member subjected
to torsion

T

$\dfrac{2G\theta}{L}$

$\dfrac{\theta}{L} =$ angle of twist
per unit length

$G =$ shearing modulus
of elasticity

FIGURE 4.17

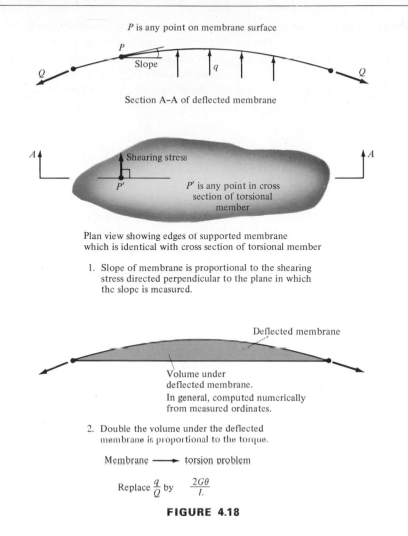

P is any point on membrane surface

Section A–A of deflected membrane

Plan view showing edges of supported membrane
which is identical with cross section of torsional member

1. Slope of membrane is proportional to the shearing
   stress directed perpendicular to the plane in which
   the slope is measured.

2. Double the volume under the deflected
   membrane is proportional to the torque.

Membrane ⟶ torsion problem

Replace $\dfrac{q}{Q}$ by $\dfrac{2G\theta}{L}$

FIGURE 4.18

Example 4.11

A series of membrane analogy experiments were performed with an elliptic
shaped boundary for the membrane with the following results:

> semimajor axis length = 4.00 in.
> semiminor axis length = 1.00 in.
> average volume under membrane = 2.95 in.^{3}
> average slope at ends of minor axis = 0.927 in./in.

A calibration membrane of circular boundary was used to determine the ratio $(2G\theta/L)/(q/Q) = 10^4$ psi. Compute the torque applied to a member of this elliptical cross section and the maximum shearing stress. Compare the experimentally determined shearing stress to a theoretical value obtained from the equations of Figure 4.16. Assume all other values to be exact in this latter calculation.

Solution

EXPERIMENTAL TORQUE AND STRESS VALUES

$$T = 10^4 \times 2 \times \text{volume} = 10^4(2)(2.95)$$

$$= \boxed{59,000 \text{ lb-in.}}$$

$$\tau = 10^4 \times \text{slope} = 10^4(0.927)$$

$$= \boxed{9270 \text{ psi}}$$

THEORETICAL STRESS VALUE

$$\tau = \frac{2T}{\pi ab^2} = \frac{2(59,000)}{\pi(4.00)(1.00)^2}$$

$$= \boxed{9390 \text{ psi}}$$

$$\text{percent difference in stresses} = \frac{9390 - 9270}{9390} \times 100$$

$$= \boxed{1.28 \text{ percent}}$$

Example 4.12
Relate the analytical solution for a circular torsional member to the membrane equations. Refer to Figure 4.19 for a diagram of the membrane.

Solution. The membrane is a paraboloid of revolution with a center ordinate h as shown in the figure. In the xz plane, a parabola through the three points $(r, 0, 0)$, $(0, 0, h)$, and $(-r, 0, 0)$ has the equation

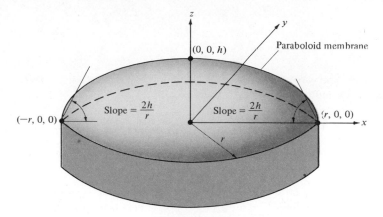

FIGURE 4.19

$$z = h - \frac{h}{r^2} x^2$$

The slope is obtained by differentiating this function with respect to x:

$$\frac{dz}{dx} = -\frac{2h}{r^2} x$$

Evaluate at $x = -r$:

$$\left(\frac{dz}{dx}\right)_{max} = \frac{2h}{r}$$

Maximum shearing stress is proportional to $2h/r$. The volume under a paraboloid of revolution is given by

$$\text{volume} = \tfrac{1}{2}\pi r^2 h$$

Torque is proportional to $2 \times \text{volume} = \pi r^2 h$. In summary:

	Analytical Solution	Membrane Solution
Maximum shearing stress	Tr/J	$\sim 2h/r$
Torque	$\tau J/r$	$\sim \pi r^2 h$

Homework Problems

4.37 A shaft has an elliptical cross section with $b = 0.06$ m, $a = 0.12$ m, and a length of 3.00 m. If the maximum shearing stress in this shaft is 250 MPa, determine the torque applied to the shaft and the angle of twist of the shaft. Refer to Figure 4.16 for appropriate equations. Use $G = 75$ GPa.

4.38 The cross section of a shaft is equilateral triangular with each side $a = 3$ in. and a length of 6 ft. If a torque of 40,000 lb-in. is applied at each end of this shaft, determine the maximum shearing stress and the angle of twist of the shaft. Use $G = 11.2 \times 10^6$ psi and obtain appropriate equations from Figure 4.16. Describe the locus of points along which the maximum shear stresses occur in this shaft.

4.39 A rectangular-cross-section shaft measures 0.5×1.5 in. If the maximum shearing stress in this shaft is 20 ksi, determine the torque T to which it is subjected. If this shaft is 4 ft long and $G = 12.0 \times 10^6$ psi, find the angle of twist of the shaft. Describe the locus of points along which the maximum shear stresses occur in this shaft. Appropriate equations may be obtained from Figure 4.16.

4.40 A shaft of elliptical cross section with $b = 0.04$ m and $a = 0.08$ m has a length of 5 m and is subjected to a torque of 20,000 N-m. Select a surface element from the shaft at a point on the end of a minor axis of a cross section and construct Mohr's circles for stress and strain for this element. State critical information obtained from these circles. Use $G = 75$ GPa.

4.41 A membrane of equilateral triangular shape with each side $a = 0.05$ m was subjected to a uniform pressure and the following results obtained:

    volume under the membrane = 3.125×10^{-6} m^3
    slope of membrane at midpoints of each of
       the sides = 1.00

calibrated ratio of $\dfrac{2G\theta/L}{q/Q} = 80.0$ MPa

Compute the torque carried by a torsional member of this cross section and the maximum shearing stress in the member.

4.42 A membrane of rectangular shape with dimensions 2.00×4.00 in. was subjected to a uniform pressure and the following results obtained:

    volume under the membrane = 2.00 in.^{3}

    slope of membrane at midpoints of longer
       sides = 0.976

    calibrated ratio of $\dfrac{2G\theta/L}{q/Q} = 20,492$ psi

Compute the torque carried by a torsional member of this cross section and the maximum shearing stress in the member.

4.43 Refer to Example 4.12 for appropriate equations in solving this problem. A membrane of circular boundary has a radius of 2 in. and central ordinate $h = 0.4$ in. Compute the maximum slope of the membrane and the volume under the membrane. If the shearing stress is 4000 psi, determine the corresponding torque applied to the shaft. What constant relates membrane measurements to shearing stresses and torques?

4.44 The constant relating membrane values to torsional member values is 50,000 psi. A membrane of circular boundary has a radius of 1 in. and a central ordinate $h = 0.15$ in. Compute the slope of the membrane and the volume under the membrane by using equations from Example 4.12. Determine the maximum shearing stress and the torque applied to the corresponding circular shaft.

*4.7   Torsion of Closed Thin-Walled Tubes

A closed thin-walled torsional member is shown in Figure 4.20(b) and the corresponding membrane is shown in Figure 4.20(a). The membrane supports a weightless plate, which accounts for the interior of the closed cross section, where shearing stresses have zero values since material to resist torques is confined to the thin-walled outer periphery. Since the weightless plate is horizontal, slopes are zero within the region occupied by the plate. Shearing stresses in this interior region are zero, since there is no material to resist torques.

Equations for shearing stress and angle of twist for such members will be developed from the membrane analogy. A review of Figures 4.17 and 4.18, together with the accompanying text, would be advisable for an understanding of the following developments.

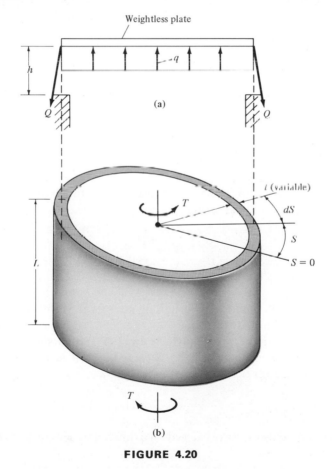

FIGURE 4.20

Recalling that double the volume under the deflected membrane is proportional to the torque and introducing a constant of proportionality C_1, we find that the equation for torque becomes

$$T = C_1(2Ah) \qquad (4.29)$$

where A is the mean of the inner and outer areas bounded by the cross-section boundaries and h is the small displacement of the deflected membrane subjected to a uniform pressure q.

Recalling that the slope of the membrane at any point is proportional to the shearing stress directed perpendicular to the plane in which the slope is measured, we can write the equation for shearing stress as follows:

$$\tau = C_1 \frac{h}{t} \qquad (4.30)$$

where t is the wall thickness, which may vary from point to point around the periphery, and C_1 and h have been defined above. Although the membrane is really curved over the depth h, an approximate slope based upon a straight membrane yields an average value for τ over the thickness of the cross section which is small enough such that there is little difference among the minimum, average, and maximum shearing stresses.

Solve Eq. 4.30 for h and substitute into Eq. 4.29, to obtain

$$T = 2A\tau t \qquad (4.31)$$

The product τt, which expresses the force per unit length, is referred to as the *shear flow*. This concept is valuable in study of the hydrodynamic analogy for the torsion problem. Only the membrane analogy is discussed in this book.

Solving Eq. 4.31 for the shearing stress yields

$$\tau = \frac{T}{2At} \qquad (4.32)$$

Equation 4.32 is most useful in determining stresses in a thin-walled closed cross-sectional member subjected to torque.

To write an equation for the angle of twist of this torsional member, vertical equilibrium of the membrane will be considered in a step-by-step fashion. The upward force exerted on the membrane is given by

$$F_U = qA \qquad (4.33)$$

The differential downward force exerted on a differential length dS of the periphery is given by

$$dF_D = \tau Q \, dS \tag{4.34}$$

The shearing stress τ appears in Eq. 4.34, since it represents the slope of the membrane and provides the appropriate vertical component of the membrane force $Q \, ds$, which acts at a small angle with the horizontal direction. To obtain the total downward force F_D, Eq. 4.34 will be integrated around the entire periphery with respect to the S coordinate, which yields the following line integral:

$$F_D = \oint_0^S \tau Q \, dS \tag{4.35}$$

Since the membrane is in equilibrium, the upward and downward forces are in balance:

$$F_U = F_D \tag{4.36}$$

Substitution from Eq. 4.33 and Eq. 4.35 into Eq. 4.36 yields

$$qA = \oint_0^S \tau Q \, dS \tag{4.37}$$

Since the membrane analogy is based upon equations derived for small displacements, the force per unit length Q in the membrane wall is assumed to be constant and may be brought outside the integral sign on the right side of Eq. 4.37. Taking this step and dividing by Q yields

$$\frac{q}{Q} A = \oint_0^S \tau \, dS \tag{4.38}$$

If we refer to Figure 4.18 and observe that q/Q may be replaced by $2G\theta/L$ as the transfer is made from the membrane to the torsion problem, Eq. 4.38 is written

$$\frac{2G\theta}{L} A = \oint_0^S \tau \, dS \tag{4.39}$$

Solve Eq. 4.39 for θ:

$$\theta = \frac{L}{2GA} \oint_0^S \tau \, dS \tag{4.40}$$

From Eq. 4.32 substitute for τ into Eq. 4.40:

$$\theta = \frac{L}{2GA} \oint_0^S \frac{T}{2At} \, dS \tag{4.41}$$

Only the thickness t varies around the periphery, and thus $T/2A$ may be taken outside the integral sign to give

$$\theta = \frac{TL}{4A^2G} \oint_0^S \frac{dS}{t} \qquad (4.42)$$

Equation 4.42 is the equation for the angle of twist θ for closed thin-walled cross-sectional members subjected to torque. Examples that follow will further clarify the meaning of Eqs. 4.32 and 4.42. If large compressive stresses develop in the walls of a very thin tube, buckling of the wall of the tube may govern rather than the shearing stress or the angle of twist given by these equations.

Example 4.13

A circular hollow tube has an outside diameter of 2.00 in. and an inside diameter of 1.80 in. The tube is 48 in. long and is subjected to a torque of 5000 lb-in. (a) Determine the average shearing stress in the wall of the tube using Eq. 4.32, and compare this value to the maximum shearing stress using Eq. 4.19. (b) Determine the angle of twist of this tube using Eq. 4.42 and compare this value with the one obtained from Eq. 4.22. Use $G = 10.0 \times 10^6$ psi. (c) Apply Eq. 4.32 to find the shear flow in this tube.

Solution

(a) Determine A, the mean of the outside and inside areas enclosed by the walls of the tube, as follows:

$$A = \left(\frac{1}{2}\right) \frac{\pi}{4} (D_o^2 + D_i^2)$$

$$= \frac{\pi}{8} [(2.00)^2 + (1.80)^2]$$

$$= 2.84 \text{ in.}^2$$

Applying Eq. 4.32 for the average shearing stress, we obtain

$$\tau = \frac{T}{2At} = \frac{5000}{2(2.84)\frac{1}{2}(2.00 - 1.80)}$$

$$= \boxed{8805 \text{ psi}}$$

Equation 4.19 yields for the maximum shearing stress with $\rho = D_o/2 = 1.00$ in.:

$$\tau_{max} = \frac{T\rho}{J} = \frac{5000(1.00)}{(\pi/32)[(2.00)^4 - (1.80)^4]}$$

$$= \boxed{9260 \text{ psi}}$$

Comparing these values on a percentage relative error basis yields

$$\text{percent error} = \frac{\tau_{max} - \tau}{\tau_{max}} \times 100$$

$$= \frac{9260 - 8805}{9260} \times 100$$

$$= \boxed{4.91 \text{ percent}}$$

The average radius to thickness ratio of this tube is 9.5, and the corresponding percentage error is 4.91 percent. For larger radius to thickness ratios, the results obtained for Eqs. 4.32 and 4.19 would differ by even smaller percentages.

(b) Since the thickness of this tube is constant, the line integral of Eq. 4.42 becomes around the complete circumference:

$$\frac{1}{t} \oint_0^S dS = \frac{1}{t} \int_0^{2\pi r_m} dS = \frac{2\pi r_m}{t}$$

where r_m is the mean radius of the tube. In this case,

$$\frac{1}{t} \oint_0^S dS = \frac{2\pi r_m}{t} = \frac{2\pi(0.95)}{0.1} = 59.7$$

The angle of twist from Eq. 4.42:

$$\theta = \frac{TL}{4A^2 G} \int_0^S \frac{dS}{t} = \frac{59.7 TL}{4A^2 G}$$

From above $A = 2.84$ in.^{2}, $G = 10 \times 10^6$ psi, $T = 5000$ lb-in., and $L = 48$ in.

$$\theta = \frac{59.7(5000)(48)}{4(2.84)^2(10 \times 10^6)}$$

$$= \boxed{0.0444 \text{ rad} \quad \text{or} \quad 2.54°}$$

From Eq. 4.22 the angle of twist:

$$\theta = \frac{TL}{JG} = \frac{5000(48)}{(\pi/32)[(2.00)^4 - (1.80)^4](10 \times 10^6)}$$

$$= \boxed{0.0444 \text{ rad} \quad \text{or} \quad 2.54°}$$

Stated to three significant figures, the angle of twist is identical from the two equations used above.

(c) Multiply Eq. 4.32 by t to find the shear flow:

$$\tau t = \frac{T}{2A} = \frac{5000}{2(2.84)} = \boxed{880 \text{ lb/in.}}$$

Example 4.14

A hollow cylindrical shaft is depicted in Figure 4.21(a). It is subjected to a torque T at each end and has a length L and is fabricated of a material with a shearing modulus of elasticity G. The membrane for this shaft is shown in Figure 4.21(b). To account for the hole in the shaft, a weightless plate of radius

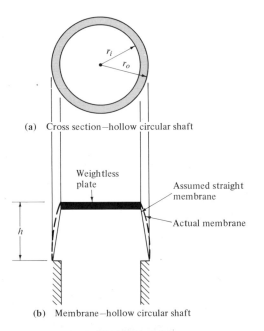

(a) Cross section—hollow circular shaft

(b) Membrane—hollow circular shaft

FIGURE 4.21

r_i is considered to be supported by the membrane. Determine (a) the maximum shearing stress τ_0 at the outside of the shaft using Eqs. 4.19 and 4.23, and (b) the average shearing stress τ using Eq. 4.32 based upon the membrane analogy for members with thin-walled cross sections. (c) Compare the answers to parts (a) and (b) on a percentage error basis using τ_0 as an exact value. Determine (d) the angle of twist θ' using Eq. 4.22, and (e) the angle of twist θ using Eq. 4.42 for members with thin-walled cross sections. (f) Compare the answers to parts (c) and (d) on a percentage error basis using θ' as an exact value. (g) Apply Eq. 4.32 to determine the shear flow in this shaft in terms of the applied torque T and cross-sectional dimensions.

Solution

(a) Equation 4.23 substituted into Eq. 4.19 gives

$$\tau_o = \frac{2Tr_o}{\pi(r_o^4 - r_i^4)}$$

(b) The mean area A enclosed by inner and outer boundaries becomes for this case:

$$A = \frac{\pi r_o^2 + \pi r_i^2}{2}$$

and the thickness

$$t = r_o - r_i$$

Substitution into Eq. 4.32 yields

$$\tau = \frac{T}{\pi(r_o^2 + r_i^2)(r_o - r_i)}$$

(c) The percentage error using τ_0 as an exact value may be written:

$$\text{percent error} = \frac{\tau_0 - \tau}{\tau_0} \times 100 \qquad (\tau_0 > \tau)$$

After algebraic manipulation, the percentage error becomes

$$\text{percent error} = 50\left(1 - \frac{r_i}{r_o}\right)$$

This linear function is plotted in Figure 4.22(a).

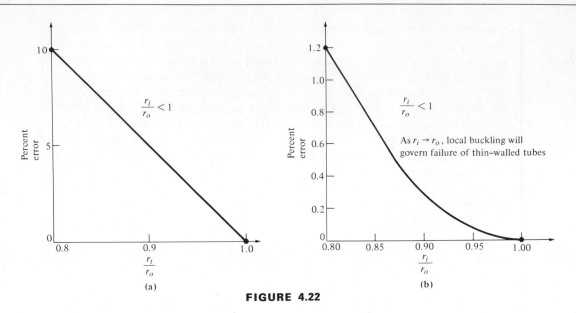

FIGURE 4.22

(d) Equation 4.22 with the appropriate polar moment of inertia J substituted yields for the angle of twist:

$$\theta' = \frac{2TL}{\pi(r_o^4 - r_i^4)G}$$

(e) Equation 4.42 for the angle of twist θ of members with thin-walled cross sections is repeated here for convenience.

$$\theta = \frac{TL}{4A^2G}\oint \frac{dS}{t}$$

The appropriate value of A is stated in part (b). The line integral is easily evaluated, since the thickness ($t = r_o - r_i$) is a constant.

$$\oint \frac{dS}{t} = \frac{1}{r_o - r_i}\oint dS$$

This line integral on the right equals the circumference of the circle with a mean radius of $(r_o + r_i)/2$:

$$\oint dS = 2\pi\left(\frac{r_o + r_i}{2}\right) = \pi(r_o + r_i)$$

Substitution into Eq. 4.42 yields

$$\theta = \frac{TL\pi(r_o + r_i)}{4((\pi r_o^2 + \pi r_i^2)/2)^2 G(r_o - r_i)}$$

Simplifying this equation for the angle of twist, we obtain

$$\theta = \frac{TL}{G\pi} \frac{r_o + r_i}{(r_o^2 + r_i^2)^2 (r_o - r_i)}$$

(f) The percentage error using θ' as an exact value may be written:

$$\text{percent error} = \frac{\theta' - \theta}{\theta'} \times 100 \qquad (\theta' > \theta)$$

After algebraic manipulation, the percentage error becomes

$$\text{percent error} = 50\left[1 - \frac{2(r_i/r_o)}{1 + (r_i/r_o)^2}\right]$$

This function is plotted in Figure 4.22(b). The approach taken in solving this problem, in which two theories are compared on a percentage basis, is widely used in engineering. Reference to Figure 4.22 enables the investigator to make meaningful statements about the percentage error involved in using the thin-walled solutions given in Eqs. 4.32 and 4.42. For example, if a 5 percent error in shearing stress calculations is to be tolerated, then $0.9 < (r_i/r_o) < 1.0$, and this range of (r_i/r_o) values corresponds to a maximum of 0.28 percent error in angle of twist calculations. For large values of (r_i/r_o) local buckling of the thin-walled cross section may govern rather than shearing stress or angle of twist.

(g) Multiply Eq. 4.32 by t to yield the shear flow:

$$\tau t = \frac{T}{2A} = \frac{T}{\pi(r_o^2 + r_i^2)}$$

Example 4.15
Four thin-walled cross sections of thickness t are shown in Figure 4.23. The length along the centerline of each of the thin walls is $2\pi r$. Each member is subjected to end torques and all four members have equal lengths L and shearing moduli of elasticity G. Determine the shearing stress and angle of twist of each of these members. Compare the answers obtained for the ratio of $t/r = 0.1$.

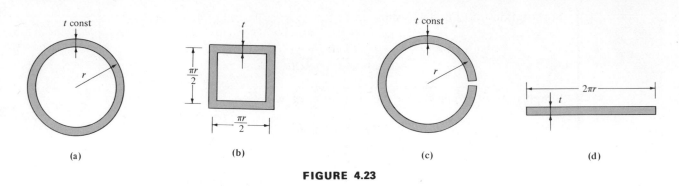

FIGURE 4.23

Solution

(a) *Closed circular thin-walled cross section.* Equation 4.32 yields for the shearing stress:

$$\tau_a = \frac{T}{2At} = \frac{T}{2\pi r^2 t}$$

Equation 4.42 yields for the angle of twist:

$$\theta_a = \frac{TL}{4A^2G} \oint \frac{dS}{t}$$

$$= \frac{TL}{4(\pi r^2)^2 G} \frac{2\pi r}{t} = \frac{TL}{2\pi r^3 tG}$$

(b) *Closed square thin-walled cross section.* Equation 4.32 yields for the shearing stress:

$$\tau_b = \frac{T}{2At} = \frac{T}{2(\pi r/2)^2 t} = \frac{2T}{\pi^2 r^2 t}$$

Equation 4.42 yields for the angle of twist:

$$\theta_b = \frac{TL}{4A^2G} \oint \frac{dS}{t}$$

$$= \frac{TL}{4(\pi r/2)^4 G} \frac{2\pi r}{t} = \frac{8TL}{\pi^3 r^3 tG}$$

(c) and (d) *Open thin-walled cross sections.* Since the tube depicted in Figure 4.23(c) has been cut lengthwise along a cylindrical element, the membranes for Figure 4.23(c) and (d) will be essentially the same and the solutions for these two problems will not differ. Refer to Figure 4.16 for the appropriate equations for the rectangular cross section.

$$\tau_c = \frac{T(3a + 1.8b)}{8a^2b^2}$$

In this case $a = \pi r$ and $b = t/2$, which upon substitution yields

$$\tau_c = \frac{T(3\pi r + 0.9t)}{2\pi^2 r^2 t^2}$$

The angle of twist is given by

$$\theta_c = \frac{TL}{KG}$$

where

$$K = ab^3 \left| \frac{16}{3} - 3.36\frac{b}{a}\left(1 - \frac{b^4}{12a^4}\right) \right|$$

Again, substitution of $a = \pi r$ and $b = t/2$ leads to

$$\theta_c = \frac{8TL}{G\pi r t^3\{\frac{16}{3} - 3.36(t/2\pi r)[1 - (t^4/192\pi^4 r^4)]\}}$$

To compare stress values, form the following ratios:

$$\frac{\tau_a}{\tau_b} = \frac{T}{2\pi r^2 t}\frac{\pi^2 r^2 t}{2T} = \frac{\pi}{4} = 0.785$$

$$\frac{\tau_a}{\tau_c} = \frac{T}{2\pi r^2 t}\frac{2\pi^2 r^2 t^2}{T(3\pi r + 0.9t)} = \frac{\pi}{3\pi(r/t) + 0.9}$$

Since t/r is given as 0.1, then $r/t = 10$ and this stress ratio becomes

$$\frac{\tau_a}{\tau_c} = \frac{\pi}{30\pi + 0.9} = 0.0330$$

Similar ratios will be formed for comparison of the angles of twist:

$$\frac{\theta_a}{\theta_b} = \frac{TL}{2\pi r^3 tG}\frac{\pi^3 r^3 tG}{8TL} = \frac{\pi^2}{16} = 0.617$$

$$\frac{\theta_a}{\theta_c} = \frac{TL}{2\pi r^3 tG}\frac{G\pi r t^3\{\frac{16}{3} - 3.36(t/2\pi r)[1 - (t^4/192\pi^4 r^4)]\}}{8TL}$$

For $t/r = 0.1$, this angle of twist ratio becomes

$$\frac{\theta_a}{\theta_c} = 0.00330$$

This comparative analysis reveals that the member with a thin-walled circular cross section has lower stresses and twists less than a member with a thin-walled square cross section of equal perimeter.

Comparative results for members of the closed cross section of Figure 4.23(a) and the open cross sections of Figure 4.23(c) and (d) are much more dramatic. Inverting the stress and angle of twist ratios obtained above reveals that the open cross section develops stresses that are 30.3 times those of the closed cross section and that the open cross-section twists through angles which are 303 times those of the closed cross section. The stress computed for the closed cross section is an average value, while the stress computed for the closed cross section is a maximum at the midpoint of each longer side. Solutions given here are for the elastic case, which means that the stresses remain below the shearing proportional limit stress discussed in Section 4.8. In general, members with closed cross sections resist torque with much lower stresses and angles of twist than do members with open cross sections.

Homework Problems

4.45 Refer to the solution of Example 4.14 and check in detail the percentage error functions plotted in Figure 4.22. If a 3 percent error in shearing stress calculations is to be tolerated, what (r_i/r_o) range of values is permissible? What is the corresponding maximum error in the angle of twist calculations? What mode of failure may invalidate these results, particularly for r_i/r_o approaching unity?

4.46 Solve Example 4.13 for a square hollow tube with outside dimensions 2.00×2.00 in. and a wall thickness of 0.10 in. Equations 4.32 and 4.42 are to be used for your solution. Also, determine the shear flow in this tube. Why are Eqs. 4.19 and 4.22 invalid for this problem?

4.47 Solve Example 4.15 for a thickness/radius ratio of 0.08 for the circular cross sections shown in Figure

4.23(a) and (c). Compare your results with the solution to Example 4.15.

4.48 A thin-walled tube 0.12 in. thick has an elliptical cross section with outside measurements as follows:

semimajor axis = 4.00 in.

semiminor axis = 3.00 in.

The 36.0-in.-long tube is subjected to a torque of 20,000 lb-in. Determine the average shearing stress in the wall of the tube. (The area bounded by an ellipse is given by $A = \pi ab$, where a and b are the semiaxes of the ellipse.)

4.49 The cross section of a torsional member is shown in Figure H4.49. A torque of 400 N-m is

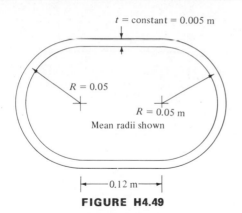

$t = \text{constant} = 0.005 \text{ m}$

$R = 0.05$

$R = 0.05 \text{ m}$

Mean radii shown

0.12 m

FIGURE H4.49

applied to each end of a member 2.5 m long. Determine the average shearing stress in the wall of this member and the angle of twist due to application of the end torques. Also, find the shear flow in this member. Use $G = 100$ GPa.

4.50 A thin-walled tube 0.0030 m thick has an elliptical cross section with outside measurements as follows:

$$\text{semimajor axis} = 0.100 \text{ m}$$

$$\text{semiminor axis} = 0.075 \text{ m}$$

This 1-m-long tube is subjected to a torque of 2250 N-m. Determine:

(a) The average shearing stress in the wall of this tube.

(b) The shear flow in this tube.

(The area bounded by an ellipse is given by $A = \pi ab$, where a and b are the semiaxes of the ellipse.)

4.51 A circular hollow tube has an outside diameter of 0.050 m and an inside diameter of 0.046 m. The tube, 1.20 m long, is subjected to a torque of 600 N-m. Determine:

(a) The average shearing stress in the wall of this tube using Eq. 4.32 and compare this value to the maximum shearing stress using Eq. 4.19.

(b) The angle of twist of this tube using Eq. 4.42 and compare this value with the one obtained from Eq. 4.22. Use $G = 80$ GPa.

4.52 A hollow, rectangular, cross-sectioned tube measures 16.0×8.0 in. outside and has a thickness of 0.40 in. It is subjected to end torques of 100,000 lb-in. and $G = 12 \times 10^6$ psi. Determine the average shearing stress in the wall of this tube and the angle of twist for a length of 8.0 ft.

4.8 Shearing Stress–Strain Properties

A typical torque versus angle of twist curve for a ductile metal specimen is shown in Figure 4.24(a). Data for such curves may be collected by subjecting torsional specimens to increasing torques and simultaneously measuring applied torques and angles of twist over a specified gage length. Solid circular specimens are often tested, although thin-walled hollow specimens have the advantage of a practically uniform shearing stress over their thin-walled annular areas. In order to construct the linear portion of the shearing stress–shearing strain curve from the experimental data for torque T and angle of twist θ, Eqs. 4.19 and 4.7 are applied as follows for solid circular specimens.

$$\tau_\rho = \frac{T\rho}{J} \tag{4.19}$$

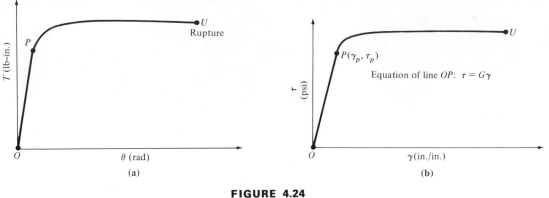

FIGURE 4.24

Substitutions of $\rho = r_o$ for the outside radius of the specimen and $J = (\pi/2)r_o^4$ yields an equation for shearing stress τ as a function of the applied torque T and the specimen size:

$$\tau = \frac{2}{\pi}\frac{T}{r_o^3} \tag{4.43}$$

Equation 4.7 for the shearing strain γ_ρ at any radial distance ρ from the center of the shaft is given by

$$\gamma_\rho = \frac{\rho\theta}{L} \tag{4.7}$$

The maximum shearing strain at the distance $\rho = r_0$ is given by

$$\gamma = \frac{r_0\theta}{L} \tag{4.44}$$

Equation 4.44 expresses the maximum shearing strain γ as a function of measurable quantities: r_o the outside specimen radius, and θ the angle of twist over the gage length L.

The linear portion of experimental torque-angle of twist data of Figure 4.24(a) may be transformed to the shearing stress–strain plot of Figure 4.24(b) by application of Eqs. 4.43 and 4.44.

The initial part of the τ–γ curve is linear over OP shown in Figure 4.24(b), and the slope of this straight line equals the shearing modulus of elasticity G. The ordinate to point P is the shearing proportional limit stress τ_p and point P is known as the shearing proportional limit because above this point the curve is no longer a straight line.

The area under the τ–γ curve up to the proportional limit P is known as the shearing modulus of resilience and is given by

$$u = \tfrac{1}{2}\tau_p \gamma_p \tag{4.45}$$

Physically, the modulus of resilience represents the elastic energy stored at the surface of the shaft per unit volume when it is stressed at the extreme fibers to the proportional limit. This elastic energy is recoverable, which means that upon release of the end torques applied to the specimen, it will return to its original condition without being permanently deformed.

Construction of the nonlinear portion of the τ–γ curve from point P to point U in Figure 4.24(b) requires an inelastic analysis of the torsion problem.

The general trend of the curve is shown in Figure 4.24(b), but inelastic torsion analysis is deferred to Section 12.4.

The modulus of rupture is obtained by calculating the shearing stress from Eq. 4.43 with the experimental rupture or ultimate torque T_u substituted for T:

$$\tau_u = \frac{2T_u}{\pi r_o^3} \tag{4.46}$$

Equation 4.46 yields a value for τ_u expressed in stress units, but this value is not truly the maximum stress when the shaft ruptures because Eq. 4.43 only provides true stresses for the linear portion OP of the plots. The modulus of rupture is an index of the torsional strength of a given material and may be used to compare static resistance of various materials under torsional loading.

The *torsional toughness* is defined as the work done per unit volume required to rupture the specimen. The area under the complete curve of Figure 4.24(b) equals the toughness of the material. A number of specimens would be tested and a statistical analysis performed in order to state mean and standard deviation values for the

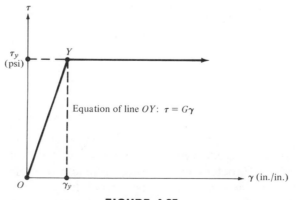

FIGURE 4.25

toughness of a given material. In fact, statistical statements are most useful for all material properties and "exact" values should be regarded as approximations.

The shearing stress–strain curve for a mild steel specimen is shown in Figure 4.25. This plot is idealized as a sloping straight line OY and a horizontal straight line to the right of point Y. Point Y is referred to as the yield point and the plot is termed the *idealized elastoplastic stress–strain curve*. The ordinate to the point Y is termed the *shearing yield stress* τ_y and the corresponding strain is termed the *shearing yield strain* γ_y. The shearing modulus of elasticity G is obtained by taking the slope of the line OY that equals τ_y/γ_y.

Example 4.16

Data were collected for a torsional specimen having a T–θ curve shaped somewhat like Figure 4.24(a).

Specimen outside diameter:	4.00 in.
Specimen inside diameter:	3.60 in.
Gage length:	10.0 in.
Ultimate torque:	200. k-in.

Torque (k-in.)	Angle of Twist (deg)
0.00	0.00
15.38	0.10
30.75	0.20
46.13	0.30
61.51	0.40
76.89	0.50
92.26	0.60
107.6	0.70
109.2	0.71
118.0	1.00
123.5	2.00
129.0	3.00
134.5	4.00
140.0	5.00
151.0	7.00
162.0	9.00
173.0	11.00

The θ values were not read beyond 11.00°. (a) Plot the T–θ curve for this torsional specimen; (b) determine the shearing modulus of elasticity for this material; (c) plot the linear portion of the τ–γ curve; (d) determine the modulus of resilience; and (e) determine the modulus of rupture.

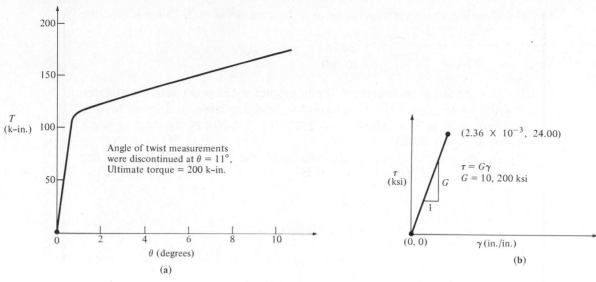

FIGURE 4.26

Solution

(a) A plot of the T–θ curve is shown in Figure 4.26(a).

(b) and (c) The point with $T = 92.26$ k-in., $\theta = 0.60°$, lies on the straight line portion of the T–θ plot. Since the specimen may be considered thin-walled, Eqs. 4.32 and 4.42 will be employed to transform torques and angles of twist to stresses and strains.

$$\tau = \frac{T}{2At}$$

$$= \frac{92.26}{2\left[\dfrac{\pi(2)^2 + \pi(1.8)^2}{2}\right]\left(\dfrac{4.00 - 3.60}{2}\right)}$$

$$= 20.28 \text{ ksi}$$

Solve Eq. 4.42 for G:

$$G = \frac{TL}{4A^2\theta}\oint_0^s \frac{dS}{t}$$

$$= \frac{92.26(10.)\,\pi\left(\dfrac{4.00 + 3.60}{2}\right)}{4\left[\dfrac{\pi(2)^2 + \pi(1.8)^2}{2}\right]^2\left(\dfrac{0.6\pi}{180}\right)\left(\dfrac{4.00 - 3.60}{2}\right)}$$

$$= \boxed{10{,}200 \text{ ksi}}$$

The shearing strain corresponding to a shearing stress of 20.28 ksi is given by

$$\gamma = \frac{\tau}{G} = \frac{20.28}{10,200} = 1.99 \times 10^{-3} \text{ in./in.}$$

The straight line portion of the T–θ plot extends to the point with coordinates: $T = 109.2$ k-in., $\theta = 0.71°$. Corresponding shearing stress and strain for this point are given by $\tau = 24.00$ ksi, $\gamma = 2.36 \times 10^{-3}$. A plot of this straight line is shown in Figure 4.26(b).

(d) The shearing modulus of resilience equals the area under the straight line shown in Figure 4.26(b). Equation 4.45:

$$u = \tfrac{1}{2}\tau_p \gamma_p$$
$$= \tfrac{1}{2}(24.00)(2.36 \times 10^{-3})$$
$$= \boxed{2.83 \times 10^{-2} \text{ in-k/in.}^3}$$

(e) The shearing modulus of rupture computed from Eq. 4.32 is given by

$$\tau_u = \frac{T}{2At}$$

$$= \frac{200}{2\left[\dfrac{\pi(2)^2 + \pi(1.8)^2}{2}\right]\left(\dfrac{4.00 - 3.60}{2}\right)}$$

$$= \boxed{43.97 \text{ ksi}}$$

Example 4.17

An idealized shearing stress–strain diagram was constructed for a mild steel specimen as shown in Figure 4.27. Determine the shearing modulus of elasticity for this specimen, the yield torque, and the corresponding angle of twist. The specimen has a diameter of 2.00 in., and the gage length, over which the angle of twist was measured, was 6.00 in.

Solution. The shearing modulus of elasticity G equals the slope of the τ–γ curve below the shearing yield point Y of Figure 4.27. This slope is given by

$$\text{slope} = G = \frac{\text{rise}}{\text{run}} = \frac{(21.0 - 0) \times 1000}{1.75 \times 10^{-3}}$$

$$= \boxed{12 \times 10^6 \text{ psi}}$$

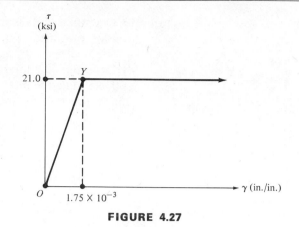

FIGURE 4.27

Equation 4.43 solved for T will provide the yield torque:

$$T = \frac{\pi r_o^3 \tau}{2}$$

$$= \frac{\pi (2.00/2)^3 (21,000)}{2} \cdot$$

$$= \boxed{32,990 \text{ lb-in.}}$$

Equation 4.44 solved for θ will provide the angle of twist corresponding to the yield torque:

$$\theta = \frac{\gamma L}{r_o}$$

$$= \frac{(1.75 \times 10^{-3})(6.00)}{2.00/2}$$

$$= \boxed{0.0105 \text{ rad} \quad \text{or} \quad 0.602°}$$

Homework Problems

4.53 Data from a torsional experiment follow.
(a) Determine the T–θ plot to scale. Angle of twist measurements were not made beyond $\theta = 30.0°$.
(b) Determine the shearing modulus of elasticity.

(c) Plot the linear portion of the shearing stress–strain curve.
(d) Determine the modulus of resilience.
(e) Determine the modulus of rupture.

Circular cylindrical specimen:

Diameter:　　　2.50 in.
Gage length:　8.00 in.

Torque (k-in.)	Angle of Twist (deg)	Torque (k-in.)	Angle of Twist (deg)
0.00	0.00	156.0	8.0
13.37	0.20	160.0	10.0
26.74	0.40	163.0	12.0
40.11	0.60	167.0	14.0
53.48	0.80	171.0	16.0
66.85	1.00	174.0	18.0
80.22	1.20	176.0	20.0
93.58	1.40	177.0	22.0
106.9	1.60	178.0	24.0
125.0	1.87	180.0	26.0
138.0	3.00	181.0	28.0
144.0	4.00	182.0	30.0
150.0	6.00	185.0	(Ultimate)

4.54　Data from a torsional experiment revealed a well-pronounced torsional yield point as shown in Figure 4.25. For the data given below:
(a) Find the shearing modulus of elasticity.
(b) Find the equation of the shearing stress–strain line below the yield point.
(c) Plot the stress–strain diagram for shear.
(d) Find the modulus of resilience.
Circular cylindrical specimen:

Diameter:　　　2.00 in.
Gage length:　6.00 in.

Torque (k-in.)	Angle of Twist (deg)	Torque (k-in.)	Angle of Twist (deg)
0.0	0.0	44.1	0.8
11.0	0.2	55.1	1.0
22.0	0.4	62.8	1.14 (yield)
33.1	0.6		

4.55　Data from a torsional experiment follow.
(a) Determine the $T–\theta$ plot to scale. (Angle of twist measurements were terminated at $\theta = 80°$.)

(b) Determine the shearing modulus of elasticity.
(c) Plot the linear portion of the shearing stress–strain curve.
(d) Determine the modulus of resilience.
(e) Determine the modulus of rupture.
Thin-walled hollow circular specimen:

Outside diameter:　0.100 m
Inside diameter:　　0.092 m
Gage length:　　　　0.250 m

Torque (N-m)	Angle of Twist (deg)	Torque (N-m)	Angle of Twist (deg)
0.	0.00	37,280.	20.0
7,100.	0.40	39,540.	30.0
14,200.	0.80	41,800.	40.0
21,300.	1.20	42,930.	50.0
24,850.	1.40	43,490.	60.0
29,930.	5.0	44,060.	70.0
33,320.	10.0	44,400.	80.0
35,580.	15.0	46,250.	(Ultimate)

4.56　Data from a torsional experiment revealed a well-pronounced torsional yield point as shown in Figure 4.25. For the data given below:
(a) Find the shearing modulus of elasticity.
(b) Find the equation of the shearing stress–strain line below the yield point.
(c) Plot the stress–strain diagram for shear.
(d) Find the modulus of resilience.
Circular cylindrical specimen:

Diameter:　　　0.05 m
Gage length:　0.15 m

Torque (N-m)	Angle of Twist (deg)	Torque (N-m)	Angle of Twist (deg)
0.	0.0	4290.	0.7
613.	0.1	4900.	0.8
1225.	0.2	5510.	0.9
1840.	0.3	6130.	1.0
2450.	0.4	6740.	1.1
3060.	0.5	7350.	1.2 (yield)
3680.	0.6		

4.57 Data from a torsional experiment are given below.

(a) Determine the T–θ plot to scale (angle of twist measurements were terminated at $\theta = 30°$).
(b) Determine the shearing modulus of elasticity.
(c) Plot the linear portion of the shearing stress–strain curve.
(d) Determine the modulus of resilience.
(e) Determine the modulus of rupture.

Circular cylindrical specimen:

> Diameter: 0.065 m
> Gage length: 0.200 m

Torque (N-m)	Angle of Twist (deg)	Torque (N-m)	Angle of Twist (deg)
0.	0.0	19,980.	12.0
2,870.	0.4	20,340.	14.0
5,750.	0.8	20,560.	16.0
8,620.	1.2	20,900.	18.0
10,050.	1.4	21,240.	20.0
12,930.	1.8	21,350.	22.0
15,800.	2.2	21,470.	24.0
16,500.	3.0	21,690.	26.0
17,060.	4.0	21,800.	28.0
18,020.	6.0	21,920.	30.0
18,750.	8.0		
19,430.	10.0	22,800.	(Ultimate)

4.58 Data from a torsional experiment are given below.

(a) Determine the T–θ plot to scale (angle of twist measurements were terminated at $\theta = 80.0°$).
(b) Determine the shearing modulus of elasticity.
(c) Plot the linear portion of the shearing stress–strain curve.
(d) Determine the modulus of resilience.
(e) Determine the modulus of rupture.

Thin-walled hollow circular specimen:

> Outside diameter: 4.00 in.
> Inside diameter: 3.60 in.
> Gage length: 10.0 in.

Torque (k-in.)	Angle of Twist (deg)	Torque (k-in.)	Angle of Twist (deg)
0.	0.0	380.	50.0
62.9	0.4	385.	60.0
125.7	0.8	390.	70.0
188.6	1.2	393.	80.0
220.	1.4		
262.	5.0	456.	(Ultimate)
293.	10.0		
315.	15.0		
330.	20.0		
350.	30.0		
370.	40.0		

Chapter **5** Stresses in Beams

5.1 Introduction

The beam is a very useful structural member that is employed in many different types of structural applications, such as floor, roof, and bridge deck systems. These members are called upon to resist bending action that is usually produced either by loads that are perpendicular to the member (transverse loads) or by pure bending moments. Regardless of the way the beam is loaded, the resulting action leads to two effects that are of major significance in the analysis of beams. These two effects are the induced stresses and the resulting deformations. Chapter 5 deals with the first of these two effects, the stresses induced in a beam, and the subject of deformations (or deflections) is dealt with in Chapter 6.

In order to discuss intelligently the question of stresses in beams, it is necessary to have a good understanding of properties of areas, including centroid location and principal centroidal moments of inertia, and their determinations. Thus, before discussing the concept of stresses in beams, it was decided to review briefly, in Section 5.2, the various concepts relating to properties of areas. This short review will ensure that the student has the background needed for proper understanding of the concepts of stresses due to symmetric and unsymmetric bending of beams.

5.2 Properties of Areas

Consider the arbitrary cross section shown in Figure 5.1 for which the centroid, point C, has been located. At point C, an xy coordinate system has been established as shown. Properties of the cross section with respect to axes through the centroid (i.e., centroidal axes) are examined in this section.

By definition, the moments of inertia with respect to the centroidal coordinate x and y axes in Figure 5.1 are given by

$$I_x = \int y^2 \, dA$$

$$I_y = \int x^2 \, dA$$

(5.1a)

where I_x and I_y represent the centroidal moments of inertia with respect to the x and y axes, respectively. Thus, by definition, the moment of inertia of an area with respect to a given axis requires that an element of area dA be multiplied by the square of its distance from this axis and the resulting product integrated over the entire area. Therefore, the unit for the moment of inertia is a length raised to the fourth power, such as in.^{4} in the British gravitational system and m^4 in the SI system.

Another quantity that is of great significance is the *polar moment of inertia*. This quantity, defined with respect to a centroidal axis, is given by

$$J_C = \int \rho^2 \, dA$$

(5.1b)

where J_C is the centroidal polar moment of inertia and the quantity ρ is defined in Figure 5.1. By using the fact that $\rho^2 = x^2 + y^2$, we can express J_C in terms of I_x and I_y. Thus

$$J_C = \int \rho^2 \, dA = \int (x^2 + y^2) \, dA$$

$$J_C = \int x^2 \, dA + \int y^2 \, dA = I_y + I_x$$

(5.1c)

Note that the concept of the polar moment of inertia has already been used in the analysis of torsional members discussed in Section 4.2.

The radii of gyration for a given area with respect to the centroidal x and y axes in Figure 5.1 are defined by the equations

$$r_x = \left(\frac{I_x}{A}\right)^{1/2}$$

$$r_y = \left(\frac{I_y}{A}\right)^{1/2}$$

(5.1d)

where r_x and r_y represent the rectangular centroidal radii of gyration with respect to the x and y axes, respectively. A polar radius of gyration may be defined in a similar manner by the equation

$$r_C = \left(\frac{J_C}{A}\right)^{1/2} \tag{5.1e}$$

where r_C is the centroidal polar radius of gyration. Substitution from Eqs. 5.1d and 5.1e into Eq. 5.1c leads to a relation among the various radii of gyration. Thus

$$r_C^2 = r_y^2 + r_x^2 \tag{5.1f}$$

Note that the unit for the radius of gyration is that of length.

The centroidal moments of inertia with respect to any set of perpendicular axes such as n and t shown in Figure 5.1 are, respectively,

$$I_n = \int t^2 \, dA$$
$$\tag{5.1g}$$
$$I_t = \int n^2 \, dA$$

If the axis n is inclined to the axis x by the angle α, assumed positive when measured in a counterclockwise direction, then from the geometry we conclude that

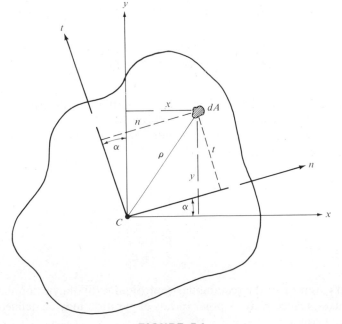

FIGURE 5.1

$$t = y \cos \alpha - x \sin \alpha$$

$$n = x \cos \alpha + y \sin \alpha \qquad \textbf{(5.1h)}$$

Substitution of the first of Eq. 5.1h into the first of Eq. 5.1g leads to

$$I_n = \int (y \cos \alpha - x \sin \alpha)^2 \, dA$$

$$= \cos^2 \alpha \int y^2 \, dA + \sin^2 \alpha \int x^2 \, dA - 2 \sin \alpha \cos \alpha \int xy \, dA \qquad \textbf{(5.1i)}$$

The mixed integral, $\int xy \, dA$, is known as the *product of inertia with respect to the orthogonal centroidal x and y axes* and is given the symbol P_{xy}. Thus, if Eq. 5.1a is substituted into Eq. 5.1i, we obtain

$$I_n = I_x \cos^2 \alpha + I_y \sin^2 \alpha - 2P_{xy} \sin \alpha \cos \alpha, \qquad \textbf{(5.1j)}$$

Substitution of appropriate trigonometric identities leads to

$$I_n = \frac{I_x + I_y}{2} + \frac{1}{2}(I_x - I_y) \cos 2\alpha - P_{xy} \sin 2\alpha \qquad \textbf{(5.1k)}$$

Equation 5.1k is mathematically similar to Eq. 2.4 previously derived for plane stress conditions. This property will be utilized in a later part of this section. Note that the centroidal moment of inertia of the area with respect to the t axis (i.e., I_t) may be found in a similar manner or more directly from Eq. 5.1k by replacing α by the quantity $\alpha + 90°$.

The centroidal product of inertia with respect to the perpendicular axes n and t can be represented by the relation

$$P_{nt} = \int nt \, dA \qquad \textbf{(5.2a)}$$

Substitution of Eq. 5.1h into Eq. 5.2a yields

$$P_{nt} = \int (x \cos \alpha + y \sin \alpha)(y \cos \alpha - x \sin \alpha) \, dA$$

$$P_{nt} = (\sin \alpha \cos \alpha)\left[\int y^2 \, dA - \int x^2 \, dA\right]$$

$$+ (\cos^2 \alpha - \sin^2 \alpha) \int xy \, dA$$

$$P_{nt} = (I_x - I_y) \sin \alpha \cos \alpha + P_{xy}(\cos^2 \alpha - \sin^2 \alpha)$$

$$P_{nt} = \tfrac{1}{2}(I_x - I_y) \sin 2\alpha + P_{xy} \cos 2\alpha \tag{5.2b}$$

Once again it is observed that Eq. 5.2b is mathematically similar to Eq. 2.6, which was obtained previously for the case of plane stress.

Equations 5.1k and 5.2b for moments of inertia can be made mathematically identical with the plane stress equations 2.4 and 2.6, respectively, by satisfying the following equalities:

$$\sigma_x \equiv I_x, \qquad \sigma_y \equiv I_y, \qquad \tau_{xy} \equiv P_{xy}$$

$$\sigma_n \equiv I_n, \qquad \tau_{nt} \equiv P_{nt} \tag{5.3}$$

Therefore, using the equalities in Eq. 5.3, one can develop, for example, expressions for the two principal moments of inertia $I_u = I_1$ and $I_v = I_2$ directly from Eq. 2.8. Thus

$$I_u = I_1 = \frac{I_x + I_y}{2} + R'' = I_A + R''$$

$$I_v = I_2 = \frac{I_x + I_y}{2} - R'' = I_A - R'' \tag{5.4}$$

where

$$R'' = \left[\left(\frac{I_x - I_y}{2} \right)^2 + (P_{xy})^2 \right]^{1/2}$$

and I_A denotes the average of the two orthogonal moments of inertia. For the sake of convenience in writing subsequent equations, the symbols I_u and I_v will, henceforth, be used in place of I_1 and I_2, respectively. Note that as in the case of principal stresses, the two principal axes of inertia u and v are perpendicular to each other.

As in the case of principal stresses, the principal moments of inertia I_u and I_v represent the algebraic maximum and algebraic minimum values, respectively. Alternatively, I_u and I_v are the moments of inertia about a set of perpendicular centroidal axes with respect to which the product of inertia is zero. Similarly, from other plane stress equations, corresponding equations for moments of inertia may be obtained directly, if needed, by using the equalities expressed in Eq. 5.3. Furthermore, because of the similarities between the plane stress equations and those for moments of inertia, Mohr's circle construction, which was developed for the solution of plane stress problems, is equally applicable to the solution of problems dealing with moments of inertia. Only minor modifications, as implied in Eq. 5.3, become necessary. The use of Mohr's circle for moments of inertia will be illustrated in Example 5.2.

It is desirable at this point to review briefly the question of determining moments and products of inertia of composite areas. If the moment of inertia of a composite area with respect to a given axis is desired, the moment of inertia for each of its component parts with respect to the given axis is found separately and the results added algebraically to obtain the moment of inertia of a composite area. Similarly, to obtain the product of inertia of a composite area with respect to a given set of orthogonal axes, the product of inertia for each component part with respect to the given set of orthogonal axes is determined separately and the results added algebraically. In order to determine the moments and products of inertia of component parts, however, the *parallel-axis theorems* are needed.

Consider the area A shown in Figure 5.2, where x and y are centroidal axes and X and Y are any axes parallel to x and y, respectively. Thus

$$I_X = \int Y^2 \, dA$$

$$I_X = \int (y + a)^2 \, dA$$

$$I_X = \int y^2 \, dA + 2a \int y \, dA + a^2 \int dA \tag{5.4a}$$

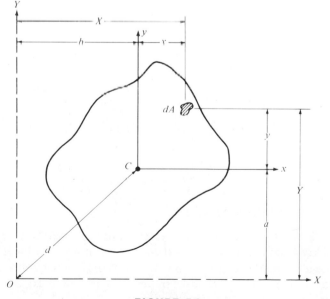

FIGURE 5.2

The integral $\int y^2\, dA$ represents the moment of inertia I_x with respect to the centroidal x axis. The integral $\int y\, dA$ represents the first moment of the area with respect to a centroidal axis and therefore must vanish; and the $\int dA$ is, of course, the area A. Thus

$$I_X = I_x + Aa^2 \tag{5.4b}$$

Similarly,

$$I_Y = I_y + Ab^2 \tag{5.4c}$$

and

$$J_O = J_C + Ad^2 \tag{5.4d}$$

Also, the product of inertia with respect to the perpendicular axes X and Y is given by

$$P_{XY} = \int XY\, dA$$

$$P_{XY} = \int (x + b)(y + a)\, dA$$

$$P_{XY} = \int xy\, dA + a\int x\, dA + b\int y\, dA + ab\int dA \tag{5.4e}$$

The integral $\int xy\, dA$ is the product of inertia P_{xy} with respect to the centroidal x and y axes. For the same reason given earlier,

$$\int x\, dA = \int y\, dA = 0 \text{ and } \int dA = A.$$

Therefore,

$$P_{XY} = P_{xy} + Aab \tag{5.4f}$$

Parallel-axis theorems are also available for radii of gyration. These can be developed quickly from the parallel-axis theorems for moments of inertia by replacing these moments of inertia by their values in terms of radii of gyration as expressed in Eqs. 5.1d and 5.1e. For example, if the moments of inertia in Eq. 5.4b are replaced by their equivalents in terms of the radii of gyration, we obtain

$$r_X^2 = r_x^2 + a^2 \tag{5.5a}$$

Similarly, from Eq. 5.4c,

$$r_Y^2 = r_y^2 + b^2 \tag{5.5b}$$

and from Eq. 5.4d,

$$r_O^2 = r_C^2 + d^2 \tag{5.5c}$$

where r_X is the rectangular radius of gyration with respect to the X axis, r_Y the rectangular radius of gyration with respect to the Y axis, and r_O the polar radius of gyration with respect to an axis through point O as shown in Figure 5.2.

Note that while moments of inertia can only assume positive values, products of inertia can be positive, negative, or zero, depending upon the choice of the coordinate axes. Also, because of the mathematical definition, P_{xy} will be zero if one or both of the x and y axes are axes of symmetry for the given area. Thus it may be concluded that an axis of symmetry is, of necessity, a principal axis of inertia. However, it should be pointed out that a principal axis of inertia is not necessarily an axis of symmetry. For example, all the cross-sectional areas shown in Figure 5.3 possess at least one axis of symmetry. This axis of symmetry is one of the two principal axes of inertia for the cross-sectional area. Consider Figure 5.3(a), which shows a rectangular area whose base is b and whose height is h. If it is assumed that $b < h$, then the moment of inertia with respect to the horizontal axis of symmetry (i.e., $bh^3/12$) is larger than that with respect to the vertical axis of symmetry (i.e., $hb^3/12$). Therefore, the horizontal axis of symmetry is the u axis and the vertical axis of symmetry is the v axis. For the circular cross section of diameter D shown in Figure 5.3(b), any pair of orthogonal centroidal axes represents the two principal axes of inertia u and v, and the principal moment of inertia has a value of $\pi D^4/64$. In the isoceles triangular cross-sectional area shown in Figure 5.3(c), the vertical axis through the centroid C is the only axis of

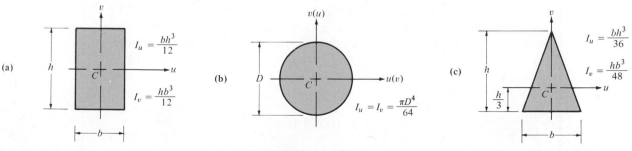

FIGURE 5.3

symmetry, and therefore it is one of the two principal centroidal axes of inertia. The second principal centroidal axis of inertia is perpendicular to the first and it is, obviously, not an axis of symmetry. If it is assumed that $b < h$, then the moment of inertia with respect to the horizontal principal centroidal axis of inertia (i.e., $bh^3/36$) is larger than that with respect to the vertical principal centroidal axis of inertia (i.e., $hb^3/48$). Therefore, the horizontal centroidal axis of inertia is the u axis and the vertical is the v axis.

In order to make the analogy of area moments of inertia with the plane stress problem expressed by Eq. 5.3 consistent, we will define P_{yx} as the negative of P_{xy}. This property is used in locating the two diametrically opposite points required for the construction of Mohr's circle, as illustrated in Example 5.2.

Selected properties of a few plane areas are shown in Appendix D.

Example 5.1

Determine (a) the centroidal principal axes and the centroidal principal moments of inertia for the inverted T section shown in Figure 5.4, and (b) its rectangular centroidal radii of gyration.

Solution

(a) The centroid C of the section is located on the vertical axis of symmetry (i.e., the y axis) at a distance above the base equal to $\bar{y}$, which is determined by dividing the area into two rectangular areas $A_1 = 16$ in.^{2} and $A_2 = 12$ in.^{2} as shown, and using the equation

$$\bar{y} = \frac{\sum A_i y_i}{\sum A_i} = \frac{A_1 y_1 + A_2 y_2}{A_1 + A_2}$$

$$= \frac{16(6) + 12(1)}{16 + 12} = 3.86 \text{ in.}$$

At the centroid C of the section, construct an x axis perpendicular to the y axis as shown. Since the y axis is an axis of symmetry, it follows that it is one of the two principal centroidal axes of inertia. Obviously, then, the x axis must be the second principal centroidal axis of inertia. Under certain conditions it is possible to determine by inspection which of the two is the u axis and which is the v axis. In complex composite sections, however, it is necessary to determine both moments of inertia before it can be ascertained which is the u and which is the v axis. In this case

$$I_x = (I_1)_x + (I_2)_x$$

where $(I_1)_x$ and $(I_2)_x$ represent the moments of inertia with respect to the x axis

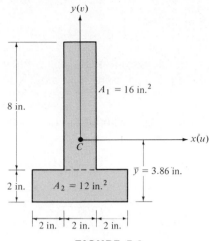

FIGURE 5.4

of rectangular areas A_1 and A_2, respectively. Thus, by using the parallel-axis theorem expressed in Eq. 5.4b,

$$(I_1)_x = \frac{2(8)^3}{12} + 16(2.14)^2 = 158.61 \text{ in.}^4$$

Similarly,

$$(I_2)_x = \frac{6(2)^3}{12} + 12(2.86)^2 = 102.16 \text{ in.}^4$$

Therefore,

$$I_x = 158.61 + 102.16 = 260.77 \text{ in.}^4$$

Also,

$$I_y = (I_1)_y + (I_2)_y$$

where $(I_1)_y$ and $(I_2)_y$ represent the moments of inertia with respect to the y axis of rectangular areas A_1 and A_2, respectively. These moments of inertia are found by using the parallel-axis theorem expressed in Eq. 5.4c. However, since the y axis coincides with the centroidal axes for both rectangular areas A_1 and A_2, the second term in Eq. 5.4c vanishes and

$$(I_1)_y = \frac{8(2)^3}{12} = 5.33 \text{ in.}^4$$

$$(I_2)_y = \frac{2(6)^3}{12} = 36.00 \text{ in.}^4$$

Therefore,

$$I_y = 5.33 + 36.00 = 41.33 \text{ in.}^4$$

Since I_x is larger than I_y, it follows that the x axis is the u principal centroidal axis and the y axis is the v principal centroidal axis. Also,

$$\boxed{I_u = I_x = 260.77 \text{ in.}^4} \qquad \text{and} \qquad \boxed{I_v = I_y = 41.33 \text{ in.}^4}$$

(b) The rectangular centroidal radii of gyration are obtained by using Eqs. 5.1d. Thus

$$r_x = \left(\frac{I_x}{A}\right)^{1/2} = \left(\frac{260.77}{28}\right)^{1/2}$$

$$= \boxed{3.05 \text{ in.}}$$

and

$$r_y = \left(\frac{I_y}{A}\right)^{1/2} = \left(\frac{41.33}{28}\right)^{1/2}$$

$$= \boxed{1.21 \text{ in.}}$$

Example 5.2
Determine the principal centroidal axes and principal centroidal moments of inertia for the cross-sectional area shown in Figure 5.5(a).

Solution. As in Example 5.1, the composite area is divided into the two rectangular areas $A_1 = 48 \times 10^{-4}$ m^2 and $A_2 = 96 \times 10^{-4}$ m^2, as shown. The centroid C of the cross section is then found as follows:

$$\bar{x} = \frac{\sum A_i x_i}{\sum A_i} = \frac{[48(6) + 96(10)] \times 10^{-6}}{(48 + 96) \times 10^{-4}} = 0.0867 \text{ m}$$

$$\bar{y} = \frac{\sum A_i x_i}{\sum A_i} = \frac{[48(26) + 96(12)] \times 10^{-6}}{(48 + 96) \times 10^{-4}} = 0.1667 \text{ m}$$

A centroidal xy coordinate system is established as shown in Figure 5.5(a). The moments of inertia I_x and I_y and the product of inertia P_{xy} are then determined. The moments of inertia I_x and I_y are found as discussed in Example 5.1. Thus

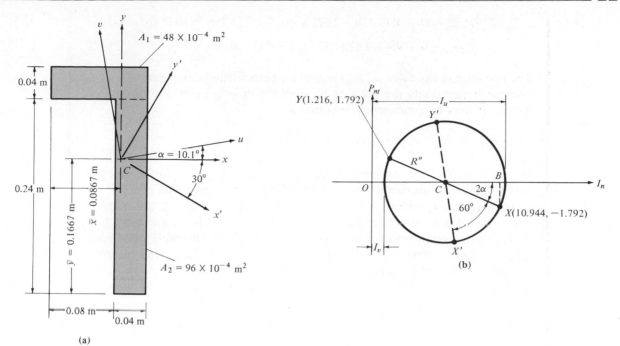

FIGURE 5.5

$$I_x = (I_1)_x + (I_2)_x$$

$$= \frac{12(4)^3 \times 10^{-8}}{12} + 48(9.33)^2 \times 10^{-8} + \frac{4(24)^3 \times 10^{-8}}{12} + 96(4.67)^2 \times 10^{-8}$$

$$= 10.944 \times 10^{-5} \ m^4$$

$$I_y = (I_1)_y + (I_2)_y$$

$$= \frac{4(12)^3 \times 10^{-8}}{12} + 48(2.67)^2 \times 10^{-8} + \frac{24(4)^3 \times 10^{-8}}{12} + 96(1.33)^2 \times 10^{-8}$$

$$= 1.216 \times 10^{-5} \ m^4$$

The product of inertia P_{xy} is found in a similar manner. Thus

$$P_{xy} = (P_1)_{xy} + (P_2)_{xy}$$

where $(P_1)_{xy}$ and $(P_2)_{xy}$ represent the products of inertia with respect to the orthogonal x and y axes of the rectangular areas A_1 and A_2, respectively. Thus, by using the parallel-axis theorem expressed in Eq. 5.4f,

$$(P_1)_{xy} = 0 + 48(9.33)(-2.67) \times 10^{-8} = -1.196 \times 10^{-5} \ \text{m}^4$$
$$(P_2)_{xy} = 0 + 96(-4.67)(1.33) \times 10^{-8} = -0.596 \times 10^{-5} \ \text{m}^4$$

The first term in Eq. 5.4f (i.e., P_{xy}) is zero for both rectangular areas A_1 and A_2 because it represents the product of inertia with respect to centroidal axes which are axes of symmetry. Therefore,

$$P_{xy} = -1.792 \times 10^{-5} \ \text{m}^4$$

Establish an I_n–P_{nt} coordinate system as shown in Figure 5.5(b). Construct Mohr's circle for moments and products of inertia by locating point X, whose coordinates are I_x and P_{xy} (i.e., 10.944×10^{-5}, -1.792×10^{-5}), and point Y, whose coordinates are I_y and $P_{yx} = -P_{xy}$ (i.e., 1.216×10^{-5}, 1.792×10^{-5}). Note that the common factor 10^{-5} has been omitted from the circle for convenience. Connect points X and Y to locate the center of the circle and complete the construction of the circle as shown in Figure 5.5(b).

From the geometry of Mohr's circle, the following values are obtained:

$$OC = \frac{I_x + I_y}{2} = \frac{(10.944 + 1.216) \times 10^{-5}}{2} = 6.080 \times 10^{-5} \ \text{m}^4$$

$$R'' = [(CB)^2 + (BX)^2] = [(4.864)^2 + (1.792)^2]^{1/2} \times 10^{-5} = 5.184 \times 10^{-5} \ \text{m}^4$$

$$I_u = OC + R'' = (6.080 + 5.184) \times 10^{-5}$$

$$= \boxed{11.264 \times 10^{-5} \ \text{m}^4}$$

$$I_v = OC - R'' = (6.080 - 5.184) \times 10^{-5}$$

$$= \boxed{0.896 \times 10^{-5} \ \text{m}^4}$$

$$2\alpha = \tan^{-1} \frac{1.792}{4.864} = \tan^{-1} 0.368 = 20.2°$$

$$\alpha = 10.1°$$

Therefore, the u axis is located at an angle of $10.1°$ ccw from the x axis and the v axis is, of course, perpendicular to the u axis. These axes (i.e., the principal centroidal axes of inertia) are shown in Figure 5.5(a).

Example 5.3
Determine the centroidal moments and the centroidal product of inertia for the section shown in Figure 5.5(a) with respect to a new set of axes obtained by rotating the x and y axes through $30°$ cw.

Solution. A new set of axes, x' and y', are shown in Figure 5.5(a) and were obtained by rotating the x and y axes through 30° cw. Point X' on Mohr's circle shown in Figure 5.5(b) is located by rotating from radius CX through an angle of 60° cw. The abscissa of this point represents the moment of inertia with respect to the x' axis, and the abscissa of point Y', diametrically opposite to point X', represents the moment of inertia with respect to the y' axis. Furthermore, the ordinates of the two points represent the product of inertia with respect to the x' and y' axes. Thus, from the geometry of Mohr's circle,

$$I_{x'} = OC + R'' \cos(2\alpha + 60) = (6.080 + 5.184 \cos 80.2) \times 10^{-5}$$

$$= \boxed{6.962 \times 10^{-5} \text{ m}^4}$$

$$I_{y'} = OC - R'' \cos(2\alpha + 60) = (6.080 - 5.184 \cos 80.2) \times 10^{-5}$$

$$= \boxed{5.198 \times 10^{-5} \text{ m}^4}$$

$$P_{x'y'} = -R'' \sin(2\alpha + 60) = -(6.080 \sin 80.2) \times 10^{-5}$$

$$= \boxed{5.991 \times 10^{-5} \text{ m}^4 = -P_{y'x'}}$$

Homework Problems

5.1–5.3 Determine the moments of inertia I_X and I_Y, and the product of inertia P_{XY} for the three sections that are shown in Figures H5.1, H5.2, and H5.3, respectively.

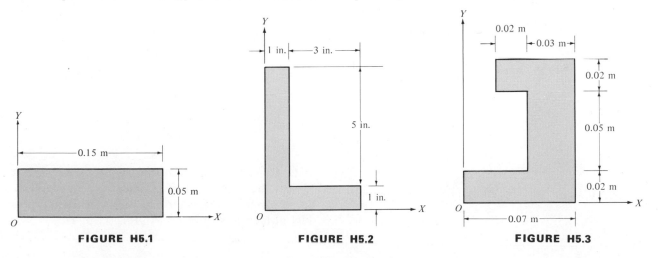

FIGURE H5.1   FIGURE H5.2   FIGURE H5.3

5.4 The moment of inertia of the area shown in Figure H5.4 with respect to the X axis is 3200 in.^{4} and with respect to the Y axis is 4700 in.^{4}. Determine the area and the centroidal moments of inertia I_x and I_y if they are known to be equal in magnitude. Find the polar moments of inertia with respect to axes through points C and O.

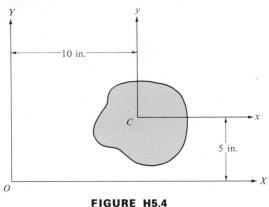

FIGURE H5.4

5.5–5.7 The sections shown in Figures H5.5, H5.6, and H5.7 have two perpendicular axes of symmetry. For each of these three sections, determine the principal centroidal axes and the principal centroidal moments of inertia.

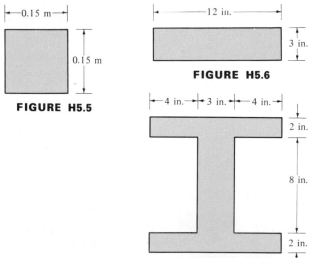

FIGURE H5.5

FIGURE H5.6

FIGURE H5.7

5.8–5.10 The sections shown in Figures H5.8, H5.9, and H5.10 have one axis of symmetry. For each of these three sections, determine the principal centroidal axes and the principal centroidal moments of inertia.

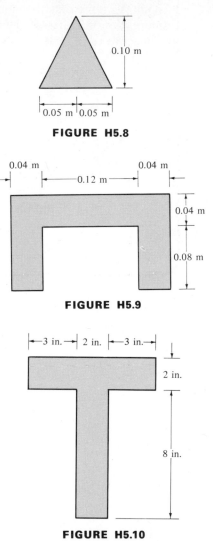

FIGURE H5.8

FIGURE H5.9

FIGURE H5.10

5.11 Consider a set of orthogonal axes inclined at 15° ccw from the axes of symmetry in Figure H5.5. Using Mohr's circle, determine the moments and product of inertia with respect to this set of axes.

5.12 Consider a set of orthogonal axes inclined at 30° cw from the axes of symmetry in Figure H5.7. Using Mohr's circle, determine the moments and product of inertia with respect to this set of axes.

5.13 Consider a set of orthogonal axes one of which is inclined at 45° ccw from the axis of symmetry in Figure H5.9. Using Mohr's circle, determine the moments and products of inertia with respect to this set of axes.

5.14–5.16 Determine the rectangular, principal centroidal radii of gyration and the centroidal polar radius of gyration for the sections shown in Figures H5.5, H5.6, and H5.7.

5.17–5.19 Determine the rectangular, principal centroidal radii of gyration and the centroidal polar radius of gyration for the sections shown in Figures H5.8, H5.9, and H5.10.

5.20–5.22 Determine, for the sections shown in Figures H5.20, H5.21, and H5.22,

(a) The centroidal principal axes and the centroidal principal moments of inertia.

(b) The rectangular centroidal radii of gyration with respect to the principal axes of inertia.

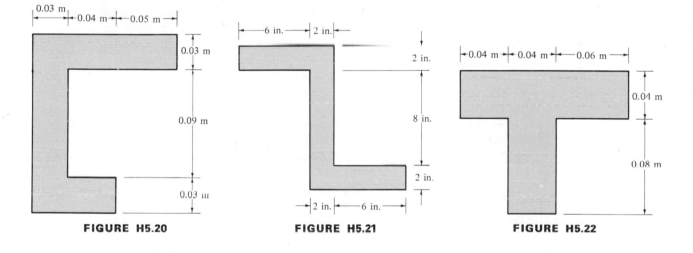

FIGURE H5.20   FIGURE H5.21   FIGURE H5.22

5.3 Flexural Stresses due to Symmetric Bending of Beams

By definition, a beam is a long and slender member that is subjected to bending action. Depending upon the position and orientation of the loads with respect to the principal axes of inertia of the beam cross section, bending may be either symmetric or unsymmetric. Symmetric bending of beams is discussed in this section and unsymmetric bending will be the subject of Section 5.5. In both types of bending, however, the assumption is made that the bending loads are so placed that they produce no twisting action in the beam. Furthermore, the assumption is made that a beam has the same cross-sectional configuration along its entire length (i.e., the beam is prismatic).

Consider two longitudinal planes, one of which contains the u principal axes of inertia and the second the v principal axes of inertia for all cross-sectional areas along the beam. Every beam, therefore, has two longitudinal principal planes which are perpendicular to each other. *Symmetric bending* of a beam occurs when all the bending loads lie in a plane that is either parallel to or coincident with one of the two longitudinal principal planes. Since, as was discussed in Section 5.2, an axis of symmetry for a cross section is a principal axis of inertia, it follows that symmetric bending results in beams with cross sections that have at least one axis of symmetry if the plane of the loads is parallel to or contains this axis of symmetry for each section.

Consider, for example, the case of a beam depicted in Figure 5.6(a), whose cross section, as shown magnified in Figure 5.6(b), has a vertical axis of symmetry. Obviously, this axis of symmetry is a principal centroidal axis of inertia (denoted as the v principal axis), and a vertical plane along the beam containing all these axes of symmetry is a longitudinal principal plane. Also, a horizontal axis through the centroid C of the section is the second centroidal principal axis of inertia (denoted as the u principal axis), and a horizontal plane along the beam containing all these horizontal centroidal principal axes is a second longitudinal principal plane. When the beam is subjected to positive bending couples M_u,* lying in the vertical longitudinal principal plane (i.e., the vertical plane of symmetry), symmetric bending occurs. The beam deforms into the configuration shown in a very exaggerated manner by the dashed curves in Figure 5.6(a). Longitudinal fibers in the upper part of the beam are shortened and those in the lower part of the beam are extended. At some position between the top and bottom of the beam, there is a surface at which the fibers are neither shortened nor extended. This surface is known as the *neutral surface* for the beam. The intersection of the neutral surface with the cross section of the beam is known as the *neutral axis for the section* and its intersection with the longitudinal principal plane of the loads is known as the *neutral axis for the beam*. As will be shown later in this section, the *neutral axis for the section* coincides with one of the two principal centroidal axes of inertia if the beam is subjected to symmetric bending (i.e., if the plane of the loads is either parallel to or is coincident with one of the two principal axes of inertia for the beam cross section, the neutral axis for the section coincides with its second principal axis of inertia). Thus in the symmetrically loaded beam shown in Figure 5.6, since the plane of the loads contains the v principal axis, the u principal axis coincides with the neutral axis for the section.

The shortening of the upper longitudinal fibers in Figure 5.6(a) is due to a system of normal compressive stresses, and the extension of its lower longitudinal fibers is caused by a system of normal tensile stresses. These normal stresses (both tension and compression) have been referred to by several names, among them "bending," "fiber," and "flexural" stresses.

---

* In this text the symbols M_u and M_v are used to signify moments that produce bending about the u and v principal centroidal axes of inertia, respectively.

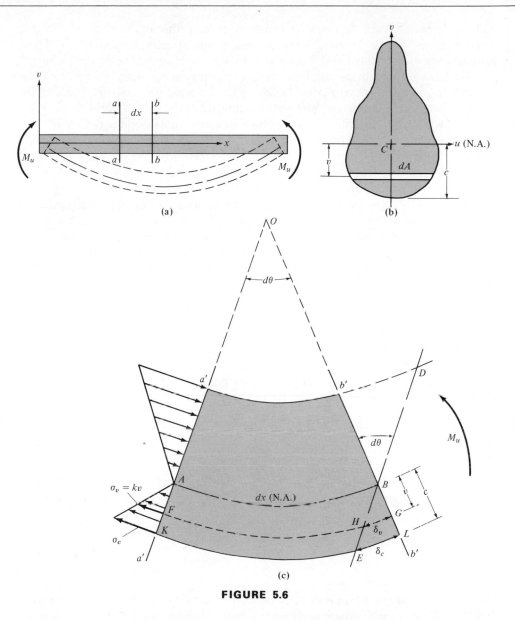

FIGURE 5.6

It should be pointed out that the flexure problem is a rather complex one, and its exact solution is beyond the scope of this book. The simple beam theory presented in this text is based upon a number of simplifying assumptions. However, the results obtained are sufficiently accurate for most engineering purposes. In addition to the two assumptions that have already been made relating to the plane of the bending loads and the uniformity of the cross section along the length of the beam, several other assumptions will be made in order to develop the analyses needed for the determination of the bending stresses in a symmetrically loaded beam. These assumptions will be stated and discussed as they are needed during the development.

The most significant assumption made in beam analysis states that plane cross sections (i.e., cross sections perpendicular to the longitudinal centroidal axis of the beam) before loading remain plane after loading. Thus the two adjacent sections a–a and b–b, which are an infinitesimal distance dx apart, remain plane but rotate with respect to each other into positions a'–a' and b'–b', respectively, as shown magnified in Figure 5.6(c). The element dx, whose ends are denoted by the symbols A and B, is coincident with the neutral axis (N.A.) for the beam, and therefore it undergoes no deformation. Furthermore, plane sections such as a–a and b–b rotate during deformation about their respective neutral axes, which appear in Figure 5.6(c) as points A and B. The rotated planes a'–a' and b'–b' form the angle $d\theta$ between them and, if extended, intersect at point O, known as the *center of curvature for the deflected beam*.

Through point B in Figure 5.6(c) construct line DBE parallel to the rotated plane a'–a'. If it is assumed that the beam is initially straight, then all longitudinal fibers contained between planes a–a and b–b have the same initial length dx. A longitudinal fiber such as FH, which is at a distance v below the neutral axis, undergoes an extension denoted by δ_v, equal to segment HG in Figure 5.6(c). Also, at the bottom of the beam, a distance c below the neutral axis, fiber KE undergoes an extension denoted by δ_c, equal to segment EL in Figure 5.6(c).

The next assumption made in the development of simplified beam theory is that deformations are relatively small. This assumption makes it possible to relate the deformations δ_v and δ_c through the similarity of triangles BGH and BLE. Thus

$$\frac{\delta_v}{\delta_c} = \frac{HG}{EL} = \frac{v}{c} \qquad (5.6a)$$

Since, by definition $\delta_v = \varepsilon_v\, dx$ and $\delta_c = \varepsilon_c\, dx$, where ε_v and ε_c are the strains corresponding to longitudinal fibers at distances v and c below the neutral axis, respectively, Eq. 5.6a leads to

$$\frac{\varepsilon_v}{\varepsilon_c} = \frac{v}{c} \qquad (5.6b)$$

which shows that the strain is directly proportional to the distance from the neutral axis. The assumption is now made that the material obeys Hooke's law (i.e., strain is proportional to stress). Therefore, ignoring Poisson's effects, one may substitute for ε_v the value σ_v/E_v and for ε_c the value σ_c/E_c in Eq. 5.6b, to obtain

$$\frac{\sigma_v/E_v}{\sigma_c/E_c} = \frac{v}{c} \qquad (5.6c)$$

where E_v and E_c are the moduli of elasticity for the material corresponding to locations defined by the distances v and c, respectively, below the neutral axis. If the material is assumed to be homogeneous, then $E_v = E_c = E$ and Eq. 5.6c becomes

$$\frac{\sigma_v}{\sigma_c} = \frac{v}{c} \tag{5.6d}$$

from which

$$\frac{\sigma_v}{v} = \frac{\sigma_c}{c} = \text{const} = k \tag{5.6e}$$

Equation 5.6e can also be written in the form

$$\sigma_v = \left(\frac{\sigma_c}{c}\right)v = kv \tag{5.6f}$$

which states the very important conclusion that the normal bending stress (i.e., fiber stress) produced by a bending moment is directly proportional to the distance of the fiber from the neutral axis. For the beam loaded as shown in Figure 5.6(a), these stresses will be compressive above and tensile below the neutral axis. Figure 5.6(c) shows the variation of the bending stress over cross section $a'-a'$ of the deflected beam. As will be shown later, this stress distribution gives rise to a bending couple which is equal in magnitude to the applied moment M_u.

It was stated earlier that in symmetrically loaded beams, the neutral axis coincides with one of the two principal centroidal axes of inertia. To prove this statement, consider the differential element of area dA at a distance v below the neutral axis of the cross section, as shown in Figure 5.6(b). The normal bending stress at this location is σ_v, which produces a differential normal force over the element of area dA equal in magnitude to $\sigma_v \, dA$ and acting along the axis of the beam (i.e., in the x direction). For any cross section along the beam, equilibrium of forces in the x direction dictates that the algebraic sum of the forces on a cross section such as $a'-a'$ in Figure 5.6(c) be equal to zero. Thus

$$\int \sigma_v \, dA = 0 \tag{5.7a}$$

Substitution of Eq. 5.6f into Eq. 5.7a leads to

$$k \int v \, dA = 0 \tag{5.7b}$$

Since k is not zero, it follows that

$$\int v \, dA = \bar{v}A = 0 \qquad \textbf{(5.7c)}$$

where A is the area of the cross section and $\bar{v}$ represents the distance from the neutral axis to the centroidal axis of the section. Since A cannot be zero, it follows that $\bar{v}$ must be zero and the neutral axis (N.A.) for the section coincides with its centroidal axis, which, in this case, is the u principal axis of inertia, as shown in Figure 5.6(b).

In order to further explain the variation of the bending (flexural or fiber) stress over a given cross section of a loaded beam, consider the beam shown in Figure 5.7(a), whose cross-sectional area shown in Figure 5.7(b) is a rectangle and which is subjected to the pure negative bending moment M_u. The plane of the applied moments is assumed to contain the vertical axis of symmetry for the cross section (i.e., the v principal centroidal axis of inertia), and therefore, as was shown earlier, the

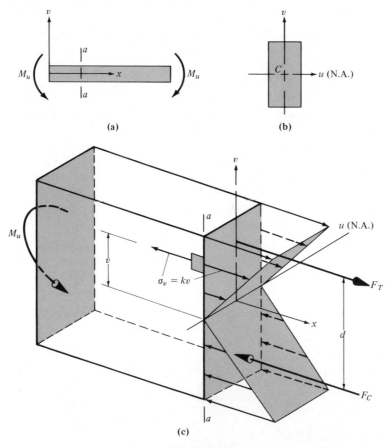

FIGURE 5.7

horizontal axis of symmetry (i.e., the u principal centroidal axis of inertia) becomes the neutral axis for the section. In this case the fibers of the beam above the neutral axis are extended and those below the neutral axis are shortened. Thus the stress distribution over any cross section for the beam will be such that the stresses are tension above and compression below the neutral axis.

A segment of the beam to the left of section a–a is isolated and a magnified three-dimensional view of its free-body diagram constructed as shown in Figure 5.7(c). Equilibrium dictates that the resisting moment on section a–a of the beam be equal in magnitude to the applied moment M_u. This resisting moment is due to the system of tensile and compressive stresses shown in section a–a in Figure 5.7(c). The neutral axis is, of course, unstressed and, as mentioned earlier, the fibers above the neutral axis are subjected to tension. Thus any stress element at a distance v above the neutral axis, such as the one shown in Figure 5.7(c) on the surface of the beam, has normal tensile stresses σ_v acting on its two planes that are perpendicular to the axis of the beam. Its other two planes, parallel to the axis of the beam, are free of any stresses. Also, since the planes on which σ_v act are free of shear stresses, it follows that they, as well as any planes perpendicular to them, are principal planes and therefore σ_v is a principal stress.

The tensile system of stresses above the neutral axis is equivalent to a tensile force F_T and the compressive system of stresses below the neutral axis is equivalent to a compressive force F_C as shown in Figure 5.7(c). Since there are no externally applied forces along the beam (i.e., along the x axis), it follows from the equilibrium of forces in the x direction that $F_T = F_C$. Therefore, forces F_T and F_C constitute a couple whose magnitude is $F_T d$ (or $F_C d$), where d is the perpendicular distance between the two forces. This couple is the resisting bending moment M_u^r at section a–a and must be equal in magnitude to the applied bending moment M_u. Thus

$$M_u^r = F_T d = F_C d = M_u \qquad (5.8)$$

Return now to Figure 5.6(b) and consider the element of area dA at a distance v below the neutral axis. As stated earlier, the normal bending stress at this location is σ_v, which produces a differential normal force acting on the element of area dA equal in magnitude to $\sigma_v\,dA$. The differential force $\sigma_v\,dA$ produces a differential resisting moment dM_u^r equal to the differential force $\sigma_v\,dA$ multiplied by its moment arm v. Thus

$$dM_u^r = \sigma_v v\,dA \qquad (5.9a)$$

and the resisting moment at section a–a of the beam is obviously the sum of all the differential moments over the entire area. Therefore,

$$M_u^r = \int dM_u^r = \int \sigma_v v\,dA \qquad (5.9b)$$

By Eq. 5.8, the resisting moment M_u^r is equal to the applied moment M_u and, by Eq. 5.6f, $\sigma_v = kv$. Making these substitutions into Eq. 5.9b leads to

$$M_u = k \int v^2 \, dA \tag{5.9c}$$

where the integral represents I_u, the moment of inertia of the area with respect to the u principal centroidal axis, which is also the neutral axis for the section. Thus

$$M_u = kI_u \tag{5.9d}$$

If the value of $k = \sigma_v / v$ is substituted in Eq. 5.9d and the resulting equation solved for σ_v, we obtain

$$\sigma_v = \frac{M_u v}{I_u} \tag{5.9e}$$

The notations σ_v and σ_c were used, for convenience, to signify the bending stresses at distances v and c from the neutral axis of the section, respectively. Since these stresses are normal to cross sections of the beam, they are oriented along the axis of the beam, which is designated in Figure 5.6(a) as the x axis. For the sake of consistency with the stress notation developed in Section 2.3, these stresses will be denoted by the symbol σ_x and Eq. 5.9e will become

$$\sigma_x = \frac{M_u v}{I_u} \tag{5.10a}$$

If the plane of the loads contained the u principal centroidal axis of inertia instead of the v principal centroidal axis of inertia, as has been assumed in developing Eq. 5.10a, the neutral axis would be the v principal centroidal axis of inertia and the bending stress σ_x would be given by the equation

$$\sigma_x = \frac{M_v u}{I_v} \tag{5.10b}$$

Equation 5.10 are the famous *flexure equations* and may be used to determine the flexural (bending or fiber) stress at any distance from the neutral axis for any location in a beam where the bending moment is known. These equations, however, were derived for the case in which the beam is subjected to a pure bending moment. Most cases that relate to the flexure of beams are such that bending is produced by loads applied transversely to the beam. Such a case is shown in Figure 5.8(a). By the methods developed in Section 1.6, any section such as a–a is subjected to a shear

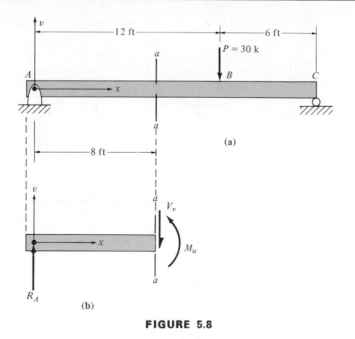

FIGURE 5.8

force V and a bending moment M, as shown in Figure 5.8(b). As will be shown in Section 5.4, the existence of a shear force is equivalent to a system of shear stresses in the plane of the section which produce a certain amount of distortion. Therefore, under these conditions, plane sections before bending do not remain plane after bending, and one of the basic assumptions made in deriving Eq. 5.10 is not satisfied. However, more refined solutions obtained for the flexure problem, including the effects of shear stresses, indicate that the flexure equations give answers that are very satisfactory for all practical purposes.

Example 5.4

Assume the cross-sectional area of the beam in Figure 5.8(a) to be rectangular, as shown in Figure 5.9(a). Let the plane of the load P be such that it contains the v principal axis of inertia and plot the flexure stress versus the v coordinate for section a–a at a distance of 8 ft from the left support.

Solution. The reactions at supports A and C for the beam in Figure 5.8(a) are computed by applying the conditions of equilibrium and are found to be

$$R_A = 10 \text{ k} \quad \text{and} \quad R_C = 20 \text{ k}$$

Consideration of the free-body diagram of that part of the beam to the left of section a–a, as shown in Figure 5.8(b), leads to the value of the bending

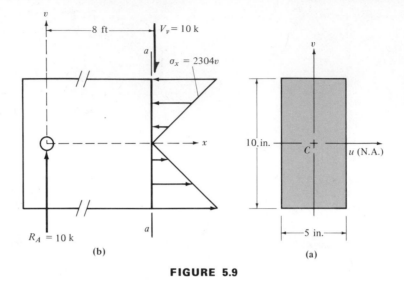

FIGURE 5.9

moment at cross section a–a. This moment is found to be

$$M_u = R_A(8) = 80 \text{ k-ft} = 960{,}000 \text{ lb-in.}$$

Note that since bending in this case is about the u principal axis of inertia, the bending moment has been designated as M_u.

By inspection it can be concluded that the horizontal axis of symmetry (i.e., the horizontal centroidal axis) is the u principal centroidal axis of inertia, and the corresponding principal moment of inertia is

$$I_u = \frac{1}{12}(5)(10)^3 = \frac{5000}{12} \text{ in.}^4$$

Thus, by Eq. 5.10a,

$$\sigma_x = \frac{960{,}000 \times 12v}{5000} = 2304v \text{ psi}$$

A plot of this equation is given in Figure 5.9(b), which shows the free-body diagram of that portion of the beam to the left of section a–a. Note the existence of the shear force V_v on section a–a. The shear stress produced by such a shear force will be dealt with in Section 5.4. Note also that the fibers of the beam above the u principal axis (i.e., the neutral axis) are in compression and that those below the neutral axis are in tension. Thus the maximum compressive value of σ_x occurs at the top of the beam, where $v = 5$ in. Therefore,

$$(\sigma_x)_{\text{top}} = 2304(5) = 11{,}520 \text{ psi}(C)$$

Furthermore, the maximum tensile stress occurs at the bottom of the beam, where v is also 5 in. Thus

$$(\sigma_x)_{bot} = 2304(5) = 11,520 \text{ psi}(T)$$

Note that only the absolute value of the coordinate v is needed in computing the flexural stress, since its sense may be determined by inspection for any point above or below the neutral axis.

Example 5.5

Assume the cross-sectional area of the beam in Figure 5.8(a) to be rectangular, as shown in Figure 5.10(a). Let the plane of the load P be such that it contains the u principal axis of inertia and plot the flexure stress versus the u coordinate for section a–a at a distance of 8 ft from the left support.

Solution. The support reactions at A and C are the same as found in Example 5.4, and therefore the bending moment at section a–a has the same value as determined in Example 5.4. However, although the cross-sectional area is identical with that in Example 5.4, the applied load in this case is such that it produces bending about the v principal centroidal axis. Therefore, the bending moment is designated as M_v and is given by

$$M_v = 960,000 \text{ lb-in.}$$

The moment of inertia with respect to the v principal centroidal axis is

$$I_v = \frac{1}{12}(10)(5)^3 = \frac{1250}{12} \text{ in.}^4$$

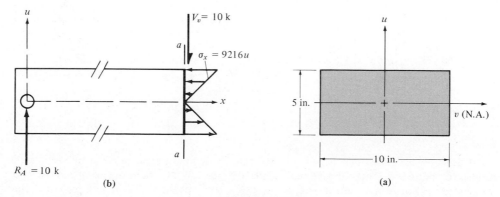

(b)

(a)

FIGURE 5.10

By Eq. 5.10b,

$$\sigma_x = \frac{960{,}000 \times 12u}{1250} = 9216u \text{ psi}$$

A plot of this equation is given in Figure 5.10(b), which shows the free-body diagram of that part of the beam to the left of section a–a. Attention is once again directed at the shear force V_v on section a–a, which will be dealt with in Section 5.4. Note that the fibers of the beam above the v principal axis (i.e., the neutral axis) are in compression and that those below the neutral axis are in tension. Thus the maximum compressive value of σ_x occurs at the top of the beam, where $u = 2.5$ in. Thus

$$(\sigma_x)_{\text{top}} = 9216(2.5) = 23{,}040 \text{ psi}(C)$$

The maximum tensile stress occurs at the bottom of the beam, where u is also 2.5 in. Therefore,

$$(\sigma_x)_{\text{bot}} = 9216(2.5) = 23{,}040 \text{ psi}(T)$$

Example 5.6
The beam depicted in Figure 5.11(a) has a cross section shown magnified in Figure 5.11(b). Assume the loads to act normal to the 0.15-m top of the T cross-sectional area. Determine (a) the maximum tensile fiber stress, and (b) the maximum compressive fiber stress. Specify the location of these stresses.

Solution. The centroid C of the T section is located at 0.1536 m above the bottom, as shown in Figure 5.11(b). By inspection, it can be concluded that the horizontal centroidal axis is the u principal axis of inertia.* The corresponding moment of inertia is found to be

$$I_u = 10.186 \times 10^{-5} \text{ m}^4$$

The reactions at the two supports A and B are found from consideration of equilibrium to be

$$R_A = 48{,}750 \text{ N} \qquad \text{and} \qquad R_B = 26{,}250 \text{ N}$$

---

* Students may convince themselves of this fact by computing the moments of inertia with respect to both horizontal and vertical centroidal axes. While the moment of inertia with respect to the horizontal centroidal axis is 10.186×10^{-5} m⁴, we find that with respect to the vertical centroidal axis, the moment of inertia is 1.615×10^{-5} m⁴.

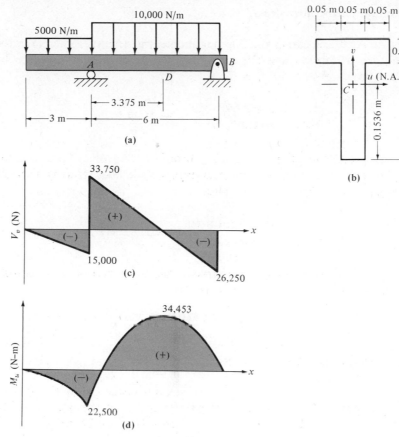

FIGURE 5.11

Since the loads on the beam are vertical (i.e., their plane contains the v principal centroidal axis of inertia), bending of the beam takes place about the u principal centroidal axis of inertia; hence the u axis is the neutral axis, and bending moments are designated by the symbol M_u.

For convenience, the shear and moment diagrams for the beam are constructed as shown in Figure 5.11(c) and (d), respectively. It is observed that there are two positions along the beam where the bending moment is a relative maximum. These two bending moments are found to be

$$(M_u)_A = -22{,}500 \text{ N-m}$$

at support A and

$$(M_u)_D = +34{,}453 \text{ N-m}$$

at point D located at 3.375 m right of support A, as shown in Figure 5.11(a).

(a) Maximum tensile fiber stress:

POSITION A: At this location the bending moment is negative. Therefore, the top fibers are in tension and the maximum tensile fiber stress is given by Eq. 5.10a, with $v = 9.64 \times 10^{-2}$ m. Thus

$$(\sigma_x)_{top} = \frac{22,500(9.64) \times 10^{-2}}{10.186 \times 10^{-5}} = 21.29 \text{ MPa}$$

POSITION D: At this location, the bending moment is positive. Therefore, the bottom fibers are in tension and the maximum tensile fiber stress is given by Eq. 5.10a with $v = 15.36 \times 10^{-2}$ m. Thus

$$(\sigma_x)_{bot} = \frac{34,453(15.36) \times 10^{-2}}{10.186 \times 10^{-5}} = 51.95 \text{ MPa}$$

Therefore, the maximum tensile fiber stress in the beam occurs at the bottom fibers of the section at position D, and its value is

$$\boxed{\sigma_x = 51.95 \text{ MPa}}$$

(b) Maximum compressive fiber stress:

POSITION A: At this location, compressive stresses occur below the neutral axis and the maximum compressive fiber stress occurs at the bottom of the section and is given by Eq. 5.10a with $v = 15.36 \times 10^{-2}$ m. Therefore,

$$(\sigma_x)_{bot} = \frac{22,500(15.36) \times 10^{-2}}{10.186 \times 10^{-5}} = 33.93 \text{ MPa}$$

POSITION D: The fibers above the neutral axis at this location are in compression. The maximum compressive fiber stress occurs at the top of the section and is given by Eq. 5.10a with $v = 9.64 \times 10^{-2}$ m. Thus

$$(\sigma_x)_{top} = \frac{34,453(9.64) \times 10^{-2}}{10.186 \times 10^{-5}} = 32.61 \text{ MPa}$$

Therefore, the maximum compressive fiber stress in the beam occurs at the bottom fibers of the section at position A, and its value is

$$\boxed{\sigma_x = 33.93 \text{ MPa}}$$

Example 5.7

The angle section shown in Figure 5.12(a) is the cross section for a cantilever beam 20 ft long, which is to be subjected to a uniform load whose intensity is 500 lb/ft acting vertically downward. (a) Rotate the section so that it offers maximum resistance to the bending action produced by the vertical load, and (b) find the maximum fiber stress induced in the beam for the orientation of the section in part (a).

Solution

(a) The centroid C of the section is located and an xy coordinate system established as shown in Figure 5.12(a). The principal centroidal axes of inertia, u and v, and the corresponding moments of inertia, I_u and I_v, are determined using the procedure discussed in Section 5.2. These principal centroidal axes of inertia are shown, properly oriented with respect to centroidal x and y axes, on the section in Figure 5.12(a). The principal centroidal moments of inertia are found to be

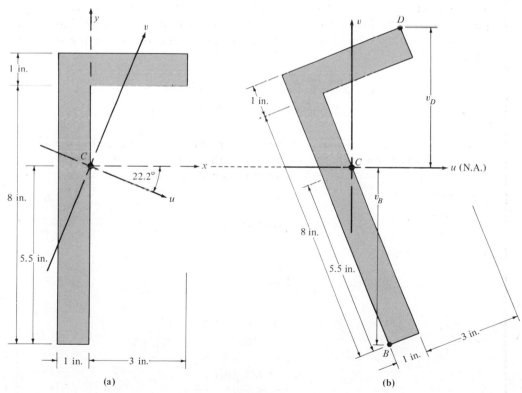

(a)   (b)

FIGURE 5.12

$$I_u = 100.7 \text{ in.}^4$$

$$I_v = 8.4 \text{ in.}^4$$

Thus, to develop maximum resistance to the bending action produced by the vertical load, the cross section must be rotated such that the u principal centroidal axis is horizontal (i.e., such that the u-principal centroidal axis is the neutral axis and is perpendicular to be applied load). The rotated cross section is shown in Figure 5.12(b).

(b) When the cross section is rotated as shown in Figure 5.12(b), the vertical load acts parallel to the v principal centroidal axis and the neutral axis coincides with the u principal centroidal axis of inertia. The fiber stress is directly proportional to the distance of the fiber from the neutral axis, as expressed by Eq. 5.10a. Thus to determine the maximum fiber stress, we need to locate the point in the cross section that is farthest from the neutral axis (i.e., the u principal centroidal axis). If the cross section is drawn to scale, the point farthest away from the neutral axis can usually be located by inspection. If the cross section is not drawn to scale, two or more points may have to be examined before deciding which is the most highly stressed fiber in the section. The cross section shown in Figure 5.12(b) is drawn approximately to scale and point B is seen, by inspection, to be the farthest away from the neutral axis. As a check, however, the distances from the neutral axis to both points B and D are computed:

$$v_B = (5.5) \cos 22.2 + (1) \sin 22.2 = 5.47 \text{ in.}$$

$$v_D = (3.5) \cos 22.2 + (3) \sin 22.2 = 4.37 \text{ in.}$$

Therefore, the maximum fiber stress occurs at point B at a cross section along the beam where the bending moment assumes its maximum value. The maximum bending moment occurs at the fixed end of the cantilever beam and is equal to

$$(M_u)_{\max} = 1{,}200{,}000 \text{ lb-in.}$$

Thus, by Eq. 5.10a,

$$(\sigma_x)_B = \frac{(M_u)_{\max} v_B}{I_u}$$

$$= \frac{1{,}200{,}000 \times 5.47}{100.7}$$

$$= \boxed{65{,}184 \text{ psi}(C)}$$

Homework Problems

5.23 A cantilever beam 20 ft long has a rectangular cross section 5 × 10 in. It carries a uniformly distributed load whose intensity is 400 lb/ft.
(a) If the plane of the load contains the v principal centroidal axis of inertia, determine the maximum tensile and compressive fiber stresses in the beam.
(b) If the plane of the load contains the u principal centroidal axis of inertia, determine the maximum tensile and compressive fiber stresses in the beam.

5.24 A simply supported beam 10 m long has a hollow circular cross-sectional area whose inside diameter is 0.15 m and outside diameter is 0.30 m. The beam is subjected to a concentrated force of 50 kN at midspan.
(a) For a location 3 m from the left support, show the variation of the flexural stress from the inner to the outer fiber.
(b) Determine the maximum tensile and compressive fiber stresses in the beam.

5.25–5.27 A cantilever beam of length L carries a concentrated force P acting vertically downward at the free end and in a plane that contains the v principal centroidal axis for its cross-sectional area. Determine the maximum tensile and compressive flexural stresses for each of the following three cases.
5.25 $L = 3$ m, $P = 20$ kN, cross section shown in Figure H5.5.
5.26 $L = 15$ ft, $P = 30$ kips, cross section shown in Figure H5.6.
5.27 $L = 25$ ft, $P = 50$ kips, cross section shown in Figure H5.7.

5.28–5.30 Repeat Problems 5.25 to 5.27 if the load P acts in a plane that contains the u principal centroidal axis for the cross-sectional area.

5.31 The beam shown in Figure H5.31 has the cross-sectional area shown in Figure H5.8 such that the

0.10 m base of the triangle is the bottom of the beam and is perpendicular to the plane of the loads.
(a) Consider a section 2 m to the right of support A and determine the variation of the flexural stress along the axis of symmetry for the triangular section.
(b) Compute the maximum tensile and compressive stresses in the beam and specify their locations.

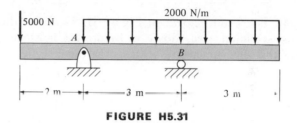

FIGURE H5.31

5.32 The beam shown in Figure H5.32 has the cross-sectional area shown in Figure H5.9 such that the 0.20-m dimension of the channel is the top of the beam and is perpendicular to the plane of the loads. Compute the values of the maximum tensile and compressive stresses in the beam and specify their locations.

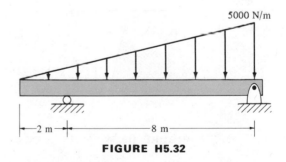

FIGURE H5.32

5.33 The beam shown in Figure H5.33 has the cross-sectional area shown in Figure H5.10 such that the 8-in. top of the T section is the top of the beam and is perpendicular to the plane of the loads. Consider the section along the beam for which the bending moment assumes its maximum value and plot the flexural stress variation along the axis of symmetry of the T section.

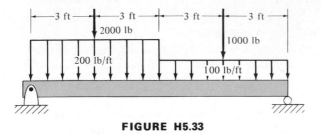

FIGURE H5.33

5.34 The beam shown in Figure H5.34(a) has the cross-sectional area shown in Figure H5.34(b) such that the 5-in. dimension of the trapezoidal section is

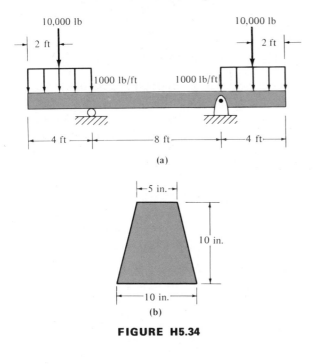

FIGURE H5.34

the top of the beam and is perpendicular to the plane of the loads. Determine the maximum tensile and compressive stresses in the beam and specify their locations.

5.35 The cross section shown in Figure H5.20 is that for a cantilever beam 3 m long that is to be subjected to a concentrated force of 8 kN acting vertically downward at its free end.
(a) Rotate the section so that it offers maximum resistance to the bending action produced by the vertical load.
(b) Determine the maximum tensile and compressive stresses induced in the beam for the rotated position of the section in part (a).

5.36 The cross section shown in Figure H5.21 is that for a simply supported beam 20 ft long which is to carry a uniform load of intensity 300 lb/ft acting vertically downward along the entire length of the beam.
(a) Rotate the section so that it develops maximum resistance to the bending action produced by the vertical load.
(b) Determine the maximum tensile and compressive stresses induced in the beam for the rotated position of the section in part (a).

5.37 The beam shown in Figure H5.31 has the cross-sectional area given in Figure H5.22. All the loads acting on the beam are assumed to act vertically downward along the entire length of the beam.
(a) Rotate the section so that it develops maximum resistance to the bending action produced by the vertical loads.
(b) Determine the maximum tensile and compressive stresses induced in the beam for the rotated position in part (a).

5.4 Shear Stresses in Symmetrically Loaded Beams

As was pointed out earlier, when bending of a beam is produced by transverse loads, any of its cross sections is subjected not only to the action of a bending moment M but also to the action of a shear force V, as illustrated in Figure 5.8(b). The shear force V represents the resultant of a system of shear stresses that are distributed in some manner over the cross-sectional area of the beam.

In order to derive an approximate distribution for the shear stresses in symmetrically loaded beams, consider a beam subjected to an arbitrary transverse loading system such as the one depicted in Figure 5.13(a). For simplicity and convenience, the cross-sectional area of the beam is assumed rectangular, as shown magnified in Figure 5.13(b). The plane of the applied loads is assumed to contain the v principal centroidal axis of inertia for the section so that symmetric bending is produced. A differential segment of the beam of length dx contained between planes a–a and b–b is isolated as a free body and shown magnified in Figure 5.13(c). The moments on planes a–a and b–b are, respectively, M_u and $M_u + dM_u$, where the quantity dM_u represents the elemental change in moment over the differential length dx. Also, the shear forces on planes a–a and b–b are V_v* and $V_v + dV_v$, respectively, where the quantity dV_v represents the elemental change in shear over the differential length dx.

The moments M_u and $M_u + dM_u$ give rise to the normal stresses depicted on planes a–a and b–b, respectively, in Figure 5.13(d). The shear forces V_v and $V_v + dV_v$ are resultants of the shear stress distributions shown schematically over planes a–a and b–b, respectively. The assumption is made that on any surface parallel to the neutral plane such as $efij$ in Figure 5.13(d), the shear stress is constant. Inherent in this assumption is the requirement that the width b of the cross-sectional area is relatively small[†] and that its sides intersecting the neutral axis are perpendicular to this axis and, of course, parallel to each other. Otherwise, the variation of the shear stress across the width of the section may be sufficiently large to render useless the following development and the resulting equation.

The determination of the shear stress τ_{vx} on any surface $efij$ located at a distance v_1 from the neutral axis is accomplished by isolating portion $efghijkl$ and constructing its free-body diagram as shown in the lower part of Figure 5.13(d). The shear stress τ_{vx} on surfaces parallel to the neutral plane must be accompanied by shear stresses τ_{xv} of equal magnitude on planes perpendicular to the neutral plane. Thus,

---

* The subscript v on the shear force V is used to emphasize the fact that this shear force is parallel to the v principal axis of inertia.

† A more refined solution of this problem by Saint Venant indicates that the width b of the section cannot be much larger than its height h if the approximate solution given by Eq. 5.12 is to yield satisfactory answers.

referring to Figure 5.13(d), we find that the shear stress on surface *efij* is τ_{vx} and that on plane *efgh* is τ_{xv}, where τ_{vx} must be equal to τ_{xv} in order to satisfy equilibrium requirements, as was shown in Section 2.3.

Consider the equilibrium of forces acting on portion *efghijkl* in a direction parallel to the longitudinal axis of the beam (i.e., along the *x* axis). There are three forces acting in the *x* direction:

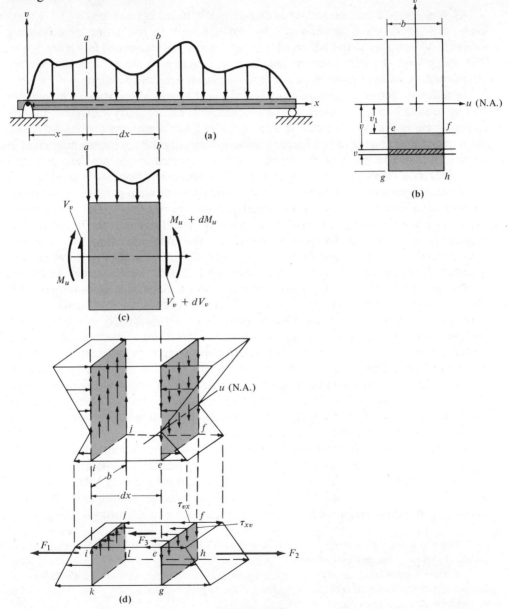

FIGURE 5.13

1. A normal force F_1 representing the resultant of the normal stresses on plane *ijkl* produced by the moment M_u. This resultant force is given by

$$F_1 = \int_{v_1}^{-h/2} \sigma_x \, dA \qquad (5.11a)$$

where dA is an element of area in plane *ijkl* at a distance v below the neutral axis and the stress σ_x is given by Eq. 5.10a. Thus substituting Eq. 5.10a into Eq. 5.11a yields

$$F_1 = \frac{M_u}{I_u} \int_{v_1}^{-h/2} v \, dA \qquad (5.11b)$$

2. A normal force F_2 representing the resultant of the normal stresses on plane *efgh* produced by the moment $M_u + dM_u$. This resultant force is also given by Eq. 5.11a, where dA is an element of area in plane *efgh* at a distance v below the neutral axis. The stress σ_x is again given by Eq. 5.10a, in which the moment is $M_u + dM_u$. Substitution of the value of σ_x as given by Eq. 5.10a into Eq. 5.11a yields

$$F_2 = \frac{M_u + dM_u}{I_u} \int_{v_1}^{h/2} v \, dA \qquad (5.11c)$$

3. A shear force F_3 representing the resultant of the shear stresses τ_{vx} acting on surface *efij*. Since the shear stress τ_{vx} is assumed constant over the thickness b of the beam, the shear force F_3 can be determined as the product of the shear stress τ_{vx} and the area over which it acts (i.e., surface area *efij*, which is equal to $b \, dx$). Thus

$$F_3 = \tau_{vx} b \, dx \qquad (5.11d)$$

Summing forces in the x direction leads to the conclusion that

$$F_3 = F_2 - F_1 \qquad (5.11e)$$

Substitution of Eqs. 5.11b, 5.11c, and 5.11d into Eq. 5.11e and simplifying leads to

$$\tau_{vx} b \, dx = \frac{dM_u}{I_u} \int_{v_1}^{-h/2} v \, dA \qquad (5.11f)$$

from which

$$\tau_{vx} = \frac{1}{bI_u} \left(\frac{dM_u}{dx} \right) \int_{v_1}^{-h/2} v \, dA \qquad (5.11g)$$

By Eq. 1.6, the quantity dM_u/dx is the shear force V_v. The integral $\int_{v_1}^{-h/2} v \, dA$ represents the first moment of the area between locations v_1 and $-h/2$ (shown shaded in Figure 5.13b) about the neutral axis. Alternatively, this integral represents the first moment about the neutral axis of the area contained between an edge of the cross section parallel to its neutral axis and the surface at which the shear stress is to be computed. This quantity is traditionally given the symbol Q. Making these substitutions in Eq. 5.11g, we obtain the equation for the horizontal shear stress τ_{vx} on any surface at a distance v from the neutral axis. Thus

$$\tau_{vx} = \frac{V_v}{bI_u} Q \qquad \text{(5.12a)}$$

As stated earlier, since $\tau_{xv} = \tau_{vx}$, Eq. 5.12 yields not only the value of the horizontal shear stress τ_{vx} at any location at a given distance from the neutral axis, but also the vertical shear stress τ_{xv} at the same location. Also, since Eq. 5.10 yielding the flexural stress was used in developing Eq. 5.12, the latter equation is subject to the same limitations dictated by the assumptions made in deriving the flexural equation.

Examination of Eq. 5.12a reveals that for a given location along a beam of a certain cross-sectional area, V_v and I_u are both constants, and therefore the value of τ_{vx} depends upon the ratio Q/b. Thus the maximum value of the horizontal shear stress occurs at a location in the cross section where the ratio Q/b is a maximum. In general, for most commonly used structural shapes, this ratio is a maximum at the neutral axis for the cross-sectional area. However, it should be noted that there are shapes for which the maximum value of the horizontal (and the vertical) shear stress does not occur at the neutral axis. Therefore, we need to examine each cross-sectional configuration on its own merit to determine the location for which Q/b is a maximum.

If the plane of the applied loads is assumed to contain the u principal instead of the v principal centroidal axis of inertia, the shear force would be V_u parallel to the u principal axis, bending would take place about the v principal axis, and the appropriate equation for the shear stress would be

$$\tau_{ux} = \frac{V_u}{hI_v} Q \qquad \text{(5.12b)}$$

The shear stress τ_{ux} is the horizontal shear stress at any location at a given distance from the neutral axis, but it is also equal to τ_{xu}, the vertical shear stress at the same location.

Example 5.8
Consider a position along the beam of Figure 5.13 where the shear force is V_v and construct the horizontal shear stress distribution as a function of the coordinate v.

Solution. The horizontal shear stress τ_{vx} is given by Eq. 5.12a, where, for a given position along the beam, the quantities V_v, b, and I_u are all constants and $I_u = \frac{1}{12}bh^3$.

To determine the horizontal shear stress τ_{vx} at any distance v_1 from the neutral axis of the rectangular cross section, Q must be evaluated for that particular distance from its basic definition. Thus

$$Q = \int_{v_1}^{-h/2} v \, dA = b \int_{v_1}^{-h/2} v \, dv = \frac{b}{2}\left(\frac{h^2}{4} - v_1^2\right)$$

Substitution of these quantities into Eq. 5.12a leads to

$$\tau_{vx} = \frac{V_u}{b(\frac{1}{12}bh^3)}\left(\frac{b}{2}\right)\left(\frac{h^2}{4} - v_1^2\right)$$

Simplification of this relation yields

$$\tau_{vx} = \frac{6V_v}{bh^3}\left(\frac{h^2}{4} - v_1^2\right)$$

which is the equation for a parabola. At the top and bottom of the rectangular cross-sectional area, where v_1 is equal to $+h/2$ and $-h/2$, respectively, τ_{vx} becomes equal to zero. The maximum value for τ_{vx} is observed to occur at the neutral axis for the rectangular area where $v_1 = 0$. Thus

$$(\tau_{vx})_{max} = \frac{3}{2}\frac{V_v}{bh} - \frac{3}{2}\frac{V_v}{A}$$

Therefore, for a rectangular cross-sectional area as shown in Figure 5.14(a), the

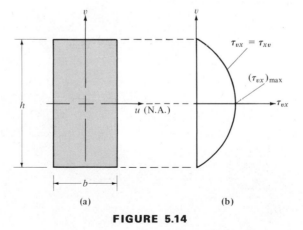

(a)   (b)

FIGURE 5.14

horizontal shear stress varies parabolically, as shown in Figure 5.14(b). Since, for any distance v_1 from the neutral axis, the vertical shear stress τ_{xv} is equal in magnitude to the horizontal shear stress τ_{vx}, the distribution shown in Figure 5.14(b) also gives the values of the vertical shear stress.

Example 5.9

Consider a position an infinitesimal distance to the right of support A in the beam shown in Figure 5.11(a). Plot the horizontal shear stress distribution at this position with respect to the v coordinate axis for the T cross section shown in Figure 5.11(b).

Solution. The shear force V_v an infinitesimal distance to the right of support A in Figure 5.11(a) was found in the solution of Example 5.6 to be $+33,750$ N. The positive sign indicates that the sum of the applied forces to the left of the section of interest is upward (i.e., the resisting shear force at the section of interest is downward). Also, from the solution of Example 5.6, $I_u = 10.186 \times 10^{-5}$ m^4. Two separate expressions have to be written for the horizontal shear stress, one for the flange and the second for the stem of the T section.

    The horizontal shear stress τ_{vx} at any location in the flange of the T section at a distance v_1 from the neutral axis is given by Eq. 5.12a. Thus

$$\tau_{vx} = \frac{V_v}{bI_u} Q$$

where

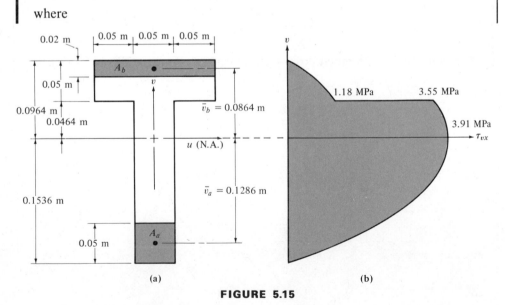

(a)   (b)

FIGURE 5.15

$$Q = \int_{v_1}^{0.0964} v \, dA = 0.15 \int_{v_1}^{0.0964} v \, dv = 7.5 \times 10^{-2}(92.93 \times 10^{-4} - v_1^2)$$

Since b for the flange is 0.15 m,

$$\tau_{vx} = \frac{33,750 \times 7.5 \times 10^{-2}}{0.15(10.186 \times 10^{-5})}(92.93 \times 10^{-4} - v_1^2)$$

$$= 16.567 \times 10^7(92.93 \times 10^{-4} - v_1^2) \, \text{N/m}^2$$

$$= 165.67(92.93 \times 10^{-4} - v_1^2) \, \text{MPa}$$

which represents the equation of a parabola applicable only in the range $0.0464 < v_1 \leq 0.0964$.

The horizontal shear stress τ_{vx} at any location in the stem of the T section a distance v_1 from the neutral axis is also given by Eq. 5.12a, where b is 0.05 m and

$$Q = \int_{v_1}^{-0.1536} v \, dA = 0.05 \int_{v_1}^{0.1536} v \, dv = 2.5 \times 10^{-2}(235.93 \times 10^{-4} - v_1^2)$$

Therefore,

$$\tau_{vx} = \frac{33,750 \times 2.5 \times 10^{-2}}{0.05(10.186 \times 10^{-5})}(235.93 \times 10^{-4} - v_1^2)$$

$$= 16.567 \times 10^7(235.93 \times 10^{-4} - v_1^2) \, \text{N/m}^2$$

$$= 165.67(235.93 \times 10^{-4} - v_1^2) \, \text{MPa}$$

which represents the equation of a parabola applicable only in the range $-0.1536 < v_1 < 0.0464$.

The T cross section is redrawn in Figure 5.15(a) for convenience. A plot of τ_{vx} versus v is obtained from the preceding two equations and shown in Figure 5.15(b). Note the abrupt increase in the value of the horizontal shear stress from 1.18 MPa at the very bottom of the flange to 3.55 MPa at the very top of the stem. This is obviously due to the abrupt decrease in the width from 0.15 m in the flange to 0.05 m in the stem of the section. It should also be pointed out that the horizontal shear stress distribution for the flange is not realistic, as these shear stresses must vanish along the bottom free edges of the flange, and we must look at the distribution shown in Figure 5.15(b) as a reasonable approximation for the stem and that part of the flange which forms the continuation of the stem. As in the case of the rectangular cross section examined in Example 5.8, the maximum value of the horizontal shear stress, which is 3.91 MPa, occurs at the neutral axis for the T section.

Example 5.10

Consider a position an infinitesimal distance to the left of support B in the beam shown in Figure 5.11(a) and (b). Determine the vertical shear stress at a distance (a) 0.05 m above the bottom of the T section, and (b) 0.02 m below the top of the T section.

Solution. The shear force V_v an infinitesimal distance to the left of support B in Figure 5.11(a) was found in the solution of Example 5.6 to be $-26{,}250$ N. The negative sign simply means that the sum of the applied forces to the left of the section of interest is downward (i.e., the resisting shear force at the section is upward). From the solution of Example 5.6, $I_u = 10.186 \times 10^{-5}$ m^4.

(a) Consider a location 0.05 m above the bottom of the T section. Obviously, this location is within the stem of the T section, where the width b is equal to 0.05 m. By definition, the quantity Q is the first moment with respect to the neutral axis of the area below or above the location at which the shear stress is to be determined. Clearly, in this case it is much easier to obtain Q from the area below the location of interest. By referring to the sketch of the T section, which has been redrawn in Figure 5.15(a), we conclude that

$$Q = A_a \bar{v}_a = (5 \times 5) \times 10^{-4} \times 0.1286 = 3.215 \times 10^{-4} \text{ m}^3$$

Therefore, the vertical shear stress at a distance of 0.05 m above the base of the T section is, by Eq. 5.12a, given by

$$\tau_{xv} = \tau_{vx} = \frac{V_v}{bI_u} Q$$

$$= \frac{26{,}250 \times 3.215 \times 10^{-4}}{0.05 \times 10.186 \times 10^{-5}}$$

$$= \boxed{1.66 \text{ MPa}}$$

(b) A location 0.02 m below the top of the T section lies in the flange of the T. It is much easier in this case to find Q from the area above the location at which the shear stress is to be determined. By reference to Figure 5.15(a), we determine that

$$Q = A_b \bar{v}_b = (2 \times 15) \times 10^{-4} \times 0.0864 = 2.592 \times 10^{-4} \text{ m}^3$$

Thus the vertical shear stress at a distance of 0.02 m below the top of the T section is, by Eq. 5.12a, given by

$$\tau_{xv} = \tau_{vx} = \frac{26{,}250 \times 2.592 \times 10^{-4}}{0.15 \times 10.186 \times 10^{-5}}$$

$$= \boxed{0.45 \text{ MPa}}$$

Example 5.11

Consider a position an infinitesimal distance to the right of support A in the beam shown in Figure 5.11(a) and (b). Determine the principal stresses and the absolute maximum shear stress at a point on the surface of the stem 0.02 m above the bottom of the T section. Show these stresses on properly oriented planes.

Solution. The bending moment at support A was found in the solution of Example 5.6 to be $M_u = -22{,}500$ N-m. The shear force V_v an infinitesimal distance to the right of support A was also found in the solution of Example 5.6 to be $+33{,}750$ N. The negative sign on the moment implies that fibers below the neutral axis (i.e., the u axis) are in compression, while those above the neutral axis are in tension. The positive sign on the shear force means that the sum of the applied forces to the left of the section of interest is upward (i.e., the resisting shear force at the section is downward). A stress element on the surface of the stem, 0.02 m above the bottom of the T section, would be acted upon by the normal and shear stresses given, respectively, by Eq. 5.10a and Eq. 5.12a. Thus

$$\sigma_x = \frac{M_u}{I_u} v = -\frac{22{,}500 \times 0.1336}{10.186 \times 10^{-5}} = -29.51 \text{ MN/m}^2$$

$$\tau_{xv} = \frac{V_v}{b I_u} Q = \frac{33{,}750 \times 143.6 \times 10^{-6}}{0.05 \times 10.186 \times 10^{-5}} = 0.95 \text{ MN/m}^2$$

This stress element is shown in Figure 5.16(a), where the X planes are planes perpendicular to the beam axis. Construction of Mohr's circle for the stress element and analysis of its geometry leads to the following values:

$$\boxed{\sigma_1 = 0.03 \text{ MN/m}^2}$$

$$\boxed{\sigma_3 = -29.54 \text{ MN/m}^2}$$

Note that this is a plane stress problem for which $\sigma_2 = 0$. Also, the plane on which σ_3 acts makes an angle of 3.7° ccw from the X plane.

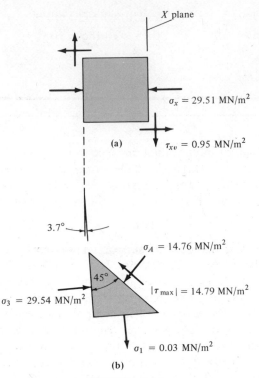

FIGURE 5.16

The absolute maximum shear stress is given by Eq. 2.13. Thus

$$|\tau_{max}| = \frac{\sigma_1 - \sigma_3}{2}$$

$$= \boxed{14.79 \text{ MN/m}^2}$$

These stresses are shown on a properly oriented element in Figure 5.16(b).

Homework Problems

5.38–5.40 A simply supported beam 20 ft long carries a uniform load of intensity 1000 lb/ft. Assume the load to act vertically downward and parallel to the v principal axis of inertia for the cross-sectional area of the beam. Find the magnitudes of the indicated shear stresses for each of the following three cases:

5.38 The cross section of the beam is shown in Figure H5.6. Determine horizontal shear stress 2 ft from left support and

(a) 3 in. above the bottom of the section.

(b) At the neutral axis for the section.

5.39 The cross section of the beam is shown in Figure H5.7. Determine the vertical shear stress 5 ft from the right support and

(a) In the web, 2 in. below the top of the section.

(b) At the neutral axis for the section.

5.40 The cross section of the beam is shown in Figure H5.10. Determine the horizontal shear stress 5 ft from the left support and

(a) In the web, 2 in. below the top of the section.

(b) At the neutral axis for the section.

5.41–5.43 A cantilever beam 4 m long carries a 50-kN concentrated force at the free end. Assume the load to act vertically downward and parallel to the v principal axis of inertia for the cross-sectional area of the beam. Find the magnitudes of the indicated shear stresses for each of the following three cases:

5.41 The cross section of the beam is shown in Figure H5.5. Determine the vertical shear stress 1 m from the free end of the beam and

(a) 0.05 m below the top of the section.

(b) At the neutral axis for the section.

5.42 The cross section of the beam is shown in Figure H5.8. Find the horizontal shear stress 2 m from the free end of the beam and

(a) 0.05 m above the bottom of the section.

(b) At the neutral axis for the section. (Note: Assume that Eqs. 5.12 are applicable to this case.)

5.43 The cross section of the beam is shown in Figure H5.9. Determine the vertical shear stress at the fixed end and

(a) 0.04 m below the top of the section an infinitesimal distance into the web.

(b) At the neutral axis for the section.

5.44–5.46 Repeat Problems 5.38 to 5.40 if the 20-ft simply supported beam carries a 25-k concentrated load at midspan instead of the 1000-lb/ft uniform load. All other conditions are the same as in Problems 5.38 to 5.40.

5.47–5.49 Repeat Problems 5.41 to 5.43 if the 4-m cantilever beam carries a uniform load of intensity 10 kN/m instead of the 50-kN concentrated force. All other conditions are the same as in Problems 5.41 to 5.43.

5.50 Assume that Eq. 5.12 are equally applicable to the case of a solid circular cross section and determine an expression for the maximum vertical shear stress in terms of the shear force V and the cross-sectional area A.

5.51 The beam shown in Figure H5.31 has the cross section depicted in Figure H5.51. Assume the plane of the loads to contain the v principal axis of inertia.

(a) Determine the magnitude of the horizontal shear stress in the webs 0.03 m below the top of the section at a position along the beam an infinitesimal distance to the right of the left support.

(b) Find the magnitude of the maximum vertical shear stress at a position along the beam an infinitesimal distance to the right of the left support.

(c) Find the magnitude of the maximum horizontal shear stress in the beam.

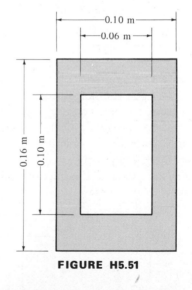

FIGURE H5.51

5.52 The beam shown in Figure H5.32 has the cross section depicted in Figure H5.52. Assume the plane of the loads to contain the v principal axis of inertia.

(a) Find the magnitude of the vertical shear stress in

the web 0.03 m above the bottom of the section at a position along the beam an infinitesimal distance to the right of the left support.

(b) Compute the magnitude of the maximum horizontal shear stress at a position along the beam an infinitesimal distance to the right of the left support.

(c) Determine the magnitude of the maximum horizontal shear stress in the beam.

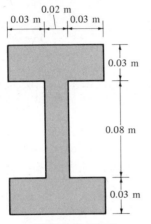

FIGURE H5.52

5.53 The beam shown in Figure H5.33 has the cross section illustrated in Figure H5.53. Assume the plane of the loads to contain the v principal axis of inertia.

(a) Compute the magnitude of the vertical shear stress 3 in. above the bottom of the section at a location along the beam 6 ft from the left support.

(b) Determine the magnitude of the maximum horizontal shear stress at a location along the beam 6 ft from the left support.

(c) Find the magnitude of the maximum horizontal shear stress in the beam.

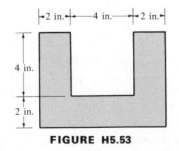

FIGURE H5.53

5.54 The beam shown in Figure H5.34(a) has the cross section illustrated in Figure H5.54. Assume the plane of the loads to contain the v principal axis of inertia.

(a) Determine the magnitude of the vertical shear stress 5 in. below the top of the section at a location along the beam an infinitesimal distance to the left of the left support.

(b) Find the magnitude of the maximum vertical shear stress at a location along the beam an infinitesimal distance to the left of the left support.

(c) Compute the magnitude of the maximum horizontal shear stress in the beam.

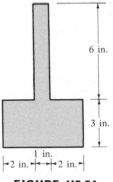

FIGURE H5.54

5.55 Repeat Problem 5.52 using the cross-sectional area shown in Figure H5.55.

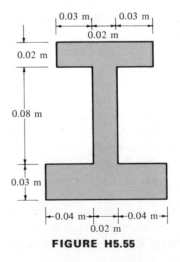

FIGURE H5.55

5.56 Repeat Problem 5.53 using the cross-sectional area shown in Figure H5.56.

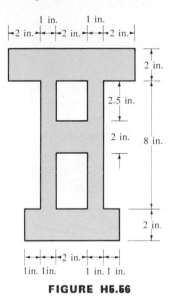

FIGURE H5.56

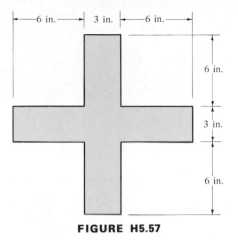

FIGURE H5.57

5.57 Repeat Problem 5.54 using the cross-sectional area shown in Figure H5.57.

5.58 Consider a point at the position defined in part (a) of Problem 5.51. Determine the principal stresses and the absolute maximum shear stress. Define the planes on which these stresses act.

5.59 Consider a point at the position defined in part (a) of Problem 5.54. Determine the principal stresses and the absolute maximum shear stress. Define the planes on which these stresses act.

*5.5 Flexural Stresses due to Unsymmetric Bending of Beams

It was stated earlier that symmetric bending of a beam takes place if the plane of the bending loads is either parallel to or contains one of the two longitudinal principal planes. *Unsymmetric bending* occurs, however, if the bending loads lie in a plane that is inclined to the longitudinal principal planes. Unlike symmetrically loaded beams, whose cross sections rotate about a neutral axis that is one of the two principal centroidal axes of inertia, cross sections of unsymmetrically loaded beams rotate about a neutral axis that is not one of the two principal centroidal axes of inertia.

Consider the arbitrary cross-sectional area shown in Figure 5.17, for which the u and v principal centroidal axes have been determined. Assume this cross-sectional area to be that of a beam for which the plane of the loads, denoted as L, makes the

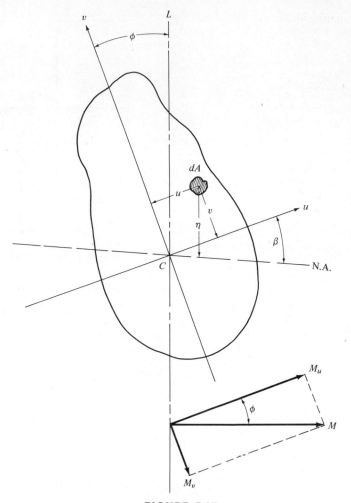

FIGURE 5.17

clockwise angle ϕ with the longitudinal principal plane that contains the v principal centroidal axis of inertia. For any position along the beam, the bending moment M may be represented by the right-hand rule, as a vector perpendicular to plane L, as shown in Figure 5.17. The sense of the moment vector M would depend upon the way the beam is loaded and supported. Without loss of generality, the moment vector M in this development is assumed to be pointed to the right. This moment vector may be decomposed into its two perpendicular components $M_u = M \cos \phi$ and $M_v = M \sin \phi$. Thus the unsymmetric bending problem may be considered as two separate cases of symmetric bending produced by M_u and M_v and occurring simultaneously. This concept will be used later in determining the flexural stress produced under unsymmetric bending conditions.

As in the case of symmetric bending, the assumption is made that bending is produced either by pure bending moments or by transverse loads. Under these conditions, the same conclusion is reached as was reached in the case of symmetric bending (see the development of Eq. 5.7 and subsequent conclusions) that the neutral axis for bending must pass through the centroid of the section. Let the neutral axis for bending make the clockwise angle β from the u principal centroidal axis, as shown in Figure 5.17. If it is considered further that all the assumptions made in the case of symmetric bending are equally applicable in the case of unsymmetric bending, then we conclude, as was concluded earlier from Eq. 5.6f, that the flexural stress is directly proportional to the distance from the neutral axis. Thus for an element of area dA at a distance η from the neutral axis, the flexural stress σ_x* can be written as

$$\sigma_x = K\eta \qquad (5.13a)$$

where K is a constant of proportionality and η can be expressed in terms of the dimensions u and v by the equation

$$\eta = v \cos \beta + u \sin \beta \qquad (5.13b)$$

Substitution of Eq. 5.13b into Eq. 5.13a yields

$$\sigma_x = K(v \cos \beta + u \sin \beta) \qquad (5.13c)$$

Equilibrium of moments with respect to the u principal centroidal axis leads to

$$M_u = \int \sigma_x v \, dA$$

$$= K \int (v \cos \beta + u \sin \beta) v \, dA$$

$$= K \left[\cos \beta \int v^2 \, dA + \sin \beta \int vu \, dA \right]$$

The first integral in this equation represents I_u, the moment of inertia of the area with respect to the u principal centroidal axis. The second integral is zero because it represents the product of inertia with respect to principal axes of inertia. Thus

$$M_u = KI_u \cos \beta \qquad (5.13d)$$

Similarly, by considering equilibrium of moments about the v principal centroidal axis, we conclude that

---

*As in previous developments, the x axis is assumed to be along the longitudinal axis of the beam.

$$M_v = KI_v \sin \beta \qquad (5.13e)$$

From the geometry in Figure 5.17 and by using Eqs. 5.13d and 5.13e, we obtain

$$\tan \phi = \frac{M_v}{M_u}$$

$$= \left(\frac{I_v}{I_u}\right) \tan \beta \qquad (5.13f)$$

Solving Eq. 5.13f for $\tan \beta$ leads to

$$\tan \beta = \frac{I_u}{I_v} \tan \phi \qquad (5.14)$$

Equation 5.14 determines the orientation, from the u principal centroidal axis of inertia, of the neutral axis for unsymmetric bending in terms of the principal centroidal moments of inertia I_u and I_v and the orientation, angle ϕ, of the plane of the loads with respect to the v principal centroidal axis of inertia. To one side of the neutral axis, fibers are stretched (tensile zone), and to the other side, fibers are compressed (compression zone). In the particular case where ϕ is zero (i.e., symmetric bending), we conclude from Eq. 5.14 that β is also zero, which means that the neutral axis coincides with the u principal centroidal axis of inertia, as was concluded earlier in analyzing symmetric bending. It is important to remember that while the angle ϕ is measured from the v axis, the angle β is measured from the u axis.

The determination of the flexural stress produced under conditions of unsymmetric bending is facilitated, as stated earlier, by considering the unsymmetric bending case as two separate cases of symmetric bending occurring simultaneously. In other words, the flexural stress at any location in a given cross section will be the algebraic sum of the stresses produced by M_u and by M_v. Thus, referring to Figure 5.17, the moment M_u, acting alone, would produce symmetric bending of the section about the u principal centroidal axis of inertia, and therefore the flexural stress at any location defined by the coordinates u and v would be given by Eq. 5.10a. Also, the moment M_v, acting alone, would produce symmetric bending of the section about the v principal centroidal axis and, consequently, the flexural stress at the same location in the section would be given by Eq. 5.10b. Therefore, the total flexural stress at a given location in the cross section would be the algebraic sum of the flexural stresses resulting from Eqs. 5.10a and 5.10b. Thus, under conditions of unsymmetric bending, the flexural stress is given by

$$\sigma_x = \frac{M_u}{I_u} v + \frac{M_v}{I_v} u \qquad (5.15)$$

The use of Eqs. 5.14 and 5.15 in the solution of unsymmetric bending problems will be illustrated in the following examples.

Example 5.12

A cantilever beam 15 ft long has a rectangular sectional area as shown in Figure 5.18. The beam carries a concentrated load $P = 10,000$ lb at the free end. This load is inclined to the v principal centroidal axis of inertia through a 30° counterclockwise angle, as shown. Locate the neutral axis and determine the maximum tensile flexural stress induced in the beam.

Solution. The principal centroidal moments of inertia are found to be

$$I_u = 170.7 \text{ in.}^4 \qquad \text{and} \qquad I_v = 42.7 \text{ in.}^4$$

The plane of the load is inclined at 30° ccw from the v principal centroidal axis of inertia. Therefore, $\phi = 30°$. Thus, by Eq. 5.14,

$$\tan \beta = \frac{I_u}{I_v} \tan \phi$$

$$= \left(\frac{170.7}{42.7}\right) \tan 30 = 2.31$$

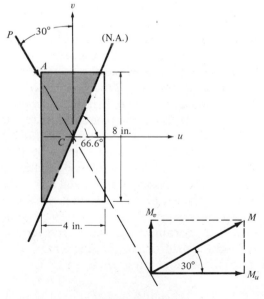

FIGURE 5.18

and

$$\boxed{\beta = 66.6°}$$

Since the plane of the load is counterclockwise from the v principal axis, the neutral axis defined by the angle β is counterclockwise from the u principal axis of inertia. This neutral axis is shown properly oriented in Figure 5.18. Fibers in the beam downward and to the right of the neutral axis are in the compression zone, and those upward and to the left of the neutral axis are in the tension zone.

The maximum tensile flexural stress occurs at the position along the beam where the bending moment is a maximum—at the fixed end of the beam. At this position

$$M = 150 \times 10^3 \text{ ft-lb} = 1800 \times 10^3 \text{ in.-lb}$$

This moment, and its two perpendicular components $M_u = 1559 \times 10^3$ in.-lb and $M_v = 900 \times 10^3$ in.-lb are shown in Figure 5.18. Since the flexural stress is directly proportional to the distance from the neutral axis, the maximum tensile flexural stress occurs in the tensile zone of the cross section (shown shaded in Figure 5.18) at the point farthest from this axis.

If the cross-sectional area and the orientation of its neutral axis are drawn approximately to scale, it is usually possible to determine, by inspection, the point in either zone that is farthest from the neutral axis. In some cases, however, the determination of such points may require a little more effort. For the case under consideration, the maximum tensile flexural stress occurs at point A, as shown in Figure 5.18. This stress is found by the use of Eq. 5.15. The first term in this equation represents the flexural stress at point A due to the moment M_u, which produces bending about the u axis. Thus, insofar as M_u is concerned, the u principal centroidal axis is the neutral axis for bending. Above this axis, fibers are in tension and below it fibers are in compression. Thus the stress at point A produced by M_u is tension. The coordinate v in the first term of Eq. 5.15 represents the distance of point A from the neutral axis (i.e., the u axis) and is equal to 4 in. The second term in Eq. 5.15 represents the flexural stress at point A due to the moment M_v, which produces bending about the v axis. Insofar as M_v is concerned, the v principal centroidal axis is the neutral axis for bending. Fibers located to the left of this axis are in tension and those to the right of this axis are in compression. Therefore, the stress at point A produced by M_v is also tension. The coordinate u in the second term of Eq. 5.15 is the distance of point A from the neutral axis (i.e., the v axis) and is equal to 2 in. Thus the maximum tensile flexural stress, by Eq. 5.15, is

$$(\sigma_x)_A = \frac{1559 \times 10^3}{170.7}(4) + \frac{900 \times 10^3}{42.7}(2)$$

$$= 36{,}532 + 42{,}155$$

$$= \boxed{78{,}687 \text{ psi}}$$

This stress is relatively high, but there are certain high-strength alloy steels that are capable of resisting such high stresses elastically.

Example 5.13

A simply supported beam 6 m long carries a uniform load of intensity 2000 N/m along its entire length and acting vertically downward. The cross-sectional area for the beam is the angle shown in Figure 5.19 and the uniform load acts normal to the 0.10-m side, which is the top of the beam. Assume that

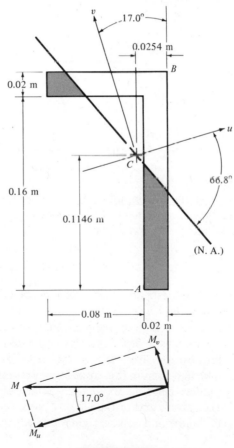

FIGURE 5.19

the load is so placed that no twisting of the section occurs and determine the maximum tensile and maximum compressive flexural stresses in the beam.

Solution. The centroid C of the section is located as shown in Figure 5.19. The principal centroidal axes of inertia, u and v, and the corresponding moments of inertia, I_u and I_v, are determined by the methods developed in Section 5.2. These principal centroidal axes of inertia are shown properly oriented in Figure 5.19. The principal moments of inertia are found to be

$$I_u = 18.219 \times 10^{-6} \text{ m}^4 \quad \text{and} \quad I_v = 2.386 \times 10^{-6} \text{ m}^4$$

Since the plane of the loads is vertical and the v principal centroidal axis is inclined to the vertical at an angle of 17.0° ccw, the angle $\phi = 17.0°$. The orientation of the neutral axis can now be determined from Eq. 5.14. Thus

$$\tan \beta = \frac{I_u}{I_v} \tan \phi$$

$$= \frac{18.219}{2.386} \tan 17.0° = 2.33$$

and

$$\beta = 66.8°$$

This neutral axis is shown properly oriented in Figure 5.19. Points in the cross section downward and to the left of the neutral axis are in the tension zone (shown shaded in Figure 5.19) and those upward and to the right of the neutral axis are in the compression zone.

The maximum tensile and maximum compressive flexural stresses occur at midspan, where the bending moment M assumes its maximum value. The maximum bending moment is found to be

$$M = 9000 \text{ N-m}$$

This moment and its two perpendicular components $M_u = 8610$ N-m and $M_v = 2630$ N-m are shown in Figure 5.19.

Since the flexural stress is directly proportional to the distance from the neutral axis, the maximum tensile flexural stress occurs at point A, which is the farthest point in the tension zone from the neutral axis. The stress at this point is given by Eq. 5.15. The first term in this equation represents the flexural stress due to M_u, which produces bending about the u principal centroidal axis. On one side of this axis (i.e., above and to the left), M_u produces compression and on the other side (i.e., below and to the right) it produces tension. Thus the flexural stress produced at point A by M_u is tension. The quantity v represents the distance from this point to the u axis and is found from the geometry of the section to be

$$v_A = (0.1146) \cos 17 + (0.0054) \sin 17 = 0.1112 \text{ m}$$

The second term in Eq. 5.15 represents the flexural stress due to M_v, which produces bending about the v principal centroidal axis. On one side of this axis (i.e., above and to the right), M_v produces compression and on the other side (i.e., below and to the left) it produces tension. Therefore, the flexural stress produced at point A by M_v is also tension. The quantity u represents the distance from this point to the v axis and is found to be

$$u_A = (0.1146) \sin 17 - (0.0054) \cos 17 = 0.0283 \text{ m}$$

Thus the maximum tensile flexural stress, by Eq. 5.15, is

$$(\sigma_x)_A - \frac{8610(0.1112)}{18.219 \times 10^{-6}} + \frac{2630(0.0283)}{2.386 \times 10^{-6}}$$

$$= (52.55 + 31.19) \times 10^6$$

$$= \boxed{83.74 \text{ MPa}}$$

Similarly, the maximum compressive flexural stress occurs at point B, the farthest point from the neutral axis in the compression zone for the section. Both M_u and M_v produce compression at this point. The quantities u_B and v_B are found to be

$$u_B = \frac{0.0254}{\cos 17} + (0.0654 - 0.0254 \tan 17) \sin 17 = 0.0434 \text{ m}$$

$$v_B = (0.0654 - 0.0254 \tan 17) \cos 17 = 0.0551 \text{ m}$$

Therefore, the maximum compressive flexural stress, by Eq. 5.15, is

$$(\sigma_x)_B = - \frac{8610(0.0551)}{18.219 \times 10^{-6}} - \frac{2630(0.0434)}{2.386 \times 10^{-6}}$$

$$= (-26.04 - 47.84) \times 10^6$$

$$= \boxed{-73.88 \text{ MPa}}$$

Homework Problems

In the following problems, assume the loads to be so placed that they produce no twisting action.

5.60 A simply supported beam 20 ft long carries a uniform load of intensity 1200 lb/ft along its entire length. The cross section for the beam is a 5 × 10 in. rectangular area oriented in such a way that the plane of the loads L is 20° cw from the v principal centroidal axis of the section, as shown in Figure H5.60. Locate the neutral axis and determine the maximum tensile and maximum compressive flexural stresses in the beam.

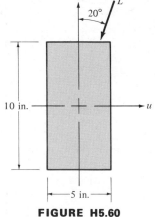

FIGURE H5.60

5.61 A cantilever beam 7 m long carries a con-

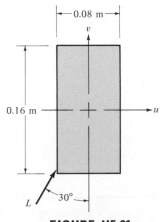

FIGURE H5.61

centrated load of 5000 N at its free end. The cross section for the beam is a rectangle and the plane of the load L is inclined to the v principal axis as shown in Figure H5.61. Locate the neutral axis and compute the maximum tensile and the maximum compressive flexural stresses in the beam.

5.62 A cantilever beam 4 m long carries a uniform load of intensity 5000 N/m along its entire length and acting vertically downward. The cross-sectional area for the beam is the angle shown in Figure H5.62 and the uniform load acts normal to the 0.15-m side, which is the top of the beam. Locate the neutral axis and determine the maximum tensile and maximum compressive flexural stresses in the beam.

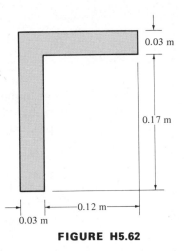

FIGURE H5.62

5.63 The cross-sectional area shown in Figure H5.2 is that for a simply supported beam 10 ft long that carries a concentrated load of 3000 lb at midspan acting vertically downward.The concentrated load acts normal to the 4-in. side of the angle, which is the bottom of the beam. Locate the neutral axis and compute the maximum tensile and maximum compressive flexural stresses in the beam.

5.64 The beam shown in Figure H5.31 has the cross-sectional area depicted in Figure H5.3. Assume the loads to act vertically downward and their plane to be normal to the 0.05-m side of the section, which is the top of the beam. Compute the maximum tensile and maximum compressive flexural stresses in the beam.

5.65 The beam shown in Figure H5.32 has the cross-sectional area illustrated in Figure H5.20. Assume the loads to act vertically downward and their plane to be normal to the 0.12-m side of the section, which is the top of the beam. Compute the maximum tensile and maximum compressive flexural stresses in the beam.

5.66 The beam shown in Figure H5.33 has the cross-sectional area depicted in Figure H5.21. Assume the loads to act vertically downward and their plane to be parallel to the stem of the Z section. Determine the maximum tensile and maximum compressive flexural stresses in the beam.

5.67 The beam shown in Figure H5.32 has the cross-sectional area illustrated in Figure H5.22. Assume the loads to act vertically downward and their plane to be normal to the 0.14-m side of the section, which is the top of the beam. Compute the maximum tensile and maximum compressive flexural stresses in the beam.

5.68 The beam shown in Figure H5.34(a) has the cross-sectional area depicted in Figure H5.34(b). Assume the 5-in. side of the trapezoidal section to be the top of the beam and that the plane of the loads makes a 60° cw angle with the beam top. Find the maximum tensile and maximum compressive flexural stresses in the beam.

Chapter **6** Deflections of Beams

6.1 Introduction

A number of practical reasons for studying beam deflections may be cited. If these deflections become excessive, plaster cracking, which is expensive to repair, may occur in buildings. Shafts acting in bending may become misaligned in their bearings due to large deflections, resulting in excessive wear and possible malfunction. High-rise structures, which contain beam components, may deflect enough to cause psychological stress among the occupants even though failure may be unlikely. Extensive glass breakage in some recently constructed tall buildings has been partially attributed to excessive deflections. The general trend toward construction and manufacture of lighter, more flexible components which are safe from a stress or load-carrying capacity standpoint has led to a number of problems at least partially attributable to deflections that are too large to be tolerated either on physical or psychological grounds or a combination of both.

In addition to these reasons, a knowledge of deflection calculations forms the basis for analysis and design of indeterminate structures. Equations of equilibrium must be supplemented by compatibility equations which involve deflection calculations, as discussed in Chapter 9. Again, modern machine and structural designs are more likely to involve indeterminate systems, and these cannot be investigated without a knowledge of deflection calculations.

Small, elastic deflections of initially straight beams are discussed in this chapter. Calculations of these deflections under service or working loads has been and is likely to remain a very important engineering consideration for optimum design.

The curvature of a bent elastic beam is directly proportional to the applied bending moment and inversely proportional to its bending stiffness. The bending stiffness is given by the product of the modulus of elasticity and the area moment of inertia (or second moment of the area) with respect to the axis of bending. Once this relationship is established, the governing differential equation for bending of beams follows by approximating the beam curvature by the second derivative of beam displacements (or deflections) with respect to a coordinate measured along the axis of the beam. This coordinate is denoted by x and has been used in Section 1.8 to write bending moment equations. The method of determining beam deflections by two successive integrations will depend upon a knowledge of these bending moment equations for proper problem formulation. This method of Section 6.3 has the advantage of yielding the equation of the elastic curve, which means that the deflection at

266

any point along the beam axis may be obtained by substitution of the appropriate x value into the resulting equation. In Section 6.4 the derivatives of the elastic curve equation are explored with an emphasis on their physical significance.

The method of superposition of Section 6.5 is of great practical importance, since it enables the engineer to combine previously solved deflection problems to obtain solutions for complex loading conditions that would be difficult to solve otherwise.

Construction of moment diagrams by parts discussed in Section 6.6 forms the basis for effective use of the area-moment method of Section 6.7, which is valuable method for finding slopes and deflections at selected points of interest along the beam axis. It is also readily adapted for finding maximum deflections.

Castigliano's second theorem of Section 6.8 provides an alternative method based upon work and energy concepts for determination of slopes and deflections at specified points along a beam axis. Study of the method will clarify the concept of work done on a beam by externally applied forces being stored internally in the beam as elastic strain energy.

In all the foregoing methods the effect of shear has been ignored compared to the effect of bending. In Section 6.9 it is shown that for large span/depth ratios the effect of shear is negligible compared to the effect of bending on elastic deflections. The chapter closes with a discussion of unsymmetric beam deflections using principal axes.

6.2 Moment–Curvature Relationship

A differential element of a beam subjected to pure bending is depicted in Figure 6.1(a). Cross-sectional centroidal axes u and v, which are also principal axes, are shown in Figure 6.1(b). To simplify the geometry, the right end of this beam element is assumed, without loss of generality, to be fixed against rotation. Before the couples M_u are applied, a cross section appears in Figure 6.1(a) as the vertical straight line $FAOE$. After the couples are applied, this plane section is assumed to remain plane and rotate to a position denoted by $F'A'OE'$. During this rotation a beam fiber denoted by AB [with cross-sectional area dA as shown in Figure 6.1(b)] shortens by an amount AA', since A moves to A'. This shortening is also given by $\varepsilon\, dx$, where ε is the bending strain at the level of this fiber and $dx(dx \approx ds)$ is the original length of the fiber AB. Application of these couples bends the differential beam element such that it has a curvature $\kappa = 1/\rho$, where ρ is the radius of curvature and G is the center of curvature. Triangles $AA'O$ and ODG are similar and thus $AA'/OA = OD/DG$. But $AA' = \varepsilon\, dx$, $OA = v$, $OD = dx$, and $DG = \rho$. Substitution of these leads to

$$\frac{\varepsilon\, dx}{v} = \frac{dx}{\rho}$$

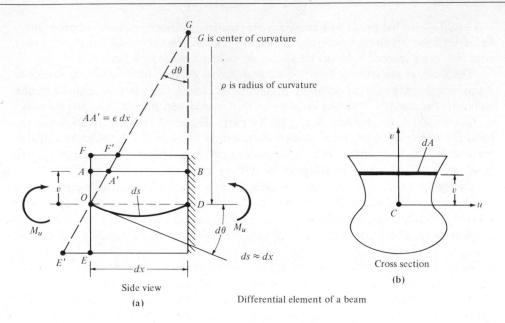

G is center of curvature

ρ is radius of curvature

Side view
(a)

Cross section
(b)

Differential element of a beam

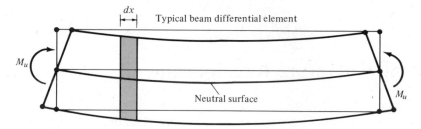

Typical beam differential element

Neutral surface

(c)

Neutral surface—bent beam of finite length

FIGURE 6.1

or

$$\frac{\varepsilon}{v} = \frac{1}{\rho} = \kappa \tag{6.1}$$

From Chapter 5,

$$\sigma_x = \frac{M_u v}{I_u} \tag{5.10a}$$

Ignoring Poisson effects, we can view the fiber AB as being uniaxially stressed, which means that Eq. 2.25 is valid and is repeated here for convenience:

$$\sigma_x = E\varepsilon_x \tag{2.25}$$

Substitution of Eq. 5.10a into Eq. 2.25 yields

$$\varepsilon_x = \frac{M_u v}{EI_u} \tag{6.2}$$

Substitution of Eq. 6.2 into Eq. 6.1 leads to

$$\kappa = \frac{1}{\rho} = \frac{M_u}{EI_u} \tag{6.3}$$

Equation 6.3 expresses the fact that the beam curvature at any beam section is directly proportional to the applied moment M_u at that section and inversely proportional to the bending stiffness EI_u, where E is the modulus of elasticity of the beam material and I_u is the moment of inertia of the cross-sectional area of the beam with respect to the u principal centroidal axis. The u axis intersects the v axis at the centroid C of the cross-sectional area and the u axis is the neutral axis of the cross section. Line OD is straight and horizontal prior to the application of couples M_u and is curved but unchanged in length after application of these couples.

The horizontal plane defined by the locus of all u axes of the initially straight beams becomes a curved cylindrical surface after application of equal moments at the ends of the beam of finite length, as shown in Figure 6.1(c). This cylindrical surface, which is unstrained, is referred to as the *neutral surface*. Rigorously, Eq. 6.3 is only valid for a beam subjected to pure bending (i.e., shear forces are not present). For beams that are long compared to their cross-sectional dimensions, the effect of shear is small compared to the effect of moments in producing deformations. Equation 6.3 will be applied, even in the presence of shears, with the reservation that short, deep beams may require special analysis. The effects of shear on deflection is considered in Section 6.9.

Example 6.1
A steel I beam is subjected to end moments M_u as shown in Figure 6.1(c). It is 12 in. deep, symmetric about both the u and v axes, and has a moment of inertia of area of its cross section I_u equal to 600 in.[4]. Determine the end moments applied to the beam when the normal stress in the beam reaches 30 ksi and find the corresponding curvature κ and radius of the curvature ρ. Use $E = 30,000$ ksi and neglect the weight of the beam.

Solution. Solve Eq. 5.10a for M_u:

$$M_u = \frac{\sigma_x I_u}{v} = \frac{30(600)}{\frac{12}{2}}$$

$$= \boxed{3000 \text{ k-in.}}$$

From Eq. (6.3):

$$\kappa = \frac{1}{\rho} = \frac{M_u}{EI_u}$$

$$= \frac{1}{\rho} = \frac{3000}{30,000(600)}$$

$$= \boxed{1.667 \times 10^{-4} \text{ 1/in.}}$$

$$\boxed{\rho = \frac{1}{\kappa} = 6000 \text{ in.}}$$

6.3 Beam Deflections—Two Successive Integrations

A beam is subjected to two equal couples M_u, as depicted in Figure 6.2. These couples are positive, as shown, and the beam deflects downward when they act on it. The deflection of the beam is denoted by v at any section a distance x from the origin O at the left end of the beam. Deflection v locates points on the neutral surface after the couples have been applied to the beam and the function $v = f(x)$ defines the curved line referred to as the *elastic curve* or the *deflected shape*. This curved line is an edge view of the neutral surface. The function $v = f(x)$, which will be determined by two successive integrations of the appropriate differential equation, is continuous and differentiable. The first derivative of this function with respect to x, dv/dx, represents the slope of the elastic curve and is a derived function $f'(x)$, which is also a function of x.

To derive the governing differential equation for beam deflections (or displacements), an approximation will be developed for beam curvature and this will be substituted into the moment–curvature relationship given by Eq. 6.3. In beginning calculus courses, the curvature of any plane curve is discussed and the equation for curvature is stated in terms of first and second derivatives as follows:

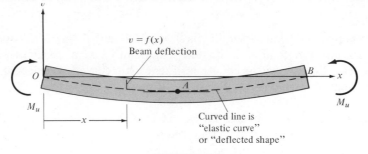

FIGURE 6.2

$$\kappa = \frac{1}{\rho} = \frac{\pm d^2v/dx^2}{[1 + (dv/dx)^2]^{3/2}} \qquad (6.4)$$

Elastic beam deflections are small compared to beam lengths and the slopes are also small. In the denominator of Eq. 6.4 the term $[1 + (dv/dx)^2]^{3/2}$ is approximately equal to 1, since the slope squared is very small compared to 1. The positive sign will be chosen for the second derivative on the right-hand side of Eq. 6.4. This choice is consistent with the positive bending moments shown in Figure 6.2. Since the first derivative varies in the following fashion as x increases—negative slope at point O, zero slope at point A, and positive slope at point B—the second derivative will be positive for the bent beam of Figure 6.2. The curvature is given approximately by

$$\kappa = \frac{1}{\rho} \approx \frac{d^2v}{dx^2} \qquad (6.5)$$

Substitution of Eq. 6.5 into Eq. 6.3 yields

$$EI_u \frac{d^2v}{dx^2} = M_u \qquad (6.6)$$

Equations 6.6 is the governing differential equation for beam deflections. It is an approximation for small elastic displacements of beams, which yields values sufficiently accurate for engineering applications provided that the beams are long compared to their cross-sectional dimensions and moment effects predominate as compared to shear effects. Applications of this equation will be considered for which EI_u remains constant over full or partial beam lengths and for which M_u will, in general, be a function of x. It should be noted that the flexure formula, Eq. 5.10a, was utilized in deriving Eq. 6.6, and thus all the assumptions made and the limitations of this former equation apply also to this governing differential equation for beam deflections.

Example 6.2

A simply supported beam of length L is shown in Figure 6.3(a). It is loaded by a downward uniform load over its full length and has a constant bending stiffness, EI_u. Determine the equation of its elastic curve (or deflected shape) and the equation for the slope of the elastic curve.

Solution. The governing differential Eq. 6.6 for this problem is a second-order one, and two integrations will be required to solve it. Each of the two integrations will require the introduction of a constant of integration. Two boundary conditions will be required to evaluate these two constants of integration:

$$x = 0 \qquad v = 0$$
$$x = L \qquad v = 0$$

These conditions state that the beam does not deflect at its supported ends. That is, the supports of the beam do not settle.

Consider the free-body diagram shown in Figure 6.3(b) and write M_u as a function of x measured from the origin at the left end of the beam.

$$M_u = R_1 x - \frac{wx^2}{2}$$

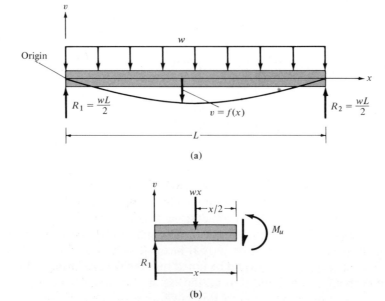

FIGURE 6.3

But $R_1 = wL/2$, yielding

$$M_u = \frac{wL}{2}x - \frac{wx^2}{2} \qquad (0 \leq x \leq L)$$

Substitute this equation for M_u into Eq. 6.6, to obtain

$$EI_u \frac{d^2v}{dx^2} = \frac{wL}{2}x - \frac{wx^2}{2}$$

To perform the first of two successive integrations used to solve this differential equation, multiply through by dx and integrate term by term as follows:

$$EI_u \int \frac{d^2v}{dx^2}\, dx = \frac{wL}{2}\int x\, dx - \frac{w}{2}\int x^2\, dx$$

$$EI_u \frac{dv}{dx} = \frac{wL}{2}\frac{x^2}{2} - \frac{w}{2}\frac{x^3}{3} + C_1$$

Again multiply by dx throughout and integrate term by term:

$$EI_u \int \frac{dv}{dx}\, dx = \frac{wL}{4}\int x^2\, dx - \frac{w}{6}\int x^3\, dx + C_1 \int dx$$

$$EI_u v = \frac{wL}{4}\frac{x^3}{3} - \frac{w}{6}\frac{x^4}{4} + C_1 x + C_2$$

The constants of integration C_1 and C_2 arise, since indefinite integrals are involved, and these constants must be evaluated by applying the boundary conditions stated above. The first condition, $x = 0$, $v = 0$, substituted into the equation for v yields

$$EI_u(0) = 0 - 0 + 0 + C_2$$

which implies that $C_2 = 0$. The second condition, $x = L$, $v = 0$, substituted into the equation for v yields

$$EI_u(0) = \frac{wL^4}{12} - \frac{wL^4}{24} + C_1 L + 0$$

Solving, we obtain

$$C_1 = -\frac{wL^3}{24}$$

Back substitution of these values for C_1 and C_2 into the equation for v and division by EI_u yields the equation of the elastic curve as a function of x for the given beam and loading. Thus

$$v = \frac{1}{EI_u}\left(\frac{wL}{12}x^3 - \frac{w}{24}x^4 - \frac{wL^3}{24}x\right) \qquad (0 \le x \le L)$$

The equation for the slope of the elastic curve may be obtained by differentiating the elastic curve equation with respect to x or substitution of the value of C_1 into the equation for dv/dx developed above after the first of the two successive integrations. This slope equation becomes

$$\frac{dv}{dx} = \frac{1}{EI_u}\left(\frac{wL}{4}x^2 - \frac{w}{6}x^3 - \frac{wL^3}{24}\right) \qquad (0 \le x \le L)$$

Example 6.3
Refer to the elastic curve and slope equations developed in Example 6.2 and obtain an equation for the maximum vertical displacement of the beam. Evaluate the maximum displacement for the following numerical values:

$$w = 4000 \text{ lb/ft}$$

$$L = 30 \text{ ft}$$

$$I_u = 3600 \text{ in.}^4$$

$$E = 30 \times 10^6 \text{ psi}$$

Solution. By symmetry, the slope of the elastic curve will be zero at midspan $(x = L/2)$ and the displacement will reach its maximum value. Substitute $x = L/2$ into the equation for v to obtain

$$v = \frac{1}{EI_u}\left[\frac{wL}{12}\left(\frac{L}{2}\right)^3 - \frac{w}{24}\left(\frac{L}{2}\right)^4 - \frac{wL^3}{24}\left(\frac{L}{2}\right)\right]$$

$$= -\frac{5wL^4}{384EI_u}$$

$$= \frac{-5(4000)(\frac{1}{12})(30 \times 12)^4}{384(30 \times 10^6)(3600)}$$

$$= \boxed{-0.675 \text{ in.}}$$

The minus sign means that the displacement is directed downward in a sense opposite to the positive sense of v assumed upward in solving Example 6.2.

Example 6.4

Write the equations of the elastic curve and its slope for the beam depicted in Figure 6.4. Locate the point of maximum downward deflection between supports A and B and compare this deflection to the upward deflection at point C.

Solution. It will be necessary to solve this problem in two parts by considering separate free-body diagrams as shown in Figure 6.4(b) and (c). Two separate moment functions are required, which are valid for segments AB and BC of the beam. These two moment functions are associated with two differential equations of the form of Eq. 6.6. Since the solution of each second-order differential equation requires the introduction of two constants of integration, a total of four constants will arise in solving this problem. Four constants require four boundary conditions for their evaluation. These boundary conditions may be stated as follows:

The beam does not deflect at point A: $x = 0, v_1 = 0$

The beam does not deflect at point B: $x = 2L, v_1 = 0$

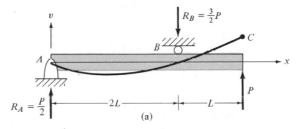

(a)

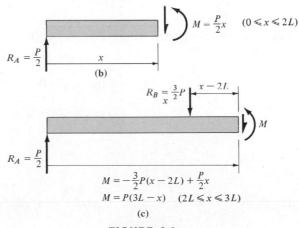

$$M = \frac{P}{2}x \quad (0 \leqslant x \leqslant 2L)$$

(b)

$$M = -\frac{3}{2}P(x - 2L) + \frac{P}{2}x$$

$$M = P(3L - x) \quad (2L \leqslant x \leqslant 3L)$$

(c)

FIGURE 6.4

The beam deflections and slopes are continuous functions at point B:

$$x = 2L \qquad v_1 = v_2 = 0$$

$$x = 2L \qquad v_1' = v_2'$$

Subscripts 1 and 2 denote displacements and slopes valid over lengths AB and BC of the beam, respectively.

SEGMENT AB $(0 \leq x \leq 2L)$: Consider the free-body diagram shown in Figure 6.4(b) and adopt the prime notation for derivatives with respect to x.

$$EI_u v_1'' = \frac{P}{2} x$$

Integrating once: $\qquad EI_u v_1' = \frac{P}{4} x^2 + C_1$

Integrating again: $\qquad EI_u v_1 = \frac{P}{12} x^3 + C_1 x + C_2$

SEGMENT BC $(2L \leq x \leq 3L)$: Consider the free-body diagram shown in Figure 6.4(c) and express the moment as a function of x.

$$M = \frac{P}{2} x - \frac{3}{2} P(x - 2L) = -Px + 3PL$$

$$EI_u v_2'' = -Px + 3PL$$

Integrating once: $\quad EI_u v_2' = -\frac{Px^2}{2} + 3PLx + C_3$

Integrating again: $\quad EI_u v_2 = -\frac{Px^3}{6} + \frac{3PL}{2} x^2 + C_3 x + C_4$

Apply the boundary conditions for determination of the four constants of integration.

$$x = 0,\ v_1 = 0: \quad \text{implies that } C_2 = 0$$

$$x = 2L,\ v_1 = 0: \quad 0 = \frac{P}{12} (2L)^3 + C_1(2L) + 0$$

$$C_1 = -\frac{PL^2}{3}$$

$$x = 2L,\ v_1' = v_2' = \frac{P}{4}(2L)^2 + C_1 = -\frac{P}{2}(2L)^2 + 3PL(2L) + C_3$$

$$PL^2 - \tfrac{1}{3}PL^2 = -2PL^2 + 6PL^2 + C_3$$

$$C_3 = -\tfrac{10}{3}PL^2$$

$$x = 2L, \; v_2 = 0: \quad 0 = -\frac{P}{6}(2L)^3 + \frac{3PL}{2}(2L)^2 + C_3(2L) + C_4$$

$$= -\tfrac{4}{3}PL^3 + 6PL^3 - \tfrac{20}{3}PL^3 + C_4$$

$$C_4 = 2PL^3$$

Substitution of these values for the constants into the elastic curve and slope equations yields

$$v_1 = \frac{1}{EI_u}\left(\frac{P}{12}x^3 - \frac{PL^2}{3}x\right) \qquad\qquad (0 \le x \le 2L)$$

$$v_1' = \frac{1}{EI_u}\left(\frac{P}{4}x^2 - \frac{PL^2}{3}\right) \qquad\qquad (0 \le x \le 2L)$$

$$v_2 = \frac{1}{EI_u}\left(-\frac{P}{6}x^3 + \frac{3PL}{2}x^2 - \frac{10}{3}PL^2x + 2PL^3\right) \qquad (2L \le x \le 3L)$$

$$v_2' = \frac{1}{EI_u}\left(-\frac{P}{2}x^2 + 3PLx - \frac{10}{3}PL^2\right) \qquad\qquad (2L \le x \le 3L)$$

To locate the point of maximum downward deflection between A and B, equate v_1' to zero and solve for x.

$$0 = \frac{1}{EI_u}\left(\frac{P}{4}x^2 - \frac{PL^2}{3}\right)$$

$$\boxed{x - \frac{2\sqrt{3}}{3}L}$$

Substitute this value of x into the v_1 equation to obtain the maximum downward deflection between A and B.

$$v_1 = \frac{1}{EI_u}\left[\frac{P}{12}\left(\frac{2\sqrt{3}L}{3}\right)^3 - \frac{PL^2}{2}\left(\frac{2\sqrt{3}}{3}L\right)\right]$$

$$= \boxed{\frac{-7\sqrt{3}}{27}\frac{PL^3}{EI_u}}$$

where the negative sign signifies a downward deflection. The maximum upward deflection occurs at point C (the right end of the beam). Substitute $x = 3L$ into v_2 to obtain this maximum upward deflection.

$$v_2 = \frac{1}{EI_u}\left[-\frac{P}{6}(3L)^3 + \frac{3PL}{2}(3L)^2 - \frac{10}{3}PL^2(3L) + 2PL^3\right]$$

$$= \boxed{\frac{PL^3}{EI_u}}$$

where the positive sign signifies an upward deflection. These deflections may be readily compared by forming their ratio.

$$\frac{|v_2|}{|v_1|} = \frac{PL^3/EI_u}{(7\sqrt{3}/27)(PL^3/EI_u)} = \frac{27}{7\sqrt{3}}$$

$$= \boxed{2.23}$$

The upward deflection at C is 2.23 times as large as the maximum downward deflection between A and B.

Homework Problems

6.1 Refer to the cantilever beam shown in Figure H6.1 and choose an origin at the left end, O. Determine the reactions at O, then write the equation of the elastic curve and the slope equation, both as functions of x measured from the left end of the beam.

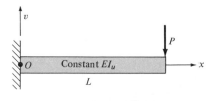

FIGURE H6.1

6.2 The flexural stress in the beam depicted in Figure H6.2 is to be limited to 20 ksi. Determine the radius of curvature of this beam if $E = 10 \times 10^6$ psi and $I_u = 2400$ in.[4] The beam is 12 in. deep.

FIGURE H6.2

6.3 A cantilever beam is loaded uniformly as shown in Figure H6.3. Select an origin at the left end, O, and write the equation of the elastic curve for this beam. Determine the deflection at the right end of this cantilever in terms of w, L, E, and I_u.

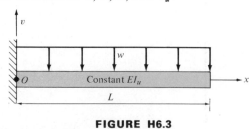

FIGURE H6.3

6.4 Refer to the beam shown in Figure H6.4, and state the four boundary conditions required to determine the equation of the elastic curve. Determine this equation and its derivative function for this beam. Express the deflection under the load as a function of P, L, E, and I_u. Is this deflection the maximum deflection? If not, express the maximum deflection as a function of P, L, E, and I_u.

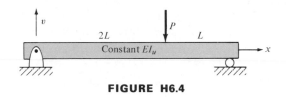

FIGURE H6.4

6.5 Choose an origin at point B for the beam depicted in Figure H6.5 and derive the equations of the elastic curve. Express the deflection at point A as a function of w, L, E, and I_u.

(HINT: You will need to consider segments AB and BC of the beam separately and use matching conditions at B.)

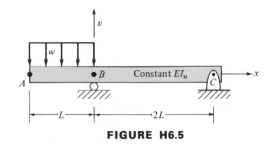

FIGURE H6.5

6.6 A couple M_1 is applied at the center of the beam shown in Figure H6.6. Express the slope at the center of the beam as a function of M_1, L, E, and I_u.

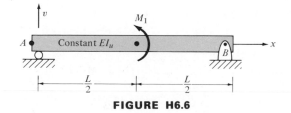

FIGURE H6.6

6.7 Choose an origin at A and write the equations of the elastic curve for the beam shown in Figure H6.7. Express the center deflection of this beam in terms of Q, L, E, and I_u.

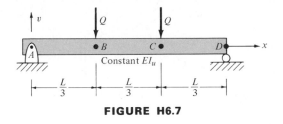

FIGURE H6.7

6.8 A beam shown in Figure H6.8 is loaded with a couple M_1 at its right end as shown. Choose an origin at A and write the equation of the elastic curve. Locate the point where the maximum deflection occurs and express this deflection in terms of M_1, L, E, and I_u.

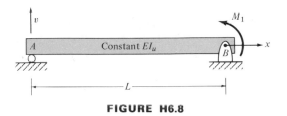

FIGURE H6.8

6.9 Choose an origin at A for the beam depicted in Figure H6.9. Write the equation of the elastic curve for each segment of the beam and express the deflection at the center of the beam in terms of w, L, E, and I_u.

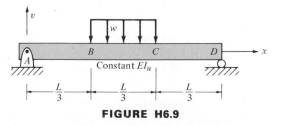

FIGURE H6.9

6.10 Refer to the beam depicted in Figure H6.10 and write the equations of the elastic curve and the slope for this beam. Determine the location of the maximum deflection and express this deflection as a function of w_0, L, E, and I_u. (The roller at B is constructed such that a downward force may be exerted on the beam.)

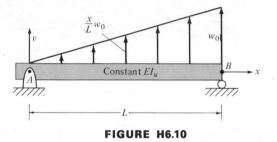

FIGURE H6.10

6.11 Write the equation of the elastic curve for each segment of the beam shown in Figure H6.11. Choose an origin at B and express the deflection at A and the deflection at the center each as functions of P, L, E, and I_u. Determine the ratio of the absolute values of these deflections. Express the slope of the beam at point A in terms of P, L, E, and I_u. Show your results on a sketch of the elastic curve.

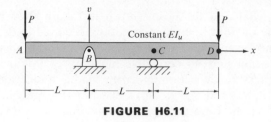

FIGURE H6.11

6.12 Write the equation of the elastic curve for each segment of the beam depicted in Figure H6.12. Locate the point of maximum downward deflection and compare the magnitude of this deflection to the magnitude of the deflection at point C. Express these deflections in terms of w, L, E, and I_u.

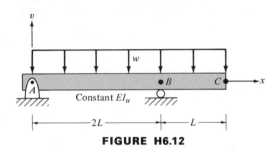

FIGURE H6.12

6.4 Derivatives of the Elastic Curve Equation and Their Physical Significance

The elastic curve equation given by the function $v = v(x)$ and its first four derivatives with respect to x are readily associated with the physical problem of the bending of a loaded beam.

The slope of the elastic curve is given by the first derivatives of $v(x)$ with respect to x:

$$\frac{dv}{dx} = \tan \theta \qquad \text{(6.7)}$$

For small angles $\tan \theta$ is approximately equal to θ, which yields

$$\frac{dv}{dx} = \theta \tag{6.8}$$

where θ is given in radians.

Equation 6.6, $EI_u(d^2v/dx^2) = M_u$, expresses the bending moment at any section in the bent beam as the product of the bending stiffness (EI_u) and the second derivative of $v(x)$ with respect to x. Differentiate Eq. 6.6 with respect to x, to yield

$$EI_u \frac{d^3v}{dx^3} = \frac{dM_u}{dx} \tag{6.9}$$

But $dM_u/dx = V$, by Eq. 1.6. Thus

$$EI_u \frac{d^3v}{dx^3} = V \tag{6.10}$$

which states that the bending stiffness (EI_u) multiplied by the third derivative of $v(x)$ with respect to x equals the shear V at any section of the bent beam.

Differentiate Eq. 6.10 with respect to x, to yield

$$EI_u \frac{d^4v}{dx^4} = \frac{dV}{dx} \tag{6.11}$$

But $dV/dx = -w$, by Eq. 1.5. Thus

$$EI_u \frac{d^4v}{dx^4} = -w \tag{6.12}$$

which states that the bending stiffness multiplied by the fourth derivative of $v(x)$ with respect to x equals the negative of the load intensity w at any section of the bent beam.

Equations 6.7 to 6.12 are applicable as long as these functions of x are continuous and differentiable. Special problems arise at points where these functions are discontinuous. For example, the shear function is discontinuous at points where concentrated loads are applied. Similarly, the moment function is discontinuous at points where concentrated couples are applied.

Example 6.5

(a) Refer to Figure 6.5(a), in which a simply supported beam is loaded by a sinusoidal loading. Determine the reactions acting on the beam, then find the shear, moment, slope, and deflection as functions of x and represent these functions graphically. (b) Without finding the reactions, start with Eq. 6.12 and derive the equation of the elastic curve.

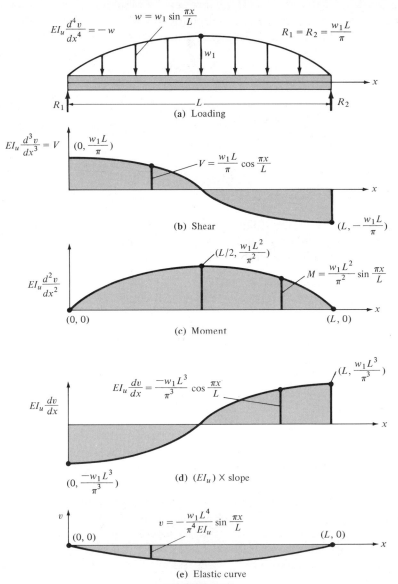

FIGURE 6.5

Solution

(a) Each reaction equals half the total downward load applied to the beam. This total downward load equals the area under the w–x curve.

$$R_1 = R_2 = w_1 \int_0^{L/2} \sin \frac{\pi x}{L} \, dx = \frac{w_1 L}{\pi} \int_0^{L/2} \sin \frac{\pi x}{L} \left(\frac{\pi}{L} \, dx \right)$$

$$= R_2 = \frac{w_1 L}{\pi} \left[-\cos \frac{\pi x}{L} \right]_0^{L/2} = \frac{-w_1 L}{\pi} (0 - 1)$$

$$= \boxed{\frac{w_1 L}{\pi} \uparrow}$$

Substitution of the sinusoidal loading intensity w into Eq. 6.12 yields

$$EI_u \frac{d^4 v}{dx^4} = -w_1 \sin \frac{\pi x}{L}$$

Multiply through by dx, adjust the differential and integrate as follows:

$$EI_u \int \frac{d^4 v}{dx^4} \, dx = \frac{-w_1 L}{\pi} \int \sin \frac{\pi x}{L} \left(\frac{\pi}{L} \, dx \right)$$

$$EI_u \frac{d^3 v}{dx^3} = \frac{w_1 L}{\pi} \cos \frac{\pi x}{L} + C_1$$

At the left end of the beam, the shear equals the reaction, and this condition will be used to find the constant of integration C_1. Thus at $x = 0$,

$$V = \frac{w_1 L}{\pi} = EI_u \frac{d^3 v}{dx^3}$$

by Eq. 6.10. Therefore,

$$\frac{w_1 L}{\pi} = \frac{w_1 L}{\pi} \cos 0 + C_1$$

and

$$C_1 = 0$$

which leads to

$$V = EI_u \frac{d^3 v}{dx^3} = \frac{w_1 L}{\pi} \cos \frac{\pi x}{L}$$

This function is shown graphically in Figure 6.5(b). It shows the variation of shear across the beam span and is the usual shear diagram. Note that the loading, which is symmetric about the center of the beam, leads to an antisymmetric shear about the center of the beam. Multiply through by dx, adjust the differential, and integrate as follows:

$$EI_u \int \frac{d^3v}{dx^3}\, dx = \frac{w_1 L^2}{\pi^2} \int \cos \frac{\pi x}{L} \left(\frac{\pi}{L}\, dx \right)$$

$$EI_u \frac{d^2v}{dx^2} = \frac{w_1 L^2}{\pi^2} \sin \frac{\pi x}{L} + C_2$$

At the left end of the beam, the moment equals zero, and this condition will be used to find the constant of integration C_2. Thus at $x = 0$, $M = EI_u(d^2v/dx^2) = 0$, by Eq. 6.6. Therefore,

$$0 = \frac{w_1 L^2}{\pi^2} \sin (0) + C_2$$

and

$$C_2 = 0$$

which leads to

$$M = EI_u \frac{d^2v}{dx^2} = \frac{w_1 L^2}{\pi^2} \sin \frac{\pi x}{L}$$

This function is shown graphically in Figure 6.5(c). It shows the variation of moment across the beam span and is the usual moment diagram. Note that symmetric loading leads to antisymmetric shear and symmetric moment about the center of the beam.

Again, multiply through by dx, adjust the differential, and integrate as follows:

$$EI_u \int \frac{d^2v}{dx^2}\, dx = \frac{w_1 L^3}{\pi^3} \int \sin \frac{\pi x}{L} \frac{\pi}{L}\, dx$$

$$EI_u \frac{dv}{dx} = \frac{-w_1 L^3}{\pi^3} \cos \frac{\pi x}{L} + C_3$$

Since the beam is loaded symmetrically, the slope at the center of the beam will vanish, and this condition will be used to find the constant of integration C_3. Thus at $x = L/2$, $dv/dx = 0$. Therefore,

$$0 = \frac{-w_1 L^3}{\pi^3} \cos \frac{\pi}{2} + C_3$$

and

$$C_3 = 0$$

which leads to

$$EI_u \frac{dv}{dx} = \frac{-w_1 L^3}{\pi^3} \cos \frac{\pi x}{L}$$

The function $EI_u(dv/dx)$ is shown graphically in Figure 6.5(d). Division of the ordinates by the constant EI_u would provide values for the slopes of the elastic curves at each point across the beam span. Slopes to the left of the center are negative and those to the right of the center are positive. Note that symmetric loading leads to antisymmetric slopes about the center of the beam. Once more, multiply through by dx, adjust the differential, and integrate as follows:

$$EI_u \int \frac{dv}{dx} \, dx = \frac{-w_1 L^4}{\pi^4} \int \cos \frac{\pi x}{L} \left(\frac{\pi}{L} \, dx \right)$$

$$EI_u v = \frac{-w_1 L^4}{\pi^4} \sin \frac{\pi x}{L} \mid C_4$$

At the left end of the beam the deflection is zero, and this condition will be used to find the constant of integration C_4.

$$0 = \frac{-w_1 L^4}{\pi^4} \sin (0) + C_4$$

$$C_4 = 0$$

$$v = \frac{-w_1 L^4}{EI_u \pi^4} \sin \frac{\pi x}{L}$$

The function $v(x)$ is shown graphically in Figure 6.5(e). It shows the variation of the deflection across the beam span and is the usual equation of the elastic curve. Note that a symmetric loading leads to a symmetric elastic curve about the center of the beam. Overall review of Figure 6.5(a) to (e) reveals that a symmetric loading function is followed by an antisymmetric shear function, and so on. This observation is a useful aid for reviewing results to check their accuracy.

(b) Reactions need not be determined, since four successive integrations of Eq. 6.12 will give v as a function of x together with four constants of integration. These four constants will be evaluated with boundary conditions stated below. The ends of the beam do not deflect vertically:

$$x = 0, v = 0 \quad \text{and} \quad x = L, v = 0$$

The ends of this simply supported beam cannot resist bending moments, and these moments are directly proportional to second derivatives of v with respect to x:

$$x = 0, \; v'' = 0 \quad \text{and} \quad x = L, \; v'' = 0$$

Equate the flexural rigidity (EI_u) multiplied by the fourth derivative of v with respect to x to the negative of the loading intensity function $(-w)$ in accord with Eq. 6.12:

$$EI_u v^{iv} = -w_1 \sin \frac{\pi x}{L}$$

Multiply through by dx, adjust the differential, and integrate as follows:

$$EI_u \int \frac{d^4 v}{dx^4} \, dx = -\frac{w_1 L}{\pi} \int \sin \frac{\pi x}{L} \frac{\pi}{L} \, dx$$

$$EI_u v''' = -\frac{w_1 L}{\pi} \left(-\cos \frac{\pi x}{L} \right) + C_1$$

Multiply through by dx, adjust the differential, and integrate to give

$$EI_u v'' = \frac{W_1 L^2}{\pi^2} \sin \frac{\pi x}{L} + C_1 x + C_2$$

Multiply by dx, adjust the differential, and integrate to give:

$$EI_u v' = \frac{-w_1 L^3}{\pi^3} \cos \frac{\pi x}{L} + C_1 \frac{x^2}{2} + C_2 x + C_3$$

Multiply by dx, adjust the differential, and integrate to give

$$EI_u v = \frac{-w_1 L^4}{\pi^4} \sin \frac{\pi x}{L} + \frac{C_1}{6} x^3 + \frac{C_2}{2} x^2 + C_3 x + C_4$$

Boundary conditions will now be substituted in these equations in order to evaluate the constants C_1 through C_4.

$x = 0, \; v = 0$:   implies that $C_4 = 0$

$x = 0, \; v'' = 0$:   implies that $C_2 = 0$

$x = L, \; v'' = 0$:   implies that $C_1 = 0$, since $\sin \pi = 0$

$x = L, \; v = 0$:   implies that $C_3 = 0$, since $\sin \pi = 0$

Since all constants of integration vanish in this particular case, the equation of the elastic curve becomes

$$v = \frac{-w_1 L^4}{EI_u \pi^4} \sin \frac{\pi x}{L}$$

This agrees with the result obtained in part (a). Loading shear, bending moment, slope, and displacement plots are shown in Figure 6.5.

Homework Problems

6.13 Refer to the cantilever beam depicted in Figure H6.13 with an origin at A. Substitute the given loading intensity function $w(x)$ into Eq. 6.12 and perform the necessary integrations required to express shear, moment, slope, and deflection of the beam as functions of x. Sketch these functions.

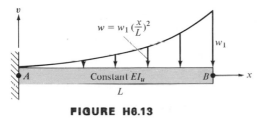

FIGURE H6.13

6.14 The cantilever beam is loaded as shown in Figure H6.14. An origin has been chosen at B to express the loading intensity as a function of x. Substitute this function $w(x)$ into Eq. 6.12 and perform the necessary integrations required to express shear, moment, slope, and deflection of the beam as functions of x. Sketch these functions and check to see that all boundary conditions are met.

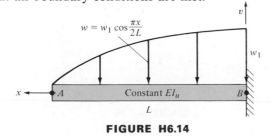

FIGURE H6.14

6.15 A simply supported beam depicted in Figure H6.15 is loaded by a sinusoidally varying load. Substitute this given loading function into Eq. 6.12 and perform the necessary integrations required to express shear, moment, slope, and deflection of the beam as functions of x. Sketch these functions and check to see that all boundary conditions are met.

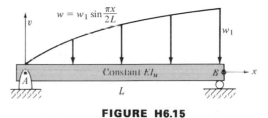

FIGURE H6.15

6.16 An origin has been chosen at point A for the simply supported beam depicted in Figure H6.16. Express the shear, moment, slope, and deflection as functions of x by substitution of the loading function $w(x)$ into Eq. 6.12. Sketch these functions and check to see that all boundary conditions are met.

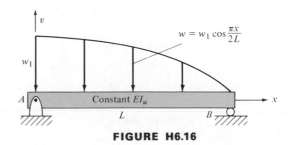

FIGURE H6.16

6.17 A linearly varying loading is applied to the simply supported beam of Figure H6.17. For an origin at point A, substitute the loading function $w(x)$ into Eq. 6.12 and derive equations for shear, moment, slope, and deflection as functions of x. Sketch these functions and check to see that all boundary conditions are satisfied.

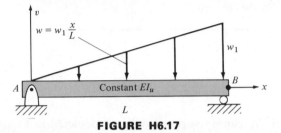

FIGURE H6.17

6.18 A cantilever beam fixed at point B is subjected to a linearly varying loading as shown in Figure H6.18. Substitute this given loading function into Eq. 6.12 and derive equations for shear, moment, slope, and deflection as functions of x. Sketch these functions and check to see that all boundary conditions are satisfied.

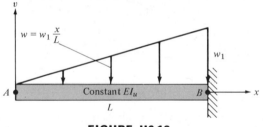

FIGURE H6.18

6.19 A parabolically distributed loading is applied to the simply supported beam shown in Figure H6.19.

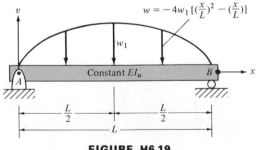

FIGURE H6.19

Substitute the given loading intensity function $w(x)$ into Eq. 6.12 and perform the necessary integrations required to express shear, moment, slope, and deflection of the beam as functions of x. Sketch these functions and check to see that all boundary conditions are satisfied. An origin has been selected at A.

6.20 Loading on the beam shown in Figure H6.20 varies as a linear function of x measured from an origin at A. Use this loading intensity function $w(x)$ in Eq. 6.12 and derive equations for shear, moment, slope, and deflection of the beam as functions of x. Sketch these functions and consider the following special cases for checking the results:

(a) $w_1 = w_2$, corresponding to a uniform loading.

(b) $w_1 = 0$, corresponding to a loading diagram of triangular shape.

Note that this is the solution for Problem 6.17, with w_1 replaced by w_2.

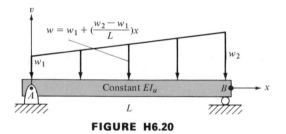

FIGURE H6.20

6.21 The simply supported beam shown in Figure H6.21 is loaded as indicated with an origin at A. Derive equations for shear, moment, slope, and deflection as functions of x by substituting the given loading intensity function $w(x)$ into Eq. 6.12. Sketch these functions and check to see that all boundary conditions are satisfied.

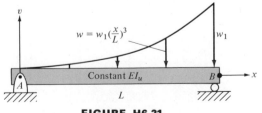

FIGURE H6.21

6.22 With an origin at the right end of the cantilever depicted in Figure H6.22, the loading varies as the cube of x/L to a maximum intensity of w_1, when $x = L$. Substitute this given loading intensity function $w(x)$ into Eq. 6.12 and perform the necessary integrations required to express shear, moment, slope, and deflection as functions of x. Sketch these functions and check to see that boundary conditions at both ends of the beam are satisfied.

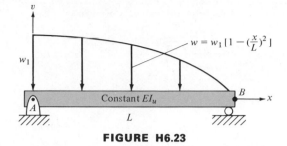

FIGURE H6.23

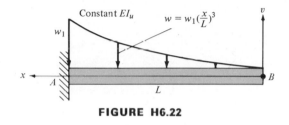

FIGURE H6.22

6.24 An origin was chosen at the left end of the simply supported beam shown in Figure H6.24 and the loading intensity functions $w(x)$ expressed as shown. Substitute this given function into Eq. 6.12 and perform the necessary integrations required to express shear, moment, slope, and deflection as functions of x. Sketch these functions and check to see that the boundary conditions are satisfied.

6.23 The simply supported beam of Figure H6.23 is subjected to the loading shown with x measured from an origin at the left end of the beam. Derive equations for shear, moment, slope, and deflection as functions of x by substituting the given loading intensity function $w(x)$ into Eq. 6.12 and performing the required integrations. Sketch these functions and check to see that the boundary conditions are satisfied.

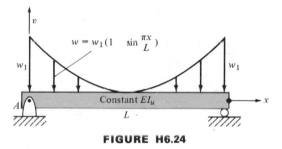

FIGURE H6.24

6.5 Beam Deflections—The Method of Superposition

Equation 6.6, $EI_u(d^2v/dx^2) = M_u$, with M_u a function of x, is an ordinary, second-order differential equation. An ordinary differential equation is said to be linear if the dependent variable and its derivatives appear only to the first power in the equation. The dependent variable v appears only in the second derivative d^2v/dx^2 of Eq. 6.6, and since this derivative is raised to the first power, the differential equation is a linear one. The coefficient, EI_u, multiplying the second derivative d^2v/dx^2, is termed the *bending stiffness* and is assumed to be constant over full or partial beam lengths. The modulus of elasticity E is a constant that relates stress to strain in the beam, and the moment of inertia I_u is a constant for prismatic beams over their lengths. Bending

TABLE 6.1
Equations for Beam Slopes and Deflections of Selected Cases

Case 1. Simply supported beam—uniformly distributed loading

$$\theta_1 = \frac{-wL^3}{24EI_u}$$

$$\theta_2 = -\theta_1$$

$$v = \frac{wx}{24EI_u}(L^3 - 2Lx^2 + x^3)$$

$$v = \frac{-5wL^4}{384EI_u}$$

Maximum at center

Case 2. Simply supported beam—concentrated load at any point

$$\theta_1 = \frac{Pb(b^2 - L^2)}{6EI_uL}$$

$$\theta_2 = \frac{Pab(2L - b)}{6EI_uL}$$

$$v = \frac{Pbx}{6EI_uL}(b^2 + x^2 - L^2)$$
$$(0 \leqslant x \leqslant a)$$

$$v = \frac{-Pb}{6EI_uL}\left[\frac{L}{b}(x-a)^3 + (L^2 - b^2)x - x^3\right]$$
$$(a \leqslant x \leqslant L)$$

$$v = \frac{-Pb(L^2 - b^2)^{3/2}}{9\sqrt{3}\,EI_uL}$$

Maximum at $x = \sqrt{\dfrac{L^2 - b^2}{3}}$

Case 3. Simply supported beam—couple C applied at left end

$$\theta_1 = \frac{-CL}{3EI_u}$$

$$\theta_2 = \frac{CL}{6EI_u}$$

$$v = \frac{-CLx}{6EI_u}\left[\left(\frac{x}{L}\right)^2 - 3\left(\frac{x}{L}\right) + 2\right]$$

$$v = \frac{-CL^2}{9\sqrt{3}\,EI_u}$$

Maximum at $x = L\left(1 - \dfrac{\sqrt{3}}{3}\right)$.

Center deflection

$$v = \frac{-CL^2}{16EI_u}$$

Case 4. Cantilever beam—uniformly distributed loading

$$\theta_2 = \frac{-wL^3}{6EI_u}$$

$$v = \frac{-wx^2}{24EI_u}(x^2 + 6L^2 - 4Lx)$$

$$v = \frac{-wL^4}{8EI_u}$$

Maximum at right end

Case 5. Cantilever beam—concentrated load at any point

$$\theta_2 = \frac{-Pa^2}{2EI_u}$$

$$v = \frac{-Px^2}{6EI_u}(3a - x)$$
$$(0 \leqslant x \leqslant a)$$

$$v = \frac{-Pa^2}{6EI_u}(3x - a)$$
$$(a \leqslant x \leqslant L)$$

$$v = \frac{-Pa^2}{6EI_u}(3L - a)$$

Maximum at right end

Case 6. Cantilever beam—couple C applied at right end

$$\theta_2 = \frac{CL}{EI_u}$$

$$v = \frac{Cx^2}{2EI_u}$$

$$v = \frac{CL^2}{2EI_u}$$

Maximum at right end

Case 7. Cantilever beam—linearly varying load

$$w\left(1 - \frac{x}{L}\right)$$

$$\theta_2 = \frac{-wL^3}{24EI_u}$$

$$v = -\frac{wx^2}{120LEI_u}(10L^3 - 10L^2x + 5Lx^2 - x^3); \quad v = \frac{-wL^4}{30EI_u}$$

Maximum at right end

Case 8. Cantilever beam—parabolically varying load

$$w\left(\frac{x}{L}\right)^2$$

$$\theta_2 = \frac{wL^3}{60EI_u}$$

$$v = \frac{-w}{12EI_u}\left[\frac{x^6}{30L^2} - \frac{L^3}{5}x + \frac{L^4}{6}\right]$$

$$v = \frac{-wL^4}{72EI_u}$$

Maximum at right end

moment M_u appearing on the right side of Eq. 6.6 is a function of x derived by applying the equations of equilibrium to beam segments of variable length x.

The equations of equilibrium are formulated for the undeformed beam. As a consequence, these equations and conclusions drawn from them are valid only as long as the small deflections do not affect the action of the applied forces.

Consider a given beam subjected to two separate loadings and use subscripts 1 and 2 to denote the two analyses. Write equations of the form of Eq. 6.6 for each separate loading:

$$EI_u \frac{d^2v_1}{dx^2} = M_{u1} \tag{6.13}$$

$$EI_u \frac{d^2v_2}{dx^2} = M_{u2} \tag{6.14}$$

If Eqs. 6.13 and 6.14 are solved separately for v_1 and v_2, the sum of these deflections, denoted by v_3, may be written

$$v_3 = v_1 + v_2 \tag{6.15}$$

Now, form the sum of the Eqs. 6.13 and 6.14 and note that d^2/dx^2 is a linear operator:

$$EI_u \frac{d^2(v_1 + v_2)}{dx^2} = M_{u1} + M_{u2} \tag{6.16}$$

or

$$EI_u \frac{d^2v_3}{dx^2} = M_{u3} \tag{6.17}$$

where

$$M_{u3} = M_{u1} + M_{u2} \tag{6.18}$$

If Eq. 6.17 is solved for v_3, the result will be the same as that given by Eq. 6.15.

The method of superposition of beam deflections is expressed by Eq. 6.15, and it is readily extended from two loadings to n loadings. In summary, superposition is valid for beam deflections because the governing differential equation is linear, the bending stiffness is constant, and the bending moments are independent of the small displacements of the beam. This method is best illustrated by examples solved with the aid of Table 6.1.

Example 6.6
Refer to Figure 6.6(a) and use the method of superposition to determine the center deflection of the beam shown. The moment of inertia of the beam cross section $I_u = 4.0 \times 10^{-5}$ m^4 and the modulus of elasticity $E = 100$ GPa. Loads are applied perpendicular to the u principal axes of inertia of the cross sections.

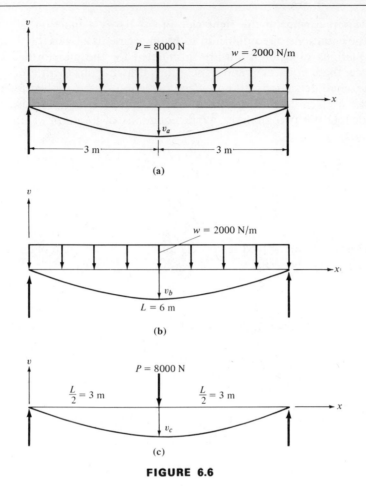

FIGURE 6.6

Solution. Decompose the problem into two cases, shown in Figure 6.6(b) and (c). Note that by the principle of superposition the total deflection at the center equals the sum of the component deflections due to the uniform and concentrated loading, respectively. Thus

$$v_a = v_b + v_c$$

Refer to Table 6.1 for the appropriate deflection formulas.

CASE 1

$$v_b = \frac{-5wL^4}{384EI_u}$$

CASE 2

$$v_c = \frac{Pbx}{6EI_u L}(b^2 + x^2 - L^2) \qquad (0 \le x \le a)$$

Since P is applied at the center of the beam and the deflection under the load is required, $x = b = L/2$, which yields

$$v_c = \frac{-PL^3}{48EI_u}$$

$$v_a = -\frac{5wL^4}{384EI_u} - \frac{PL^3}{48EI_u}$$

Substitute the appropriate numerical values using consistent units:

$$v_a = \frac{-5(2000)(6)^4}{384(100 \times 10^9)(4 \times 10^{-5})} - \frac{8000(6)^3}{48(100 \times 10^9)(4 \times 10^{-5})}$$

$$= -8.44 \times 10^{-3} - 9.00 \times 10^{-3}$$

$$= \boxed{-17.44 \times 10^{-3} \text{ m}}$$

The negative sign means that the deflection is directed downward.

Example 6.7

Refer to Figure 6.7(a) and use the method of superposition to determine the end deflection and slope of the cantilever beam shown. The moment of inertia of the beam cross section $I_u = 160$ in.^{4} and the modulus of elasticity $E = 30 \times 10^3$ ksi.

Solution. Decompose the problem into the two cases shown in Figure 6.7(b) and (c). Note that the end deflection and slope equals the sum of the component deflections and slopes due to the concentrated load and the couple, respectively. In writing the superposition equation, consider all deflections to be positive and the correct sign will follow from the values of Table 6.1.

$$v_a = v_b + v_c$$

Refer to Table 6.1 for the appropriate deflection formulas.

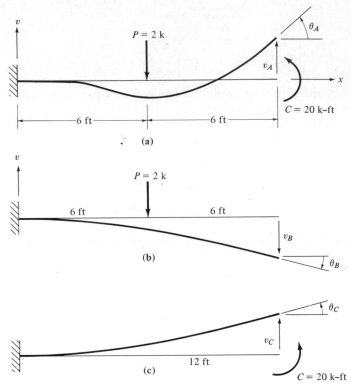

FIGURE 6.7

CASE 3

$$v_b = \frac{-Pa^2}{6EI_u}(3x - a) \qquad (a \le x \le L)$$

Since P is applied at the midpoint of the cantilever beam and the deflection at the right end is required, $a = L/2$ and $x = L$, which yields

$$v_b = \frac{-P(L/2)^2}{6EI_u}\left(3L - \frac{L}{2}\right)$$

$$= \frac{-5PL^3}{48EI_u}$$

CASE 4

$$v_c = \frac{CL^2}{2EI_u}$$

$$v_a = \frac{-5PL^3}{48EI_u} + \frac{CL^2}{2EI_u}$$

Substitute the appropriate numerical values using consistent units:

$$v_a = \frac{-5(2)(12 \times 12)^3}{48(30 \times 10^3)(160)} + \frac{20(12)(12 \times 12)^2}{2(30 \times 10^3)(160)}$$

$$= -0.130 + 0.518$$

$$= \boxed{0.388 \text{ in.}}$$

The positive sign means that the deflection is directed upward.
Similarly, for slopes, the superposition equation becomes

$$\theta_a = \theta_b + \theta_c$$

Again referring to Table 6.1 for the appropriate slope formulas from cases 5 and 6, we see that

$$\theta_a = \frac{-P(L/2)^2}{2EI_u} + \frac{CL}{EI_u}$$

Substitute the appropriate numerical values using consistent units:

$$\theta_a = \frac{-2(6 \times 12)^2}{2(30 \times 10^3)(160)} + \frac{20 \times 12(12 \times 12)}{30 \times 10^3(160)}$$

$$= -0.00108 + 0.00720$$

$$- \boxed{0.00612 \text{ rad} \quad \text{or} \quad 0.351°}$$

The positive sign means that the tangent has rotated counterclockwise from the unloaded to the loaded configuration.

Homework Problems

6.25 Determine the deflection at the right end of the cantilever beam shown in Figure H6.25. Use the method of superposition and obtain the appropriate formulas from Table 6.1.

FIGURE H6.25

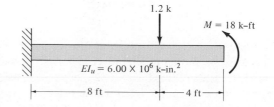

1.2 k

$M = 18$ k–ft

$EI_u = 6.00 \times 10^6$ k–in.^{2}

8 ft

4 ft

6.26 Determine the deflection at the center of the simply supported beam depicted in Figure H6.26. Refer to Table 6.1 for the appropriate formulas and use the method of superposition.

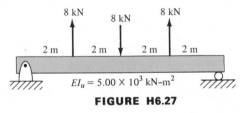

FIGURE H6.26

6.27 Find the deflection of the simply supported beam at the center of the span due to the loadings shown in Figure H6.27. Use the method of superposition and obtain the appropriate formulas from Table 6.1.

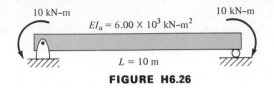

FIGURE H6.27

6.28 Equal and oppositely directed couples are applied to the ends of the beam depicted in Figure H6.28. Determine the deflection at the center of this beam by applying the appropriate formulas from Table 6.1. Use the method of superposition.

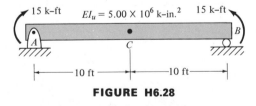

FIGURE H6.28

6.29 Determine the deflection at the center of the beam loaded as shown in Figure H6.29. Also find the slopes at the ends of this beam by superposition using the appropriate formulas from Table 6.1.

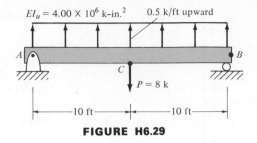

FIGURE H6.29

6.30 Apply formulas from Table 6.1 to determine by superposition the center deflection and end slopes of the beam depicted in Figure H6.30.

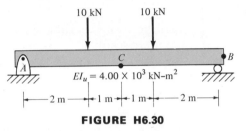

FIGURE H6.30

6.31 A cantilever beam is loaded as shown in Figure H6.31. Determine the slope and deflection at the free end B using the method of superposition. Appropriate formulas are given in Table 6.1.

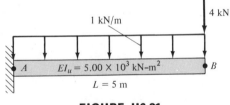

FIGURE H6.31

6.32 Use the method of superposition and Table 6.1 to find the slope and deflection at the free end B of the cantilever shown in Figure H6.32.

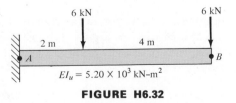

FIGURE H6.32

6.33 Find the slope and deflection at the free end B of the cantilever shown in Figure H6.33. Use Table 6.1 and the method of superposition.

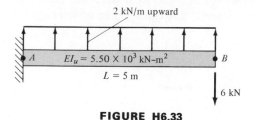

FIGURE H6.33

6.34 A cantilever beam is loaded as shown in Figure H6.34. Determine the slope and deflection at the free end using the method of superposition and appropriate formulas from Table 6.1.

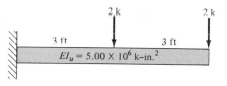

FIGURE H6.34

6.35 Determine the slope and deflection at the right end of the cantilever depicted in Figure H6.35. Use the method of superposition and appropriate formulas from Table 6.1.

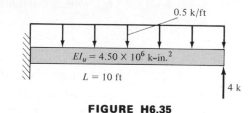

FIGURE H6.35

6.36 A simply supported beam is subjected to two concentrated forces as shown in Figure H6.36. Determine the center deflection of this beam using the method of superposition and appropriate formulas from Table 6.1.

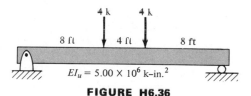

FIGURE H6.36

6.6 Construction of Moment Diagrams by Cantilever Parts

Writing equations for bending moments as a function of x and plotting the associated moment diagram was discussed in Section 1.8, but it is convenient to develop a new approach in this section which will prove quite useful in applying the area-moment method developed in Section 6.7. This method for construction of moment diagrams considers the original beam to be replaced by a series of cantilever beams that are loaded by the system of applied forces and reactions of the original beam. Each cantilever is loaded by a single force or couple. Location of the fixed end of these cantilevers is arbitrary since bending moments M_u in a beam are physical quantities that are independent of the choice of an origin for the dependent variable x

in terms of which the moments may be expressed. Individual moment diagrams are constructed for each cantilever, and if these diagrams are superimposed, the resulting diagram is the correct and complete moment diagram as constructed by the methods of Section 1.8. The method is illustrated in Examples 6.8 and 6.9. Reference will be made to Table 6.2, which contains four cases of moment diagrams for cantilever beams.

TABLE 6.2
Moment Diagrams for Selected Cantilever Beams

Case 1. Concentrated load at any point

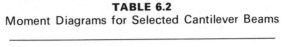

Case 2. Uniform loading over length a from fixed end

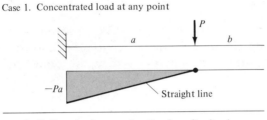

Case 3. Uniform loading over length b from free end

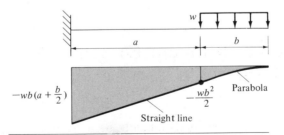

Case 4. Couple M applied at a distance a from fixed end

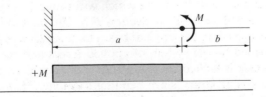

Example 6.8

Refer to Figure 6.8(a) and construct the moment diagram for this beam by three different approaches as follows: (a) Draw free-body diagrams, write the moment equations as a function of x, and plot the functions; (b) use the method of cantilever parts by choosing the fixed ends of the cantilevers at point B; and (c) use the method of cantilever parts by choosing the fixed ends at point C.

Refer to Table 6.2 for cantilever moment diagrams.

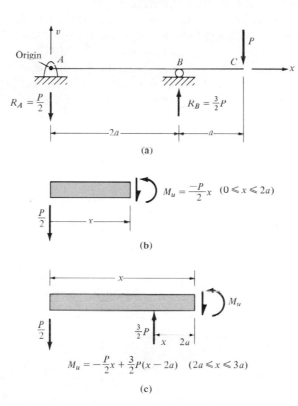

$$M_u = \frac{-P}{2}x \quad (0 \leqslant x \leqslant 2a)$$

(b)

$$M_u = -\frac{P}{2}x + \frac{3}{2}P(x - 2a) \quad (2a \leqslant x \leqslant 3a)$$

(c)

(d)

FIGURE 6.8

Solution

(a) The solution is depicted in Figure 6.8(d). Appropriate free-body diagrams are shown in Figure 6.8(b) and (c). It is left as an exercise for the student to check these equations and the moment diagram.

(b) The solution is depicted in Figure 6.9. Appropriate cantilevers and their moment diagrams are shown in Figure 6.9(b) and (c). Cantilever AB shown in Figure 6.9(b) is loaded by the reaction $R_A = P/2$. Cantilever BC, loaded by the force P applied downward at C, is shown in Figure 6.9(c). If cantilever moment diagrams of Figure 6.9(b) and (c) are superimposed (or added algebraically), the resulting diagram will be that of Figure 6.8(d).

(c) The solution is depicted in Figure 6.10. Appropriate cantilevers and their moment diagrams are shown in Figure 6.10(b) and (c). Cantilever AC, shown in Figure 6.10(b), is loaded by the reaction $R_A = P/2$. Cantilever AC, shown in Figure 6.10(c), is loaded by reaction $R_B = \frac{3}{2}P$ at B and is free of loading between A and B. If cantilever moment diagrams of Figure 6.10(b) and (c) are superimposed (or added algebraically), the resulting diagram will be that of Figure 6.8(d).

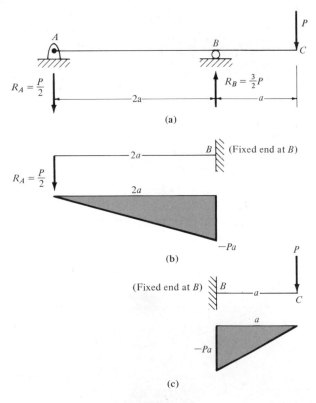

FIGURE 6.9

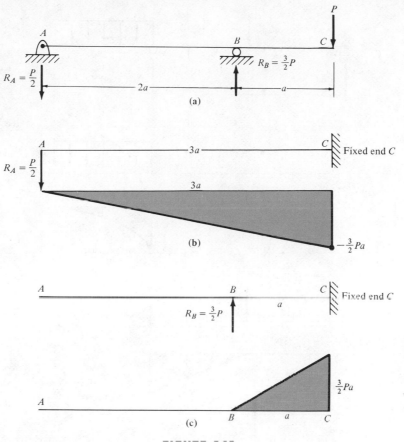

FIGURE 6.10

Any one of the three methods used above for drawing the moment diagram has yielded correct results, but there are distinct advantages in using the method of cantilever parts when the area-moment method of Section 6.7 is utilized to determine beam deflections.

Example 6.9

Construct moment diagrams by cantilever parts for the beam depicted in Figure 6.11(a). Choose the fixed ends of the cantilevers at point C. Refer to Table 6.2 for cantilever moment diagrams.

Solution. The three cantilevers and their corresponding moment diagrams are shown in Figure 6.11(b), (c), and (d). Cantilever CD, shown in Figure 6.11(b), is loaded with a uniform load w and the corresponding moment function is

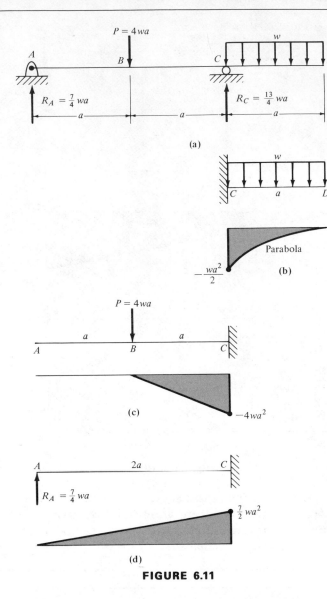

FIGURE 6.11

parabolic. Cantilever AC, shown in Figure 6.11(c), is loaded by a downward load P at point B and is free of loading between A and B. The corresponding moment function is linear. Cantilever AC, shown in Figure 6.11(d), is loaded by the upward reaction $R_A = \frac{7}{4}wa$ at point A and the corresponding moment function is linear. If the cantilever moment diagrams of Figure 6.11(b), (c), and (d) are superimposed (or added algebraically), the resulting diagram would coincide with the one obtained by the method of Section 1.8 for constructing moment diagrams.

6.7 Beam Deflections—The Area-Moment Method

Two theorems form the basis for the *area-moment method*, and these depend upon the moment curvature relationship expressed by Eq. 6.6:

$$EI_u \frac{d^2v}{dx^2} = M_u$$

Theorem I is developed by integrating this equation between limits x_A and x_B, as shown in Figure 6.12.

$$\int_{x_A}^{x_B} \frac{d^2v}{dx^2} \, dx = \int_{x_A}^{x_B} \frac{M_u}{EI_u} \, dx \tag{6.19}$$

$$\left.\frac{dv}{dx}\right|_{x_B} - \left.\frac{dv}{dx}\right|_{x_A} = \int_{x_A}^{x_B} \frac{M_u}{EI_u} \, dx \tag{6.20}$$

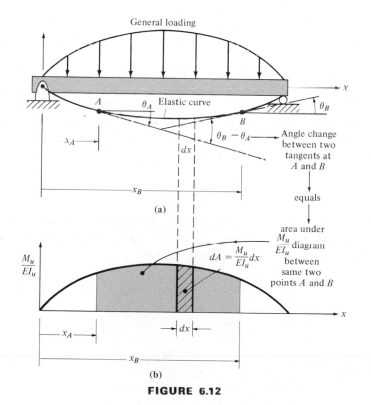

FIGURE 6.12

By Eq. 6.8, the left side of Eq. 6.20 may be written as $\theta_B - \theta_A$, to yield

$$\theta_B - \theta_A = \int_{x_A}^{x_B} \frac{M_u}{EI_u} \, dx \qquad (6.21)$$

Equation 6.21 is the mathematical form of the first area-moment theorem. In order to interpret the integral on the right side of this equation, refer to Figure 6.12(b), which contains a plot of the moment diagram, whose ordinates have been divided by the bending stiffness EI_u at each point along the beam span. A differential area under this curve equals $(M_u/EI_u) \, dx$. Thus the integral on the right side of Eq. 6.21 equals the area under the M_u/EI_u diagram between the limits x_A and x_B.

THEOREM I. *The angle change between two tangents to the elastic curve [such as A and B of Figure 6.12(a)] equals the area under the M_u/EI_u diagram between the same two points A and B.*

Figure 6.13 depicts the geometric and other information involved in the derivation of the second theorem. In Figure 6.13(a), a tangent is drawn at any point A a distance x_A from the origin at the left support, and the second theorem will provide a means of calculating the tangential deviation t_{BA} at any point B a distance x_B from the origin. In general, tangential deviations are measured normal to the original beam axis from certain points on the elastic curve to tangents drawn to the elastic curve at other points.

Points C and D are located a differential distance $dx = dx_1$ apart on the elastic curve of Figure 6.13(a). This differential segment is located a distance x_1 to the left of point B, which has been chosen as the origin for the variable x_1. Consider the differential angle $d\theta$ between the tangents to the elastic curve at points C and D and the differential tangential deviation dt, which is a differential part of the total deviation t_{BA} and relate $d\theta$ and dt as follows:

$$dt = x_1 \, d\theta$$

This differential equation is valid because beam displacements and slope angles are small even though they are shown relatively large in the figure for convenience. Equation 6.8, $dv/dx = \theta$, substituted into Eq. 6.6,

$$\frac{d^2v}{dx^2} = \frac{d}{dx}\left(\frac{dv}{dx}\right) = \frac{M_u}{EI_u}$$

yields

$$\frac{d\theta}{dx} = \frac{M_u}{EI_u} \qquad (6.22)$$

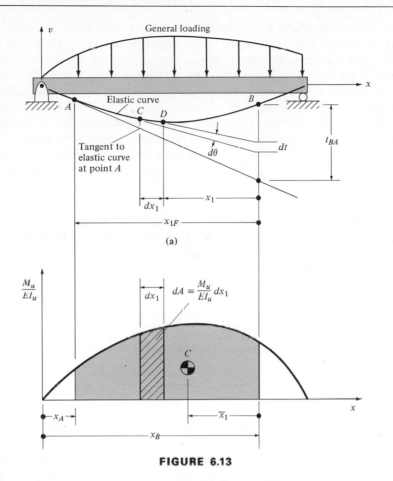

FIGURE 6.13

which may be written $d\theta = (M_u/EI_u)\,dx$. Substitute this expression for $d\theta$ into $dt = x_1\,d\theta$ with $dx = dx_1$ to obtain

$$dt = \left(\frac{M_u}{EI_u}\right)x_1\,dx_1 \tag{6.23}$$

Integrate Eq. 6.23 with limits on t from zero to t_{BA} and limits on x_1 from zero to x_{1F}.

$$\int_0^{t_{BA}} dt = t_{BA} = \int_0^{x_{1F}} \frac{M_u}{EI_u} x_1\,dx_1 \tag{6.24}$$

Equation 6.24 is the mathematical form of the second theorem. Subscripts of the tangential deviation t refer to the point where the deviation is to be determined (i.e.,

B) and to the point where the tangent is drawn (i.e., *A*). Thus t_{BA} refers to the tangential deviation at *B* with respect to a tangent drawn at *A*. Note that the sequence in which the subscripts are written is very important. In order to interpret the integral on the right side of Eq. 6.24, refer to Figure 6.13(b), which contains a plot of a moment diagram whose ordinates have been divided by the bending stiffness EI_u at each point along the beam span. A differential area under this curve equals $(M_u/EI_u)\,dx_1$, and when this differential area is multiplied by x_1 to produce $(M_u/EI_u)x_1\,dx_1$, the result is the differential first area moment with respect to point *B*, the origin for x_1. Upon integration from zero to x_{1F}, the right side of Eq. 6.24 represents the first area moment of the M/EI_u diagram between *A* and *B* with respect to *B*.

THEOREM II. *The tangential deviation t_{BA} at any point B with respect to a tangent drawn to the elastic curve at any point A equals the first area moment of the M/EI_u diagram between A and B with respect to point B.*

Equation 6.24 may be written in another form by noting that the integral on the right side of the equation equals $A_1\bar{x}_1$, where A_1 equals the area under the M/EI_u diagram between *A* and *B* and $\bar{x}_1$ locates the centroid of this area A_1 with respect to point *B*, where the tangential deviation t_{BA} is to be determined. A generalization to *n* component areas under the M/EI_u diagram is readily written as follows:

$$t_{BA} = \sum_{i=1}^{n} A_i\bar{x}_{1i} \qquad (i = 1, 2, \ldots, n) \qquad \textbf{(6.25)}$$

Area and centroid information for a selected number of simple geometric shapes is provided in Appendix D.

It is important to realize that, in general, angle changes and tangential deviations obtained from the two theorems are not usually of direct interest but are an indirect means of obtaining beam deflections. This will be illustrated in Examples 6.10 and 6.11. The difference between beam deflections and tangential deviations is shown qualitatively in Figure 6.14. Positive deviations are those for which the elastic curve lies above the tangent, and negative deviations are those for which the tangent lies above the elastic curve.

Slopes θ are positive when the tangent to the elastic curve has rotated counterclockwise from the undeformed to the deformed position. This is consistent with the mathematical definition of positive slope for a coordinate system with an origin located at the left of the beam with *x* positive to the right and *v* positive upward. The first area-moment theorem refers to changes of slope, and these will be considered positive counterclockwise with increasing values of *x*, which is consistent with the positive sign convention for M/EI_u diagram ordinates.

Selected properties of a few plane areas are shown in Appendix D. These properties will facilitate the use of the area-moment method in the solution of beam deflection problems.

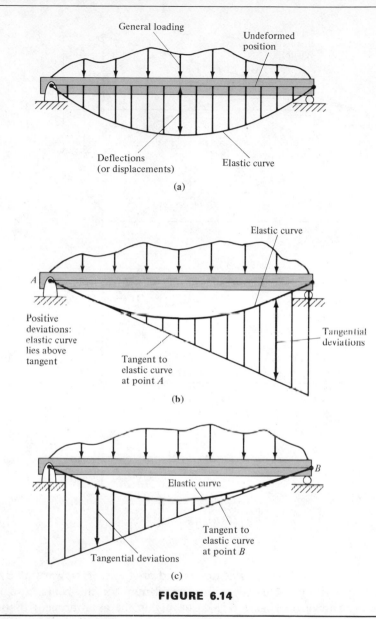

FIGURE 6.14

Example 6.10

Use the second area-moment theorem to determine the slope at the left end θ_A and the deflection at the right end v_C for the beam depicted in Figure 6.15(a). The beam is loaded by a concentrated load P at its right end and is supported by a pin at A and a roller at B. The bending stiffness EI_u is a constant. Express the answers in terms of P, L, E, and I_u.

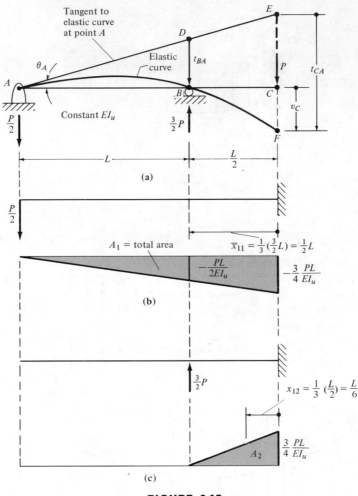

FIGURE 6.15

Solution. Correct reactions of $P/2$ downward at A and $\frac{3}{2}P$ upward at B were computed by writing the equations of equilibrium for the beam, and these reactions are shown on Figure 6.15(a). Applying the area-moment theorems depends upon a precise understanding of the geometry of a given problem. Students should refer to Figure 6.15(a) as they study the following statements as a prelude to solving the problem.

1. The beam is straight prior to application of the load P. Straight line ABC represents this initial position of the beam.
2. The load P is gradually applied to the right end of the beam, and it bends

until an edge view of the neutral surface takes on the shape of the curved line *ABF* referred to as the elastic curve.

3. Points *A* and *B* do not deflect because the rigid pin and roller at these reactions, respectively, are assumed to rest on nondeformable surfaces. In other words, settlement of the supports is assumed to be prevented.

4. A tangent drawn to the elastic curve at point *A* is the straight line *ADE*. Tangents may be drawn to the elastic curve at any point, but only two points *A* and *B*, which do not deflect, are practical choices for solving this problem.

5. Two tangential deviations t_{BA} and t_{CA} are shown in the figure. Tangential deviations are measured perpendicular to the undeformed beam axis from points on the tangent drawn to the elastic curve. (Point *B* also lies on the initial position given by the straight line *ABC*, since the support at *B* does not settle.) The tangential deviation t_{CA} extends perpendicular to the undeformed beam axis from point *E* on the tangent to point *F* on the elastic curve.

6. The slope at the left end is θ_A, which is the angle between the straight line *ABC* and the tangent *ADE*. The deflection at *C* is v_C, which is measured perpendicular to the undeformed beam axis from *C* to *F*. Deflections and slopes shown in Figure 6.15(a) are greatly exaggerated compared to the beam length. Actual slopes and deflections are small when the stresses are elastic, and the theory of this chapter is a small-deflection theory.

Refer to triangle *ADB* and recall that for small angles $\tan \theta = \theta$:

$$\theta_A = \frac{|t_{BA}|}{L}$$

Triangles *ADB* and *AEC* are similar triangles; thus

$$\frac{EC}{CA} = \frac{DB}{BA}$$

where

$$EC = EF - CF = t_{CA} - v_C$$

$$CA = \tfrac{3}{2}L$$

$$DB = t_{BA}$$

$$BA = L$$

Substituting yields

$$\frac{t_{CA} - v_C}{\tfrac{3}{2}L} = \frac{t_{BA}}{L}$$

Solve for v_C:

$$v_C = t_{CA} - \tfrac{3}{2} t_{BA}$$

The slope θ_A and deflection v_C have been expressed in terms of tangential deviations by referring to the geometry of this particular problem depicted in Figure 6.15(a). Success in applying the two area-moment theorems is closely related to making a correct sketch similar to Figure 6.15(a) and expressing the desired unknowns, such as θ_A and v_C, symbolically in terms of tangential deviations, such as t_{BA} and t_{CA}.

To apply the second area-moment theorem, construct the moment diagrams by cantilever parts as shown in Figure 6.15(b) and (c). In this case the fixed ends of the cantilever were chosen at the right end C. Quite often it will be advisable to choose the fixed ends at the end opposite to the end where the tangent is drawn. For example, in this solution the tangent was drawn at the left end, A, and the fixed ends of the cantilevers chosen at the right end, C. If a tangent is drawn at a point other than the extremities of the beam, it may be advisable to choose the fixed ends for the cantilevers at the point where the tangent is drawn.

To find the tangential deviation t_{BA} by the second theorem, take the moment of the M/EI_u diagram between A and B with respect to B. In this particular problem only the cantilever moment diagram shown in Figure 6.15(b) is involved, since the moment in the cantilever of Figure 6.15(c) vanishes between A and B; that is, only that part of the M_u/EI_u diagram between points A and B should be considered.

$$t_{BA} = \underbrace{\frac{1}{2}\left(\frac{-PL}{2EI_u}\right)(L)}_{\text{area}} \underbrace{\left(\frac{1}{3}L\right)}_{\text{arm}} = \frac{-PL^3}{12EI_u}$$

The negative sign means that point B on the elastic curve lies below the tangent. But

$$\theta_A = \frac{|t_{BA}|}{L}$$

$$= \frac{|(-PL^3/12EI_u)|}{L} = \boxed{\frac{PL^2}{12EI_u}}$$

The rotation is counterclockwise from the undeformed to the deformed position of the beam.

To find the tangential deviation t_{CA} by the second theorem, take the moment of the M_u/EI_u diagram between A and C with respect to C. The complete triangle moment diagrams shown in Figure 6.15(b) and (c) are involved and their moments are to be taken with respect to point C. From Eq. 6.25,

$$t_{CA} = A_1 \bar{x}_{11} + A_2 \bar{x}_{12}$$

$$= \underbrace{\frac{1}{2}\left(\frac{-3PL}{4EI_u}\right)\left(\frac{3}{2}L\right)}_{\text{area}} \underbrace{\left(\frac{1}{3} \cdot \frac{3}{2}L\right)}_{\text{arm}} + \underbrace{\frac{1}{2}\left(\frac{3PL}{4EI_u}\right)\frac{L}{2}}_{\text{area}} \underbrace{\left(\frac{1}{3} \cdot \frac{L}{2}\right)}_{\text{arm}}$$

$$= \frac{PL^3}{EI_u}\left(\frac{-9}{32} + \frac{1}{32}\right) = -\frac{PL^3}{4EI_u}$$

But

$$v_C = t_{CA} - \tfrac{3}{2}t_{BA}$$

$$= \frac{-PL^3}{4EI_u} - \frac{3}{2}\left(\frac{-PL^3}{12EI_u}\right)$$

$$= \boxed{\frac{-PL^3}{8EI_u}} \quad \text{(a downward deflection at point } C\text{)}$$

Example 6.11

Refer to Figure 6.16(a) and use the area-moment method to determine the rotation of the tangent at A, the deflection at the center B, and the maximum deflection of the beam together with its location. Express results in terms of w, a, E, and I_u.

Solution. Refer to Figure 6.16(b) and note the tangent drawn to the elastic curve at point A. Since the angle θ_A is small,

$$\theta_A = \frac{t_{CA}}{2a}$$

From similar triangles ABE and ACF,

$$\frac{BE}{AB} = \frac{CF}{AC}$$

or

$$\frac{v_B + t_{BA}}{a} = \frac{t_{CA}}{2a}$$

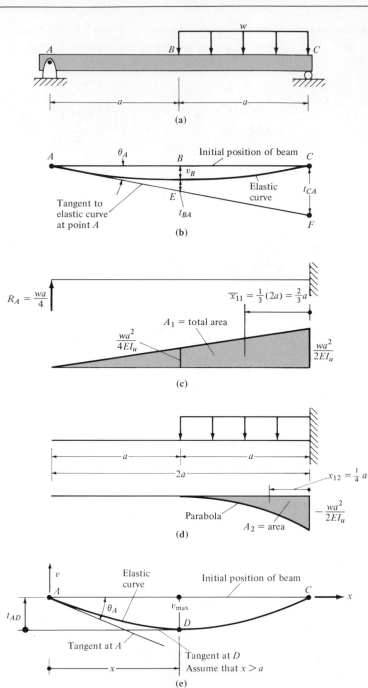

FIGURE 6.16

Solving for v_B, we obtain

$$v_B = \tfrac{1}{2}t_{CA} - t_{BA}$$

Construct moment diagrams by cantilever parts and divide ordinates by EI_u as shown in Figure 6.16(c) and (d). Apply the second area-moment theorem to determine t_{CA} and t_{BA}.

The tangential deviation t_{BA} equals the moment of the area under the M_u/EI_u diagram between A and B with respect to B. The diagram of Figure 6.16(d) is not involved, since $M_u = 0$ for this cantilever between A and B. Referring to Figure 6.16(c) and noting that the moment at B in this cantilever is $wa^2/4$, we find that the tangential deviation becomes

$$t_{BA} = \underbrace{\frac{1}{2}\frac{wa^2}{4EI_u}}_{\text{area}} \underbrace{a\,\frac{1}{3}a}_{\text{arm}} = \frac{wa^4}{24EI_u}$$

The tangential deviation t_{CA} equals the moment of the area under the M_u/EI_u diagram between A and C with respect to C. The complete diagrams shown in Figure 6.16(c) and (d) are both involved in computing t_{CA}. From Eq. 6.25,

$$t_{CA} = A_1\bar{x}_{11} + A_2\bar{x}_{12}$$

$$= \underbrace{\frac{1}{2}\frac{wa^2}{2EI_u}(2a)}_{\text{area}}\underbrace{\frac{1}{3}\cdot 2a}_{\text{arm}} + \underbrace{\frac{1}{3}\left(\frac{-wa^2}{2EI_u}\right)a}_{\text{area}}\underbrace{\left(\frac{1}{4}a\right)}_{\text{arm}}$$

$$= \frac{wa^4}{EI_u}\left(\frac{1}{3} - \frac{1}{24}\right)$$

$$= \frac{7}{24}\frac{wa^4}{EI_u}$$

But

$$\theta_A = \frac{t_{CA}}{2a} = \frac{1}{2a}\frac{7wa^4}{24EI_u}$$

$$= \boxed{\frac{7wa^3}{48EI_u}}$$

The rotation at A is clockwise, as shown in Figure 6.16(b), and

$$v_B = \frac{1}{2}t_{CA} - t_{BA}$$

$$= \frac{1}{2}\left(\frac{7}{24}\frac{wa^4}{EI_u}\right) - \frac{wa^4}{24EI_u}$$

$$= \boxed{\frac{5wa^4}{48EI_u}} \quad \text{downward.}$$

In order to locate and determine the maximum deflection, refer to Figure 6.16(e). At point D, where the deflection is a maximum, the slope of the tangent is zero or the tangent is horizontal. Apply the first area-moment theorem to locate the distance x from A to D. If we assume that $x > a$ and recall that $(\theta_D - \theta_A)$ equals the area under the M_u/EI_u diagram between A and D, the appropriate equation for x becomes

$$\theta_D - \theta_A = \frac{1}{2}\left(\frac{x}{2a}\frac{wa^2}{2EI_u}\right)x - \frac{1}{3}\frac{w(x-a)^2}{2EI_u}(x-a)$$

But $\theta_D = 0$ and $\theta_A = -7wa^3/48EI_u$. The negative sign is required for θ_A since the slope of the elastic curve for v positive upward and x positive to the right is negative.

$$0 - \left(-\frac{7}{48}\frac{wa^3}{EI_u}\right) = \frac{1}{8}\frac{wa}{EI_u}x^2 - \frac{1}{6}\frac{w(x-a)^3}{EI_u}$$

Substitution of $x = \alpha a$, where $(1 \leq \alpha \leq 2)$ and algebraic simplification leads to the following cubic equation in α:

$$8\alpha^3 - 30\alpha^2 + 24\alpha - 1 = 0$$

Newton's method, after several iterations, yields a value of $\alpha = 1.080444$, which will be rounded off to 1.08 to give a value of $x = 1.08a$.

In order to find the maximum deflection, refer to Figure 6.16(e) and note that $v_{max} = t_{AD}$. The tangential deviation t_{AD} equals the moment of the area under the M_u/EI_u diagram between A and D with respect to A.

$$v_{max} = t_{AD} = \underbrace{\frac{1}{2}\frac{wax}{4EI_u}x}_{\text{area}}\underbrace{\left(\frac{2}{3}x\right)}_{\text{arm}} - \underbrace{\frac{1}{3}\frac{w}{EI_u}\frac{(x-a)^2}{2}(x-a)}_{\text{area}}\underbrace{\left[a + \frac{3}{4}(x-a)\right]}_{\text{arm}}$$

where $x = 1.08a$.

$$v_{max} = \frac{1}{12}\frac{wa^4}{EI_u}(1.08)^3 - \frac{wa^3}{6EI_u}(0.08)^3(1.06a)$$

$$= \boxed{0.105\frac{wa^5}{EI_u}}$$

For this particular problem, v_{max} is only slightly larger than the deflection at the center, which equals $0.104(wa^4/EI_u)$.

Example 6.12

Use the second area-moment theorem to determine the deflection at point C of the cantilever beam depicted in Figure 6.17. This beam has a varying moment of inertia as shown in the figure. Express the deflection at C in terms of P, a, E, and I_u.

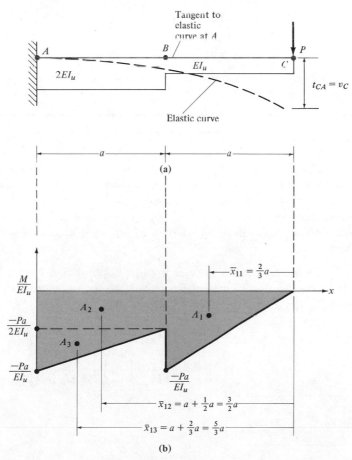

FIGURE 6.17

Solution. Draw the tangent to the elastic curve at point A, the fixed end of the cantilever beam as shown in Figure 6.17(a). This tangent to the elastic curve at the fixed end of the cantilever is also the undeformed axis of the beam and thus, in this special case, the deflections will equal the tangential deviations. For point C this may be expressed as follows:

$$v_C = t_{CA}$$

Construct the M_u/EI_u diagrams as shown in Figure 6.17(a). The tangential deviation t_{CA} equals the moment of the entire M_u/EI_u diagram with respect to the right end C. From Eq. 6.25

$$t_{CA} = A_1 \bar{x}_{11} + A_2 \bar{x}_{12} + A_3 \bar{x}_{13}$$

$$= -\frac{1}{2}\frac{Paa}{EI_u}\left(\frac{2}{3}a\right) - \frac{Pa}{2EI_u}(a)\left(\frac{3}{2}a\right) - \frac{1}{2}\frac{Pa}{2EI_u}a\left(\frac{5}{3}a\right)$$

$$= \boxed{v_C = -\frac{3}{2}\frac{Pa^3}{EI_u}}$$

The negative sign means that the tangent lies above the elastic curve, which corresponds to a downward deflection at point C.

Example 6.13
Use the second area-moment theorem to determine the center deflection of the symmetric beam, which is symmetrically loaded as shown in Figure 6.18. This beam has a varying moment of inertia as shown in the figure. Express the deflection in terms of P, a, E, and I_1. Compare this deflection to the deflection of a beam with constant bending stiffness EI_1 of total length $4a$ subjected to the same loading.

Solution. Draw the tangent to the elastic curve at point B in Figure 6.18(a). A tangent drawn to the elastic curve at B is parallel to the undeformed beam axis. In this case the tangential deviation t_{CB} equals the deflection v_B. By Eq. 6.25

$$t_{CB} = A_1 \bar{x}_{11} + A_2 \bar{x}_{12} + A_3 \bar{x}_{13}$$

$$= \frac{1}{2}\frac{Pa}{2EI_u}a\left(\frac{2}{3}a\right) + \frac{1}{2}\frac{Pa}{8EI_u}a\left(\frac{5}{3}a\right) + \frac{Pa}{8EI_u}(a)\left(\frac{3}{2}a\right)$$

$$\boxed{v_B = t_{CB} = \frac{11}{24}\frac{Pa^3}{EI_1}}$$

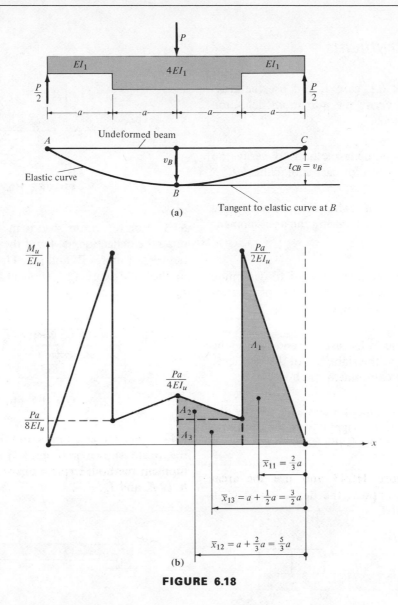

FIGURE 6.18

The center deflection of a beam with constant bending stiffness EI_1 of total length $4a$ subjected to a center loading P equals $\frac{4}{3}(Pa^3/EI_1)$. The ratio of these deflections is $\frac{32}{11}$, which means that the beam of constant bending stiffness EI_1 deflects almost three times as much at the center as the beam shown in Figure 6.18(a).

Homework Problems

6.37 Refer to Table 6.1, case 1, and use the area-moment method to verify the maximum deflection and end slope equations.

6.38 Use the area-moment method to verify the center deflection for $a > b$ for case 2 of Table 6.1.

6.39 Refer to case 3 of Table 6.1 and verify the center deflection equation by using the area-moment method.

6.40 Apply the area-moment method to determine the deflection and slope at the right end of the cantilever shown in case 4 of Table 6.1.

6.41 Refer to Table 6.1, case 5, and verify the deflection and slope at the right end of the cantilever beam by using the area-moment method.

6.42 Use the area-moment method to verify the slope and deflection at the right end of the cantilever beam depicted in case 6 of Table 6.1.

6.43 Refer to Figure H6.43 and use the area-moment method to express the deflection at C in terms of w, L, E, and I_u.

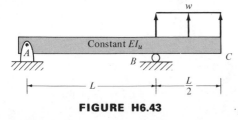

FIGURE H6.43

6.44 Determine the deflections at points B and D of the beam depicted in Figure H6.44 in terms of P, L, E, and I_u using the area-moment method.

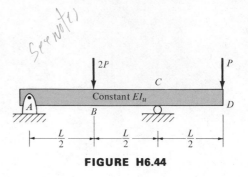

FIGURE H6.44

6.45 For the beam shown in Figure H6.45, determine the deflection at A and the maximum deflection between points B and C. Use the area-moment method and express answers in terms of M_A, L, E, and I_u.

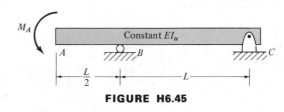

FIGURE H6.45

6.46 Find the end slopes and the deflection at B for the beam shown in Figure H6.46, using the area-moment method. Express answers in terms of M_B, a, b, L, E, and I_u.

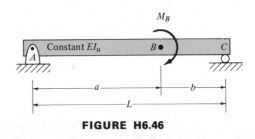

FIGURE H6.46

6.47 Apply the area-moment method to determine the deflection at A and the maximum deflection between B and C for the beam depicted in Figure H6.47. Express results in terms of w, a, E, and I_u.

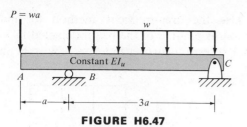

FIGURE H6.47

6.48 Apply the area-moment method to determine the following deflections and slopes of the beam depicted in Figure H6.48:
(a) Deflection at A.
(b) Slope at B.
(c) Maximum deflection between B and C.
(d) Deflection $1.5a$ to the right of B.
(e) Slope at C.
(f) Deflection at D.

Express answers in terms of P, a, E, and I_u.

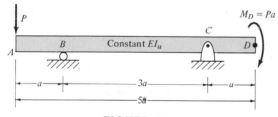

FIGURE H6.48

6.49 Apply the area-moment method to determine the following deflections and slopes of the beam depicted in Figure H6.49:
(a) Slope at A.
(b) Deflection at B.
(c) Deflection at C.
(d) Slope at D.

Express answers in terms of P, b, E, and I_u.

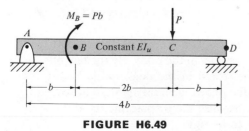

FIGURE H6.49

6.50 Apply the area-moment method to determine the following deflections and slopes of the beam depicted in Figure H6.50:
(a) Slope at A.
(b) Deflection at B.
(c) Maximum deflection of the beam.
(d) Deflection at C.
(e) Slope at D.

Express answers in terms of w, a, E, and I_u.

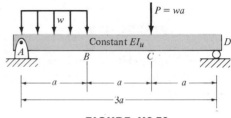

FIGURE H6.50

6.51 Refer to Figure H6.51 and use the area-moment method to determine the deflection at point C as a function of P, a, E, and I_1.

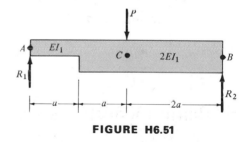

FIGURE H6.51

6.52 Refer to Figure H6.52 and use the area-moment method to determine the deflection at point C as a function of P, a, E, and I_1.

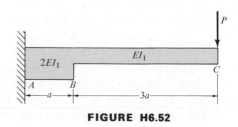

FIGURE H6.52

6.53 Use the area-moment method to find the deflection at point A of the beam depicted in Figure H6.53. Express the answer in terms of P, a, E, and I_1.

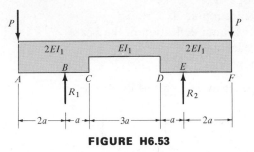

FIGURE H6.53

6.54 Refer to Figure H6.54 and use the area-moment method to find the deflection of point C in terms of w, a, E, and I_1.

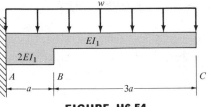

FIGURE H6.54

6.55 Use the area-moment method to find the deflection at point B of the beam depicted in Figure H6.55. Express the answer in terms of w, a, E, and I_1.

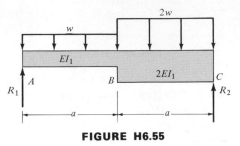

FIGURE H6.55

6.56 Refer to Figure H6.56 and use the area-moment method to find the deflection at point D in terms of P, a, E, and I_1.

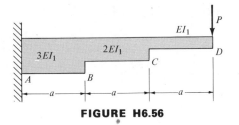

FIGURE H6.56

★**6.8** Beam Deflections—Castigliano's Second Theorem

Alberto Castigliano (1847–1884), when he was 26 years old, presented his thesis for the engineer's degree to the Turin Polytechnical Institute in Italy, and it contained a statement of his famous theorems. His second theorem will be derived for any body fabricated of a linearly elastic material with unyielding supports and then applied to obtain beam deflections.

The system of forces P_i in Figure 6.19(a) (where $i = 1, 2, 3, \ldots, n$) do external work on the body which is stored as internal strain energy in the body. By the law of conservation of energy the external work must be stored in the deformable body since losses (e.g., temperature change in the body due to heat generated) are ignored. No work is done at the supports by the reactions because the displacements at the reactions are zero since the supports are assumed to be rigid.

The effect on the strain energy U of changing any one of the forces P_i by an

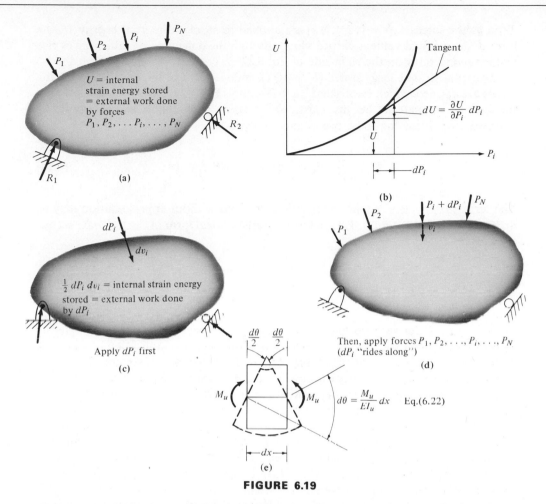

FIGURE 6.19

amount dP_i is depicted in Figure 6.19(b). This change dU in the internal energy stored when added to U yields

$$U + dU = U + \frac{\partial U}{\partial P_i} dP_i \tag{6.26}$$

Now reverse the order in which the forces are applied. First apply dP_i as shown in Figure 6.19(c) and then apply the forces $P_i(i = 1, 2, \ldots, n)$ as depicted in Figure 6.19(d). This operation leads to

$$U + dU = \tfrac{1}{2} dP_i \, dv_i + dP_i v_i + U \tag{6.27}$$

The factor of one-half in the first term on the right side of Eq. 6.27 is required, since dP_i builds up from a zero value to a final value of dP_i with an average value of $\tfrac{1}{2} dP_i$.

Then as the forces $P_i (i = 1, 2, \ldots, n)$ are applied to store an internal energy U, the force dP_i at its final value is carried along through the deflection v_i, which gives rise to the second term on the right side of Eq. 6.27.

The total strain energy stored $(U + dU)$ is independent of the order in which the forces are applied, which means that Eq. 6.26 equals Eq. 6.27. Order of application of the forces is immaterial because elastic deformation is reversible and energy losses are neglected. Equating these two equations leads to

$$U + \frac{\partial U}{\partial P_i} dP_i = \frac{1}{2} dP_i \, dv_i + dP_i v_i + U \tag{6.28}$$

The term $\frac{1}{2} dP_i \, dv_i$ is a second-order differential and without approximation may be dropped from the equation. This yields the mathematical form of *Castigliano's second theorem*:

$$\frac{\partial U}{\partial P_i} = v_i \tag{6.29}$$

The first partial derivative of the strain energy with respect to any one of the external forces of the system is equal to the deflection of the point of application of that force P_i, and in the direction of that force. The body to which the system of forces is applied is fabricated of linearly elastic material, the supports are unyielding, and the principal of superposition has been employed in the derivation. Equation 6.29 may also be interpreted to apply to moments and slopes as well as to forces and deflections. If P_i is replaced by M_i and v_i is replaced by θ_i, then the equation may be used to obtain rotations. This use of the equation will be covered in the examples.

To apply Castigliano's second theorem to the problem of beam deflections, an equation for the strain energy stored in a bent beam must be developed. The effects of shear will be ignored in comparison to bending in development of this equation. Refer to Figure 6.19(e), which shows a differential length dx of beam subjected to bending moments, and recall that the work done by a couple equals the average value of the couple multiplied by the angle of rotation of the couple. The couples M_u build up from an initial value of zero to a final value M_u as the loads are increased gradually from zero to their final static values, which means that the average couple is $M_u/2$ acting on each face of the differential beam segment and each face rotates through an angle $d\theta/2$, as shown. Thus

$$\underbrace{dU}_{\substack{\text{differential} \\ \text{internal} \\ \text{strain} \\ \text{energy}}} = \underbrace{2}_{\substack{\text{two couples,} \\ \text{one on each} \\ \text{face of the} \\ \text{element}}} \times \underbrace{\frac{M_u}{2}}_{\substack{\text{average} \\ \text{couple}}} \times \underbrace{\frac{d\theta}{2}}_{\substack{\text{angle of} \\ \text{rotation} \\ \text{of each} \\ \text{couple}}} = \frac{1}{2} M_u \, d\theta \tag{6.30}$$

From Eq. 6.22, $d\theta = M_u/EI_u \, dx$, and when substituted into Eq. 6.30, the differential strain energy stored in the beam element of differential length dx becomes

$$dU = \frac{M_u^2 \, dx}{2EI_u} \tag{6.31}$$

Integrate both sides of Eq. 6.31 to obtain the equation for the bending strain energy stored in a beam. The integration on the right-hand side of Eq. 6.32 extends over the complete length of the beam, as indicated by the symbol L.

$$U = \int_L \frac{M_u^2 \, dx}{2EI_u} \tag{6.32}$$

Three examples will clarify the application of Eqs. 6.29 and 6.32 to the problem of determining beam slopes and deflections.

Example 6.14

Use Castigliano's second theorem to find the deflection of the cantilever beam at its free end A as shown in Figure 6.20(a). Then determine the rotation of the beam at point A.

Solution. Refer to the free-body diagram shown in Figure 6.20(b) and the moment equation $M_u = -Px$, which is valid over the entire beam length. Express the total internal strain energy stored in the beam as follows:

$$U = \int \frac{M_u^2 \, dx}{2EI_u}$$

$$= \frac{1}{2EI_u} \int_0^L (-Px)^2 \, dx = \frac{P^2 L^3}{6EI_u}$$

By Eq. 6.29,

$$\boxed{\frac{\partial U}{\partial P} = v_A = \frac{PL^3}{3EI_u}}$$

The positive result means that the senses of the force and deflection are in agreement. In this case the deflection is directed downward.

To obtain the end rotation of the cantilever beam, a fictional couple M_1 is applied at the left end of the beam as shown in Figure 6.20(c). Once the internal strain energy U is differentiated partially with respect to M_1, then M_1 is as-

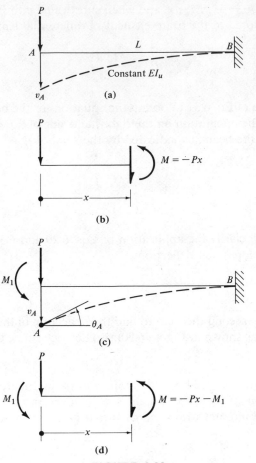

FIGURE 6.20

signed the value of zero, which results in the correct value for the rotation. From Figure 6.20(d), $M_u = -Px - M_1$ will be utilized to determine U in Eq. 6.32:

$$U = \frac{1}{2EI_u} \int_0^L (-Px - M_1)^2 \, dx$$

$$= \frac{1}{2EI_u} \int_0^L (P^2 x^2 + 2PM_1 x + M_1^2) \, dx$$

$$= \frac{1}{2EI_u} \left(\frac{P^2 L^3}{3} + PM_1 L^2 + M_1^2 L \right)$$

$$\frac{\partial U}{\partial M_1} = \frac{1}{2EI_u} (PL^2 + 2M_1 L)$$

Equate M_1 to zero to obtain the correct equation for θ_A:

$$\boxed{\frac{\partial U}{\partial M_1} = \theta_A = \frac{PL^2}{2EI_u}}$$

This rotation has a positive sign, which indicates agreement between θ_A and M_1. Since M_1 is ccw, θ_A is ccw, as indicated in Figure 6.20(c). The foregoing solution could have been shortened by interchanging the order of integration and differentiation as shown below, since the loadings P and M_1 are both independent of x, which is the variable of integration.

$$U = \frac{1}{2EI_u} \int_0^L (-Px - M_1)^2 \, dx$$

$$\frac{\partial U}{\partial M_1} = \frac{1}{EI_u} \int_0^L (-Px - M_1)(-1) \, dx$$

$$= \frac{1}{EI_u} \int_0^L (Px + M_1) \, dx$$

Since the differentiation has been performed, it is permissible to let $M_1 = 0$ prior to integration.

$$\theta_A = \frac{\partial U}{\partial M_1} = \frac{1}{EI_u} \int_0^L Px \, dx = \frac{PL^2}{2EI_u}$$

θ_A is ccw, in agreement with the ccw sense of M_1. The result is the same as that obtained previously.

Example 6.15
Use Castigliano's second theorem to find the rotation at the left end of the simply supported beam depicted in Figure 6.21(a) loaded by a linearly varying loading.

Solution. Refer to Figure 6.21(b) for the appropriate free-body diagram and the bending moment function, where M_1 is a fictitious couple that will be set equal to zero after differentiation partially with respect to M_1.

$$M_u = R_1 x + M_1 - \frac{w_1}{6L} x^3$$

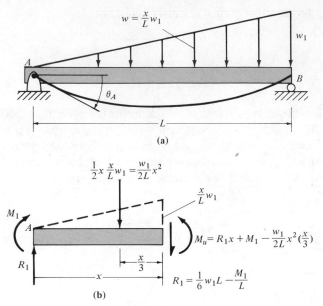

FIGURE 6.21

where

$$R_1 = \frac{w_1 L}{6} - \frac{M_1}{L}$$

$$M_u = \left(\frac{w_1 L}{6} - \frac{M_1}{L}\right)x + M_1 - \frac{w_1}{6L}x^3$$

$$U = \frac{1}{2EI_u}\int_0^L \left[\left(\frac{w_1 L}{6} - \frac{M_1}{L}\right)x + M_1 - \frac{w_1}{6L}x^3\right]^2 dx$$

$$\frac{\partial U}{\partial M_1} = \frac{1}{EI_u}\int_0^L \left[\left(\frac{w_1 L}{6} - \frac{M_1}{L}\right)x + M_1 - \frac{w_1}{6L}x^3\right]\left(-\frac{x}{L} + 1\right) dx$$

Let $M_1 = 0$ in the foregoing equation.

$$\frac{\partial U}{\partial M_1} = \frac{1}{EI_u}\int_0^L \left(\frac{w_1 L}{6}x - \frac{w_1}{6L}x^3\right)\left(1 - \frac{x}{L}\right) dx$$

$$= \frac{1}{EI_u}\left[\frac{w_1 L}{12}x^2 - \frac{w_1}{24L}x^4 - \frac{w_1}{18}x^3 + \frac{w_1}{30L^2}x^5\right]_0^L$$

$$= \theta_A = \boxed{\frac{7w_1 L^3}{360EI_u}}$$

θ_A and M_1 are both cw, as indicated in Figure 6.21(a) and (b).

Example 6.16

Use Castigliano's second theorem to find the deflection at the center of the beam depicted in Figure 6.22(a) in terms of P, a, E, and I_u.

Solution. Introduce the fictitious force Q at the center of the beam where the deflection is sought, since a real force is not applied at this location. Refer to Figure 6.22(b) and write the following moment functions by drawing the appropriate free-body diagrams. The reader is urged to draw the necessary diagrams and check the following equations. The origin for x is located at the left end of the beam.

$$M_u = \left(\tfrac{2}{3}P + \tfrac{1}{2}Q\right)x \qquad\qquad\qquad (0 \le x \le 2a)$$

$$M_u = \left(\tfrac{2}{3}P + \tfrac{1}{2}Q\right)x - P(x - 2a) = -\tfrac{1}{3}Px + 2Pa + \tfrac{1}{2}Qx \qquad (2a \le x \le 3a)$$

$$M_u = \left(\tfrac{2}{3}P + \tfrac{1}{2}Q\right)x - P(x - 2a) - Q(x - 3a)$$

$$M_u = \left(-\frac{P}{3}\right)x + 2Pa - \frac{1}{2}Qx + 3Qa \qquad\qquad (3a \le x \le 6a)$$

$$U = \frac{1}{2EI_u}\int_0^{2u}\left[\left(\frac{2}{3}P + \frac{1}{2}Q\right)x\right]^2 dx + \frac{1}{2EI_u}\int_{2a}^{3a}\left(-\frac{1}{3}Px + 2Pa + \frac{1}{2}Qx\right)^2 dx$$

$$+ \frac{1}{2EI_u}\int_{3a}^{6a}\left(-\frac{1}{3}Px + 2Pa - \frac{1}{2}Qx + 3Qu\right)^2 dx$$

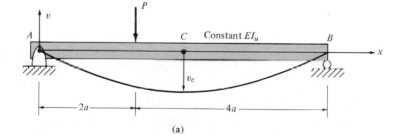

(a)

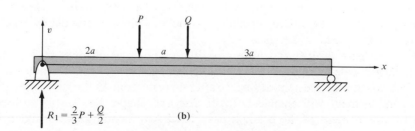

$$R_1 = \frac{2}{3}P + \frac{Q}{2}$$

(b)

FIGURE 6.22

$$\frac{\partial U}{\partial Q} = \frac{1}{EI_u} \int_0^{2a} \left(\frac{2}{3}P + \frac{1}{2}Q \right) \frac{x^2}{2} \, dx + \frac{1}{EI_u} \int_{2a}^{3a} \left(-\frac{1}{3}Px + 2Pa + \frac{1}{2}Qx \right) \left(\frac{1}{2}x \, dx \right)$$

$$+ \frac{1}{EI_u} \int_{3a}^{6a} \left(-\frac{1}{3}Px + 2Pa - \frac{1}{2}Qx + 3Qa \right) \left(-\frac{1}{2}x + 3a \right) dx$$

Let $Q = 0$ in the preceding equation, since the differentiation partially with respect to Q has been completed.

$$\frac{\partial U}{\partial Q} = \frac{1}{EI_u} \int_0^{2a} \left(\frac{2}{3}P \right) \frac{x^2}{2} \, dx + \frac{1}{EI_u} \int_{2a}^{3a} \left(-\frac{1}{3}Px + 2Pa \right) \frac{1}{2}x \, dx$$

$$+ \frac{1}{EI_u} \int_{3a}^{6a} \left(-\frac{1}{3}Px + 2Pa \right) \left(-\frac{1}{2}x + 3a \right) dx$$

Carrying out the integrations above and substitution of the limits yields

$$\frac{\partial U}{\partial Q} = v_c = \boxed{\frac{23}{6} \frac{Pa^3}{EI_u}}$$

The sense of v_c and Q are both downward as shown in Figure 6.22(a) and (b), which means that the deflection is directed downward.

★**6.9** Deflections Associated with Shearing Deformations

Methods developed in the previous sections of this chapter are all based upon bending deformations and ignore shearing deformations. In this section an approximate method for determining deflections associated with shearing deformations will be developed and two examples will illustrate the fact that for large beam length/depth ratios, the effects of shear on deflections are practically negligible compared to the effects of bending.

Refer to Figure 6.23(a) and note any differential-sized element at P, which is drawn to larger scale in Figure 6.23(c). This element is selected from the neutral plane before deformation or the neutral surface after deformation. As the general loading is applied, this element will change both its size and shape. The size or volumetric change of any element of the beam is associated with bending or normal stresses, while the shape change of any element of the beam is associated with shearing stresses. Since the objective of this development is to study deflections associated with shearing deformations, only the shape change depicted in Figure 6.23(c) need be considered. Geometrically, the tangent of the angle γ is given by

$$\tan \gamma = -\frac{dv}{dx} \tag{6.33}$$

where the negative sign has been introduced since positive shearing stresses shown produce a downward deflection, which is negative. The angle γ is small for elastic deformations, which leads to

$$\gamma = -\frac{dv}{dx} \tag{6.34}$$

But

$$\gamma = \frac{\tau}{G}$$

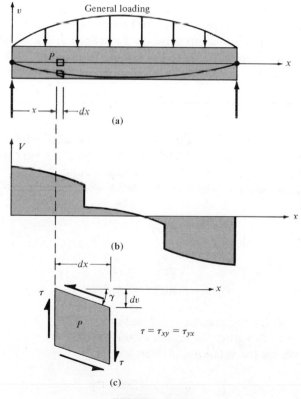

FIGURE 6.23

and the shearing stress may be written

$$\tau = \beta_1 \frac{V}{A} \qquad (6.35)$$

with V being the shear, which is a function of x. The form of this variation of shear force with x is provided by the shear diagram. The quotient of V and the total area A of the cross section is the average shearing stress over the cross-sectional area. The multiplier β_1 transforms the average shearing stress to the shearing stress at the neutral surface, where the deflections are calculated. For a rectangular cross section, $\beta_1 = 1.50$. For I-shaped beams it can be shown that the web carries almost all the vertical shear, and an approximate equation for this beam becomes $\tau \approx V/A_{\text{web}}$, in which case $\beta_1 = 1.00$. Substitute Eq. 6.35 into $\gamma = \tau/G$ and equate to Eq. 6.34 to yield

$$\frac{dv}{dx} = \frac{-\beta_1 V}{GA} \qquad (6.36)$$

Equation 6.36 is the differential equation to be integrated in order to determine v as a function of x. This deflection v supplements the deflection due to bending. Examples 6.17 and 6.18 illustrate applications of Eq. 6.36 for finding deflections associated with shearing deformations.

Example 6.17

Refer to the cantilever of Figure 6.24(a) and determine the deflection under the load due to shear deformation. Determine the ratio of the bending deflection to the shear deflection as a function of the length/depth ratio of the beam. Plot and discuss this function. The E/G ratio is 2.5, and the weight of the beam is to be neglected.

Solution. Refer to Eq. 6.36.

$$\frac{dv}{dx} = \frac{-\beta_1 V}{GA}$$

where $\beta_1 = 1.50$ for rectangular cross sections
$V = -P$ along the full length of the beam
$A = bh$ for the rectangular cross section

$$\frac{dv}{dx} = \frac{-1.50(-P)}{Gbh}$$

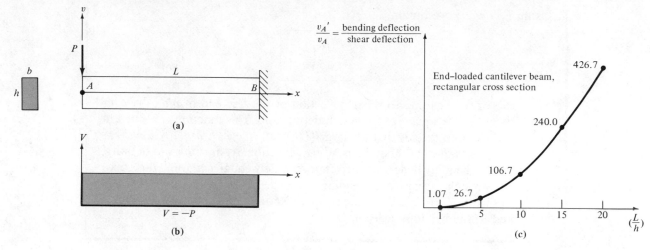

$$\frac{v_A'}{v_A} = \frac{\text{bending deflection}}{\text{shear deflection}}$$

End-loaded cantilever beam, rectangular cross section

426.7

240.0

106.7

1.07 26.7

$\left(\frac{L}{h}\right)$

(a)

$V = -P$

(b)

(c)

FIGURE 6.24

Multiply by dx and integrate observing that $v = 0$ for $x = L$.

$$\int_{v_A}^{0} dv = \frac{1.50P}{Gbh} \int_{0}^{L} dx$$

$$0 - v_A = \frac{1.5PL}{Gbh}$$

$$\boxed{v_A = \frac{-1.5PL}{Gbh}}$$ (downward deflection at A due to shear deformation)

The deflection under load due to bending deformation is given by

$$v_A' = \frac{-PL^3}{3EI_u}$$

and

$$I_u = \tfrac{1}{12}bh^3$$

Form the ratio v_A'/v_A,

$$\frac{v_A'}{v_A} = \frac{-PL^3}{3(\tfrac{1}{12}bh^3)E} \frac{-Gbh}{1.5PL} = \frac{8}{3}\frac{G}{E}\left(\frac{L}{h}\right)^2$$

Using $E/G = 2.5$, we obtain

$$\boxed{\dfrac{v'_A}{v_A} = \dfrac{16}{15}\left(\dfrac{L}{h}\right)^2}$$

The ratio of the bending to shear deflection at the free end of the cantilever beam varies as the square of the length/depth ratio. This function is plotted in Figure 6.24(c) and reveals that for large (L/h) ratios (e.g., $L/h \geq 10$), the shear deflection is less than $1/100$ of the bending deflection. At an L/h ratio of unity, the shear and bending deflections are approximately equal. Although this curve is for a particular beam and loading, it is generally true that for large span/depth ratios, say $L/h \geq 10$, the shear deflections are practically negligible compared to the bending deflections.

Example 6.18

Refer to the simply supported beam of Figure 6.25(a) and determine the center deflection due to shear deformation. Determine the ratio of bending deflection to the shear deflection as a function of the length/depth ratio. Plot and discuss this function. Neglect the weight of the beam and use an E/G ratio of 2.5.

Solution. Refer to Eq. 6.26,

$$\frac{dv}{dx} = -\frac{\beta_1 V}{GA}$$

where $\beta_1 = 1.50$ for the rectangular cross section; refer to the shear diagram of
         Figure 6.25(b) for the variation of V along the beam length
   $A = bh$ for the rectangular cross section

$$\frac{dv}{dx} = \frac{-1.50V}{Gbh}$$

where V is regarded as a function of x, which may be represented as follows:

$$V = P \qquad (0 < x < a)$$
$$V = 0 \qquad (a < x < 2a)$$
$$V = -P \qquad (2a < x < 3a)$$

Since the center deflection is requested, the integration will proceed with x varying from zero to $1.50a$. At $x = 0$, $v = 0$

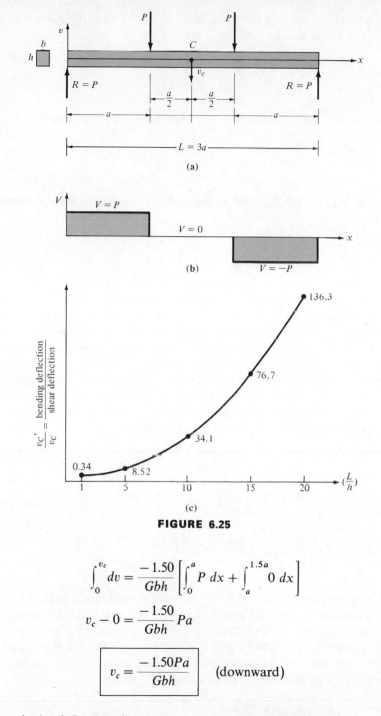

FIGURE 6.25

$$\int_0^{v_c} dv = \frac{-1.50}{Gbh}\left[\int_0^a P\,dx + \int_a^{1.5a} 0\,dx\right]$$

$$v_c - 0 = \frac{-1.50}{Gbh}Pa$$

$$\boxed{v_c = \frac{-1.50Pa}{Gbh}} \quad \text{(downward)}$$

where v_c is the deflection due to shear at the center of the beam.

It is left as an exercise for the student to prove that the deflection at the center of the beam due to bending is given by

$$v'_c = -\frac{23}{24}\frac{Pa^3}{EI_u}$$

Form the ratio of v'_c/v_c using $E/G = 2.5$ and $I_u = \frac{1}{12}bh^3$, to obtain

$$\frac{v'_c}{v_c} = \frac{46}{15}\left(\frac{a}{h}\right)^2$$

In order to express this function in terms of the length/depth ratio, recall that $a = L/3$.

$$\boxed{\frac{v'_c}{v_c} = \frac{46}{135}\left(\frac{L}{h}\right)^2}$$

Once again the ratio of the bending to shear deflection varies as the square of the length/depth ratio. This function is plotted in Figure 6.25(c) and reveals that for large depth/span ratios (say, $L/h > 10$), the shear deflection is less than $\frac{3}{100}$ of the bending deflection. At an L/h ratio of unity, the shear deflection is approximately three times the bending deflection. Although this curve is for a particular beam and loading, it is generally true that for the large span/depth ratios, say $L/h > 10$, the shear deflections are practically negligible compared to the bending deflections. Conversely, shear deflections should be investigated for beams with relatively small length/depth ratios.

*6.10 Unsymmetric Beam Deflections

Consider a cantilever beam such as the one depicted in Figure 6.26(a) subjected to a vertical load as shown. The load P is applied through the shear center C_s of the end cross section, which assures that the beam will bend without twisting. In the cross section depicted in Figure 6.26(b), the centroid C and the shear center C_s are assumed to coincide.* If these two points do not coincide, the loading must be applied to prevent twisting (i.e., through the shear center).

* Shear center is discussed in Section 14.3.

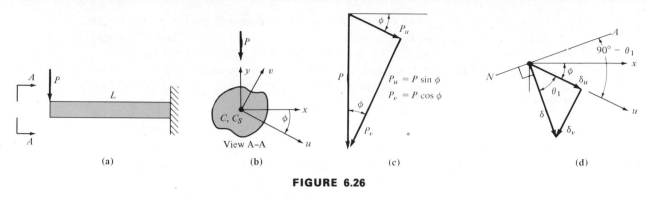

FIGURE 6.26

Axes u and v, which are the principal axes, are determined from Mohr's circle, as discussed in Section 5.2. Once these principal axes of inertia are known, the applied force is resolved into components parallel to these axes, as shown in Figure 6.26(c), and the deflection components parallel to the u and v axes are determined from the following equations, based upon Table 6.1 case 5 with $a = L$:

$$\delta_u = \frac{P \sin \phi L^3}{3EI_v} \tag{6.37}$$

$$\delta_v = \frac{P \cos \phi L^3}{3EI_u} \tag{6.38}$$

where P is the applied load, ϕ defines the direction of the principal axis u with respect to the horizontal x axis or the principal v axis with respect to the vertical y axis, L is the length of the cantilever beam, E is the modulus of elasticity of the material, and I_u and I_v are principal moments of inertia of the cross-sectional area with respect to axes through the centroid. The perpendicular deflection components δ_u and δ_v may be combined vectorially to obtain the total deflection δ, which is oriented perpendicular to the neutral axis (N.A.) shown in Figure 6.26(d). The total deflection is given by

$$\delta = \sqrt{\delta_u^2 + \delta_v^2} \tag{6.39}$$

This total deflection makes an angle of $(\theta_1 + \phi)$ with the horizontal x axis, where θ_1 is given by

$$\theta_1 = \arctan \frac{\delta_v}{\delta_u} \tag{6.40}$$

Since the total deflection δ is perpendicular to the neutral axis, the angle between the u axis and N.A. will be $(90° - \theta_1)$, as shown in Figure 6.26(d).

Although our discussion has focused on an end-loaded cantilever beam, the approach is quite general. For example, the divisor of 3 in Eqs. 6.37 and 6.38 would be replaced by 48 if the problem involved computation of the center deflection of a simply supported beam of total length L loaded by a concentrated force at its center. Example 6.19 will clarify the procedure for computing unsymmetric beam deflections.

Example 6.19

A cantilever beam of length $L = 10$ ft has the cross section shown in Figure 6.27, which is an angle $9 \times 4 \times 1$ in. The beam is fabricated of steel for which $E = 30 \times 10^6$ psi and the following moments of inertia are given:

$$I_x = 97 \text{ in.}^4$$

$$I_y = 12 \text{ in.}^4$$

$$I_{xy} = 18 \text{ in.}^4$$

The beam is subjected to a concentrated load $P = 1000$ lb at the free end acting vertically downward such that the beam bends without twisting. Determine the deflection of the cantilever beam at its free end.

Solution. The Mohr's circle construction is shown in Figure 6.27(c) together with the necessary calculations. From this construction

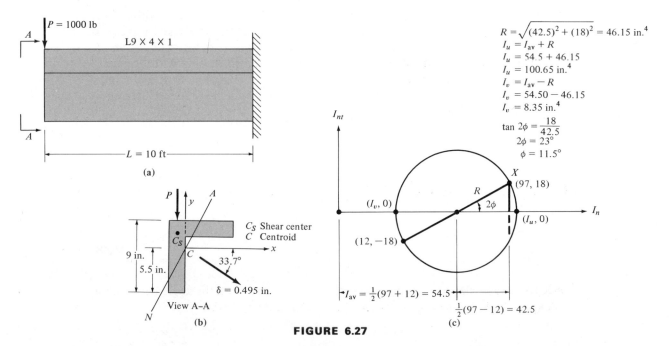

FIGURE 6.27

$$I_u = I_{max} = 100.65 \text{ in.}^4$$

$$I_v = I_{min} = 8.35 \text{ in.}^4$$

$$\phi = 11.5° \quad \text{(cw from } x \text{ axis to } u \text{ axis)}$$

The deflection at the end of a cantilever beam subjected to a concentrated load at the free end is given by the equation

$$\delta = \frac{PL^3}{3EI}$$

Applying Eqs. 6.37 and 6.38, we obtain

$$\delta_u = \frac{P_u L^3}{3EI_v} = \frac{(1000 \sin 11.5°)(10 \times 12)^3}{3(30 \times 10^6)(8.35)} = 0.458 \text{ in.}$$

$$\delta_v = \frac{P_v L^3}{3EI_u} = \frac{(1000 \cos 11.5°)(10 \times 12)^3}{3(30 \times 10^6)(100.65)} = 0.187 \text{ in.}$$

Applying Eqs. 6.39 and 6.40, we find that

$$\delta = \sqrt{\delta_u^2 + \delta_v^2} = \sqrt{(0.458)^2 + (0.187)^2} = 0.495 \text{ in.}$$

$$\theta_1 = \arctan \frac{\delta_v}{\delta_u} = \arctan \frac{0.187}{0.458} = 22.2°$$

$$\theta_1 + \phi = 22.2° + 11.5° = 33.7°$$

The total deflection of the free end of the cantilever is shown in Figure 6.27(b).

Homework Problems

Text sections and problems are to be associated as follows:

Section 6.8, Problems 6.57 to 6.64.
Section 6.9, Problems 6.65 to 6.69.
Section 6.10, Problems 6.70 to 6.76.

6.57 Refer to Figure H6.43 and use Castigliano's second theorem to find the deflection at point C of this beam in terms of w, L, E, and I_u.

6.58 Refer to Figure H6.44 and find the rotation at point A in terms of P, L, E, and I_u by applying Castigliano's second theorem.

6.59 Refer to Figure H6.45 and use Castigliano's second theorem to find the rotation at point A of this beam in terms of M_A, L, E, and I_u.

6.60 Apply Castigliano's second theorem to find the

rotation at point B of the beam shown in Figure H6.46 in terms of M_B, a, b, E, and I_u.

6.61 Find the deflection of point A of the beam shown in Figure H6.47 using Castigliano's second theorem. Avoid replacing P by wa until the final step when you are ready to express the deflection in terms of w, a, E, and I_u.

6.62 Refer to Figure H6.48 and use Castigliano's second theorem to express the deflection at point A in terms of P, a, E, and I_u. Replace M_D by Pa after differentiating partially with respect to P.

6.63 Use Castigliano's second theorem to determine the deflection of point C of the beam shown in Figure H6.49 in terms of P, b, E, and I_u. Replace M_B by Pb after differentiating partially with respect to P.

6.64 Refer to Figure H6.50 and determine the deflection of point C in terms of w, a, E, and I_u by applying Castigliano's second theorem. Replace P by wa after differentiating partially with respect to P.

6.65 Refer to case 1 of Table 6.1 and determine the center deflection of this beam associated with shearing deformations. Consider a rectangular cross section $(b \times h)$ and an $E/G = 2.50$. Form the ratio of the bending deflection to the shearing deflection at the center as a function of the beam length/depth ratio squared. Plot this deflection versus L/h equal to 1, 5, 10, 15, 20 similar to Figure 6.24(c).

6.66 Repeat Problem 6.65 for case 2 of Table 6.1 with $a = b = L/2$.

6.67 Repeat Problem 6.65 for case 3 of Table 6.1.

6.68 Repeat Problem 6.65 for case 4 of Table 6.1 using the free end deflections.

6.69 Repeat Problem 6.65 for case 5 of Table 6.1 using the free end deflections with $a = b = L/2$.

6.70 Refer to Example 6.19 and replace the concentrated force by a uniform loading of $w = 100$ lb/ft and compute the end deflection of this cantilever beam. $E = 30 \times 10^6$ psi.

6.71 Refer to Example 6.19, replace P by an upward force of 500 lb applied at the same point, and replace the cross section by an angle $6 \times 6 \times 1$ in., then determine the end deflection of this cantilever beam. $E = 30 \times 10^6$ psi.

6.72 Refer to Figure H6.6, let $M_1 = 10,000$ lb-ft, and $L = 8$ ft. Determine the center deflection of this beam if the angle cross section of Example 6.19 is used. $E = 30 \times 10^6$ psi. Assume freedom to rotate about the u and v axes at the ends of this beam.

6.73 Refer to Figure H6.7, let $Q = 200$ lb, and $L = 9$ ft. Determine the center deflection of this beam if the angle cross section of Example 6.19 is used. $E = 30 \times 10^6$ psi. Assume freedom to rotate about the u and v axes at the ends of this beam.

6.74 Refer to Figure H6.8, let $M_1 = 6000$ lb-ft, and $L = 6$ ft. Determine the center deflection of this beam if the angle cross section of Example 6.19 is used. $E = 30 \times 10^6$ psi. Assume freedom to rotate about the u and v axes at the ends of this beam.

6.75 Refer to Problem 6.71 and change the cross section to a rectangle 6×4 in., with the 6-in. side rotated 45 clockwise with respect to the vertical. Then determine the end deflection of the cantilever beam. In this case the shear center coincides with the centroid and the load is applied through this common point.

6.76 Solve Problem 6.74 with a rectangular cross section. The rectangle measures 5×2.50 in. with the 5-in. side rotated 30° clockwise with respect to the vertical. Determine the center deflection of this beam with M_1 reduced by a factor of 10 to 600 lb-ft. In this case the shear center coincides with the centroid. The couple M_1 is applied in the vx plane and the x axis passes through the centroids of all cross sections.

Chapter 7 Combined Stresses and Theories of Failure

7.1 Introduction

In the preceding chapters methods were developed and discussed for the analysis of members subjected to one type of load only. For example, Chapter 3 dealt with members subjected only to axial loads, Chapter 4 with members subjected only to torsional loads, and Chapter 5 with members subjected only to flexural loads. There are numerous engineering situations relating to structural and machine applications, however, in which members are subjected to the simultaneous action of two or more types of loads.

Four different load combinations are considered in this chapter. These are (1) axial and torsional loads, (2) axial and flexural loads, (3) torsional and flexural loads, and (4) torsional, flexural, and axial loads. In all instances, the material is assumed not to be stressed beyond the elastic limit, to justify the use of the principle of superposition. This principle then leads to the development of analytical methods that make it possible to examine the stress condition induced by any one of the four load combinations mentioned here.

When a member is subjected to uniaxial tension or compression, one can predict, with a certain degree of confidence, the magnitude of the uniaxial load at which failure by yielding or by inelastic action would occur. However, when a member is subjected to combined loads, the cause of failure cannot be accurately ascertained, and resort is made to one of the many theories that have been proposed to predict failure under combined loads. The four most commonly used theories of failure are developed and discussed, and they are then compared to each other under a specific state of stress to emphasize the differences that exist among these various theories.

7.2 Axial and Torsional Stresses

There are many engineering situations in which a member is called upon to resist the simultaneous action of axial and torsional loads. Familiar examples of such situations include the case of the screw stem and that of the drill shaft. For example, as the screw is driven by a screw driver, turning and pushing occur simultaneously. The turning effect results in a system of resisting shearing forces between the screw and the material into which it is being driven that gives rise to a resisting torque. The pushing effect leads to compression action in the screw. Thus the stress analysis in the stem of the screw should account for the action of both the torque and the compression load.

Analysis of combined axial and torsional loads is accomplished by the use of the principle of superposition, which was introduced and used in Section 2.10. For simplicity and convenience, only members with solid or hollow circular cross sections will be analyzed in this section, although the method of superposition is equally

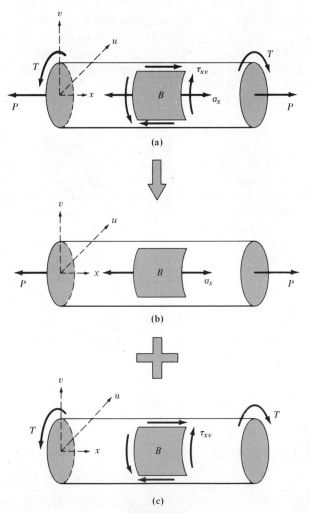

FIGURE 7.1

applicable to any arbitrary cross-sectional area as long as the material is not stressed beyond the proportional limit.

Consider, for example, the circular shaft shown in Figure 7.1(a), which is subjected to the combined effects of the tensile axial force P and the torque T. By the principle of superposition, this member can be considered to consist of the two members shown in Figure 7.1(b) and (c). The member shown in Figure 7.1(b) is acted upon by the action of the axial force P acting alone, and therefore the stress system induced in the shaft, at any cross section along its length, is a simple uniform normal stress distribution whose intensity σ_x is given by Eq. 3.1. The member shown in Figure 7.1(c) is subjected to the torque T acting alone and, at any cross section along the shaft, there is induced a shear stress distribution, τ_{xv}, given by Eq. 4.19, that varies linearly from zero at the center to a maximum value at the outside surface of the circular cross section. Thus to investigate the most severe stress condition existing in the shaft as a result of the combined action of P and T, one should consider any stress element on the surface of the shaft such as the one depicted at point B in Figure 7.1. This stress element has two planes parallel and two planes perpendicular to the axis of the shaft. For convenience, the xv coordinate system is so chosen that the x axis is along the shaft. The stress condition at point B due to the axial load P acting alone is shown in Figure 7.1(b), and that at the same point due to the torque T acting alone is shown in Figure 7.1(c). The superposition of these two stress conditions results in the stress element depicted in Figure 7.1(a). Note that this is a plane stress condition, since the stress in the u direction [perpendicular to the cylindrical surface of the shaft but not shown in Figure 7.1(a)] is zero. Having determined the combined stress condition, we can compute the principal stresses and the maximum shear stress by the use of Mohr's circle. This procedure is further illustrated in Example 7.1.

Example 7.1
Let the diameter of the shaft shown in Figure 7.1 be 4 in. Let the load P be 80 k tension and the torque T be 60 k-in. Determine the principal stresses and the absolute maximum shear stress induced in the shaft. Show these stresses on properly oriented planes.

Solution. The stress element at point B on the surface of the shaft of Figure 7.1 is isolated and shown in Figure 7.2(a) in relation to the xv coordinate plane. Remember that the x axis was chosen parallel to the longitudinal axis of the shaft. By Eq. 3.1, the normal stress σ_x is found to be

$$\sigma_x = \frac{P}{A} = \frac{80}{12.57} = 6.37 \text{ ksi}$$

Also, by Eq. 4.19, the shear stress τ_{xv} is found to be

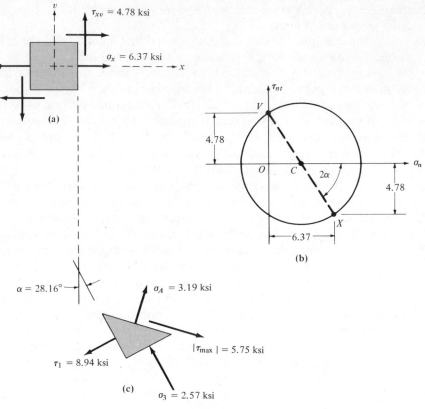

FIGURE 7.2

$$\tau_{xv} = \frac{T\rho}{J} = \frac{60(2)}{25.13} = 4.78 \text{ ksi}$$

These stresses are shown on the stress element in Figure 7.2(a). Mohr's circle for this stress condition is constructed as shown in Figure 7.2(b). From its geometry, the following values are determined:

$$OC = 3.185$$

$$R = 5.75$$

$$\sigma_1 = OC + R$$

$$= \boxed{8.94 \text{ ksi}}$$

$$\sigma_2 = \boxed{0}$$

$$\sigma_3 = OC - R$$

$$= \boxed{-2.57 \text{ ksi}}$$

$$|\tau_{max}| = \frac{\sigma_1 - \sigma_3}{2}$$

$$= \boxed{5.75 \text{ ksi}}$$

Note that σ_2 acts in the u direction, which is perpendicular to the shaft surface and is zero because this is a force-free surface. Note also that in this case, the absolute maximum shear stress is equal to the radius of Mohr's circle. Also,

$$\tan 2\alpha = \tan^{-1} \frac{4.78}{3.185} = 56.32°$$

and

$$\boxed{\alpha = 28.16°}$$

The principal stresses and the absolute maximum shear stress are shown on a properly oriented stress element in Figure 7.2(c). Thus it can be concluded that the principal planes on which $\sigma_1 = 8.94$ ksi acts make 28.16° ccw angles with cross sections of the shaft (i.e., with the X planes), and the principal planes on which $\sigma_3 = -2.57$ ksi acts make 28.16° ccw angles with the longitudinal axis of the shaft (i.e., with the V plane). Obviously, $\sigma_2 = 0$ would be normal to the surface of the shaft, as stated earlier.

Homework Problems

7.1 Solve Example 7.1 if the load $P = 80$ kips is compressive and all other conditions in the example remain the same.

7.2 A hollow steel shaft 5 m long, 0.15 m outside diameter, and 0.06 m inside diameter is fixed rigidly at one end and subjected to an axial tensile force of 300 kN and a torque of 5 kN-m at the other end. Determine:

(a) The principal stresses and the absolute maximum shear stress induced in the member and specify the planes on which these stresses act.

(b) The change in length and the angle of twist of the shaft.

Assume that $E = 200 \times 10^9$ N/m² and $\mu = 0.25$.

7.3 A 4-in.-diameter solid shaft is subjected to the torques and to the axial force shown in Figure H7.3.

Construct the torque and force diagrams and compute the principal stresses at the most severely stressed point or points in the member.

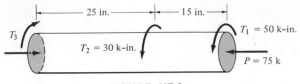

FIGURE H7.3

7.4 A composite steel shaft is subjected to the torques and to the axial load shown in Figure H7.4. Determine:

(a) The principal stresses and the absolute maximum shear stress at the most severely stressed point or points in the member.

(b) The total deformation and total angle of twist of the composite member.

Assume that $E_{st} = 30 \times 10^6$ psi and $E_{al} = 10 \times 10^6$ psi and let $G = (\frac{4}{10})E$ for both materials.

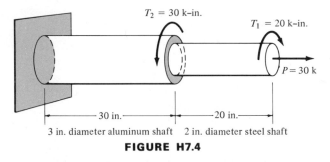

FIGURE H7.4

7.5 A hollow steel shaft 0.10 m outside diameter is rigidly welded to a 0.06-m solid steel shaft and subjected to the torques and to the compressive axial force as shown in Figure H7.5. Construct the torque

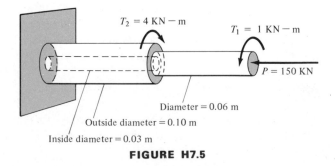

FIGURE H7.5

and force diagrams and determine the principal stresses and the absolute maximum shear stress at the most severely stressed point or points in the member.

7.6 The solid steel shaft shown in Figure H7.6 is subjected to a torque $T = 3$ kN-m and an axial force P. The diameter of the shaft is 0.07 m, and if the principal stress of largest magnitude at point B is known to have a magnitude of 120 MPa, what is the value of the force P?

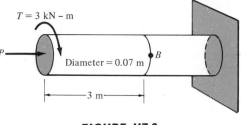

FIGURE H7.6

7.7 Solve Problem 7.6 if the absolute maximum shear stress instead of the principal stress of largest magnitude at point B is known to be 100 MPa. All other conditions in the problem remain the same.

7.8 The solid steel shaft shown in Figure H7.8 is subjected to a torque T and an axial force $P = 75$ k. The diameter of the shaft is 3 in., and if the principal stress of largest magnitude at point B is known to have a magnitude of 14 ksi, what is the value of the torque T?

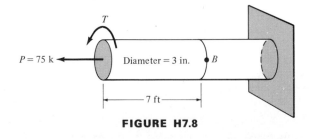

FIGURE H7.8

7.9 Solve Problem 7.8 if the absolute maximum shear stress instead of the principal stress of largest magnitude at point B is known to be 12 ksi. All other conditions in the problem remain the same.

7.3 Axial and Flexural Stresses

The simultaneous action of axial and flexural loads occurs frequently in machine and structural members. A very common example of such occurrence is the case of certain types of traffic light frames. An idealized model of such a frame is shown in Figure 7.3(a), in which, for simplicity, the weight of the traffic light fixture and that of the horizontal member AB have been grouped together into one single force P acting at point A. Equilibrium dictates that, at any cross section in the vertical member BE, there exists a normal force $F_n = P$ and a bending moment $M = Pd$ as shown in the free-body diagrams of Figure 7.3(b) and (c). The combined effects of F_n and M may be determined by the principle of superposition.

Consider the free-body diagram of vertical member DE of Figure 7.3(c), which is repeated magnified in Figure 7.4(a). By the principle of superposition, this vertical member may be assumed to consist of the two members shown in Figure 7.4(b) and (c). The member shown in Figure 7.4(b) is acted upon by the compressive axial force F_n acting alone, and therefore the induced stress system at any cross section of the member is a uniform normal stress distribution (compression in this case) whose intensity σ_y is given by Eq. 3.1 (i.e., $\sigma_y = F_n/A$). This uniform stress distribution is shown for an arbitrary cross section a–a in the lower part of Figure 7.4(b). The member shown in Figure 7.4(c) is subjected to the action of the bending moment M_u acting alone, and at any cross section in the member, the induced normal (flexural) stress distribution, σ_y, is given by Eq. 5.10 [i.e., $\sigma_y = (M_u/I_u)v$] if it is assumed that symmetric bending exists. As a matter of fact, a rectangular cross section as shown in the top part of Figure 7.4(a) is assumed, and all problems in this section are con-

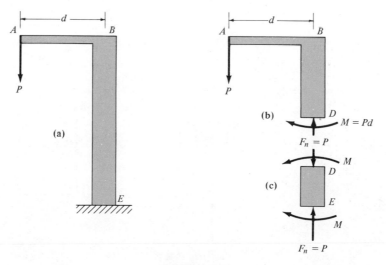

FIGURE 7.3

sidered to deal with symmetric bending cases. The flexural stresses resulting from the action of the moment M_u vary linearly from zero at the neutral axis (i.e., the u axis) to maximum compression on the left edge and maximum tension on the right edge of the member. This linearly varying stress distribution is shown for the arbitrary cross section a–a in the lower part of Figure 7.4(c). Since the stresses shown in Figure 7.4(b) and (c) are both normal stresses acting over the same cross section of the member, they may be superposed (added algebraically). Examination of the stress distribution shown in Figure 7.4(b) and (c) reveals that those fibers to the left of the neutral axis for bending [point N.A. in Figure 7.4(c)] are subjected to compression due to F_n and also due to M_u. Therefore, the stresses to the left of the neutral axis are additive. On the other hand, the fibers to the right of the neutral axis are subjected to compression due to F_n and tension due to M_u. Thus the stresses to the right of the neutral axis subtract from each other. Therefore, the superposition of the two stress systems in Figure 7.4(b) and (c) yields the stress distribution shown in the lower part

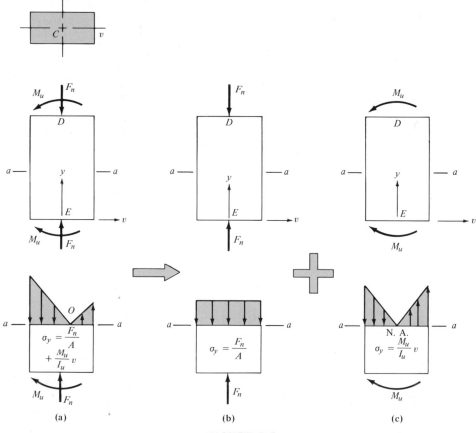

FIGURE 7.4

of Figure 7.4(a), which is the result of the simultaneous action of the normal force F_n and the bending moment M_u. Thus the resultant stress, σ_y, at any point in the cross section of the member subjected to the combined action of F_n and M_u is given by

$$\sigma_y = \frac{F_n}{A} + \frac{M_u}{I_u} v \qquad\qquad (7.1)$$

where the addition of the two terms is to be interpreted in an algebraic sense (i.e., each of the two terms may be either positive or negative). Note that the fiber of zero stress for the combined action of F_n and M_u [point O in the lower part of Figure 7.4(a)] does not coincide with the neutral axis for bending [i.e., point N.A. in Figure 7.4(c), which is the projection of the u principal centroidal axis] but is shifted slightly to the right of this axis. The extent of the shift in a given situation would depend upon the relative magnitudes of the normal stresses produced by F_n and M_u. The fiber of zero stress for the combined action of F_n and M_u is sometimes referred to as the *neutral axis*, but, in this text, to avoid confusion, the term *neutral axis* is reserved for bending only and the expression *fiber* or *axis of zero stress* for the simultaneous combined effects of axial forces and bending moments.*

Application of the foregoing principles will be illustrated in Example 7.2.

Example 7.2
A short compression block is subjected to an eccentric load (i.e., a load that does not pass through the centroidal axis of the member) $P = 500$ kN, as shown in Figure 7.5(a). However, the load is applied in such a way that it is contained within the v–x principal longitudinal plane of the member, as shown in Figure 7.5(b). Construct the normal stress distribution in the member at any section, such as a–a.

Solution. Isolate that portion of the compression member to the right of section a–a and construct its free-body diagram as shown in Figure 7.5(c). Note that equilibrium dictates that at any section a–a there be present a compressive force $F_n = P = 500$ kN and a bending moment $M_u = Pe = 500(0.04) = 20$ kN-m.

It is evident that, in this case, bending takes place about the u principal axis of inertia, and therefore symmetric bending occurs. The principal moment of inertia with respect to the u principal controidal axis is found to be $I_u = 3.5 \times 10^{-5}$ m^4.

* The location of point O may be determined by setting the value of σ_y in Eq. 7.1 equal to zero. The resulting equation yields the value of v from the u axis for which the combined stress vanishes.

The bending (flexural) stress σ_x due to the moment M_u is given by Eq. 5.10. Thus

$$\sigma_x = \frac{M_u}{I_u} v = 571.4v \ \text{MN/m}^2$$

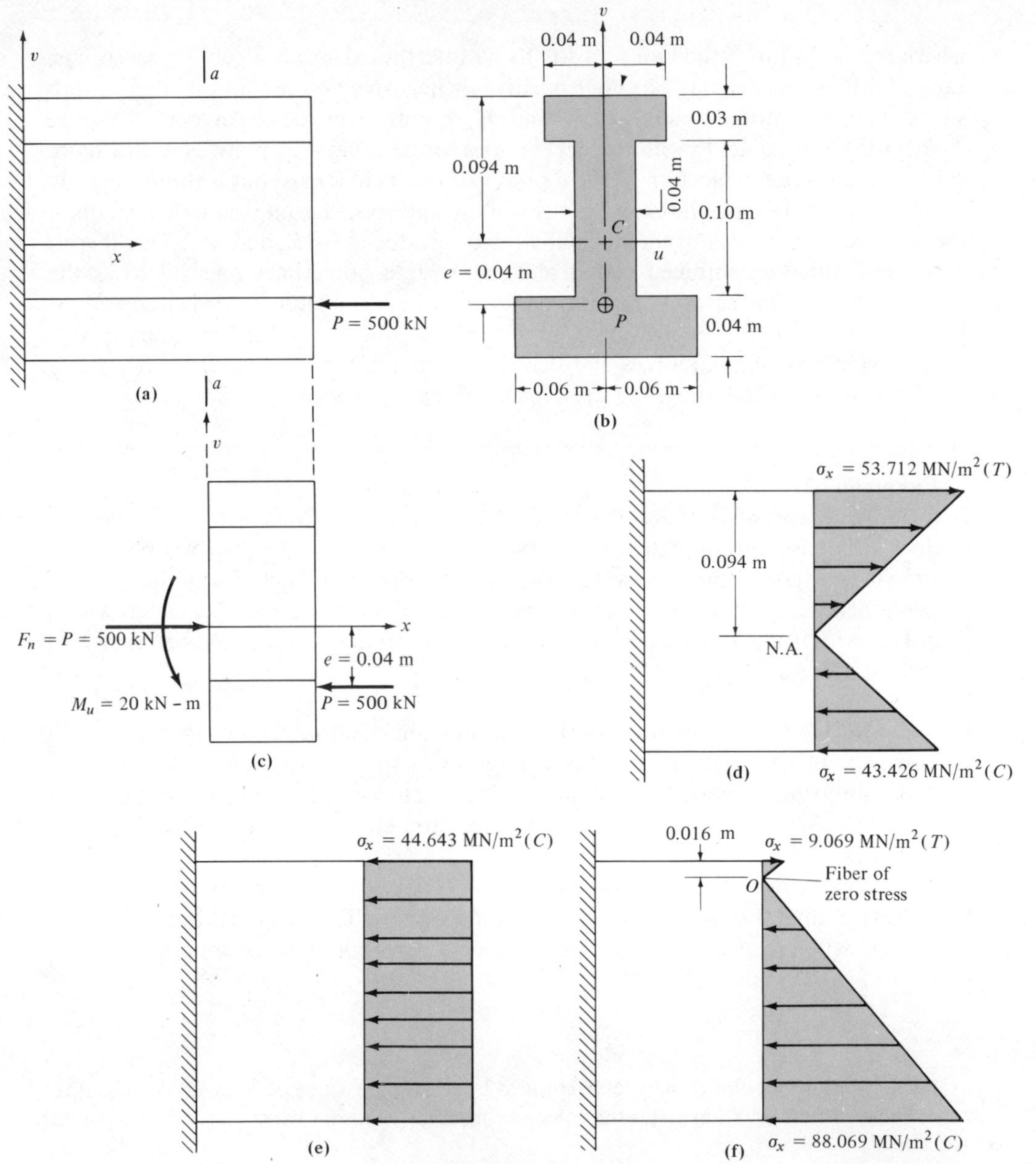

FIGURE 7.5

where v represents the distance, in meters, from the neutral axis (the u axis) to any point in the cross section. The top fibers are in tension and the bottom fibers in compression. This stress distribution is plotted in Figure 7.5(d). The normal stress σ_x due to the compressive force P is given by Eq. 3.1. Thus

$$\sigma_x = \frac{F_n}{A} = -\frac{P}{A} = -44.643 \text{ MN/m}^2$$

which is a constant compressive stress across the section and is represented in Figure 7.5(e). Superposition of these two stress distributions (i.e., algebraic addition according to Eq. 7.1) yields the composite stress distribution shown in Figure 7.5(f). Note that the point of zero stress, point O, is a distance of 0.016 m from the upper edge of the section. Thus the eccentric load P induces the normal stress distribution shown in Figure 7.5(f) such that fibers in the cross section above point O are in tension and those below point O are in compression.

Example 7.3

A simply supported beam AD is subjected to a 20-k force at point D inclined 30° to its longitudinal axis as shown in Figure 7.6(a). The cross section for the beam is a rectangle, as shown magnified in Figure 7.6(b). The load is so applied that symmetric bending about the u axis of the section takes place. Consider a section along the beam an infinitesimal distance to the left of support B and determine the principal stresses and the absolute maximum shear stress at (a) a point at the top of the section, (b) at a point at the centroid of the section, and (c) at a point 1.5 in. below the top of the section.

Solution. The support reactions are found by the methods of statics to be

$$A_v = -3.33 \text{ k} \qquad A_x = 17.32 \text{ k} \qquad B_v = 13.33 \text{ k}$$

These reactions are shown on the free-body diagram of the entire beam in Figure 7.6(c). The free-body diagram of that portion of the beam to the left of the section of interest is shown in Figure 7.6(d). Equilibrium requires that at the section of interest there be a bending moment $M_u = 239.76$ k-in., a normal tensile force $F_n = 17.32$ k, and a shear force $V_v = 3.33$ k, as shown in Figure 7.6(d). The existence of a shear force makes this type of problem slightly different from that discussed in Example 7.2 and results in a system of shear stresses in the cross sections of the beam as discussed in Section 5.4. These shear stresses have to be included in the analysis along with the normal stresses produced by F_n and M_u.

(a) A stress element at the top of the section is isolated in such a way that two of its planes are parallel and the other two perpendicular to the longitudinal

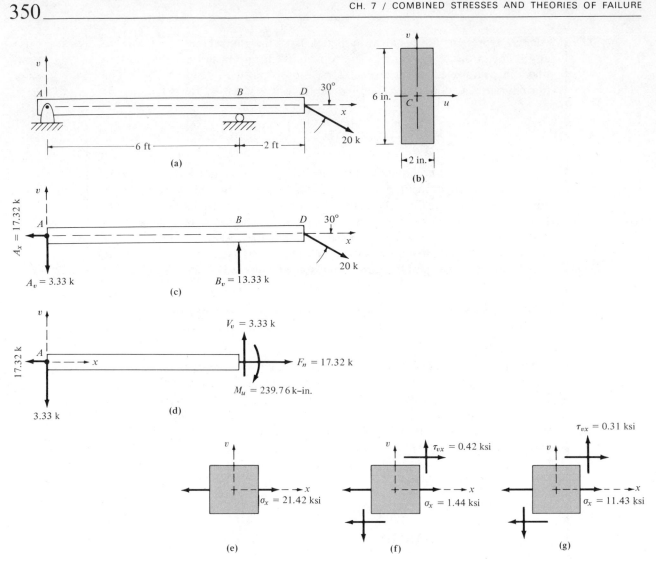

FIGURE 7.6

axis of the beam (i.e., to the x axis). This stress element is shown in Figure 7.6(e). Since the shear stress due to V_v at the top of the beam is zero (see Section 5.4), the only stress acting on this element is a normal stress σ_x on the X plane due to the action of both F_n and M_u and is given by Eq. 7.1. Thus

$$\sigma_x = \frac{F_n}{A} + \frac{M_u}{I_u} v = \frac{17.32}{12} + \frac{239.76(3)}{36} = 1.44 + 19.98 = 21.42 \text{ ksi}$$

Thus this stress element represents a uniaxial stress condition which is a special case of the plane stress problem. Since there are no shear stresses on either of the two planes, we conclude that these two planes are principal planes and that

$$\sigma_1 = 21.42 \text{ ksi} \qquad \text{and} \qquad \sigma_2 = \sigma_3 = 0$$

Also, by Eq. 2.13b, the absolute maximum shear stress at the point is found to be

$$|\tau_{max}| = \frac{\sigma_1 - \sigma_3}{2}$$

$$= \boxed{10.71 \text{ ksi}}$$

(b) A stress element is selected at the centroid of the section with planes as described in part (a). This stress element is shown in Figure 7.6(f). At the centroid of the section, the flexural stress due to M_u is zero (see Section 5.3). Therefore, this stress element is subjected to a normal stress σ_x due to F_n and a shear stress τ_{vx} due to V_v. These values are computed below and are shown in Figure 7.6(f). Thus, by Eq. 3.1,

$$\sigma_x = \frac{F_n}{A} = \frac{17.32}{12} = 1.44 \text{ ksi}$$

and, by Eq. 5.12a,

$$\tau_{vx} = \frac{V_v Q}{b I_u} = \frac{3.33(9)}{2(36)} = 0.42 \text{ ksi}$$

The principal stresses may now be obtained from a Mohr's circle solution. Such a solution leads to

$$\sigma_1 = 1.55 \text{ ksi} \qquad \sigma_2 = 0 \qquad \sigma_3 = -0.11 \text{ ksi}$$

Note that since this is a plane stress condition, one of the three principal stresses is zero, labeled here as σ_2 to satisfy the requirement that $\sigma_1 > \sigma_2 > \sigma_3$. The absolute maximum shear stress is determined from Eq. 2.13b. Thus

$$|\tau_{max}| = \frac{\sigma_1 - \sigma_3}{2}$$

$$= \boxed{0.83 \text{ ksi}}$$

(c) A stress element with planes as defined in part (a) is isolated at a point 1.5 in. below the top of the section. This stress element is shown in Figure 7.6(g). The stresses acting on this element are computed as follows:

By Eq. 7.1,

$$\sigma_x = \frac{F_n}{A} + \frac{M_u}{I_u} v$$

$$= \frac{17.32}{12} + \frac{239.76}{36}(1.5)$$

$$= 1.44 + 9.99 = 11.43 \text{ ksi}$$

By Eq. 5.12a,

$$\tau_{vx} = \frac{V_v Q}{b I_u} = \frac{3.33(6.75)}{2(36)} = 0.31 \text{ ksi}$$

From a Mohr's circle solution, the following principal stresses are determined:

$$\boxed{\sigma_1 = 11.44 \text{ ksi}} \qquad \boxed{\sigma_2 = 0} \qquad \boxed{\sigma_3 = -0.008 \doteq 0}$$

The value of σ_3 is sufficiently small to justify the assumption that it is equal to zero.

The absolute maximum shear stress is given by Eq. 2.13b,

$$|\tau_{max}| = \frac{\sigma_1 - \sigma_3}{2}$$

$$= \boxed{5.72 \text{ ksi}}$$

Homework Problems

7.10 Determine the normal stress distribution at section *a–a* of the machine link shown in Figure H7.10. What is the principal stress of largest magnitude at this section?

7.11 A compression member is subjected to an eccentric load as shown in Figure H7.11. Plot the normal stress distribution at section *a–a* of the member. What is the value of the absolute maximum shear stress at this section?

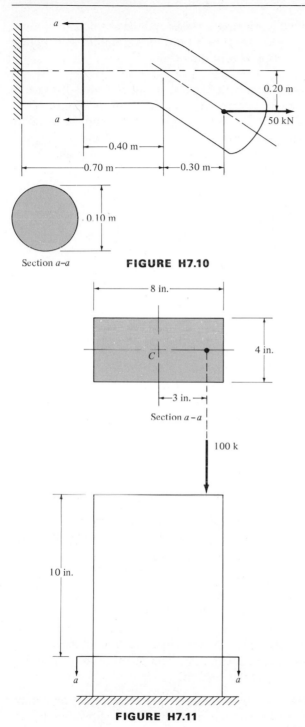

Section a–a

FIGURE H7.10

FIGURE H7.11

7.12 The frame of a machine press is shown in Figure H7.12. The maximum tensile normal stress at section a–a is known to be 200 MPa. Determine the force P.

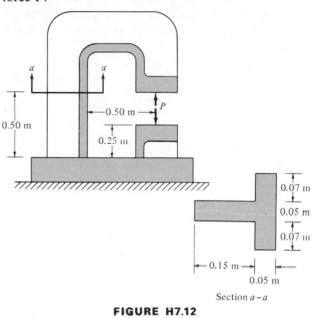

Section a–a

FIGURE H7.12

7.13 Determine the maximum tensile and maximum compressive stresses at any section such as a–a of the member shown in Figure H7.13.

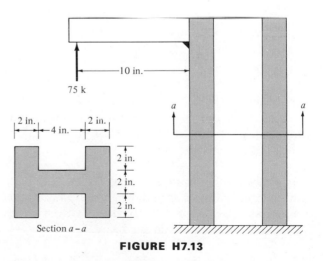

Section a–a

FIGURE H7.13

7.14 Assume the two loads acting on the cantilever beam shown in Figure H7.14 to be applied at the centroid of the T section and to be contained in the plane of symmetry of the beam. Compute the principal stresses and the absolute maximum shear stress at a point on top of the T section at location *a–a* in the beam.

pute the principal stresses and the absolute maximum shear stress at a point at the bottom of the T section at location *a–a* in the beam.

7.18 Consider a location an infinitesimal distance to the right of support *C* in the beam of Figure H7.18 and determine the principal stresses and the absolute maximum shear stress at the top of the rectangular section. Assume the 50-k load to act through the centroid of the cross section at point *A*.

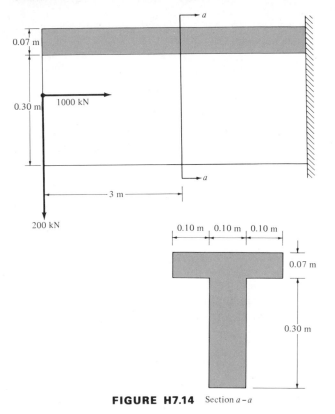

FIGURE H7.14   Section *a–a*

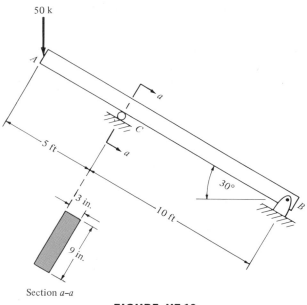

FIGURE H7.18

7.15 Refer to the beam in Figure H7.14 and compute the principal stresses and the absolute maximum shear stress at a point at the centroid of the T section at location *a–a* in the beam.

7.16 Refer to the beam in Figure H7.14 and compute the principal stresses and the absolute maximum shear stress at a point 0.10 m above the bottom of the T section at location *a–a* in the beam.

7.17 Refer to the beam in Figure H7.14 and com-

7.19 Refer to the beam of Figure H7.18 and for the same location along the beam defined in Problem 7.18, determine the principal stresses and the absolute maximum shear stress at the bottom of the rectangular section.

7.20 Refer to the beam of Figure H7.18 and for the same location along the beam defined in Problem 7.18, determine the principal stresses and the absolute value of the maximum shear stress at the centroid of the rectangular section.

7.21 Compute the principal stress of largest magnitude in section *a–a* of the frame shown in Figure H7.21.

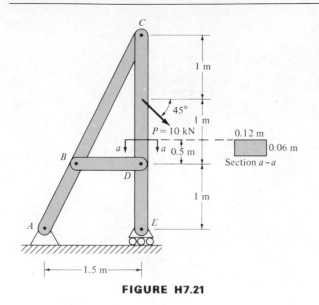

FIGURE H7.21

7.22 The compression block shown in Figure H7.22 is subjected to the load $P = 50$ k, which is eccentric with respect to both the u and v principal centroidal axes of the rectangular cross section, as shown. Determine the normal stresses at the four corners of the member at any location such as a–a.

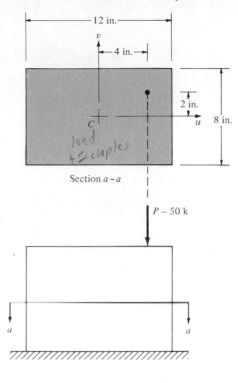

FIGURE H7.22

7.4 Torsional and Flexural Stresses

Many mechanical and structural components are called upon to resist the simultaneous action of torsional and flexural loads. A typical example of such a member is the crank shaft of an automobile engine. An idealized model of such a member is shown in Figure 7.7(a). The force $F_1 = F$ is transmitted from the piston through the connecting rod and is eccentric to the axis of the crank shaft so as to produce the required rotary motion. This force, $F_1 = F$, creates three types of action in the crank shaft: twisting, bending, and direct shear.

The eccentric force $F_1 = F$ may be transformed into a force $F_3 = F$ and a torque $T = Fe$ by applying two equal, collinear, and opposite forces $F_2 = F$ and $F_3 = F$ at point O. Point O is the origin of an x–u–v coordinate system in which the x axis is coincident with the longitudinal centroidal axis of the crank shaft. Note that the force system consisting of $F_1 = F$ and $F_2 = F$ represents a couple (torque) equal in magnitude to Fe and produces twisting of the crank shaft about its longitudinal centroidal

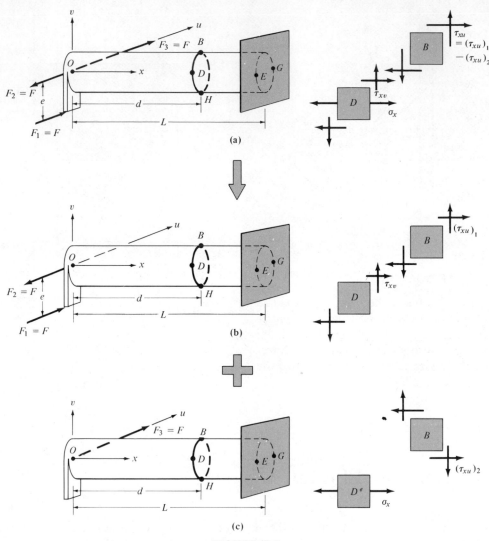

FIGURE 7.7

axis (i.e., about the x axis). The remaining force $F_3 = F$ produces bending of the crank shaft about axes parallel to the v axis. The magnitude of the bending moment M_v varies linearly from zero at point O to a value equal to FL at the fixed support.

The principle of superposition is again used to solve this problem. Under the assumption that the crank shaft is not stressed beyond the proportional limit of the material, the principle of superposition makes it possible to consider the member shown in Figure 7.7(a) as consisting of the two members shown in Figure 7.7(b) and (c). The member shown in Figure 7.7(b) is subjected to the pure torque $T = Fe$ and that shown in Figure 7.7(c) to a bending moment M_v and to the direct shear V_u caused by the force $F_3 = F$.

Consider the two points B and D on the surface of the crank shaft at a distance d from the origin of the coordinate system. Under the action of the pure torque in Figure 7.7(b), a stress element at point B is subjected to the shear stress $(\tau_{xu})_1 = T\rho/J = Fec/J$, in which c is the radius of the crank shaft. This same element is not stressed as a result of the action of the bending moment M_v in Figure 7.7(c), because it lies on the neutral axis for bending. However, the force $F_3 = F$ gives rise to a direct shear stress as explained in Section 5.4, and at point B the value of this shear stress $(\tau_{xu})_2$ is approximately equal to $\frac{4}{3}F/A$, where A is the cross-sectional area of the crank shaft. This value, $(\tau_{xu})_2 = \frac{4}{3}F/A$, follows from the application of Eq. 5.12b to a circular cross section. It should be pointed out, however, that for members that are long in comparison to their cross-sectional dimensions, the direct shear stress is relatively small in comparison to the shear stress produced by the twisting couple.

A stress element at point D is subjected to a shear stress $\tau_{xv} = (\tau_{xu})_1 = Fec/J$ as a result of the torque in Figure 7.7(b), and to a tensile flexural stress $\sigma_x = (M_v/I_v)u = Fdc/I_v$ as a result of the bending moment M_v in Figure 7.7(c). However, at point D the effect of the direct shear V_u due to $F_3 = F$ is zero. The stress elements for points B and D are depicted for the separate actions of the torque and of the bending moment in Figure 7.7(b) and (c), respectively. The superposition of these two separate actions leads to the composite stress elements shown in Figure 7.7(a). Note that the shear stress τ_{xu} on element B is the difference between $(\tau_{xu})_1$ and $(\tau_{xu})_2$, since these shear stresses at this point are opposite to each other in sense. However, at point H, diametrically opposite to point B on the surface of the crank shaft, these two shear stresses would have the same sense, and the resultant shear stress τ_{xu} would be the sum of $(\tau_{xu})_1$ and $(\tau_{xu})_2$.

Once the composite stress element at a given point on the crank has been established, Mohr's circle solution is used to determine the principal stresses and the maximum shear stress acting at the point. This procedure is illustrated in Example 7.4.

Example 7.4

Let the force F acting on the crank shaft of Figure 7.7(a) be 8000 lb. Assume the diameter of the crank shaft to be 4 in. and e, d, and L to be 6, 10, and 15 in., respectively. Determine (a) the principal stresses and the absolute maximum shear stress at point D, and (b) the principal stresses and the absolute maximum shear stress at the point where the crank shaft is most highly stressed. Show all of these stresses on properly oriented planes.

Solution

(a) The stress element at point D on the surface of the crank shaft is redrawn, for convenience, in Figure 7.8(a) in relation to x–v coordinate system of Figure 7.7. The stresses σ_x and $\tau_{xv} = \tau_{vx}$ are computed as follows:

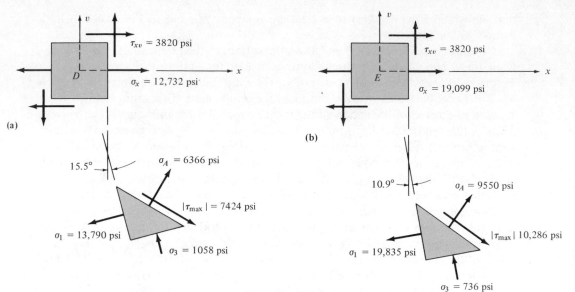

FIGURE 7.8

By Eq. 5.10,

$$\sigma_x = \frac{M_v}{I_v} u = \frac{Fd}{I_v} u = \frac{(8000)(10)(2)}{(\pi/64)(4)^4} = 12{,}732 \text{ psi}$$

By Eq. 4.19,

$$\tau_{xv} = \tau_{vx} = \frac{T\rho}{J} = \frac{Fec}{J} = \frac{(8000)(6)(2)}{(\pi/32)(4)^4} = 3820 \text{ psi}$$

A Mohr's circle solution leads to the following principal stresses:

$$\boxed{\sigma_1 = 13{,}790 \text{ psi}} \qquad \boxed{\sigma_2 = 0} \qquad \boxed{\sigma_3 = -1058 \text{ psi}}$$

By Eq. 2.13b,

$$|\tau_{max}| = \frac{\sigma_1 - \sigma_3}{2}$$

$$= \boxed{7424 \text{ psi}}$$

The principal stresses σ_1 and σ_3, as well as the absolute maximum shear stress, are shown in Figure 7.8(a) on an element properly oriented with respect to the

X plane (i.e., with respect to the plane perpendicular to the axis of the crank shaft). In other words, the plane on which the principal stress σ_1 acts is inclined at 15.5° ccw from the normal cross sections of the crank shaft.

(b) The torque is constant along the length of the crank shaft and, consequently, the shear stress $\tau_{xv} = \tau_{vx}$ on the surface of the crank shaft is the same along the entire length. As indicated previously, however, the bending moment varies linearly from zero at point O to $FL = (8000)(15) = 120,000$ lb-in. at the right support. Thus the most severely stressed points in the crank shaft are points E and G at the right support, where the flexural stresses assume their maximum values. Note that while the flexural stress at point E is tensile, it is equal in magnitude to the compressive flexural stress at point G. Note also that the effect of the direct shear V_u at these two points is zero. Thus the principal stresses and the absolute maximum shear stress, while opposite in sense, would have identical magnitudes at both points, and therefore only the stresses at point E will be investigated in this solution.

The stress element at point E is shown in Figure 7.8(b), in which $\tau_{xv} = \tau_{vx}$ has the same value (3820 psi) as that determined in part (a) and the flexural stress is given by Eq. 5.10. Thus

$$\sigma_x = \frac{M_v}{I_v} u$$

$$= \frac{(120,000)(2)}{(\pi/64)(4)^4} = 19,099 \text{ psi}$$

A Mohr's circle solution leads to the following principal stresses

$$\boxed{\sigma_1 = 19,835 \text{ psi}} \qquad \boxed{\sigma_2 = 0} \qquad \boxed{\sigma_3 = -736 \text{ psi}}$$

By Eq. 2.13b,

$$\left|\tau_{\max}\right| = \frac{\sigma_1 - \sigma_3}{2}$$

$$= \boxed{10,286 \text{ psi}}$$

The principal stresses σ_1 and σ_3 as well as the absolute maximum shear stress are shown in Figure 7.8(b) on an element properly oriented with respect to the X plane (i.e., with respect to the plane perpendicular to the axis of the crank shaft). In other words, the plane on which the principal stress σ_1 acts is inclined at 10.9° ccw from the normal cross sections of the crank shaft.

Homework Problems

7.23 The following data apply to the crank shaft shown in Figure 7.7(a): $F = 10$ kN, diameter = 0.10 m, $e = 0.18$ m, $d = 0.25$ m, and $L = 0.40$ m. Determine:

(a) The principal stresses and the absolute maximum shear stress at point B.

(b) The principal stresses and the absolute maximum shear stress at the point where the crank shaft is most highly stressed.

Show all these stresses on properly oriented planes.

7.24 The following data apply to the crank shaft shown in Figure 7.7(a): diameter = 5 in., $e = 10$ in., $d = 20$ in. The maximum value of the tensile stress at point B is known to be 15,000 psi. Determine the force F.

7.25 Repeat Problem 7.24 if the absolute value of the shear stress at point B, instead of the maximum value of the tensile stress at this point, is known to be 12,000 psi. All other conditions remain the same.

7.26 Compute the principal stresses and the absolute maximum shear stress at point B on the outside surface of the hollow shaft shown in Figure H7.26.

Show these stresses on properly oriented planes. The pulley is fastened rigidly to the shaft.

7.27 Compute the principal stresses and the absolute maximum shear stress at point D on the outside surface of the hollow shaft shown in Figure H7.26. Show these stresses on properly oriented planes.

7.28 Compute the principal stresses and the absolute maximum shear stress at the point where the hollow shaft in Figure H7.26 is most severely stressed. Show these stresses on properly oriented planes.

7.29 Determine the principal stresses and the absolute maximum shear stress at point B on the surface of the 5-in.-diameter shaft shown in Figure H7.29. Show these stresses on properly oriented planes. The two 18-in.-diameter pulleys are fastened rigidly to the solid shaft, which is supported in frictionless bearings at its two ends. These two bearings may be assumed to act as simple beam supports.

7.30 Compute the principal stresses and the absolute maximum shear stress at point D on the surface of the shaft described in Problem 7.29. Show all these stresses on properly oriented planes.

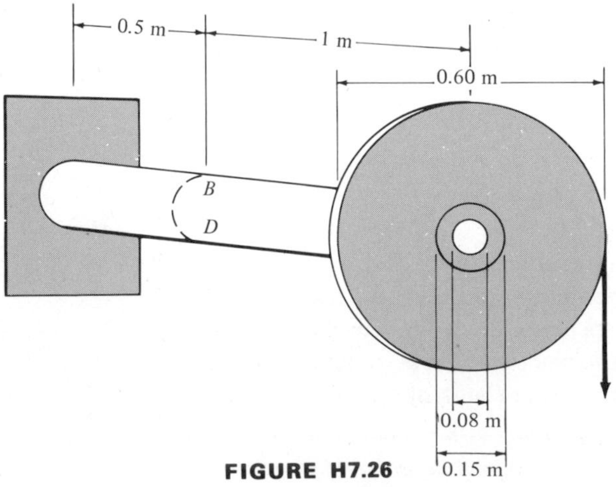

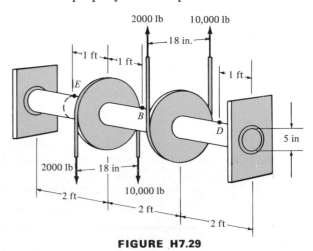

FIGURE H7.26

FIGURE H7.29

7.31 Compute the principal stresses and the absolute maximum shear stress at point E on the surface of the shaft described in Problem 7.29. Show all these stresses on properly oriented planes.

7.32 The following data apply to the crank shaft shown in Figure 7.7(a): $F = 12$ kN, $e = 0.30$ m, $d = 0.50$ m, and the maximum tensile stress at point D is known to be 100 MPa. Determine the diameter of the crank shaft.

7.33 Repeat Problem 7.32 if the absolute maximum shear stress at point D, instead of the maximum tensile stress at this point, is known to be 70 MPa. All other conditions remain unchanged.

7.34 The following data apply to the crank shaft shown in Figure 7.7(a): $F = 20$ k, $e = 15$ in., $d = 30$ in., and the maximum compressive stress at point B is known to be 2 ksi. Compute the diameter of the crank shaft.

7.5 Torsional, Flexural, and Axial Stresses

In numerous cases, structural and machine members are subjected to the simultaneous action of torsion, flexure, and direct axial loads. This type of loading occurs, for example, in shafts that are keyed to bevel gears. The geometry of the bevel gear is such that, in general, the interacting force between two mating gears has components that cause extension or contraction of the shaft in addition to twisting and bending, which may occur in more than one plane. A simplified model of such a member bent in one plane only is depicted in Figure 7.9(a), which differs from the member shown in Figure 7.7(a) in that an axial tensile force P has been added. The eccentric force $F_1 = F$ is transformed into a force $F_2 = F$ that produces bending, direct shear, and a torque $T = Fe$ as was discussed in Section 7.4. Assuming that the shaft is not stressed beyond the proportional limit, one may use the principle of superposition to separate the composite action shown in Figure 7.9(a) into the three separate actions shown in Figure 7.9(b), (c), and (d). As indicated in Section 7.4, the member shown in Figure 7.9(b) is subjected to the pure torque $T = Fe$ and that shown in Figure 7.9(c) to the direct shear V_u and the bending moment M_v produced by the force $F_3 = F$. The member shown in Figure 7.9(d) is subjected to the pure tensile axial force P that produces extension of the shaft fibers.

Once again, consider the two points B and D on the surface of the shaft at a distance d from the origin of the coordinate system. An element at point B as shown in Figure 7.9(b) is subjected to the shear stress $(\tau_{xu})_1 = T\rho/J = Fec/J$, in which c is the radius of the shaft. If we consider the action of the bending moment M_v produced by force $F_3 = F$ as shown in Figure 7.9(c), the element at point B is unstressed, since it lies on the neutral axis for bending. As explained earlier, the direct shear stress $(\tau_{xu})_2$ at point B due to $F_3 = F$ is, approximately, $\frac{4}{3}F/A$, where A is the cross-sectional area of the shaft. This shear stress is shown on the element at point B in Figure 7.9(c). The tensile force P in Figure 7.9(d) produces a uniform tensile stress distribution on any

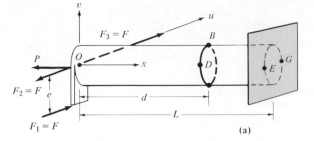
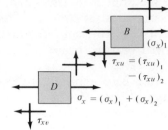

(a)

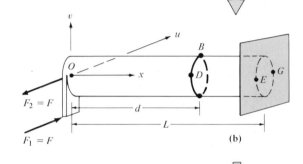
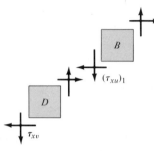

(b)

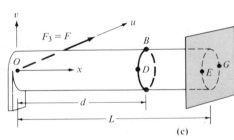
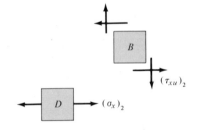

(c)

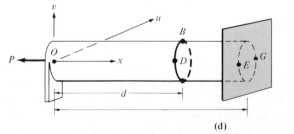
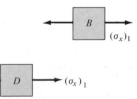

(d)

FIGURE 7.9

cross section of the shaft. Thus, owing to the force P, an element at point B is subjected to the tensile stress $(\sigma_x)_1 = F_n/A = P/A$ as shown in Figure 7.9(d).

An element at point D in Figure 7.9(b) is acted upon by the shear stress $\tau_{xv} = (\tau_{xu})_1 = Fec/J$. This same element in Figure 7.9(c) is subjected to the flexural tensile stress

$$(\sigma_x)_2 = \frac{M_v}{I_v} u = \frac{Fdc}{I_v}$$

Finally, this element in Figure 7.9(d) is subjected to the tensile stress $(\sigma_x)_1 = P/A$.

The stresses developed at points B and D from the separate actions depicted in Figure 7.9(b), (c), and (d) are superposed to yield the composite stress elements shown for these two points in Figure 7.9(a). The Mohr's circle solution is then applied to obtain the principal stresses at these points. Such a solution is illustrated in Example 7.5.

Example 7.5

Let the force F acting on the shaft of Figure 7.9(a) be 10 kN and the tensile force P be 50 kN. Assume the diameter of the shaft to be 0.10 m and e, d, and L to be 0.20, 0.30, and 0.50 m, respectively. Determine (a) the principal stresses and the absolute maximum shear stress at point D, and (b) the principal stresses and the absolute maximum shear stress at the point where the shaft is most highly stressed. Show all these stresses on properly oriented planes.

Solution

(a) The stress element at point D is redrawn, for convenience, in Figure 7.10(a) in relation to the x–v coordinate system of Figure 7.9. The stresses σ_x and $\tau_{xv} = \tau_{vx}$ are determined as follows:

$$\sigma_x = (\sigma_x)_1 + (\sigma_x)_2$$

where, by Eq. 3.1,

$$(\sigma_x)_1 = \frac{P}{A} = \frac{50 \times 10^3}{(\pi/4)(0.10)^2} = 6.37 \text{ MPa}$$

and, by Eq. 5.10,

$$(\sigma_x)_2 = \frac{M_v}{I_v} u = \frac{Fd}{I_v} u$$

$$= \frac{10 \times 10^3 (0.30)(0.05)}{\pi/64 (0.10)^4} = 30.56 \text{ MPa}$$

Therefore,

$$\sigma_x = 6.37 + 30.56 = 36.93 \text{ MPa}$$

By Eq. 4.19,

$$\tau_{xv} = \tau_{vx} = \frac{T\rho}{J} = \frac{Fec}{J}$$

$$= \frac{10 \times 10^3 (0.20)(0.05)}{\pi/32(0.10)^4} = 10.19 \text{ MPa}$$

A Mohr's circle solution leads to the following principal stresses:

$$\boxed{\sigma_1 = 39.56 \text{ MPa}} \qquad \boxed{\sigma_2 = 0} \qquad \boxed{\sigma_3 = -2.62 \text{ MPa}}$$

By Eq. 2.13b,

$$|\tau_{\max}| = \frac{\sigma_1 - \sigma_3}{2}$$

$$= \boxed{21.09 \text{ MPa}}$$

All these stresses are depicted on properly oriented planes in Figure 7.10(a).

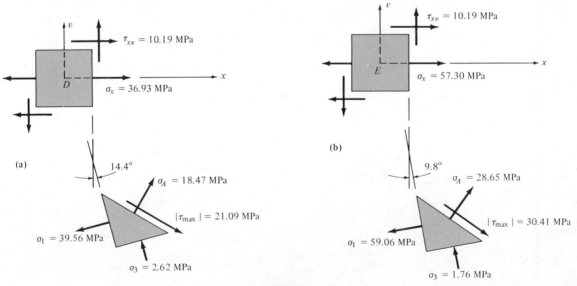

FIGURE 7.10

(b) Both the torque and the axial load are constants along the entire length of the shaft. Therefore, both the shear stress $\tau_{xv} = \tau_{vx}$ and the tensile stress $(\sigma_x)_1$ on the surface of the shaft are the same along its entire length. The bending moment, however, increases to a maximum value at the fixed support. At this location, the bending moment produces a flexural stress $(\sigma_x)_2$, which is tension at point E and compression at point G. Since the direct normal stress at both points is tensile, the most severely stressed point in the shaft is point E, where $(\sigma_x)_1$ and $(\sigma_x)_2$ are additive. The stress element at this point is shown in Figure 7.10(b), in which $\tau_{xv} = \tau_{vx}$ has the same magnitude (10.19 MPa) as found in part (a) of the solution, and the normal stress σ_x is determined as follows:

$$\sigma_x = (\sigma_x)_1 + (\sigma_x)_2$$

where, by Eq. 3.1,

$$(\sigma_x)_1 = \frac{P}{A} = \frac{50 \times 10^3}{(\pi/4)(0.10)^2} = 6.37 \text{ MPa}$$

and, by Eq. 5.10,

$$(\sigma_x)_2 = \frac{M_v}{I_v} u = \frac{FL}{I_v} u$$

$$= \frac{10 \times 10^3 (0.50)(0.05)}{\pi/64 (0.10)^4} = 50.93 \text{ MPa}$$

Therefore,

$$\sigma_x = 6.37 + 50.93 = 57.30 \text{ MPa}$$

A Mohr's circle solution yields the following principal stresses:

$$\boxed{\sigma_1 = 59.06 \text{ MPa}} \qquad \boxed{\sigma_2 = 0} \qquad \boxed{\sigma_3 = -1.76 \text{ MPa}}$$

By Eq. 2.13b,

$$|\tau_{max}| = \frac{\sigma_1 - \sigma_3}{2}$$

$$= \boxed{30.41 \text{ MPa}}$$

All of these stresses are shown on properly oriented planes in Figure 7.10(b).

Homework Problems

7.35 The following data apply to the shaft shown in Figure 7.9(a): $F = 20$ kN, $P = 60$ kN, diameter $= 0.08$ m, $e = 0.30$ m, $d = 0.50$ m, $L = 0.65$ m. Determine the principal stresses and the absolute maximum shear stress at point B. Show all of these stresses on properly oriented planes.

7.36 Determine the principal stresses and the absolute maximum shear stress at the point where the shaft described in Problem 7.35 is most highly stressed. Show all these stresses on properly oriented planes.

7.37 The following data apply to the shaft shown in Figure 7.9(a): $P = 7$ F, diameter $= 5$ in., $e = 45$ in., $d = 40$ in. The maximum value of the tensile stress at point B is known to be 15,000 psi. Determine P and F.

7.38 Repeat Problem 7.37 if the maximum value of the compressive stress at point D, instead of the maximum value of the tensile stress at point B, is known to be 10,000 psi. All other conditions remain the same.

7.39 Repeat Problem 7.37 if the absolute maximum shear stress at point B, instead of the maximum value of the tensile stress at this point, is known to be 12,000 psi. All other conditions remain the same.

7.40 Repeat Problem 7.37 if the absolute maximum shear stress at point D, instead of the maximum value of the tensile stress at point B, is known to be 20,000 psi. All other conditions remain the same.

7.41 In addition to the 15-kN load, the hollow shaft shown in Figure H7.26 is subjected to a compressive axial force of 30 kN. Compute the principal stresses and the absolute maximum shear stress at point B on the outside surface of the hollow shaft. Show these stresses on properly oriented planes.

7.42 Repeat Problem 7.41 for point D on the outside surface of the hollow shaft shown in Figure H7.26.

7.43 Compute the principal stresses and the absolute maximum shear stress at the point where the hollow shaft described in Problem 7.41 is most highly stressed. Show these stresses on properly oriented planes.

7.44 In addition to the given loads, the shaft shown in Figure H7.29 is subjected to a tensile axial force of 50,000 lb. Compute the principal stresses and the absolute maximum shear stress at point B on the surface of the 5-in.-diameter shaft. Show these stresses on properly oriented planes.

7.45 Repeat Problem 7.44 for point D on the surface of the shaft shown in Figure H7.29.

7.46 Repeat Problem 7.44 for point E on the surface of the shaft shown in Figure H7.29.

7.47 The following data apply to the shaft shown in Figure 7.9(a): $F = 15$ kN, $P = 100$ kN, $e = 0.30$ m, $d = 0.60$ m, and the absolute maximum shear stress at point B is known to be 60 MN/m^2. Determine the diameter of the shaft.

7.48 The following data apply to the shaft shown in Figure 7.9(a): $F = 15$ k, $P = 75$ k, $e = 15$ in., $d = 50$ in., and the maximum tensile stress at point D is known to be 50 ksi. Compute the diameter of the shaft.

7.49 The following data apply to the shaft shown in Figure 7.9(a): $F = 20$ kN, $e = 0.45$ m, $d = 0.80$ m and the diameter of the shaft is 0.10 m. If the maximum tensile stress at point D is known to be 200 MPa, determine the value of the tensile axial force P.

*7.6 Theories of Failure

For purposes of this section, the failure of a member is defined as one of two conditions:

1. Fracture of the material of which the member is made. This type of failure is characteristic of brittle materials.
2. Initiation of inelastic (plastic) behavior in the material. This type of failure is the one generally exhibited by ductile materials.

When a member is subjected to simple tension or compression, failure, as defined earlier, occurs when the applied axial stress reaches a limiting value σ_0. If the member is ductile, σ_0 represents the yield point or the yield strength of the material; if the member is brittle, it represents the ultimate or the fracture strength. However, when a member is subjected to a biaxial or triaxial state of stress, the cause of failure, whether by fracture or by inelastic action, is unknown and failure is predicted by one of several theories of failure.

A failure theory is a criterion that is used in an effort to predict the failure of a given material when subjected to a complex stress condition, from knowledge of the properties of this material as obtained from the simple tension, compression, or torsion tests. Several theories have been proposed, but only the four most widely used theories will be presented and discussed briefly.

Statement of Theories of Failure

The Maximum Principal Stress Theory

According to this theory, failure of a member subjected to any state of stress occurs when the principal stress of largest magnitude, $|\sigma_{max}|$, in the member reaches the limiting value σ_0, as obtained from the simple tension or compression test. Thus, according to this theory, it is immaterial how complex or how simple the state of stress in a member is. This member will fail when the principal stress of largest magnitude in the member reaches the critical value σ_0, which is assumed in this section to be numerically the same for tension and for compression. Since, in a given situation, the principal stress of largest magnitude $|\sigma_{max}|$ may be either σ_1 or σ_3, the *maximum principal stress theory* may be represented mathematically as follows:

$$|\sigma_{max}| = |\sigma_1| = \sigma_0 \qquad \text{or} \qquad |\sigma_{max}| = |\sigma_3| = \sigma_0 \qquad \textbf{(7.2)}$$

For ductile materials σ_0 in Eq. 7.2 is the yield point or the yield strength, and for brittle materials it is the ultimate or the fracture strength as obtained from the simple tension or compression test.

Experimental evidence shows that the maximum principal stress theory does not predict the behavior of ductile material satisfactorily. However, it has been found that this theory of failure is in close agreement with the results of experiments for brittle materials. Thus the maximum principal stress theory is generally used to predict failure of brittle materials such as cast iron when subjected to complex stress conditions.

The Maximum Principal Strain Theory

The *maximum principal strain theory* predicts failure of a member subjected to any combination of loads, when the principal strain of largest magnitude, $|\varepsilon_{max}|$, in the member reaches a limiting value ε_0. This limiting or critical value ε_0 is obtained from the simple tension or compression test and represents the strain corresponding to the yield point or the yield strength (i.e., $\varepsilon_0 = \sigma_0/E$) in the case of a ductile material. Thus since Hooke's law is assumed in the determination of ε_0, this theory of failure is applicable only when the stresses are within the proportional range for the material. Furthermore, the quantity $\varepsilon_0 = \sigma_0/E$ is assumed in this section to have the same numerical value in tension and in compression.

The principal strain of largest magnitude, $|\varepsilon_{max}|$, could be one of the two principal strains ε_1 or ε_3, depending upon the stress conditions existing in the member. Thus the maximum principal strain theory may be represented by the following mathematical expression:

$$|\varepsilon_{max}| = |\varepsilon_1| = \varepsilon_0 \quad \text{or} \quad |\varepsilon_{max}| = |\varepsilon_3| = \varepsilon_0 \qquad \textbf{(7.3a)}$$

where ε_1 and ε_3 are defined by Eq. 2.30′ in terms of the three principal stresses σ_1, σ_2, and σ_3. In the special plane stress case in which one of the three principal stresses, say σ_2, is zero, the following relations may be written to represent the maximum principal strain theory. Thus, if

$$|\varepsilon_{max}| = |\varepsilon_1| = \left| \frac{\sigma_1}{E} - \mu \frac{\sigma_3}{E} \right| = \frac{\sigma_0}{E}$$

it follows that

$$|\sigma_1 - \mu\sigma_3| = \sigma_0 \qquad \textbf{(7.3b)}$$

Also, if

$$|\varepsilon_{max}| = |\varepsilon_3| = \left| \frac{\sigma_3}{E} - \mu \frac{\sigma_1}{E} \right| = \frac{\sigma_0}{E}$$

it follows that

$$|\sigma_3 - \mu\sigma_1| = \sigma_0 \qquad \textbf{(7.3c)}$$

As in the case of the maximum principal stress theory, the maximum principal strain theory has been found by experiment to be unsatisfactory for the prediction of ductile material failure. However, it is reasonably adequate to predict failure of brittle materials if for ε_0 we use the strain at fracture.

Maximum Shear Stress Theory

The *maximum shear stress theory* states that a member subjected to any state of stress fails when the absolute maximum shear stress $|\tau_{\max}|$ in the member reaches the critical value τ_0 as obtained from the simple tension or compression test. The critical or limiting value τ_0 is the maximum shear stress in the tension or compression test when failure by yielding (inelastic action) occurs. For a uniaxial stress condition, the shear stress is equal to one-half the value of the normal stress, as may be concluded from Mohr's circle. Therefore, in the simple tension or compression test, when yielding takes place, $\tau_0 = \sigma_0/2$.

The absolute maximum shear stress for any combination of stresses may be found in terms of the principal stress by Eq. 2.13b to be $(\sigma_1 - \sigma_3)/2$. Thus a mathematical expression may be written to represent the maximum shear stress theory of failure as follows:

$$\left| \frac{\sigma_1 - \sigma_3}{2} \right| = \frac{\sigma_0}{2}$$

from which

$$|\sigma_1 - \sigma_3| = \sigma_0 \tag{7.4}$$

Thus for a given state of stress, the principal stresses σ_1 and σ_3 are determined by the methods of Sections 2.5 and 2.6. The absolute value of the difference between these two principal stresses is then equated to the property σ_0 as obtained from the simple tension or compression test. As stated earlier, the quantity σ_0 is assumed in this section to have the same numerical value in tension and in compression.

Experimental evidence indicates that the maximum shear stress theory is in reasonable agreement with the performance of ductile materials, and it is, therefore, commonly used to predict the behavior of these materials under complex states of stress. Attempts to predict behavior of brittle materials by this theory have not been successful.

The Energy of Distortion Theory

According to the *energy of distortion theory*, failure of a member subjected to combined loads takes place when the strain energy per unit volume used up in changing the shape of the member (energy of distortion) becomes equal to the energy of distortion per unit volume absorbed at failure by inelastic action in the uniaxial tension or compression test. Thus this theory assumes that only a part of the total strain energy stored in a member (the energy of distortion part) is responsible for failure of the member by inelastic action. The remaining part, known as the *energy of volume change*, is responsible for the change in volume of the member and does not contribute to failure by inelastic action. Some experimental evidence exists that supports this assumption: Several materials were subjected to large hydrostatic pressure, resulting in appreciable changes in volume but no changes in shape, and no failure by yielding was detected.

To develop a mathematical relation that expresses the energy of distortion theory, we need to decompose the total strain energy per unit volume, u, imparted to a member into its two component parts. As mentioned earlier, these two component parts are the energy of distortion per unit volume, u_d, and the energy of volume change per unit volume, u_v. Thus

$$u = u_v + u_d \tag{7.5a}$$

and

$$u_d = u - u_v \tag{7.5b}$$

Therefore, to determine the energy of distortion per unit volume u_d for any state of stress, one needs to determine the total strain energy per unit volume, u, and subtract from it the energy of volume change per unit volume, u_v.

The total strain energy per unit volume, u, stored in a member subjected to a uniaxial state of stress in which the only nonzero principal stress is σ_1 is given by Eq. 3.5 to be $u = \sigma_1 \varepsilon_1 / 2$. The strain energy per unit volume stored in a member subjected to a triaxial state of stress represented by the three principal stresses σ_1, σ_2, and σ_3 may be obtained by superposition if it is assumed that the material is linearly elastic and not stressed beyond the elastic limit. Thus

$$u = \frac{\sigma_1 \varepsilon_1}{2} + \frac{\sigma_2 \varepsilon_2}{2} + \frac{\sigma_3 \varepsilon_3}{2} \tag{7.6a}$$

Substitution of the values of the three principal strains in terms of the principal stresses from Eq. 2.30' leads, after simplification, to the total strain energy per unit volume stored in a member subjected to a triaxial state of stress. Thus

$$u = \frac{1}{2E} \left[(\sigma_1^2 + \sigma_2^2 + \sigma_3^2) - 2\mu(\sigma_1 \sigma_2 + \sigma_2 \sigma_3 + \sigma_1 \sigma_3) \right] \tag{7.6b}$$

Consider a unit volume of material subjected to the three principal stresses σ_1, σ_2, and σ_3. The principal strains associated with σ_1, σ_2, and σ_3 are, respectively, $\varepsilon_1, \varepsilon_2$, and ε_3 and may be expressed in terms of these stresses by Eq. 2.30'. Thus the change in volume per unit volume, Δ_v, becomes

$$\Delta_v = (1 + \varepsilon_1)(1 + \varepsilon_2)(1 + \varepsilon_3) - 1 \tag{7.7a}$$

Since the strains ε_1, ε_2, and ε_3 are very small quantities, their products may be neglected in comparison to unity, and the change in volume per unit volume may be written as

$$\Delta_v = \varepsilon_1 + \varepsilon_2 + \varepsilon_3 \tag{7.7b}$$

The total strain energy per unit volume given by Eq. 7.6b may be separated into the two component parts u_v and u_d by resolving the three principal stresses σ_1, σ_2, and σ_3 acting on the unit volume shown in Figure 7.11(a) into the two states of stress shown in Figure 7.11(b) and (c). The state of stress shown in Figure 7.11(b) represents a hydrostatic stress condition is which all three principal stresses are equal to the quantity σ_A. The quantity σ_A needs to be adjusted so that the stress condition shown in Figure 7.11(b) produces the entire volume change in the unit element. It follows, then, that the stress condition shown in Figure 7.11(c) would produce all of the distortion in the unit element without producing any volume change. If the stress condition shown in Figure 7.11(c) is to produce no volume change in the unit element, then, by Eq. 7.7b, the algebraic sum of the three principal strains produced by the three principal stresses $\sigma_1 - \sigma_A$, $\sigma_2 - \sigma_A$, and $\sigma_3 - \sigma_A$ must be equal to zero. Thus using Eqs. 2.30' and 7.7b, we obtain

$$\Delta_v = \frac{1}{E}\{(\sigma_1 - \sigma_A) - \mu[(\sigma_2 - \sigma_A) + (\sigma_3 - \sigma_A)]\}$$

$$+ \frac{1}{E}\{(\sigma_2 - \sigma_A) - \mu[(\sigma_1 - \sigma_A) + (\sigma_3 - \sigma_A)]\} \tag{7.8a}$$

$$+ \frac{1}{E}\{(\sigma_3 - \sigma_A) - \mu[(\sigma_1 - \sigma_A) + (\sigma_2 - \sigma_A)]\} = 0$$

When Eq. 7.8a is solved for σ_A, we obtain

$$\sigma_A = \tfrac{1}{3}(\sigma_1 + \sigma_2 + \sigma_3) \tag{7.8b}$$

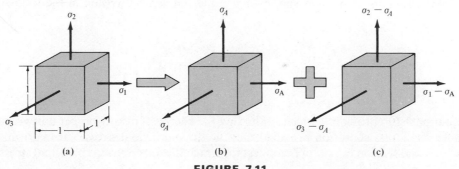

FIGURE 7.11

which indicates that in order for the state of stress in Figure 7.11(c) to produce no volume change in the unit element, the hydrostatic stress σ_A must be equal to the average of the three principal stresses. Under these conditions, the entire volume change in the unit element is produced by the state of stress shown in Figure 7.11(b), and the energy of volume per unit volume, u_v, may be obtained by adding the energies produced by the three stresses σ_A in a manner analogous to Eq. 7.6a. Thus

$$u_v = \frac{1}{2}\sigma_A \frac{1}{E}(\sigma_A - \mu\sigma_A - \mu\sigma_A)$$

$$+ \frac{1}{2}\sigma_A \frac{1}{E}(\sigma_A - \mu\sigma_A - \mu\sigma_A)$$

$$+ \frac{1}{2}\sigma_A \frac{1}{E}(\sigma_A - \mu\sigma_A - \mu\sigma_A)$$

$$= \frac{3}{2}\frac{1 - 2\mu}{E}\sigma_A^2 \tag{7.9a}$$

When the value of σ_A is substituted from Eq. 7.8b, we obtain the following relation for the energy of volume change per unit of volume:

$$u_v = \frac{1 - 2\mu}{6E}(\sigma_1 + \sigma_2 + \sigma_3)^2 \tag{7.9b}$$

The energy of distortion per unit volume, u_d, may now be obtained from Eq. 7.5b by substituting for u its value from Eq. 7.6b and for u_v its value from Eq. 7.9b. Thus

$$u_d = \frac{1}{2E}[(\sigma_1^2 + \sigma_2^2 + \sigma_3^2) - 2\mu(\sigma_1\sigma_2 + \sigma_2\sigma_3 + \sigma_1\sigma_3)]$$

$$- \frac{1 - 2\mu}{6E}(\sigma_1 + \sigma_2 + \sigma_3)^2 \tag{7.10a}$$

After rearranging terms and simplifying, Eq. 7.10a can be written in the following form:

$$u_d = \frac{1 + \mu}{6E}[(\sigma_1 - \sigma_2)^2 + (\sigma_2 - \sigma_3)^2 + (\sigma_3 - \sigma_1)^2] \tag{7.10b}$$

Equation 7.10b provides a means of finding the energy of distortion per unit volume under any stress condition provided that the three principal stresses are known. In the uniaxial tension or compression test for which the only nonzero principal stress is

σ_1, failure by inelastic action occurs when this principal stress reaches the limiting value σ_0. Thus the energy of distortion per unit volume stored in the material during the uniaxial tension or compression test is given by the expression

$$(u_d)_{\text{uniaxial}} = \frac{1+\mu}{3E}\sigma_0^2 \tag{7.11}$$

which is obtained from Eq. 7.10b by setting σ_2 and σ_3 equal to zero and σ_1 equal to σ_0. The quantity expressed in Eq. 7.11 represents a unique property for a given material and is assumed, in this section to be numerically the same in tension and in compression.

The energy of distortion theory may now be stated mathematically as follows: failure by inelastic action of a member subjected to a complex state of stress occurs when the energy of distortion per unit volume in this member as given by Eq. 7.10b reaches the critical value of the material given by Eq. 7.11. Thus

$$\frac{1+\mu}{6E}[(\sigma_1 - \sigma_2)^2 + (\sigma_2 - \sigma_3)^2 + (\sigma_3 - \sigma_1)^2] = \frac{1+\mu}{3E}\sigma_0^2 \tag{7.12a}$$

which, when simplified, reduces to

$$(\sigma_1 - \sigma_2)^2 + (\sigma_2 - \sigma_3)^2 + (\sigma_3 - \sigma_1)^2 = 2\sigma_0^2 \tag{7.12b}$$

Equation 7.12b represents the energy of distortion theory of failure by yielding (i.e., by inelastic action)—which has been found to provide the best agreement, among all other theories of failure, with the behavior of ductile materials.

The relation expressed in Eq. 7.12b may also be obtained on the basis of what has come to be known as the *octahedral shear stress theory*. This theory states that failure by inelastic action of a member subjected to combined loads occurs when the octahedral shear stress in this member reaches a critical value equal to the octahedral shear stress at yielding in the uniaxial tension or compression test.

The octahedral plane is defined as one whose normal makes equal angles with the three principal directions. For a triaxial state of stress represented by σ_1, σ_2, and σ_3, the octahedral shear stress, τ_{oct}, is given by

$$\tau_{\text{oct}} = \tfrac{1}{3}[(\sigma_1 - \sigma_2)^2 + (\sigma_2 - \sigma_3)^2 + (\sigma_3 - \sigma_1)^2]^{1/2} \tag{7.12c}$$

In the uniaxial tension or compression test for which the only nonzero principal stress is σ_1, failure by inelastic action occurs when this principal stress reaches the limiting value σ_0. Thus the octahedral shear stress at failure by yielding in the uniaxial tension or compression case is obtained from Eq. 7.12c by setting $\sigma_2 = \sigma_3 = 0$ and $\sigma_1 = \sigma_0$ and is found to be

$$(\tau_{\text{oct}})_{\text{uniaxial}} = \frac{\sqrt{2}}{3}\sigma_0 \qquad \textbf{(7.12d)}$$

Therefore, a mathematical expression for the octahedral shear stress theory of failure results when Eq. 7.12c is set equal to Eq. 7.12d. Thus

$$\frac{1}{3}[(\sigma_1 - \sigma_2)^2 + (\sigma_2 - \sigma_3)^2 + (\sigma_3 - \sigma_1)^2]^{1/2} = \frac{\sqrt{2}}{3}\sigma_0$$

from which

$$(\sigma_1 - \sigma_2)^2 + (\sigma_2 - \sigma_3)^2 + (\sigma_3 - \sigma_1)^2 = 2\sigma_0^2 \qquad \textbf{(7.12e)}$$

which is identical to Eq. 7.12b.

Comparison of Theories of Failure

Each of the theories of failure that has been presented agrees with results of experiments only under certain conditions of stress and for a specific type of material (i.e., ductile or brittle). In other words, none of the various theories of failure predicts failure successfully under all possible load combinations and for both ductile and brittle materials. As has already been pointed out, experimental evidence has shown that, in general and as a rule of thumb, the maximum principal stress and maximum principal strain theories of failure are successful in predicting failure of brittle materials, while the maximum shear stress and the energy of distortion theories (or the octahedral shear stress theory) are successful in predicting failure of ductile materials.

In addition to the type of material (ductile or brittle), theories of failure may be compared to each other under an infinite number of states of stress. Their comparison under one state of stress will not, in general, be the same as their comparison under other states of stress. Therefore, in comparing theories of failure, one has to be careful in specifying not only the type of material, but also the state of stress under which the comparison is being made. In this text a theoretical comparison is made of the four theories of failure that were presented earlier on the basis of the plane stress condition shown in Figure 7.12(a). Note that as discussed in Sections 7.2, 7.4, and 7.5, such a stress condition results from the application of a torque and an axial load; a torque and a bending moment; or a torque, an axial load, and a bending moment to a circular shaft.

The first step in the application of any one of the theories of failure is to determine the three principal stresses, since all of the four theories have been expressed in terms

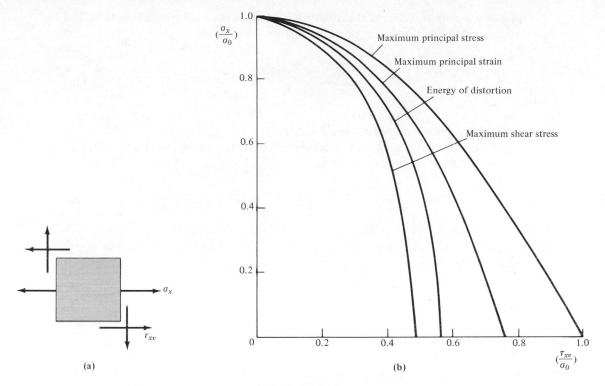

(a)

(b)

FIGURE 7.12

of these stresses. Thus by the use of a Mohr's circle solution, the principal stresses are found to be as follows:

$$\sigma_1 = \frac{\sigma_x}{2} + \sqrt{\left(\frac{\sigma_x}{2}\right)^2 + \tau_{xv}^2} \qquad \sigma_3 = \frac{\sigma_x}{2} - \sqrt{\left(\frac{\sigma_x}{2}\right)^2 + \tau_{xv}^2}$$

and, of course, $\sigma_2 = 0$, since Figure 7.12(a) represents a plane stress condition.

Application of the four theories of failure to the stress condition depicted in Figure 7.12(a) is accomplished as follows.

The Maximum Principal Stress Theory
Since the principal stress of largest magnitude is σ_1, Eq. 7.2 expressing this theory may be written

$$\frac{\sigma_x}{2} + \sqrt{\left(\frac{\sigma_x}{2}\right)^2 + \tau_{xv}^2} = \sigma_0$$

Simplification and division by the quantity σ_0^2 leads to the following dimensionless expression for the maximum principal stress theory of failure:

$$\frac{\sigma_x}{\sigma_0} + \left(\frac{\tau_{xv}}{\sigma_0}\right)^2 = 1 \tag{7.13a}$$

The Maximum Principal Strain Theory

Equation 7.3b is the applicable relation, since for the stress condition shown in Figure 7.12(a), the principal strain of largest magnitude is ε_1. Thus

$$\frac{\sigma_x}{2} + \sqrt{\left(\frac{\sigma_x}{2}\right)^2 + \tau_{xv}^2} - \mu\left(\frac{\sigma_x}{2} - \sqrt{\left(\frac{\sigma_x}{2}\right)^2 + \tau_{xv}^2}\right) = \sigma_0$$

If the value of μ is assumed to be 0.3 and the relation is simplified and divided by σ_0^2, one obtains the following approximate dimensionless expression for this theory of failure:

$$0.3\left(\frac{\sigma_x}{\sigma_0}\right)^2 + 0.7\left(\frac{\sigma_x}{\sigma_0}\right) + 1.7\left(\frac{\tau_{xv}}{\sigma_0}\right)^2 = 1 \tag{7.13b}$$

The Maximum Shear Stress Theory

Equation 7.4 expressing this theory of failure becomes

$$2\sqrt{\left(\frac{\sigma_x}{2}\right)^2 + \tau_{xv}^2} = \sigma_0$$

Simplification and division by σ_0^2 yields the following dimensionless relation for the maximum shear stress theory of failure:

$$\left(\frac{\sigma_x}{\sigma_0}\right)^2 + 4\left(\frac{\tau_{xv}}{\sigma_0}\right)^2 = 1 \tag{7.13c}$$

The Energy of Distortion Theory

Substitution of the values of the three principal stresses into Eq. 7.12b yields

$$\left[\frac{\sigma_x}{2} + \sqrt{\left(\frac{\sigma_x}{2}\right)^2 + \tau_{xv}^2}\right]^2 + \left[\frac{\sigma_x}{2} - \sqrt{\left(\frac{\sigma_x}{2}\right)^2 + \tau_{xv}^2}\right]^2 \cdot$$

$$+ \left[2\sqrt{\left(\frac{\sigma_x}{2}\right)^2 + \tau_{xv}^2}\right]^2 = 2\sigma_0^2$$

Simplification of this relation and division by the quantity σ_0^2 results in the following dimensionless expression for this theory of failure:

$$\left(\frac{\sigma_x}{\sigma_0}\right)^2 + 3\left(\frac{\tau_{xv}}{\sigma_0}\right)^2 = 1 \qquad \text{(7.13d)}$$

Equations 7.13 are plotted in Figure 7.12(b). It can be seen that, for the stress condition shown in Figure 7.12(a), the four theories of failure are in perfect agreement only for the case where $\tau_{xv}/\sigma_0 = 0$ (i.e., for the uniaxial case where the stress element is subjected to σ_x only). For any other combination of σ_x and τ_{xv} acting on the element, the four theories of failure provide different answers. For example, for a stress condition in which $\sigma_x/\sigma_0 = 0.5$, failure of the member would occur when $\tau_{xv}/\sigma_0 = 0.433, 0.500, 0.582$ and 0.707, according to the maximum shear stress theory, the energy of distortion theory, the maximum principal strain theory, and the maximum principal stress theory, respectively. However, for any combination of σ_x and τ_{xv}, the maximum shear stress theory of failure is the most conservative, and the maximum principal stress theory of failure is the least conservative.

Example 7.6
A solid circular shaft is subjected to a torque $T = 40,000$ lb-in. and a bending moment $M = 80,000$ lb-in. Assume the material to be steel, for which the yield strength in tension and compression $\sigma_0 = 30,000$ psi. What should be the diameter of the shaft so that failure does not take place by yielding according to the energy of distortion theory?

Solution. A stress element similar to the one shown in Figure 7.12(a) is selected on the surface of the circular shaft, and the stresses σ_x and τ_{xv} are computed in terms of the unknown diameter d as follows.

By Eq. 5.10,

$$\sigma_x = \frac{M_u}{I_u} v$$

$$= \frac{80,000(d/2)}{(\pi/64)d^4} = \frac{814,873.3}{d^3} \text{ psi}$$

By Eq. 4.19,

$$\tau_{xv} = \frac{T\rho}{J}$$

$$= \frac{40,000(d/2)}{(\pi/32)d^4} = \frac{203,718.3}{d^3} \text{ psi}$$

A Mohr's circle solution yields the principal stresses as follows:

$$\sigma_1 = \frac{862{,}964.7}{d^3} \text{ psi} \qquad \sigma_2 = 0 \qquad \sigma_3 = \frac{-48{,}091.4}{d^3} \text{ psi}$$

Application of the energy of distortion theory, Eq. 7.12b, yields

$$\left(\frac{862{,}964.7}{d^3}\right)^2 + \left(\frac{48{,}091.4}{d^3}\right)^2 + \left(\frac{911{,}056.1}{d^3}\right)^2 = 2(30{,}000)^2$$

Solution of the preceding equation leads to the following value for diameter d:

$$\boxed{d = 3.09 \text{ in.}}$$

(Note that this solution may also be obtained by use of Eq. 7.13d.)

Example 7.7

Solve Example 7.6 using the maximum shear stress theory of failure instead of the energy of distortion theory.

Solution. The three principal stresses as determined in terms of the unknown diameter d in Example 7.6 are repeated here for convenience. Thus

$$\sigma_1 = \frac{862{,}964.7}{d^3} \text{ psi} \qquad \sigma_2 = 0 \qquad \sigma_3 = \frac{-48{,}091.4}{d^3} \text{ psi}$$

The maximum shear stress theory, Eq. 7.4, yields the following relation:

$$\frac{911{,}056.1}{d^3} = 30{,}000$$

Therefore,

$$\boxed{d = 3.12 \text{ in.}}$$

(Note that this solution may also be obtained by use of Eq. 7.13c.) Thus for the specific stress condition in this problem, the diameter required by the maximum shear stress theory of failure is only slightly larger than that needed by the energy of distortion theory. It should be noted, however, that the difference in the values of the diameter predicted by these two theories depends, to a large extent, upon the state of stress acting on the member (i.e., upon the relative magnitudes of σ_x and τ_{xv}).

Example 7.8

A circular pipe made of a brittle material has a 0.10-m inside diameter and a 0.15-m outside diameter. It is subjected to a torque $T = 70$ kN-m and a compressive axial force P kN. The material is known to have an ultimate compressive strength $\sigma_0 = 250$ MPa and an ultimate tensile strength $\sigma_0 = 100$ MPa. Determine the maximum value of the compressive force P so that the member does not fail according to the maximum principal stress theory.

Solution. A stress element similar to the one shown in Figure 7.12(a) is selected on the surface of the hollow shaft and the stresses σ_x and τ_{xv} are computed. Note that σ_x is expressed in terms of the unknown force P. Thus

$$\tau_{xv} = \frac{T\rho}{J} = \frac{(70 \times 10^3)(0.075)}{(\pi/32)(0.15^4 - 0.10^4)} = 131.63 \text{ MPa}$$

$$\sigma_x = -\frac{P}{A} = -\frac{P \times 10^3}{(\pi/4)(0.15^2 - 0.10^2)} = -0.102P \text{ MPa}$$

The principal stresses are now computed using a Mohr's circle solution and are found to be as follows:

$$\sigma_1 = [\sqrt{(131.63)^2 + (0.051P)^2} - 0.051P] \text{ MPa}$$

$$\sigma_2 = 0$$

$$\sigma_3 = -[\sqrt{(131.63)^2 + (0.051P)^2} + 0.051P] \text{ MPa}$$

Since the ultimate strengths in tension and compression are considerably different from each other, both properties have to be investigated as follows.

TENSILE PROPERTY: The maximum principal stress theory, Eq. 7.2, yields the following relation:

$$\sqrt{(131.63)^2 + (0.051P)^2} - 0.051P = 100 \tag{a}$$

Solution of this relation leads to a value of $P = 718.3$ kN. Note that any compressive load P greater in magnitude than 718.3 kN leads to a tensile stress value σ_1 [given by the left-hand side of Eq. (a)] which is less than the tensile property of 100 MPa. Thus, according to the tensile property of this brittle material, any compressive load larger than 718.3 kN is acceptable.

COMPRESSIVE PROPERTY: Again, the maximum principal stress theory, Eq. 7.2, yields the following relation when applied to the compressive property of this material:

$$\sqrt{(131.63)^2 + (0.051P)^2} + 0.051P = 250 \qquad \textbf{(b)}$$

Solution of Eq. (b) yields a value for $P = 1771.8$ kN. Any compressive load greater than 1771.8 kN would lead to a compressive stress σ_3 [left-hand side of Eq. (b)], which is greater than the compressive property of 250 MPa for this brittle material.

Thus the compressive property is the limiting factor, and both tensile and compressive properties are satisfied by ensuring that the compressive load has a maximum value of

$$\boxed{P = 1771.8 \text{ kN}}$$

Example 7.9

Refer to the hollow shaft described in Example 7.8 and assume the brittle material to be one for which the ultimate strength in tension $\sigma_0 = 250$ MPa and the ultimate strength in compression $\sigma_0 = 700$ MPa. Determine the maximum value of the compressive load P so that the member does not fail according to the maximum principal strain theory. Assume that $\mu = 0.28$.

Solution. The three principal stresses, in terms of the unknown force P, as determined in Example 7.8 are repeated here for convenience. Thus

$$\sigma_1 = [\sqrt{(131.63)^2 + (0.051P)^2} - 0.051P] \text{ MPa}$$

$$\sigma_2 = 0$$

$$\sigma_3 = -[\sqrt{(131.63)^2 + (0.051P)^2} + 0.051P] \text{ MPa}$$

As in Example 7.8, both tensile and compressive characteristics of the material need to be investigated, as follows.

TENSILE PROPERTY: Application of the maximum principal strain theory, Eq. 7.3b, using the tensile ultimate strength $\sigma_0 = 250$ MPa, yields

$$(1 + \mu)\sqrt{(131.63)^2 + (0.051P)^2} - (1 - \mu)(0.051P) = 250$$

Substituting $\mu = 0.28$ and simplifying yields the following quadratic equation:

$$P^2 - 6.30 \times 10^3 P - 21.45 \times 10^6 = 0$$

Solution of this quadratic equation leads to $P = 8750$ kN. Note that any compressive load smaller in magnitude than 8750 kN would satisfy the tensile criterion.

COMPRESSIVE PROPERTY: Application of the maximum principal strain theory, Eq. 7.3c, using the compressive ultimate strength $\sigma_0 = 700$ MPa, yields

$$(1 + \mu)\sqrt{(131.63)^2 + (0.051P)^2} + (1 - \mu)(0.051P) = 700$$

Substituting $\mu = 0.28$ and simplifying leads to the following quadratic equation:

$$P^2 + 17.66 \times 10^3 P - 158.46 \times 10^6 = 0$$

The solution of this quadratic equation leads to $P = 6545$ kN. Any compressive load of lesser magnitude than 6545 kN would obviously satisfy the compressive criterion.

Therefore, the compressive property of the material is the controlling factor, and both tensile and compressive properties are satisfied by ensuring that the compressive load has a maximum value of

$$\boxed{P = 6545 \text{ kN}}$$

Homework Problems

7.50 A solid circular shaft is subjected to a torque $T = 50,000$ lb-in. and a tensile axial force $P = 75,000$ lb. Assume the material to be steel for which the yield strengths, σ_0, in tension and compression are numerically identical and equal to 30,000 psi. Determine the diameter of the shaft so that it does not fail by yielding, using the maximum shear stress theory of failure.

7.51 Assume the same conditions as in Problem 7.50 and find the diameter of the shaft using the energy of distortion theory.

7.52 A hollow circular shaft of 0.05 m inside diameter and 0.09-m outside diameter is subjected to

a bending moment $M = 7.2$ kN-m and a torque T. Assume the material to be aluminum for which the yield strengths, σ_0, in tension and compression are numerically identical and equal to 200 MN/m^2. Use the maximum shear stress theory and determine the value of the torque T so that the shaft does not fail by yielding.

7.53 Solve Problem 7.52 using the energy of distortion theory instead of the maximum shear stress theory.

7.54 A solid circular shaft is subjected to the simultaneous action of a torque $T = 60,000$ lb-in., a bending moment $M = 100,000$ lb-in., and a compressive axial for $P = 80,000$ lb. Assume the material to be steel for which the yield strengths, σ_0, in tension and compression are numerically identical and equal to 25,000 psi. Determine the diameter of the shaft by the maximum shear stress theory so that it does not fail by yielding.

7.55 Solve Problem 7.54 using the energy of distortion theory instead of the maximum shear stress theory.

7.56 Assume the shaft described in Problem 7.50 to be made of a brittle material for which the ultimate or fracture strengths, σ_0, in tension and compression are assumed identical and equal to 10,000 psi and for which Poisson's ratio $\mu = 0.30$. Determine the diameter of the shaft using the maximum principal stress theory so that the shaft does not fail by fracture.

7.57 Repeat Problem 7.56 using the maximum principal strain theory.

7.58 The hollow shaft described in Problem 7.52 is made of a brittle material for which the ultimate or fracture strengths, σ_0, in tension and compression are assumed identical and equal to 125 MPa and for which Poisson's ratio $\mu = 0.30$. Determine the value of the torque T by the maximum principal stress theory so that it does not fail by fracture.

7.59 Repeat Problem 7.58 using the maximum principal strain theory instead of the maximum principal stress theory.

Chapter 8 Column Theory and Analyses

8.1 Introduction

Members that resist axial or eccentric compression forces are found in many structures and machines, and the overall behavior of such prismatic columns or struts will be studied in this chapter. Local, torsional, and lateral buckling and the like, although quite important in engineering, are beyond the scope of this text.

Fundamental concepts of the stability of equilibrium are developed and studied in detail with the aid of a column model. Equilibrium of a system is stable, unstable, or neutral, depending upon whether the potential energy of the system is a minimum, maximum, or constant, respectively. Mathematical criteria enable the engineer to classify the state of equilibrium of a system as stable, unstable, or neutral. This background is essential for understanding buckling phenomena.

Rational understanding of column behavior began about 200 years ago with the study of ideal or perfect columns by Leonard Euler, a Swiss mathematician, who developed his now famous equation for the critical (or Euler) buckling load applicable to long columns. Relaxation of idealizing assumptions leads to a study of eccentrically loaded and initially curved columns.

Empirical equations developed to fit experimental data for allowable column loads are studied by means of examples, with emphasis placed upon material for which the equations were developed, relationship of buckling loads to allowable loads, validity of equations related to slenderness ratio ranges, and column support conditions.

Finally, the tangent modulus of elasticity is used to extend the Euler equation to the full range of slenderness ratios. As recently as 1946, F. R. Shanley, an American professor of engineering, made important contributions to understanding the generalized Euler equation.

Eleven examples show applications of the theory to practical analysis of column behavior. Column design will be studied in Section 10.6.

8.2 Stability of Equilibrium

Equilibrium of a given system is stable, unstable, or neutral, and it is desirable to have criteria for deciding which of these three classifications describes a given position of equilibrium. Figure 8.1(a) depicts the three classifications and provides a

383

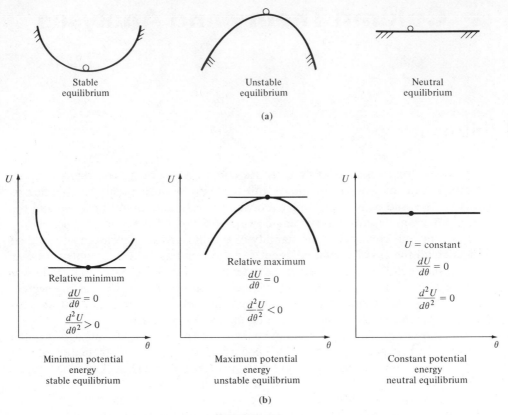

FIGURE 8.1

visual means of recalling the criteria. For *stable equilibrium*, imagine a small sphere at rest at the lowest point of a concave cylindrical container. If the small sphere is displaced slightly from this position of equilibrium and released, it will oscillate about the low point under the action of gravity. This oscillation will be damped out by frictional forces and atmospheric resistance until the sphere comes to rest in the original equilibrium position at the lowest point of the container. The behavior of the sphere characterizes stable equilibrium. If a system in equilibrium is displaced slightly from a position of equilibrium by applying small disturbing forces and upon removal of these disturbing forces the system returns to the same position of equilibrium, the system is said to be in a state of stable equilibrium

For *unstable equilibrium*, imagine a small sphere at rest at the highest point of a convex cylindrical body. If the small sphere is displaced slightly from this position of equilibrium and released, it will not return to its original equilibrium position. The behavior of the sphere characterizes unstable equilibrium. If a system in equilibrium is displaced slightly from a position of equilibrium by applying small disturbing forces and upon removal of these disturbing forces the system does not return to the same position of equilibrium, the system is said to be in a state of unstable equilibrium.

For *neutral equilibrium*, imagine a small sphere at rest in a horizontal surface. If the small sphere is displaced slightly from this position of equilibrium and released, it will remain in equilibrium in the new position to which it was displaced. The behavior of the sphere characterizes neutral equilibrium. If a system in equilibrium is displaced slightly from a position of equilibrium by applying small disturbing forces and upon removal of these disturbing forces the system remains in the new equilibrium position to which it was moved, the system is said to be in a state of neutral equilibrium.

Stable equilibrium is associated with a minimum value of the potential energy function discussed in physics and introductory mechanics courses. Considering the potential energy U to be a function of a single variable θ, then the conditions for stable equilibrium follow. For equilibrium

$$\frac{dU}{d\theta} = 0 \tag{8.1}$$

For a minimum value of the potential energy function, and therefore for stable equilibrium,

$$\frac{d^2U}{d\theta^2} > 0 \tag{8.2}$$

For unstable equilibrium, in addition to Eq. 8.1, the potential energy function must be a maximum, which means that

$$\frac{d^2U}{d\theta^2} < 0 \tag{8.3}$$

For neutral equilibrium, in addition to Eq. 8.1, the potential energy function must be a constant, which means that

$$\frac{d^2U}{d\theta^2} = 0 \tag{8.4}$$

It is necessary to examine higher-ordered derivatives if both the first and second derivatives vanish. If all derivatives vanish, the system is in a state of neutral equilibrium since the potential energy is then a constant. If the first nonzero derivative is of even order and positive, the equilibrium is stable. In all other cases, unstable equilibrium results.

These criteria for stable, unstable, and neutral equilibrium are shown pictorially in Figure 8.1(b). Only single-degree-of-freedom systems are considered in this section.

Examples 8.1 and 8.2 apply these concepts of the stability of equilibrium to the column model shown in Figure 8.2. This system is referred to as a *column model*, since it is a one-degree-of-freedom simplification of a column such as the one depicted in Figure 8.8(c). The real column fixed at its base and free at the top with a load applied

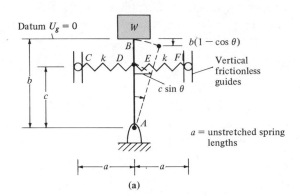

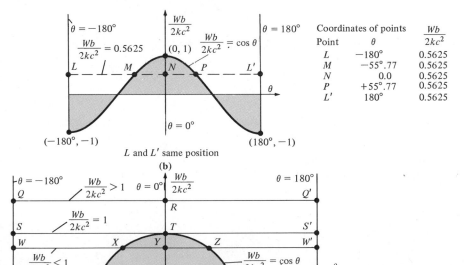

Coordinates of points

Point	θ	$\dfrac{Wb}{2kc^2}$
L	$-180°$	0.5625
M	$-55°.77$	0.5625
N	0.0	0.5625
P	$+55°.77$	0.5625
L'	$180°$	0.5625

FIGURE 8.2

at its top shown in Figure 8.8(c) has an infinite number of degrees of freedom, since a function is required to describe its deflected shape. Since the rod AB of Figure 8.2 is assumed to be perfectly rigid, only one generalized coordinate (i.e., θ) is required to describe its position. This simplification enables the investigator to focus on the concepts of stability of equilibrium while studying the behavior of the column model. This procedure should be viewed as an aid to understanding the concepts involved and not as a means of solving real column problems, which will be studied in the remainder of this chapter.

Example 8.1

The rigid rod AB shown in Figure 8.2(a) is free to rotate about the frictionless pin at A and is restrained by the two springs CD and EF, each with spring constant equal to k. Ends C and F of the springs are constrained to move in vertical frictionless guides which do not interfere with rotation of the rod. A weight W is supported at the top of the rigid rod. The unstretched length of each spring is a and each spring remains horizontal as the rod moves from one position to another. Ignore the weight of the rigid rod AB in comparison to W. (a) For the following numerical values, determine the positions of equilibrium of the rod and state whether these positions are stable, unstable or neutral equilibrium positions: $W = 150$ N, $b = 0.6$ m, $c = 0.4$ m, and $k = 500$ N/m. (b) Provide a general discussion of the equilibrium of the rod AB for all possible numerical inputs for $W = mg$, b, c, and k.

Solution

(a) The potential energy of the system U is given by: $U = U_g + U_e$. Where the datum for the gravitational potential energy U_g has been chosen through the top of the rod B, when $\theta = 0$ and U_e is the elastic potential strain energy stored in the two springs. Therefore, $U = -Wb(1 - \cos \theta) + 2(\frac{1}{2}k)(c \sin \theta)^2$. To determine the positions of equilibrium of the system, differentiate U with respect to θ and equate to zero.

$$\frac{dU}{d\theta} = -Wb \sin \theta + 2kc^2 \sin \theta \cos \theta$$

$$\sin \theta(-Wb + 2kc^2 \cos \theta) = 0$$

$$\sin \theta = 0$$

$$\boxed{\theta = 0^\circ, \pm 180^\circ}$$

$$-Wb + 2kc^2 \cos \theta = 0$$

$$\frac{Wb}{2kc^2} = \cos \theta$$

Substitution of the given numerical values, $W = 150$ N, $b = 0.6$ m, $c = 0.4$ m, and $k = 500$ N/m yields

$$\cos \theta = \frac{150(0.6)}{2(500)(0.4)^2} = 0.5625$$

$$\boxed{\theta = 55°.77, \; -55°.77}$$

There are five positions of equilibrium, given by $0°$, $\pm 55.77°$, $\pm 180°$. The $\theta = -180°$ position repeats the $\theta = +180°$ position, and both need not be considered.

In order to determine whether these positions are stable, unstable, or neutral, the second derivative of U with respect to θ will be examined.

$$\frac{d^2 U}{d\theta^2} = -Wb \cos \theta + 2kc^2(-\sin^2 \theta + \cos^2 \theta)$$

Replace $\sin^2 \theta$ by $1 - \cos^2 \theta$ and simplify to yield

$$\frac{d^2 U}{d\theta^2} = 4kc^2 \cos^2 \theta - Wb \cos \theta - 2kc^2$$

Substitute numerical values from above for k, c, b, and W:

$$\frac{d^2 U}{d\theta^2} = 320 \cos^2 \theta - 90 \cos \theta - 160$$

Next, we tabulate the results of investigating the sign of the second derivative for each of the equilibrium positions. Recall that a positive second derivative corresponds to a minimum potential energy and stable equilibrium, while a negative second derivative corresponds to a maximum potential energy and unstable equilibrium. A graphical summary of results is shown in Figure 8.2(b).

Equilibrium Angle θ	Sign of Second Derivative	Point in Figure 8.2(b)	Equilibrium Is
−180.00	+	L	Stable
−55.77	−	M	Unstable
0.00	+	N	Stable
+55.77	−	P	Unstable
+180.00	+	L'	Stable

(b) Refer to Figure 8.2(c) in order to understand the general discussion of equilibrium of the given system. The dimensionless quantity $Wb/2kc^2$ must be greater than zero because each quantity, W, b, and so on, must be positive on physical grounds. A given physical system may be classified according to whether $Wb/2kc^2$ is less than, equal to, or greater than unity. Equating the first derivative of the potential energy U with respect to the angle θ to zero yields the following general solutions for equilibrium positions: $\theta = 0°$ [the positive vertical axis of Figure 8.2(c)], $\theta = \pm 180°$ (the vertical marked lines shown in the figure), $\theta = \arccos(Wb/2kc^2)$ [associate with the portion of the cosine function shown in Figure 8.2(c)].

The quantity $Wb/2kc^2 = $ constant denotes a line parallel to the θ axis, which characterizes a given physical system. Equilibrium positions are associated with intersection of the horizontal lines $Wb/2kc^2 = $ constant with $\theta = 0°$, $\theta = \pm 180°$, and the cosine function from $-90°$ to $+90°$. The tabulation below summarizes all possible equilibrium positions of the system. Stability of equilibrium was established in each case by considering $Wb/2kc^2$ to be less than, equal to, or greater than unity substituted into the second derivative $d^2U/d\theta^2$ in the following form:

$$\frac{d^2U}{d\theta^2} = 2kc^2\left(2\cos^2\theta - \frac{Wb}{2kc^2}\cos\theta - 1\right)$$

Higher derivatives were investigated to establish point T as corresponding to unstable equilibrium.

Equilibrium Angle θ	$\dfrac{Wb}{2kc^2}$	Sign of Second Derivative	Point in Figure 8.2(c)	Equilibrium Is
$-180°$	< 1	$+$	W	Stable
$-90° < -\theta_1 < 0°$	< 1	$-$	X	Unstable
$0°$	< 1	$+$	Y	Stable
$0° < \theta_1 < 90°$	< 1	$-$	Z	Unstable
$+180°$	< 1	$+$	W'	Stable
$-180°$	$= 1$	$+$	S	Stable
$0°$	$= 1$	0	T	Unstable
$+180°$	$= 1$	$+$	S'	Stable
$-180°$	> 1	$+$	Q	Stable
$0°$	> 1	$-$	R	Unstable
$+180°$	> 1	$+$	Q'	Stable

$\theta = +180°$ and $\theta = -180°$ refer to identical positions of the system for each value of $Wb/2kc^2$.

Example 8.2

Refer to Example 8.1 and discuss the $\theta = 0°$ equilibrium positions associated with points R, T, and Y of Figure 8.2(c). Then derive and discuss the small-displacement-theory solution for the system.

Solution. A convenient way to consider variations of the dimensionless quantity $Wb/2kc^2$ is to imagine all quantities except W to be held constant. Large values of W, corresponding to a heavy weight attached to the rigid rod of Figure 8.2(a), are associated with the large values of the quantity $Wb/2kc^2$. Imagine the system in equilibrium in the $\theta = 0°$ or the upright position, and let W take on a relatively large value associated with point R on the vertical axis of Figure 8.2(c). Point R corresponds to an unstable equilibrium position. Physically, W is so large that the springs are unable to pull the rod back to the vertical position if it is slightly displaced from this position. Next, imagine the system in equilibrium in the $\theta = 0°$ or upright position and let W take on a relatively small value associated with point Y on the vertical axis of Figure 8.2(c).

Point Y corresponds to a stable equilibrium position. Physically, W is so small that the springs are able to pull the rod back to the vertical position if it is slightly displaced from this position. Finally, if W takes on a precise value such that $Wb/2kc^2$ equals unity, point T on the vertical axis of Figure 8.2(c) is associated with unstable equilibrium of the system. Decreasing W values are associated with points R, T, and Y, in that order, from unstable to stable equilibrium positions.

Small-displacement theory applied to the system of Figure 8.2(a) refers to restricting θ to small values. It will be shown that this theory leads to the same results as those discussed above for the $\theta = 0°$ equilibrium position except for point T. In Example 8.1, θ was not restricted to small values and all possible equilibrium positions were revealed in its solution, shown pictorially in Figure 8.2(c). Small-displacement-theory solutions are of great interest in the study of the elastic behavior in this chapter because large displacements are quite often associated with inelastic behavior.

From the solution of Example 8.1, the potential energy U of the system is given by $U = -Wb(1 - \cos \theta) + 2(\frac{1}{2}k)(c \sin \theta)^2$. Use the following approximate replacements for the cosine and sine functions in U which are satisfactory for small angles θ:

$$\cos \theta \approx 1 - \frac{\theta^2}{2}$$

$$\sin \theta \approx \theta$$

This yields U for small-displacement theory:

$$U = -Wb\frac{\theta^2}{2} + kc^2\theta^2$$

To determine the positions of equilibrium of the system, differentiate U with respect to θ and equate to zero.

$$\frac{dU}{d\theta} = -Wb\theta + 2kc^2\theta$$

Equating $dU/d\theta$ to zero,

$$\theta(-Wb + 2kc^2) = 0$$

and equating each factor to zero, we obtain

$$\boxed{\theta = 0°}$$

$$-Wb + 2kc^2 = 0$$

$$\boxed{\frac{Wb}{2kc^2} = 1}$$

These values duplicate those obtained from the large-deflection-theory solution to Example 8.1, which is represented graphically in Figure 8.2(c). That is, $\theta = 0°$ refers to the vertical axis, which contains points R, T, and Y, while $Wb/2kc^2 = 1$ refers specifically to point T.

In order to investigate the stability of equilibrium, write out the second derivative of the potential energy U with respect to the angle θ.

$$\frac{d^2U}{d\theta^2} = -Wb + 2kc^2$$

Factor $2kc^2$ on the right-hand side of the preceding equation:

$$\frac{d^2U}{d\theta^2} = 2kc^2\left(\frac{-Wb}{2kc^2} + 1\right)$$

Since $2kc^2 > 0$ on physical grounds, the sign of the second derivative depends upon the term in parentheses. Consider point R of Figure 8.2(c), for which $Wb/2kc^2 > 1$. If $Wb/2kc^2 > 1$, then $d^2U/d\theta^2 < 0$, which is associated with a maximum value of U and unstable equilibrium. Consider point Y of Figure 8.2(c) for which $Wb/2kc^2 < 1$. If $Wb/2kc^2 < 1$, then $d^2U/d\theta^2 > 0$, which is associated with a minimum value of U and stable equilibrium. Consider point T of Figure 8.2(c), for which $Wb/2kc^2 = 1$. If $Wb/2kc^2 = 1$, then $d^2U/d\theta^2 = 0$. Note also that all higher-ordered derivatives of U with respect to θ are identically zero. Point T is associated with a neutral equilibrium position. Statements made previously about points R, T, and Y of Figure 8.2(c) are fully confirmed by the small-displacement-theory solution except for point T. A change in the function for U from large to small displacements has shifted point T from unstable to neutral equilibrium. The associated precise value of W is referred to as the *critical load*. In other words, T corresponds to transition from unstable to stable equilibrium as W decreases.

Homework Problems

8.1 Refer to Example 8.1 and Figure 8.2 and verify each of the equilibrium positions for $Wb/2kc^2$ equal to 0.5625 by writing the moment equilibrium equation with respect to the pin at A. Carefully construct free-body diagrams for each position of equilibrium.

8.2 Refer to Example 8.2 and verify the small displacement theory solutions by drawing free-body diagrams and writing the moment equilibrium equations with respect to the pin at A.

8.3 The rigid rod AB shown in Figure H8.3 has a length L and is attached to a torsional spring at its lower end A and supports a weight $W = mg$ at its upper end B. Ignore the weight of the rod in comparison to the weight W, which is supported by the rod. Determine the positions of equilibrium of the rod and state whether these positions are stable, unstable, or

neutral equilibrium positions. Solve by expressing the potential energy as a function of θ, and make a dimensionless plot of WL/K versus θ for the equilibrium positions. Determine the equilibrium position corresponding to the particular values: $W = mg = 250$ lb, $L = 1.00$ ft, and $K = 55.0$ lb-ft/rad ($0 \leq \theta < \pi$).

8.4 The column model shown in Figure H8.4 may be described as follows. A weight W applies the load to the model consisting of two rigid rods AB and BC.

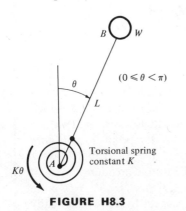

FIGURE H8.3

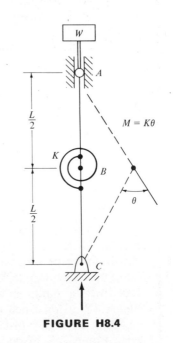

FIGURE H8.4

These rods are free to rotate at A and C but are connected together by a torsional spring at B. The moment resistance of this spring is given by $K\theta$, where K is the torsional spring constant and θ is the angle between the two rods. Express the critical value

of W, corresponding to neutral equilibrium of the model, in terms of K and L, where $L/2$ is the length of each of the rods.

(HINT: As θ approaches zero, the system is in neutral equilibrium.)

8.3 Ideal Column Theory—The Euler Critical Load

    Leonard Euler (1707–1783), a Swiss mathematician, more than two hundred years ago, laid the foundations for the study of column behavior. His name appears frequently in the literature of mathematics, science, and engineering, primarily because he holds the all-time record for mathematical productivity. Euler wrote over 80 volumes of mathematics, many of enduring interest and usefulness.

    Ideal column theory is based upon the following assumptions, which are related to the fundamental case shown in Figure 8.3(a):

1. Loads are applied at the ends of the column and without eccentricity. That is, the line of action of the applied forces coincides with the longitudinal axis of the prismatic column that passes through the centroid of all cross sections of the member.
2. The member is perfectly straight before the loads are applied.
3. The pins (or hinges) at the ends are frictionless, which implies that no resistance to

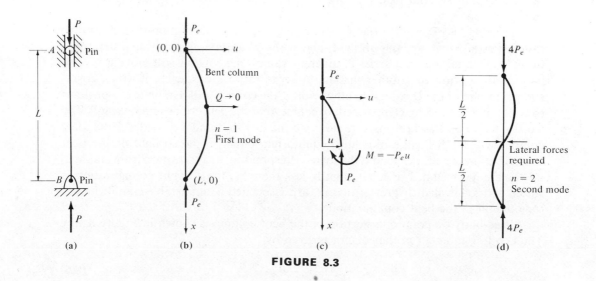

FIGURE 8.3

rotation is encountered at the ends of the column. End A of the column is free to move toward end B of the column as the load increases and the column compresses.

4. The column is fabricated of homogeneous and isotropic material.
5. The column weight is ignored in the analysis.
6. The end pins are constructed such that the column is free to buckle about any axis of the cross sections.

Since the column is assumed to be prismatic and fabricated of homogeneous material, the flexural rigidity EI_v is a constant. The moment of inertia I_v is the minimum moment of inertia of the column cross section, and the column will buckle about the axis of least resistance to bending. Displacements from the straight configuration to the bent configuration will be directed perpendicular to the principal centroidal axis v or in the direction of the u axis.

If it were possible to construct and load an ideal column as described above, it would not buckle as depicted in Figure 8.3(b) but would simply continue to shorten axially under increasing loads applied at the ends A and B. To assure that the ideal column will buckle, a very small lateral force, such as Q of Figure 8.3(b), may be introduced and allowed to approach zero in the limit. This lateral load is so small compared to critical Euler load P_e, derived below, that it need not be considered in the analysis.

The load P is gradually built up from a zero initial value to a value denoted by P_e, the Euler load. If the load applied to the column is less than P_e and a small lateral force Q is applied to the column, it will bend, as indicated in Figure 8.3(b), but upon removal of Q the column will return to the straight configuration, indicating stable equilibrium. If P is further increased until P_e is reached, then when Q is applied and removed, the column will remain in the slightly bent configuration. In other words, P_e is the smallest axial load for which the column will remain in the slightly bent configuration.

If the load is exactly P_e, the column is in a state of indifferent or neutral equilibrium. If straight, it remains straight, and if the small lateral disturbing force Q is thought of as taking on a series of different values, the column will move from one bent configuration to another through a series of positions, each of which corresponds to one of the Q values. In other words, the column will remain in the position to which it is moved by Q, provided, of course, that the deflections remain small. The word *bifurcation* has been used to describe the fact that under a load P_e, the ideal column will be in a state of neutral equilibrium either in the straight or the bent configuration. The Euler load is the load corresponding to a transition from stable to unstable equilibrium. For loads slightly less than P_e, the straight configuration is stable and loads slightly greater than P_e are associated with unstable equilibrium of the column in the bent configuration.

A free-body diagram of a portion of the bent column is shown in Figure 8.3(c). The bending moment in the column is given by

$$M_v = -P_e u \tag{8.5}$$

In this case the bending moment is a function of the small lateral deflections of the column, which contrasts with beam bending moments, studied in Chapters 1 and 6, which are functions of the longitudinal coordinate x.

Equation 6.3 relates bending moments to curvature provided that the column deflections are small enough that the corresponding stress and strain are related with the constant E, the modulus of elasticity of the material.

$$M_v = EI_v \frac{d^2u}{dx^2}$$

which is Eq. 6.6 repeated with subscripts u and v interchanged.

Substitution of Eq. 8.5 into Eq. 6.6 yields

$$-P_e u = EI_v \frac{d^2u}{dx^2} \tag{8.6}$$

Transpose $-P_e u$ to the right side of Eq. 8.6 and divide by EI_v, to obtain

$$\frac{d^2u}{dx^2} + \frac{P_e}{EI_v} u = 0 \tag{8.7}$$

Let

$$\frac{P_e}{EI_v} = k^2 \tag{8.8}$$

and rewrite Eq. 8.7 as follows:

$$\frac{d^2u}{dx^2} + k^2u = 0 \tag{8.9}$$

Equation 8.9 is an ordinary, second-order, linear, homogeneous differential equation with constant coefficients. The general solution of this equation is given by

$$u = A \sin kx + B \cos kx \tag{8.10}$$

The reader is urged to differentiate Eq. 8.10 twice and substitute u and its second derivative into Eq. 8.9 to check that the general solution does satisfy the differential equation, regardless of the values of A and B.

The values of the constants A and B will be determined from the boundary conditions. The boundary conditions are stated by observing that the end points of the column do not deflect or do not leave the x axis during loading of the column.

These boundary conditions are

$$x = 0, \quad u = 0 \tag{8.11}$$

$$x = L, \quad u = 0 \tag{8.12}$$

Substitution of condition 8.11 into Eq. 8.10 yields

$$0 = A \sin 0 + B \cos 0$$

$$0 = 0 + B(1)$$

$$B = 0$$

Substitution of condition 8.12 into Eq. 8.10 yields

$$0 = A \sin kL + 0 \cos kL$$

$$A \sin kL = 0$$

If $A = 0$, the column has not deflected (since $B = 0$), and this refers to the straight or unbuckled configuration. For the deflected configuration $A \neq 0$, $\sin kL$ must equal zero. This implies that

$$kL = n\pi \quad (n = 0, \pm 1, \pm 2, \ldots) \tag{8.13}$$

Squaring both sides of Eq. 8.13 gives

$$k^2 L^2 = n^2 \pi^2 \tag{8.14}$$

But

$$k^2 = \frac{P_e}{EI_v} \tag{8.8}$$

$$P_e = \frac{n^2 \pi^2 EI_v}{L^2} \tag{8.15}$$

The first nonzero value of n, which equals unity, yields the equation for the Euler column load:

$$P_e = \frac{\pi^2 EI_v}{L^2} \tag{8.16}$$

This equation reveals that the Euler critical load varies directly as the product of the modulus of elasticity and the minimum moment of inertia and inversely as the length of the column squared.

Returning to Eq. 8.10, we can write the solution to the governing differential equation as follows:

$$u = A \sin\left(\sqrt{\frac{P_e}{EI_v}}\, x \right) \tag{8.17}$$

Since both boundary conditions have been utilized, the constant A remains indeterminate in Eq. 8.17. The ideal column of Figure 8.3(b) has a deflected shape which is half a sine wave, but the amplitude A has not been expressed in terms of known quantities. This is a limitation of the linear theory, which is based upon an approximation for the curvature. If the curvature is not approximated, it is possible to determine the amplitude of the deflected shape, but this is of less interest than Eq. 8.16 for the critical load.

If $n = 2$ is chosen, Eq. 8.15 provides the second-mode solution depicted in Figure 8.3(d), for which the critical load is four times P_e and lateral forces are required at the midpoint of the column in order for it to buckle in this full sine wave. Since the critical load for the second and higher modes is larger than the critical load for the first mode, only the first mode is of practical interest unless the lateral forces are applied in order to have the column buckle in these higher modes.

To write Eq. 8.16 in terms of a critical stress, divide the equation throughout by the column cross-sectional area, A:

$$\frac{P_e}{A} = \frac{\pi^2 E}{L^2}\frac{I_v}{A} \tag{8.18}$$

Let $\sigma_{cr} = P_e/A$, the critical stress, and note that $I_v/A = r_v^2$, where r_v is the radius of gyration of the cross-sectional area with respect to the v principal axis. Equation 8.18 then takes the form

$$\sigma_{cr} = \frac{\pi^2 E}{(L/r_v)^2} \tag{8.19}$$

where (L/r_v) is known as the *slenderness ratio* for the column; Eq. 8.19 is sketched in Figure 8.4. The critical stress varies directly with the modulus of elasticity and inversely with $(L/r_v)^2$, the slenderness ratio squared. Since E is a material property, a different curve is required for each different material of which columns may be fabricated. As long as σ_{cr} does not exceed the proportional limit stress of the material, Eq. 8.19 is valid as indicated in Figure 8.4. The critical stress falls off rapidly in value as the slenderness ratio increases, since an inverse-square relationship holds. The critical stress concept may be deceptive to beginners unless they understand that elastic buckling is reversible; that is, if the Euler load (or critical stress) is removed from the column, it will return to its undeformed or straight shape. In other words, the column material has not been harmed or destroyed. Of course, if the deflections

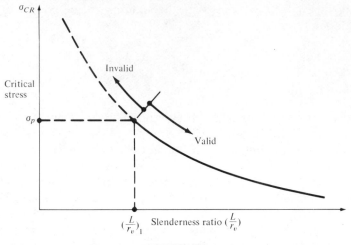

FIGURE 8.4

become too large, the material will be permanently deformed. Very small increases in the load above the critical value will result in large increases in the deflection as revealed by large-deflection-theory analysis.

Example 8.3

An axially loaded aluminum column is 5.00 m long and has the cross section shown in Figure 8.5. It is pin-connected at the ends. Use an $E = 71$ GPa and a unit weight of 27,150 N/m³. Determine (a) the critical Euler load for this column, (b) the critical stress for this column, and (c) the ratio of the Euler load to the weight of the column. (The slenderness ratio is such that the Euler equation applies in this case.)

Solution

(a) The principal axes are u and v and $I_v < I_u$. The column will buckle about the v axis and deflect in the u direction.

$$I_v = \frac{1(0.40)(0.20)^3}{12} - \frac{1(0.32)(0.12)^3}{12} = 22.0587 \times 10^{-5} \text{ m}^4$$

$$A = 0.40(0.20) - 0.32(0.12) = 0.0416 \text{ m}^2$$

$$r_v = \sqrt{\frac{I_v}{A}} = \sqrt{\frac{22.0587 \times 10^{-5}}{0.0416}} = 0.0728 \text{ m}$$

The slenderness ratio is $5/0.0728 = 68.7$.

The Euler critical load is given by Eq. 8.16:

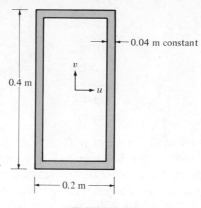

FIGURE 8.5

$$P_e = \frac{\pi^2 E I_v}{l_r^2}$$

$$= \frac{\pi^2 (71 \times 10^9)(22.0587 \times 10^{-5})}{(5)^2}$$

$$= \boxed{6.183 \times 10^6 \text{ N}}$$

(b) The critical stress

$$\sigma_{\text{cr}} = \frac{P_e}{A} = \frac{6.183 \times 10^6}{0.0416}$$

$$= \boxed{148.6 \times 10^6 \text{ N/m}^2}$$

$$= \boxed{148.6 \text{MPa}}$$

(c) Column weight equals the volume multiplied by the weight per unit volume.

$$W = 0.0416 \times 5 \times 27{,}150$$

$$= \boxed{5650 \text{ N}}$$

The ratio of P_e to W is

$$\frac{P_e}{W} = \frac{6.183 \times 10^6}{5650}$$

$$= \boxed{1094}$$

or the weight W is less than 0.1 percent of P_e.

At an L/r_v ratio of 68.7, the Euler equation applies for this aluminum column and the weight of the column is very small compared to the applied critical loading. Of course, the design load for this column would be considerably less than the critical load because this latter load implies elastic buckling of the column. A buckled column would be undesirable as a member of a structure because of the large resulting lateral deflections. A slight increase in the load above P_e could result in a very large deflection and probable inelastic deformation of the column.

Example 8.4

An axially loaded, wide-flanged steel column whose cross section is shown in Figure 8.6 is pin-ended and has the following geometric and material properties: length = 40.0 ft, $A = 59.39$ in.^{2}, $I_u = 2538.8$ in.^{4}, $I_v = 979.7$ in.^{4}, weight = 202 lb/ft, $E = 30 \times 10^3$ ksi, and $\sigma_y = 33.0$ ksi. (Assume that the yield stress for steel approximates the proportional limit stress.) Determine (a) the critical Euler column load, (b) the critical stress for this column, (c) the shortest length of this column for which the Euler equation applies, and (d) the ratio of the critical Euler load to the weight of the column.

Solution

(a) Equation 8.16:

$$P_e = \frac{\pi^2 E I_v}{L^2}$$

$$= \frac{\pi^2 (30 \times 10^3)(979.7)}{(40 \times 12)^2}$$

$$= \boxed{1259 \text{ k}}$$

(b)

$$\sigma_{cr} = \frac{P_e}{A} = \frac{1259}{59.39}$$

$$= \boxed{21.2 \text{ ksi}}$$

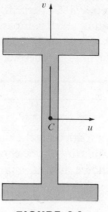

FIGURE 8.6

(c) Equate σ_{cr} to $\sigma_y = 33$ ksi to find the smallest length for which the Euler equation applies. The minimum radius of gyration yields the maximum slenderness ratio.

$$r_v = \sqrt{\frac{I_v}{A}} = \sqrt{\frac{979.7}{59.39}} = 4.06 \text{ in.}$$

From Eq. 8.19:

$$\sigma_{cr} = \frac{\pi^2 E}{(L/r_v)^2}$$

Solving for L yields

$$L = \sqrt{\frac{\pi^2 E r_v^2}{\sigma_{cr}}} = \pi r_v \sqrt{\frac{E}{\sigma_{cr}}}$$

$$L = \pi(4.06)\sqrt{\frac{30 \times 10^3}{33}}$$

$$= \boxed{384.6 \text{ in.} \quad \text{or} \quad 32.1 \text{ ft}}$$

The Euler equation would not apply for columns shorter than 32.1 ft with all other quantities given for this example.

(d) $W = 202$ lb/ft $\times$ 40 ft $= 8080$ lb or 8.08 k.

$$\frac{P_e}{W} = \frac{1259}{8.08}$$

$$= \boxed{155.8}$$

or the weight of the column is only 0.64 percent of P_e, the critical applied load. The weight of this steel column is small compared to the applied critical loading. Again, the design load for this column would be considerably less than the critical load because this latter load implies elastic buckling of the column.

Example 8.5

The cross section of an axially loaded steel column is shown in Figure 8.7(a). Although this shape would ordinarily not be chosen for a column, architectural requirements governed the choice and an analysis is needed. The column is

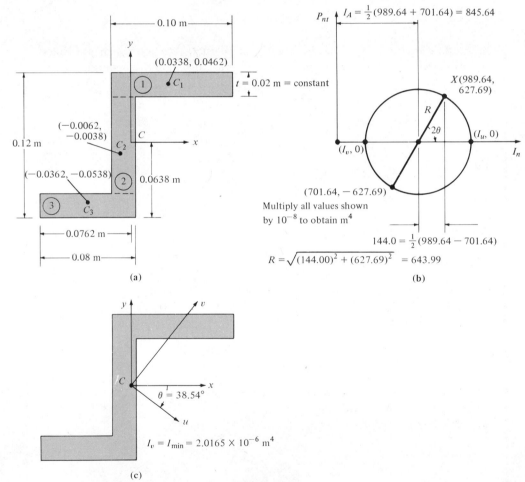

(a)

$$I_A = \tfrac{1}{2}(989.64 + 701.64) = 845.64$$

$X(989.64, 627.69)$

$(I_v, 0)$ $(I_u, 0)$

$(701.64, -627.69)$

Multiply all values shown by 10^{-8} to obtain m^4

$$144.0 = \tfrac{1}{2}(989.64 - 701.64)$$

$$R = \sqrt{(144.00)^2 + (627.69)^2} = 643.99$$

(b)

$\theta = 38.54°$

$$I_v = I_{min} = 2.0165 \times 10^{-6} \ m^4$$

(c)

FIGURE 8.7

pin-connected at each end and is free to buckle about its axis of minimum moment of inertia. This pin-connected column is 4.00 m long, $E = 200$ GPa, and the unit weight of steel is 77,090 N/m³. Use a yield stress of 228 MPa to approximate the proportional limit stress. Determine (a) the principal moments of inertia of the column cross section, (b) the critical Euler load for this column, (c) the critical stress for this column, (d) the shortest length of this column for which the Euler equation applies, and (e) the ratio of the critical Euler load to the weight of the column.

Solution

(a) Subdivide the area into three rectangles, as shown in Figure 8.7(a).

$$A = 0.10(0.02) + 0.08(0.02) + 0.08(0.02) = 0.0052 \text{ m}^2$$

$$\begin{aligned}
I_x &= \tfrac{1}{12}(0.10)(0.02)^3 + 0.10(0.02)(0.0462)^2 \\
&\quad + \tfrac{1}{12}(0.02)(0.08)^3 + 0.08(0.02)(-0.0038)^2 \\
&\quad + \tfrac{1}{12}(0.08)(0.02)^3 + 0.08(0.02)(-0.0538)^2 \\
&= 9.8964 \times 10^{-6} \text{ m}^4
\end{aligned}$$

$$\begin{aligned}
I_y &= \tfrac{1}{12}(0.02)(0.10)^3 + (0.10)(0.02)(0.0338)^2 + \tfrac{1}{12}(0.08)(0.02)^3 \\
&\quad + (0.08)(0.02)(-0.0062)^2 + \tfrac{1}{12}(0.02)(0.08)^3 + (0.08)(0.02)(-0.0362)^2 \\
&= 7.0164 \times 10^{-6} \text{ m}^4
\end{aligned}$$

$$\begin{aligned}
P_{xy} &= (0.10)(0.02)(0.0338)(0.0462) + (0.08)(0.02)(-0.0062)(-0.0038) \\
&\quad + (0.08)(0.02)(-0.0362)(-0.0538) \\
&= 6.2769 \times 10^{-6} \text{ m}^4
\end{aligned}$$

Refer to the Mohr's circle shown in Figure 8.7(b) and complete the calculations as follows:

$$\begin{aligned}
I_u &= I_A + R \\
&= (845.64 + 643.99) \times 10^{-8} \\
&= \boxed{1489.63 \times 10^{-8} \text{ m}^4}
\end{aligned}$$

$$\begin{aligned}
I_v &= I_A - R \\
&= (845.64 - 643.99) \times 10^{-8} \\
&= \boxed{201.65 \times 10^{-8} \text{ m}^4}
\end{aligned}$$

This column will buckle about the v axis or the axis with respect to which the moment of inertia is a minimum.

$$\tan 2\theta = \frac{627.69}{144.00} = 4.359$$

$$2\theta = 77.08°$$

$$\theta = 38.54°$$

Figure 8.7(c) shows the principal axes for this cross section.

(b) From Eq. 8.16, determine P_e:

$$P_e = \frac{\pi^2 E I_v}{L^2}$$

$$= \frac{\pi^2 (200 \times 10^9)(201.65 \times 10^{-8})}{(4^2)10^3}$$

$$= \boxed{248.78 \text{ kN}}$$

(c)
$$\sigma_{cr} = \frac{P_e}{A} = \frac{248.78 \times 10^3}{52 \times 10^{-4} \times 10^6}$$

$$= \boxed{47.8 \text{ MPa}}$$

(d) Equate σ_{cr} to $\sigma_y = 228$ MPa, in order to find the smallest length for which the Euler equation applies. The minimum radius of gyration yields the maximum slenderness ratio which is associated with the smaller critical stress.

$$r_v = \sqrt{\frac{I_v}{A}} = \sqrt{\frac{201.65 \times 10^{-8}}{52 \times 10^{-4}}} = 0.0197 \text{ m}$$

Solve Eq. 8.19 for L:

$$L = \pi r_v \sqrt{\frac{E}{\sigma_{cr}}}$$

$$= \pi(0.0197) \sqrt{\frac{200 \times 10^9}{228 \times 10^6}}$$

$$= \boxed{1.83 \text{ m}}$$

The Euler equation would not apply for columns shorter than 1.83 m with all other quantities given for this example.

(e) $W = 52 \times 10^{-4} \ \text{m}^2 \times 4.00 \ \text{m} \times 77.09 \ \text{kN/m}^3$

$$= \boxed{1.603 \ \text{kN}}$$

$$\frac{P_e}{W} = \frac{248.78}{1.603}$$

$$= \boxed{155.2}$$

or the weight of the column is 0.64 percent of P_e, the critical applied load. The weight of this steel column is small compared to the applied critical loading. Again, the design load for this column would be considerably less than the critical load because this latter load corresponds to elastic buckling of the column.

Homework Problems

8.5 Refer to Example 8.3 and Figure 8.5. Change the overall dimensions from 0.40 m × 0.20 m to 0.20 m × 0.10 m and reduce the wall thickness from 0.04 m to 0.02 m, then answer the questions posed in Example 8.3. All other given quantities remain unchanged.

8.6 Refer to Example 8.4 and answer the questions posed after changing the cross-sectional properties to the following wide-flanged steel section. $A = 62.1 \ \text{in.}^2$, $I_u = 2670 \ \text{in.}^4$, $I_v = 1030 \ \text{in.}^4$, and weight = 211 lb/ft. All other given quantities are to remain unchanged.

8.7 Refer to Example 8.5 and Figure 8.7(a) and answer the questions posed after decreasing the width of the top flange (i.e., rectangle 1) from 0.10 m to 0.08 m and decreasing the total depth of the section from 0.12 m to 0.10 m. All other quantities are to remain unchanged.

8.8 An axially loaded, welded steel column has the cross section shown in Figure H8.8. It is 25 ft long and pin-connected at each end. $E = 29 \times 10^3$ ksi and $\sigma_p = 30$ ksi. Steel weighs 0.284 lb/in.^{3}. Determine:
(a) The critical Euler column load.
(b) The critical stress for this column.
(c) The shortest length of this column for which the Euler equation applies.
(d) The ratio of the critical Euler load to the weight of the column.
(Neglect weld metal in your calculations.)

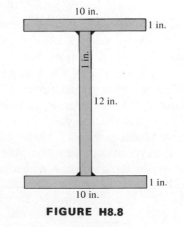

FIGURE H8.8

8.9 Figure H8.9 shows the cross section of a welded steel column which was designed to meet special framing conditions in a building addition. It is axially loaded and fabricated of welded steel and is 16 ft long. If $E = 29 \times 10^6$ psi, $\sigma_p = 32,000$ psi, and the column may be assumed pinned at both ends, find:

(a) The critical Euler column load.

(b) The critical stress for this column.

(c) The shortest length of this column for which the Euler equation applies.

(Neglect weld metal in your calculations.)

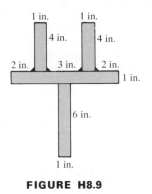

FIGURE H8.9

8.10 Construct a critical stress versus slenderness ratio plot similar to Figure 8.4 for a material for which $E = 20 \times 10^3$ ksi and $\sigma_p = 25$ ksi. Assign slenderness ratio values and compute the corresponding critical stresses. Clearly note the range of validity of the Euler equation for this material.

8.11 Experiments have shown that the modulus of elasticity of steel is about 29×10^6 psi regardless of the alloying elements. Compute the minimum slenderness ratio for validity of the Euler equation for three steel alloys with yield stresses of 36, 46, and 50 ksi. (Assume that the proportional limit stress is approximately equal to the yield stress.) Consider a maximum slenderness ratio of 200 for steel columns for building applications and compute a range of slenderness ratios for which the critical stress will be independent of the alloy of steel chosen. (For short or intermediate-length columns there will be an advantage in choosing higher-strength alloys.)

8.12 An aluminum solid cylindrical rod 0.50 in. in diameter acts as a compression member in a machine. Determine the minimum length of this member if its critical stress is not to exceed 10 ksi and $E = 10 \times 10^3$ ksi. Both ends of the member are assumed to be pinned. Euler's equation is valid.

8.13 Solve Problem 8.12 if the member is square in cross section with each side equal to 0.50 in. All other given quantities are to remain the same.

8.14 Solve Problem 8.12 if the member is rectangular in cross section 0.25×0.50 in. All other given quantities are to remain the same.

8.15 An axially loaded aluminum column is 6.00 m long and has a square box shape which measures, on the outside, 0.20 m $\times 0.20$ m with a wall thickness of 0.016 m. It is pin-connected at the ends with $E = 71$ GPa and a unit weight of 27.15 kN/m^3. Determine:

(a) The critical Euler load for this column.

(b) The critical stress for this column.

(c) The ratio of this Euler load to the weight of the column.

(The slenderness ratio is such that the Euler equation applies in this case.)

8.16 A truss is fabricated of welded steel pipe members connected to thin gusset plates at their ends. A preliminary analysis is being made with the assumption that all members are pin-connected at their ends and loaded axially. A 22-ft-long steel compression member has the following cross-sectional properties: $A = 8.40$ in.^{2} and $I = 40.5$ in.^{4}. Use $E = 29 \times 10^6$ psi and the yield stress of the steel is 36,000 psi.

(a) Determine the critical Euler load for this member of the truss.

(b) Justify the use of the Euler equation in this case.

8.4 Effect of End Conditions on Behavior of Columns

End conditions other than pinned or hinged ones described in assumption 3 of Section 8.3 may be treated by rewriting Eq. 8.1 in terms of an effective length L_e as follows:

$$\sigma_{cr} = \frac{\pi^2 E}{(L_e/r_v)^2} \tag{8.20}$$

where appropriate L_e values are shown in Figure 8.8. A column with pins or hinges at both ends is referred to as the *fundamental case*, which is shown in Figure 8.8(a) with an effective length L_e equal to its actual length L.

If both ends of the column are fixed as shown in Figure 8.8(b), end moment restraints result in a deflected shape which has reversals of curvature (or inflection points) at points A and B. Since the second derivative of the deflection with respect to the lengthwise coordinate x will vanish at inflection points and the bending moments are proportional to the second derivatives, the moments vanish at points A and B. Between these two points the column will behave as though it were pinned, and since the length of AB is one-half the actual length of the column, the effective length equals one-half the actual length for this case. In terms of loading, return to Eq. 8.16 for P_e and substitute L_e for L as follows:

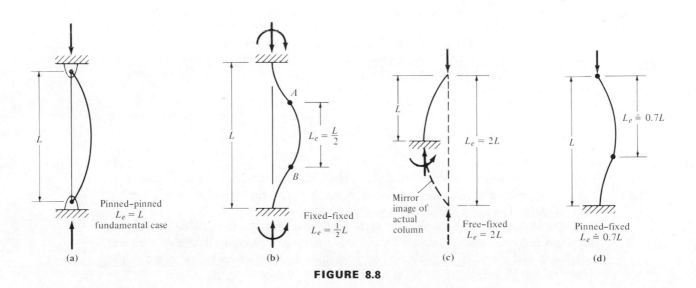

FIGURE 8.8

$$P_e = \frac{\pi^2 EI_v}{L_e^2} \tag{8.21}$$

But $L_e = L/2$ for the fixed-fixed case, which yields

$$P_e = \frac{\pi^2 EI_v}{(L/2)^2} = \frac{4\pi^2 EI_v}{L^2} \tag{8.22}$$

The column with perfectly fixed ends will buckle elastically at a load that is four times the load required to buckle it if the ends were pinned. This increased resistance is provided by the moment restraints at the ends of the column depicted in Figure 8.8(b).

If one end of the column is free and the other end is fixed, as shown in Figure 8.8(c), a mirror image of the actual column must be constructed in order to reveal that the effective length equals twice the actual length. Substitution of this effective length into Eq. 8.21 yields a critical load:

$$P_e = \frac{1}{4} \frac{\pi^2 EI_v}{L^2} \tag{8.23}$$

The column with one end free and the other end fixed will buckle elastically at a load that is one-fourth as large as the load required to buckle it if both ends were pinned. Even though moment restraint is provided at the lower end of the column, the fact that the upper end is free of any restraint greatly reduces the buckling load for this case, shown in Figure 8.8(c).

If one end of the column is pinned and the other end is fixed, as shown in Figure 8.8(d), an inflection point is located at a distance of approximately seven-tenths the actual length from the upper end, which is pinned. The effective length is approximately seven-tenths the actual length. Substitution of this effective length into Eq. 8.21 yields an approximate critical load which is satisfactory for engineering purposes.

$$P_e \approx 2.04 \frac{\pi^2 EI_v}{L^2} \tag{8.24}$$

Since moment restraint is provided at the lower end while the upper end is pinned, as shown in Figure 8.8(d), the critical load lies between the values obtained for the fundamental case and the fixed–fixed case.

The reader should understand that these effective lengths are valid for both elastic column behavior and inelastic column behavior. Effective lengths discussed here and depicted in Figure 8.8 may also be derived from first principles as indicated in the Homework Problems.

*8.5 Eccentrically Loaded and Initially Curved Columns

To investigate imperfect columns, the first two assumptions of ideal column theory will be relaxed and the consequences investigated. The problem of eccentrically loading a column and the problem of loading an initially curved column will be formulated and the results discussed.

All assumptions of Section 8.3 except the first one apply to the column depicted in Figure 8.9(a), since this column is loaded by forces applied with equal eccentricities at each end. This eccentricity e is measured in the u direction. To write the governing differential equation, refer to the free-body diagram shown in Figure 8.9(b). From Eq. 6.6 with u and v interchanged:

$$M_v = EI_v \frac{d^2u}{dx^2}$$

where

$$M_v = -P(e + u). \tag{8.25}$$

Equation 8.25 into Eq. 6.6 yields

$$EI_v \frac{d^2u}{dx^2} + P(e + u) = 0 \tag{8.26}$$

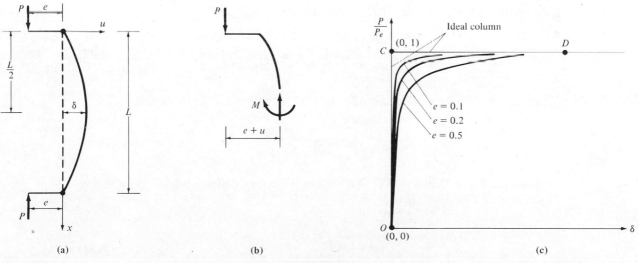

(a)   (b)   (c)

FIGURE 8.9

Divide by EI_v and transpose the constant term:

$$\frac{d^2u}{dx^2} + \frac{P}{EI_v}u = -\frac{P}{EI_v}e \qquad (8.27)$$

Let

$$k^2 = \frac{P}{EI_v} \qquad (8.8)$$

$$\frac{d^2u}{dx^2} + k^2u = -k^2e \qquad (8.28)$$

The boundary conditions are as follows:

$$x = 0, \qquad u = 0 \qquad (8.29)$$

$$x = L, \qquad u = 0 \qquad (8.30)$$

Completing the solution of the differential equation (8.28) subject to the boundary conditions of Eqs. 8.29 and 8.30 is left as a homework assignment (see Problem 8.20). This solution is given by

$$u = e\left(\cos kx + \frac{1 - \cos kL}{\sin kL}\sin kx - 1\right) \qquad (8.31)$$

Arbitrarily multiply Eq. 8.8 by L^2/π^2 as follows:

$$\frac{k^2L^2}{\pi^2} = \frac{PL^2}{\pi^2 EI_v} = \frac{P}{P_e} \qquad (8.32)$$

Multiply Eq. 8.32 by π^2 and take the square root of both sides to obtain

$$kL = \pi\sqrt{\frac{P}{P_e}} \qquad (8.33)$$

Evaluate Eq. 8.31 at $x = L/2$ and denote the center deflection by δ as shown in Figure 8.9(a). The result is

$$\delta = e\left[\sec\left(\frac{\pi}{2}\sqrt{\frac{P}{P_e}}\right) - 1\right] \qquad (8.34)$$

A plot of Eq. 8.34 is shown in Figure 8.9(c). First, consider ideal column behavior shown on this plot. The vertical line OC and the horizontal line CD are to be associated with ideal column behavior. As the force P applied to the ideal column is increased, there is no deflection of the column until $P = P_e$. At point C, the ideal column is in neutral equilibrium and line CD indicates that the column is either straight (corresponding to point C) or the column is bent with any small deflection (corresponding to any point of line CD). Three eccentricities ($e = 0.1, 0.2,$ and 0.5) were arbitrarily chosen to illustrate the effects of nonconcentric loading of the column. Note that for eccentric loading the column begins to bend from the beginning of load application rather than remaining straight, as indicated by the fact that the three curves for various e values depart from the vertical line OC (i.e., nonzero P/P_e values, however small, are associated with nonzero δ values). As the eccentricity increases, the load corresponding to a given δ decreases. All curves approach the horizontal line CD as δ becomes infinite, but the column will behave inelastically long before this condition is reached.

All assumptions of Section 8.3 except the second one apply to the column depicted in Figure 8.10(a), since this column is initially bent. For convenience the bent shape is assumed to be sinusoidal and expressed mathematically as follows:

$$u_0 = A_0 \sin \frac{\pi x}{L} \tag{8.35}$$

where A_0 is the amplitude or midpoint deflection of the column.

To write the governing differential equation, refer to the free-body diagram shown in Figure 8.10(b). From Eq. 6.6 with u and v interchanged:

$$M_v = EI_v \frac{d^2u}{dx^2}$$

where

$$M_v = -P(u_0 + u) \tag{8.36}$$

Substitution of Eq. 8.36 into Eq. 6.6 yields

$$EI_v \frac{d^2u}{dx^2} + P(u_0 + u) = 0 \tag{8.37}$$

Next, divide by EI_v and transpose the term containing u_0:

$$\frac{d^2u}{dx^2} + \frac{P}{EI_v} u = -\frac{P}{EI_v} u_0 \tag{8.38}$$

Substitute $P/EI_v = k^2$ and $u_0 = A_0 \sin \pi x/L$ into Eq. 8.38, to yield

$$\frac{d^2u}{dx^2} + k^2u = -k^2 A_0 \sin \frac{\pi x}{L} \qquad (8.39)$$

The boundary conditions are as follows:

$$x = 0, \qquad u = 0 \qquad (8.40)$$

$$x = L, \qquad u = 0 \qquad (8.41)$$

Completing the solution of the differential equation 8.39 subject to the boundary conditions of Eqs. 8.40 and 8.41 is left as a homework assignment (see Problem 8.21). This solution is given by

$$u = \frac{A_0}{1 - P/P_e} \sin \frac{\pi x}{L} \qquad (8.42)$$

It should be noted that Eq. 8.33 will prove useful in writing the solution in this form.

Evaluate Eq. 8.42 at $x = L/2$ and denote the center deflection by δ as shown in Figure 8.10(a). The result is

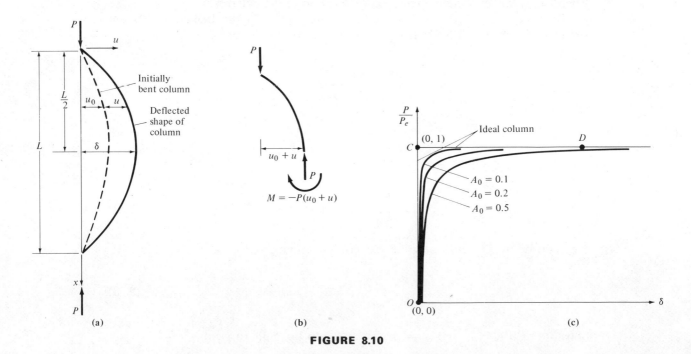

FIGURE 8.10

$$\delta = \frac{A_0}{1 - P/P_e} \qquad\qquad \textbf{(8.43)}$$

A plot of Eq. 8.43 is shown in Figure 8.10(c). Once again, ideal column behavior is associated with lines OC and CD, and the discussion of concepts for Figure 8.9(c) applies here as well. Once more, the column begins to bend from the beginning of load application rather than remaining straight, as indicated by the fact that the three curves for various initial amplitudes ($A_0 = 0.1$, 0.2, and 0.5) depart from the vertical line OC (i.e., nonzero P/P_e values, however small, are associated with nonzero δ values). As the value of A_0 increases, the load corresponding to a given δ decreases. All curves approach the horizontal line CD as δ become infinite, but the column will behave inelastically long before this condition is reached. Obviously, an infinite displacement is not possible.

All columns that are parts of machines or members of structures are imperfect and the discussion above shows that the load-deflection behavior of these imperfect columns depends upon the nature and magnitude of these imperfections. Imperfect columns do not remain straight until the Euler load is reached but begin to bend as soon as load is applied to them. Results shown in Figures 8.9(c) and 8.10(c) are only valid as long as the column stresses are equal to or less than the proportional limit stress.

Homework Problems

8.17 Refer to Figure 8.8(b) and derive the equation for the critical Euler load for a column fixed at both ends which will enable you to verify the effective length for this case. Formulate and solve the appropriate differential equation subject to the boundary conditions for this case. Carefully show the free-body diagram required for formulation of the differential equation.

8.18 Refer to Figure 8.8(c) and repeat Problem 8.17 for this case of a column fixed at one end and free at the other end.

8.19 Refer to Figure 8.8(d) and repeat Problem 8.17 for this case of a column pinned at one end and fixed at the other end.

8.20 Refer to Section 8.5 and Figure 8.9 and complete the solution of the differential equation for an eccentrically loaded column.

8.21 Refer to Section 8.5 and Figure 8.10 and complete the solution of the differential equation for an initially curved column.

8.22 Refer to Section 8.5 and Figure 8.9(c) and compute and plot the function P/P_e versus δ for $e = 0.35$. Assign P/P_e values arbitrarily in the range 0.0 to 1.0 and compute corresponding δ values for plotting. Discuss the results.

8.23 Refer to Section 8.5 and Figure 8.10(c) and compute and plot the function P/P_e versus δ for

$A_0 = 0.40$. Assign P/P_e values arbitrarily in the range 0.0 to 1.0 and compute corresponding δ values for plotting. Discuss the results.

8.24 An eccentrically loaded column is subjected to a loading equal to 0.6 of the Euler critical loading with an eccentricity of 0.1 in. What is the deflection of the column? You are to assume that the column behaves elastically.

8.25 A column is initially bent in a sinusoidal shape with an amplitude $A_0 = 0.002$ m. It is then subjected to an axial loading equal to 0.8 of the Euler critical loading. What is the deflection of the column? You are to assume elastic behavior of the column.

8.26 Formulate the following problem by deriving the governing differential equation and stating the boundary conditions: A pin-ended column of length L is initially bent in a sinusoidal shape of amplitude A_0 and simultaneously loaded with an eccentric load of eccentricity e.

8.6 Empirical Equations for Column Allowable Loads

If a series of pin-ended columns of varying slenderness ratio L_e/r_v are tested and critical stresses are computed for each column, a set of experimental values, as shown in the upper part of Figure 8.11, are obtained. All tested columns are assumed to have been fabricated of the same material and care taken to eliminate eccentricity of the axial loading and to eliminate end restraint for pinned ends. These experimental values could be fitted with curves of known functional form, but the ordinates would correspond to failure of the columns. Of greater interest are the curves corresponding to failure values divided by a factor of safety, which are shown in the lower part of Figure 8.11. These smaller ordinates are known as the *allowable* or *working stresses* and form the basis of column analysis and design. At this stage the factor of safety may be viewed as a positive number by which the failure stress is divided to obtain the allowable stress. More detailed discussion of the factor of safety is given in Chapter 10. The term *failure* as employed here has different meanings, which depend upon the slenderness ratios of the columns. For long columns, *failure* refers to the fact that a small increase in the load above the critical Euler load will result in a large increase in the lateral displacements of the column. If the Euler load is removed, the material of which the column is fabricated is unharmed and the column will return to its original unbent shape. Thus for long columns the term *failure* refers to large lateral elastic displacements associated with buckling of the columns. For very short columns *failure* refers to crushing or compressive yielding of the material of which the column is fabricated, and this phenomenon is irreversible. Axial displacements of a permanent nature characterize the failure of short columns. Columns of intermediate length fail by inelastic buckling, which is also irreversible. In this case both axial and lateral displacements will be permanent, and upon removal of the load the column will remain in the bent shape.

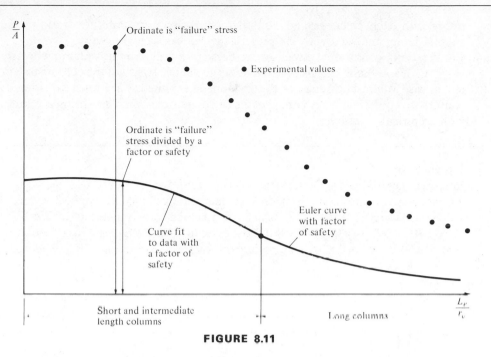

FIGURE 8.11

A very large number of empirical column equations have been devised and presented by various investigators since the time of Euler. Examples will be studied to illustrate the use of several sets of empirical column equations for analysis of column allowable loads. Design of columns will be covered in Section 10.6.

Certain fundamental considerations should be borne in mind by those who use empirical column equations:

1. Empirical column equations vary with the material from which the columns are fabricated. Equations suitable for timber columns are useless for high-strength-steel columns.

2. Many empirical column equations have a factor of safety incorporated into them, which means that the computed P/A stresses are allowable stresses which refer to safe loadings without danger of failure. However, some empirical equations do not incorporate a factor of safety, which means that the computed P/A stresses are to be associated with failure of the columns. In this case the engineer is responsible for incorporation of a factor of safety that will either be prescribed or be left to personal judgment. Ordinarily, the engineer's choice of a factor of safety would have to be approved by a governmental body or other authority.

3. Practically all empirical column equations are valid for restricted values of slenderness ratios L_e/r_v. Outside the prescribed L_e/r_v range of values, the equations are not valid. Problems may be avoided by associating a range of slenderness

ratios with each equation and carefully comparing the slenderness ratio of the column being analyzed with the prescribed slenderness ratio range.

4. Column end conditions discussed in Section 8.4 determine column effective lengths L_e, and these lengths differ from the geometric length of the column except in the case where both ends of the columns are pinned. Care must be taken to conservatively estimate the nature of the end conditions when analyzing a column with empirical equations.

Example 8.6

Empirical equations for Douglas fir columns are shown in Figure 8.12. Use this information to determine the safe axial loading that may be applied to the following Douglas fir columns, which are assumed to be pinned at each end. ($E = 1.70 \times 10^6$ psi). (a) A 6.00-ft-long column of rectangular cross section 5.50 × 9.50 in. (b) A 25.0-ft-long column of square cross section 9.50 × 9.50 in.

Solution

(a) Compute the slenderness ratio of the column in order to decide which equation is appropriate for the analysis. The radius of gyration r_v is given by

$$r_v = \sqrt{\frac{I_v}{A}} = \sqrt{\frac{\frac{1}{12}bd^3}{bd}} = \frac{d}{\sqrt{12}}$$

where d is the least dimension of the rectangular column. In this case $d = 5.50$ in.

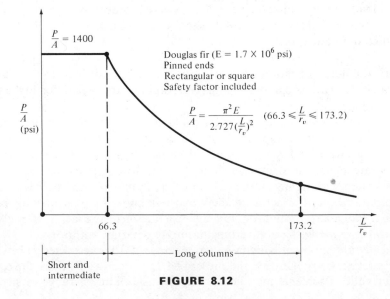

$\dfrac{P}{A} = 1400$

Douglas fir ($E = 1.7 \times 10^6$ psi)
Pinned ends
Rectangular or square
Safety factor included

$$\frac{P}{A} = \frac{\pi^2 E}{2.727\left(\frac{L}{r_v}\right)^2} \quad \left(66.3 \leqslant \frac{L}{r_v} \leqslant 173.2\right)$$

$\dfrac{P}{A}$ (psi)

66.3

173.2

$\dfrac{L}{r_v}$

Short and intermediate

Long columns

FIGURE 8.12

$$r_v = \frac{5.50}{\sqrt{12}} = 1.588 \text{ in.}$$

The slenderness ratio is given by

$$\frac{L}{r_v} = \frac{6.00 \times 12}{1.588} = 45.3$$

From Figure 8.12, $P/A = 1400$ psi provided that $L/r_v < 66.3$. Since $45.3 < 66.3$, this column is short or intermediate in length and the allowable load may be obtained as follows:

$$\frac{P}{A} = 1400 \text{ psi}$$

$$A = 5.50 \times 9.50 = 52.25 \text{ in.}^2$$

The allowable load for this Douglas fir column is

$$\boxed{P = 52.25 \times 1400 = 73{,}150 \text{ lb}}$$

(b) Again, compute the slenderness ratio.

$$r_v = \frac{d}{\sqrt{12}} = \frac{9.50}{\sqrt{12}} = 2.74 \text{ in.}$$

$$\frac{L}{r_v} = \frac{25.0 \times 12}{2.74} = 109.5$$

Since $66.3 < 109.5 < 173.2$, this column is a long column and the Euler equation with a factor of safety of 2.727 will be used from Figure 8.12.

$$\frac{P}{A} = \frac{\pi^2 E}{2.727(L/r_v)^2}$$

$$= \frac{\pi^2 (1.70 \times 10^6)}{2.727(109.5)^2} = 513.1 \text{ psi}$$

$$A = 9.50 \times 9.50 = 90.25 \text{ in.}^2$$

The allowable load for this Douglas fir column is

$$\boxed{P = 90.25 \times 513.1 = 46{,}307 \text{ lb}}$$

The reader should refer to Figure 8.12 and the solution above and note that the allowable stresses for short and intermediate columns are given by ordinates to a horizontal straight line and that the allowable stresses for long columns are given by ordinates to the Euler equation with a factor of safety. In summary, the following fundamental considerations were carefully made:

1. Material: Douglas fir.
2. Factors of safety are incorporated in the equations.
3. L/r ranges:
   (a) Short and intermediate: < 66.3.
   (b) Long: 66.3 to 173.2.
4. End conditions: pinned ends (commonly assumed for timber columns with flat ends).

Further reductions of the loads may be necessary for grading and moisture variation. Such factors are beyond the scope of this analysis.

Example 8.7

(a) A welded steel column is fixed at both ends and is 4.00 m long. It has the cross section shown in Figure 8.13 and is fabricated from structural steel with a yield stress of 344.8 MPa, for which the appropriate column allowable stress

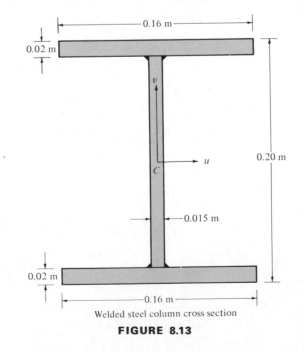

Welded steel column cross section

FIGURE 8.13

information is provided in Figure 8.14(a). Determine the safe load that may be applied to this column. Secondary members are those members whose failure would not result in overall failure of the structure in question. Maximum slenderness ratios for the information provided in Figure 8.14(a) are as follows: Secondary members: 140; main members: 120. Assume that the column is a secondary member and use $E = 200$ GPa. (b) If the length of the column of part (a) is changed to 10.80 m and all other information is unchanged, determine the safe load for this longer column.

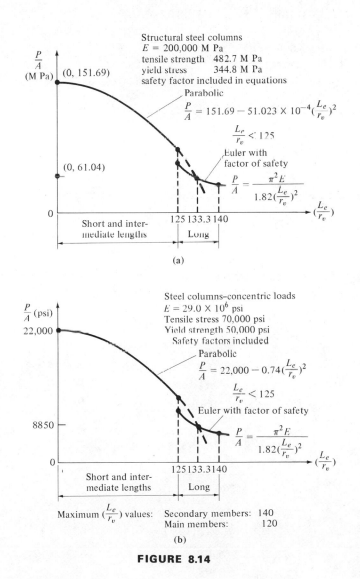

FIGURE 8.14

Solution

(a) Calculate the principal moments of inertia with respect to the u–v set of axes through the centroid C of the welded steel column shown in Figure 8.13. Ignore the weld metal in all calculations.

$$I_u = \tfrac{1}{12}(0.015)(0.16)^3 + 2[\tfrac{1}{12}(0.16)(0.02)^3 + (0.16)(0.02)(0.09)^2]$$

$$= \boxed{57.173 \times 10^{-6} \text{ m}^4}$$

$$I_v = \tfrac{1}{12}(0.16)(0.015)^3 + 2(\tfrac{1}{12})(0.02)(0.16)^3$$

$$= \boxed{13.698 \times 10^{-6} \text{ m}^4}$$

$I_v < I_u$,   column will buckle about the v axis.

$$A = 0.16(0.015) + 2(0.16)(0.02) = 88 \times 10^{-4} \text{ m}^2$$

$$r_v = \sqrt{\frac{I_v}{A}} = \sqrt{\frac{13.698 \times 10^{-6}}{88 \times 10^{-4}}} = 0.0395 \text{ m}$$

Since both ends of the column are fixed, $L_e = \tfrac{1}{2}L = \tfrac{1}{2}(4) = 2$ m.

$$\frac{L_e}{r_v} = \frac{2}{0.0395} = 50.6$$

Refer to Figure 8.14 and note that $50.6 < 125.1$, which means that the following equation is applicable for finding the allowable or safe stress:

$$\frac{P}{A} = 151.69 - 51.023 \times 10^{-4}\left(\frac{L_e}{r_v}\right)^2$$

$$= 151.69 - 51.023 \times 10^{-4}(50.6)^2 = 138.6 \text{ MPa}$$

Since $A = 88 \times 10^{-4}$ m^2, the safe load is given by

$$P = 88 \times 10^{-4} \times 138.6 \times 10^6 \times \frac{1}{10^3}$$

$$= \boxed{1220 \text{ kN}}$$

(b) $L_e = \tfrac{1}{2}L = \tfrac{1}{2}(10.8) = 5.4$ m

$$\frac{L_e}{r_v} = \frac{5.4}{0.0395} = 136.7$$

The result, 136.7, is less than 140, which is the maximum L_e/r_v for secondary members. Again, refer to Figure 8.14 and note that $125 < 136.7 < 140$, which means that the Euler equation with a factor of safety of 1.82 is applicable with $E = 200$ MPa and $A = 88.0 \times 10^{-4}$ m^2.

$$\frac{P}{A} = \frac{\pi^2 E}{1.82(L_e/r_v)^2}$$

$$\frac{P}{88 \times 10^{-4}} = \frac{\pi^2(200 \times 10^9)}{1.82(136.7)^2(10)^6} = 58.04 \text{ MPa}$$

$$P = \frac{(88.0 \times 10^{-4})(58.04 \times 10^6)}{10^3}$$

$$= \boxed{510.8 \text{ kN}}$$

which is the safe load for this longer column. Once again, care was taken to use equations for the proper material, slenderness ratios were computed so as to select the appropriate equations, the fixed-end conditions were associated with effective lengths equal to one-half the geometric lengths, and appropriate safety factors are incorporated in the equations from Figure 8.14(a).

Example 8.8

Column allowable stresses versus L_e/r_v are shown graphically in Figure 8.15 for an aluminum alloy. A 30-ft-long pipe column is fabricated of this alloy and has end connections such that $L_e = 0.75L$. Determine the allowable axial load

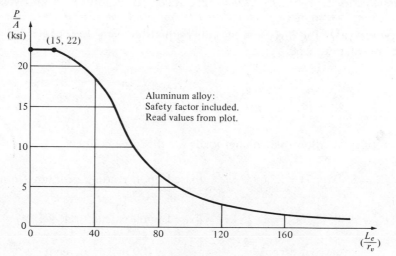

Aluminum alloy:
Safety factor included.
Read values from plot.

FIGURE 8.15

for this column. Cross-sectional properties are: outside diameter = 6.625 in., wall thickness = 0.432 in.

Solution. Calculate the inside diameter of the pipe:

$$D_i = D_o - 2t = 6.625 - 2(0.432) = 5.761 \text{ in.}$$

The area is given by

$$A = \frac{\pi}{4}(D_o^2 - D_i^2) = \frac{\pi}{4}[(6.625)^2 - (5.761)^2]$$

$$= 8.405 \text{ in.}^2$$

The moments of inertia are equal for all diametral axes:

$$I_v = \frac{\pi}{64}(D_o^4 - D_i^4) = \frac{\pi}{64}[(6.625)^4 - (5.761)^4]$$

$$= 40.491 \text{ in.}^4$$

The radius of gyration is given by

$$r_v = \sqrt{\frac{I_v}{A}} = \sqrt{\frac{40.491}{8.405}} = 2.195 \text{ in.}$$

The effective length $L_e = 0.75L = 0.75(30 \times 12) = 270$ in. Compute the slenderness ratio $L_e/r_v = 270/2.195 = 123$. Refer to Figure 8.15 and read an approximate ordinate of $P/A = 2.7$ ksi. Since this L_e/r_v is large, it is feasible to compute a value for P/A from the Euler equation using $E = 10,600$ ksi and a factor of safety of 2.50.

$$\frac{P}{A} = \frac{\pi^2 E}{2.5(L_e/r_v)^2} = \frac{\pi^2(10,600)}{2.5(123)^2} = 2.766 \text{ ksi}$$

Computing the allowable column loads for this pipe column:

$P_1 = 8.405(2.7) = 22.69$ k   (based upon reading ordinate from Figure 8.15)

$P_2 = 8.405(2.766) = 23.25$ k   (based upon calculation of P/A)

$$\text{percent error} = \frac{23.25 - 22.69}{23.25} \times 100 = 2.41 \text{ percent}$$

A slight underestimate by reading the curve value would be acceptable for most engineering applications. The allowable load on the pipe column:

$$P = 23.25 \text{ k}$$

Example 8.9

A cast iron column is 10.0 ft long and has a hollow circular cross section with an outside diameter of 5.00 in. and an inside diameter of 4.50 in. A straight line equation that contains a factor of safety is suitable for the analysis of this cast iron column.

$$\frac{P}{A} = 12,000 - 60\frac{L}{r}$$

(provided that $35 < L/r < 100$). Determine the allowable load P that may be applied safely to this column.

Solution. The radius of gyration is given by

$$r = \sqrt{\frac{I}{A}} = \sqrt{\frac{\pi/64(D_o^4 - D_i^4)}{\pi/4(D_o^2 - D_i^2)}} = \frac{1}{4}\sqrt{\frac{5^4 - 4.5^4}{5^2 - 4.5^2}}$$

$$= 1.68 \text{ in.}$$

The slenderness ratio is

$$\frac{L}{r} = \frac{10 \times 12}{1.68} = 71.43$$

This ratio lies between the defined limits of validity:

$$35 < 71.43 < 100$$

$$\frac{P}{A} = 12,000 - 60\frac{L}{r}$$

$$A = \frac{\pi}{4}(5^2 - 4.5^2) = 3.73 \text{ in.}^2$$

$$P = 3.73[12,000 - 60(71.43)]$$

$$= \boxed{28,770 \text{ lb}}$$

Example 8.10

The allowable load P on a longleaf yellow pine column is to be determined from the following equation, which contains a factor of safety:

$$\frac{P}{A} = 1450\left[1 - 1.62 \times 10^{-6}\left(\frac{L}{d}\right)^4\right]$$

Provided that $L/d < 21$, where d is the least dimension of a rectangular cross section. The column is 12 ft long and has dressed cross-sectional dimensions 9.5 × 11.5 in.

Solution. The least cross-sectional dimension d equals 9.5 in. and $L/d = (12 \times 12)/9.5 = 15.16$. The area $A = 9.5 \times 12.5 = 118.75$ in.^{2}. Since $L/d = 15.16 < 21$, the equation is applicable and yields a value for P:

$$P = 118.75 \times 1450[1 - 1.62 \times 10^{-6}(15.16)^4]$$

$$= \boxed{157,450 \text{ lb}}$$

Homework Problems

8.27 A steel column has the W (wide-flanged) shape shown in Figure H8.27 and is fabricated of steel with a yield stress of 50 ksi. Curves and equations for analysis of the column are shown in Figure 8.14(b). Determine the slenderness ratio of the column if it is 10.0 ft long and is pinned at both ends. Determine the safe axial load for the column and compare this load to the weight of the column on a percentage basis.

8.28 Rectangular structural steel tubing is to be used for a column that is 20.0 ft long and fixed at both ends. Properties of the tubing cross section are shown in Figure H8.28, and the appropriate column curves and equations are shown in Figure 8.14(b). Determine the effective length of the column and its associated slenderness ratio. What is the safe load that may be applied to this column? What percentage of the safe load is the weight of this column?

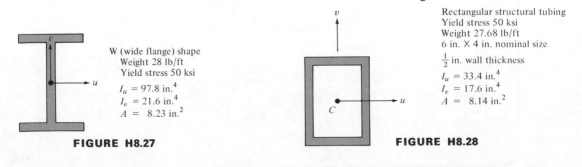

W (wide flange) shape
Weight 28 lb/ft
Yield stress 50 ksi
I_u = 97.8 in.^{4}
I_v = 21.6 in.^{4}
A = 8.23 in.^{2}

FIGURE H8.27

Rectangular structural tubing
Yield stress 50 ksi
Weight 27.68 lb/ft
6 in. × 4 in. nominal size
$\frac{1}{2}$ in. wall thickness
I_u = 33.4 in.^{4}
I_v = 17.6 in.^{4}
A = 8.14 in.^{2}

FIGURE H8.28

8.29 A column cross section, referred to as a *double-angle strut*, is shown in Figure H8.29. This steel column is 8.00 ft long and is pinned at both ends. Determine the critical slenderness ratio for this column and find the safe load for it using the column equations of Figure 8.14(b). Determine the ratio of this safe load to the weight of the column.

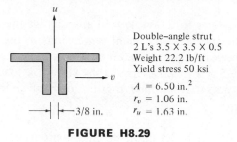

Double–angle strut
2 L's 3.5 × 3.5 × 0.5
Weight 22.2 lb/ft
Yield stress 50 ksi

$A = 6.50$ in.^{2}
$r_v = 1.06$ in.
$r_u = 1.63$ in.

FIGURE H8.29

8.30 A steel pipe column cross section is shown in Figure H8.30. This column is welded to very stiff members at its ends and thus may be considered to have fixed ends. It is 30.0 ft long and fabricated of steel with a 36-ksi yield stress. Determine the safe axial load for this pipe column and compare its weight to the safe load. The appropriate equation for the allowable stress (psi) for this column is

$$\frac{P}{A} = 15{,}840 - 0.533\left(\frac{L_e}{r_v}\right)^2$$

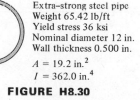

Extra–strong steel pipe
Weight 65.42 lb/ft
Yield stress 36 ksi
Nominal diameter 12 in.
Wall thickness 0.500 in.

$A = 19.2$ in.^{2}
$I = 362.0$ in.^{4}

FIGURE H8.30

8.31 A Douglas fir column has a rectangular cross section 11.5 × 13.5 in. and is 22.0 ft long. Appropriate equations are shown in Figure 8.12. End conditions are such that the given length is equal to the effective length. Determine the slenderness ratio of this timber column. Find the safe load that may be applied to this column. $E = 1.70 \times 10^6$ psi.

8.32 A welded steel column is 8.00 m long and has the box shape shown in Figure H8.32. It is fabricated of structural steel with a yield stress of 344.8 MPa. The appropriate column equations are shown in Figure 8.14. Determine the safe load that this welded pin-ended column will support. Neglect weld metal in evaluating the section area properties.

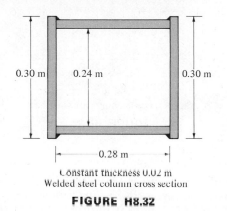

0.30 m   0.24 m   0.30 m

|— 0.28 m —|

Constant thickness 0.02 m
Welded steel column cross section

FIGURE H8.32

8.33 A welded steel column is 5.25 m long and has the cross-sectional shape depicted in Figure H8.33. It is fabricated of structural steel with a yield stress of 344.8 MPa. The appropriate column equations are shown in Figure 8.14. Determine the safe load that this pin-ended column will support. Neglect weld metal in evaluating the section area properties.

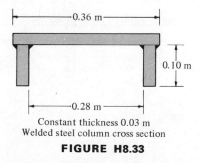

|————— 0.36 m —————|

0.10 m

|————— 0.28 m —————|

Constant thickness 0.03 m
Welded steel column cross section

FIGURE H8.33

8.34 A small machine element must act as a strut that is axially loaded and pinned at both ends. Its cruciform cross section is shown in Figure H8.34 and it has a length of 0.30 m. It is fabricated of steel, for which the appropriate equations are shown in Figure 8.14. Determine the safe load that this strut will support.

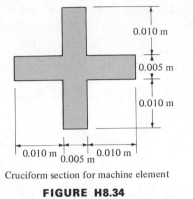

Cruciform section for machine element

FIGURE H8.34

8.35 A bridge member carries axial compressive forces and is built up of two channels laced together as shown in Figure H8.35. Determine the back-to-back distance *d* such that the moment inertia of the column section will be equal for both the *u* and *v* axes through the centroid. Neglect the lacing in making these calculations. This bridge compression member is 45.0 ft long and is fabricated of structural steel, for which the appropriate column equations are shown in Figure 8.14(b). The member is assumed to be pinned at its ends and the lacing is such that the channels act together as a solid member. Determine the safe axial load that this member will resist.

8.36 Refer to Example 8.9 and change the length of the column from 10.0 to 14.0 ft. Determine the allowable load *P* that may be safely applied to this cast iron column of hollow circular cross section.

8.37 Refer to Example 8.10 and change the column length and cross-sectional dimensions to the following values: length = 13.0 ft and rectangular cross section 7.5 × 9.5 in. Determine the allowable load *P* that may be safely applied to this longleaf yellow pine column.

8.38 Refer to Example 8.9 and compare the hollow circular cast iron column with a change of shape to a hollow square cross section. This new shape is to be 5.00 × 5.00 in. outside and is to have a wall thickness such that its area will be the same as the hollow circular cross section of Example 8.9. All other given quantities are to remain the same. Assuming that manufacturing and erection costs are the same, which cross section is to be preferred on the basis of safe load-carrying capacity? Provide numerical justification for your answers.

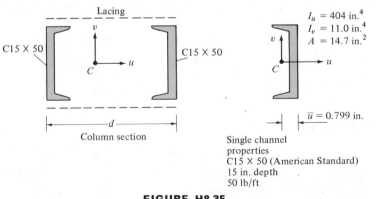

Single channel
properties
C15 × 50 (American Standard)
15 in. depth
50 lb/ft

FIGURE H8.35

★8.7 Generalized Euler Equation Using the Tangent Modulus of Elasticity

In 1744, Euler first derived his column buckling equation but not until over 200 years later was it fully understood in a general sense for the full range of slenderness ratios. In 1899, the conclusion was drawn by Considère and Engesser that the equation applied only to long columns. Prior to this time it was thought to apply to short as well as to long columns. Test results in the short and intermediate ranges of slenderness ratios compared so poorly with the theory that many held the theory to be incorrect.

Engesser proposed the *tangent modulus theory*, which replaces the modulus of elasticity (or Young's modulus) by the tangent modulus of elasticity:

$$P_a = \frac{\pi^2 E_t I_0}{L_e^2} \qquad \textbf{(8.44)}$$

where

$$E_t = \frac{d\sigma}{d\varepsilon} \qquad \textbf{(8.45)}$$

is the slope of the stress–strain diagram at any point, as shown in Figure 8.16.

Considère advanced the idea that when the column begins to bend at the critical load, the appropriate modulus for the concave side is the tangent modulus E_t and the

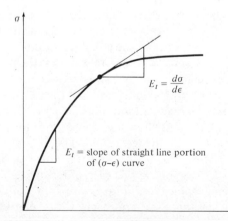

$E_t = \dfrac{d\sigma}{d\epsilon}$

E_t = slope of straight line portion of $(\sigma-\epsilon)$ curve

FIGURE 8.16

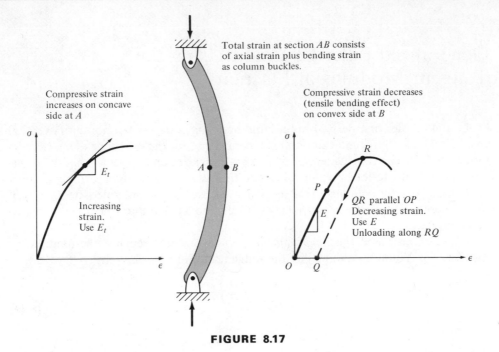

Total strain at section AB consists of axial strain plus bending strain as column buckles.

Compressive strain increases on concave side at A

Compressive strain decreases (tensile bending effect) on convex side at B

σ

E_t

Increasing strain. Use E_t

ϵ

$A \bullet \quad \bullet B$

σ

R

P

E

QR parallel OP Decreasing strain. Use E Unloading along RQ

$O \quad Q$

ϵ

FIGURE 8.17

appropriate modulus for the convex side is Young's modulus E. This idea is presented pictorially in Figure 8.17. From this idea developed the *double modulus theory*, which gained widespread acceptance after von Kármán derived it independently in 1910.

In 1947, F. R. Shanley reexamined these theories and concluded that the tangent modulus theory and not the double modulus theory is correct. The double modulus theory is based upon the assumption that the axial load remains constant as the column moves from the straight to the bent configuration. Shanley observed that it is possible for the axial load to increase, rather than remain constant, as the column begins to bend, which means that no strain reversal will occur at any point in the cross section. If strain reversals do not occur, the tangent modulus is the appropriate one to use. The load deflection curve in Figure 8.18 reveals that the maximum load P_s for the Shanley column model (using an actual stress–strain relationship) lies between the load based upon the tangent modulus load and the reduced modulus load.

Many engineers have accepted the tangent modulus theory as the basis for design, since it leads to a lower buckling load than the double modulus theory and the tangent modulus load agrees better with the test results. The theory as presented here is useful in dealing with aluminum alloys but is not applicable, without modifications, to steel alloys, because of the presence of residual stresses in steel.

Although, in practice, semigraphical techniques are quite often employed to obtain P/A versus L_e/r_v plots, an analytical approach will be employed here which clearly focuses on the fundamental concepts involved. Ramberg and Osgood proposed the following function for fitting stress–strain data, and this is the starting

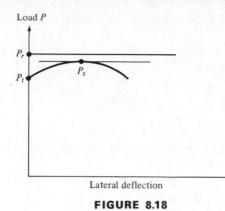

FIGURE 8.18

point for a discussion of the tangent modulus theory used to obtain column allowable stresses versus slenderness ratio curves. The Ramberg–Osgood function contains three parameters, E, n, and B, as shown in the following equation:

$$\varepsilon = \frac{\sigma}{E} + \left(\frac{\sigma}{B}\right)^n \tag{8.46}$$

where E is Young's modulus and n and B are chosen such that the ε–σ function fits the experimental data; n is a nondimensional exponent and B has stress units in order that the equation for $\varepsilon = f(\sigma)$ be dimensionally homogeneous. If $E = 10 \times 10^6$ psi, $n = 12$, and $B = 80{,}000$ psi, the Ramberg–Osgood equation becomes

$$\varepsilon = 10^{-7}\sigma + 1.455 \times 10^{-59}\sigma^{12} \tag{8.47}$$

If values of σ are assigned arbitrarily, varying from 0 to 70,000 psi, and the corresponding values of ε calculated, the resulting plot of the stress–strain curve is shown in Figure 8.19. The plot is not essential for this development but graphically shows

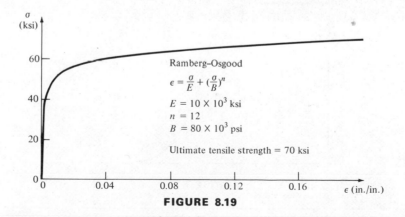

FIGURE 8.19

that fitting stress–strain data with such a function is feasible, particularly for aluminum alloys.

To develop an equation for the tangent modulus as a function of stress σ, differentiate Eq. 8.46 with respect to σ:

$$\frac{d\varepsilon}{d\sigma} = \frac{1}{E} + \frac{n}{B^n}\sigma^{n-1}$$

But $E_t = d\sigma/d\varepsilon$ from Eq. 8.45, and the tangent modulus may be expressed as follows:

$$E_t = \frac{1}{d\varepsilon/d\sigma} = \frac{1}{1/E + (n/B^n)\sigma^{n-1}} \tag{8.48}$$

which may be written in the form

$$E_t = \frac{E}{1 + (nE/B^n)\sigma^{n-1}} \tag{8.49}$$

Choosing, as before, $E = 1.0 \times 10^7$ psi, $n = 12$, and $B = 80{,}000$ psi, we find that the equation for the tangent modulus becomes

$$E_t = \frac{10^7}{1 + 1.746 \times 10^{-51}\sigma^{11}} \tag{8.50}$$

If values of σ are assigned, as before, from 0 to 70,000 psi and the corresponding values of E_t calculated, the resulting plot of the E_t versus σ curve is shown in Figure 8.20. This plot is similar to those obtained for aluminum alloys.

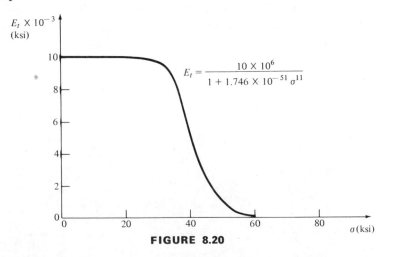

$$E_t = \frac{10 \times 10^6}{1 + 1.746 \times 10^{-51}\sigma^{11}}$$

FIGURE 8.20

Introducing a factor of safety N into the Euler equation and using the tangent modulus E_t results in

$$\frac{P}{A} = \frac{\pi^2 E_t}{N(L_e/r_v)^2} \qquad (8.51)$$

The allowable stress P/A and σ are related through the factor of the safety as follows:

$$\sigma = N\frac{P}{A} \qquad (8.52)$$

Substitution of Eq. 8.52 into Eq. 8.49 for E_t, together with replacement of E_t with Eq. 8.51, leads to

$$\frac{P}{A} + \frac{nE}{B^n}N^{n-1}\left(\frac{P}{A}\right)^n = \frac{\pi^2 E}{N(L_e/r_v)^2} \qquad (8.53)$$

This equation relates the allowable stress P/A to the slenderness ratio L_e/r_v. Once more, choosing $E = 1.0 \times 10^7$ psi, $n = 12$, and $B = 80{,}000$ psi, together with a factor of safety $N = 2.5$, and then solving for L_e/r_v yields

$$\frac{L_e}{r_v} = \frac{6283.2}{\sqrt{P/A + 4.163 \times 10^{-47}(P/A)^{12}}} \qquad (8.54)$$

Assign values of P/A varying from 2000 to 27,000 and calculate corresponding slenderness ratios L_e/r_v to obtain points required to plot Figure 8.21. This plot is the column analysis and design curve, which extends over the entire L_e/r_v range from

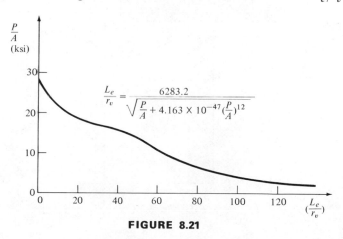

FIGURE 8.21

very short to long columns. Once the stress–strain curve for the given material has been fitted with an appropriate function, the remainder of the analysis proceeds as shown above. This rational analysis, using the tangent modulus E_t, extends the range of validity of the Euler equation to all slenderness ratios of interest.

A major advantage of this rational theory employing the tangent modulus when compared to the empirical-equations approach of Section 8.6 is the simplification in experimentation and the inherently lower cost of a testing program. Once the tangent modulus theory has successfully been applied to a given material, only simple tension and compression tests need be run to obtain the stress–strain properties of new alloys that may be developed. The theory requires only this information, which enables an engineering analyst to develop the column analysis and design curves. The empirical approach of Section 8.6 requires the testing of a large number of columns of various slenderness ratios. These tests would usually be run with full-scale specimens, which means that an extensive and expensive testing program would be required for each new alloy developed.

Example 8.11

A T-shaped cross section, as shown in Figure 8.22, is to be used as a 4.06-m-long column cross section. It is to be fabricated from the material for which the column analysis information is shown in Figure 8.21. Determine the safe load for this column, which is to be fixed at both ends.

Solution. First, compute the area properties of the cross section.

$$A = 0.20(0.03) + 0.20(0.03) = 0.0120 \text{ m}^2$$

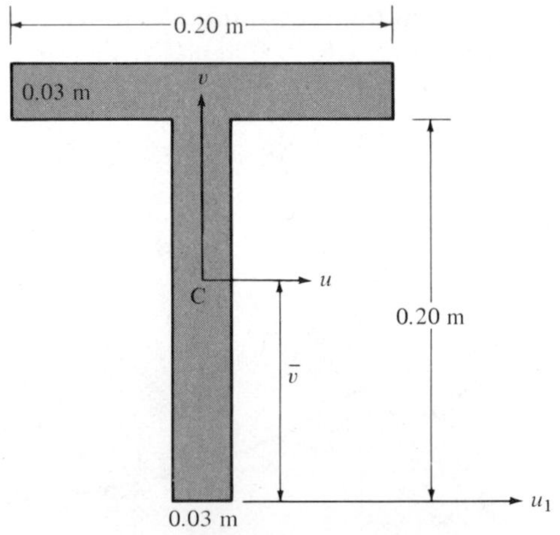

FIGURE 8.22

Locate the centroid C with respect to the u_1 axis:

$$\bar{v} = \frac{A_1 v_1 + A_2 v_2}{A_1 + A_2}$$

$$= \frac{0.20(0.03)(0.215) + 0.20(0.03)(0.10)}{0.0120}$$

$$= 0.1575 \text{ m}$$

Compute the area moments of inertia as follows:

$$I_u = \bar{I}_1 + A_1 d_1^2 + \bar{I}_2 + A_2 d_2^2$$

$$= \tfrac{1}{12}(0.20)(0.03)^3 + (0.20)(0.03)(0.215 - 0.1575)^2 + \tfrac{1}{12}(0.03)(0.20)^3$$

$$+ 0.20(0.03)(0.10 - 0.1575)^2$$

$$= 60.125 \times 10^{-6} \text{ m}^4$$

$$I_v = \bar{I}_1 + A_1 d_1^2 + \bar{I}_2 + A_2 d_2^2$$

$$= \tfrac{1}{12}(0.03)(0.20)^3 + 0 + \tfrac{1}{12}(0.20)(0.03)^3 + 0$$

$$= 20.45 \times 10^{-6} \text{ m}^4$$

$I_v < I_u,$ column will buckle about the v axis.

$$r_v = \sqrt{\frac{I_v}{A}} = \sqrt{\frac{20.45 \times 10^{-6}}{0.0120}} = 0.0413 \text{ m}$$

$$L_e = \tfrac{1}{2}L = \tfrac{1}{2}(4.06) = 2.03 \text{ m}$$

$$\frac{L_e}{r_v} = \frac{2.03}{0.0413} = 49.15$$

From Figure 8.21, $P/A = 14.0$ ksi. Note the introduction of different units, which is legitimate, since L/r is a nondimensional ratio. Convert P/A to the appropriate units:

$$\frac{P}{A} = 14.0 \text{ ksi} \times \frac{6.895 \text{ MPa}}{1 \text{ ksi}} = 96.53 \text{ MPa}$$

$$P = 0.0120 \text{ m}^2 \times 96.53 \times 10^6 \frac{\text{N}}{\text{m}^2} \times \frac{1 \text{ kN}}{10^3 \text{ N}}$$

$$= \boxed{1158.4 \text{ kN}}$$

As for empirical column equations, note should be taken of the following:

1. The allowable stress was obtained from the curve devised for the material of which the column is to be fabricated.
2. The factor of safety $N = 2.5$ is included in the plotted values.
3. The slenderness ratio was computed using the effective length of a column fixed at both ends. One curve and equation serves for the entire range of slenderness ratios in this case, since the tangent modulus theory generalizes the Euler theory for all slenderness ratios of interest.
4. Fixed-ended conditions were noted which reduced the effective length to one-half of the geometric length.

Homework Problems

8.39–8.43 These problems require information for Ramberg–Osgood stress–strain plots and safety factors provided in the accompanying tabulation. In each case, the student is to make the calculations required to construct the following plots:

(a) Stress–strain (see Figure 8.19).
(b) Tangent modulus of elasticity–stress (see Figure 8.20).
(c) Allowable stress–slenderness ratio (see Figure 8.21).

Neat plots, together with well-organized calculations sheets, are to be submitted by the student.

Problem Number	$E \times 10^{-6}$ psi	n	$B \times 10^{-3}$ psi	N
8.39	10	14	84	2.5
8.40	10	16	86	2.0
8.41	10	18	88	3.0
8.42	10	20	90	2.5
8.43	6.5	10	50	2.75

8.44 A 12.0-ft-long column is fabricated of an aluminum alloy. Appropriate column analysis equations are provided in Figure H8.44(a), and cross-sectional properties of this S (standard I) shape are given in

Figure H8.44(b). The effective length L_e of the column is obtained by multiplying its geometric length by 0.75. Determine the safe load that may be applied to this aluminum alloy column.

(HINT: Assume that $\sigma < 23$ ksi and use $E_t = 10,000$ ksi. Then check the validity of this assumption.) Form the ratio of the safe load to the weight of this aluminum column.

Generalized Euler equation

$$\frac{P}{A} = \frac{\pi^2 E_t}{2.5\left(\frac{L_e}{r_v}\right)^2} \quad \text{but not greater than 14.0 ksi}$$

Safety factor of 2.5, $\sigma = 2.5\dfrac{P}{A}$

Tangent modulus of elasticity E_t
(data fitted with two straight lines)

$E_t = 10,000$ ksi $\sigma \leqslant 23$ ksi

$E_t = -750\sigma + 27,250$ $23 \leqslant \sigma \leqslant 35$ ksi

Aluminum alloy column analysis

(a)

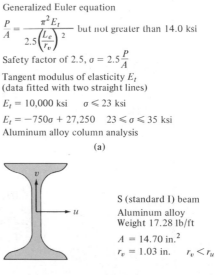

S (standard I) beam
Aluminum alloy
Weight 17.28 lb/ft

$A = 14.70$ in.^{2}
$r_v = 1.03$ in. $r_v < r_u$

(b)

FIGURE H8.44

8.45 A wide-flange aluminum column is 8.00 ft long and has a cross-sectional area of 7.81 in.^{2} with a least radius of gyration of $r_v = 1.41$ in. Refer to Figure H8.44(a) for the appropriate column analysis equations. Effective length L_e of the column is obtained by multiplying its geometric length by 0.75. Determine the safe load that may be applied to this aluminum alloy column.
(HINT: In this case $\sigma > 23$ ksi.)

8.46 A solid cylindrical aluminum alloy rod acts as a compression component in a machine. It has a diameter of 0.50 in. and a length of 10.0 in. Column analysis equations are given in Figure H8.44(a). Assume the rod to be pinned at both ends and determine the safe load that it will support.

8.47 An aluminum-alloy column is 18.0 ft long and may be assumed to have fixed ends. The cross section is square, 4 × 4 in. outside, with a wall thickness of 0.50 in. Using the information contained in Figure H8.44(a) for this alloy, find the safe load that may be applied to this column.

8.48 Refer to Problem 8.47 and devise two hollow circular cross sections to replace the square one given such that the carrying capacities of the columns will remain the same. Consider two separate replacements:
(a) Equal wall thicknesses (i.e., 0.50 in.).
(b) Equal areas.
All other given information is to remain the same. Discuss the merits of these replacements.

8.49 An aluminum column 10.0 ft long is pinned at both ends and has the same cross section as the steel column analyzed in Example 8.5. Appropriate equations for this aluminum alloy column are provided in Figure H8.44(a). Determine the safe load for this aluminum column.

8.50 Refer to Figure H8.44(a) and use the tangent modulus information provided there together with the equation $E_t = d\sigma/d\varepsilon$ to write stress–strain functions covering the stress range for σ from 0 to 35 ksi. Make a careful plot of your results.

Chapter 9 Statically Indeterminate Members

9.1 Introduction

The problems encountered in all the preceding chapters dealt with members subjected to force systems known as *statically determinate* force systems. The reason for this name is the fact that in all of these problems, the force system (including applied as well as known and unknown reactive forces, torques, and moments) could be fully ascertained by satisfying the applicable equations of equilibrium. Under certain conditions, however, members are subjected to force systems that cannot be completely defined by satisfying only the conditions of equilibrium. For these members the equations of equilibrium lead to systems of equations in which the number of unknown quantities exceeds the number of equations. Thus the system of equations is not solvable unless supplemented, according to the principle of *consistent deformations*, by additional equations that are based upon member deformations. Such members are said to be subjected to *statically indeterminate* force systems.

The analysis of statically indeterminate members is discussed in this chapter. Three classes of statically indeterminate members are distinguished depending upon the type of load to which members are subjected. Section 9.2 deals with statically indeterminate members under axial loads, Section 9.3 with statically indeterminate members under torsional loads, and Sections 9.4 to 9.7 with statically indeterminate members under flexural loads. Also, the effects of uniform temperature changes on the behavior of structural and machine members is dealt with in certain Examples and Homework Problems.

9.2 Statically Indeterminate Members Under Axial Loads

The procedure used in the solution of statically indeterminate problems containing axially loaded members is best illustrated by examining an actual problem. Thus consider the frame shown in Figure 9.1(a). The horizontal member is assumed to be rigid and is supported by three rods as shown. The two outside rods have the same

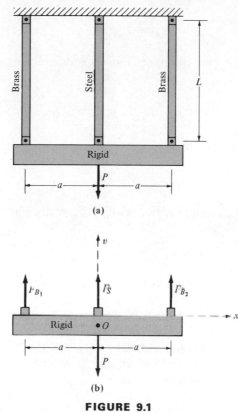

FIGURE 9.1

cross-sectional area A_B and are made of brass for which the modulus of elasticity is E_B. The center rod has a cross-sectional area A_S and is made of steel for which the modulus of elasticity is E_S. Assume that the frame is subjected to the vertical load P and that the rigid member remains horizontal after the load is applied. It is required to determine the normal stresses induced in the brass and steel rods and the vertical movement of the rigid member.

A free-body diagram of the rigid member is shown in Figure 9.1(b). Note that this free-body diagram contains the three unknown forces F_{B1}, F_{B2}, and F_S. Since only two equations of equilibrium are available, the given frame is statically indeterminate. Viewed differently, the frame is said to contain one more constraint than is required for equilibrium. For example, the steel rod may be removed without affecting the equilibrium of the frame. The steel bar, therefore, is said to be *redundant* or unnecessary for the equilibrium of the system.

Application of the two available equations of equilibrium to the forces shown in Figure 9.1(b) leads to the following relations:

$$\circlearrowleft \quad \sum M_0 = 0$$

$$F_{B2}(a) - F_{B1}(a) = 0$$

Therefore,

$$F_{B2} = F_{B1} = F_B \tag{a}$$

which indicates that the forces in the two brass rods are identical.

$\uparrow \ \Sigma F_v = 0$

$$F_{B1} + F_{B2} + F_S - P = 0 \tag{b}$$

Substituting Eq. (a) into Eq. (b) yields

$$2F_B + F_S - P = 0 \tag{c}$$

Since $F_B = \sigma_B A_B$ and $F_S = \sigma_S A_S$, Eq. (c) may be written in terms of the required stresses as follows:

$$2\sigma_B A_B + \sigma_S A_S - P = 0 \tag{d}$$

Equation (d) contains the two unknown quantities σ_B and σ_S and is, by itself, insufficient for a unique determination of their values. Thus a second equation containing the same two unknown quantities needs to be developed to complete the solution of the problem. This second equation is derived by applying the principle of *consistent deformations*. For this particular frame, the principle of consistent deformation requires that the deformations of the three rods are compatible with the geometric requirements of the problem. Since the rigid member is to remain horizontal, the extension of the brass rods, δ_B, must be exactly the same as the extension of the steel rod, δ_S. This physical condition is expressed mathematically by the relation

$$\delta_B = \delta_S \tag{e}$$

According to Eq. 2.16a, $\delta_B = \varepsilon_B L$ and $\delta_S = \varepsilon_S L$, where ε_B and ε_S are the normal (axial) strains in the brass and steel rods, respectively. Thus, by Eq. (e), $\varepsilon_B L = \varepsilon_S L$, and therefore

$$\varepsilon_B = \varepsilon_S \tag{f}$$

The assumption is now made that the materials (both brass and steel) obey Hooke's law and, by Eq. 2.25, $\varepsilon_B = \sigma_B / E_B$ and $\varepsilon_S = \sigma_S / E_S$. Thus Eq. (f) may also be expressed in terms of the desired stresses as follows:

$$\frac{\sigma_B}{E_B} = \frac{\sigma_S}{E_S} \tag{g}$$

Therefore, the solution of the given statically indeterminate problem has been reduced to the solution of the two simultaneous equations (d) and (g), which leads to the normal stresses σ_B and σ_S. Since Hooke's law was used in the solution, the resulting values of stress should be carefully compared to the proportional limit of the materials to ensure that those limits have not been exceeded. Should the proportional limit of the material be exceeded, the solution is obviously not valid. Under such conditions another type of solution based upon inelastic analysis is required. Such a solution is beyond the scope of this chapter. However, certain procedures based upon inelastic behavior are discussed in Chapter 12. Once the stress values are determined, the vertical movement of the rigid member is found by computing the deformation of the brass rod, δ_B, or the deformation of the steel rod, δ_S. Thus

$$\delta_B = \delta_S = \varepsilon_S L = \frac{\sigma_S}{E_S} L \qquad \textbf{(h)}$$

Example 9.1

The following data apply to the frame shown in Figure 9.1(a): $P = 13{,}750$ lb, $a = 3$ in., $L = 5$ ft, $A_S = 1.000$ in.^{2}, $A_B = 0.375$ in.^{2}, $E_S = 30 \times 10^6$ psi, and $E_B = 15 \times 10^6$ psi. Determine the normal stresses induced in the steel and brass rods and the movement of the rigid member.

Solution. Substitution of the given data into Eqs. (d) and (g) developed earlier leads to the following two simultaneous equations:

$$0.75\sigma_B + \sigma_S - 13{,}750 = 0$$

$$2.00\sigma_B - \sigma_S = 0$$

A simultaneous solution of these two equations yields

$$\boxed{\sigma_S = 10{,}000 \text{ psi}}$$

$$\boxed{\sigma_B = 5000 \text{ psi}}$$

The steel and brass used in this frame would have to have proportional limits at least equal to 10,000 psi and 5000 psi, respectively, if the solution is to be valid.

The vertical movement of the rigid member is found from Eq. (h). Thus

$$\delta_B = \delta_S = \frac{10{,}000}{30 \times 10^6} \times 60$$

$$= \boxed{0.02 \text{ in.}}$$

Example 9.2

The rigid bar ABD shown in Figure 9.2(a) is hinged at A and supported by the aluminum rod BC and the concrete support DE. The system is subjected to the 500-kN force. The cross-sectional areas of the aluminum rod and concrete support are $A_A = 7 \times 10^{-4}$ m^2 and $A_C = 70 \times 10^{-4}$ m^2, respectively. The modulii of elasticity for aluminum and concrete are $E_A = 70 \times 10^9$ N/m^2 and $E_C = 30 \times 10^9$ N/m^2, respectively. Compute the normal stresses developed in the aluminum and concrete. Assume linearly elastic behavior.

Solution. The free-body diagram of the rigid member ABD is shown in Figure 9.2(b), in which F_A is the force in the aluminum rod and F_C is the force in the

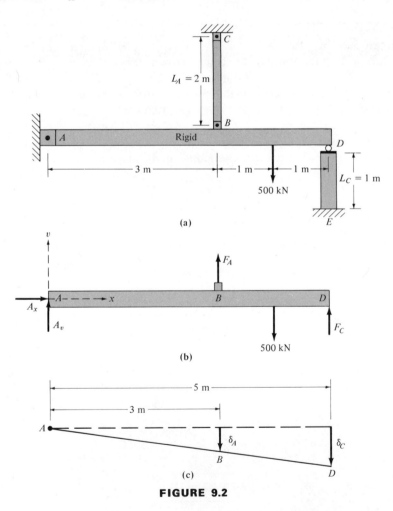

FIGURE 9.2

concrete support. The horizontal and vertical components of the reaction at hinge A are designated by A_x and A_v, respectively. Clearly, this force system is statically indeterminate, since there are four unknown forces and only three available equations of equilibrium. In other words, the structure shown in Figure 9.2(a) has one redundant member. This redundant member, for example BC, could be eliminated and still have a structure capable of sustaining the applied force. Application of the conditions of equilibrium to the force system in Figure 9.2(b) yields

$$\rightarrow \sum F_x = 0$$

$$A_x = 0 \tag{a}$$

$$\uparrow \sum F_v = 0$$

$$A_v + F_C + F_A - 500 = 0 \tag{b}$$

$$\circlearrowleft \quad \sum M_A = 0$$

$$3F_A + 5F_C - 2000 = 0 \tag{c}$$

Thus the conditions of equilibrium reduce to two equations [i.e., Eqs. (b) and (c)] with the three unknown F_A, F_C, and A_v. Another equation, containing two of the three unknowns, is needed to complete the solution. This equation is obtained from the principal of consistent deformations. In this case this principle implies that the extension, δ_A, of the aluminum rod BC must be compatible with the contraction, δ_C, of the concrete support DE. Since the member ABD is rigid, this compatibility condition is expressed, geometrically, by the line diagram shown in Figure 9.2(c). Thus from similar triangles,

$$\frac{\delta_C}{5} = \frac{\delta_A}{3}$$

and

$$\delta_A = \tfrac{3}{5}\delta_C \tag{d}$$

Since the materials are assumed to behave in a linearly elastic manner, Hooke's law applies and

$$\delta_A = \left(\frac{\sigma_A}{E_A}\right)L_A = \frac{F_A L_A}{A_A E_A} = \frac{2 \times 10^{-6} F_A}{49} \text{ m}$$

and

$$\delta_C = \left(\frac{\sigma_C}{E_C}\right)L_C = \frac{F_C L_C}{A_C E_C} = \frac{1 \times 10^{-6} F_C}{210} \text{ m}$$

Therefore, Eq. (d) may be expressed in terms of the two unknown quantities F_A and F_C. Thus

$$\frac{2 \times 10^{-6} F_A}{49} = \frac{\frac{3}{5} \times 10^{-6} F_C}{210}$$

and

$$F_C = 14.29 F_A \qquad \text{(e)}$$

Equations (c) and (e) may now be solved simultaneously to obtain the values of F_A and F_C. Thus $F_A = 26.86$ kN and $F_C = 383.88$ kN. The vertical component of the force in the hinge at A, if needed, can be computed from Eq. (b). The normal stresses σ_A and σ_C in the aluminum rod and concrete support, respectively, are determined as follows:

$$\sigma_A = \frac{F_A}{A_A} = \frac{26.86 \times 10^3}{7 \times 10^{-4}}$$

$$= \boxed{38.4 \text{ MN/m}^2}$$

$$\sigma_C = \frac{F_C}{A_C} = \frac{383.88 \times 10^3}{70 \times 10^{-4}}$$

$$= \boxed{54.8 \text{ MN/m}^2}$$

Example 9.3

A vertical steel member is fixed between the two unyielding supports A and B, as shown in Figure 9.3(a). When the temperature is 70°F, there is no stress in the steel member. Determine the axial stresses developed in segments AC and CB of the member when a load $P = 50$ k is applied as shown and the temperature is increased to 650°F. Assume linear elastic behavior and that the distance between A and B (i.e., $L = a + b$) remains unchanged. The following data apply: the modulus of elasticity for steel, $E = 30 \times 10^6$ psi, $a = 2$ ft, $b = 3$ ft, the cross-sectional area of the steel member $A = 6$ in.^{2}, and the coefficient of thermal expansion for steel, $\alpha = 6.5 \times 10^{-6}$ in./in./°F.

Solution. The solution of this problem is accomplished most conveniently by considering it in two stages. Stage 1 would consider the effect of the load P only and stage 2 the effects of the temperature change (thermal effects) only. Once these two effects are analyzed separately, they can be superimposed using the

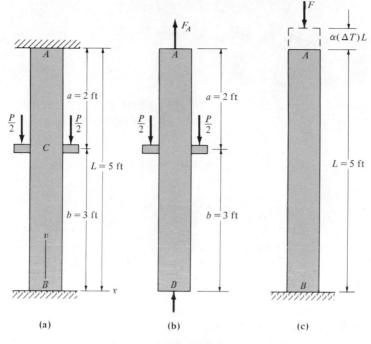

FIGURE 9.3

principle of superposition, which is valid since linear elastic behavior is assumed.

STAGE 1: A free-body diagram of the steel member when only the load P is acting is shown in Figure 9.3(b), where the reactive forces F_A and F_B at supports A and B, respectively, represent those due only to the applied force P. Thus there are two unknown quantities and only one available equation of equilibrium. Therefore,

$$\uparrow \sum F_v = 0$$

$$F_A + F_B - 50 = 0 \tag{a}$$

A second relationship among the forces is obtained from the principal of consistent deformations. Since the length L is to remain unchanged, the lengthening of segment AC, δ_{AC}, must be numerically equal to the shortening of segment BC, δ_{BC}. Thus

$$\delta_{AC} = \delta_{BC} \tag{b}$$

in which $\delta_{AC} = (F_A)(a)/AE$ and $\delta_{BC} = (F_B)(b)/AE$. Substitution of these quantities into Eq. (b) yields

$$(F_A)(a) = (F_B)(b) \qquad\qquad\qquad\qquad\text{(c)}$$

where $a = 2$ ft and $b = 3$ ft. Therefore,

$$F_A = 1.5F_B \qquad\qquad\qquad\qquad\text{(d)}$$

A simultaneous solution of Eqs. (a) and (d) leads to $F_A = 30$ k and $F_B = 20$ k. Thus the axial stresses induced by the force P in segments AC and BC are

$$\sigma_{AC} = \frac{F_A}{A} = \frac{30}{6} = 5 \text{ ksi} = 5000 \text{ psi}(T)$$

$$\sigma_{BC} = \frac{F_B}{A} = \frac{20}{6} = 3.33 \text{ ksi} = 3330 \text{ psi}(C)$$

STAGE 2: Engineering materials, when not restrained, experience a change in dimensions if subjected to a change in temperature. The change in length per unit length per degree change in temperature is referred to as the *coefficient of thermal expansion* and denoted by the symbol α. The coefficient of thermal expansion is a unique property for a given material, and representative values of this property are given in Appendix C for a few materials.

Consider, for example, the case of an unrestrained prismatic bar of length L. If subjected to a temperature change ΔT, the bar would experience a change in length equal to $\alpha(\Delta T)L$. If this prismatic bar, however, were restrained so that the change in length could not occur in part or in whole, forces would be induced in it giving rise to a system of stresses known as *thermal stresses*. Obviously, similar effects may be developed in an unrestrained prismatic bar if it were subjected to a nonuniform temperature distribution. However, this and other examples as well as the Homework Problems in this chapter deal only with prismatic members under uniform temperature changes.

Thus when the temperature of the steel member AB in Figure 9.3 is increased by the amount $\Delta T = 75.8 - 70 = 5.8°F$, it would have the tendency to increase in length by $\alpha(\Delta T)L$. Obviously, since the length of the member L remains unchanged, this tendency to increase in length cannot materialize and a compressive restraining force due to thermal effects is induced in the supports at A and B. Determination of this compressive restraining force is most conveniently accomplished by assuming that one of the two supports, say support A, is removed as in Figure 9.3(c) to allow the member to undergo free expansion by the amount $\alpha(\Delta T)L = 6.5 \times 10^{-6}(5.8)60 = 22.62 \times 10^{-4}$ in. This free expansion would have to be eliminated by the application of the compressive force F as shown in Figure 9.3(c). The compressive force would need to be of sufficient magnitude to produce a contraction in the steel member equal to 22.62×10^{-2} in. The decrease in length resulting from the compressive force F in the steel member is $FL/AE = 3.33 \times 10^{-7}F$ in. Therefore,

$$3.33 \times 10^{-7}F = 22.62 \times 10^{-4}$$

and

$$F = -6793 \text{ lb}$$

Thus the compressive axial stress σ, produced by thermal effects throughout the steel member, becomes

$$\sigma = \frac{F}{A} = \frac{-6793}{6} = -1132 \text{ psi} = 1132 \text{ psi}(C)$$

SUPERPOSITION OF STAGES 1 AND 2: The resultant stresses in segments AB and BC of the steel member are obtained by superposing the stresses due to the applied load P and to the change in temperature. Therefore, the resultant stress in segment AB is

$$\sigma'_{AB} = 5000 - 1132$$

$$= \boxed{3868 \text{ psi}(T)}$$

and the resultant stress in segment BC is

$$\sigma'_{BC} = -3330 - 1132 = -4462 \text{ psi}$$

$$= \boxed{4462 \text{ psi}(C)}$$

Homework Problems

In all the following problems, assume that the materials obey Hooke's law.

9.1 A composite compression member consisting of a concentric steel bar and an aluminum tube is loaded as shown in Figure H9.1. Determine the axial stresses in the steel bar and in the aluminum tube if $E_S = 30 \times 10^6$ psi, $E_A = 10 \times 10^6$ psi, $A_S = A_A = 0.2$ in.^{2}.

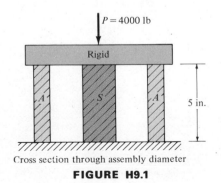

Cross section through assembly diameter

FIGURE H9.1

9.2 The frame shown in Figure H9.2 consists of two magnesium outside bars marked M and a center aluminum bar marked A. If the aluminum bar is 0.001 in shorter than the two magnesium bars, determine the load P so that the axial stresses induced in the three bars are identical. Each bar has a cross-sectional area of 2 in.^{2}. Assume that $E_A = 10 \times 10^6$ psi and $E_M = 6.5 \times 10^6$ psi. Compute also the shortening of the magnesium bars.

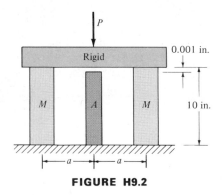

FIGURE H9.2

9.3 An aluminum cylinder is filled with concrete as shown in Figure H9.3. The assembly is then subjected to the compressive load $P = 750$ kN. Compute the axial stress induced in the aluminum cylinder and in the concrete material as well as the shortening of the assembly. Assume that $E_A = 70$ GN/m^2 and $E_C = 30$ GN/m^2.

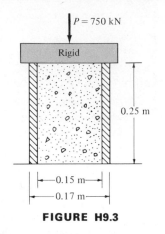

FIGURE H9.3

9.4 The rigid horizontal member is supported by the two rods A and B as shown in Figure H9.4. The 100-kN load is so placed that the rigid member remains horizontal. Assume the following data: $A_A = 4 \times 10^{-4}$ m^2, $A_B = 3 \times 10^{-4}$ m^2, $E_A = 100 \times 10^9$ N/m^2, and $E_B = 120 \times 10^9$ N/m^2. Determine:
(a) The angle θ for equilibrium.
(b) The position of the 100-kN force on the horizontal rigid member.
(c) The axial stress developed in rods A and B.
(d) The vertical movement of the rigid member.

9.5 The rigid horizontal member is supported by the aluminum rod A and brass rod B, as shown in Figure H9.5. When the 50-kip load is applied, the rigid

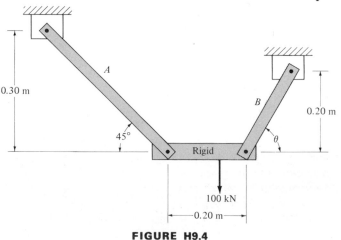

FIGURE H9.4

member remains horizontal and the axial stress induced in the brass rod is $\sigma_B = 15$ ksi. If the moduli of elasticity for aluminum and brass are, respectively, $E_A = 10 \times 10^3$ ksi and $E_B = 15 \times 10^3$ ksi, determine the necessary cross-sectional areas of the two rods.

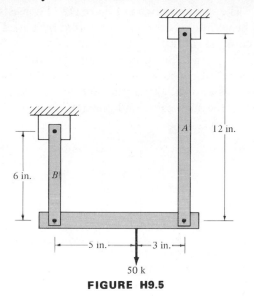

FIGURE H9.5

9.6 A compression block is made up of two materials as shown in Figure H9.6. The magnesium portion measures 0.06 m × 0.12 m in cross section while the wooden portion measures 0.12 m × 0.12 m in cross section. If $E_M = 50 \times 10^9$ N/m^2 and $E_W = 10 \times 10^9$ N/m^2, and if the 250-kN load is so placed that the rigid platen remains horizontal, determine:

(a) The position of the 250-kN load.

(b) The compressive axial stresses in the magnesium and in the wood.

(c) The shortening of the compression block.

9.7 The assembly shown in Figure H9.7 consists of an aluminum block 0.10 m long placed on top of a steel block that is 0.15 m long. The assembly is subjected to a load P which deforms it by 5×10^{-4} m. If both blocks have the same cross-sectional area $A = 6 \times 10^{-4}$ m^2, determine the value of the load P. Assume that $E_S = 210$ GN/m^2 and $E_A = 70$ GN/m^2.

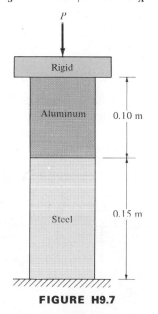

FIGURE H9.7

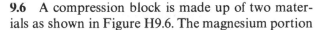
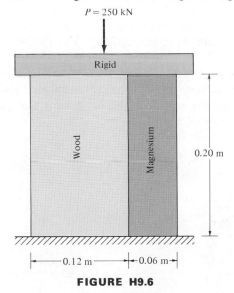

FIGURE H9.6

9.8 The assembly shown in Figure H9.8 consists of an outer cylinder of material A, a central solid core of material B, and filling the concentric space between A and B is material C. A compressive load P is applied through the rigid plate deforming the entire assembly uniformly by an amount $\delta = 0.002$ in. Determine the load P and the axial stresses in the various components of the assembly. Assume that $E_A = 25 \times 10^6$ psi, $E_B = 15 \times 10^6$ psi, and $E_C = 8 \times 10^6$ psi.

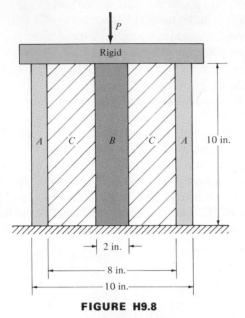

FIGURE H9.8

9.9 The entire assembly shown in Figure H9.9 is made of aluminum for which $E = 10 \times 10^6$ psi. If the load $P = 30$ k, determine the axial stresses and deformations in both parts of the assembly.

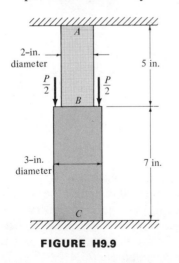

FIGURE H9.9

9.10 Refer to the assembly shown in Figure H9.9 and assume segment AB to be steel for which

$E_S = 30 \times 10^6$ psi and segment BC to be aluminum for which $E_A = 10 \times 10^6$ psi. If the load $P = 40$ k, find the axial stresses and deformations in both parts of the assembly.

9.11 The system shown in Figure H9.11 consists of a rigid beam hinged at A and supported by a steel flexible cable at B ($E_S = 200$ GN/m^2 and $A_S = 0.7 \times 10^{-4}$ m^2) and a concrete support at C($E_C = 30$ GN/m^2, $A_C = 50 \times 10^{-4}$ m^2). A load of 50 kN is applied as shown. Determine the axial stresses induced in the steel cable and concrete support.

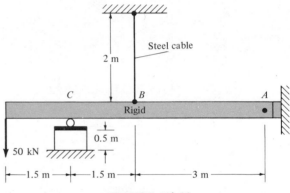

FIGURE H9.11

9.12 The rigid beam shown in Figure H9.12 is hinged at A and supported at B and C by a steel and an aluminum rod, respectively. The following data apply: $E_S = 30 \times 10^6$ psi, $A_S = 0.2$ in.^{2}, $E_A = 10 \times 10^6$ psi, $A_A = 1.0$ in.^{2}. Determine the load P to produce an axial stress of 15,000 psi in the steel rod. Also determine the stress in the aluminum rod and the deformation of both rods.

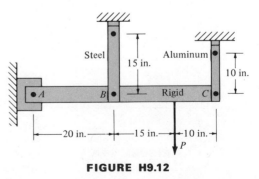

FIGURE H9.12

9.13 The system shown in Figure H9.13 consists of a rigid beam hinged at A and supported at B by means of a flexible cable made of material X. A load

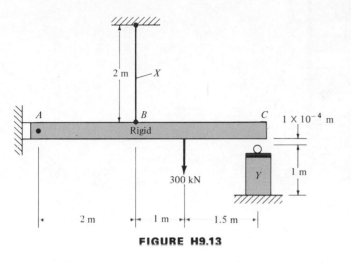

FIGURE H9.13

$P = 300$ kN is applied as shown. Before the load is applied, a gap of 1×10^{-4} m exists between the rigid beam and the support at C, which is made of material Y. If $E_X = 250 \times 10^9$ N/m², $A_X = 1 \times 10^{-4}$ m², $E_Y = 25 \times 10^9$ N/m², and $A_Y = 60 \times 10^{-4}$ m², determine the axial stresses and deformations in materials X and Y.

9.14 Determine the axial stresses in Problem 9.9 if in addition to the 30-k load, there is a temperature drop $\Delta T = 40°F$. Assume the coefficient of thermal expansion for aluminum, $\alpha = 12.5 \times 10^{-6}$ in./in./°F.

9.15 Determine the axial stresses in Problem 9.10 if in addition to the 40-k load, there is a temperature increase $\Delta T = 100°F$. Assume the coefficients of thermal expansions for steel and aluminum to be, respectively, $\alpha_S = 6.5 \times 10^{-6}$ in./in./°F and $\alpha_A = 12.5 \times 10^{-6}$ in./in./°F.

9.3 Statically Indeterminate Members Under Torsional Loads

The procedure used in the solution of problems dealing with statically indeterminate members under torsional loads is essentially the same as that used for statically indeterminate axially loaded members. As was done in Section 9.2, the conditions of equilibrium have to be supplemented by the principle of consistent deformations. In the case of torsionally loaded members, the principal of consistent deformations relates to the distortions (angles of twist) of these members. Only torsional members with circular cross sections will be considered.

Consider, for example, the case of the two material shaft shown in Figure 9.4, which is fixed at its left end and subjected to the torque T at its right end. The hollow shaft is made of material X and the concentric solid shaft of material Y. Both shafts are fastened to a rigid plate at the right end and the entire assembly may be assumed to act as one single unit. The maximum shear stresses developed in both materials are to be determined.

The applied torque T is resisted partially by each of the two shafts. Thus consider the free-body diagram of that portion of the composite shaft to the right of section a–a, as shown in the three-dimensional view of Figure 9.4(c). The resisting torque in

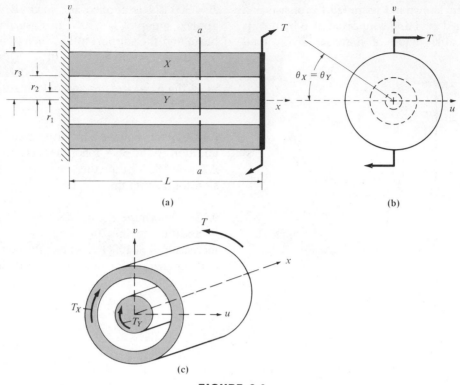

FIGURE 9.4

hollow shaft X is T_X and that in solid shaft Y is T_Y. The only available and applicable equation of equilibrium relates to the sum of the torques about the x axis. Thus

 $\sum T_x = 0$

$$T_X + T_Y - T = 0 \qquad \textbf{(a)}$$

Equation (a) contains the two unknowns T_X and T_Y and, obviously, their determination requires the development of a second relation containing these same two unknowns. This second relation is developed by observing that the total distortion (i.e., the total angle of twist θ) of the solid shaft is the same as that of the hollow shaft. In other words, over the entire length L of the assembly, the angle of twist θ_X of hollow shaft X is identical with the angle of twist θ_Y of solid shaft Y as shown schematically in Figure 9.4(b). Thus

$$\theta_X = \theta_Y \qquad \textbf{(b)}$$

Equation (b) may be expressed in terms of the torques T_X and T_Y by using Eq. 4.22. Therefore,

$$\left(\frac{TL}{JG}\right)_X = \left(\frac{TL}{JG}\right)_Y \qquad\qquad \textbf{(c)}$$

The use of Eq. 4.22 requires that Hooke's law be obeyed by both materials. Since the length L for shafts X and Y is the same, Eq. (c) may be written in the form

$$T_X = \frac{J_X G_X}{J_Y G_Y} T_Y \qquad\qquad \textbf{(d)}$$

A simultaneous solution of Eqs. (a) and (d) leads to the values of the two torques T_X and T_Y from which the maximum shear stresses in both shafts are determined by the use of Eq. 4.19.

Example 9.4

Refer to the composite shaft of Figure 9.4 and let $r_1 = 0.10$ m, $r_2 = 0.20$ m, $r_3 = 0.30$ m, and $L = 8$ m. The applied torque $T = 670$ kN-m and the materials are such that $G_X = 50$ GN/m^2 and $G_Y = 100$ GN/m^2. Determine (a) the maximum shear stresses in both shafts, and (b) the total angle of twist of the entire assembly.

Solution

(a) Substitution of the given data into Eqs. (a) and (d) leads to the following two simultaneous equations in T_X and T_Y:

$$T_X + T_Y - 670 = 0$$

$$T_X = 32.5 T_Y$$

A simultaneous solution of these two equations yields

$$T_X = 650 \text{ kN-m} \qquad \text{and} \qquad T_Y = 20 \text{ kN-m}$$

Therefore,

$$\tau_X = \frac{T\rho}{J} = \frac{650 \times 10^3 (0.30)}{(\pi/32)(0.60^4 - 0.40^4)}$$

$$= \boxed{19.1 \text{ MN/m}^2}$$

$$\tau_Y = \frac{T\rho}{J} = \frac{20 \times 10^3 (0.10)}{(\pi/32)(0.20^4)}$$

$$= \boxed{12.7 \text{ MN/m}^2}$$

(b)

$$\theta_X = \theta_Y = \left(\frac{TL}{JG}\right)_X$$

$$= \frac{650 \times 10^3 \times 8}{(\pi/32)(0.60^4 - 0.40^4)(50 \times 10^9)}$$

$$= \boxed{0.0102 \text{ rad}}$$

Example 9.5

The system shown in Figure 9.5(a) consists of a steel shaft 3 in. in diameter fixed at both ends D and F and fastened rigidly to the rigid vertical member BC, which in turn is attached to the two aluminum rods at B and C as shown. The length l of the aluminum rods is 20 in. and their cross-sectional area A is 1 in.^{2}. Other needed dimensions are shown in Figure 9.5(a). Assume the coefficient of thermal expansion for aluminum, $\alpha = 12.5 \times 10^6$ in./in./°F, the modulus of elasticity for aluminum $E = 10 \times 10^6$ psi, the modulus of rigidity for steel $G = 12 \times 10^6$ psi, and that the two materials behave in a linearly elastic manner. If the temperature of the two aluminum rods is dropped by the amount $\Delta T = 400$°F, determine the magnitude of torque developed in both segments of the steel shaft.

Solution. When the temperature of the aluminum rods drops through ΔT, each of the two rods will have a tendency to shorten by an amount $\delta_T = \alpha(\Delta T)l$. However, because of the restraining force F induced in each aluminum rod through the rigid member BC, each of the two aluminum rods will shorten only through the amount $\delta = \delta_T - \delta_F = \alpha(\Delta T)l - Fl/AE$. A free-body diagram of the rigid member BC is shown in Figure 9.5(b). As a result of the deformation δ of the aluminum rods, the rigid member BC rotates through the angle θ, which is also the angle of twist of the two portions of the steel shaft at section E. The geometric relation between θ and δ is shown in Figure 9.5(c). For small deformations,

$$\theta = \frac{\delta}{c} = \frac{1}{c}\left[\alpha(\Delta T)l - \frac{Fl}{AE}\right]$$

$$= \frac{1}{20}\left[12.5 \times 10^{-6}(400)(20) - \frac{20F}{1 \times 10 \times 10^6}\right]$$

$$= (5 \times 10^{-3}) - (10^{-7})F \tag{a}$$

As mentioned earlier, the angles of twist of the two portions of the steel shaft at section E are identical, to satisfy the law of consistent deformations. Thus

$$\theta_{DE} = \theta_{EF}$$

$$a\left(\frac{T}{JG}\right)_{DE} = b\left(\frac{T}{JG}\right)_{EF}$$

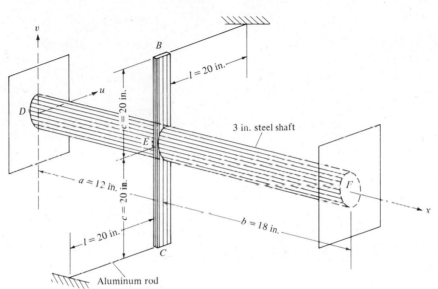

(a)

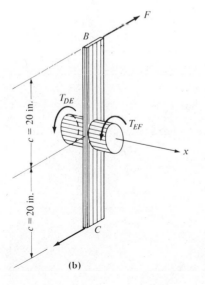

(b)

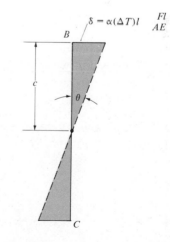

(c)

FIGURE 9.5

Since J and G for segments DE and EF of the steel shaft are identical, this relation reduces to

$$aT_{DE} = bT_{EF}$$

$$T_{DE} = \left(\frac{b}{a}\right)T_{EF}$$

$$= 1.5T_{EF} \qquad \textbf{(b)}$$

Also, since $\theta_{DE} = \theta_{EF} = \theta$, the following relations may be established:

$$a\left(\frac{T}{JG}\right)_{DE} = \theta$$

$$1.26 \times 10^{-7}T_{DE} = 5(10^{-3}) - (10^{-7})F \qquad \textbf{(c)}$$

and

$$b\left(\frac{T}{JG}\right)_{EF} = \theta$$

$$1.89 \times 10^{-7}T_{EF} = 5(10^{-3}) - (10^{-7})F \qquad \textbf{(d)}$$

Referring now to the free-body diagram shown in Figure 9.5(b) and summing torques about the axis of the steel shaft (i.e., about the x axis), we obtain

$$\sum T_x = 0$$

$$T_{DE} + T_{EF} - 2Fc = 0$$

$$T_{DE} + T_{EF} - 40F = 0 \qquad \textbf{(e)}$$

Substituting Eq. (b) into Eq. (e) yields

$$T_{EF} = 16F \qquad \textbf{(f)}$$

A simultaneous solution of Eqs. (d) and (f) leads to the following value for the torque in portion EF:

$$\boxed{T_{EF} = 25,600 \text{ lb-in.}}$$

Finally, Eq. (b) yields

$$\boxed{T_{DE} = 38,400 \text{ lb-in.}}$$

Homework Problems

In all the following problems, assume that the materials behave in a linearly elastic manner.

9.16 The solid shaft shown in Figure H9.16 is fixed at both ends A and C and is subjected to the forces $F = 25,000$ lb creating a torque at location B. The material of the shaft is steel ($G = 12 \times 10^6$ psi) and its diameter is 4 in. Determine the absolute maximum shearing stress and the maximum angle of twist induced in the shaft.

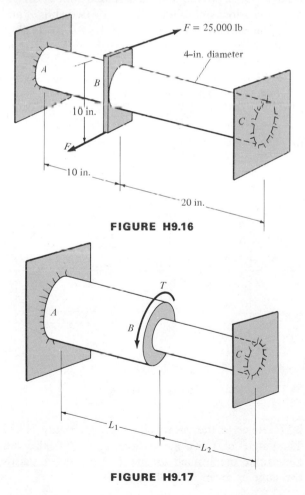

FIGURE H9.16

FIGURE H9.17

9.17 The two-diameter shaft shown in Figure H9.17 consists of a 0.10-m-diameter segment AB and a 0.06-m-diameter segment BC rigidly fixed at its two ends A and C. The material is aluminum for which $G = 30$ GN/m^2. Let $L_1 = 2.4$ m and $L_2 = 0.8$ m. Assume that the torque $T = 8$ kN-m and determine:
(a) The absolute maximum shearing stress developed in the system.
(b) The maximum angle of twist.

9.18 The composite shaft shown in Figure H9.17 consists of a 0.10-m solid aluminum shaft ($G = 30 \times 10^9$ N/m^2) securely fastened to a 0.06-m solid steel shaft ($G = 80 \times 10^9$ N/m^2), so that the entire assembly acts as one single unit. A torque $T = 10$ kN-m is applied as shown. Assume that $L_1 = 1$ m and $L_2 = 2$ m. Determine:
(a) The torques developed in the steel and aluminum shafts.
(b) The rotation of the section at B with respect to the sections at A and at C.
(c) The absolute maximum shearing stress in the aluminum shaft.

9.19 A sectional view of a steel shaft is shown in Figure H9.19. The shaft is fixed at its two ends A and B and subjected to a torque T at section C. Determine the absolute maximum shearing stress to which the shaft is subjected. The following data apply: $T = 150,000$ lb-in., $L = 1.5$ ft, $D_1 = 2$ in., $D_2 = 3$ in., and $G = 12 \times 10^6$ psi.

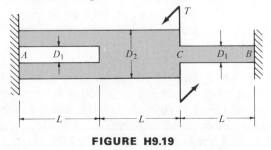

FIGURE H9.19

9.20 Refer to the shaft in Figure H9.19. Segment AC is titanium ($G = 6 \times 10^6$ psi) and segment CB is aluminum ($G = 4 \times 10^6$ psi). Determine the absolute maximum shearing stress in the shaft and the angle of twist at section C. The following data apply: $T = 100,000$ lb-in., $D_1 = 3$ in., $D_2 = 4$ in., and $L = 2$ ft.

9.21 The shaft shown in Figure H9.21 is fixed at its two ends A and B. A concentrated torque T is applied at section C. The following data are provided: $a = 1$ m, $b = 1.5$ m, $D = 0.06$ m, $G = 30$ GN/m^2, and the angle of twist at section C, $\theta_C = 0.02$ rad. Determine the magnitude of the applied torque T.

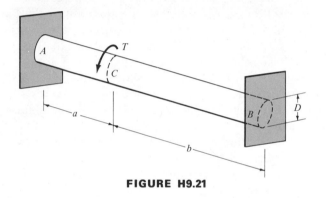

FIGURE H9.21

9.22 Refer to the shaft shown in Figure H9.21. The following data apply: $a = 2$ ft, $b = 4$ ft, $D = 4$ in, $G = 6 \times 10^6$ psi, and the absolute maximum shearing stress in segment AC, $|\tau_{max}|_{AC} = 10,000$ psi. Determine the magnitude of the applied torque T.

9.23 Refer to the shaft shown in Figure H9.21. The following data apply: $T = 700$ N-m, $a = 1.5$ m, $b = 2.0$ m, $G = 50$ GN/m^2, and the angle of twist at section C, $\theta_C = 0.015$ rad. Determine the diameter of the shaft.

9.24 Refer to the shaft shown in Figure H9.21. The following data apply: $T = 20,000$ lb-ft, $a = 1.5$ ft, $b = 3.0$ ft, $G = 8 \times 10^6$ psi, and the absolute maximum shearing stress in segment CB, $|\tau_{max}|_{CB} = 15,000$ psi. Determine the diameter of the shaft.

9.25 A composite shaft of length L consists of an aluminum ($G = 4 \times 10^6$ psi) core surrounded by a magnesium ($G = 2.5 \times 10^6$ psi) shell with an outside diameter of 6 in. If the outside diameter of the aluminum core, d, is exactly equal to the inside diameter of the magnesium shell, determine the value of d so that the torque in the aluminum core is twice that in the magnesium shell. Find the absolute maximum shear stresses in each part of the composite shaft if the applied torque is 9000 lb-ft.

9.26 A sectional view of a composite shaft made up of hollow shafts X and Z and solid shaft Y is shown in Figure H9.26. Member C is assumed thin and rigid and the entire assembly is fixed at its two ends A and B. The following data apply: $L_1 = 0.20$ m, $L_2 = 0.30$ m, $d_1 = 0.04$ m, $d_2 = 0.05$ m, $d_3 = 0.06$ m, $d_4 = 0.06$ m, $d_5 = 0.07$ m, $G_X = 40 \times 10^9$ N/m^2, $G_Y = 10 \times 10^9$ N/m^2, $G_Z = 60 \times 10^9$ N/m^2, and $T = 400$ N-m. Determine:

(a) The absolute maximum shearing stresses in the three shafts.

(b) The angle of twist at C.

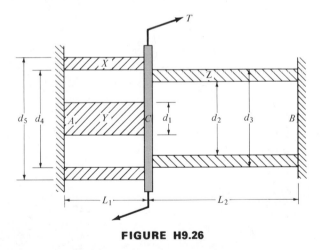

FIGURE H9.26

9.27 Refer to the composite shaft in Figure H9.26. The angle of twist at section C, $\theta_C = 0.0025$ rad. Determine the applied torque T if all other data are the same as given in Problem 9.26.

9.28 The 5-in.-diameter aluminum shaft shown in Figure H9.28 is fixed at its two ends A and B and is subjected to the two torques T_1 and T_2 as shown. Assume that $G = 4 \times 10^6$ psi, $L_1 = 2$ ft, $L_2 = 4$ ft, $L_3 = 3$ ft, $T_1 = 10,000$ lb-ft, and $T_2 = 15,000$ lb-ft. Determine the absolute maximum shearing stress in the shaft.

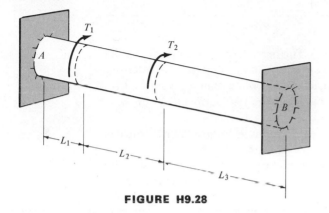

FIGURE H9.28

9.29 The 4-in.-diameter brass ($G = 5.5 \times 10^6$ psi) shaft shown in Figure H9.29 is fixed at end A and fastened securely to a rigid horizontal member at end B. The two ends of the rigid member are attached to identical magnesium ($\alpha = 14.5 \times 10^{-6}$ in./in./°F) rods whose cross-sectional area is 0.5 in.^{2}. Other needed information is shown in Figure H9.29. If the temperature of the magnesium rods is dropped by the amount $\Delta T = 250$°F, determine the absolute maximum shearing stress in the brass shaft and the rota-

tion of the rigid horizontal member. Assume that $E_M = 6.5 \times 10^6$ psi.

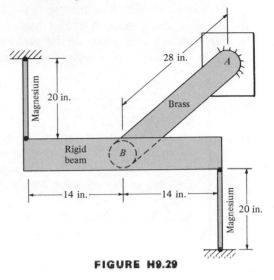

FIGURE H9.29

9.30 Refer to the assembly shown in Figure 9.5(a). Segment DE is a 5-in-diameter aluminum ($G = 4 \times 10^6$ psi) shaft and segment EF is a 3-in-diameter steel ($G = 12 \times 10^6$ psi) shaft. The two identical rods attached to the ends of rigid member BC are aluminum ($E = 10 \times 10^6$ psi, $\alpha = 12.5 \times 10^{-6}$ in./in./°F) and their cross-sectional area A is 0.75 in.^{2}. The following data are provided: $a = 15$ in., $b = 20$ in., $c = 30$ in., and $l = 30$ in. Determine the temperature drop that would produce a rotation of 0.03 rad in the rigid member BC.

9.4 Statically Indeterminate Members Under Flexural Loads—Two Successive Integrations

Statically indeterminate members under flexural loads are analyzed using the same basic procedure discussed in Sections 9.2 and 9.3. The applicable equations of equilibrium need to be supplemented by additional relations satisfying the principle of consistent deformations. When using the method of two successive integrations, the moment equation for the flexural member (i.e., the differential equation of the elastic curve for the beam) is written in terms of unknown forces and/or moments. Thus, in addition to the two constants of integration, the slope and deflection equations would contain one or more unknown quantities that are evaluated from the boundary conditions of the problem. The following examples illustrate the use of the method of two successive integrations in the solution of statically indeterminate beams.

Example 9.6

Consider the beam shown in Figure 9.6(a). The beam is simply supported at its left end A and fixed at its right end B. It carries a load that varies in intensity from zero at support A to w at support B according to the relation $w_x = (x/L)w$. Determine the unknown forces and moments at the two supports.

Solution. The free-body diagram of the beam shown in Figure 9.6(b) indicates that there are four unknown reactive components, A_v, B_x, B_v, and M_B. Since there are only three applicable equilibrium equations, the beam is statically indeterminate. These three equations of equilibrium are applied as follows:

$\rightarrow \sum F_x = 0$ yields

$$\boxed{B_x = 0} \tag{a}$$

$\uparrow \sum F_v = 0$ yields

$$A_v + B_v - \tfrac{1}{2}wL = 0 \tag{b}$$

$\circlearrowright \sum M_B = 0$ yields

$$\tfrac{1}{6}wL^2 - A_v L - M_B = 0 \tag{c}$$

Equations (b) and (c) represent two simultaneous algebraic equations with the three unknown A_v, B_v, and M_B. One more equation is needed to complete the solution. This equation is obtained from knowledge of the deformation characteristics and support conditions of the beam (i.e., by applying the principle of consistent deformations).

Using the method of two successive integrations and the free-body diagram shown in Figure 9.6(c), the differential equation for the elastic curve of the beam is written as follows:

$$EI_u v'' = M_u = A_v x - \frac{1}{6}\frac{w}{L}x^3$$

After two successive integrations, the following two equations are obtained:

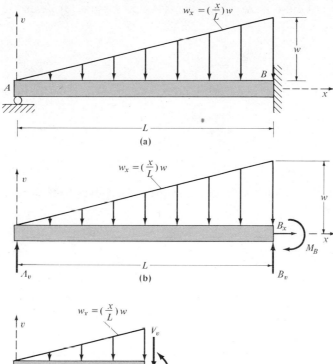

FIGURE 9.6

$$EI_u v' = \frac{A_v}{2} x^2 - \frac{1}{24} \frac{w}{L} x^4 + C_1 \qquad \textbf{(d)}$$

$$EI_u v = \frac{A_v}{6} x^3 - \frac{1}{120} \frac{w}{L} x^5 + C_1 x + C_2 \qquad \textbf{(e)}$$

where C_1 and C_2 are constants of integration. These two constants, along with A_v, are unknown quantities that can be determined from the existing boundary conditions. These boundary conditions are

1. $x = 0$, $v = 0$.
2. $x = L$, $v = 0$.
3. $x = L$, $v' = 0$.

Substitution of these three conditions into Eqs. (d) and (e) yields the following values:

$$C_1 = -\frac{wL^3}{120}, \quad C_2 = 0$$

and

$$\boxed{A_v = \frac{wL}{10}}$$

Substituting this value for A_v into Eqs. (b) and (c) leads to the values of B_v and M_B as follows:

$$\boxed{B_v = \frac{4wL}{10}} \quad \text{and} \quad \boxed{M_B = \frac{wL^2}{15}}$$

Example 9.7

The beam shown in Figure 9.7(a) is fixed at both ends and carries a load that varies in intensity from zero at support A to w at support B according to the relations $w_x = (x/L)^2 w$. Determine the reactive forces and moments at the two supports.

Solution. The free-body diagram of the beam is shown in Figure 9.7(b), and it shows that there are a total of four unknown reactive components, A_v, B_v, M_A, and M_B. Reactive components in the x direction are zero, since there are no applied forces in this direction. Only two equilibrium equations are applicable as follows:

$$\circlearrowleft \quad \sum M_B = 0$$

$$\tfrac{1}{12}wL^2 - A_v L + M_A - M_B = 0 \qquad \textbf{(a)}$$

$$\uparrow \quad \sum F_v = 0$$

$$A_v + B_v - \tfrac{1}{3}wL = 0 \qquad \textbf{(b)}$$

Equations (a) and (b) are two simultaneous algebraic equations containing the four unknowns A_v, B_v, M_A, and M_B. Therefore, two additional relations are needed among the unknowns in order to complete the solution. These two additional relations are obtained, as in Example 9.6, by considering the boundary conditions and the principle of consistent deformations. Thus we can write the differential equation for the elastic curve of the beam as follows:

$$EI_u v'' = M_u = A_v x - M_A - \frac{1}{12} \frac{w}{L^2} x^4$$

Performing two successive integrations, we obtain the following two equations:

$$EI_u v' = \frac{A_v}{2} x^2 - M_A x - \frac{1}{60} \frac{w}{L^2} x^5 + C_1 \qquad \textbf{(c)}$$

$$EI_u v = \frac{A_v}{6} x^3 - \frac{M_A}{2} x^2 - \frac{1}{360} \frac{w}{L^2} x^6 + C_1 x + C_2 \qquad \textbf{(d)}$$

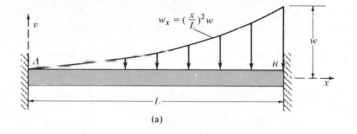

(a)

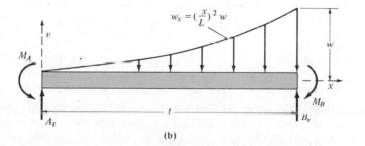

(b)

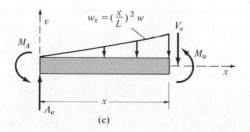

(c)

FIGURE 9.7

in which C_1 and C_2 are constants of integration. Equations (c) and (d) contain four unknowns, and therefore four boundary conditions are needed for their determination. These boundary conditions are

1. $x = 0, v = 0.$
2. $x = 0, v' = 0.$
3. $x = L, v = 0.$
4. $x = L, v' = 0.$

These boundary conditions, when substituted into Eqs. (c) and (d), lead to four simultaneous equations in $C_1, C_2, A_v,$ and M_A. Their solution yield the following values:

$$C_1 = C_2 = 0 \qquad \boxed{M_A = \frac{wL^2}{60}} \qquad \boxed{A_v = \frac{wL}{15}}$$

Using the values just determined for M_A and A_v in Eqs. (a) and (b) leads to the values of M_B and B_v as follows:

$$\boxed{M_B = \frac{wL^2}{30}} \quad \text{and} \quad \boxed{B_v = \frac{4wL}{15}}$$

Example 9.8

The steel beam shown in Figure 9.8(a) is simply supported at end B by the aluminum rod BC whose length is l and whose cross-sectional area is A. The beam is fixed at end D and carries a uniform load of intensity w. Assume the coefficient of thermal expansion for aluminum to be α, the modulus of elasticity for aluminum to be E_A and that for steel to be E_S, and the moment of inertia for the beam to be I_u. Before the load is applied, rod BC is unstressed. If, in addition to the load on the steel beam, the temperature of the aluminum rod is dropped by the amount ΔT, determine the force induced in rod BC and the reaction components at support D.

Solution. The free-body diagram of the steel beam BD is shown in Figure 9.8(b), which also shows the free-body diagram of the aluminum rod BC. Since there are no applied forces in the x direction, reactive forces in this direction do not exist and the free-body diagram of beam BD has three unknown reaction components F, D_v, and M_D. Note that the unknown F is also the force acting on rod BC.

Only two equations of equilibrium are applicable as follows:

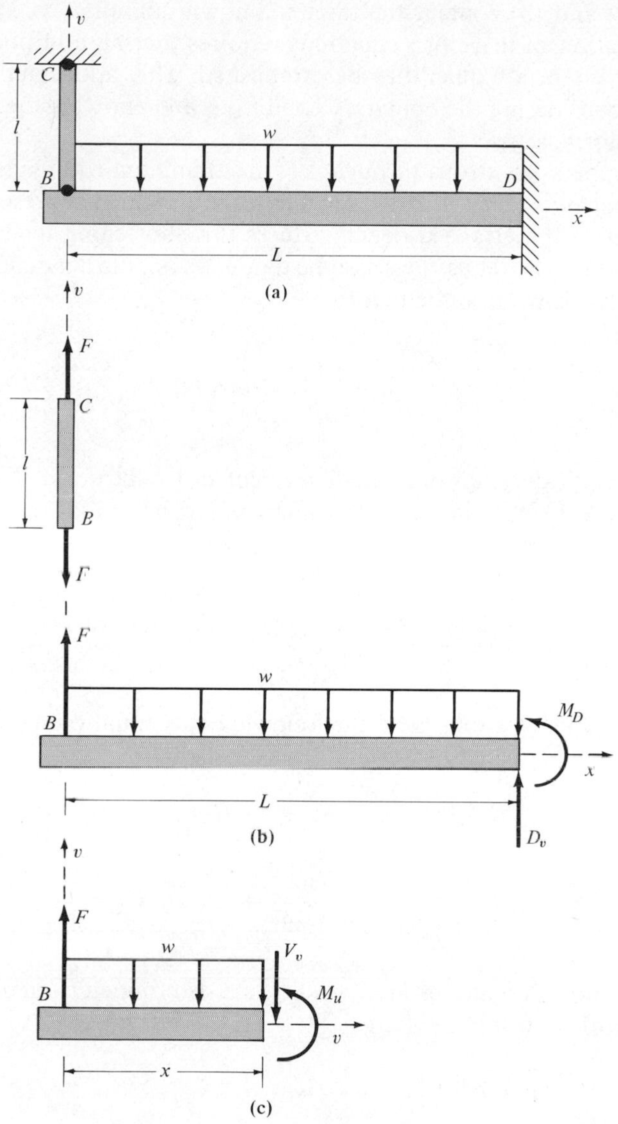

FIGURE 9.8

↺ $\sum M_D = 0$

$$\frac{wL^2}{2} - FL - M_D = 0 \qquad \textbf{(a)}$$

↑ $\sum F_v = 0$

$$F + D_v - wL = 0 \qquad \textbf{(b)}$$

Equations (a) and (b) contain the three unknown quantities F, M_D, and D_v. Thus the solution of these two equations requires that an additional relation among these unknown quantities be established. This additional relation is obtained by considering the boundary conditions and applying the principle of consistent deformations.

As the temperature drops through ΔT, the aluminum rod tends to shorten by the amount $\delta_T = \alpha(\Delta T)l$. However, the force F, which represents the restraining effect of the attached beam, reduces this shortening tendency by the amount $\delta_F = Fl/AE_A$. Thus, if we assume that $\delta_T < \delta_F$, the net deflection of the beam at B is downward and equal to

$$\delta_F - \delta_T = \frac{Fl}{AE_A} - \alpha(\Delta T)l$$

The free-body diagram of a small segment of the beam at the left end is shown in Figure 9.8(c), from which the differential equation for the elastic curve is written as follows:

$$E_S I_u v'' = M_u = Fx - \frac{wx^2}{2}$$

Two successive integrations yield the following two equations:

$$E_S I_u v' = \frac{Fx^2}{2} - \frac{wx^3}{6} + C_1 \qquad \textbf{(c)}$$

$$E_S I_u v = \frac{Fx^3}{6} - \frac{wx^4}{24} + C_1 x + C_2 \qquad \textbf{(d)}$$

The three unknown quantities F, C_1, and C_2 are determined from the boundary conditions, which, in this case, are

1. $x = 0$, $v = \alpha(\Delta T)l - Fl/AE_A$.
2. $x = L$, $v = 0$.
3. $x = L$, $v' = 0$.

Substitution of these boundary conditions into Eqs. (c) and (d) leads to the following value for F:

$$\boxed{F = \left(\frac{3AE_A}{8}\right)\left(\frac{wL^4 + 8E_S I_u \alpha(\Delta T)l}{AE_A L^3 + 3I_u E_S l}\right)} \qquad \textbf{(e)}$$

If this value for F is now used in Eqs. (a) and (b), we obtain the unknown quantities D_v and M_D as follows:

$$D_v = wL - \left(\frac{3AE_A}{8}\right)\left(\frac{wL^4 + 8E_S I_u \alpha(\Delta T)l}{AE_A L^3 + 3I_u E_S l}\right) \tag{f}$$

$$M_D = \frac{wL^2}{2} - \left(\frac{3LAE_A}{8}\right)\left(\frac{wL^4 + 8E_S I_u \alpha(\Delta T)l}{AE_A L^3 + 3I_u E_S l}\right) \tag{g}$$

Note that the quantity ΔT represents a temperature drop and should be treated as a positive quantity in all of the above relations.

Homework Problems

Assume linear elastic behavior in all the following problems.

9.31
(a) Derive Eqs. (e), (f), and (g) in Example 9.8.
(b) The following data are provided in Example 9.8:
$l = 0.5$ m, $L = 4$ m, $w = 100$ kN/m, $\alpha = 10 \times 10^{-6}$ m/m/°C, $A = 4 \times 10^{-4}$ m², $E_A = 70 \times 10^9$ N/m², $E_S = 200 \times 10^9$ N/m², temperature drop, $\Delta T = 200$°C, and $I_u = 30 \times 10^{-6}$ m⁴. Determine the axial stress and axial deformation in rod BC.

9.32 The beam shown in Figure H9.32 is simply supported at end A and fixed at end B. It carries a uniform load of intensity w. Find the reaction com-

ponents at A and B. Express answers in terms of w and L.

9.33 The beam shown in Figure H9.33 is fixed at both ends and carries a load that varies linearly from zero at end A to w at end B according to the relation $w_x = (x/L)w$. Determine the reaction components at both ends. Express answers in terms of w and L.

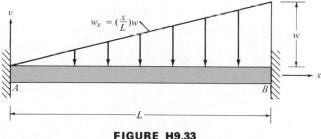

FIGURE H9.33

9.34 Refer to the beam shown in Figure H9.32. Before the uniform load is applied, it was discovered that there was a gap equal to δ between the beam and the support at A. Determine the reaction components at the two supports and compare them to those found in Problem 9.32.

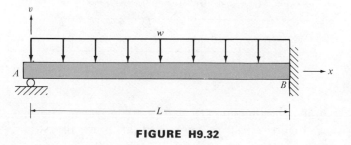

FIGURE H9.32

9.35 The aluminum beam shown in Figure H9.35 is simply supported by a magnesium compression block at B and fixed at C. It carries a load that varies from zero at B to w at C according to the relation $w_x = (x/L)w$. The following data are provided: $L = 12$ ft, $E_A = 10 \times 10^6$ psi, $I_u = 70$ in.^{4}, $w = 2000$ lb/ft, $l = 1$ ft, $E_M = 6 \times 10^6$ psi, and the cross-sectional area for the magnesium compression block $A = 20$ in.^{2}. Determine the axial stress in the magnesium member and the reaction components at the fixed support.

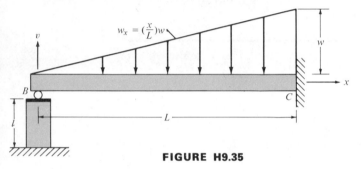

FIGURE H9.35

9.36 The beam shown in Figure H9.36 is simply supported at B and fixed at C. It carries a uniform load over the span between B and C. A moment M_A is applied at the free end of the overhang AB as shown. Determine the value of M_A if:
(a) The slope at B is zero.
(b) The slope at B is $-0.01wL^3/EI_u$.
(c) The slope at B is $+0.02wL^3/EI_u$.
Express your answers in terms of w and L.

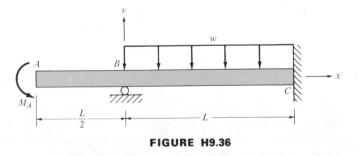

FIGURE H9.36

9.37 The beam shown in Figure H9.37 is simply supported at B and fixed at C. It carries a load that varies in intensity from zero at B to w at C according to the relation $w_x = (x/L)^2w$. A downward load P is applied at the free end of the overhang AB. Determine in terms of w, P, and L the reaction components at B and C.

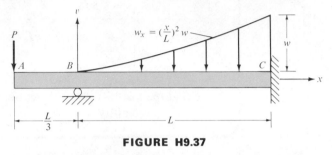

FIGURE H9.37

9.38 Refer to Figure H9.35 and assume that when the load is applied, the temperature of the magnesium block increases by $\Delta T = 300°F$. If the coefficient of thermal expansion for magnesium is $\alpha = 15 \times 10^{-6}$ in./in./°F, determine the axial stress in the magnesium block and the components of the reaction at support C.

9.39 The beam shown in Figure H9.39 is fixed at A and simply supported at B. It carries a uniform load of intensity w over the overhang BC. Determine:
(a) The reaction components at A and B in terms of w and L.
(b) The location of the maximum deflection between A and B.
(c) The slope at B.
(d) The deflection halfway between A and B.

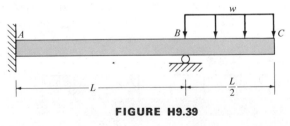

FIGURE H9.39

9.40 The aluminum ($E = 10 \times 10^6$ psi) beam shown in Figure H9.40 is supported at B by the steel

($E = 30 \times 10^6$ psi) rod BD of length $l = 12$ ft and cross-sectional area $A = 0.5$ in.^{2} and is fixed at C. It carries a uniform load of intensity $w = 800$ lb/ft over the entire length. If $L = 10$ ft and the moment of inertia for the aluminum beam is $I_u = 1000$ in.^{4}, determine:

(a) The axial stress in the steel rod.
(b) The deformation of the steel rod.
(c) The reaction components at C.

9.41 Repeat Problem 9.40, if in addition to the load w, the temperature of the steel rod drops by 400°F. Assume the coefficient of thermal expansion for steel to be $\alpha = 6.5 \times 10^{-6}$ in./in./°F.

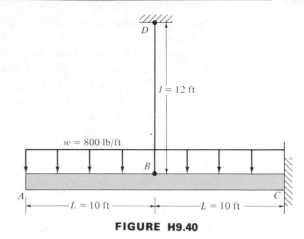

FIGURE H9.40

9.5 Statically Indeterminate Members Under Flexural Loads—Superposition

The concept of superposition has already been introduced and used in previous chapters. In finding the deflections and rotations produced by several loads acting on flexurally loaded members, one determines the deflections and rotations due to each load acting separately and determines the resultant deflections and rotations by superposing (i.e., combining algebraically) these separate effects. The basic concept underlying the method of superposition requires the decomposition of the given statically indeterminate beam into two or more statically determinate ones for which solutions may be easily determined, or which have already been obtained and are available in tabulations similar to the one shown in Table 6.1. In a given statically indeterminate situation, the method of superposition would be used to fulfill the principle of consistent deformations. This procedure would lead to one or more relations among the unknown quantities that would be used to supplement the applicable equations of equilibrium. The use of the method of superposition is illustrated by the following examples.

Example 9.9
Repeat Example 9.6 using the method of superposition.

Solution. The beam of Example 9.6 is repeated, for convenience, in Figure 9.9(a). The effect of the simple support at A is to provide a vertical reactive force A_v that keeps the beam from deflecting downward at that location. If this

redundant support were removed momentarily, the remaining structure would
be a statically determinate cantilever fixed at its right end B and subjected to
the load $w_x = (x/L)w$ as shown in Figure 9.9(b). This cantilever beam would
deflect downward at its free end through the amount v_1 as shown in Figure
9.9(b). Obviously, the deflection v_1 does not occur because of the redundant
support at A, which develops the reactive force A_v. Thus this reactive force,
acting at the end of the cantilever beam, would have to produce an upward
deflection v_2 as shown in Figure 9.9(c). Of necessity, this upward deflection v_2
would have to be equal in magnitude to the downward deflection v_1 in order
for the resultant deflection at point A to be zero. In other words, the algebraic
sum of v_1 and v_2 must be zero. Therefore,

$$v_1 + v_2 = 0 \qquad\qquad \textbf{(a)}$$

The whole process of analyzing this problem may be summed up by stating
that the statically indeterminate beam shown in Figure 9.9(a) is equivalent to
the superposition (algebraic sum) of the two statically determinate beams
shown in Figure 9.9(b) and (c).

From the information contained in Table 6.1, one may determine the values
of v_1 and v_2 as follows:

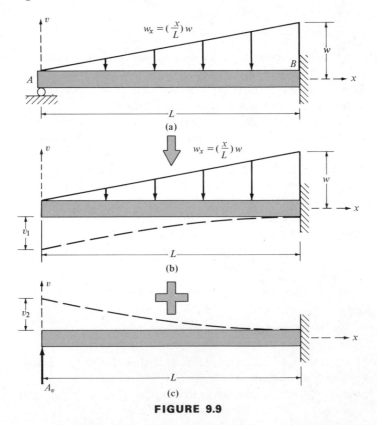

FIGURE 9.9

$$v_1 = -\frac{wL^4}{30EI_u} \tag{b}$$

and

$$v_2 = \frac{A_v L^3}{3EI_u} \tag{c}$$

Substitution of Eqs. (b) and (c) into Eq. (a) yields the value for A_v. Thus

$$\boxed{A_v = \frac{wL}{10}}$$

which is the same as that obtained by using the method of two successive integrations in Example 9.6. The determination of the reaction components B_v and M_B is identical to what was done in Example 9.6.

Example 9.10
Solve Example 9.8 by superposition.

Solution. For convenience, the beam of Example 9.8 is repeated in Figure 9.10(a). As in the case of Example 9.9, this beam may be decomposed into two statically determinate cantilever beams as shown in Figure 9.10(b) and (c). The condition that needs to be satisfied is that the algebraic sum of the deflection, v_1 and v_2, be equal to the deflection v at the end of the beam shown in Figure 9.10(a). Therefore,

$$v_1 + v_2 = v \tag{a}$$

where, from Table 6.1,

$$v_1 = -\frac{wL^4}{8E_S I_u} \tag{b}$$

$$v_2 = \frac{FL^3}{3E_S I_u} \tag{c}$$

The deflection v is also the deflection of rod BC and, as explained in the solution of Example 9.8, it is given by

$$v = -\left[\frac{Fl}{AE_A} - \alpha(\Delta T)l\right] \tag{d}$$

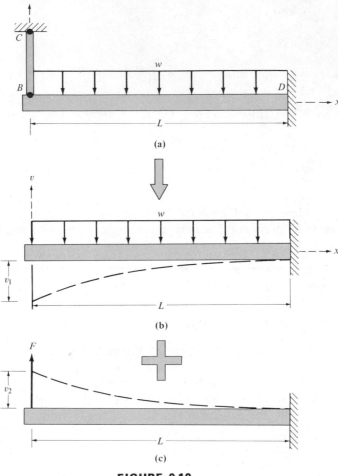

FIGURE 9.10

where the negative sign is due to the fact that the deflection is downward. Substitution of Eqs. (b), (c), and (d) into Eq. (a) leads to the value of the force F as follows:

$$F = \left(\frac{3AE_A}{8}\right)\left(\frac{wL^4 + 8E_S I_u \alpha(\Delta T)l}{AE_A L^3 + 3I_u E_S l}\right)$$

This is obviously identical to the value found by the method of two successive integrations used in Example 9.8. The determination of the reaction components D_v and M_D at the fixed support D is accomplished in exactly the same manner as was done in the solution of Example 9.8.

Example 9.11

Beam AC is simply supported at its two ends and at midspan, point B, as shown in Figure 9.11(a). Before the load P is applied, there is a gap δ between the beam and the midspan support at B. After the load P is applied, however, contact is made between support B and the beam so that support B carries part of the load. Determine the amount of the load carried by each of the three supports A, B, and C.

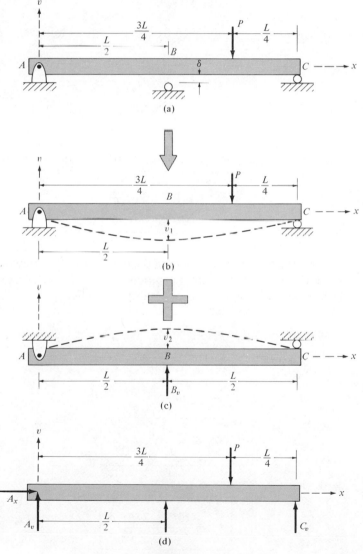

FIGURE 9.11

Solution. As explained in the previous two examples, the given statically indeterminate beam may be considered as the superposition of the two statically determinate beams shown in Figure 9.11(b) and (c). The load P produces a downward deflection v_1 at the midspan of the beam shown in Figure 9.11(b). The unknown reaction B_v sustained by support B produces an upward deflection v_2 at the midspan of the beam shown in Figure 9.11(c). The magnitude of the deflection v_2 is less than that of v_1 by the amount δ. In other words, the algebraic sum of v_1 and v_2 must be equal to the negative of δ. Thus

$$v_1 + v_2 = -\delta \tag{a}$$

The deflections v_1 and v_2 are obtained from Table 6.1 as follows:

$$v_1 = -\frac{11PL^3}{768EI_u} \tag{b}$$

$$v_2 = +\frac{B_v L^3}{48EI_u} \tag{c}$$

Substitution of Eqs. (b) and (c) into Eq. (a) yields the value of B_v as follows:

$$\boxed{B_v = \frac{11P}{16} - \frac{48EI_u\delta}{L^3}}$$

Considering now the free-body diagram of the entire beam as shown in Figure 9.11(d), we can determine the reaction components A_x, A_v, and C_v by application of the three available equations of equilibrium. Thus

$\rightarrow \sum F_x = 0$ yields

$$A_x = 0$$

$\circlearrowleft \sum M_c = 0$, yields

$$\boxed{A_v = \frac{24EI_u\delta}{L^3} - \frac{3P}{32}}$$

$\uparrow \ \sum F_v = 0$, yields

$$\boxed{C_v = \frac{13P}{32} + \frac{24EI_u\delta}{L^3}}$$

Note that the gap δ influences the value of all the vertical reactions. Note also that the equations that have been obtained are valid for any value of δ, including the case when $\delta = 0$ (i.e., when there is no gap between the beam and the center support).

Homework Problems

Assume linear elastic behavior in all the following problems.

9.42 Solve Problem 9.32 by the method of superposition.

9.43 Solve Problem 9.33 by the method of superposition.

9.44 Solve Problem 9.34 by the method of superposition.

9.45 Solve Problem 9.35 by the method of superposition.

9.46 Solve Problem 9.36 by the method of superposition.

9.47 Solve part (a) of Problem 9.39 by the method of superposition.

9.48 Solve Problem 9.40 by the method of superposition.

9.49 Solve Problem 9.41 by the method of superposition.

9.50 The steel beam shown in Figure H9.50 is fixed at its right end B and supported by a steel rod at its left end C. The modulus of elasticity for steel is E, the moment of inertia for the beam is I_u, and the cross-sectional areas for the steel rod is A. Use the method of superposition to determine in terms of w, l, L, E, I_u,

and A, the force in the steel rod, and the reaction components at the fixed support B.

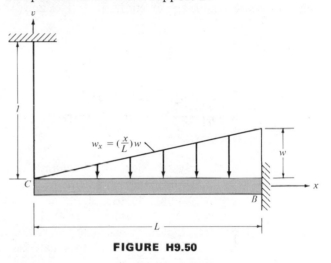

FIGURE H9.50

9.51 Refer to the system shown in Figure H9.51. Beam AB is fixed at end A and supported at end B at

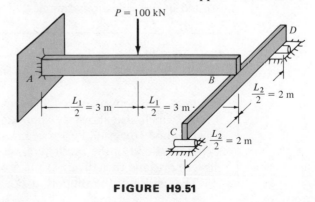

FIGURE H9.51

the midpoint of cross beam *CD*, which is simply supported. Both beams are steel for which $E = 200 \text{ GN/m}^2$ and have the same moment of inertia $I_u = 22.5 \times 10^{-6} \text{ m}^4$. Use the method of superposition to determine the force acting on beam *CD* and the reaction components at support *A*. Also find the maximum deflection of beam *CD*.

9.52 The system shown in Figure H9.52 consists of beam *AB* carrying a uniform load of intensity *w* along the entire span. It is simply supported at its two ends *A* and *B* and at midspan *E* by means of a second simply supported beam *CD*. Use the method of superposition to determine the reactions at *A*, *B*, *C*, and *D*.

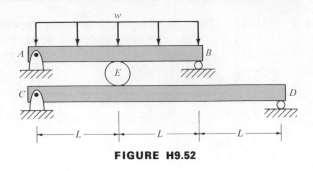

FIGURE H9.52

9.53 Repeat Example 9.7 using the method of superposition.

9.6 Statically Indeterminate Members Under Flexural Loads—Area Moment

In solving a statically indeterminate flexural problem, the applicable equations of equilibriums may be supplemented by satisfying the principle of consistent deformations through the use of the area-moment method. This procedure requires the development of geometric relations, in terms of tangential deviations and/or slope changes, that are consistent with the boundary conditions of the problem. Therefore, it is essential that an approximate but neat representation be made of the elastic curve of the beam so that the necessary geometric relations may be developed. The determination of tangential deviations and slope changes is facilitated if the M/EI_u diagrams are constructed by cantilever parts. The use of these various concepts is illustrated in the following examples.

Example 9.12
Solve Example 9.6 using the area-moment method.

Solution. For convenience, the beam of Example 9.6 is repeated in Figure 9.12(a), which also shows an approximation for the elastic curve of the beam. The free-body diagram for the beam is shown in Figure 9.12(b). Using point *B* as the fixed end of the cantilevers, we construct the moment diagram by cantilever parts and then divide by the product EI_u, as shown in Figure 9.12(c) and (d). The nature of the support at *B* is such that a tangent drawn to the elastic curve at this point passes through point A. Therefore, by the second theorem of the area-moment method, we conclude that the tangential deviation of point A with respect to point B is equal to zero. Thus

$$t_{AB} = 0 \qquad \textbf{(a)}$$

Determination of the tangential deviation t_{AB} from the information in Figure 9.12(c) and (d) yields

$$\frac{1}{2}\frac{A_v L}{EI_u}(L)\left(\frac{2}{3}L\right) - \frac{1}{4}\frac{wL^2}{6EI_u}(L)\left(\frac{4}{5}L\right) = 0 \qquad \textbf{(b)}$$

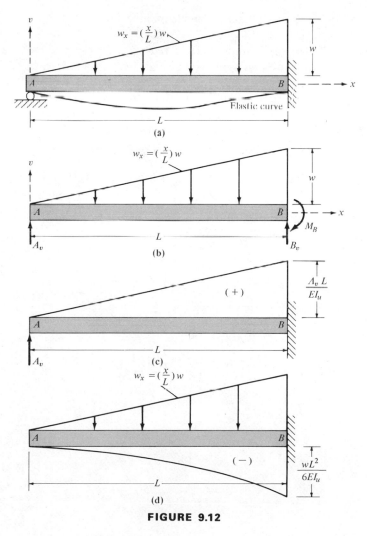

FIGURE 9.12

Solution of Eq. (b) for the value of A_v leads to

$$A_v = \frac{wL}{10} \qquad\qquad \textbf{(c)}$$

Determination of the reaction components B_v and M_B from the equilibrium conditions is accomplished in the same manner as was done in Example 9.6.

Example 9.13
Solve Example 9.11 by the area-moment method.

Solution. The beam of Example 9.11 is redrawn, for convenience, in Figure 9.13(a), which also shows an approximate elastic curve. The free-body diagram for the beam is shown in Figure 9.13(b). Using point A as the fixed end of the cantilevers, we construct the M/EI_u diagram by cantilever parts as shown in Figure 9.13(c), (d), and (e). A tangent to the elastic curve is drawn at C as shown in Figure 9.13(a), which also shows the tangential deviations t_{AC} and t_{BC}. The geometric relations shown in Figure 9.13(a) lead to the conclusion that

$$\tfrac{1}{2}|t_{AC}| = |t_{BC}| + |\delta| \qquad\qquad \textbf{(a)}$$

The tangential deviations t_{AC} and t_{BC} are determined from the M/EI_u diagrams in Figure 9.13(c), (d), and (c) as follows:

$$
\begin{aligned}
t_{AC} &= \frac{1}{2}\frac{C_v L}{EI_u}(L)\left(\frac{L}{3}\right) - \frac{1}{2}\frac{3PL}{4EI_u}\left(\frac{3L}{4}\right)\left(\frac{L}{4}\right) + \frac{1}{2}\frac{B_v L}{2EI_u}\left(\frac{L}{2}\right)\left(\frac{L}{6}\right) \\
&= \frac{C_v L^3}{6EI_u} + \frac{B_v L^3}{48EI_u} - \frac{9PL^3}{128EI_u} \qquad\qquad \textbf{(b)}
\end{aligned}
$$

$$
\begin{aligned}
t_{BC} &= \frac{1}{2}\frac{C_v L}{2EI_u}\left(\frac{L}{2}\right)\left(\frac{L}{6}\right) - \frac{1}{2}\frac{PL}{4EI_u}\left(\frac{L}{4}\right)\left(\frac{L}{12}\right) \\
&= \frac{C_v L^3}{48EI_u} - \frac{PL^3}{384EI_u} \qquad\qquad \textbf{(c)}
\end{aligned}
$$

Substitution of Eqs. (b) and (c) into Eq. (a) yields

$$16C_v + 8B_v - P = \frac{768EI_u\delta}{L^3} \qquad\qquad \textbf{(d)}$$

Equation (d) contains the two unknown quantities C_v and B_v. A second equation containing the same two unknowns can be obtained by summing moments about point A. Thus, referring to the free-body diagram of Figure 9.13(b), we can write

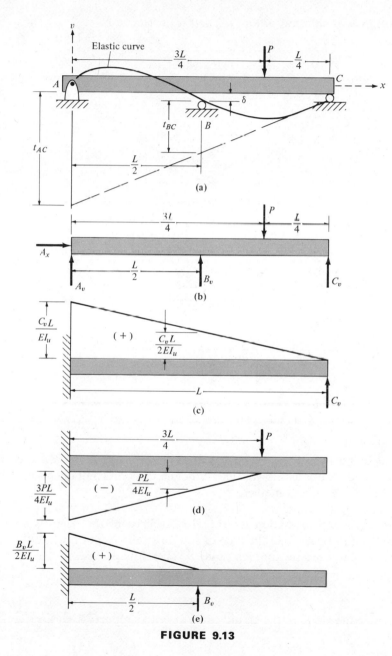

FIGURE 9.13

$\circlearrowleft \sum M_B = 0$ yields

$$C_v L + B_v \frac{L}{2} - P \frac{3L}{4} = 0 \qquad \text{(e)}$$

A simultaneous solution of Eqs. (d) and (e) yields

$$\boxed{B_v = \frac{11P}{16} - \frac{48EI_u \delta}{L^3}}$$

and

$$\boxed{C_v = \frac{13P}{32} + \frac{24EI_u \delta}{L^3}}$$

Application of two additional equilibrium equations lead to the values of A_x and A_v as follows:

$\rightarrow \sum F_x = 0$ yields

$$\boxed{A_x = 0}$$

$\uparrow \sum F_v = 0$ yields

$$\boxed{A_v = \frac{24EI_u \delta}{L^3} - \frac{3P}{32}}$$

Example 9.14

The beam shown in Figure 9.14(a) is fixed at both ends A and B and carries a concentrated load P as shown. Determine the reaction components at A and B using the area-moment method.

Solution. An approximation for the elastic curve of the beam is shown in Figure 9.14(a). Because of the type of supports, a tangent drawn to the elastic curve at point A passes through point B. Therefore,

$$t_{BA} = 0 \qquad \text{(a)}$$

Also, since the slope of the elastic curve at both supports is zero, we conclude that

$$\theta_{AB} = 0 \qquad \textbf{(b)}$$

Since there are no forces applied in the x direction, the free-body diagram as shown in Figure 9.14(b) shows only four reaction components, A_v, B_v, M_A, and M_B. Using point B as the fixed end of the cantilevers, we construct the M/EI_u diagram for the beam by cantilever parts as shown in Figure 9.14(c), (d),

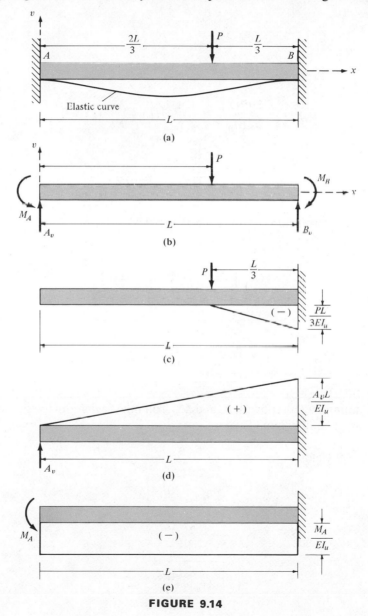

FIGURE 9.14

and (e). These M/EI_u diagrams are used to satisfy the requirements of Eqs. (a) and (b) as follows:

$$t_{BA} = -\frac{1}{2}\frac{PL}{3EI_u}\left(\frac{L}{3}\right)\left(\frac{L}{9}\right) + \frac{1}{2}\frac{A_vL}{EI_u}(L)\left(\frac{L}{3}\right) - \frac{M_A}{EI_u}(L)\left(\frac{L}{2}\right) = 0$$

Therefore,

$$\frac{A_vL}{3} - M_A - \frac{PL}{81} = 0 \qquad\qquad \textbf{(c)}$$

Also,

$$\theta_{AB} = -\frac{1}{2}\frac{PL}{3EI_u}\frac{L}{3} + \frac{1}{2}\frac{A_vL}{EI_u}(L) - \frac{M_A}{EI_u}(L) = 0$$

Therefore,

$$\frac{A_vL}{2} - M_A - \frac{PL}{18} = 0 \qquad\qquad \textbf{(d)}$$

A simultaneous solution of Eqs. (c) and (d) yields

$$\boxed{A_v = \frac{7P}{27}}$$

and

$$\boxed{M_A = \frac{2PL}{27}}$$

The remaining two unknowns, B_v and M_B, can now be determined by using the conditions of equilibrium. Thus

$\uparrow \sum F_v = 0$ yields

$$\boxed{B_v = \frac{20P}{27}}$$

$\circlearrowleft \sum M_B = 0$ yields

$$\boxed{M_B = \frac{4PL}{27}}$$

Homework Problems

Assume linear elastic behavior in all the following problems.

9.54 Solve Problem 9.32 by the area-moment method.

9.55 Solve Problem 9.33 by the area-moment method.

9.56 Solve Problem 9.34 by the area-moment method.

9.57 Solve Problem 9.35 by the area-moment method.

9.58 Solve Problem 9.36 by the area-moment method.

9.59 Solve part (a) of Problem 9.39 by the area-moment method.

9.60 Solve Problem 9.40 by the area-moment method.

9.61 Solve Problem 9.41 by the area-moment method.

9.62 Solve Example 9.8 by the area-moment method.

9.63 Solve Problem 9.50 by the area-moment method.

9.64 Solve Problem 9.51 by the area-moment method.

9.65 Solve Problem 9.52 by the area-moment method.

9.66 Determine the reaction components at supports A and B for the beam shown in Figure H9.66. Use the area-moment method.

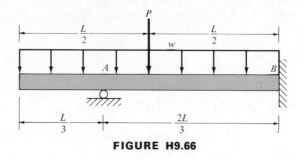

FIGURE H9.66

9.67 Determine the reaction components at supports A and B for the beam shown in Figure H9.67. Use the area-moment method.

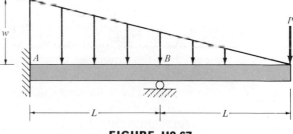

FIGURE H9.67

9.68 The steel beam shown in Figure H9.68 is fixed at B and restrained by two steel rods of length l which are connected to the ends of rigid arms securely

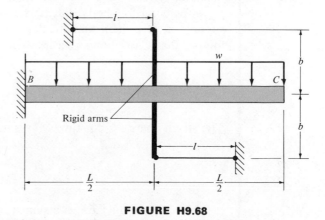

FIGURE H9.68

fastened to the beam at midspan. The beam carries a uniform load of intensity w over the entire span. Assume the cross-sectional area of the steel rods to be A, the modulus of elasticity for steel to be E, and the moment of inertia for the beam to be I_u, and determine the force induced in the rods and the reaction components at support B. Use the area-moment method.

⋆9.7   Statically Indeterminate Members Under Flexural Loads—Castigliano's Second Theorem

Castigliano's second theorem as expressed in Eq. 6.29 may be used to satisfy the principle of consistent deformations and therefore to supplement the equations of equilibrium as they apply to a given statically indeterminate beam. This procedure requires the determination of the strain energy in the beam in terms of one or more unknown reaction components. According to Eq. 6.29, the strain energy is then differentiated partially with respect to one or more of the unknown reaction components and the resulting expressions equated to known deflections or rotations as dictated by the boundary conditions of the problem. This procedure leads to simultaneous equations that may be used to determine some of the unknown reaction components. The remaining unknown reaction components may then be determined from the applicable equations of equilibrium. The use of Castigliano's second theorem is illustrated by the following examples.

Example 9.15
Solve Example 9.6 using Castigliano's second theorem.

Solution. Refer to the beam of Example 9.6 shown in Figure 9.6(a) and the accompanying free-body diagrams shown in Figure 9.6(b) and (c). Since the support at A is such that the deflection of the beam at this point is zero, it follows that Castigliano's second theorem expressed in Eq. 6.29 leads to

$$\frac{\partial U}{\partial A_v} = 0 \tag{a}$$

The strain energy U for the beam is given by Eq. 6.32 as follows:

$$U = \int_0^L \frac{M_u^2}{2EI_u}\, dx \tag{b}$$

in which the product EI_u is assumed to be a constant. Therefore,

$$\frac{\partial U}{\partial A_v} = \frac{1}{EI_u} \int_0^L M_u \frac{\partial M_u}{\partial A_v} \, dx \qquad \text{(c)}$$

From the free-body diagram of Figure 9.6(c),

$$M_u = A_v x - \frac{1}{6}\frac{w}{L} x^3 \qquad \text{(d)}$$

Therefore,

$$\frac{\partial M_u}{\partial A_v} = x \qquad \text{(e)}$$

Substitution of Eqs. (d) and (e) into Eq. (c) leads to the following expression:

$$\frac{\partial U}{\partial A_v} = \frac{1}{EI_u}\left(\frac{A_v L^3}{3} - \frac{1}{30} w L^4 \right) \qquad \text{(f)}$$

Using Castigliano's second theorem as expressed in Eq. (a), we conclude that

$$\boxed{A_v = \frac{wL}{10}}$$

which, of course, is identical to the value obtained by the three previous methods. The determination of the reaction components B_v and M_B is accomplished by the application of the conditions of equilibrium as was done in Example 9.6.

Example 9.16
Repeat Example 9.8 by the use of Castigliano's second theorem.

Solution. Refer to the beam of Example 9.8 shown in Figure 9.8(a) and the free-body diagrams shown in Figure 9.8(b) and (c). At its end B, the beam deflects through the amount v, which is also the deflection of rod BC. The determination of v was thoroughly explained in the solution of Example 9.8 and was found to be given by the expression

$$v = -\left[\frac{Fl}{AE_A} - \alpha(\Delta T)l \right] \qquad \text{(a)}$$

Therefore, Castigliano's second theorem as given by Eq. 6.29 yields

$$\frac{\partial U}{\partial F} = v = \alpha(\Delta T)l - \frac{Fl}{AE_A} \qquad \text{(b)}$$

Since the strain energy for the beam is given by

$$U = \int_0^L \frac{M_u^2}{2 E_S I_u}\, dx \qquad \textbf{(c)}$$

in which $E_S I_u$ is a constant, it follows that

$$\frac{\partial U}{\partial F} = \frac{1}{E_S I_u} \int_0^L M_u \left(\frac{\partial M_u}{\partial F}\right) dx \qquad \textbf{(d)}$$

From the free-body diagram of Figure 9.8(c),

$$M_u = Fx - \frac{wx^2}{2} \qquad \textbf{(e)}$$

Thus

$$\frac{\partial M_u}{\partial F} = x \qquad \textbf{(f)}$$

Using the values given by Eqs. (e) and (f) in Eq. (d), we obtain the following expression:

$$\frac{\partial U}{\partial F} = \left(\frac{1}{E_S I_u} \frac{F L^3}{3} - \frac{w L^4}{8}\right) \qquad \textbf{(g)}$$

Substituting Eq. (g) into Eq. (b), we obtain the relation needed to determine the value of F. Thus

$$F = \left(\frac{3 A E_A}{8}\right)\left(\frac{w L^4 + 8 E_S I_u \alpha (\Delta T) l}{A E_A L^3 + 3 I_u E_S l}\right)$$

which is identical to the value found previously by other methods. The reactive components D_v and M_D are determined in exactly the same manner as was done previously by the application of the conditions of equilibrium.

Example 9.17
Solve Example 9.14 by Castigliano's second theorem.

Solution. Refer to the beam of Example 9.14 shown in Figure 9.14(a) and the free-body diagram shown in Figure 9.14(b). Because of the nature of the support at A, the slope and the deflection at this point are both equal to zero. Therefore, Castigliano's second theorem may be written for these conditions as follows:

$$\frac{\partial U}{\partial A_v} = v_A = 0 \tag{a}$$

$$\frac{\partial U}{\partial M_A} = \theta_A = 0 \tag{b}$$

The strain energy U is given by the equation

$$U = \int \frac{M_u^2}{2EI_u} \, dx \tag{c}$$

in which EI_u is a constant. Therefore,

$$\frac{\partial U}{\partial A_v} = \frac{1}{EI_u} \left[\int_0^{2L/3} M_u \left(\frac{\partial M_u}{\partial A_v} \right) dx + \int_{2L/3}^{L} M_u \left(\frac{\partial M_u}{\partial A_v} \right) dx \right] \tag{d}$$

$$\frac{\partial U}{\partial M_A} = \frac{1}{EI_u} \left[\int_0^{2L/3} M_u \left(\frac{\partial M_u}{\partial M_A} \right) dx + \int_{2L/3}^{L} M_u \left(\frac{\partial M_u}{\partial M_A} \right) dx \right] \tag{e}$$

The moment M_u in the equations above requires definition for sections to the left, and to the right, of the load P as follows:

$$M_u = A_v x - M_A \qquad\qquad 0 \le x \le \tfrac{2}{3}L \tag{f}$$

$$M_u = A_v x - M_A - P(x - \tfrac{2}{3}L) \qquad \tfrac{2}{3}L \le x \le L \tag{g}$$

Therefore,

$$\frac{\partial M_u}{\partial A_v} = x \quad \text{and} \quad \frac{\partial M_u}{\partial M_A} = -1 \qquad 0 \le x \le \tfrac{2}{3}L \tag{h}$$

$$\frac{\partial M_u}{\partial A_v} = x \quad \text{and} \quad \frac{\partial M_u}{\partial M_A} = -1 \qquad \tfrac{2}{3}L \le x \le L \tag{i}$$

Substitution of Eqs. (f), (g), (h), and (i) into Eqs. (d) and (e) leads to the following relations:

$$\frac{\partial U}{\partial A_v} = \frac{1}{EI_u} \left(\frac{A_v L^3}{3} - \frac{M_A L^2}{2} - \frac{4PL^3}{81} \right) \tag{j}$$

$$\frac{\partial U}{\partial M_A} = \frac{1}{EI_u} \left(-\frac{A_v L^2}{2} + M_A L + \frac{PL^2}{18} \right) \tag{k}$$

Substitution of Eqs. (j) and (k) into Eqs. (a) and (b), respectively, leads to two simultaneous equations in A_v and M_A as follows:

$$54A_vL - 81M_A - 8PL = 0 \qquad \text{(l)}$$
$$-9A_vL + 18M_A + PL = 0 \qquad \text{(m)}$$

A simultaneous solution of Eqs. (1) and (m) yields the following values:

$$A_v = \frac{7P}{27}$$

and

$$M_A = \frac{2PL}{27}$$

The reaction components B_v and M_B are determined by the conditions of equilibrium in exactly the same manner as was done in Example 9.14.

Homework Problems

Assume linear elastic behavior in all the following problems.

9.69 Solve Problem 9.32 by Castigliano's second theorem.

9.70 Solve Problem 9.33 by Castigliano's second theorem.

9.71 Solve Problem 9.34 by Castigliano's second theorem.

9.72 Solve Problem 9.35 by Castigliano's second theorem.

9.73 Solve Problem 9.36 by Castigliano's second theorem.

9.74 Solve Problem 9.40 by Castigliano's second theorem.

9.75 Solve Problem 9.50 by Castigliano's second theorem.

9.76 Solve Problem 9.51 by Castigliano's second theorem.

9.77 Solve Problem 9.66 by Castigliano's second theorem.

9.78 Solve Problem 9.67 by Castigliano's second theorem.

9.79 Solve Problem 9.68 by Castigliano's second theorem.

Chapter 10 Introduction to Component Design

10.1 Introduction

The first nine chapters in this text dealt exclusively with the development and use of the tools required to analyze members subjected to axial and transverse forces, to torques, and to bending moments as well as to combinations of these effects. In all instances, methods were developed to compute the stresses and deformations resulting from a given set of loading conditions.

One major objective of the analytical methods of the preceding nine chapters is the ability to produce structural and machine systems that are capable of performing given functions without failure. This concept forms the basis for the design process in which the student is required to integrate knowledge from the preceding chapters to arrive at a practical solution to a given problem. Chapter 10 deals with an introductory treatment of the concepts of design as they relate to individual component parts and Chapter 11 extends this introductory treatment to the design of simple systems. Thus Section 10.2 discusses briefly some of the definitions and fundamentals of the design process. Section 10.3 concentrates on the design of axially loaded members, Section 10.4 on the design of torsional members, Section 10.5 on the design of flexural members, Section 10.6 on the design of columns, and finally, Section 10.7 on the design of components subjected to combined loads.

10.2 Basic Concepts of Design

Design differs from analysis in three major ways:

1. Design solutions are not unique. Many different solutions would be acceptable from a strength standpoint but other factors, such as those discussed under (3), are important in choosing a final design.
2. Design often requires trial and error or iterative methods to obtain solutions. An informal approach is taken in this text, but methods such as the Newton–Raphson, linear interpolation, and so on, could readily be applied for solving the resulting equations.

487

3. Final design involves a consideration of numerous factors other than strength and deformation characteristics. Availability of materials and shapes, fabrication or manufacturing costs, erection or assembly procedures, and so on, would be examples of such factors.

The ultimate objective of a design process is to select suitable material and to choose proper dimensions for the components of a structure or machine so that they perform their functions without failure when subjected to the design loads. As pointed out above, however, a complete and thorough design process includes not only analytical considerations dealing with stress and deformation levels as well as material properties, but also such considerations as economic, weight, and fabrication factors. Selection of materials involves a knowledge of metallurgy which most engineering students study in courses dealing with materials science. Although the Examples of this chapter will deal with components fabricated of preselected materials, it should be understood that these are only preliminary selections and would have to be modified, particularly when environmental factors such as high or low temperatures, exposure to corrosive atmospheric conditions, and so on, are encountered.

The limited definition provided in Section 7.6 for the term *failure* will be broadened in this chapter to include the several types that may occur in a given member, structure, or machine.

In general, failure of a member, a structure, or a machine may be defined as any action that prevents it from fulfilling the functions for which it was designed. Several types of failure are discussed briefly in the following paragraphs.

1. Failure by excessive elastic action. Excessive elastic deformations sometimes lead to undesirable effects resulting in failure. For example, large elastic deflections in the components of a high-rise building may lead to plaster cracks, glass breakage, and excessive movement that result in undesirable psychological effects. Also, as discussed in Chapter 8, columns may buckle (fail) by purely elastic action which may lead to the complete collapse of the structure of which these columns are members.

2. Failure by inelastic (plastic) action. Most designs today are based upon the premise that members behave elastically. This obviously means that the member stresses are kept well within the elastic limit of the material. If, for any reason, the service conditions cause the member stresses to exceed the elastic limit, inelastic (plastic) action of the members ensues and the resulting permanent deformations are usually sufficiently large to prevent the structure or machine from properly performing its intended function. This type of failure is referred to as failure by yielding.

Yielding in a given material may occur at ordinary temperatures if the loads are sufficiently high to cause the stresses to exceed the yield point or yield stress of the material. However, yielding can also occur at stresses well below the yield point or yield stress if the material is subjected to these stresses at elevated temperatures for sufficiently long periods of time. Such yielding and the resulting permanent deformations are referred to by the name *creep*. Thus under certain combinations of temperature and time, a given member, structure, or machine may fail by yielding through the phenomenon of creep.

3. *Failure by fracture*. Under certain conditions, failure may be caused by fracture of one or more members of a structure or machine. Fracture may occur suddenly or progressively. For example, a brittle material such as cast iron, if subjected to sufficiently high loads at ordinary temperature, will fracture abruptly without exhibiting appreciable plastic deformations. Also, a normally ductile material such as steel, if subjected to very low temperatures, may behave in a brittle manner and fail by sudden fracture if impact loads are applied. Progressive fracture, on the other hand, is caused by the repeated application of loads. For example, members of a bridge are repeatedly stressed every time traffic passes over it. This repeated application of loads results in the initiation and propagation of cracks that ultimately lead to failure by fracture. This progressive type of failure is known as *fatigue*, a topic that is discussed in more detail in Section 13.4.

Thus a major task faced by the designer would be to anticipate the type of failure that may occur under certain conditions and produce a design that would avoid that type of failure. However, it is not always possible to anticipate the type of failure that may occur in a given situation because all the necessary conditions relating to a given design may not be accurately known. Therefore, the designer may have to estimate some of the pertinent parameters, which include such factors as service loads, service temperatures, and material properties on the basis of past experience and sound judgment. Consider, for example, the way properties of materials are obtained in the laboratory. As discussed in Section 3.4, small samples (specimens) are prepared from a given material, and it is assumed that the average properties obtained from the few tests that are performed are a true representation of the behavior of that material. Furthermore, the properties obtained from a small sample are not necessarily the same as would be obtained from a full-size member of the same material. And testing of full-size members is economically unfeasible.

Thus the designer is faced with a situation that is shrouded with many uncertainties. Because of these uncertainties, the designer tries to ensure that the system (member, structure, or machine) being designed is subjected to stresses well below the failure stresses. Thus, if failure is judged to be by inelastic action, the failure stress would be the yield point or yield stress of the material. Therefore, for the system to be safe from failure by inelastic action, the design stresses (stresses to which the system is subjected during service) should be considerably less than the yield point or yield strength of the material. The design is then said to contain a margin of safety, which accounts for the many uncertainties and which, of course, provides the designer with a feeling of security.

The term *allowable stress* is used to signify the design stress, the largest stress that a given material can safely sustain in service. Values of allowable stresses for various materials have been arrived at through vast experience and sound engineering judgment. These values are normally specified in city ordinances and other building, machine, and structural codes.

As implied earlier, the allowable stress is appreciably less than the failure-causing stress in a given material. This failure-causing stress is a unique property for a given material and is either the yield point, σ_y (yield strength, σ_s), or the ultimate strength, σ_u, depending upon the type of failure expected. The ratio of the failure-causing stress to the allowable stress provides a measure of the margin of safety in a given design and is given the name "factor of safety" and denoted in this text by the symbol N. Thus

$$N = \frac{\text{failure-causing stress}}{\text{allowable stress}}$$ (10.1)

For example, if the failure-causing stress is the yield point or yield strength for the material, the factor of safety is designated as N_y and if the failure-causing stress is the ultimate strength, the symbol would be N_u.

Terms such as "allowable," "working," and "service" when used as adjectives with terms such as "stress," "strain," "deformation," and "rotation" refer to permissible values of these quantities associated with satisfactory behavior of structural or machine components under the actual loadings to which they may be safely subjected during their useful lifetimes. Use of this terminology implies that a factor of safety has already been incorporated in the stated value. For example, an allowable rotation for a shaft may be stated as some part of a radian over a specified length, and this would imply that a factor of safety has already been incorporated in the value and that the loadings to be used in the design are those which might reasonably be expected to be applied to the shaft during its useful service life.

It should be pointed out that other definitions of the factor of safety, based upon load and energy, instead of stress are in use under certain conditions. However, the only definition used for the factor of safety in this text is the one given by Eq. 10.1.

10.3 Design of Axially Loaded Members

The design of axially loaded members, whether under tensile or compressive loads, is accomplished by the use of the relations that were developed in Sections 3.2 and 3.3 between the applied loads and the resulting stresses and deformations. In the case of members subjected to compressive loads, however, only short members are considered, to avoid the possibility of column action. Long compression members that are subject to column action are designed according to the procedures developed in Section 10.6.

As stated in Section 10.2, a complete design of any member requires consideration of many factors, which include material selection as well as proper dimensions of the member. In the introductory treatment developed in this section, the material is preselected and the design process is limited to the analytical steps that lead to the dimensions necessary to ensure with some degree of confidence that the member does not fail when subjected to the design loads. The application of these concepts to the design of axially loaded members is illustrated in the following examples.

Example 10.1

A tension member is to be designed such that it can resist a load of 75 kN with a factor of safety $N_y = 1.5$ based upon failure by yielding. The member is to be 4 m long and its deformation is not to exceed a maximum allowable value of 0.0075 m. For weight reasons, the member is to be made of an aluminum alloy for which the yield strength is 300 MPa. Determine the safe cross-sectional dimensions for the member if it is to have (a) a solid circular cross section, and (b) a solid square cross section. Assume that $E = 70$ GPa.

Solution. By Eq. 3.1,

$$\sigma = \frac{F_n}{A}$$

where σ is the allowable stress, which in this case is $\sigma_s/N_y = 300/1.5 = 200$ MPa and $F_n = 75$ kN. Therefore,

$$A = \frac{F_n}{\sigma} = \frac{75,000}{200 \times 10^6} = 3.75 \times 10^{-4} \text{ m}^2$$

Thus, on the basis of strength alone, the cross-sectional area should be at least 3.75×10^{-4} m². Whether or not this area is adequate depends upon the deformation characteristics of the tension member. This deformation consideration is investigated as follows. By Eqs. 2.16a and 2.25,

$$\delta = \varepsilon L = \frac{\sigma}{E} L = \frac{F_n L}{AE}$$

Thus

$$\delta = \frac{75,000(4)}{(3.75 \times 10^{-4})(70 \times 10^9)} = 0.01143 \text{ m}$$

Since $\delta > 0.0075$ m, a redesign is necessary on the basis of the deformation requirements. Thus

$$A = \frac{F_n L}{\delta E} = \frac{75,000(4)}{0.0075(70 \times 10^9)} = 5.71 \times 10^{-4} \text{ m}^2$$

(a) The diameter D of a solid circular cross section needed would be given by

$$D = 2\sqrt{\frac{A}{\pi}} = 2\sqrt{\frac{5.71 \times 10^{-4}}{\pi}}$$

$$= \boxed{2.70 \times 10^{-2} \text{ m}}$$

(b) The dimension b of the side of a solid square cross section needed would be given by

$$b = \sqrt{A} = \sqrt{5.71 \times 10^{-4}}$$

$$= \boxed{2.39 \times 10^{-2} \text{ m}}$$

Availability and other practical considerations may dictate some increase in these dimensions.

Example 10.2

A two-part member of solid circular cross sections is to carry the axial loads shown in Figure 10.1(a). The entire member is to be made of structural steel for which the yield strength is 36,000 psi and the ultimate strength is 66,000 psi. Design the member so that it will safely resist the applied loads with a factor of safety $N = 2.5$ based upon (a) failure by yielding, and (b) failure by fracture.

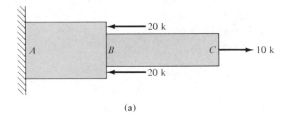

(a)

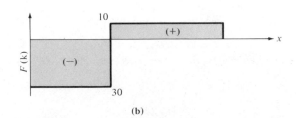

(b)

FIGURE 10.1

Assume that the deformations of the member are not significant design considerations.

Solution. A force analysis reveals that part AB is subjected to a compressive normal force of 30 k and part BC to a tensile normal force of 10 kips, as shown in Figure 10.1(b). By Eq. 3.1,

$$\sigma = \frac{F_n}{A} = \frac{F_n}{(\pi/4)D^2}$$

in which σ is the allowable stress and D is the diameter of the solid member. This relation may now be solved for the diameter D to yield

$$D = 2\sqrt{\frac{F_n}{\pi\sigma}}$$

(a) Failure by yielding: by Eq. 10.1, the allowable stress σ is

$$\sigma = \frac{\sigma_s}{N_y} = \frac{36,000}{2.5} = 14,400 \text{ psi}$$

Therefore,

$$D_{AB} = 2\sqrt{\frac{30,000}{\pi(14,400)}}$$

$$= \boxed{1.63 \text{ in.}}$$

and

$$D_{BC} = 2\sqrt{\frac{10,000}{\pi(14,400)}}$$

$$= \boxed{0.94 \text{ in.}}$$

Thus, if failure is to be by yielding of the material, the diameters would have to be at least 1.63 in. for part AB and 0.94 in. for part BC. However, practical considerations may dictate some increase in these diameters.

(b) Failure by fracture: the allowable stress in this case is

$$\sigma = \frac{\sigma_u}{N_u} = \frac{66,000}{2.5} = 26,400$$

Therefore,

$$D_{AB} = 2\sqrt{\frac{30,000}{\pi(26,400)}}$$

$$= \boxed{1.20 \text{ in.}}$$

and

$$D_{BC} = 2\sqrt{\frac{10,000}{\pi(26,400)}}$$

$$= \boxed{0.69 \text{ in.}}$$

Again, practical considerations may necessitate that these diameters be increased somewhat.

Note that the abrupt change in diameter between parts AB and BC results in stress concentration effects, consideration of which will be postponed until Section 13.2.

Example 10.3

An axially loaded member is to carry the loads shown in Figure 10.2. The entire member is to be made of a magnesium alloy ($E = 45$ GPa) for which the yield strength is 250 MPa. The deformation of the entire member is to be limited to an allowable value of 0.001 m. Determine the required cross-sectional areas of parts AB and BC using a factor of safety $N_y = 3$ based upon failure by yielding.

Solution. In this simple case shown in Figure 10.2, inspection shows that the force in AB is 100 kN(C) and the force in BC is 300 kN(T). The allowable stress based upon a factor of safety of 3 would be $250/3 = 83.3$ MPa. Therefore, by Eq. 3.1, the cross-sectional areas of the two parts may be determined as follows:

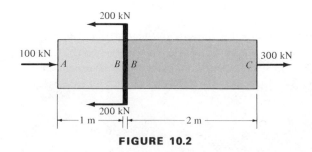

FIGURE 10.2

$$A_{AB} = \frac{100,000}{83.3 \times 10^6} = 12 \times 10^{-4} \text{ m}^2$$

and

$$A_{BC} = \frac{300,000}{83.3 \times 10^6} = 36 \times 10^{-4} \text{ m}^2$$

These values need to be checked to ensure that the allowable deformation has not been exceeded. Thus

$$\delta_{\text{total}} = \sum \frac{F_n L}{AE} = -\frac{100,000(1)}{12 \times 10^{-4}(45 \times 10^9)} + \frac{300,000(2)}{36 \times 10^{-4}(45 \times 10^9)}$$

$$= 18.5 \times 10^{-4} \text{ m}$$

Since $\delta_{\text{total}} > 0.001$ m, a redesign is needed.

Examination of the deformation equation reveals that the total deformation may be decreased from its value of 18.5×10^{-4} m in one of two ways:

1. Increase the negative term. This can only be accomplished by decreasing the cross-sectional area of part AB. This approach, however, would lead to a stress value in part AB which would exceed the allowable, and therefore it is impractical.
2. Decrease the positive term by increasing the cross-sectional area of member BC. This is the practical approach to the redesign problem, because any increase in the cross-sectional area of part BC would result in a stress value lower than the allowable and would allow a reduction in the total deformation of the member. Thus, since the total deformation is to be limited to 0.001 m,

$$0.001 = -\frac{100,000(1)}{A_{AB} \times 45 \times 10^9} + \frac{300,000(2)}{A_{BC} \times 45 \times 10^9} = \frac{10^{-5}}{4.5}\left(\frac{6}{A_{BC}} - \frac{1}{A_{AB}}\right)$$

In order to maintain the stress in part AB within the allowable value, the cross-sectional area of this member is maintained at 12×10^{-4} m². The cross-sectional area of part BC is, however, adjusted in order to decrease the positive deflection so that the total deflection does not exceed 0.001 m. Thus

$$0.001 = \frac{10^{-5}}{4.5}\left(\frac{6}{A_{BC}} - \frac{1}{12 \times 10^{-4}}\right)$$

Solution of this relation shows that the proper value for A_{BC} is 46.75×10^{-4} m².

Therefore, to meet both stress and deformation limitations, the member shown in Figure 10.2 would have to have the following minimum cross-sectional dimensions:

$$A_{AB} = 12 \times 10^{-4} \text{ m}^2$$

and

$$A_{BC} = 46.75 \times 10^{-4} \text{ m}^2$$

Note that the stress in part BC of the member would be only 64.17 MPa, considerably less than the allowable value of 83.3 MPa. Thus the factor of safety in part BC is $250/64.17 \doteq 4$, which is larger than that for part AB, where the factor of safety is 3.

Homework Problems

10.1 A tension member is to be fabricated from an aluminum alloy ($E = 10 \times 10^6$ psi) for which the yield strength is 40,000 psi and the ultimate strength is 60,000 psi. The member is to carry a load of 40 kips and is to be designed using a factor of safety of $N_u = 3$ based upon failure by fracture and an allowable deformation of 0.25 in. Determine:
(a) The minimum cross-sectional area and the maximum length dictated by the requirements of the problem.
(b) The factor of safety based upon failure by yielding.

10.2 A short compression concrete member ($E = 20$ GPa) is to have a hollow circular cross section such that its inside diameter is one-third of the outside diameter and its length is to be 2 m. The member is to carry a compressive load of 200 kN and its shortening should not exceed an allowable value of 0.002 m. Neglect any possibility of column action and determine the diameters of the hollow cross section. Assume that the concrete has an ultimate strength of

20 MPa and use a factor of safety $N_u = 4$ based upon failure by fracture.

10.3 An S* shape member with a length of 20 ft is to carry a tension load of 400 kips and have an allowable deformation of 0.2 in. If the material has a yield strength of 36,000 psi and an ultimate strength of 62,000 psi, determine the proper specification for the S section if the member is to be designed using a factor of safety $N_y = 2$ based upon yielding. What would be the factor of safety based upon fracture? Use $E = 30 \times 10^6$ psi.

10.4 A two-part member having square cross sections is to resist the loads shown in Figure H10.4. The

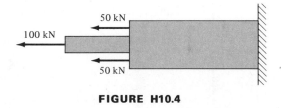

FIGURE H10.4

---

* The S designation (see Appendix E) is now used for what had been known as American Standard steel I sections.

entire member is to be made of a material for which the yield strength is 300 MPa and the ultimate strength is 400 MPa. Design the member so that it will safely carry the applied loads with a factor of safety $N = 2$ based upon:
(a) Failure by yielding.
(b) Failure by fracture.
Assume that the deformations of the member are not significant design considerations.

10.5 A three-part member having circular cross sections is to resist the loads shown in Figure H10.5. Material A has an ultimate strength of 50,000 psi, material B 40,000 psi, and material C 30,000 psi. Construct the force diagram for the member. Design this member so that it will safely carry the applied loads with a factor of safety $N_u = 2.5$ based upon failure by fracture. Assume that the deformations of the member are not significant design considerations.

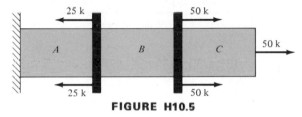

FIGURE H10.5

10.6 An axially loaded member is to resist the action of the loads shown in Figure H10.6. The entire member is to be made of structural steel ($E = 30 \times 10^6$ psi) for which the yield strength is 36,000 psi. The deformation of the entire member is limited to an allowable value of 0.15 in. Find the necessary cross-sectional areas of parts AB and BC using a factor of safety $N_y = 2$ based upon failure by yielding. If S sections are to be used for the member, select proper S sizes to meet the needs of the problem.

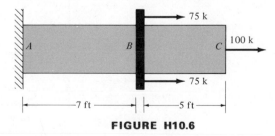

FIGURE H10.6

10.7 An axially loaded member is to resist the action of the loads shown in Figure H10.7. Material S is a steel ($E = 200$ GPa) for which the yield strength is 200 MPa and material M is a magnesium alloy ($E = 40$ GPa) for which the yield strength is 150 MPa. The deformation of the entire member is to be within an allowable value of 0.003 m. Determine the cross-sectional areas of the two parts using a factor of safety $N_y = 2.5$ based upon failure by yielding. If hollow circular cross sections in which the outside diameter is twice the inside diameter are to be used for both parts of the member, determine the proper values of these diameters.

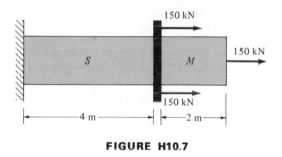

FIGURE H10.7

10.8 An axially loaded member is to carry the loads shown in Figure H10.8. The entire member is to be made of an aluminum alloy ($E = 10.5 \times 10^6$ psi) for which the yield strength is 24,000 psi and the ultimate strength is 36,000 psi. The deformation of the entire member is to be restricted to an allowable value of 0.2 in. Construct the force diagram for the member. Determine the required cross-sectional areas of the two parts of this member using a factor of safety $N_y = 1.5$ based upon failure by yielding. What would be the factor of safety based upon failure by fracture?

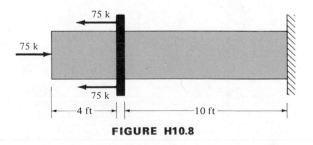

FIGURE H10.8

10.4   Design of Torsional Members

Shafts are designed to resist applied torques. The first step in design is to construct a diagram showing the variation of applied torques along the member length as discussed in Sections 1.4 and 1.5. Next, a working stress and an allowable angle of twist per unit length are selected by the designer or are specified in a code. Critical torque values together with the allowable stress and allowable angle of twist are substituted into Eqs. 4.19, 4.22, and 4.23 in order to design solid or hollow circular cylindrical shafts. Design procedure is best learned by studying the following examples.

Example 10.4

Refer to Figure 1.8 and design a solid steel shaft of constant diameter to resist these applied torques. Use a shearing yield stress of 16,000 psi with a factor of safety of $N_y = 2$ and an allowable rotation of 2.00×10^{-4} rad/in. Comment on the design if steel were to be replaced by (a) a relatively brittle material, or (b) a thin-walled tube. Use $G = 10 \times 10^6$ psi.

Solution. The critical design torque is 20 k-in., selected from the T versus x diagram of Figure 1.8. Equation 4.19 is applicable: choose $\rho = D/2$ to maximize the shearing stress. For a solid shaft $J = (\pi/32)D^4$ and τ_ρ is the allowable shearing stress. Substitute these values to obtain

$$\tau_\rho = \frac{\tau_y}{N_y} = \frac{16,000}{2} = \frac{20,000(D/2)}{(\pi/32)D^4}$$

Solve for the diameter, D:

$$\boxed{D = 2.34 \text{ in.}}$$

Use Eq. 4.22 to check the rotation of this shaft:

$$\theta = \frac{TL}{JG}$$

$$\frac{\theta}{L} = \frac{20,000}{(\pi/32)(2.34)^4(10 \times 10^6)} = 6.79 \times 10^{-4} \text{ rad/in.}$$

Since this value exceeds the allowable rotation of 2.00×10^{-4} rad/in., the shaft must be redesigned using Eq. 4.22 as follows:

$$\frac{\theta}{L} = \frac{T}{JG}$$

$$2.00 \times 10^{-4} = \frac{20,000}{(\pi D^4/32)(10 \times 10^6)}$$

$$\boxed{D = 3.18 \text{ in.}}$$

The shaft must be at least 3.18 in. in diameter to meet the stated stress and rotation design criteria. Other factors, such as the availability of material stock of this diameter and machining capabilities, would need to be considered before deciding upon a final diameter.

To fully understand the following comments, the reader should review Section 4.4 thoroughly. Although applied torques give rise to shearing stresses, the Mohr's circle construction reveals that tensile and compressive principal stresses are always associated with torsion. These principal stresses are each equal in magnitude to the shearing stress computed from the torsion formula, Eq. 4.19 with $\rho = r$. For pure torsion: $\sigma_1 = \tau$, $\sigma_2 = 0$, and $\sigma_3 = -\tau$.

COMMENTS

(a) If a given shaft were fabricated of a brittle material, the tensile normal stress σ_1 would be critical, since such a shaft is likely to fail in tension by brittle fracture due to these stresses.

(b) If a thin-walled tube were to be used as a shaft, the compressive normal stress σ_3 may be critical, since such a shaft may fail due to local buckling associated with high compressive stresses.

Example 10.5

Design a hollow circular steel shaft to resist the maximum torque of Figure 1.11. Use a shearing yield stress of 30,000 psi with a factor of safety of $N_y = 2$ and an allowable rotation of 5.00×10^{-4} rad/in. Select the inside diameter equal to 0.6 of the outside diameter. This diametral ratio will assure that local buckling is not critical in this case and that shearing stress or rotation per unit length will govern the design. Use $G = 10 \times 10^6$ psi.

Solution. The design torque from Figure 1.11 is 100 k-ft $= 1,200,000$ lb-in. Equation 4.23 becomes

$$J_1 = \frac{\pi}{32}[D_o^4 - (0.6D_o)^4] = 0.08545D_o^4$$

Equation 4.19 is applicable with J_1 from Eq. 4.23 replacing J:

$$\frac{\tau_y}{N_y} = \frac{30,000}{2} = \frac{1,200,000(D_o/2)}{0.08545D_o^4}$$

Solve for the outside diameter, D_o:

$$D_o = 7.76 \text{ in.}$$

$$D_i = 0.6D_o = 4.66 \text{ in.}$$

Use Eq. 4.22 to check the rotation of this shaft:

$$\frac{\theta}{L} = \frac{T}{JG}$$

$$= \frac{1,200,000}{0.08545(7.76)^4(10 \times 10^6)} = 3.87 \times 10^{-4} \text{ rad/in.}$$

Since this value is less than the allowable of 5.00×10^{-4} rad/in, the diameters determined above are acceptable. Hollow circular shaft diameters are

$$\boxed{D_o = 7.76 \text{ in.}} \quad \text{and} \quad \boxed{D_i = 4.66 \text{ in.}}$$

Practical considerations may dictate a small increase in the outside diameter or a decrease in the inside diameter. If the torsional loading is known to be distributed as shown in Figure 1.11, it may be economically feasible to design a tapered shaft rather than one of constant diameter. Tapered shafts would probably not be economically feasible unless a relatively large quantity is required.

Example 10.6

Design a solid circular cylindrical shaft to transmit 1600 kW at 500 rev/min. The endurance limit (refer to Section 13.4) of the material is 210 MPa and a factor of safety of $N_e = 3$ is to be used with respect to fatigue failure. An allowable static rotation of 5.00° is specified for this 4.00-m-long shaft. Use $G = 77.5$ GPa.

Solution. Solve Eq. 4.28 for the applied torque T in terms of the power P and the angular velocity of the shaft, ω:

$$T = \frac{P}{\omega}$$

Substitute values to obtain

$$T = \frac{1600 \text{ kW} \times \dfrac{1000 \text{ W}}{1 \text{ kW}} \times \dfrac{1 \text{ N-m/sec}}{1 \text{ W}}}{500 \dfrac{\text{rev}}{\text{min}} \times \dfrac{1 \text{ min}}{60 \text{ sec}} \times \dfrac{2\pi \text{ rad}}{1 \text{ rev}}}$$

$$= 30{,}560 \text{ N-m}$$

Equation 4.19 is applicable: $\tau_\rho = T\rho/J$:

$$\frac{\sigma_e}{N_e} = \frac{210 \times 10^6}{3} = \frac{(30{,}560)D/2}{(\pi/32)D^4}$$

$$D = 0.1305 \text{ m}$$

Use Eq. 4.22 to check the rotation angle of this shaft:

$$\theta = \frac{TL}{JG} = \frac{30{,}560(4)}{(\pi/32)(0.1305)^4(77.5 \times 10^9)} = 0.0554 \text{ or } 3.17°$$

Since $3.17° < 5.00°$, the shaft has satisfactory rotation characteristics. Choose:

$$\boxed{D = 0.1305 \text{ m}}$$

Practical considerations may dictate a slight increase in this shaft diameter.

Example 10.7
Design a solid shaft of rectangular cross section to resist a torque of 1500 N-m. The shearing yield stress of the material is 300 MPa. Use a factor of safety of $N_y = 2$ with respect to first yield in shear. An allowable rotation of 10.00° is specified for this 2-m shaft. Use $G = 77.5$ GPa. Choose a length/width ratio of 2.00 for this rectangular cross section.

Solution. Appropriate equations are given in Figure 4.16 for shafts of rectangular cross section. Note that a and b are the half-length and half-width, respectively.

$$\frac{a}{b} = 2 \qquad \text{or} \qquad a = 2b$$

$$\tau = \frac{T(3a + 1.8b)}{8a^2b^2}$$

Substitute as follows:

$$\frac{\tau_y}{N_y} = \frac{300 \times 10^6}{2} = \frac{1500[3(2b) + 1.8b]}{8(2b)^2b^2}$$

Solving for b gives

$$b = 0.0134 \text{ m}$$
$$a = 2b = 0.0268 \text{ m}$$

Use the following equations to check the rotation:

$$K = ab^3\left[\frac{16}{3} - 3.36\frac{b}{a}\left(1 - \frac{b^4}{12a^4}\right)\right]$$

$$= 2bb^3\left[\frac{16}{3} - 3.36\frac{b}{2b}\left(1 - \frac{b^4}{12(2b)^4}\right)\right]$$

$$= 7.324b^4 = 7.324(0.0134)^4 = 23.61 \times 10^{-8} \text{ m}^4$$

$$\theta = \frac{TL}{KG} = \frac{1500(2)}{(23.61 \times 10^{-8})77.5 \times 10^9} = 0.164 \text{ rad} \quad \text{or} \quad 9.40°$$

Since the rotation of 9.40° is less than the allowable value of 10.00°, the rectangular cross section is given by

$$\text{length} = 2a = 2(0.0268)$$

$$= \boxed{0.0536 \text{ m}}$$

$$\text{width} = 2b = 2(0.0134)$$

$$= \boxed{0.0268 \text{ m}}$$

Practical considerations may dictate small increases in these cross-sectional dimensions. A circular or hollow circular shaft could be designed with less material than the rectangular one to meet the stated requirements. A rectangular shaft would only be chosen if other conditions dictated this choice.

Homework Problems

Refer to Section 13.2 for a discussion of stress concentrations.

10.9 Design a solid cylindrical shaft of circular cross section to resist an applied torque of 100,000 lb-in. The allowable shearing stress is 10,000 psi and the angle of twist per unit length is not to exceed 5.00×10^{-4} rad/in. Use $G = 10 \times 10^6$ psi.

10.10 Design a solid cylindrical shaft of circular cross section to resist an applied torque of 12,000 N-m. The allowable shearing stress is 70 MPa, and the angle of twist per unit length is not to exceed 2.00×10^{-2} rad/m. Use $G = 72$ GPa.

10.11 A cylindrical shaft is to be designed with a hollow circular cross section such that the inside diameter is 0.80 of the outside diameter. If the allowable shearing stress is 12,000 psi and the shaft is to carry a torque of 120,000 lb-in., determine the cross-sectional dimensions of the shaft. Assume that the angle of twist of this shaft is satisfactory.

10.12 A shaft of hollow circular cross section is cylindrical and is required to transmit a torque of 14,000 N-m. Choose the inside diameter to be 0.75 of the outside diameter. If the allowable shearing stress is 80 MPa and the angle of twist per unit length is not to exceed 1.80×10^{-2} rad/m, determine the cross-sectional dimensions of this shaft. Use $G = 74$ GPa.

10.13 Refer to Figure H10.13 and draw the torque versus x diagram for this stepped shaft. The shaft is solid and is to be fabricated of a material with a shearing yield stress of 160 MPa. Use a factor of safety of $N_y = 2$ with respect to yield in shear and determine the diameters D_1, D_2, and D_3. Assume that the shafts have minimal stress concentrations and satisfactory rotation characteristics.

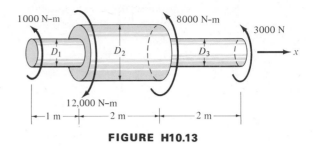

FIGURE H10.13

10.14 Refer to Figure H10.14 and draw the torque versus x diagram for this stepped shaft. The shaft is solid and is to be fabricated of a material with a shearing yield stress of 22,000 psi. Use a factor of safety of $N_y = 2$ with respect to yield in shear and determine the diameters D_1, D_2, and D_3. Assume that the shafts have minimal stress concentrations and satisfactory rotation characteristics.

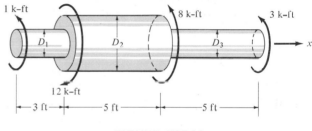

FIGURE H10.14

10.15 Refer to Example 4.7 and Figure 4.12. This shaft is to be fabricated of a material with a shearing endurance limit of 30,000 psi with a factor of safety of $N_e = 3$ with respect to fatigue failure. Determine the diameter of this shaft to transmit power as given. Assume that the shaft has minimal stress concentrations which are allowed for in the factor of safety. Static rotation characteristics are satisfactory. Refer to Section 13.4 for a discussion of endurance limit.

10.16 Refer to Problem 4.33 and Figure H4.15. The given diameters D_1 and D_2 are to be revised to meet the conditions stated below. The shafts are to be fabricated of a material with a shearing endurance limit of 220 MPa with a factor of safety of $N_e = 3$ with respect to fatigue failure. Determine the diameters of these shafts to transmit power as given. Assume that the shaft has minimal stress concentrations which are allowed for in the factor of safety. Static rotation characteristics are satisfactory. Refer to Section 13.4 for a discussion of endurance limit.

10.17 Refer to Problem 4.25 and Figure H4.13. The given diameter is to be revised to meet conditions stated below. The shafts AB and BC are to be fabricated of a material with a shearing endurance limit of 27,000 psi with a factor of safety of $N_e = 3$ with respect to fatigue failure. Determine the diameter of these shafts to transmit power as given. Assume that the shaft has minimal stress concentrations which are allowed for in the factor of safety. Static rotation characteristics are satisfactory. Refer to Section 13.4 for a discussion of endurance limit.

10.18 Design a solid shaft of rectangular cross section to resist an applied torque of 25,000 lb-in.

Choose the length of the rectangle to be twice the width. The allowable shearing stress is 10,000 psi and the angle of twist per unit length is not to exceed 4.00×10^{-4} rad/in. Use $G = 10 \times 10^6$ psi and refer to Figure 4.16 for appropriate equations.

10.19 Design a solid shaft of rectangular cross section to resist an applied torque of 4000 N-m. Choose the length of the rectangle to be twice the width. The allowable shearing stress is 70 MPa and the angle of twist is not to exceed 1.80×10^{-2} rad/m. Use $G = 75$ GPa and refer to Figure 4.16 for appropriate equations.

10.20 Refer to Section 4.7 for appropriate equations to design a thin-walled tube of rectangular cross section. The cross-sectional length/width ratio is 2.00 and the wall thickness is 0.40 in. It is subjected to end torques of 50,000 lb-in. Use $G = 12 \times 10^6$ psi. The shaft has a length of 8.00 ft. The allowable shearing stress is 6000 psi and the angle of twist over the total length is not to exceed 0.05 rad. Choose a length and width for the cross section of this tube to meet the stated requirements.

10.5 Design of Flexural Members

The design of flexural members is based upon the analytical procedures that were developed in Chapter 5 for stresses and in Chapter 6 for deformations. Thus the reader is urged to review these analytical procedures prior to embarking on the design of members subjected to flexural or bending loads.

As in previous sections of this chapter, flexural members are designed to safely carry loads without exceeding certain stress and deformation limits. Generally, the procedure consists of first designing the member to satisfy the allowable stress requirement. Then the deformation (deflection) is checked to see if it is within the allowable limits. If not, a redesign becomes necessary to ensure that both stress and deformation requirements are satisfied.

Except for beams that are very short in comparison with their cross-sectional dimensions, the shear stresses that are developed are usually relatively small in comparison to normal stresses. Thus, in general, the design process, in addition to satisfying the deformation requirement, is limited to satisfying the normal stress requirement. However, there are some cases of long beams where the shearing stress needs to be evaluated. These cases include beams that are fabricated by fastening two or more sections together, or timber beams, to make sure that the fastening between the various sections of the beam or the bonding between the timber grains, respectively, is sufficiently strong to carry the applied loads.

The various concepts mentioned above are illustrated in the following examples.

Example 10.8
A structural steel ($E = 30 \times 10^6$ psi) cantilever beam of rectangular cross section is to carry the loads as shown in Figure 10.3(a). The yield strength in

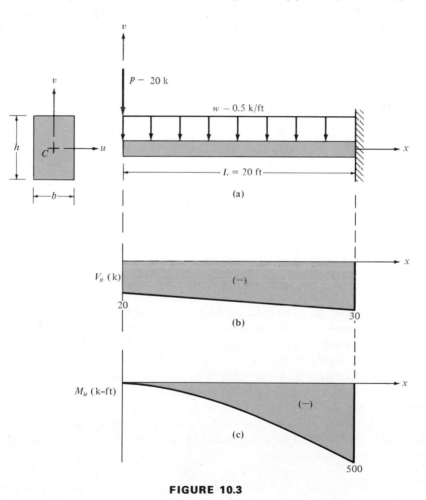

FIGURE 10.3

tension or compression for the material is given as 50,000 psi and the allowable deflection for the beam is to be 1.25 in. The span of the beam is 20 ft. Determine a suitable size if the width b of the rectangular cross section is one-half of its height h. Use a factor of safety $N_y = 2.5$ based upon failure by yielding and assume that the shearing stresses in the beam are not significant design factors.

Solution. It is good design procedure, as a first step in the solution, to construct both the shear and moment diagrams for the beam under consideration. These diagrams are constructed for the beam of Figure 10.3(a) in Figure 10.3(b) and (c). Thus, in this case, it is obvious that the maximum flexural stress occurs at the fixed end of the beam. If we ignore effects of stress concentration, which are discussed in Section 13.2, the flexural stress at the fixed end of the beam is given by Eq. 5.10a, $\sigma_x = M_u v / I_u$, in which $M_u = 500$ k-ft and in which σ_x is the allowable stress, which in this case is $\sigma_y / N_y = 50,000/2.5 = 20,000$ psi. If both v and I_u are expressed in terms of the height of the section h, we are able to relate the known quantities to the unknown section height h. Thus

$$\sigma_x = \frac{M_u v}{I_u}$$

$$20,000 = \frac{500 \times 10^3 \times 12(h/2)}{\frac{1}{12}(h/2)(h)^3} = \frac{72 \times 10^6}{h^3}$$

from which

$$h = 15.3 \text{ in.}$$

and

$$b = \frac{15.3}{2} \doteq 7.7 \text{ in.}$$

Therefore, on the basis of stress alone, the rectangular cross section should be at least 7.7×15.3 in. in order that the beam is able to carry the applied loads safely. However, the maximum deflection of the beam needs to be examined to ensure that the allowable deflection of 1.25 in. is not exceeded.

Equations for the deflection of beams may be obtained from tabulations similar to the one given in Table 6.1. If such tabulations are not available or if they do not provide the answers to the specific beam under consideration, a deflection solution becomes necessary using one of the several methods discussed in Chapter 6. In this particular case the deflection equation may be obtained from Table 6.1 by superposing cases 4 and 5. Thus the deflection v at the free end of the cantilever is

$$v = -\left(\frac{wL^4}{8EI_u} + \frac{PL^3}{3EI_u}\right)$$

where the minus sign signifies that the deflection is downward. Note that the second term in this equation was obtained from case 5 by setting $a = L$ in the equation for maximum deflection. Thus

$$v = - \left[\frac{(500/12)(20 \times 12)^4}{8(30 \times 10^6) \times \frac{1}{12}(7.7)(15.3)^3} + \frac{(20,000)(20 \times 12)^3}{3(30 \times 10^6) \times \frac{1}{12}(7.7)(15.3)^3} \right]$$

$$= -1.587 \text{ in.}$$

Since $|v| > 1.25$ in., a redesign becomes necessary on the basis of the deflection requirements. Therefore, using the deflection equation above with $v = -1.25$ in., and expressing I_u in terms of h, we obtain

$$1.25 = \frac{24L^3}{h^4 E} \left(\frac{wL}{8} + \frac{P}{3} \right)$$

and

$$h^4 = \frac{19.2L^3}{E} \left(\frac{wL}{8} + \frac{P}{3} \right)$$

$$= \frac{19.2(20 \times 12)^3}{30 \times 10^6} \left[\frac{(500/12)(20 \times 12)}{8} + \frac{20,000}{3} \right]$$

$$= 70,041.6$$

from which

$$\boxed{h = 16.3 \text{ in.}}$$

and therefore

$$\boxed{b \doteq 8.2 \text{ in.}}$$

Availability and other practical considerations may dictate some increase in these dimensions.

Example 10.9

A timber ($E = 1.5 \times 10^6$ psi) beam of rectangular cross section is to serve as a simply supported beam as shown in Figure 10.4(a). The allowable normal stress for the timber is given as 2000 psi and the allowable shearing stress as 150 psi. Assume the grain of the wood to run parallel to the axis of the beam and that the allowable deflection at midspan is 1.00 in. Determine a suitable size if the width b of the rectangular cross section is one-third of its height h.

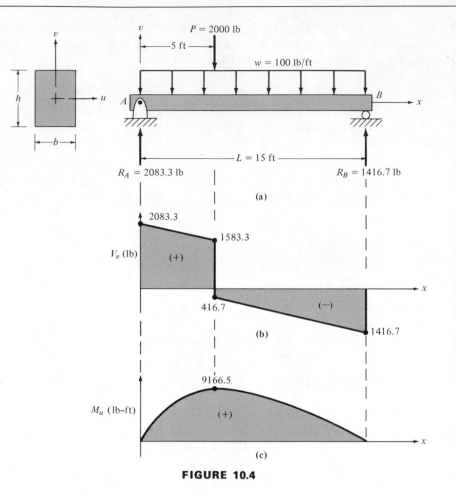

FIGURE 10.4

Solution. The reactive forces at supports A and B are determined and the shear and moment diagrams constructed as shown in Figure 10.4(b) and (c). It is evident that the maximum flexural stress occurs at a section 5 ft from the left support, where $M_u = 9166.5$ lb-ft, while the maximum shearing stress (neglecting stress concentration effects) occurs at the left support, where $V_v = 2083.3$ lb. Thus, by Eq. 5.10a, the maximum flexural stress is

$$\sigma_x = \frac{M_u v}{I_u}$$

$$2000 = \frac{9166.5 \times 12(h/2)}{\frac{1}{12}(h/3)(h)^3} = \frac{1,979,964}{h^3}$$

from which

$$h = 9.97 \text{ in.} \doteq 10.0 \text{ in.}$$

and

$$b = \frac{9.97}{3} \doteq 3.3 \text{ in.}$$

By Eq. 5.12a, the maximum shearing stress is

$$\tau_{vx} = \frac{V_v}{bI_u} Q = \frac{2083.3(3.3)(5)(2.5)}{3.3(\frac{1}{12})(3.3)(10)^3} = 94.7 \text{ psi}$$

Therefore, the shearing stress allowable of 150 psi is satisfied.

The deflection of the beam under consideration may be obtained by super-posing cases 1 and 2 in Table 6.1. Thus, from case 1, due to the uniform load, the deflection at midspan is

$$v = -\frac{5wL^4}{384EI_u}$$

From case 2, due to the concentrated load, the deflection at midspan is

$$v = -\frac{Pb}{48EI_u}(3L^2 - 4b^2)$$

where $b = \frac{1}{3}L$. Making this substitution, we obtain

$$v = -\frac{23PL^3}{1296EI_u}$$

Therefore, the total deflection at midspan would be

$$v = -\frac{L^3}{EI_u}\left(\frac{5wL}{384} + \frac{23P}{1296}\right)$$

$$= -\frac{(15 \times 12)^3}{1.5 \times 10^6(\frac{1}{12})(3.3)(10)^3}\left[\frac{5(100/12)(15 \times 12)}{384} + \frac{23(2000)}{1296}\right]$$

$$= -0.778 \text{ in.}$$

where the negative sign indicates that the deflection is downward. Since this deflection is within the allowable of 1.00 in., the design is proper and the rectangular cross section should be such that

$$\boxed{b = 3.3 \text{ in.}}$$

and

$$h = 10.0 \text{ in.}$$

Of course, practical considerations may dictate that these sizes be increased somewhat.

Example 10.10

The structural steel ($E = 200$ GPa) beam shown in Figure 10.5(a) is to have an S section. If the structural steel has a yield strength in tension and compression of 250 MPa and if the beam is to have an allowable deflection of 0.03 m at the right end, determine a suitable S section. Use a factor of safety $N_y = 2$ and assume that the shearing stresses are not significant design factors.

Solution. The reactive forces at supports A and B are computed and are shown in Figure 10.5(a). The shear diagram is shown in Figure 10.5(b) and the composite moment diagram in Figure 10.5(c). For convenience in the computations of deflections, the moment diagram is also shown by cantilever parts in Figure 10.5(d) using a section at C as the fixed end of the cantilevers.

Thus the maximum moment occurs at support B and has a value of 60 kN-m. Therefore, by Eq. 5.10a, the maximum flexural stress is

$$\sigma_x = \frac{M_u v}{I_u} = \frac{M_u}{I_u/v} = \frac{M_u}{Z}$$

where the quantity $Z = I_u/v$ is known as the section modulus for the section and σ_x is the allowable stress, which in this case is $\sigma_y/N_y = 250/2 = 125$ MPa. Thus

$$125 \times 10^6 = \frac{60{,}000}{Z}$$

$$Z = \frac{60{,}000}{125 \times 10^6} = 4.8 \times 10^{-4} \text{ m}^3$$

Since structural steel sections have not yet been standardized in SI units, this value of Z will be transformed into inch units in order to be able to use the data in Appendix E. Therefore,

$$Z = 4.8 \times 10^{-4} \text{ m}^3 = 29.3 \text{ in.}^3$$

Examination of Appendix E reveals that a standard section with a designation of S12 × 31.8 has a section modulus $Z = 36.0$ in.^{3} and therefore is more than

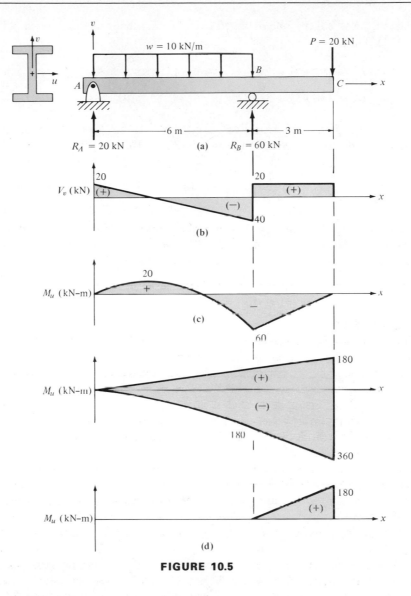

FIGURE 10.5

adequate to carry the loads as given. The next smaller size (i.e., S10 × 35) has a section modulus $Z = 29.2$ in.^{3}, which is a little less than the required value.

The deflection at the right end of the beam could be obtained by superposing several simpler beam cases, but in this solution it was decided to obtain this deflection by use of the area-moment method. In terms of tangential deviations, the deflection v_c at the right end of the beam would be

$$v_c = t_{CA} - \tfrac{3}{2} t_{BA}$$

where t_{CA} and t_{BA} may be obtained from the moment diagram shown in Figure 10.5(d). Thus

$$EI_u t_{CA} = [810(3) + 270(1) - 360(4.5) - 270(2) - 540(1)] \times 10^3 = 0$$

$$EI_u t_{BA} = [360(2) - 360(1.5)] \times 10^3 = 180 \times 10^3 \text{ N-m}^3$$

The value of I_u may be read from Appendix E to be 215.8 in.^{4}. If this value is transformed into *SI* units, it becomes 89.823×10^{-6} m^4. Therefore,

$$v_c = \frac{1}{200 \times 10^9 (89.823 \times 10^{-6})} [0 - \tfrac{3}{2}(180) \times 10^3] = -0.015 \text{ m}$$

where the negative sign signifies that the deflection of the beam at point C is downward. Since the value found for the deflection is well within the allowable value of 0.030 m, the S12 × 31.8 section is suitable for the application.

Example 10.11

The aluminum beam shown in Figure 10.6(a) is to be fabricated by fastening two rectangular sections as shown in Figure 10.6(b) and (c). The threaded fasteners are to be aluminum for which the yield strength in shear is 105 MPa and are to have a diameter of 0.015 m. Determine the appropriate spacing d for a single row of fasteners as shown in Figure 10.6(b) using a factor of safety $N_y = 1.5$ so that the beam can carry the loads safely. Assume that flexural stresses and beam deflection are not significant design considerations.

Solution. The reactive forces at supports A and B are first computed and are shown in Figure 10.6(a). Even though only the shear diagram is needed in this solution, both the shear and moment diagrams are shown in Figure 10.6(d) and (e), respectively. It is obvious that the maximum value of the shear force V_v occurs an infinitesimal distance to the right of support A and has a value of 33,750 N.

The centroid of the T section is computed and its principal centroidal axes of inertia are located as shown in Figure 10.6(c). The moment of inertia I_u with respect to the u principal axis of inertia is found to be 10.186×10^{-5} m^4. Therefore, the horizontal shear stress at the junction between the two rectangular sections making up the T is given by Eq. 5.12a. Thus

$$\tau_{vx} = \frac{V_v}{bI_u} Q$$

$$= \frac{33,750}{(0.05)(10.186 \times 10^{-5})} (0.05 \times 0.20 \times 0.0536) = 3.552 \text{ MPa}$$

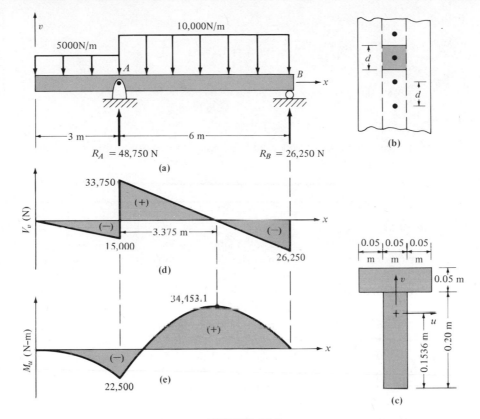

FIGURE 10.6

Therefore, the horizontal shear force V_x that has to be transmitted from the top rectangle to the bottom rectangle by each one of the fasteners may be determined by multiplying $\tau_{vx} = 3.552$ MPa by the shaded area shown in Figure 10.6(b). Thus

$$V_x = 3.552 \times 10^6 (0.05d) = 1.776 \times 10^5 d \text{ N}$$

The allowable shear stress in the fasteners is given as $\sigma_y/N_y = 105/1.5 = 70$ MPa. Therefore, the allowable shear force V_A per fastener is obtained by multiplying the allowable shear stress by the cross-sectional area of the fastener. Thus

$$V_A = 70 \times 10^6 \times \frac{\pi}{4}(0.015)^2 = 1.237 \times 10^4 \text{ N}$$

If V_A is equated to V_x, the required value of the fastener spacing d is obtained.

Thus

$$1.776 \times 10^5 \, d = 1.237 \times 10^4$$

and

$$\boxed{d = 0.0697 \text{ m}}$$

In practice d may be varied along the span to accommodate variation in the shear ordinates.

Example 10.12

A steel beam ($E = 30 \times 10^6$ psi) is to be simply supported over a span of 10 ft and is to carry a uniform load of intensity 400 lb/ft over its entire length, acting vertically downward as shown in Figure 10.7(a). Certain design requirements dictate that the cross-sectional area of the beam be an equal-legged angle with the load acting perpendicularly to one of its sides, as shown in Figure 10.7(b). The material has a yield strength in tension or compression of 36,000 psi. Select a suitable equal-legged angle section using a factor of safety $N_y = 2$ and an allowable deflection in the beam equal to 0.20 in. Assume that shear stresses and localized buckling problems are not significant design considerations and that the loads are so placed that they produce no twisting action of the beam.

Solution. Since the plane of the load is not parallel to one of the two principal centroidal axes of inertia, the beam is loaded unsymmetrically and the solution requires a trial-and-error procedure as described below.

The maximum moment in the beam occurs at midspan and has a value equal to 5000 lb-ft = 60,000 lb-in. This moment may be decomposed into its two components, M_u and M_v as shown in Figure 10.7(b), in which $M_u = M_v = 42,420$ lb-in.

FIRST TRY: Angle L5 $\times$ 5 $\times$ $\frac{3}{4}$. This angle is shown in Figure 10.7(c) approximately to scale. From Appendix E the following values are obtained:

$$I_x = I_y = 15.7 \text{ in.}^4$$

$$\bar{x} = \bar{y} = 1.52 \text{ in.}$$

$$r_v = 0.97 \text{ in.} \qquad A = 6.94 \text{ in.}^2$$

Therefore,

$$I_v = r_v^2 A = 6.5 \text{ in.}^4$$

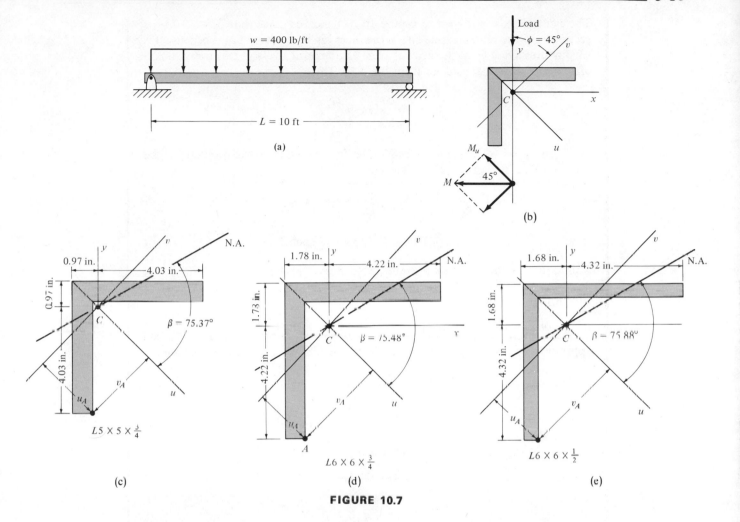

FIGURE 10.7

Also, since $I_x + I_y = I_u + I_v$, it follows that

$$I_u = 2(15.7) - 6.5 = 24.9 \text{ in.}^4$$

The neutral axis for bending is located by using Eq. 5.14. Thus

$$\beta = \tan^{-1}\left(\frac{I_u}{I_v} \tan \varphi\right)$$

$$= \tan^{-1}\left(\frac{24.9}{6.5} \times 1\right) = 75.37°$$

The neutral axis is shown in Figure 10.7(c) and indicates that point A in the section is in the tension zone and is the most highly stressed point because it is the farthest away from the neutral axes. The flexural stress at this point, therefore, needs to be determined from Eq. 5.15. Thus

$$\sigma_A = \frac{M_u}{I_u} v_A + \frac{M_v}{I_v} u_A$$

in which $v_A = 3.0$ in. and $u_A = 2.69$ in., as determined from the geometry of the cross section in Figure 10.7(c). Thus

$$\sigma_A = \frac{42,420}{24.9}(3.00) + \frac{42,420}{6.5}(2.69)$$

$$= 5110.8 + 17,555.4 = 22,666.2 \text{ psi}$$

This value is larger than the allowable stress, which is given as $\sigma_s/N_y = 36,000/2 = 18,000$ psi. Therefore, this section has to be rejected on the basis of the allowable stress.

SECOND TRY: Angle L6 × 6 × $\frac{3}{4}$. This angle is shown in Figure 10.7(d) approximately to scale. As in the first try, the following values are obtained:

$$I_x = I_y = 28.2 \text{ in.}^4$$

$$\bar{x} = \bar{y} = 1.78 \text{ in.}$$

$$r_v = 1.17 \text{ in.} \qquad A = 8.44 \text{ in.}^2$$

$$I_v = (1.17)^2(8.44) = 11.6 \text{ in.}^4$$

$$I_u = 2(28.2) - 11.6 = 44.8 \text{ in.}^4$$

$$\beta = \tan^{-1}\left(\frac{44.8}{11.6} \times 1\right) = 75.48°$$

The neutral axis for bending is shown in Figure 10.7(d) and again indicates that point A is the most highly stressed. Again, the geometry of the section is used to find v_A and u_A to be 3.72 in. and 2.26 in., respectively. Thus

$$\sigma_A = \frac{42,420}{44.8}(3.72) + \frac{42,420}{11.6}(2.26)$$

$$= 3522.4 + 8264.6 = 11,787.0 \text{ psi}$$

This value is well within the allowable stress of 18,000 psi. Another try, however, is made to ensure that the section is not overdesigned on the basis of flexural stress. Once that is done, the deflection of the beam needs to be examined.

THIRD TRY: Angle L6 × 6 × $\frac{1}{2}$. This angle is shown in Figure 10.7(e) approximately to scale. The following values are obtained:

$$I_x = I_y = 19.9 \text{ in.}^4$$

$$\bar{x} = \bar{y} = 1.68 \text{ in.}$$

$$r_v = 1.18 \text{ in.} \qquad A = 5.75 \text{ in.}^2$$

$$I_v = (1.18)^2(5.75) = 8.0 \text{ in.}^4$$

$$I_u = 2(19.9) - 8.0 = 31.8 \text{ in.}^4$$

$$\beta = \tan^{-1}\left(\frac{31.8}{8.0} \times 1\right) = 75.88°$$

The neutral axis for bending is shown in Figure 10.7(e). Point A is again the one that is most highly stressed and the distances v_A and u_A are found to be 3.89 in. and 2.22 in., respectively. Thus

$$\sigma_A = \frac{42{,}420}{31.8}(3.89) + \frac{42{,}420}{8.0}(2.22)$$

$$= 5189.1 + 11{,}771.6 = 16{,}960.7 \text{ psi}$$

This value is obviously less than the allowable stress of 18,000 psi. Therefore, the L6 × 6 × $\frac{1}{2}$ section is acceptable from the standpoint of allowable normal stress. The deflection of the beam, using this cross section, needs to be examined.

Case 1 in Table 6.1 reveals that the midspan (maximum) deflection δ for a simply supported beam of length L subjected to a uniform load of intensity w is equal to $5wL^4/384EI$. Therefore, the components of deflection at midspan in the u and v directions are

$$\delta_v = \frac{5L^4}{384E}\frac{w_v}{I_u} = \frac{5(10 \times 12)^4}{384 \times 30 \times 10^6}\frac{0.707(400/12)}{I_u} = \frac{2.12}{I_u}$$

$$= \frac{2.12}{31.8} = 0.067 \text{ in.}$$

and

$$\delta_u = \frac{5L^4}{384E}\frac{w_u}{I_v} = \frac{5(10 \times 12)^4}{384 \times 30 \times 10^6}\frac{0.707(400/12)}{I_v} = \frac{2.12}{I_v}$$

$$= \frac{2.12}{8.0} = 0.265 \text{ in.}$$

The resultant midspan deflection is, therefore,

$$\delta = \sqrt{\delta_u^2 + \delta_v^2} = 0.27 \text{ in.}$$

This value of deflection is excessive in terms of the allowable of 0.20 in. and, consequently, the L6 × 6 × $\frac{1}{2}$ angle is rejected. This conclusion prompts a further examination of the L6 × 6 × $\frac{3}{4}$ angle section that was attempted during the second try. Recall that the flexural stress was found to be 11,787.0 psi, well within the allowable limit. However, the deflection still remains to be examined to ensure that it is within the allowable value. Thus

$$\delta_v = \frac{2.12}{I_u} = \frac{2.12}{44.8} = 0.047 \text{ in.}$$

$$\delta_u = \frac{2.12}{I_v} = \frac{2.12}{11.6} = 0.183 \text{ in.}$$

and

$$\delta = \sqrt{\delta_u^2 + \delta_v^2} = 0.19 \text{ in.}$$

This value is less than the allowable deflection of 0.20 in. and therefore the equal-legged angle that is appropriate for the application is the

$$\boxed{\text{L6} \times 6 \times \tfrac{3}{4}}$$

Homework Problems

10.21 An aluminum alloy ($E = 10 \times 10^6$ psi) cantilever beam having a circular cross section and a span of 15 ft is to carry a load that varies linearly from zero at the free end to a maximum value of 3 k/ft at the fixed end. The yield strength for the material in tension and compression is given as 40,000 psi and the allowable deflection for the beam is to be 2.0 in. Determine a suitable diameter for the beam using a factor of safety $N_y = 2$ based upon failure by yielding. Assume that the shearing stresses are not significant design factors.

10.22 A simply supported beam is to have a span of 6 m and a rectangular cross section whose base b is $\frac{1}{4}$

of its height h. The material is to be structural steel ($E = 200$ GPa) and the beam is to support a uniform load of intensity 4 kN/m over the entire length and a concentrated load at midspan of 10 kN. The material has a yield strength of 250 MPa and the beam is to have an allowable deformation of 0.01 m. Determine suitable values of b and h using a factor of safety $N_y = 2.5$ based upon failure by yielding. Assume that the shearing stresses are not significant design considerations.

10.23 A timber ($E = 1.5 \times 10^6$ psi) beam of rectangular cross section is to serve as a cantilever beam with a span of 12 ft. It is to carry a uniform load of intensity 75 lb/ft over the entire length and two concentrated forces of 1500 lb each, one acting at the free end and the other at midspan. The allowable normal stress for the timber is 1500 psi and the allowable shearing stress is 100 psi. Assume the grain of the wood to run parallel to the axis of the beam and that the allowable deflection is 1.5 in. Determine a suitable size if the width b of the rectangular cross section is one-half of its height h.

10.24 A timber ($E = 10$ GPa) beam is simply supported and subjected to the loads shown in Figure H10.24(a). The beam is to be constructed by gluing two rectangular pieces to form a T section, as shown in Figure H10.24(b). The allowable normal stress for timber is 14 MPa and the allowable shearing stress for the glue is 0.4 MPa. Determine a suitable value for the dimension t if the allowable deflection in the beam is 0.04 m.

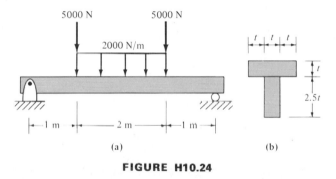

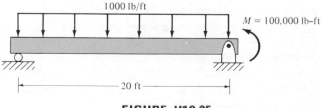

(a)   (b)

FIGURE H10.24

10.25 The structural steel ($E = 30 \times 10^6$ psi) beam shown in Figure H10.25 is to have an S section. The structural steel has a yield strength in tension and compression of 36 ksi. If the beam is to have an allowable deflection of 0.5 in., determine a suitable S section. Use a factor of safety $N_y = 2.5$ based upon

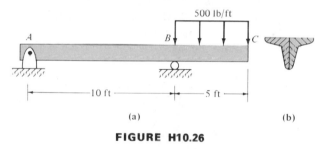

FIGURE H10.25

failure by yielding and assume that the shearing stresses are not significant design factors.

10.26 The structural steel ($E = 30 \times 10^6$psi) beam shown in Figure H10.26(a) is to be constructed by welding two standard equal-legged angles together to form a T section as shown in Figure H10.26(b). The structural steel has a yield strength in tension and compression of 36 ksi. If the beam is to have an allowable deflection of 0.75 in at its right end (at point C), select a suitable equal-legged angle for the application. Use a factor of safety $N_y = 2$ based upon failure by yielding and assume that the shearing stresses are not significant design considerations.

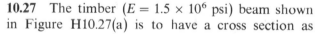

(a)   (b)

FIGURE H10.26

10.27 The timber ($E = 1.5 \times 10^6$ psi) beam shown in Figure H10.27(a) is to have a cross section as

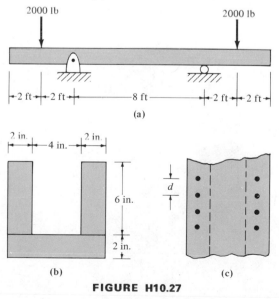

(b)   (c)

FIGURE H10.27

shown in Figure H10.27(b). This cross section is to be fabricated by nailing three rectangular timber members together by using steel spikes for which the yield strength in shear is 12,000 psi. The diameter of each spike is $\frac{3}{16}$ in. and they are arranged in the fashion shown in Figure H10.27(c). Determine the appropriate spacing d so that the beam can carry the loads safely. Use a factor of safety $N_y = 2.0$ and assume that flexural stresses and beam deflections are not significant design considerations.

10.28 A structural steel ($E = 200$ GPa) cantilever beam 6 m long is to carry a uniform load of intensity 4 kN/m over the entire span. The beam is to be fabricated by fastening three rectangular sections together to form an I section as shown in Figure H10.28(a) and (b). The steel fasteners are 0.015 m in diameter and are arranged in a single row (both top

and bottom) as shown in Figure H10.28(b). Determine the appropriate spacing d for the fasteners if the material has a yield strength in shear of 120 MPa and if a factor of safety $N_y = 2.25$ is to be used. Assume that flexural stresses and beam deflections are not significant design considerations.

10.29 An aluminum alloy cantilever beam 5 m long is to carry a uniform load of intensity 4 kN/m over the entire length. The cross section for the beam is to be a rectangle whose width b is one-half of its height h and the load is inclined at 45° to the v principal centroidal axis of the cross section as shown in Figure H10.29. The material has a yield strength in tension and compression equal to 200 MPa and the beam is to have an allowable deflection of 0.04 m. Use a factor of safety $N_y = 1.5$ and determine a suitable size for the rectangular section. Assume that shearing stresses are not significant design factors.

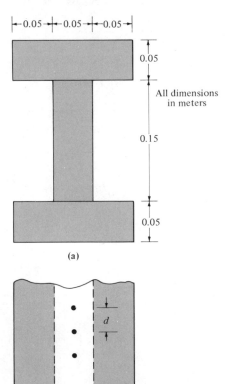

(a)

(b)

FIGURE H10.28

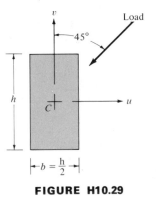

FIGURE H10.29

10.30 A simply supported beam with a span of 21 ft is to carry two concentrated loads, 20,000 lb each, acting at the third points along the span. The material is to be structural steel ($E = 30 \times 10^6$ psi) and an S section is to be used for the application. Design requirements are such that both loads have to be applied at 30° clockwise from the vertical axis of the S section. The allowable stress for the material is 20,000 psi and the allowable deflection for the beam is 0.75 in. Select a suitable S section for this application. Assume that shear stresses are not significant design considerations.

10.6 Design of Columns

A knowledge of the concepts developed in Chapter 8 is required for the design of columns or struts. Students are urged to review the Examples and their own solutions to the Homework Problems in Chapter 8. Section 8.6 should be studied again in detail prior to continuing with a study of this section in which four examples develop design approaches for columns. Except in special cases, the design of columns involves iterative procedures as shown in these examples.

Example 10.13

A Douglas fir timber column is to be designed to carry an axial force of 11,000 lb using the information of Figure 8.12. The column is 16 ft long and has a rectangular cross section. Column ends may be assumed pinned and the factor of safety is included in the recommended equations. Use $E = 1.1 \times 10^6$ psi.

Solution. For a rectangular cross section ($h \times b$) it is possible to express the slenderness ratio in terms of the least dimension (h) of the cross section.

$$r_v = \sqrt{\frac{I_v}{A}} = \sqrt{\frac{\frac{1}{12}bh^3}{bh}} = \frac{h}{\sqrt{12}} \qquad \text{(where } h < b)$$

$$\frac{L}{r_v} = \frac{16 \times 12}{h/\sqrt{12}} = \frac{665}{h}$$

From Figure 8.12, assuming that $66.3 \leq L/r_v \leq 173.2$,

$$\frac{P}{A} = \frac{\pi^2 E}{2.727(665/h)^2}$$

Substitute values:

$$\frac{11,000}{bh} = \frac{\pi^2(1.1 \times 10^6)}{2.727(665/h)^2}$$

$$bh^3 = 1222$$

Assign $b = 7.5$ in., then $h = 5.46$ in.

Dressed dimensions of a 6 × 8 are 5.5 × 7.5, which would provide

$$bh^3 = 7.5(5.5)^3 = 1248 \text{ in.}^4$$

This size is satisfactory, since $1248 > 1222$, but the actual slenderness ratio must be checked:

$$\frac{L}{r_v} = \frac{665}{h} = \frac{665}{5.5} = 121$$

Since $66.3 < 121 < 173.2$, the assumption is valid and the equation has been correctly applied. The factor of safety is $N = 2.727$ with respect to long column elastic buckling. This may appear to be a relatively large factor of safety, but timber is not homogeneous and isotropic as assumed in Euler column theory. Furthermore, moisture content and timber grade influence the load-carrying capacity of such a column. These considerations lead to the conclusion that a relatively high factor of safety is reasonable.

Example 10.14

An aluminum alloy push rod is a component of a machine. This rod is 15 in. long and solid circular in cross section. Refer to Figure 8.15 for allowable stress values. The rod is required to carry a force of 2000 lb under operating conditions. Design the rod, assuming that both ends are fixed.

Solution. In order to obtain the effective length of this column, refer to Figure 8.8(b).

$$L_e = \tfrac{1}{2}L = \tfrac{1}{2}(15) = 7.50 \text{ in.}$$

Express the radius of gyration r_v in terms of the rod diameter, D:

$$r_v = \sqrt{\frac{I_v}{A}} = \sqrt{\frac{\pi D^4/64}{\pi D^2/4}} = \frac{D}{4}$$

The slenderness ratio becomes

$$\frac{L_e}{r_v} = \frac{7.50}{D/4} = \frac{30}{D}$$

Assume that $D = 0.5$ in., $L_e/r_v = 30/0.5 = 60$. Read $P/A = 11.5$ ksi from Figure 8.15 for $L_e/r_v = 60$, or

$$A = \frac{P}{11.5} = \frac{2.0}{11.5} = 0.174 \text{ in.}^2$$

$$\frac{\pi D^2}{4} = 0.174 \quad \text{and} \quad D = 0.47 \text{ in.}$$

The calculated value of 0.47 is nearly equal to the assumed value of 0.50 in. Choose

$$\boxed{D = 0.50 \text{ in.}}$$

In general, several iterative steps would be required for convergence, but the calculations would proceed as shown above.

Example 10.15

A large number of struts are to be fabricated of the material for which column design information is given in Figure 8.21. The cross section is to be a hollow square such that the inside dimension is to equal 0.8 of the outside dimension. Each strut is to support an axial compressive force of 40 k and has a length of 12 ft with pinned ends. Figure 8.21 is a plot of Eq. 8.54 and contains a factor of safety of $N = 2.50$.

Solution. Assume that $P/A = 10$ ksi or $A = P/10$. Then

$$A = \frac{40}{10} = 4.00 \text{ in.}^2$$

Let b be the outside dimension of the square cross section; then $0.8b$ is the inside dimension and $A - b^2 - (0.8b)^2 = 0.36b^2$.

$$I_v = \tfrac{1}{12}[b^4 - (0.8b)^4] = 0.0492b^4$$

$$r_v = \sqrt{\frac{I_v}{A}} = \sqrt{\frac{0.0492b^4}{0.36b^2}} = 0.370b$$

Equate the area A to 4.00 in.^{2} and solve for b: $0.36b^2 = 4.00$, $b = 3.33$ in. The radius of gyration, $r_v = 0.370b = 0.370(3.33) = 1.23$ in. The slenderness ratio, $L/r_v = (12 \times 12)/1.23 = 117$. From Figure 8.21, read $P/A = 3.1$ ksi. Since the assumed P/A of 10 ksi does not equal the value of 3.1 ksi, read from the curve, another iteration is required.

For a second iteration, average these two values and assume:

$$\frac{P}{A} = \frac{1}{2}(10 + 3.1) = 6.55 \text{ ksi}$$

$$A = \frac{P}{6.55} = \frac{40}{6.55} = 6.11 \text{ in.}^2$$

$$A = 0.36b^2 = 6.11, \qquad b = 4.12 \text{ in.}$$

$$r_v = 0.370b = 0.370(4.12) = 1.52 \text{ in.}$$

$$\frac{L}{r_v} = \frac{144}{1.52} \doteq 94.7$$

From Figure 8.21, read $P/A = 4.50$ ksi. Since the assumed value of 6.55 ksi does not equal the value of 4.50 ksi read from the curve, another iteration is required.

For a third iteration, average these two values and assume:

$$\frac{P}{A} = \frac{1}{2}(6.55 + 4.50) = 5.53 \text{ ksi}$$

$$A = \frac{P}{5.53} = \frac{40}{5.53} = 7.23 \text{ in.}^2$$

$$A = 0.36b^2 = 7.23, \qquad b = 4.48 \text{ in.}$$

$$r_v = 0.370b = 0.370(4.48) = 1.66 \text{ in.}$$

$$\frac{L}{r_v} = \frac{14.4}{1.66} = 86.7$$

From Figure 8.21, read $P/A = 5.25$ ksi.

For a fourth iteration, assume:

$$\frac{P}{A} = \frac{1}{2}(5.53 + 5.25) = 5.39 \text{ ksi}$$

$$A = \frac{P}{5.39} = \frac{40}{5.39} = 7.42 \text{ in.}^2$$

$$A = 0.36b^2 = 7.42, \qquad b = 4.54 \text{ in.}$$

$$r_v = 0.370b = 0.370(4.54) = 1.68 \text{ in.}$$

$$\frac{L}{r_v} = \frac{144}{1.68} = 85.7$$

From Figure 8.21, read $P/A = 5.38$ ksi. This value of 5.38 ksi is approximately equal to the assumed value of 5.39 ksi. Choose $b = 4.54$ in. for an outside dimension of the hollow square and $0.8(4.54) = 3.63$ in. for an inside dimension. This provides a wall thickness of 0.455 in. It would be advisable to check for possible local buckling of the walls of this member before approving the final design. Since a large number of these struts are to be ordered, it may be feasible to fabricate special hollow square cross sections of the specified dimensions. Otherwise, the nearest standard available section would be selected from data provided by manufacturers.

Homework Problems

10.31 A Douglas fir column is 20 ft long and is to support an allowable load of 50,000 lb. The column ends are pinned and the data of Figure 8.12 are to be used to design this column. A square cross section is to be chosen. Assume that the difference between nominal and dressed dimensions is 0.5 in. (e.g., an 8.00-in. dimension is 7.50 in., dressed). Use a modulus of elasticity of 1.2×10^6 psi.

10.32 A cast iron column of hollow circular cross section is to be designed using the following allowable stress equation:

$$\frac{P}{A} = 12,000 - 60\frac{L}{r} \quad \left(\text{provided that } 35 \le \frac{L}{r} \le 100\right)$$

The column is to carry an axial load of 20,000 lb. with a length of 12.0 ft. Use an inside diameter equal to 0.8 of the outside diameter.

10.33 Solve Problem 10.32 using a solid circular cross section and compare this solution with the solution for a hollow circular cross section with an inside diameter equal to 0.8 of the outside diameter. Compare weights of the two columns using a unit weight

of 450 lb/ft^3 for cast iron. For a slenderness ratio greater than 100, use Euler's equation for pinned ends with $E = 15 \times 10^6$ psi and a factor of safety of 2.

10.34 An aluminum compression rod is to be designed using the information provided in Figure 8.15. The rod is of square cross section and is to carry a force of 3000 lb. during operation of the machine of which it is a component. This member is 20 in. long and is fixed at one end and pinned at the other end.

10.35 Redesign the compression rod of Problem 10.34 using a hollow circular cross section with the inside diameter equal to 0.80 of the outside diameter.

10.36 A machine component is required to carry a force of 15,000 N. It is 0.80 m long and may be assumed to be pinned at both ends. Euler's equation is valid for designing this component with a factor of safety of 3. The modulus of elasticity of the material is 200 GPa. Determine the dimensions of a rectangular cross section of this machine component if the width is to be twice the thickness.

10.37 A large number of compression members are to be fabricated of the material, for which column design information is given in Figure 8.21. The cross section is to be a hollow square such that the inside dimension is equal to 0.75 of the outside dimension. Each member is to support 20,000 lb. and has a length of 24 ft. with both ends fixed. Figure 8.21 is a plot of Eq. 8.54 and contains a factor of safety of $N = 2.50$.

10.38 Solve Problem 10.37 if the load is increased to 40,000 lb.

10.39 Solve Problem 10.37 if one end of each compression member is fixed and the other end is hinged. Effective length information is given in Figure 8.8.

10.40 Choose a structural steel pipe section to carry an allowable load of 300 kN. This column is assumed to be pinned at both ends and is 2.50 m long. Appropriate design equations are shown in Figure 8.14(a).

*10.7 Design of Components to Resist Combined Loadings

Review of Section 1.9 will improve the reader's understanding of design of components to resist such loadings. General review of Chapter 7 and intensive review of Section 7.6 is also suggested, particularly the maximum principal stress theory and the maximum shear stress theory. These two theories will be employed in the examples and Homework Problems.

In each of the following design examples, member weights have been neglected in comparison to the applied loadings. In practice, estimates are made of the effect of component weights in the form of what are termed *dead loads*.

Stress concentrations are also not considered here, but it is relatively simple to incorporate factors to allow for stress raisers, as discussed in Section 13.2.

Designs presented here are characterized as preliminary, since many additional factors would have to be considered before making the decisions required for a final design. The primary lesson to be learned from a study of this section is that combined loadings are frequently encountered in practice and that solutions for component geometric variables are usually iterative.

Example 10.16
Refer to Figure 1.23 and assign the following allowable design values for the axial force and torque applied to the circular cylindrical member shown: $P = 100$ kN and $T = 20$ kN-m. This component is 3.00 m long and is to be machined from a material for which the allowable stresses are 200 MPa in tension or compression and 115 MPa in shear. Select a diameter for this component to meet these stress criteria, and then check to see whether the following deformation criteria are satisfied. Allowable axial deformation: 1×10^{-3} m.

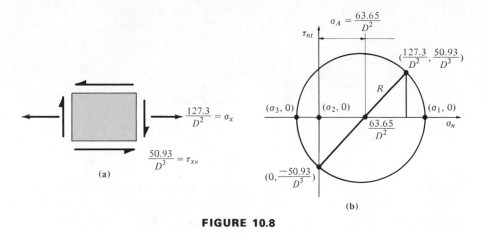

FIGURE 10.8

Allowable angle of twist: 5.00°. Revise the design, if required, to meet these deformation criteria. ($E = 206$ GPa and $\mu = 0.3$ for this material.)

Solution. Consider a stress element taken from the surface of this member and calculate the normal and shear stresses acting on the element from Eqs. 3.1 and 4.19.

$$\sigma_x = \frac{P}{A} = \frac{100}{\pi D^2/4} = \frac{127.3}{D^2} \quad \text{kN/m}^2$$

$$\tau_{xv} = \frac{T\rho}{J} = \frac{(20/2)D/2}{(\pi/32)D^4} = \frac{50.93}{D^3} \quad \text{kN/m}^2$$

This stress element is shown in Figure 10.8(a). Mohr's circle is shown in Figure 10.8(b). The radius R is given by

$$R = \sqrt{\left(\frac{63.65}{D^2}\right)^2 + \left(\frac{50.93}{D^3}\right)^2}$$

The maximum normal stress will be equated to the allowable value of 200 MPa: $\sigma_1 = \sigma_A + R$. The factor 10^3 is introduced on the right-hand side of the following equation, since the axial force and torque are both expressed in kN.

$$200 \times 10^6 = \left[\frac{63.65}{D^2} + \sqrt{\left(\frac{63.65}{D^2}\right)^2 + \left(\frac{50.93}{D^3}\right)^2}\right] \times 10^3$$

Use an iterative procedure to solve this equation for D.

D (m)	Right-Hand Side of Equation (MPa)
0.0400	836.6
0.0500	433.7
0.0600	254.1
0.0640	210.4
0.0645	205.7
0.0650	201.1
0.0655	196.7

0.0650 m $< D <$ 0.0655 m will satisfy the normal stress criterion. Next, determine the diameter required to satisfy the shearing stress criterion. The absolute maximum shearing stress given by the radius R of the Mohr's circle will now be equated to the allowable value of 115 MPa:

$$115 \times 10^6 = \left[\sqrt{\left(\frac{63.65}{D^2}\right)^2 + \left(\frac{50.93}{D^3}\right)^2} \right] \times 10^3$$

Again, use an iterative procedure to solve this equation for D.

D (m)	Right-Hand Side of Equation (MPa)
0.0800	100.0
0.0780	107.8
0.0750	121.3
0.07625	115.4

A shaft of $D = 0.0763$ m will satisfy both the normal and shearing stress criteria. A check of the deformation will now be made using a diameter of 0.0763 m.

The axial deformation is given by

$$\delta_x = \frac{\sigma_x L}{E} = \frac{127.3}{(0.0763)^2} \frac{(3)10^3}{206 \times 10^9} = 3.2 \times 10^{-4} \text{ m}$$

Since this elongation of 3.2×10^{-4} is less than the allowable value of 1×10^{-3} m, only a check of the rotation is required before deciding upon a final design diameter. The rotation is given by Eq. 4.22: $\theta = TL/JG$.

Reference to Figure 1.23(a) reveals that one-half of 3.00 m is the appropriate length to use in this equation and $0.5T = 10$ kN-m is the appropriate torque. From Eq. 4.23:

$$G = \frac{E}{2(1 + \mu)} = \frac{206}{2(1 + 0.3)} = 79.2 \text{ GPa}$$

$$\theta = \frac{TL}{JG} = \frac{0.5(20 \times 10^3)(3/2)}{(\pi/32)(0.0763)^4 79.2 \times 10^9} = 0.057 \text{ rad} \quad \text{or} \quad 3.27°$$

Since this value of 3.27° is less than the allowable angle of twist of 5.00°, the rotation criterion is met. *A component of 0.0763-m diameter will meet all the criteria.* Other practical factors not discussed here may require a revision of this diameter before the final design drawings are forwarded to the manufacturer for fabrication of the required number of these components.

Example 10.17

Refer to Figure 1.24 and note that the component shown is subjected to axial force, shear, and bending moment. A box-shaped cross section of constant thickness of 0.01 m is welded of four equal-sized plates as shown in Figure 10.9. The 0.01-m thickness was chosen to meet a minimum thickness requirement due to environmental conditions. Selection of thinner plates may lead to corrosion failure. A large number of these box shapes are to be fabricated by welding the plates together as depicted in the figure. Neglect weld metal in computing section properties and note that the centroid lies at the geometric center of the shape.

Design forces for a 6.00-m-long member are $P_1 = 200$ kN applied centroidally and $P = 100$ kN. Allowable stresses are 300 MPa in tension or compression and 170 MPa in shear. Determine the width w of each flange plate, which equals the depth d of each web plate. In this preliminary design, deformations need not be considered.

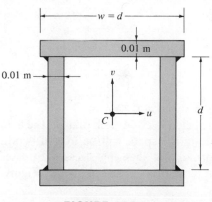

FIGURE 10.9

Solution. Geometric properties for this box section are given by

$$A = 0.02w + 0.02d = 0.04d$$

$$I_u = \frac{2}{12}(0.01)\,d^3 + 2\left(\frac{1}{12}\right)d(0.01)^3 + 2d(0.01)\left(\frac{d}{2} + 0.005\right)^2$$

$$I_u = \tfrac{2}{3} \times 10^{-2}d^3 + 1 \times 10^{-4}d^2 + \tfrac{2}{3} \times 10^{-6}d$$

A section of the beam taken just to the left of point B must resist an axial force P_1, a shearing force P, and a bending moment $PL/4$. This section is clearly the critical one to consider for this preliminary design. Keep in mind that machine and structural components are rarely designed to resist a single set of loads. Usually, a number of different loadings must be considered in design of components.

An element taken from the bottom of the beam will be subjected to normal stresses associated with axial and bending forces but will be free of shearing stresses. Recall that an element free of shearing stresses is subjected to principal stresses and express the stresses as a function of d.

$$\sigma_x = \frac{P_1}{A} + \frac{M_u v}{I_u}$$

$$300 \times 10^6 = \left[\frac{200}{0.04d} + \frac{\left(\dfrac{100 \times 6}{4}\right)\left(\dfrac{d}{2} + 0.01\right)}{\tfrac{2}{3} \times 10^{-2}d^3 + 1 \times 10^{-4}d^2 + \tfrac{2}{3} \times 10^{-6}d}\right] \times 10^3$$

Solve this equation by iteration:

d (m)	Right-Hand Side of Equation (MPa)
0.200	312.1
0.205	297.6
0.204	300.4
0.2041	300.1

Temporarily, assign d a value of 0.2041 m.

A stress element selected in either web plate at the neutral axis, again just to the left of point B along the span, will be free of bending normal stress but subjected to axial normal and shearing stresses. Stresses on this element are computed below as a function of d and shown in Figure 10.10(a).

$$\sigma_x = \frac{P_1}{A} = \frac{200}{0.04d} = \frac{5000}{d} \qquad \text{kN/m}^2$$

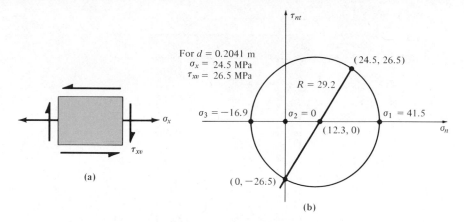

FIGURE 10.10

$$\tau_{xv} = \frac{V_v Q}{b I_u} = \frac{100[(0.01)d(d/2 + 0.005) + 2(0.01)(d/2)(d/4)]}{0.02[(\frac{2}{3} \times 10^{-2} d^3 + 1 \times 10^{-4} d^2 + \frac{2}{3} \times 10^{-6} d)]}$$

which reduces to

$$\tau_{xv} = \frac{5000(0.0075 d^2 + 5 \times 10^{-5} d)}{\frac{2}{3} \times 10^{-2} d^3 + 1 \times 10^{-4} d^2 + \frac{2}{3} \times 10^{-6} d} \qquad \text{kN/m}^2$$

Before constructing Mohr's circle shown in Figure 10.10(b), assign d the value 0.2041 m. Compare values from Mohr's circle to allowable stresses as follows:

Maximum normal stress: 41.5 < 300 MPa

Absolute maximum shearing stress: 29.2 < 170 MPa

In fact, the stresses shown on the element of Figure 10.10(a) are so small compared to the allowables that construction of the circle is not required for this design.

Select $d = w = 0.2041$ m for this preliminary design. *This box section could be fabricated of four plates each* 0.2041 × 0.01 × 6.00 m. A revised selection for the final design would depend upon available plate sizes, welding procedures, and other factors not considered in this preliminary design.

Example 10.18
Refer to Figure 1.25 and note that the member shown is subjected to axial force, torque, shear, and bending moment. A box-shaped cross section as shown in Figure 10.11 is to be fabricated by welding plates together as shown.

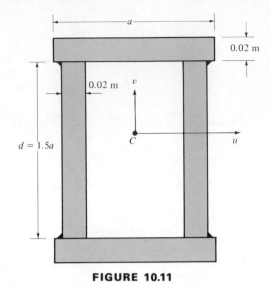

FIGURE 10.11

Initial selection of plate thickness was made as shown, although this may need to be revised in the final design in order to prevent local buckling due to large compressive forces developed in the flanges. Available thicknesses of rolled plates may also require a revision of the value 0.02 m shown. The box shape was chosen because this closed shape resists torsional loads much better than an I or other open shapes. The depth/flange width ratio was chosen to provide good bending resistance.

Design forces for a 8.00-m-long member are: $Q_1 = 400$ kN applied centroidally (use the symbol Q_1 to avoid confusion with the symbol Q in the shearing stress equation), $T = 400$ kN-m, and $P = 200$ kN. Allowable stresses are 250 MPa in tension or compression and 140 MPa in shear. Determine the width of each flange plate a. In this preliminary design, deformations need not be considered.

Solution. Geometric properties for this box section are given by

$$A = 0.04a + 0.04d = 0.10a$$

$$I_u = 2\left(\frac{1}{12}\right)(0.02)d^3 + 2\left[\frac{1}{12}a(0.02)^3 + 0.02a\left(\frac{d}{2} + 0.01\right)^2\right]$$

Substitute for $d = 1.5a$ and collect terms:

$$I_u = 3.375 \times 10^{-2}a^3 + 6.000 \times 10^{-4}a^2 + 5.333 \times 10^{-6}a$$

(provided that $a \geq 0.04$)

Inspection of the diagrams of Figure 1.25 reveals that two cross sections should be investigated. Just to the left and just to the right of B are the two sections of interest. Although the internal torque is zero just to the left of B, the direct shear has a larger magnitude at this section than for the section just to the right of B. It is likely that the section just to the right of B is the critical one, and it will be considered first. At this section, axial force $= Q_1 = 400$ kN, torque $= \frac{3}{4}T = 300$ kN-m, shear $= \frac{3}{4}P = 150$ kN, and bending moment magnitude $= PL/4 = 200(8)/4 = 400$ kN-m.

An element taken from the top of the member will be subjected to normal stresses associated with axial and bending effects as well as torsion effects but will be free of direct shear effects.

$$\sigma_x = \frac{Q_1}{A} + \frac{M_u v}{I_u}$$

$$= \frac{400}{0.10a} + \frac{400[(1.5a/2) + 0.02]}{3.375 \times 10^{-2}a^3 + 6.000 \times 10^{-4}a^2 + 5.333 \times 10^{-6}a}$$

$$= \frac{4000}{a} + \frac{300a + 8}{3.375 \times 10^{-2}a^3 + 6.000 \times 10^{-4}a^2 + 5.333 \times 10^{-6}a}$$

The section is closed and will be assumed to be thin-walled at this stage, which means that Eq. 4.32 applies.

$$\tau = \frac{T}{2At}$$

Note that A refers to the mean of the outside and inside areas for the cross section.

$$\tau_{vx} = \frac{300}{2\frac{1}{2}[a(1.5 + 0.04) + (a - 0.04)(1.5a)](0.02)}$$

$$= \frac{15,000}{a(3a - 0.02)} \qquad \text{(provided that } a \geq 0.04\text{)}$$

A Mohr's circle construction will provide the following equation for maximum normal stress:

$$\sigma_1 = \frac{\sigma_x}{2} + \sqrt{\left(\frac{\sigma_x}{2}\right)^2 + \tau_{vx}^2} = 250 \times 10^6$$

Upon substitution from above, this equation would contain a single unknown a. Stresses must be multiplied by 10^3 to account for the fact that forces, moments, and torques are expressed in kN units. Solve this equation by iteration as follows:

a	σ_x	τ_{vx}	Left-Hand Side of Equation (MPa)
0.20	250.5	129.3	305.3 > 250. ∴ a must be increased
0.25	162.6	82.2	196.9 < 250. ∴ a must be decreased
0.225	199.3	101.8	242.1 < 250. ∴ a must be decreased
0.224	201.0	102.7	244.2
0.221	206.3	105.6	250.8
0.222	204.5	104.6	248.5
0.2214	205.6	105.2	249.9

Temporarily, let $a = 0.2214$ m. Note that $\tau_{vx} = 105.2$ MPa is well below the allowable value of 140.0 MPa. Next, a check of a stress element at the neutral axis will be made at this cross section for which: axial force $= Q_1 = 400$ kN, torque $= 300$ kN-m, and shear $= 150$ kN.

$$\sigma_x = \frac{Q_1}{A}$$

$$= \frac{400}{0.1(0.2214)} = 1.81 \times 10^4 \text{ kN/m}^2$$

$$\tau_{vx} = \frac{V_v Q}{b I_u} + \frac{T}{2At}$$

where

$$I_u = 3.375 \times 10^{-2}(0.2214)^3 + 6.000 \times 10^{-4}(0.2214)^2$$
$$+ 5.333 \times 10^{-6}(0.2214) = 3.9687 \times 10^{-4} \text{ m}^4$$

$$Q = 0.02a\left(\frac{1.5a}{2} + 0.01\right) + 2\left(\frac{1.5a}{2}\right)(0.02)\left(\frac{1.5a}{4}\right)$$
$$= 2.625 \times 10^{-2}a^2 + 2 \times 10^{-4}a = 1.33 \times 10^{-3} \text{ m}^3$$

$$b = 0.04 \text{ m}$$

$$A = \frac{a}{2}(3a - 0.02) = \left(\frac{0.2214}{2}\right)[3(0.2214) - 0.02] = 713.1 \times 10^{-4} \text{ m}^2$$

$$\tau_{vx} = \frac{150(1.33 \times 10^{-3})}{0.04(3.9687 \times 10^{-4})} + \frac{300}{2(713.1 \times 10^{-4})(0.02)}$$

$$= 11.78 \times 10^4 \text{ kN/m}^2$$

A Mohr's circle construction will not be required in this case, since $\sigma_x = 18.1$ MPa and $\tau_{vx} = 117.8$ MPa would not result in maximum values approaching the allowable stresses. Furthermore, observation of all values computed above leads to the conclusion that investigation of stresses on a section to the left of B will not be required.

Select $a = 0.2214$ m for this preliminary design. *This box section will be fabricated of two flange plates each* $0.2214 \times 0.02 \times 8.00$ m *and two web plates each* $0.3321 \times 0.02 \times 8.00$ m. A revised selection for the final design would depend upon available plate sizes, welding procedures, and other factors not considered in this preliminary design. The a/t ratio is $0.2214/0.02 = 11.07$, which is large enough to justify the use of the equation for torsion of thin-walled members. Further analyses would be required before approval of the final design.

Homework Problems

Neglect the weight of the member in each of the following design problems.

10.41 Refer to Example 10.16 and redesign the member as a hollow circular cylindrical one such that the inside diameter is 0.8 of the outside diameter.

10.42 Refer to Example 10.16, increase the axial force to $P = 200$ kN and the torque to $T = 40$ kN-m and redesign the member. Use a solid circular cylindrical member.

10.43 Refer to Example 10.16 and redesign the member for allowable stress of 300 MPa in tension or compression and 175 MPa in shear. Assume that deformations are not critical in this case.

10.44 Refer to Example 10.17 and redesign the member as a hollow circular cylindrical one such that the inside diameter is 0.75 of the outside diameter.

10.45 Refer to Example 10.17 and redesign the member for allowable stresses of 200 MPa in tension or compression and 140 MPa in shear.

10.46 Refer to Example 10.17, increase the design axial force P_1 to 400 kN and the transverse force P to 200 kN, and redesign the component using a plate thickness of 0.02 m. Retain the square box cross section.

10.47 Refer to Example 10.18 and redesign the member for allowable stresses of 300 MPa in tension or compression and 200 MPa in shear.

10.48 Refer to Example 10.18, increase the axial force $Q = Q_1$ to 800 kN, the torque to $T = 800$ kN-m, and the transverse force to $P = 400$ kN and redesign the member using a plate thickness of 0.04 m.

Chapter 11 Introduction to System Design

★11.1 Introduction

Chapter 11 serves as a means of integrating all the analytical tools developed in Chapters 1 to 9 with the basic component design concepts introduced in Chapter 10. It develops an introductory treatment of the concepts and techniques underlying the design of relatively simple two- and three-dimensional structural systems. Since structural systems consist of structural components that are interconnected in some manner, a discussion is given in Section 11.2 for the design of two types of connections or joints commonly used to transmit axial loads. These are the high-strength bolted connection and the fillet welded connection. Sections 11.3 and 11.4 then discuss the analytical tools as well as the design concepts needed, respectively, for two- and three-dimensional statically determinate structural systems. Finally, some computer-aided design concepts are given in Section 11.5 to provide the student with added capability, particularly as it relates to the handling of statically indeterminate structural systems.

★11.2 Design of Connections to Transmit Axial Loads

In this section, high-strength bolted and fillet welded joints will be designed for axially loaded members. Concepts used in designing these connections form the basis for design of other more complex connections. These concepts are relatively easy to understand and apply and the resulting connections are safe and economical. The allowable stresses have been carefully chosen after a thorough review of extensive statistical analyses of experimental results from a large number of tests of various types of connections. Design of connections is more conservative than the design of members of structures by a substantial margin. Fastener failure loads are still well above the maximum yield capacity of the connected material.

Numerous high-strength bolts or many lineal feet of welding are required to connect the members of a structure together. For example, the Verrazano-Narrows Bridge, connecting Brooklyn and Staten Island, is one of the largest structures ever

536

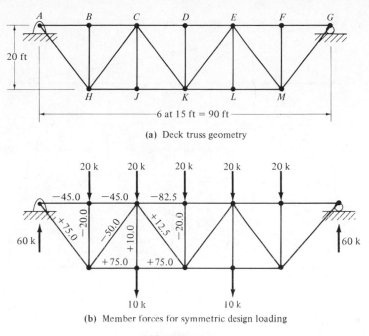

(a) Deck truss geometry

(b) Member forces for symmetric design loading

FIGURE 11.1

high-strength bolted. Bethlehem Steel Corporation supplied over 2 million high-strength bolts for the two 70-story-high towers alone.

The deck truss shown in Figure 11.1 has been analyzed for the design loading shown and will be connected together with high-strength bolts. Figure 11.2 depicts a truss analyzed for the design loading shown and will be connected together by welding.

Two basic types of high-strength bolts are those specified under ASTM (American Society for Testing Materials) designations A325 and A490. All high-strength bolts are heat-treated by quenching and tempering. The A325 Type 1 bolt is produced from medium-carbon steel and the A490 bolt is produced from alloy steel.

High-strength bolted connections to transmit axial forces are either friction- or bearing-type connections. High-strength bolts are able to transmit large forces in friction between the connected parts because the bolts are torqued until they carry large tensile forces which are transmitted as large compression forces which act normal to the connected parts. Figure 11.3 shows typical curves for bolt tension versus bolt elongation for a $\frac{7}{8}$-in.-diameter A325 bolt. Axial tension tests of full-size bolts are normally used to determine mechanical properties of high-strength bolts. The proof load and tensile strength are specified for different-size bolts. In normal use, each bolt is preloaded by turning the nut against the resistance of the connection materials. The tightening induces shear stresses due to torsion as well as tensile stresses, and this reduces the ultimate stress up to a maximum of 30 percent. After a bolt is installed by tightening the nut and then loaded by an applied tension force, the

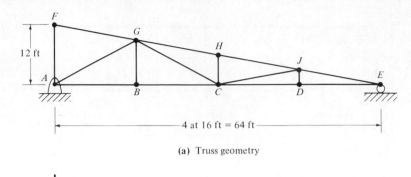

(a) Truss geometry

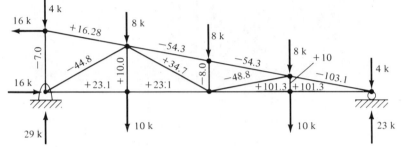

(b) Member forces for design loading

FIGURE 11.2

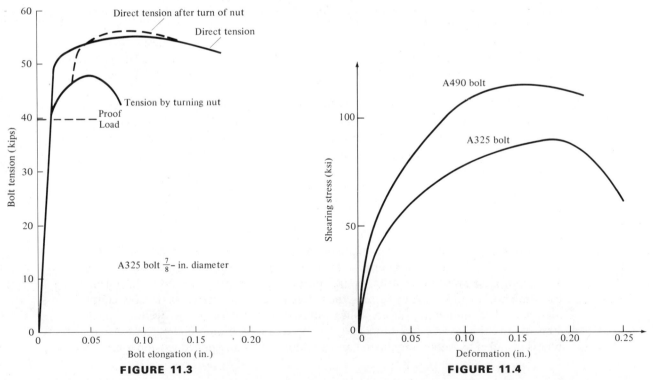

FIGURE 11.3

FIGURE 11.4

bolt is able to resist the full tensile strength, as shown in Figure 11.3. This supports the long-accepted conclusion that a bolt which does not break while being tightened is a "good" bolt. In friction-type joints, the bolts are not actually loaded in shear. However, American design practice equates slip resistance to an equivalent shear stress. Shearing stress deformation curves are shown in Figure 11.4 for an A325 and an A490 high-strength bolt. Increased shear strength of the A490 bolt accompanies increased tensile strength of the A490 bolt when compared to the A325 bolt. A slight decrease in ductility of the A490 bolt compared to A325 bolt is evident in Figure 11.4. Fastener shearing strength is also influenced by the location of the shearing plane. Failure through the threads results in a lower shearing strength due to the reduced area.

A high-strength bolted joint will be designed in Example 11.1 and a fillet welded joint will be designed in Example 11.2. Fillet welds transmit forces through shearing stresses, and appropriate methods for determining fillet weld lengths will be discussed in the example.

Example 11.1

Refer to the deck truss and the design member forces shown in Figure 11.1 and design a high-strength bolted joint to resist the forces at joint H. Consider a bearing-type connection with the bolt threads excluded from the shear planes. Use $\frac{7}{8}$-in.-diameter A325 bolts for which the shearing stress allowable is 22 ksi for this application. All double-angle members and $\frac{1}{2}$-in.-connection plates are to be fabricated of steel with a yield stress of 36 ksi. Determine the number of high-strength bolts required to transmit the design forces in each member and draw a sketch of the joint that would be helpful in preparing the design drawings for this structure. Use an allowable bearing stress on the projected area of bolts in bearing-type connections equal to $1.35\sigma_y$, where σ_y is the yield stress of the connected member.

Solution. Determine the design values for a single high-strength bolt. Since double angles are used back to back with a $\frac{1}{2}$-in. connection plates between them, the high-strength bolts are in double shear.

$$\text{Double shear resistance of each bolt} = 2 \times \frac{\pi}{4}(0.875)^2 \times 22 = 26.5 \text{ k}$$

$$\text{Bearing stress allowable} = 1.35\sigma_y = 1.35 \times 36 = 48.6 \text{ ksi}$$

Use the projected area of a nominal-bolt-diameter bearing on a $\frac{1}{2}$-in. connection plate to find the bearing resistance per bolt.

$$\text{Bearing resistance on connection plate} = 48.6 \times 0.875 \times 0.5 = 21.3 \text{ k}$$

Since $21.3 < 26.5$, the bearing is critical.

From Figure 11.1 at joint H, member HA resists a tensile force of 75.0 k, member HB resists a compressive force of 20.0 k, member HC resists a compressive force of 50.0 k, and member HJ resists a tensile force of 75.0 k. Division of these forces by the critical bearing force of 21.3 k/bolt will give the required number of A325 high-strength bolts required to transmit the member forces at joint H.

Member HA: $\dfrac{75.0}{21.3} = 3.52$, say four bolts

Member HB: $\dfrac{20}{21.3} = 0.94$, say two bolts

Member HC: $\dfrac{50}{21.3} = 2.35$, say three bolts

Member HJ: $\dfrac{75}{21.3} = 3.52$, say four bolts

A design sketch of joint H for this deck truss is shown in Figure 11.5. Although one bolt would connect member HB satisfactorily, when the design

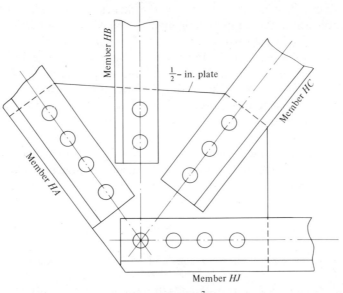

All high–strength bolts $\frac{7}{8}$– in. diameter

All members consist of two angles

FIGURE 11.5
Joint H of Deck Truss.

sketch was prepared the decision was made to use two bolts for member *HB*. It is good practice to avoid one-bolt connections since failure of the single connector leads to loss of the connected member and perhaps to failure of the entire structure. Some revision of this joint design may be required before the final detail drawings are prepared for fabrication and erection of this truss. Bolt spacings and edge distances would be shown on these final drawings.

Example 11.2

Refer to the truss and the design member forces shown in Figure 11.2. Design the connection at joint *H* as a welded one using an allowable shearing stress of 18.0 ksi on the throat of fillet welds to be used. Assume the top chord of the truss is to be spliced at this point rather than being continuous through the joint. Truss members consist of two angles back to back separated by $\frac{1}{2}$-in.-thick spacers and small gusset plates at the joints. Determine the size and lengths of welds required and draw a sketch of the joint which would be helpful in preparing the design drawings of the structure.

Solution. The basic equation $\tau = F/A$ will be used to design the welds for the members *GH*, *HJ*, and *HC* which meet a joint *H*. The allowable shearing stress $\tau = 18$ ksi is to be used regardless of the direction of the applied force F with respect to the cross-sectional area A. An analysis of the truss was performed, using the method of joints, in order to find the axial forces F in each member, as shown in Figure 11.2(b). Members *GH* and *HJ* must each resist a force of 54.3 kips, and member *HC* must resist a force of 8 k. The theoretical throat dimension of the fillet weld is 0.707t, as depicted in the lower left of Figure 11.6. The area A equals the theoretical throat dimension multiplied by the required length of the fillet weld:

$$A = 0.707tL$$

where t is the size of the fillet weld. Assume that $t = \frac{3}{8}$ in. for the fillet welds at this joint and apply the equation $\tau = F/A$.

MEMBERS *GH* AND *HJ*:

$$18 = \frac{54.3}{0.707(\frac{3}{8})L}$$

Solving for L: $L = 11.38$ in., say 12 in. Two angles composing member *GH* will have four legs, which gives $\frac{12}{4} = 3$ in. for each leg. Likewise, each of the four legs of the two angles composing member *HJ* would be connected with 3 in. of

Member	$\frac{3}{8}$- in. fillet weld length
GH	6 in. each side
HG	6 in. each side
HC	1 in. each side

All members—double angles
$\frac{1}{2}$- in. connection plate

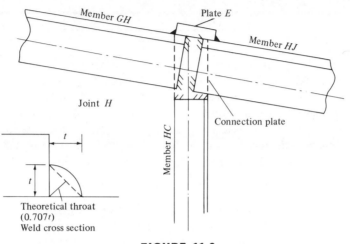

FIGURE 11.6

$\frac{3}{8}$-in. fillet weld to the connection plate and plate E of Figure 11.6. Plate E would be proportioned such that it and the $\frac{1}{2}$-in. connection plate would safely transmit the full force of 54.3 k in the top chord.

MEMBER HC: Again applying the equation $\tau = F/A$:

$$18 = \frac{8}{0.707(\frac{3}{8})L_1}$$

Solving for L_1: $L_1 = 1.67$ in., say 2 in. Two angles composing member HC will have four legs, which gives $\frac{2}{4} = 0.5$ in. for each leg if all four legs are connected. If only two legs are connected, as shown in Figure 11.6, each connected leg would be welded along a 1-in. length. The $\frac{1}{2}$-in. connection plate would also be checked for stress to see that it could safely carry the 8-k design load of member HC delivered through the welds shown at the top of member HC.

The welded connection shown in Figure 11.6 would be revised prior to its inclusion on the design drawing and then would appear fully detailed on the detail drawings from which the truss is fabricated and erected. A decision would be made as to which welds would be made in the shop and which ones would be made in the field when the truss is erected. Plate E and the connection plate would be proportioned to transmit the forces across the joint, and additional plate(s) may be required on the outstanding legs of member *HC*. Completion of these design details is not possible until the members have been designed and the problem is reviewed in the overall perspective of fabrication and erection of the complete structure. In fact, the idea of using fillet welds may be discarded and the top chord splice made by using butt welds. Selection of appropriate electrodes and welding procedures would also be part of a complete design of this welded truss.

Homework Problems

11.1 Refer to Figure 11.1 and use the method of sections to verify that member *CD* must carry a compressive design force of 82.5 k. Consider that the top chord of this deck truss will be spliced at joint *D* and find the number of A325 high-strength bolts required to transmit the force in member *CD* at joint *D*. Assume this is to be a friction-type connection for which bearing stress is not restricted. The allowable shearing stress on the $\frac{3}{4}$-in.-diameter high-strength bolts is 15.0 ksi and the bolts are in double shear.

11.2 Solve Problem 11.1 for a bearing-type connection. Consider a connection plate $\frac{1}{2}$ in. thick with a yield stress of 36 ksi. Assume that the combined thickness of the two angles of the connected member exceeds the plate thickness of $\frac{1}{2}$ in., which means that the bearing on the plate is critical. Use an allowable bearing stress of 1.35 times the yield stress of the connected material and assume shear conditions as for Problem 11.1.

11.3 Refer to Figure 11.1 and use the method of sections to verify that member *HC* must carry a compressive force of 50 kips. Find the number of A325 high-strength bolts required to transmit the force in member *HC* at joint *C*. Assume this to be a friction-type connection for which bearing stress is not restricted. The allowable shearing stress on the $\frac{5}{8}$-in.-diameter high-strength bolts is 15.0 ksi and the bolts are in double shear.

11.4 Solve Problem 11.3 for a bearing-type connection. Consider a connection plate $\frac{1}{2}$ in. thick with a yield stress of 36 ksi. Assume that the combined thickness of the two angles of the connected member exceeds the plate thickness of $\frac{1}{2}$ in., which means that bearing on the plate is critical. Use an allowable bearing stress of 1.35 times the yield stress of the connected material and assume shear conditions as for Problem 11.3.

11.5 Consider joint *A* of the deck truss of Figure 11.1 and verify that member *AB* must carry a compressive force of 45 k. Find the number of A325 high-strength bolts required to transmit the force in member *AB* at joint *A*. Assume this to be a friction-type connection for which bearing stress is not restricted. The allowable stress on the $\frac{3}{4}$-in.-diameter high-strength bolts is 15.0 ksi, and the bolts are in double shear.

11.6 Solve Problem 11.5 for a bearing-type connection. Consider a connection plate $\frac{1}{2}$ in. thick with a yield stress of 36 ksi. Assume that the combined thickness of the two angles of the connected member exceeds the plate thickness of $\frac{1}{2}$ in., which means that bearing on the plate is critical. Use an allowable bearing stress of 1.35 times the yield stress of the connected material and assume shear conditions as for Problem 11.5.

11.7 Refer to Figure 11.2 and use the method of sections to verify that member *CD* carries an axial tension force of 101.3 k. Assume that the bottom chord is to be spliced at joint *C* and determine the total length of $\frac{3}{8}$-in. fillet weld required to transmit the force in member *CD*. Use an allowable shearing stress of 18 ksi on the fillet welds regardless of the direction of the applied forces.

11.8 Consider joint *A* of the truss shown in Figure 11.2 and verify by the method of sections that member *AG* carries a compressive force of 44.8 k.

Determine the total length of $\frac{3}{8}$-in. fillet required to transmit the force in member *AG*. Use an allowable shearing stress of 18 ksi on the fillet welds regardless of the direction of the applied forces.

11.9 Refer to Figure 11.2 and use the method of sections to verify that member *BC* carries a force of 23.1 k. Assume that the bottom chord is to be spliced at joint *C* and determine the total length of $\frac{3}{8}$-in. fillet weld required to transmit the force in member *BC*. Use an allowable shearing stress of 18 ksi on the fillet welds regardless of the direction of the applied forces.

★**11.3** Design of Two-Dimensional Systems

Many structural systems can be treated as two-dimensional systems in determining the forces acting on their various components. In this section relatively simple statically determinate two-dimensional structural systems are analyzed to determine the forces acting on certain components of these structures. These components are then designed using the procedures that have been developed in Chapter 10. No attempt will be made in this section at designing the connections between the various components of a given system. The design of connections, both bolted and welded, has already been discussed in Section 11.2.

The concepts underlying the design of statically determinate two-dimensional structural systems will be illustrated in the following examples.

Example 11.3
Members *EC* and *BC* of the plane truss shown in Figure 11.7(a) are to be made of structural steel and are to be designed to carry the loads as shown. These two members of the truss are to have S (standard I) sections. Select suitable sizes for these S sections assuming that the deformations of the truss are not critical design considerations. Assume an allowable stress of 100 MPa.

Solution. Before members *CE* and *CB* can be properly designed, the forces acting on them have to be determined. This is accomplished by a force analysis of the truss. Generally, the force analysis on a truss is performed with the assumption that the various members of the truss are two-force members joined together at their ends by frictionless pins, even though the joints may actually be bolted or welded.

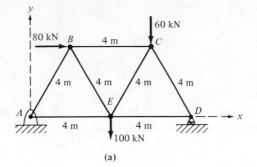

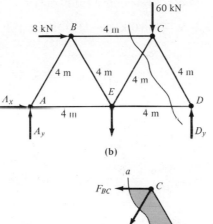

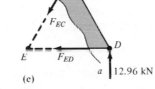

FIGURE 11.7

Consider the free-body diagram of the entire truss as shown in Figure 11.7(b). The external reactions are determined as follows:

$\circlearrowleft \quad \sum M_A = 0$

$$D_y(8) - 100(4) - 60(6) - 80(4 \cos 30) = 0$$

$$D_y = 129.6 \text{ kN}\uparrow$$

$\rightarrow \quad \sum F_x = 0$

$$80 + A_x = 0$$

$$A_x = -80 = 80 \text{ kN}\leftarrow$$

$$\uparrow \quad \sum F_y = 0$$

$$A_y + 129.6 - 100 - 60 = 0$$

$$A_y = 30.4 \text{ kN}\uparrow$$

The internal axial forces acting on members BC and EC may be determined by considering the equilibrium of the free-body diagram of that section of the truss to the right of plane a–a, as shown in Figure 11.7(c). Thus

$$\circlearrowleft \quad \sum M_E = 0$$

$$F_{BC}(4 \cos 30) - 60(2) + 129.6(4) = 0$$

$$F_{BC} = -115.0 \text{ kN} = 115.0 \text{ kN}(C)$$

$$\uparrow \quad \sum F_y = 0$$

$$129.6 - 60 - F_{EC} \cos 30 = 0$$

$$F_{EC} = 80.4 \text{ kN}(T)$$

DESIGN OF MEMBER EC: By Eq. 3.1,

$$A = \frac{F_n}{\sigma} = \frac{80.4 \times 10^3}{100 \times 10^6} = 8.04 \times 10^{-4} \text{ m}^2$$

Since structural steel sections have not yet been standardized in SI units, this value of A will be transformed into inch units in order to be able to use the data in Appendix E. Therefore,

$$A = 8.04 \times 10^{-4} \text{ m}^2 = 1.25 \text{ in.}^2$$

Examination of Appendix E reveals that the smallest standard S section satisfying the cross-sectional area requirement is the S3 × 5.7, whose area is 1.67 in.^{2} = 10.77 × 10^{-4} m^2. Thus member EC is 4 m long, as shown in Figure 11.7(a), and would have to have a standard S section whose designation is

$$\boxed{\text{S3} \times 5.7}$$

DESIGN OF MEMBER BC: Member BC is subjected to a compressive force of 115.0 kN, and therefore it would have to be designed as a column. The ends of this column may be assumed to be pinned. Therefore, the effective length of the column is 4 m. Let $E = 200 \times 10^9$ N/m^2.

Assume an S3 × 5.7 standard section. From Appendix E, the following properties are obtained:

$$A = 1.67 \text{ in.}^2 = 10.77 \times 10^{-4} \text{ m}^2$$

$$r_v = 0.522 \text{ in.} = 1.33 \times 10^{-2} \text{ m}$$

Therefore,

$$\frac{L_e}{r_v} = \frac{4}{1.33 \times 10^{-2}} = 300.75$$

Thus, according to the design information in Figure 8.14(a), this is a long column and the applicable equation is

$$\frac{P}{A} = \frac{\pi^2 E}{1.82(L_e/r_v)^2} = \frac{1084.57 \times 10^9}{(L_e/r_v)^2} \text{ N/m}^2$$

For $L_e/r_v = 300.75$,

$$\frac{P}{A} = \frac{1084.57 \times 10^9}{(300.75)^2} = 11.99 \text{ MPa}$$

Therefore, $P = 11.99(A) = 12.91$ kN, which is considerably less than the required load-carrying capacity.

Assume an S12 × 50 standard section.

$$A = 14.7 \text{ in.}^2 = 94.84 \times 10^{-4} \text{ m}^2$$

$$r_v = 1.03 \text{ in.} = 2.62 \times 10^{-2} \text{ m}$$

$$\frac{L_e}{r_v} = \frac{4}{2.62 \times 10^{-2}} = 152.67$$

which is also a long column and

$$\frac{P}{A} = \frac{1084.57 \times 10^9}{(152.67)^2} = 46.53 \text{ MPa}$$

Therefore, $P = 46.53(A) = 441.3$ kN, which is much larger than needed.

Assume an S8 × 23 standard section.

$$A = 6.77 \text{ in.}^2 = 43.68 \times 10^{-4} \text{ m}^2$$

$$r_v = 0.798 \text{ in.} = 2.03 \times 10^{-2} \text{ m}$$

$$\frac{L_e}{r_v} = \frac{4}{2.03 \times 10^{-2}} = 197.04$$

which is also a long column and

$$\frac{P}{A} = \frac{1084.57 \times 10^9}{(197.04)^2} = 27.94 \text{ MPa}$$

Therefore, $P = 27.94(A) = 122.0$ kN, which is only slightly higher than the 115.0-kN carrying capacity that is required. However, another try may be made at the next-smaller section, the S8 × 18.4 standard section, to ensure against overdesign. Thus

$$A = 5.41 \text{ in.}^2 = 34.90 \times 10^{-4} \text{ m}^2$$

$$r_v = 0.831 \text{ in.} = 2.11 \times 10^{-2} \text{ m}$$

$$\frac{L_e}{r_v} = \frac{4}{2.11 \times 10^{-2}} = 189.57$$

which is still a long column and

$$\frac{P}{A} = \frac{1084.57 \times 10^9}{(189.57)^2} = 30.18 \text{ MPa}$$

Therefore, $P = 30.18(A) = 105.3$ kN, which is inadequate. Therefore, the needed section for member BC is

$$\boxed{\text{S8} \times 23}$$

Example 11.4

The frame shown in Figure 11.8(a) is to be constructed from structural steel ($E = 30 \times 10^6$ psi) for which the yield strength is 50,000 psi in tension or compression and 30,000 psi in shear. Design member EC, which is to be fabricated by welding two American Standard channels as shown in the sectional view in Figure 11.8(a). Also design member DF, which is to have an S section. Assume that the deformations of the frame are not significant design factors. Use a factor of safety $N_y = 2$ based upon failure by yielding for members of the frame other than columns.

Solution. A force analysis of the frame is needed to determine the forces acting on members CE and DF. Consider the free-body diagram of the entire frame as shown in Figure 11.8(b).

$$\circlearrowleft \quad \sum M_B = 0$$

$$F_y(12) - 50(4) - 30(12) = 0$$

$$F_y = 46.667 \text{ k} \uparrow$$

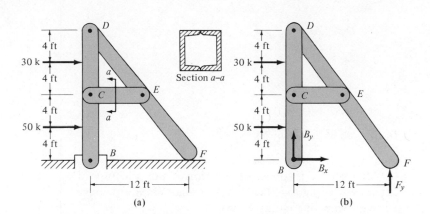

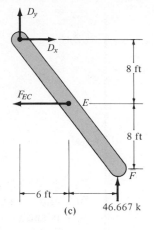

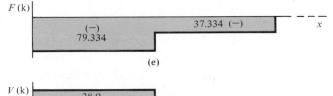

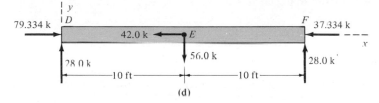

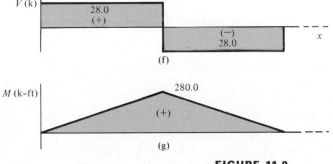

FIGURE 11.8

Consider now the free-body diagram of member DF, as shown in Figure 11.8(c).

$$\circlearrowleft \ \sum M_D = 0$$

$$46.667(12) - F_{EC}(8) = 0$$

$$F_{EC} = 70 \ k(T)$$

$$\rightarrow \ \sum F_x = 0$$

$$D_x - 70 = 0$$

$$D_x = 70 \text{ k} \rightarrow$$

$$\uparrow \ \sum F_y = 0$$

$$46.667 + D_y = 0$$

$$D_y = -46.667 = 46.667 \text{ k} \downarrow$$

DESIGN OF MEMBER EC: Member EC is to resist the action of a tensile force of 70 k and is to be fabricated by welding together two American Standard channels as shown in the sectional view of Figure 11.8(a).

The allowable stress for the material is

$$\frac{\sigma_y}{N_y} = \frac{50,000}{2} = 25,000 \text{ psi}$$

Therefore, the total (i.e., for both channels) required cross-sectional area is given by the relation

$$A = \frac{F_n}{\sigma} = \frac{70 \times 10^3}{25,000} = 2.80 \text{ in.}^2$$

Thus the cross-sectional area per channel is $2.80/2 = 1.40$ in.^{2}. Reference to Appendix E reveals that a $C3 \times 5$ channel provides the needed properties with a cross-sectional area of 1.47 in.^{2}. Thus the required American Standard channel is

$$\boxed{C3 \times 5}$$

DESIGN OF MEMBER DF: The various forces acting on member DF have been decomposed into components parallel or perpendicular to this member and the results are shown in Figure 11.8(d), on a sketch of the member which has been rotated into a horizontal position for convenience. The force, shear, and moment diagrams of this member are shown, respectively, in Figure 11.8(e), (f), and (g).

Member DF will be designed to resist the combined action of axial and flexural stresses and the resulting cross section, then checked to ensure that it can safely resist the shear and compressive buckling loads. The pin hole at point E will be ignored in the design process.

By Eq. 7.1,

$$\sigma_x = \frac{F_n}{A} + \frac{M_u}{I_u}v = \frac{F_n}{A} + \frac{M_u}{I_u/v} = \frac{F_n}{A} + \frac{M_u}{Z}$$

The most highly stressed points in the member are at the top of the member an infinitesimal distance to the left of point E, where $F_n = 79.334$ k $= 79,334$ lb, and $M_u = 280$ k-ft $= 3360 \times 10^3$ lb-in.

Assume an S24 × 100 section. From Appendix E, one obtains the following properties:

$$A = 29.4 \text{ in.}^2$$
$$Z = 199 \text{ in.}^3$$

Therefore,

$$\sigma_x = -\frac{79,334}{29.4} - \frac{3360 \times 10^3}{199} = -19,583 \text{ psi}$$

Since the allowable stress is $50,000/2 = 25,000$ psi, the stress computed for the S24 × 100 section is well within the allowable. However, one needs to examine a few smaller sections to ensure that the member is not overdesigned. Thus assume an S20 × 95 section.

$$A = 27.9 \text{ in.}^2$$
$$Z = 161 \text{ in.}^3$$
$$\sigma_x = \frac{-79,334}{27.9} - \frac{3360 \times 10^3}{161} = -23,713 \text{ psi}$$

Next, assume an S20 × 85 section.

$$A = 25.0 \text{ in.}^2$$
$$Z = 152 \text{ in.}^3$$
$$\sigma_x = -\frac{79,334}{25.0} - \frac{3360 \times 10^3}{152} = -25,279 \text{ psi}$$

This stress is slightly above the allowable of 25,000 psi. Therefore, the required S section for this application is the S20 × 95 for which the normal stress σ_x at the top of the member was found to be 23,713 psi compression. By constructing Mohr's circle for the stress element at this location, we conclude that the three principal stresses at the top of the member are $\sigma_1 = \sigma_2 = 0$ and $\sigma_3 = -23,713$ psi. Therefore, by Eq. 2.13b,

$$|\tau_{max}| = \frac{\sigma_1 - \sigma_3}{2} = 11{,}857 \text{ psi}$$

which is well within the allowable shear stress, $30{,}000/2 = 15{,}000$ psi.

The S20 × 95 section will also be checked to ensure that it is adequate to carry the compressive buckling loads as well as the shear forces.

Member DF, as shown in Figure 11.8(d), may be assumed to consist of two columns, DE and EF, each with pinned ends and having a length of 10 ft. The most severely loaded column is DE, which carries a load of 79.334 kips. Using the S20 × 95 section, we determine the following properties from Appendix E.

$$A = 27.9$$

$$r_v = 1.33 \text{ in.}$$

$$\frac{L_e}{r_v} = \frac{10 \times 12}{1.33} = 90.23$$

and according to Figure 8.14(b), member DE is a short or intermediate column for which the applicable equation is

$$\frac{P}{A} = 22{,}000 - 0.74\left(\frac{L_e}{r_v}\right)^2$$

Therefore,

$$\frac{P}{A} = 22{,}000 - 0.74(90.23)^2 = 15{,}980 \text{ psi} = 15.98 \text{ ksi}$$

and

$$P = 15.98 \times 27.9 = 445.8 \text{ k}$$

which is more than adequate to resist the applied compressive buckling load of 79.334 k.

By Eq. 5.12a,

$$\tau_{xv} = \frac{V_v}{bI_u}Q$$

Here V_v is the maximum shear force, which in this case is 28.0 k; b is the thickness of the flange and I_u is the moment of inertia about the u principal centroidal axis, which from Appendix E are found to be 0.800 in. and 1610 in.[4], respectively. Since the maximum value of τ_{xv} is desired, the quantity Q is the first moment of the area to one side of the u axis about this axis. The value of Q, however, can only be determined approximately from the data provided in

Appendix E. Nevertheless, this approximate value of Q will be sufficiently adequate for the purpose of checking the adequacy of the S20 × 95 section. Thus

$$\tau_{xv} = \frac{28.0 \times 10^3}{0.800 \times 1610} \left[(10.0 - 0.916)(0.800)\left(\frac{1}{2}\right)(10.0 - 0.916) \right.$$
$$\left. + (0.916)(7.200)(10.0 - 0.458) \right] \doteq 2086 \text{ psi}$$

Thus the maximum horizontal or vertical shear stress occurring in the member is well within the allowable 15,000 psi.

Therefore, the required section for the member shown in Figure 11.8(d) is

$$\boxed{\text{S20} \times 95}$$

Homework Problems

11.10 Design members AB of the truss in Figure 11.7(a). Assume that this member is to be fabricated from a structural steel pipe for which the design information contained in Figure 8.14 is applicable if the member is subjected to compressive loads. Use an allowable stress of 200 MPa if the member is subjected to tensile instead of compressive loads.

11.11 Repeat Problem 11.10 for member AE of the truss in Figure 11.7(a).

11.12 Repeat Problem 11.10 for member BE of the truss of Figure 11.7(a). Assume, however, that this member is to be fabricated by welding two American Standard channels together as shown in the sectional view of Figure 11.8(a).

11.13 Repeat Problem 11.10 for member CD of the truss of Figure 11.7(a). Assume the member to have a rectangular cross section whose width b is one-half of its height h.

11.14 Refer to Example 11.4 and Figure 11.8 and design member BD. Assume the member to have an S (Standard I) section. All other information in Example 11.4 is applicable.

11.15 Refer to the frame shown in Figure H3.6. The two-force member BE is to be designed of an aluminum alloy ($E - 10 \times 10^6$ psi) for which the allowable stress in tension or compression is 15,000 psi and the allowable deformation is 0.1 in. Determine the necessary cross-sectional area.

11.16 Refer to the frame shown in Figure H3.7, and design member DC. Assume the cross section of this member to be a pipe whose inside diameter is three-fourths of its outside diameter. Assume also that the material is steel for which the allowable stress is 200 MPa in tension or compression and 150 MPa in shear. Assume that the deformations of the frame are not significant design considerations. Check for column action using the information in Figure 8.14.

11.17 Refer to the frame shown in Figure H3.7 and design member *BE*. Assume the material to be a magnesium alloy $(E = 45 \times 10^9 \text{ N/m}^2)$ for which the yield strength is 250 MPa in tension or compression and 150 MPa in shear. Assume a factor of safety $N_y = 2$ based upon failure by yielding and a solid circular cross section for the member. Ignore any column action that may be present in the member and assume the deformations of the member to be insignificant design factors.

11.18 Design member *AB* of the frame shown in Figure H11.18. The material is to be structural steel for which the allowable stress is 20 ksi in tension or compression and 12 ksi in shear. Assume an S (Standard I) section for this member and ignore any deformations of the member.

11.19 Repeat Problem 11.18 for member *BC*. All other information remains the same.

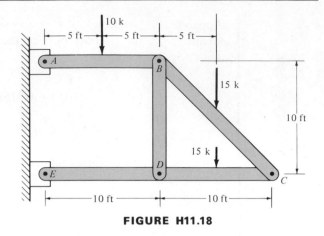

FIGURE H11.18

11.20 Repeat Problem 11.18 for member *BD*. Assume the cross section of this member to be a structural steel pipe for which design information is available in Appendix E. All other information in Problem 11.18 remains the same.

★**11.4** Design of Three-Dimensional Systems

In Section 11.3 two-dimensional structural systems were discussed. The force analysis for such systems was performed, using the scalar-algebra method, prior to undertaking the design process for each individual member of the system.

The scalar-algebra method has serious limitations when applied to the force analysis of three-dimensional structural systems. Some of these limitations are eliminated by the use of the vector-algebra method, which is a much more effective and much more elegant formulation of the three-dimensional force problem.

Only simple statically determinate three-dimensional structural systems are treated in this section. In general, three-dimensional structural systems consist of members subjected to axial loads and/or to bending and in some cases to torsion. However, since the design of such members was discussed in detail in Chapter 10 and reviewed in Section 11.3, the process will not be repeated here. The examples that follow will be limited to undertaking the three-dimensional force analysis that yields the loads acting on the various members of the system. The Homework Problems, however, will require the student to design individual components as was done in previous sections.

Example 11.5

Consider the frame shown in Figure 11.9(a). The vertical member AC is supported at A by a ball and socket and at C by two flexible steel rods CD and CE which are symmetrically located with respect to the x axis. The force P applied at point C is in the xy plane and makes a 45° angle with the y axis. Design cables CD and CE as well as member AC.

Solution. The solution that follows is limited to the force analysis required to obtain the loads acting on the three members of the frame. The design of these three members forms the basis for some of the Homework Problems.

A free-body diagram of member AC is shown in Figure 11.9(b). The three numbers shown parenthetically adjacent to the letter defining a given point in the structure are, respectively, the x, y, and z coordinates of that point. Note that there are five unknown quantities, and therefore five independent equations are needed for a complete solution. These five equations are obtained using the vector-algebra method.

The first step in a solution, using the vector-algebra method, should be to express all forces in vector form. Thus

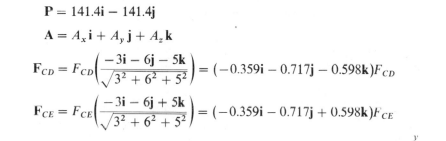

$$\mathbf{P} = 141.4\mathbf{i} - 141.4\mathbf{j}$$

$$\mathbf{A} = A_x\mathbf{i} + A_y\mathbf{j} + A_z\mathbf{k}$$

$$\mathbf{F}_{CD} = F_{CD}\left(\frac{-3\mathbf{i} - 6\mathbf{j} - 5\mathbf{k}}{\sqrt{3^2 + 6^2 + 5^2}}\right) = (-0.359\mathbf{i} - 0.717\mathbf{j} - 0.598\mathbf{k})F_{CD}$$

$$\mathbf{F}_{CE} = F_{CE}\left(\frac{-3\mathbf{i} - 6\mathbf{j} + 5\mathbf{k}}{\sqrt{3^2 + 6^2 + 5^2}}\right) = (-0.359\mathbf{i} - 0.717\mathbf{j} + 0.598\mathbf{k})F_{CE}$$

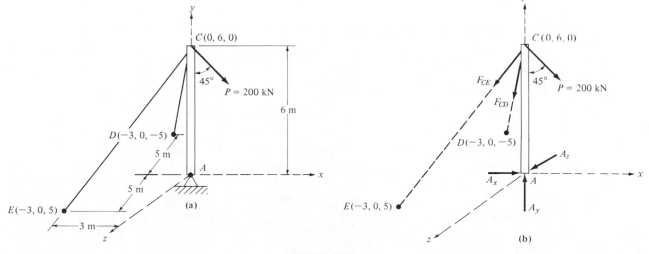

FIGURE 11.9

where boldface type indicates a vector quantity. The symbols $\mathbf{i}$, $\mathbf{j}$, and $\mathbf{k}$ are the unit vectors in the x, y, and z directions, respectively. Letters in italic type represent scalar quantities. Equilibrium requires that the forces acting on the structure form a system equivalent to zero. Thus

$$\sum \mathbf{F} = 0 \quad \text{and} \quad \sum \mathbf{M} = \sum (\mathbf{r} \times \mathbf{F}) = 0$$

where $\mathbf{F}$, $\mathbf{M}$, and $\mathbf{r}$ are symbolic representations of vector forces, vector moments, and vector radii, respectively. The first of the two equilibrium conditions above yields

$$141.4\mathbf{i} - 141.4\mathbf{j} + A_x\mathbf{i} + A_y\mathbf{j} + A_z\mathbf{k} - 0.359\mathbf{i}F_{CD} - 0.717\mathbf{j}F_{CD}$$
$$- 0.589\mathbf{k}F_{CD} - 0.359\mathbf{i}F_{CE} - 0.717\mathbf{j}F_{CE} + 0.598\mathbf{k}F_{CE} = 0$$

Grouping $\mathbf{i}$, $\mathbf{j}$, and $\mathbf{k}$ terms and setting their sum individually equal to zero, we obtain the following three scalar equations:

$$141.4 + A_x - 0.359F_{CD} - 0.359F_{CE} = 0 \qquad \textbf{(a)}$$

$$-141.4 + A_y - 0.717F_{CD} - 0.717F_{CE} = 0 \qquad \textbf{(b)}$$

$$A_z - 0.598F_{CD} + 0.598F_{CE} = 0 \qquad \textbf{(c)}$$

Equations (a), (b), and (c) contain five unknown quantities, A_x, A_y, A_z, F_{CD}, and F_{CE}, and therefore additional equations containing the same unknown quantities are required for a solution. Thus, using the second of the foregoing equilibrium conditions (i.e., $\sum \mathbf{M}_A = 0$), we obtain the following vector equation:

$$6\mathbf{j} \times (141.4\mathbf{i} - 141.4\mathbf{j}) + (6\mathbf{j}) \times (-0.359\mathbf{i} - 0.717\mathbf{j} - 0.598\mathbf{k})F_{CD}$$
$$+ (6\mathbf{j}) \times (-0.359\mathbf{i} - 0.717\mathbf{j} + 0.598\mathbf{k})F_{CE} = 0$$

Performing the indicated vector operations yields the following:

$$-848.4\mathbf{k} + 0 + 2.154\mathbf{k}F_{CD} + 0 - 3.588\mathbf{i}F_{CD} + 2.154\mathbf{k}F_{CE} + 0 + 3.588\mathbf{i}F_{CE} = 0$$

Grouping $\mathbf{i}$ and $\mathbf{k}$ terms and setting their sums individually equal to zero, we obtain the following two additional scalar equations:

$$-3.588F_{CD} + 3.588F_{CE} = 0 \qquad \textbf{(d)}$$

$$2.154F_{CD} + 2.154F_{CE} - 848.4 = 0 \qquad \textbf{(e)}$$

Simultaneous solution of Eqs. (a) to (e) yields the values of the unknowns, A_x, A_y, A_z, F_{CD}, and F_{CE}, as follows:

$$A_x = -0$$

$$A_y = 423.8 \text{ kN}$$

$$A_z = 0$$

$$F_{CD} = 196.9 \text{ kN}$$

$$F_{CE} = 196.9 \text{ kN}$$

As stated earlier, the design of the three members CD, CE, and AC will be treated in Problems 11.21 and 11.22.

Example 11.6

The system shown in Figure 11.10(a) consists of the rigid bent ABD, which is supported at point A by a ball and socket, at point B by a flexible rod BF, and at point D by a compression member DE. Assume the connection at D to be a universal joint and the support at E to be a ball and socket and design member DE.

Solution. As in Example 11.5, only the force analysis required to obtain the loads in member DE will be developed, while the actual design of this member as well as member BF will form the basis for Problems 11.23 and 11.24, respectively.

Consider the free-body diagram of member ABD as shown in Figure 11.10(b). This free-body diagram contains the five unknown quantities A_x, A_y, A_z, F_{BF}, and F_{DE}. The same procedure used in Example 11.5 may be used here to determine these unknown quantities. However, since only the force on member DE (i.e., F_{DE}) is needed, it is possible to obtain a single moment equation that contains this quantity as the only unknown. This equation may be obtained by summing the moments of all forces about line AF. This would obviously eliminate the unknowns A_x, A_y, A_z, and F_{BF} and would contain F_{DE} as the only unknown.

Proceeding as in Example 11.5, we express the forces F_{DE} and P in vector form. Thus

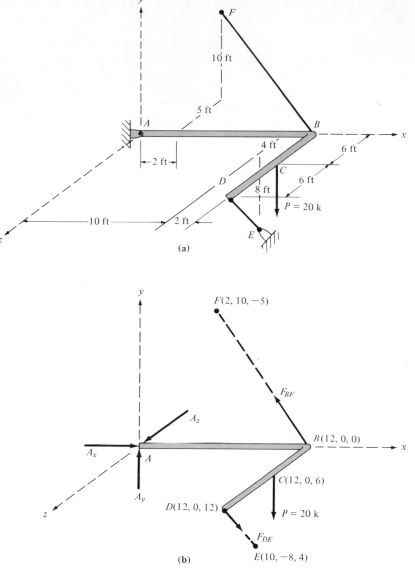

FIGURE 11.10

$$\mathbf{P} = -20\mathbf{j}$$

$$\mathbf{F}_{DE} = F_{DE}\left(\frac{-2\mathbf{i} - 8\mathbf{j} - 8\mathbf{k}}{\sqrt{2^2 + 8^2 + 8^2}}\right) = (-0.174\mathbf{i} - 0.696\mathbf{j} - 0.0696\mathbf{k})F_{DE}$$

The sum of moments of all forces about line AF may be expressed symbolically in vector form as follows:

$$\sum M_{AF} = \sum \lambda_{AF} \cdot \mathbf{M}_A = \sum \lambda_{AF} \cdot (\mathbf{r} \times \mathbf{F}) = 0$$

where λ_{AF} represents a unit vector along line AF and the symbols $\mathbf{r}$, $\mathbf{F}$, and $\mathbf{M}$ represent quantities defined previously. Thus

$$\lambda_{AF} = \frac{2\mathbf{i} + 10\mathbf{j} - 5\mathbf{k}}{\sqrt{2^2 + 10^2 + 5^2}} = 0.176\mathbf{i} + 0.880\mathbf{j} - 0.440\mathbf{k}$$

Therefore,

$$\sum \lambda_{AF} \cdot (\mathbf{r} \times \mathbf{F}) = (0.176\mathbf{i} + 0.880\mathbf{j} - 0.440\mathbf{k})$$
$$\cdot [(12\mathbf{i} + 12\mathbf{k}) \times (-0.174\mathbf{i} - 0.696\mathbf{j} - 0.696\mathbf{k})F_{DE}$$
$$+ (12\mathbf{i} + 6\mathbf{k}) \times (-20\mathbf{j})] = 0$$

and

$$(0.176\mathbf{i} + 0.880\mathbf{j} - 0.440\mathbf{k})$$
$$\cdot [(12\mathbf{i} + 12\mathbf{k}) \times (-0.174\mathbf{i} - 0.696\mathbf{j} - 0.696\mathbf{k})F_{DE}] = -126.72$$

The left-hand side of this equation represents the mixed triple product of three vectors (a scalar quantity) and therefore may be expressed in determinant form. Thus

$$\begin{vmatrix} 0.176 & 0.880 & -0.440 \\ 12 & 0 & 12 \\ -0.174 & -0.696 & -0.696 \end{vmatrix} (F_{DE}) = -126.72$$

which leads to an algebraic equation in F_{DE} as follows:

$$10.65 F_{DE} = -126.72$$

Therefore,

$$F_{DE} = -11.90 \text{ k}$$

The negative sign signifies that the force F_{DE} is opposite to the sense assumed in the free-body diagram of Figure 11.10(b). This fact leads to the conclusion that member DE is in compression and should be designed as a compression member. The design of this compression member will be treated in Problem 11.23.

Homework Problems

11.21 Refer to Example 11.5 and Figure 11.9 and design member *AC*, which is to be made of structural steel and is to have an **S** (Standard I) section. Use the information contained in Figure 8.14(a).

11.22 Refer to Example 11.5 and Figure 11.9 and design members *CE* and *CD*. As stated in Example 11.5, these two members are flexible steel rods. Assume an allowable stress of 200 MPa and an allowable deformation of 4×10^{-2} m. Let $E = 200$ GN/m².

11.23 Refer to Example 11.6 and Figure 11.10. Design member *DE* to resist the compressive load determined in the solution of Example 11.6. Assume that this member is to be a structural steel pipe for which the design information contained in Figure 8.14(b) is applicable.

11.24 Refer to Example 11.6 and Figure 11.10. Determine the force in the flexible rod *BF*. Assume the rod to be steel for which the allowable stress and

deformation are, respectively, 20 ksi and 0.2 in. Determine a suitable diameter for the rod. Let $E = 30 \times 10^6$ psi.

11.25 The frame shown in Figure H11.25 consists of the boom *AC* supported at *A* by a ball and socket and at *B* by two flexible steel rods *BD* and *BE*. The load *P* has a magnitude of 150 kN and is applied in the *xy* plane such that it makes a 20° angle with the *x* axis. Design the boom if it is to be an aluminum pipe whose inside diameter is one-half of its outside diameter. Assume the material to have a yield strength of 300 MPa in tension or compression and 200 MPa in shear. Use a factor of safety of $N_y = 2.5$ based upon failure by yielding. Assume deformations to be insignificant design factors.

11.26 Refer to Problem 11.25 and Figure H11.25 and design the flexible rods *BE* and *BD*. The rods are steel for which the allowable stress and deformation are, respectively, 400 MPa and 1×10^{-2} m. Let $E = 200$ GN/m².

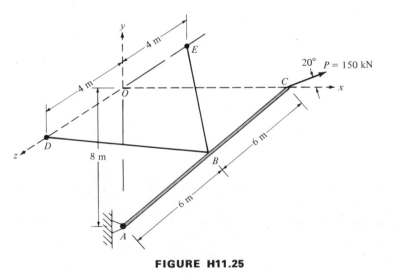

FIGURE H11.25

11.27 The space frame shown in Figure H11.27 may be assumed to be pin-connected at joint A. The supports at B, C, and D are balls and sockets. Design member AB of the frame assuming it to be structural steel whose yield strength is 36 ksi, using a factor of

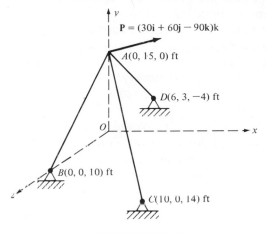

FIGURE H11.27

safety $N_y = 2$. The member is to have an S section and the deformations of the frame are to be negligible design factors.

11.28 Refer to the frame of Figure H11.27 and design member AC assuming it to be an aluminum pipe whose allowable stress is 10 ksi. Let the outside diameter of the pipe be four-thirds times its inside diameter and assume that the deformations of the frame are negligible design factors. Use information contained in Figure 8.15 if needed.

11.29 Refer to the frame shown in Figure H11.27 and design member AD. Assume that this member is to be fabricated from structural steel pipe for which the design information contained in Figure 8.14(b) is applicable if the member is subjected to compressive loads. On the other hand, if the member is subjected to tensile loads, use an allowable stress of 20 ksi. Assume the deformations of the frame to be negligible design factors.

★11.5 Computer-Aided Design Using the STRESS Language

A large number of problem-oriented languages are now available for programming digital computers. They have been developed to enable engineers and scientists to readily communicate with digital computers. IBM STRESS was developed in the Civil Engineering Department of The Massachusetts Institute of Technology. STRESS and STRUDL are two problem-oriented languages widely employed by designers, but only STRESS will be discussed here. STRESS is an acronym for *STRuctural Engineering System Solver* and STRUDL is an acronym for *STRUctural Design Language*. An introduction to the use of STRESS will be given here by providing complete programs for an indeterminate plane truss, an indeterminate plane frame, and a hingeless arch. STRESS employs matrix stiffness methods to analyze two- and three-dimensional structures. An introduction to these methods is provided in Section 14.10.

Design of indeterminate structures composed of members of varying stiffness presents the following dilemma. To determine the internal forces in a statically indeterminate structure, the stiffnesses of the members of the structure must be known. These stiffnesses depend upon the cross-sectional geometry and material of the various members of the structure. But the cross-sectional geometry cannot be determined until the internal forces are known. The dilemma is resolved by assuming initial cross-sectional dimensions, computing member stiffnesses, and then proceeding with the analysis for the internal forces. Once the internal forces are known, the various member sizes are selected and their stiffnesses are compared to those assumed initially. Usually, further analyses and revisions of member sizes will be required. In other words, design of statically indeterminate structures is an iterative procedure. Digital computers very rapidly and accurately carry out these iterations for structures composed of large numbers of members.

Only the first iterative step will be taken for the structures of this section. STRESS provides output of internal forces, reactions, and displacements. This information would be used to determine member sizes and additional iterations would be required. Determination of member sizes has been discussed in Chapter 10 and earlier sections of Chapter 11, so this aspect of design will not be repeated here. Additional STRESS analyses would generally be required for other loadings, since a number of loadings and combinations of loadings are considered in the design of most structures. STRUDL would be more useful in design practice, since it includes the design step of selecting member sizes to meet various criteria. Learning the STRUDL language is no more difficult than learning the STRESS language. Good judgment based upon a knowledge of fundamental concepts and experience is required for proper use of these languages. Output from any digital computer should always be viewed with scepticism. Order-of-magnitude appraisals and check calculations should always be a vital part of assessing the validity of any computer output.

Any structure composed of components, such as those discussed in this book, may be analyzed for internal forces once the following information is specified:

1. The geometry of the structure.
2. The properties of the material of which the structure is fabricated.
3. The loads applied to the structure.

A STRESS program provides this information in compact form. The digital computer has been programmed to formulate and solve the equations of equilibrium and compatibility and then to write out the results in readily useable form. Each of the three programs given here may be punched one line to a punch card and run at any computer center where the STRESS language is available. Only the JOB card and the control cards must be added to those shown in Figures 11.12, 11.15, and 11.18. After these adjustments are made, the card deck will be ready for the card reader at a computer center other than the one at Bradley University, where these programs were run.

Example 11.7

Write a STRESS program for the statically indeterminate plane truss shown in Figure 11.11. External loads are applied at joints 2 and 3 as shown in the figure. Geometric data are provided in the figure and the truss is fabricated of steel for which $E = 29,000$ ksi.

Solution. Study of Figure 11.12, showing the STRESS program, together with Figure 11.11 will enable the reader to understand the program and the straightforward nature of the language. The hinged joints of the truss are numbered 1, 2, 3, and 4 and the members are numbered 1 to 6. An origin is chosen at joint 1 for the global coordinate system XY shown in Figure 11.11. The global coordinate system is considered right-handed and thus the Z axis (not shown) is directed outward and perpendicular to the plane of the figure. JOINT COORDINATES are given in the global or overall, coordinate system. These coordinates are given in XY order and Z is omitted because the structure lies in the XY plane. The letter S following the coordinates of joints 1 and 4 signifies that these are supports for the structure where it is attached to the earth or to

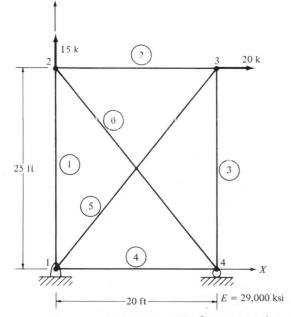

Cross–section properties: $AX = 7.8$ in.^{2}, $IZ = 17.7$ in.^{4}

Statically indeterminate plane truss

FIGURE 11.11

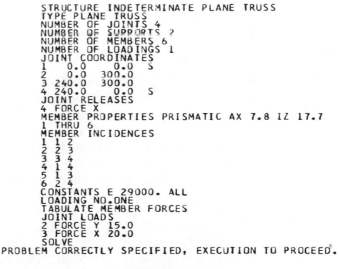

```
STRUCTURE INDETERMINATE PLANE TRUSS
TYPE PLANE TRUSS
NUMBER OF JOINTS 4
NUMBER OF SUPPORTS 2
NUMBER OF MEMBERS 6
NUMBER OF LOADINGS 1
JOINT COORDINATES
1    0.0      0.0    S
2    0.0    300.0
3  240.0    300.0
4  240.0      0.0    S
JOINT RELEASES
4 FORCE X
MEMBER PROPERTIES PRISMATIC AX 7.8 IZ 17.7
1 THRU 6
MEMBER INCIDENCES
1 1 2
2 2 3
3 3 4
4 1 4
5 1 3
6 2 4
CONSTANTS E 29000. ALL
LOADING NO.ONE
TABULATE MEMBER FORCES
JOINT LOADS
2 FORCE Y 15.0
3 FORCE X 20.0
SOLVE
PROBLEM CORRECTLY SPECIFIED, EXECUTION TO PROCEED.
```

FIGURE 11.12

another structure assumed to be nondeformable. The roller at joint 4 is assumed to be frictionless, which means that a reaction does not act in the global X direction at joint 4. This is stated under JOINT RELEASES in the program.

MEMBER PROPERTIES are stated in the local coordinate system for each member. These local coordinate systems are not shown, but each member has a local coordinate system. Consider member 5, which extends from joint 1 to joint 3. (Refer to MEMBER INCIDENCES and the entry 5   1   3 for this order.) The local coordinate system for member 5 has an origin at the first-named joint (i.e., 1). The local x axis is directed positively from the origin at 1 along the member toward joint 3. (This choice of origin and local x axis positive direction is always made consistent with the MEMBER INCIDENCES statement of the program.) For planar structures, the local z axis is chosen positive in the same direction as the global Z, which is directed outward and perpendicular to the plane of the figure. Finally, the local y axis is chosen positive such that the local coordinate system xyz is a right-handed orthogonal coordinate system. All members of the truss have equal cross-sectional areas and bending resistances. AX 7.8 of the program refers to the cross-sectional area of each member, which is oriented perpendicular to the local x axis of each member. IZ 17.7 refers to the area moment of inertia of each member with respect to the local z principal axis of each member. Actually, the members of this pin-connected truss are not bent, because of the fact that loads and reactions are only applied at the joints. If loads were applied in the XY plane at points between the joints, the bending resistance IZ would be required. In any analysis involving bending, the yz set of axes is assumed to comprise the principal axes.

All members of the truss are fabricated of steel for which the modulus of elasticity is 29,000 ksi. One statement of the program: CONSTANT E 29000. ALL states this information on material properties of the members of this truss.

The truss is loaded at joints 2 and 3, and this information is given under JOINT LOADS in the program. Joint loads directions are stated in terms of global coordinates. For example, the 20-k load applied at joint 3 of the truss is applied in the positive X direction and the corresponding program statement is 3 FORCE X 20.0. If the load were to act leftward or in the negative X direction, -20.0 would be the appropriate program entry.

The TABULATE command specifies the output requested. In this case only member forces are desired as output. If both forces and joint displacements are requested, the appropriate program statement becomes TABULATE ALL.

SOLVE and STOP commands complete the STRESS program. These are followed by control cards, which complete the card deck. It is convenient, but not required, to start STRESS commands in card column 7. Control card information must be punched in card columns as specified by systems programmers at each computer center. This control card information is programmed in another language, referred to as JCL (*Job Control Language*).

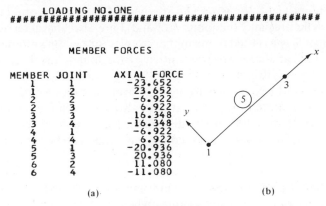

```
             STRUCTURE  INDETERMINATE  PLANE  TRUSS
        ###########################################################


             LOADING  NO.ONE
        ###########################################################


                  MEMBER  FORCES

        MEMBER JOINT      AXIAL FORCE
           1     1          -23.652
           1     2           23.652
           2     2           -6.922
           2     3            6.922
           3     3           16.348
           3     4          -16.348
           4     1           -6.922
           4     4            6.922
           5     1          -20.936
           5     3           20.936
           6     2           11.080
           6     4          -11.080
```

(a)   (b)

FIGURE 11.13

The last statement of Figure 11.12, **PROBLEM CORRECTLY SPECIFIED, EXECUTION TO PROCEED,** is not a part of the input card deck but is printed out by the high-speed printer to indicate that there are no errors in the STRESS language programming and that the calculations will be made. However, the programmer must still carefully check the program and input values to see that he or she is solving the desired problem. Digital computers will readily provide correct solutions to problems that the programmer did not intend to solve if the programmer is careless about providing proper input values.

Figure 11.13(a) provides output for the statically indeterminate plane truss. Axial forces are given for each member in terms of the local coordinate system for each member. For example, consider member 5 and the corresponding output:

MEMBER	JOINT	AXIAL FORCE
5	1	−20.936
5	3	20.936

Refer to the local coordinate system for member 5 shown in Figure 11.13(b). At joint 1 the axial force is negative, which means that it acts in the negative local x direction, and at joint 5 the axial force is positive, which means that it acts in the positive local x direction. Taken together, these statements mean that member 5 is in tension and carries a force of 20.936 k or 20,936 lb. The remainder of the output may be interpreted in a similar fashion.

Example 11.8

Write a STRESS program for the statically indeterminate plane rigid frame shown in Figure 11.14. Joints 1 and 4 are rigidly fastened to ground (or another very rigid structure) and the joints at 2 and 3 are also characterized as rigid. The term *rigid frame* is used to mean that couples (or bending moments) as well as axial and shear forces are transmitted at the four joints 1, 2, 3, and 4. A horizontal force of 10 kN is applied to the right at joint 3, and a horizontal uniform load of 0.1 kN/cm acts to the right over the entire length of member 1. Geometric and material property data are also shown in the figure. Although the centimeter (1 m = 100 cm) is discouraged as a unit of length in the SI system of units, it is sometimes convenient and will be employed in this example.

Solution. The STRESS program for this plane rigid frame is shown in Figure 11.15. This program has many features which are very similar to those discussed in Example 11.7 and shown in Figure 11.12. Only a brief discussion of different and new features shown in Figure 11.15 will be discussed here. The statement **TYPE PLANE FRAME** provides for transmission of couples as well as forces at the joints. The global coordinate system origin is chosen 300 cm

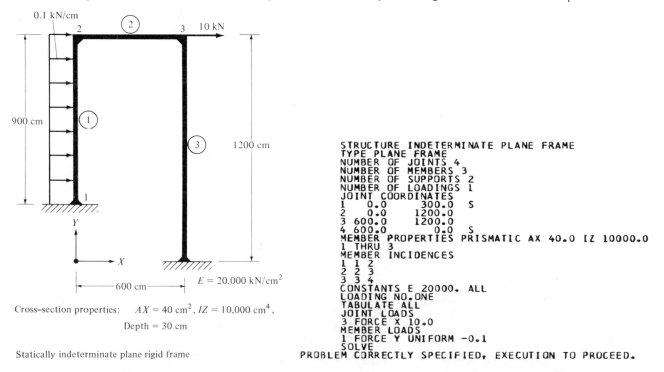

Cross-section properties: $AX = 40 \text{ cm}^2$, $IZ = 10{,}000 \text{ cm}^4$,

Depth = 30 cm

Statically indeterminate plane rigid frame

FIGURE 11.14

```
STRUCTURE INDETERMINATE PLANE FRAME
TYPE PLANE FRAME
NUMBER OF JOINTS 4
NUMBER OF MEMBERS 3
NUMBER OF SUPPORTS 2
NUMBER OF LOADINGS 1
JOINT COORDINATES
1    0.0      300.0    S
2    0.0     1200.0
3  600.0     1200.0
4  600.0        0.0    S
MEMBER PROPERTIES PRISMATIC AX 40.0 IZ 10000.0
1 THRU 3
MEMBER INCIDENCES
1  1 2
2  2 3
3  3 4
CONSTANTS E 20000. ALL
LOADING NO.ONE
TABULATE ALL
JOINT LOADS
3 FORCE X 10.0
MEMBER LOADS
1 FORCE Y UNIFORM -0.1
SOLVE
PROBLEM CORRECTLY SPECIFIED, EXECUTION TO PROCEED.
```

FIGURE 11.15

vertically below joint 1 in order to place this planar structure in the first quadrant (XY), as indicated. The statement TABULATE ALL requests joint displacement as well as force and reaction outputs. The uniform load is a load applied to member 1, and this information is provided by the statements

MEMBER LOADS
1 FORCE Y UNIFORM −0.1

Here Y of the program statement refers to the local y axis for member 1. The local x axis for member 1 is directed positively along the member from joint 1 (the origin) toward joint 2, which is consistent with

MEMBER INCIDENCES
1 1 2

```
          STRUCTURE INDETERMINATE PLANE FRAME
  ############################################################


          LOADING NO.ONE
  ############################################################



          MEMBER FORCES

    MEMBER    JOINT    AXIAL FORCE      SHEAR FORCE          MOMENT
      1        1        -30.903          83.904            25423.43
      1        2         30.903           6.096             9590.22
      2        2          6.101         -30.903            -9590.23
      2        3         -6.101          30.903            -8951.40
      3        3         30.903          16.111             8951.40
      3        4        -30.903         -16.111            10381.90


          APPLIED JOINT LOADS,    FREE JOINTS

    JOINT    FORCE X        FORCE Y        MOMENT Z
      2       0.005         -0.000          -0.01
      3      10.010          0.0            -0.00


       REACTIONS,APPLIED LOADS SUPPORT JOINTS

    JOINT    FORCE X        FORCE Y        MOMENT Z
      1      -83.904        -30.903        25423.43
      4      -16.111         30.903        10381.90


          FREE JOINT DISPLACEMENTS

    JOINT   X-DISPLACEMENT   Y-DISPLACEMENT    ROTATION
      2        14.1795          0.0348         -0.0052
      3        14.1749         -0.0464         -0.0043



          STRUCTURE INDETERMINATE PLANE FRAME
  ==========================================================
```

FIGURE 11.16

In order for the local z axis for member 1 to coincide with the global Z axis and for the local coordinate system to be right-handed, the local y axis for member 1 must be directed positively to the left and perpendicular to the member. Since the uniform load acting on member 1 is directed to the right, which is the negative y direction in the local coordinate system, the loading is programmed as a negative intensity with the given magnitude of 0.1 kN/cm. In general, member loads are expressed in terms of local coordinates and joint loads are expressed in terms of global coordinates.

Force and displacement output is shown in Figure 11.16 for this plane rigid frame. Units of the output are consistent with the input information. In general, the programmer must provide inputs in consistent units in order for the output to be consistent. STRESS is independent of the system of units employed by the programmer. In this particular case, forces are in kN, moments are in kN-cm, linear displacements are in cm, and rotational displacements are in rad. MEMBER FORCES are given in the local coordinate system for each member. APPLIED JOINT LOADS, REACTIONS, and FREE JOINT DISPLACE- MENTS are given in the global coordinate system. MOMENTS are positive when acting counterclockwise on a member or on a joint, which is consistent with a vector representation of these couples along the positive Z or z axes, which are directed positively outward and perpendicular to the XY or xy planes of Figure 11.14.

Example 11.9

Write a STRESS program for the parabolic hingeless arch depicted in Figure 11.17. The arch is rigidly supported at its end joints, numbered 1 and 21, and is subjected to a uniform temperature increase of 30°F. The equation of the parabolic arch axis is given by $Y = \frac{1}{120}X^2$, where X and Y are global coordinates for an origin chosen at the crown of the arch. Other geometric and material data are shown in the figure.

Solution. Since the arch axis is curved, it was divided into 20 straight members or segments, as indicated in Figure 11.17. Such subdividing of members is widely employed by engineers for computer analyses, and this approach is termed the *finite-element method*. The STRESS program for this parabolic arch is shown in Figure 11.18. This program has many features that are very similar to those discussed in Examples 11.7 and 11.8 and shown in Figures 11.12 and 11.15, respectively. The single new statements of this arch program are

      MEMBER TEMPERATURE CHANGE 0.0000065
      1 THROUGH 20 30.0

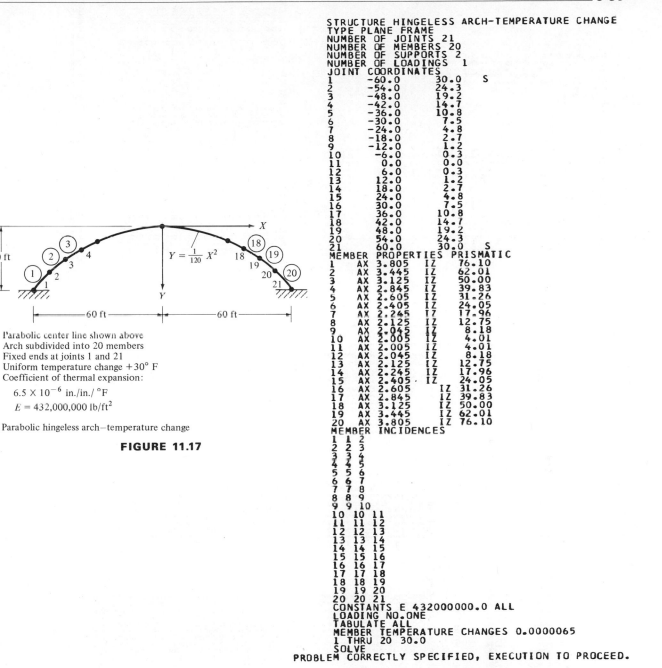

30 ft

$Y = \frac{1}{120} X^2$

— 60 ft — — 60 ft —

Parabolic center line shown above
Arch subdivided into 20 members
Fixed ends at joints 1 and 21
Uniform temperature change $+30° \text{ F}$
Coefficient of thermal expansion:

6.5×10^{-6} in./in./ °F

$E = 432{,}000{,}000 \text{ lb/ft}^2$

Parabolic hingeless arch—temperature change

FIGURE 11.17

```
STRUCTURE HINGELESS ARCH-TEMPERATURE CHANGE
TYPE PLANE FRAME
NUMBER OF JOINTS 21
NUMBER OF MEMBERS 20
NUMBER OF SUPPORTS 2
NUMBER OF LOADINGS  1
JOINT COORDINATES
1         -60.0         30.0      S
2         -54.0         24.3
3         -48.0         19.2
4         -42.0         14.7
5         -36.0         10.8
6         -30.0          7.5
7         -24.0          4.8
8         -18.0          2.7
9         -12.0          1.2
10         -6.0          0.3
11          0.0          0.0
12          6.0          0.3
13         12.0          1.2
14         18.0          2.7
15         24.0          4.8
16         30.0          7.5
17         36.0         10.8
18         42.0         14.7
19         48.0         19.2
20         54.0         24.3
21         60.0         30.0      S
MEMBER PROPERTIES PRISMATIC
1     AX  3.805   IZ    76.10
2     AX  3.445   IZ    62.01
3     AX  3.125   IZ    50.00
4     AX  2.845   IZ    39.83
5     AX  2.605   IZ    31.26
6     AX  2.405   IZ    24.05
7     AX  2.245   IZ    17.96
8     AX  2.125   IZ    12.75
9     AX  2.045   IZ     8.18
10    AX  2.005   IZ     4.01
11    AX  2.005   IZ     4.01
12    AX  2.045   IZ     8.18
13    AX  2.125   IZ    12.75
14    AX  2.245   IZ    17.96
15    AX  2.405   IZ    24.05
16    AX  2.605   IZ    31.26
17    AX  2.845   IZ    39.83
18    AX  3.125   IZ    50.00
19    AX  3.445   IZ    62.01
20    AX  3.805   IZ    76.10
MEMBER INCIDENCES
1   1  2
2   2  3
3   3  4
4   4  5
5   5  6
6   6  7
7   7  8
8   8  9
9   9  10
10  10 11
11  11 12
12  12 13
13  13 14
14  14 15
15  15 16
16  16 17
17  17 18
18  18 19
19  19 20
20  20 21
CONSTANTS E 432000000.0 ALL
LOADING NO.ONE
TABULATE ALL
MEMBER TEMPERATURE CHANGES 0.0000065
1 THRU 20 30.0
SOLVE
PROBLEM CORRECTLY SPECIFIED, EXECUTION TO PROCEED.
```

FIGURE 11.18

##
STRUCTURE HINGELESS ARCH-TEMPERATURE CHANGE
##

##
LOADING NO. ONE
##

MEMBER FORCES

MEMBER	JOINT	AXIAL FORCE	SHEAR FORCE	MOMENT
1	1	-18410.188	17466.563	655047.06
1	2	-18410.188	-17466.563	-510496.00
2	2	-19344.750	-16424.250	-510497.00
2	3	-19344.750	-16424.250	-381162.00
3	3	-20310.125	-15215.500	381156.13
3	4	-20310.125	-15215.500	-267039.94
4	4	-21287.938	-13831.063	267071.63
4	5	-21287.938	-13831.063	-168095.00
5	5	-22246.438	-12234.125	168139.56
5	6	-22246.438	-12234.125	-84364.88
6	6	-23154.688	-10417.926	84379.75
6	7	-23154.688	-10417.926	-15834.94
7	7	-23955.000	-8385.043	15816.35
7	8	-23955.000	-8385.043	-37486.40
8	8	-24618.375	-6163.637	-37453.83
8	9	-24618.375	-6163.637	75573.81
9	9	-25096.750	-3773.063	75583.88
9	10	-25096.750	-3773.063	98475.56
10	10	-25346.375	-1270.125	-98462.38
10	11	-25346.375	-1270.125	-106092.69
11	11	-25346.375	-1263.375	-106104.75
11	12	-25346.375	-1263.375	-98515.06
12	12	-25098.750	-3754.250	98500.88
12	13	-25098.750	-3754.250	-75723.38
13	13	-24626.875	-6149.406	-75572.88
13	14	-24626.875	-6149.406	-37540.94
14	14	-23956.125	-8392.953	-37452.39
14	15	-23956.125	-8392.953	-15900.64
15	15	-23147.625	-10419.883	-15887.31
15	16	-23147.625	-10419.883	-84445.06
16	16	-22247.063	-12237.125	84398.75
16	17	-22247.063	-12237.125	-168194.00
17	17	-21288.688	-13825.813	-168172.44
17	18	-21288.688	-13825.813	-261111.50
18	18	-20312.563	-15202.063	-261125.56
18	19	-20312.563	-15202.063	-381140.94
19	19	-19348.500	-16411.313	-381194.94
19	20	-19348.500	-16411.313	-510428.00
20	20	-18412.688	-17457.625	-510437.88
20	21	-18412.688	-17457.625	-654915.00

APPLIED JOINT LOADS, FREE JOINTS

JOINT	FORCE X	FORCE Y	MOMENT Z
2	-0.801	0.492	-1.00
3	-0.738	0.625	-5.88
4	9.102	8.555	31.69
5	2.035	-3.875	44.56
6	1.871	-0.301	14.88
7	-10.277	2.293	-18.59
8	-1.898	8.020	-32.57
9	0.551	-0.317	-10.06
10	-0.605	-5.654	-13.19
11	-0.336	1.506	-12.06
12	-0.133	-6.506	-14.19
13	0.055	-3.351	150.50
14	0.777	-4.941	88.54
15	5.055	-4.703	-13.33
16	5.758	2.055	-46.31
17	-6.258	-3.941	-21.56
18	-12.980	15.957	14.06
19	-0.152	11.012	54.00
20	-1.980	-1.594	9.88

REACTIONS, APPLIED LOADS SUPPORT JOINTS

JOINT	FORCE X	FORCE Y	MOMENT Z
1	-25377.461	-16.785	655047.06
21	-25373.117	-24.988	-654915.00

FREE JOINT DISPLACEMENTS

JOINT	X-DISPLACEMENT	Y-DISPLACEMENT	ROTATION
2	-.0007	-.0015	-.0001
3	-.0007	-.0037	-.0003
4	-.0002	-.0066	-.0004
5	.0000	-.0099	-.0005
6	-.0011	-.0136	-.0006
7	-.0011	-.0174	-.0006
8	-.0016	-.0212	-.0006
9	-.0018	-.0247	-.0005
10	-.0009	-.0275	-.0004
11	.0000	-.0286	-.0000
12	.0009	-.0275	.0004
13	.0016	-.0247	.0005
14	.0018	-.0212	.0006
15	.0016	-.0174	.0006
16	.0011	-.0136	.0006
17	.0011	-.0099	.0005
18	.0004	-.0066	.0006
19	-.0007	-.0037	.0003
20	-.0007	-.0015	.0001

==
STRUCTURE HINGELESS ARCH-TEMPERATURE CHANGE
==

FIGURE 11.19

This subjects all 20 members of the arch to a 30°F increase in temperature, and since the arch is monolithic and rigidly constrained at its end joints (1 and 21), internal forces and couples will be developed all along the arch axis. The arch is fabricated of a material for which the coefficient of thermal expansion is 0.0000065 in./in./°F or ft/ft/°F.

Force and displacement output is shown in Figure 11.19 for this parabolic arch. MEMBER FORCES are given in the local coordinate system for each member. APPLIED JOINT LOADS, REACTIONS, and FREE JOINT DISPLACEMENTS are given in the global coordinate system. MOMENTS are positive when acting clockwise on a member or a joint, which is consistent with a vector representation of these couples along the positive Z or z axes, which are directed positively inward and perpendicular to the XY or xy planes of Figure 11.17. Pound-foot units have been used consistently in the program. Theoretically, the APPLIED JOINT LOADS, FREE JOINTS should all equal zero, since external forces are not applied to the arch at these joints. All of these output values, while not zero, are very small when compared to the member forces on a percentage basis and hence may be neglected. These nonzero values arise in practice because digital computers, like pocket calculators, make approximate calculations rather than exact ones. Single-precision calculations were made for the STRESS programs presented in this section. For systems that involve large numbers of members it may be necessary to employ double-precision calculations to solve the resulting matrix equations.

Homework Problems

Problems 11.35 to 11.37 require the use of a digital computer with the STRESS language available on the system. Consult computer center personnel for appropriate control cards to use with STRESS.

11.30 Draw free-body diagrams of members 1 and 2 of the statically indeterminate plane truss of Figure 11.11. Refer to MEMBER INCIDENCES statements of Figure 11.12 for these members and clearly show the local coordinate systems on your free-body diagrams. MEMBER FORCES are shown in Figure 11.13.

11.31 Draw a complete free-body diagram of member 1 of the plane rigid frame of Figure 11.14.

Show the local coordinate system on this diagram and the forces applied to the member at joints 1 and 2 by referring to Figures 11.15 and 11.16. Show that this member is in equilibrium under the action of the applied uniform load and the forces applied to it at the joints 1 and 2.

11.32 Refer to Figures 11.17, 11.18, and 11.19 for the parabolic hingeless arch and construct a complete free-body diagram of member 3 of the arch. Show that the equations of equilibrium are satisfied by the forces applied to this member at joints 3 and 4. Clearly show the local coordinate system for this member, which is straight between joints 3 and 4.

11.33   Refer to Figures 11.17, 11.18, and 11.19 for the parabolic hingeless arch and construct a complete free-body diagram of the entire arch. REACTIONS, APPLIED LOADS SUPPORT JOINTS for joints 1 and 21 of Figure 11.19 will provide the end reactions. Show that the equations of equilibrium are satisfied by this *self-equilibrating system.*

11.34   Refer to Figures 11.14, 11.15, and 11.16 for the plane rigid frame and construct a free-body diagram of member 3, which is subjected to forces at joints 3 and 4. Clearly show the local coordinate system for this member and write the equations of equilibrium for it to show that the equations are satisfied.

11.35   Write a STRESS program and run it at the computer center for the structure shown in Figure 11.11. Remove the loads shown at joints 2 and 3 and apply the following loads at joint 3: a force of 15 k directed horizontally to the left and a force of 10 k directed vertically downward. Interpret the output by showing all member forces on a diagram of the truss and clearly show whether each member is in tension or compression.

11.36   Write a STRESS program and run it at the computer center for the plane frame of Figure 11.14. Remove uniform loading applied to member 1 and the concentrated load applied at joint 3. Apply a downward uniform loading to member 2 of 0.1 kN/cm intensity over its full length. Interpret the output by showing complete free-body diagrams of all three members of the frame.

11.37   Write a STRESS program and run it at the computer center for the hingeless arch of Figure 11.17. Remove the uniform temperature loading and apply a downward concentrated load of 2000 lb at the crown joint 11. Interpret the output by drawing a free-body diagram of the entire arch and show that it is in equilibrium.

Chapter 12 Analysis and Design for Inelastic Behavior

12.1 Introduction

Increased understanding of the behavior of materials and structural and machine components has led to a change in design emphasis from stress levels to load capacity. In Chapters 10 and 11, computed stresses associated with service or working loads were compared to allowable stresses. These allowable stresses were determined by applying a factor of safety to some "failure" stress. Traditional designs of the past as well as much current engineering practice has made use of these concepts. However, recent trends have refocused the attention of design engineers on the load-carrying capacity of components and overall structures. Service or working loads are multiplied by load factors greater than unity, and materials and sizes of components are selected to "fail" under the action of these factored loads. The term *failure* is defined in this chapter to mean that the component after being subjected to a maximum load continues to deflect under practically constant load. This concept restricts the discussion to ductile metals under static loading at ordinary temperatures. Once the calculations are completed, the component should safely withstand the service or working loads, which are much smaller than the ultimate or factored loads.

This chapter provides an introduction to analysis and design for inelastic behavior, which has come to be referred to as *plastic analysis and design* or *limit analysis and design*. Axial, torsional, and bending components are analyzed and designed on the basis of the load-carrying capacity. In each case the reader should consider the designs as preliminary, since a number of other factors, not discussed in this introductory chapter, must be considered in arriving at final component design.

*12.2 Inelastic Behavior of Axially Loaded Members

An idealized elastoplastic stress–strain curve is shown in Figure 12.1. Members fabricated of such an ideal material would, upon being axially loaded, have uniform internal stresses defined as follows:

573

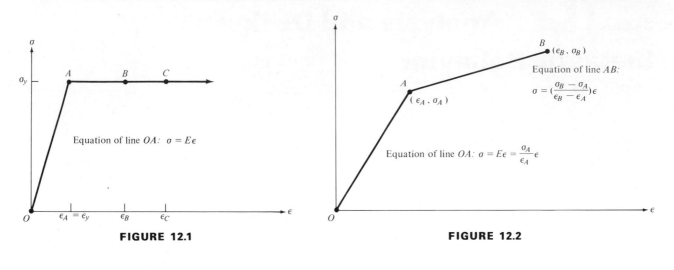

FIGURE 12.1   **FIGURE 12.2**

$$\sigma = E\varepsilon \qquad (0 \le \varepsilon \le \varepsilon_y) \tag{12.1}$$

$$\sigma = \sigma_y \qquad (\varepsilon \ge \varepsilon_y) \tag{12.2}$$

It is important to note that regardless of the value of the unit strain ε, when it exceeds ε_y, the stress equals the yield stress, σ_y. Observe that $\varepsilon_C > \varepsilon_B > \varepsilon_A$, but the corresponding stress for each of these strains is σ_y. In other words, there are an infinite number of strains corresponding to the yield stress of the material. This is a convenient simplification of the behavior of ductile metals, which ignores the effect of strain handling.

A second idealized stress–strain diagram is shown in Figure 12.2, which may be described as a piecewise linear stress–strain diagram. Members fabricated of such an ideal material would, upon being axially loaded, have uniform internal stresses defined as follows:

$$\sigma = E\varepsilon = \frac{\sigma_A}{\varepsilon_A}\varepsilon \qquad (0 < \varepsilon < \varepsilon_A) \tag{12.3}$$

$$\sigma = \left(\frac{\sigma_B - \sigma_A}{\varepsilon_B - \varepsilon_A}\right)\varepsilon \qquad (\varepsilon_A < \varepsilon < \varepsilon_B) \tag{12.4}$$

Of course, the coordinates of points A and B must be determined by experiment and the straight lines connecting O to A and A to B approximate the experimental stress–strain curve for a material of interest.

The following examples will clarify the inelastic behavior of axially loaded members fabricated of materials whose stress–strain diagrams are depicted in Figures 12.1 and 12.2.

Example 12.1

Rods A and B support a relatively stiff beam, which may be assumed to behave as a rigid member, as shown in Figure 12.3(a). A downward force P is applied at point D and the hinge at C is assumed to be frictionless. The stress–strain diagram for each rod is depicted in Figure 12.1. Determine (a) which rod (A or B) yields first. (What is the corresponding value of $P = P_y$?); (b) the largest load $P = P_p$ that may be applied to the right end of the rigid horizontal member; and (c) the deflections associated with P_y and P_p and construct a load-deflection plot for the structure.

Solution

(a) Refer to Figure 12.3(b), which indicates the relative displacements of the ends of the two rods. Since the horizontal member is assumed to be rigid and the angle is small, the relationship between the displacements becomes

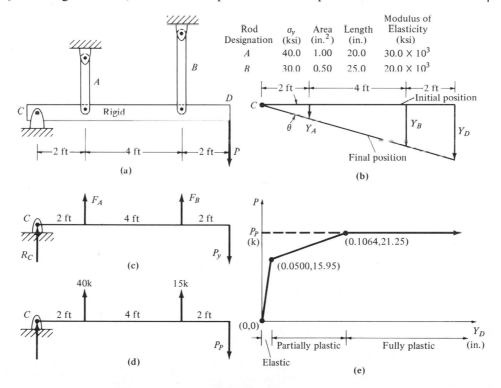

Rod Designation	σ_y (ksi)	Area (in.²)	Length (in.)	Modulus of Elasticity (ksi)
A	40.0	1.00	20.0	30.0×10^3
B	30.0	0.50	25.0	20.0×10^3

(a)

(b)

(c)

(d)

(e)

FIGURE 12.3

$$\frac{Y_A}{2} = \frac{Y_B}{2 + 4}$$

or

$$Y_A = \tfrac{1}{3} Y_B$$

From the definition of unit strain,

$$Y_A = \varepsilon_A L_A \qquad \text{and} \qquad Y_B = \varepsilon_B L_B$$

Substitution of these expressions into the equation relating the displacements gives

$$\varepsilon_A L_A = \tfrac{1}{3} \varepsilon_B L_B$$

or

$$\varepsilon_A = \frac{1}{3} \frac{L_B}{L_A} \varepsilon_B$$

Substitution of $L_B = 25.0$ and $L_A = 20.0$ gives

$$\varepsilon_A = \tfrac{1}{3} \tfrac{25}{20} \varepsilon_B$$
$$= 0.417 \varepsilon_B$$

The yield strains of each rod equals the yield stress of each divided by its corresponding modulus of elasticity:

$$\varepsilon_{yA} = \frac{40}{30 \times 10^3} = 1.33 \times 10^{-3}$$

$$\varepsilon_{yB} = \frac{30}{20 \times 10^3} = 1.50 \times 10^{-3}$$

Assume that rod B yields first and substitute its yield strain of 1.50×10^{-3} into the strain compatibility equation $\varepsilon_A = 0.417 \varepsilon_B$:

$$\varepsilon_A = 0.417(1.50 \times 10^{-3}) = 0.626 \times 10^{-3}$$

since

$$\varepsilon_A < \varepsilon_{yA} \qquad \text{(i.e., } 0.626 \times 10^{-3} < 1.33 \times 10^{-3})$$

Then rod B yields first, as assumed above.
     In order to determine the force $P = P_y$ associated with yielding of rod B, the

forces F_A and F_B in the rods must be determined from the known strains. In general, for the axially loaded rods,

$$\sigma = E\varepsilon, \quad F = A\sigma, \quad F = AE\varepsilon$$

$$F_A = (1.00)(30 \times 10^3)(0.626 \times 10^{-3}) = 18.78 \text{ k}$$

$$F_B = (0.50)(20 \times 10^3)(1.50 \times 10^{-3}) = 15.0 \text{ k}$$

Refer to Figure 12.3(c) and write the moment equilibrium equation with respect to point C.

$$\circlearrowleft \quad \sum M_C = 0: P_y(8) - F_A(2) - F_B(6) = 0$$

$$8P_y - 2(18.78) - 6(15.0) = 0$$

$$\boxed{P_y = 15.95 \text{ k}}$$

(b) The largest load $P = P_p$ that may be applied to the right end of the rigid horizontal member is associated with yielding of both bars over their entire cross sections. Rod B carries a force of 15.0 k and rod A carries a force equal to its area multiplied by its yield stress, or 40.0 k, as shown in Figure 12.3(d). As before, write the moment equation of equilibrium with respect to point C.

$$\circlearrowleft \quad \sum M_C = 0: P_p(8) - 40(2) - 15(6) = 0$$

$$\boxed{P_p = 21.25 \text{ k}}$$

(c) The deflection associated with P_y may be obtained by first noting that the elongation of rod B when it yields is given by

$$Y_B = \varepsilon_B L_B = 1.50 \times 10^{-3}(25) = 0.0375 \text{ in.}$$

Refer to Figure 12.3(b), and from similar triangles,

$$\frac{Y_D}{8} = \frac{Y_B}{6}$$

Solve for the deflection Y_D at the applied load:

$$Y_D = \tfrac{8}{6} Y_B = \tfrac{8}{6}(0.0375)$$

$$\boxed{Y_D = 0.0500 \text{ in.}} \quad \text{and} \quad \boxed{P_y = 15.95 \text{ k}}$$

The deflection associated with P_p may be obtained by noting that rod A yields last and that just prior to yielding, the usual elastic equations are applicable:

$$Y_A = \varepsilon_{YA} L_A = (1.33 \times 10^{-3})(20) = 0.0266 \text{ in.}$$

Refer to Figure 12.3(b), and from similar triangles,

$$\frac{Y_D}{8} = \frac{Y_A}{2}$$

Solve for the deflection at the applied load:

$$Y_D = 4Y_A = 4(0.0266)$$

$$\boxed{Y_D = 0.1064 \text{ in.}} \qquad \text{and} \qquad \boxed{P_p = 21.25 \text{ k}}$$

The load-deflection plot for this structure is shown in Figure 12.3(e). Elastic, partially plastic, and fully plastic regions are noted in the plot. Once the fully plastic load P_p is reached, the deflections increase without further increase in the applied loading.

Example 12.2

A rigid vertical member is supported by rods A and B and the frictionless pin at C as shown in Figure 12.4(a). The force P is applied horizontally at point D as shown and the rod B is supported to prevent it from buckling as it is compressed. Figure 12.2 depicts the stress–strain diagram for rods A and B with point coordinates specified as follows:

Rod	ε_A	σ_A (MPa)	ε_B	σ_B (MPa)
A	0.0010	100	0.0020	140
B	0.0012	80	0.0018	120

Assuming that the structure will fail when the first rod reaches stress–strain levels corresponding to point B in Figure 12.2, determine the failure load P and the corresponding deflection for this structure.

Solution. Refer to Figure 12.4(b) in order to relate the deformations of the rods by considering similar triangles.

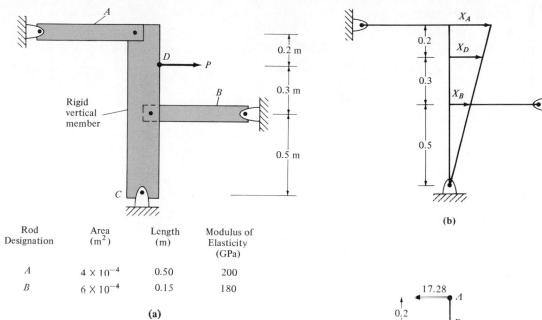

Rod Designation	Area (m^2)	Length (m)	Modulus of Elasticity (GPa)
A	4×10^{-4}	0.50	200
B	6×10^{-4}	0.15	180

(a)

(b)

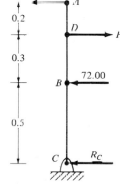

(c)

FIGURE 12.4

$$\frac{X_A}{1.0} = \frac{X_B}{0.5}$$

$$X_A = 2X_B$$

From the definition for axial strain,

$$X_A = \varepsilon_A L_A \qquad X_B = \varepsilon_B L_B$$

and

$$X_A = 0.5\varepsilon_A \qquad X_B = 0.15\varepsilon_B$$

Substitute into $X_A = 2X_B$:

$$0.5\varepsilon_A = 2(0.15\varepsilon_B)$$

$$\varepsilon_A = 0.6\varepsilon_B$$

This relationship, based upon the gometry of the given structure, must hold throughout the loading range.

Assume that rod B fails first at an ultimate strain of 0.0018. The corresponding strain in rod A will be $\varepsilon_A = 0.6(0.0018) = 0.00108$. Since $0.00108 < 0.00200$, rod B will fail first.

From the known strains in the rods, calculate the corresponding stresses. At failure, the stress in rod B is 120 MPa, corresponding to its failure strain of 0.0018. To determine the stress in rod A, write the equation of line AB of Figure 12.2 using the point coordinates specified in this problem for rod A:

$$\sigma = \left(\frac{\sigma_B - \sigma_A}{\varepsilon_B - \varepsilon_A}\right)\varepsilon = \left(\frac{140 - 100}{0.0020 - 0.0010}\right)\varepsilon$$

$$= 40{,}000\varepsilon$$

Substitute the known final strain in rod A:

$$\sigma_A = 40{,}000(0.00108) = 43.2 \text{ MPa}$$

Calculate the forces in the rods:

$$F_A = \sigma_A A = 43.2 \times 10^6 \times 4 \times 10^{-4} = 17.28 \text{ kN}$$

$$F_B = \sigma_B A = 120 \times 10^6 \times 6 \times 10^{-4} = 72.00 \text{ kN}$$

Refer to Figure 12.4(c) and sum the moments of the forces with respect to an axis perpendicular to the plane through point C.

$$\circlearrowleft \ \sum M_C = 0: -17.28(1.00) + P(0.8) - 72.00(0.5) = 0$$

Solving, we obtain

$$\boxed{P = 66.6 \text{ kN}}$$

In order to find the corresponding deflection at point D, refer to Figure 12.4(b), and from similar triangles,

$$\frac{X_D}{0.8} = \frac{X_B}{0.5}$$

$$X_D = 1.6X_B$$

From above,

$$X_B = 0.15\varepsilon_B$$

$$X_D = 1.6(0.15\varepsilon_B) = 0.24\varepsilon_B$$

Rod B fails when $\varepsilon_B = 0.0018$:

$$X_D = 0.24(0.0018)$$

$$= 0.000432 \text{ m}$$

In summary, the structure will fail when P reaches 66.6 kN when D has deflected 0.000432 m to the right.

*12.3 Design for Inelastic Behavior of Axially Loaded Members

Plastic design and *limit design* are synonymous terms for *design for inelastic behavior*. Service or working loads, discussed in Chapter 10, are multiplied by load factors and the structure is proportioned to resist these factored loads at the fully plastic stage of deformation. Stresses are allowed to reach yield levels because the service or working loads have been increased by load factors exceeding unit values by substantial amounts. Under actual expected loadings (i.e., service loadings) the stresses will be well below the yield values and failure or collapse of the structure would not occur. In fact, the focus of the designer changes from stress levels to load levels, which is more attuned to physical reality, since structures are designed to resist applied forces. Historically, engineers have devoted a great deal of study to determination of elastic stress levels, but recent progress in the analysis of inelastic behavior of structures has enabled engineers to redirect their attention to the load-carrying capacity of structures. Examples 12.3 and 12.4 illustrate this approach to design for axially loaded members.

Example 12.3
Design the members of the three-bar truss depicted in Figure 12.5(a) to resist a service load $P = 100$ kN. Use a load factor of 2 and consider all three members to have equal areas. The yield stress of the material of which the members are fabricated is 400 MPa and Figure 12.1 shows the stress–strain diagram.

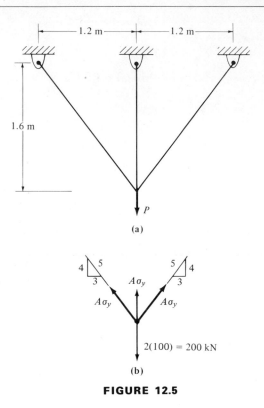

FIGURE 12.5

Solution. The design load is $2 \times 100 = 200$ kN, and each resisting force equals the area A multiplied by the yield stress σ_y as shown in Figure 12.5(b). When this condition is reached, the loaded joint of the truss will continue to deform under the constant load of 200 kN.

$$\downarrow \quad \sum F_y = 0$$

$$200 \times 10^3 - A\sigma_y - 2A\sigma_y\left(\tfrac{4}{5}\right) = 0$$

$$\tfrac{13}{5}A\sigma_y = 200 \times 10^3$$

$$A = \frac{200 \times 10^3}{\tfrac{13}{5}(400 \times 10^6)}$$

$$= \boxed{1.92 \times 10^{-4} \text{ m}^2}$$

Other considerations, such as elastic deflections and design of connections, are not discussed here so that we can focus on design for inelastic behavior.

Example 12.4

Redesign rods A and B (i.e., determine new areas for each rod) of the structure depicted in Figure 12.3(a) for a working applied load P of 40 k. Use a load factor of 2 and yield stresses of 40.0 and 30.0 ksi for rods A and B, respectively. Arbitrarily, choose the area of rod B to be equal to one-half of the area of rod A.

Solution. The appropriate free-body diagram is shown in Figure 12.6. The rod forces are given by

$$F_A = \sigma_A A_A$$

$$F_B = \sigma_B A_B$$

But $A_B = 0.5A_A$, as given in the problem statement. Sum the moments of the forces with respect to an axis through C perpendicular to the plane of the forces:

$$\circlearrowleft \quad \sum M_C = 0$$

$$F_A(2) + F_B(6) - 80(8) = 0$$

Substitute for F_A and F_B in terms of A_A:

$$2\sigma_A A_A + 6\sigma_B(0.5A_A) - 640 = 0$$

Substitute for the yield stresses, to obtain

$$2(40.0)A_A + 6(30.0)(0.5A_A) = 640$$

$$\boxed{A_A = 3.76 \text{ in.}^2}$$

$$A_B = 0.5(3.76)$$

$$\boxed{A_B = 1.88 \text{ in.}^2}$$

Other considerations, such as elastic deflections and design of connections, are not discussed here so that we can focus on design for inelastic behavior.

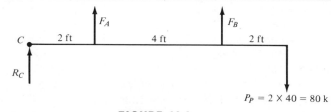

FIGURE 12.6

Homework Problems

12.1–12.3 These problems refer to Figure 12.3, with given quantities as shown in the accompanying table. In each case determine:

(a) Which rod (A or B) yields first.
(b) The largest load $P = P_P$ that may be applied to the right end of the rigid horizontal member.
(c) The deflections associated with P_y and P_P.
(d) Construct a load-deflection plot for the structure.

Problem Number	Rod Desig- nation	Yield Stress (ksi)	Area (in.²)	Length (in.)	E (ksi)
12.1	A	36	2.00	30.0	30×10^3
	B	36	0.75	40.0	30×10^3
12.2	A	50	3.00	20.0	20×10^3
	B	40	2.00	20.0	30×10^3
12.3	A	60	1.75	60.0	10×10^3
	B	40	1.00	20.0	10×10^3

12.4–12.6 These problems refer to Figure 12.4, with given quantities as shown in the accompanying table. In each case use the stress–strain information for rods A and B as given in Example 12.2. Assuming that the structure will fail when the first rod reaches stress–strain levels corresponding to point B of Figure 12.2, determine the failure load P and the corresponding deflection for this structure.

Problem Number	Rod Desig- nation	Area (m²)	Length (m)
12.4	A	8×10^{-4}	0.4
	B	6×10^{-4}	0.8
12.5	A	10×10^{-4}	1.2
	B	5×10^{-4}	1.8
12.6	A	6.80×10^{-4}	0.75
	B	3.05×10^{-4}	0.75

12.7–12.9 These problems refer to Figure 12.5, with given quantities as shown in the accompanying table.

Consider all three members of the truss to have equal areas. Use a load factor of 1.8 and determine the bar areas required for the truss members. Figure 12.1 shows the stress–strain diagram for the material of which the members are to be fabricated.

Problem Number	Yield Stress (MPa)	Working Load, P (kN)
12.7	360	50
12.8	480	400
12.9	660	600

12.10–12.12 These problems refer to Figure H12.10, with given quantities as shown in the accompanying table. Design the cable (i.e., determine its area) to resist working load P with a load factor of 3.2. Figure 12.1 shows the stress–strain diagram for the material of which the cable is to be fabricated.

Problem Number	Yield Stress (MPa)	Working Load, P (kN)	Length, a (m)
12.10	200	100	0.50
12.11	240	200	0.80
12.12	320	175	0.64

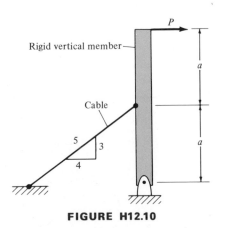

FIGURE H12.10

⋆**12.4** Inelastic Behavior of Torsional Members

Figure 12.7(a) depicts the cross section of a hollow circular shaft fabricated of a ductile material subjected to a torque T_y associated with first shearing yield at the outermost fiber. This yield torque, from Eq. 4.18, is given by

$$T_y = \frac{\tau_y(\pi/2)(r_o^4 - r_i^4)}{r_o} \qquad (12.5)$$

The fully plastic condition is depicted in Figure 12.7(b). The shearing stress/shearing strain curve is assumed to have the shape of Figure 12.1. Consider an annular differential area on which the stress τ_y is a constant and write the equation for the differential force acting on this area:

$$dF = \tau_y 2\pi\rho \, d\rho \qquad (12.6)$$

Multiplication of this differential force by the moment arm yields an expression for the differential torque:

$$dT_p = 2\pi\tau_y\rho^2 \, d\rho \qquad (12.7)$$

Integration of both sides of this differential equation gives the equation for T_p.

$$T_p = 2\pi\tau_y \int_{r_i}^{r_o} \rho^2 \, d\rho = \frac{2\pi\tau_y}{3}(r_0^3 - r_i^3) \qquad (12.8)$$

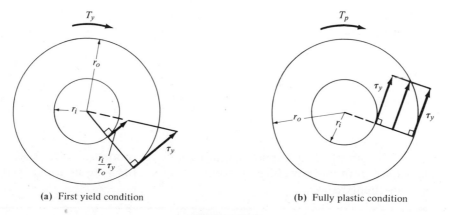

(a) First yield condition   (b) Fully plastic condition

FIGURE 12.7

Equations 12.5 and 12.8, when specialized for a solid circular shaft by letting $r_i = 0$ and $r_o = r$, give the following equations:

$$T_y = \frac{\pi \tau_y r^3}{2} \qquad\qquad (12.9)$$

$$T_p = \frac{2\pi \tau_y r^3}{3} \qquad\qquad (12.10)$$

Example 12.5

A solid circular shaft is fabricated of a material with a shearing yield stress of 500 MPa and has a radius of 0.10 m. Determine the first yield torque and the fully plastic torque for this shaft and the ratio of these torques. Is this ratio a function of material and geometric properties of the shaft? An idealized shearing stress/shearing strain curve is assumed to have the shape of Figure 12.1.

Solution. Equations 12.9 and 12.10 are the appropriate equations to use in this case.

$$T_y = \frac{\pi \tau_y r^3}{2} = \frac{\pi (500 \times 10^6)(0.1)^3}{2}$$

$$= \boxed{785.4 \text{ kN-m}}$$

$$T_p = \frac{2\pi \tau_y r^3}{3} = \frac{2\pi (500 \times 10^6)(0.1)^3}{3}$$

$$= \boxed{1047 \text{ kN-m}}$$

Form the ratio of T_p to T_y:

$$\frac{T_p}{T_y} = \frac{2\pi \tau_y r^3}{3} \frac{2}{\pi \tau_y r^3} = \frac{4}{3}$$

$$= \boxed{1.33}$$

This ratio is independent of material and geometric properties, since both τ_y and r cancel from the preceding equation. Of course, the analysis applies only to a solid circular shaft.

Example 12.6
A hollow circular shaft fabricated of a material with a shearing yield stress of 500 MPa has an outside radius $r_o = 0.10$ m and an inside radius $r_i = 0.06$ m. Determine the first yield torque and the fully plastic torque for this shaft and the ratio of these torques. Construct a plot of this ratio of torques (T_p/T_y) versus the ratio of the inner to the outer radii (r_i/r_o).

Solution. Equations 12.5 and 12.8 are the appropriate equations to use in this case.

$$T_y = \frac{\tau_y(\pi/2)(r_o^4 - r_i^4)}{r_o} = \frac{500 \times 10^6 (\pi/2)[(0.10)^4 - (0.06)^4]}{0.10}$$

$$= \boxed{683.6 \text{ kN-m}}$$

$$T_p = \frac{2\pi\tau_y}{3}(r_o^3 - r_i^3) = \frac{2\pi(500 \times 10^6)}{3}[(0.10)^3 - (0.06)^3]$$

$$= \boxed{821.0 \text{ kN-m}}$$

Form the ratio of T_p to T_y:

$$\frac{T_p}{T_y} = \frac{2\pi\tau_y}{3}(r_o^3 - r_i^3)\frac{r_o}{\tau_y(\pi/2)(r_o^4 - r_i^4)}$$

$$= \frac{4}{3}\frac{1 - (r_i/r_o)^3}{1 - (r_i/r_o)^4}$$

A plot of this equation is shown in Figure 12.8. In this example $r_i/r_o = 6/10 = 0.6$ and $T_p/T_y = 1.20$. Note that when $r_i/r_o = 0$ the torque ratio is 4/3,

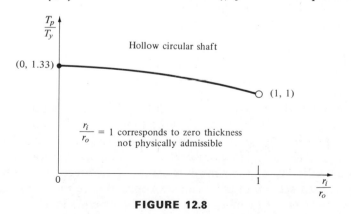

FIGURE 12.8

which checks the result for a solid circular shaft. As $r_i/r_o \to 1$, the torque ratio approaches unity, but this result must not be accepted uncritically because as the ratio of the radii approaches unity, the wall thickness of the shaft approaches zero and local buckling of the tube wall will govern rather than the stress conditions analyzed here. Note that the torque ratio is independent of material properties, since τ_y cancels from the equation, but the ratio does depend upon the geometric ratio of the radii.

★12.5 Inelastic Behavior of Bending Members

Assumptions of the simple theory of inelastic behavior of bending members may be stated as follows:

1. An idealized elastoplastic stress–strain curve for the material is shown in Figure 12.1. Strain hardening is neglected and the material is ductile under static loading at ordinary temperatures.
2. Tensile and compressive properties of the material are assumed to be identical.
3. The effects of axial force and shear force are neglected in comparison to the effects of bending moments. Many structures (e.g., single-story rigid frames) are proportioned such that this assumption is valid.
4. Joints are assumed capable of transferring fully plastic moments from member to member. Proper design of connections will ensure that this assumption is satisfied.

Figure 12.9 depicts the stress distribution under increasing bending moments for a member subjected to pure bending. The moment associated with first yield is given by the flexure formula (refer to Eq. 5.10)

$$M_y = \sigma_y \frac{I_u}{v} \tag{12.11}$$

As the applied moment increases, the stress distribution corresponding to M_1, where $M_y < M_1 < M_p$, is as shown in Figure 12.9(b). Finally, the distribution corresponding to the fully plastic moment M_p is shown in Figure 12.9(c). At this stage the entire cross section has yielded and stresses over the entire cross section are at the yield value. This same distribution is shown in Figure 12.10, together with the resultant forces and the cross section of the member. Application of the equations of equilibrium to a segment of the member gives the following equations:

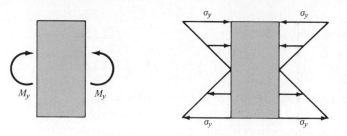

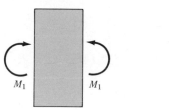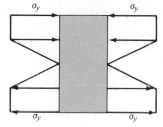

(a) First yield condition

(b) Partially plastic condition

(c) Fully plastic condition

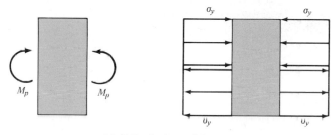

FIGURE 12.9

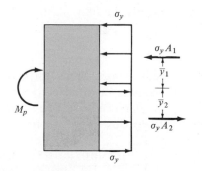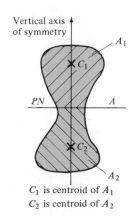

C_1 is centroid of A_1
C_2 is centroid of A_2

FIGURE 12.10

$$\rightarrow \ \sum F_x = 0$$

$$-\sigma_y A_1 + \sigma_y A_2 = 0 \qquad\qquad (12.12)$$

$$A_1 = A_2 \qquad\qquad (12.13)$$

Equation 12.13 reveals that the plastic neutral axis (PNA) of Figure 12.10 is located by dividing the total area into two equal parts. In general, the PNA need not pass through the centroid of the entire cross section, which is true of the elastic neutral axis. This means that, in general, there will be a shifting of the neutral axis as the applied moments are increased.

Sum the moments with respect to the plastic neutral axis as follows:

$$\circlearrowleft \ \sum M_{\mathrm{PNA}} = 0$$

$$M_p - \sigma_y A_1 \bar{y}_1 - \sigma_y A_2 \bar{y}_2 = 0 \qquad\qquad (12.14)$$

Solve for M_p:

$$M_p = \sigma_y (A_1 \bar{y}_1 + A_2 \bar{y}_2) \qquad\qquad (12.15)$$

The shape factor f, which is a cross-sectional property, is defined as the ratio of M_p to M_y. Equations 12.15 and 12.11 provide the appropriate equation:

$$f = \frac{M_p}{M_y} = \frac{A_1 \bar{y}_1 + A_2 \bar{y}_2}{I_u / v} \qquad\qquad (12.16)$$

Example 12.7
The cross section of a beam is depicted in Figure 12.11. The yield stress of the material of which the beam is fabricated is 400 MPa. Determine (a) the location of the elastic neutral axis, (b) the location of the plastic neutral axis, (c) the first yield moment, (d) the fully plastic moment, and (e) the shape factor for the cross section.

Solution
(a) The elastic neutral axis passes through the centroid of the entire cross section, and this axis is oriented perpendicular to the plane of the applied forces. Choosing a reference axis along the top of the cross section and measuring y values positively downward, we find that the necessary equation is

$$\bar{y} = \frac{0.15(0.03)(0.015) + 0.03(0.15)(0.105)}{0.15(0.03) + 0.03(0.15)}$$

$$= \boxed{0.06 \ \mathrm{m}}$$

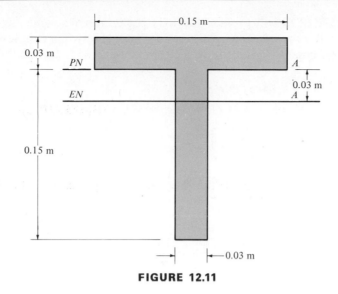

FIGURE 12.11

which locates the elastic neutral axis as a horizontal line 0.06 m below the top of the cross section.

(b) Equation 12.13 is applicable for locating the plastic neutral axis of this T section:

$$A_1 = A_2$$

Since $0.15(0.03) = 0.03(0.15)$, the plastic neutral axis is located 0.03 m below the top of the cross section, or at the junction of the flange and the web. Clearly, the neutral axis shifts upward as the loading increases from elastic to fully plastic levels.

(c) The first yield moment requires computation of the moment of inertia of the total cross-sectional area with respect to the elastic neutral axis.

$$I_{NA} = \tfrac{1}{12}(0.15)(0.03)^3 + 0.15(0.03)(0.06 - 0.015)^2 + \tfrac{1}{12}(0.03)(0.15)^3$$
$$+ 0.03(0.15)(0.105 - 0.06)^2$$
$$= 27.0 \times 10^{-6} \ m^4$$

Equation 12.11 with $v = 0.18 - 0.06 = 0.12$ m will provide the first yield moment:

$$M_y = \frac{400 \times 10^6 (27 \times 10^{-6})}{0.12}$$

$$= \boxed{90.0 \ kN\text{-}m}$$

(d) Equation 12.15 will provide the fully plastic moment and the reader is urged to review the solution in terms of an internal resisting couple consisting of equal and opposite compressive and tensile forces.

$$M_p = \sigma_y(A_1 \bar{y}_1 + A_2 \bar{y}_2)$$
$$= 400 \times 10^6 [0.15(0.03)(0.015) + 0.03(0.15)(0.075)]$$
$$= \boxed{162 \text{ kN-m}}$$

(e) The shape factor f equals the ratio of the fully plastic to the first yield moment:

$$\boxed{f = \frac{M_p}{M_y} = \frac{162}{90} = 1.80}$$

This factor is a function of the geometry of the cross section and is independent of the material yield stress. In other words, if this flexural member were fabricated of another ductile material with a different yield stress, the shape factor would still be 1.80, since it depends only upon the geometry of the member.

Example 12.8
Refer to Figure 12.12 for a square hollow cross section of constant thickness for a flexural member. The member is fabricated of a ductile material with a yield stress σ_y. Show that the shape factor f may be expressed as a function of the ratio of b to a as follows:

$$f = \left(\frac{3}{2}\right)\frac{1 - (b/a)^3}{1 - (b/a)^4}$$

Solution. Equation 12.16 is appropriate if the areas are interpreted as the net areas, since the cross section is a hollow one.

$$f = \frac{M_p}{M_y} = \frac{A_1 \bar{y}_1 + A_2 \bar{y}_2}{I/v}$$

$$\frac{I}{v} = \frac{\frac{1}{12}(2a)^4 - \frac{1}{12}(2b)^4}{a} = \left(\frac{4}{3}\right)\frac{a^4 - b^4}{a}$$

The numerator of the f equation may be expressed as follows:

$$2a(a)a - 2b(b)(b) = 2a^3 - 2b^3 = 2(a^3 - b^3)$$

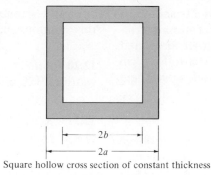

Square hollow cross section of constant thickness

FIGURE 12.12

Substitution to obtain f yields

$$f = 2(a^3 - b^3) \cdot \frac{3}{4} \frac{a}{a^4 - b^4}$$

$$= \boxed{\left(\frac{3}{2}\right) \frac{1 - (b/a)^3}{1 - (b/a)^4}}$$

Homework Problems

12.13 A solid circular shaft fabricated of a material with a shearing yield stress of 600 MPa has a radius of 0.05 m. Determine the first yield torque and the fully plastic torque for this shaft and the ratio of these torques. Explain why this ratio is independent of the material and geometric properties of the shaft. Of course, the ratio only applies to a solid circular shaft.

12.14 A solid circular shaft fabricated of a material with a shearing yield stress of 50 ksi has a radius of 3 in. Determine the first yield torque and the fully plastic torque for this shaft and the ratio of these torques. Explain why this ratio is independent of the material and geometric properties of the shaft.

12.15 A beam of hollow circular cross section is bent about a diameter. Show that its shape factor f in bending is given by

$$f = \frac{16}{3\pi} \frac{\beta^3 + \beta^2 + \beta}{\beta^3 + \beta^2 + \beta + 1}$$

where $\beta = D_o/D_i$, the ratio of the outside diameter to the inside diameter.

12.16 A beam of hollow rectangular cross section is bent about its neutral axis. Show that its shape factor f is given by

$$f = \frac{3}{2} \frac{[1 - (b/a)(d/c)^2]}{[1 - (b/a)(d/c)^3]}$$

where $2a$ is the outside width, $2c$ is the outside depth, $2b$ is the inside width, and $2d$ is the inside depth.

12.17 Show that the shape factor for a beam of rectangular cross section is 1.50. Construct stress distribution diagrams, and determine resultant forces and couples in order to compute f. Then refer to Problem 12.16 and specialize the result for a hollow rectangular cross section in order to verify that $f = 1.50$ for the special case of a rectangular cross section.

12.18 Refer to the equation for f given in Problem 12.15 and determine f for the special case where β becomes infinite (i.e., D_i approaches zero or the section is solid-circular).

12.19 Refer to the equation for f given in Problem 12.15 and determine f for the special case where β

approaches unity. What is the physical meaning of this solution, and does it have any limitation?

12.20 Refer to Example 12.7, increase the flange width to 0.20 m, and solve the same problem again.

12.21 Refer to Example 12.8 and prepare a scaled plot of f versus b/a. Comment on any special case(s).

12.22 A square, hollow cross section for which $b = 2$ in. and $a = 8$ in. is to be used as a beam cross section. Determine f for this cross section by applying fundamental concepts, and check your answer by reference to the equation presented in Example 12.8.

*12.6 The Plastic Hinge Concept

If a unit length of beam fabricated of a ductile material is subjected to increasing bending moments and a plot constructed of the variation of the bending moment versus the curvature per unit length (ϕ), the actual variation is as shown in Figure 12.13. In the simplified theory the idealized curve, consisting of two straight lines, is used to represent what is termed *plastic hinge* behavior. The terminology is used to denote the fact that once M_p is reached at the section, further rotation takes place at

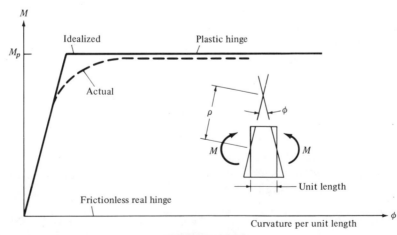

FIGURE 12.13

the section without any increase in bending moment. This behavior is best under-stood by reflecting on the behavior of a frictionless real hinge that rotates freely with zero resistance to applied bending moments as shown in Figure 12.13. In other words, once a beam has yielded over a full cross section, the beam curvature (or rotation) per unit length increases freely under a constant moment equal to the fully plastic moment.

*12.7 Pseudo-Mechanism Formation and Moment Redistribution for Beams

A simply supported beam loaded by a uniform load as shown in Figure 12.14(a) will resist increasing loads until a plastic hinge forms at the center under the action of the fully plastic load, w_p. A larger force cannot be applied to the beam and it will continue to deflect under a constant load. Since free rotation may occur at the ends of this beam and a plastic hinge has formed at the center, the beam behaves much as does a mechanism, and since the plastic hinge at the center is not a real hinge constructed in the beam, the beam is referred to as a *pseudo-mechanism*. The simply supported beam is statically determinate and a single plastic hinge is sufficient to form a pseudo-mechanism.

Statically indeterminate beams require the formation of more than one plastic hinge if they are to behave as pseudo-mechanisms. These plastic hinges do not form simultaneously but form first at sections where the moments have their largest abso-lute values, and subsequently at other sections, where the moments are relatively large in magnitude. For example, the fixed-end beam shown in Figure 12.14(b) is loaded with a uniform load w_1 and plastic hinges have formed at the fixed ends A and B, where the bending moments have their largest magnitudes. If the force per unit length w_1 is increased to the fully plastic value w_p, a final plastic hinge forms at the center C and the beam begins to act as a pseudo-mechanism, as depicted in Figure 12.14(c). An idealized sketch of the pseudo-mechanism is shown in Figure 12.14(d).

Although yielded zones of material, referred to as plastic hinges, extend over finite lengths of beams as shown in Figure 12.14(a), (b), and (c), it is assumed in simple plastic theory, as discussed here, that the plastic hinges form at points as shown in Figure 12.14(d). As the fixed-end beam changes from the condition shown in Figure 12.14(b) to the condition shown in Figure 12.14(c), the applied force per unit length increases from w_1 to w_p. This increasing force per unit length can no longer be resisted at the ends because plastic hinges have already formed at the ends and further increase in moment resistance beyond M_p is not possible for this idealized elasto-plastic material. Moment resistance to increases in load beyond w_1 is provided by the beam at sections other than those at the ends, until the final plastic hinge is

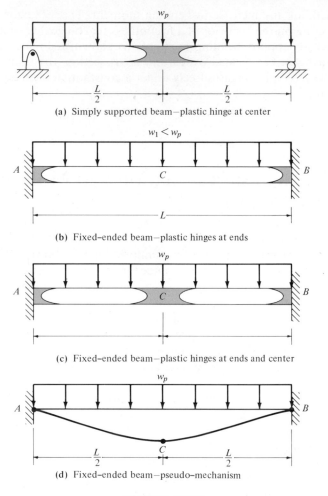

(a) Simply supported beam—plastic hinge at center

(b) Fixed–ended beam—plastic hinges at ends

(c) Fixed–ended beam—plastic hinges at ends and center

(d) Fixed–ended beam—pseudo–mechanism

FIGURE 12.14

formed at the center. This progressive increase in moment resistance after the end plastic hinges have formed is referred to as *redistribution of the moments in the beam.* Examples 12.9, 12.10, and 12.11 will further clarify the plastic hinge and moment redistribution concepts.

Example 12.9

A beam fixed at both ends is subjected to a concentrated load P at its center, as shown in Figure 12.15(a). For $L = 6$ m, $M_y = 160$ kN-m, and $f = 1.20$, determine the center force P corresponding to first yield and the fully plastic load P_p corresponding to formation of a pseudo-mechanism.

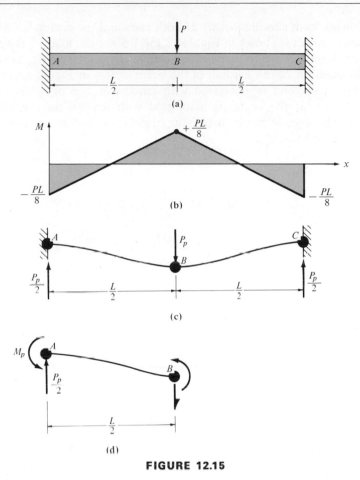

FIGURE 12.15

Solution. Refer to the elastic moment diagram for this loading, which is shown in Figure 12.14(b). In this case the moment resistances at both ends and the center are equal, which means that first yielding will occur on the top and bottom of the beam at both ends and the center. The load P_y, corresponding to first yielding, is calculated as follows:

$$M_y = \frac{P_y L}{8}$$

or

$$P_y = \frac{8M_y}{L} = \frac{8(160)}{6}$$

$$= \boxed{213 \text{ kN}}$$

Plastic hinges form simultaneously at both ends and the center, for which the pseudo-mechanism is shown in Figure 12.15(c). Since the load P_p is applied at the center, the end shears are each equal to $P_p/2$. To determine P_p, consider the free-body diagram of the left half of the beam shown in Figure 12.15(d). The couple M_p at the end is associated with tension on top of the beam, while the couple M_p at the center is associated with tension on the bottom of the beam. The pseudo-mechanism is in equilibrium just as the load reaches the fully plastic value P_p.

$$\circlearrowleft \quad \sum M_B = 0$$

$$M_p + M_p - \frac{P_p}{2}\frac{L}{2} = 0$$

$$P_p = \frac{8M_p}{L}$$

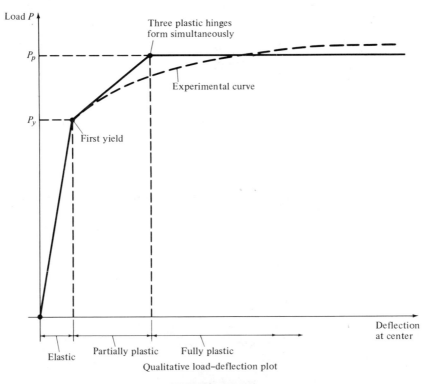

FIGURE 12.16

where

$$M_p = fM_y = 1.20(160) = 192 \text{ kN-m}$$

$$P_p = \frac{8(192)}{6}$$

$$= \boxed{256 \text{ kN}}$$

In this particular case, the ratio of P_p to P_y equals f, as it would for the statically determinate simply supported beam with a center load. Since the resisting moments at the ends and center are equal for the elastic solution, and all three plastic hinges form simultaneously for this example, there is no redistribution of moments. This is not true in general, and the following example will illustrate the advantage of redistribution of moments in indeterminate structures. A qualitative load-deflection plot is shown in Figure 12.16.

Example 12.10

A two-span continuous beam subjected to proportional loading is shown in Figure 12.17(a). Determine the value for P for formation of (a) mechanism ABC, and (b) mechanism CDE. Use $M_p = 250$ kN-m. (c) Which pseudo-mechanism is critical? State the load required to form this pseudo-mechanism.

Solution

(a) Figure 12.17(b) depicts pseudo-mechanism ABC and appropriate free-body diagrams for determination of the force required for collapse of this span.

$\uparrow$ $\sum F_y = 0$:   for ABC:

$$V_A + V_{CL} = 2P$$

$\circlearrowleft$ $\sum M_B = 0$:   for AB:

$$V_A(3) - M_p = 0$$

$$V_A = \frac{M_p}{3}$$

$\circlearrowleft$ $\sum M_B = 0$:   for BC:

$$M_p + M_p - V_{CL}(3) = 0$$

$$V_{CL} = \frac{2M_p}{3}$$

Substitute for V_A and V_{CL} into the vertical-forces summation equation:

$$\frac{M_p}{3} + \frac{2M_p}{3} = 2P$$

$$P = \frac{M_p}{2} = \frac{250}{2}$$

$$= \boxed{125 \text{ kN}} \quad \text{(for } ABC\text{)}$$

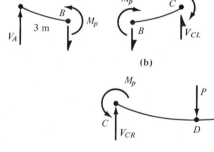

(a)

(b)

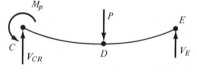

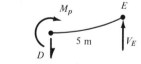

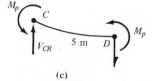

(c)

FIGURE 12.17

(b) Figure 12.17(c) depicts pseudo-mechanism CDE and appropriate free-body diagrams for determination of the force required for collapse of this span.

$$\uparrow \sum F_y = 0: \quad \text{for } CDE:$$

$$V_{CR} + V_E = P$$

$$\circlearrowleft \ \sum M_D = 0: \quad \text{for } DE:$$

$$M_p - V_E(5) = 0$$

$$V_E = \frac{M_p}{5}$$

$$\circlearrowleft \ \sum M_D = 0: \quad \text{for } CD:$$

$$-M_p - M_p + V_{CR}(5) = 0$$

$$V_{CR} = \frac{2M_p}{5}$$

Substitute for V_E and V_{CL} into the vertical-forces summation equation:

$$\tfrac{2}{5}M_p + \tfrac{1}{5}M_p = P$$

$$P = \tfrac{3}{5}M_p = \tfrac{3}{5}(250)$$

$$= \boxed{150 \text{ kN}} \qquad \text{(for } CDE)$$

Since $125 < 150$, mechanism ABC will form first under a total load of $2P = 250$ kN and the right span will be subjected to $P = 125$ kN, which is well below the load of 150 kN required to form mechanism CDE.

The fact that the loading in the left span $2P$ is twice the loading in the right span P is referred to as *proportional loading*, and is an assumption commonly made for simple plastic theory.

Example 12.11
Refer to Figure 12.14, let $M_p = 600$ k-in., $f = 1.15$, and $L = 30$ ft, and determine the following: (a) uniform load associated with first yield of the simply supported beam; (b) uniform load associated with collapse of the simply supported beam; (c) ratio of the answer to part (b) to the answer to part (a); (d) uniform load associated with first yield of the fixed-end beam; (e) uniform load associated with collapse of the fixed-end beam; and (f) ratio of the answer to part (e) to the answer to part (d), and explain why this ratio is different from the answer to part (c).

Solution

(a) For a uniformly loaded simply supported beam, the maximum moment is $wL^2/8$.

$$\frac{wL^2}{8} = M_y$$

$$M_y = \frac{M_p}{f} = \frac{600}{1.15} \times \frac{1}{12} = 43.48 \text{ k-ft}$$

$$w = \frac{8M_y}{L^2} = \frac{8(43.48)}{30^2}$$

$$= \boxed{0.386 \text{ k/ft}, \qquad \text{first yield}}$$

(b) Collapse occurs when a plastic hinge forms at the center. This means that the beam will continue to deflect under constant load.

$$\frac{wL^2}{8} = M_p$$

$$w = \frac{8M_p}{L^2} = \frac{8}{(30)^2} \times 600 \times \frac{1}{12}$$

$$= \boxed{0.444 \text{ k/ft}, \qquad \text{collapse}}$$

(c) Since the beam is determinate, this ratio equals the shape factor, f.

$$\boxed{\frac{0.444}{0.386} = 1.15 = f}$$

(d) Elastic analysis of a fixed-end beam yields the following critical moment magnitudes for the ends and center of the beam:

$$\text{End moment magnitude:} \quad wL^2/12$$

$$\text{Center moment magnitude:} \quad wL^2/24$$

The first yield occurs at the ends of the beam:

$$\frac{wL^2}{12} = M_y = 43.48$$

$$w = \frac{12(43.48)}{(30)^2}$$

$$= \boxed{0.580 \text{ k/ft}, \quad \text{first yield}}$$

(e) Plastic hinges form first at both ends and then at the center. Refer to Figure 12.14(c) and (d). Free-body diagrams will be described later and the reader is urged to sketch them as he or she follows the discussion, since a figure has not been provided. $\sum F_y = 0$ (free-body diagram of the entire beam). There are equal shears at each end, by symmetry:

$$V_L = V_R = \frac{wL}{2}$$

↻ $\sum M_c = 0$ (free-body diagram of segment AC of the beam)

$$M_p + M_p - V_L\left(\frac{L}{2}\right) + w\frac{L}{2}\frac{L}{4} = 0$$

Substitute for V_L and solve for w, to obtain

$$w = \frac{16M_p}{L^2} = \frac{16(600/12)}{(30)^2}$$

$$= \boxed{0.889 \text{ k/ft}, \quad \text{collapse}}$$

(f) Since the beam is indeterminate, there is a redistribution of moments after the end hinges form, which means that additional moment resistance is possible until the third hinge forms at the center.

$$\boxed{\frac{0.889}{0.580} = 1.53 \neq f}$$

This ratio is considerably larger than $f = 1.15$, which reflects the redistribution of moments possible in an indeterminate structure. Of course, this ratio differs for different structures and loadings.

Homework Problems

12.23 The fixed-end beam shown in Figure H12.23 will form a pseudo-mechanism when plastic hinges form at A, B, and C. Determine the fully plastic load P_p for $M_p = 800$ k-in.

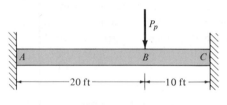

FIGURE H12.23

12.24 A two-span continuous beam is loaded with proportional loads as shown in Figure H12.24. Determine the proportionality constant K_1 such that psuedo-mechanisms ABC and CDE form simultaneously. Use $M_p = 900$ k-in.

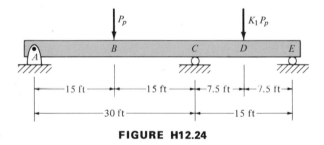

FIGURE H12.24

12.25 Refer to Problem 12.23 and position the load 12 ft from the right end C and keep the span at 30 ft. Determine the fully plastic load P_p for $M_p = 800$ k-in.

12.26 Refer to Problem 12.24 and assign K_1 a value of 1.50. Determine which pseudo-mechanism (ABC or CDE) forms first and the corresponding value of P_p.

12.27 Refer to Figure 12.14; let $M_p = 500$ kN-m, $f = 1.20$, and $L = 10$ m, and determine the following:

(a) The uniform load associated with the first yield of the simply supported beam.
(b) The uniform load associated with collapse of the simply supported beam.
(c) The ratio of the answer to part (b) to the answer to part (a).
(d) The uniform load associated with the first yield of the fixed-end beam.
(e) The uniform load associated with collapse of the fixed-end beam.
(f) The ratio of the answer to part (e) to the answer to part (d). Explain why this ratio is different from the answer to part (c).
(HINT: Refer to Example 12.11.)

12.28 Refer to Example 12.10 and solve the same problem using a yield moment $M_y = 800$ k-in. and a shape factor $f = 1.18$.

12.29 Refer to the beam depicted in Figure H12.29. Determine the proportionality constant K_2 such that pseudo-mechanisms ABC and CDE form simultaneously. Use $M_p = 400$ kN-m.

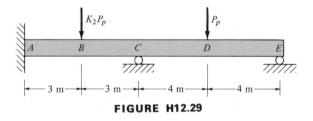

FIGURE H12.29

12.30 Refer to the beam depicted in Figure H12.29 and assign K_2 a value of 1.25. For $M_p = 400$ kN-m, determine which pseudo-mechanism (ABC or CDE) forms first and the corresponding critical value of P_p.

12.31 For the fixed-end beam shown in Figure H12.31, let $M_y = 750$ kN-m, $f = 1.10$, and determine P_p required for the formation of pseudo-mechanism $ABCD$.

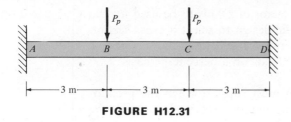

FIGURE H12.31

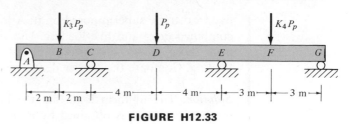

FIGURE H12.33

12.32 Refer to Figure H12.31 and imagine the fixed end D replaced by a roller support that permits free rotation. Let $M_y = 900$ kN-m, $f = 1.20$, and determine P_p for collapse after plastic hinges have formed at sections A, B, and C.

12.33 Refer to Figure H12.33 and note the proportional loading system applied to this three-span continuous beam. Determine the proportionality

constants K_3 and K_4 such that the three pseudo-mechanisms (ABC, CDE, and EFG) form simultaneously. Use $M_p = 600$ kN-m.

12.34 Refer to the three-span continuous beam depicted in Figure H12.33 and assign values to the proportionality constants as follows: $K_3 = 2$ and $K_4 = 1.25$. For $M_p = 500$ kN-m, determine which pseudo-mechanism (ABC, CDE, or EFG) forms first and the corresponding critical value of P_p.

★**12.8** Design for Inelastic Behavior of Torsional and Bending Members

This introduction to plastic design of shafts and beams provides for member selection based upon torque and moment resistance to factored static loadings. The reader should be aware that other requirements not considered here may govern the final design of such members. For example, fatigue failure is often of paramount concern in designing members to resist torque, and erection conditions may require larger beam sections than an "in-place" analysis would indicate. Final design of machine and structural components requires consideration of many factors. Some factors are amenable to analytical methods, and other factors are resolved by judgment based upon experience. The general design philosophy, expressed in Section 12.3, which focuses on load-carrying capacity rather than stress levels, will be followed in the design of shafts and beams.

Example 12.12
A 30-ft-long beam is fixed at both ends and subjected to a uniform load of 5.0 k/ft. It is to be fabricated of steel with a yield stress of 36 ksi. The uniform load is a service loading which allows for both the dead load of the beam and a

live load to be superimposed on it. A load factor of 2.00 is to be used for the combined dead and live loading. Use plastic design to select a cross section from those shown in Appendix F and determine the permissible service live loading. Compare the service dead and live loads on a percentage basis.

Solution. The uniform load that the beam must support when the pseudo-mechanism forms is obtained by multiplying the total service loading by the load factor. In this case

$$w_u = 2.00 \times 5.0 = 10.0 \text{ k/ft}$$

The loaded beam and the pseudo-mechanism are depicted in Figure 12.18(a). Each reaction equals half the total downward load or 150 k acting upward. To determine the required plastic moment M_p, write the equilibrium equation for moments with respect to an axis through B as shown in Figure 12.18(b).

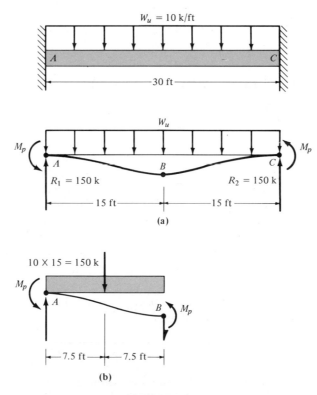

FIGURE 12.18

$$\circlearrowleft \quad \sum M_B = 0$$

$$M_p + 150(7.5) + M_p - 150(15) = 0$$

$$\boxed{M_p = 562.5 \text{ k-ft}}$$

Refer to Appendix F and *select a steel beam W24 × 76*, which provides an $M_p = 603$ k-ft for a yield stress of 36 ksi. Although a steel beam W21 × 82 provides an $M_p = 576$ k-ft for a yield stress of 36 ksi, it weighs more per foot and is therefore not as economical as the W24 × 76. If for architectural or other reasons a shallower beam than the W24 × 76 were required, the W21 × 82 would be selected.

The total service load is given as 5.0 k/ft, and the beam selected weighs 76 lb/ft = 0.076 k/ft. The service live load equals $5.000 - 0.076 = 4.924$ k/ft. The service dead load is a small percentage of the service live load: $(0.076/4.924) \times 100 = 1.54$ percent. For long-span structures such as arches or suspension bridges, the dead load may become a major consideration for the design engineer. In practice, dead loads are practically always estimated or known by the designer. In this example the dead load is a small percentage of the live load, but this is not universally true.

Example 12.13

Design a hollow circular shaft with an inside diameter equal to 0.8 of the outside diameter to resist a service maximum torque of 500 k-in. applied at the center of a 30-ft-long shaft which is torsionally fixed at both ends. Use a load factor of 1.8 and a tensile yield stress of 36 ksi. Assume that the shearing yield stress equals the tensile yield stress divided by $\sqrt{3}$. Select a standard pipe cross section for this shaft.

Solution. The ultimate torque is obtained by multiplying the service torque by the load factor:

$$T_u = 500 \times 1.8 = 900 \text{ k-in.}$$

This ultimate torque will be shared equally at each end of the shaft, since it is applied at the center and the end conditions are identical.

$$\tau_y = \frac{\sigma_y}{\sqrt{3}} = \frac{36}{\sqrt{3}} = 20.8 \text{ ksi}$$

Equation 12.8 solved for $r_o^3 - r_i^3$ becomes

$$\left(r_o^3 - r_i^3\right) = \frac{3T_p}{2\pi\tau_y} = \frac{3(900/2)}{2\pi(20.8)}$$

Required: $\left(r_o^3 - r_i^3\right) = 10.33$ in.^{3}. But $r_i = 0.8r_o$:

$$r_o^3 - (0.8r_o)^3 = 10.33$$

$$r_o = 2.76 \text{ in.}$$

$$D_o = 5.52 \text{ in.} \quad \text{and} \quad D_i = 4.42 \text{ in.}$$

From Appendix E, 5-in.-nominal-diameter double-extra-strong pipe with $D_o = 5.563$ in. and $D_i = 4.063$ in. would provide

$$\left(r_o^3 - r_i^3\right) = \frac{(5.563)^3 - (4.063)^3}{2^3} = 13.14 \text{ in.}^3$$

Since $13.14 > 10.33$, the pipe is satisfactory provided that it is fabricated of a material with a tensile yield stress of 36 ksi or better. Of course, selection of larger-diameter or stronger pipe for this application would be uneconomical.

Homework Problems

12.35 Refer to Example 12.12 and increase the total service loading from 5.0 to 6.5 k/ft. Design the beam for this increased loading and determine the permissible service live loading. Compare the service dead and live loads on a percentage basis.

12.36 Refer to Example 12.12 and redesign the beam using first yielding as a criterion (refer to Chapter 10). Use the 36-ksi steel with a factor of safety of 2. The total service loading is to remain at 5.0 k/ft. Compare this design with the solution to Example 12.12. (HINT: First yielding will occur at both ends at the top and bottom of the beam under a moment of $wL^2/12$.)

12.37 Refer to Example 12.12 and solve the problem with the following changes:

(a) Change the span from 30 to 40 ft.
(b) Change the steel yield stress from 36 to 50 ksi.

12.38 Refer to Example 12.13 and change the ratio of the inside to the outside diameter from 0.8 to 0.9 and select a pipe cross section for this hollow shaft from Appendix E.

12.39 Design a hollow circular shaft with an inside diameter equal to 0.85 of the outside diameter to resist a service maximum torque of 2000 N-m applied at the center of a 4-m-long shaft which is torsionally fixed at both ends. Use a load factor of 1.7. The shaft is to be fabricated of a material for which the shearing yield stress is 300 MPa. If the pipe sections shown in Appendix E are fabricated of a material for which the shearing yield stress is 300 MPa, select a pipe section for this shaft.

12.40 Design a solid circular shaft to resist a service maximum torque of 2500 N-m applied at the center of a 3-m-long shaft which is torsionally fixed at both ends. Use a load factor of 2.0. The shaft is to be fabricated of a material for which the shearing yield stress is 200 MPa.

12.41 A fixed-end beam subjected to two concentrated forces is shown in Figure H12.41. These forces are service loadings which include a small allowance for the beam dead load. Use a load factor of 2.0 and select a beam from Appendix F. Carefully convert units when you select the lightest beam from the table for a yield stress of 50 ksi

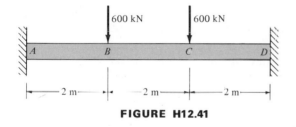

FIGURE H12.41

12.42 A fixed-end beam shown in Figure H12.42 is subjected to a service live load of 220 k, which includes a small allowance for the beam dead load. Use a load factor of 2.0 and select a beam fabricated of steel with a yield stress of 50 ksi from Appendix F.

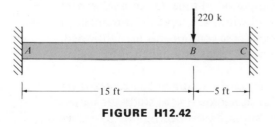

FIGURE H12.42

12.43 A two-span continuous beam depicted in Figure H12.43 is to be fabricated of steel having a

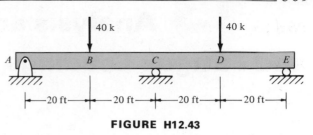

FIGURE H12.43

yield stress of 36 ksi. It is to carry two symmetrically placed service live loads of 40 k which have been increased by a small percentage to account for the dead load of the beam. Use a load factor of 2 and refer to Appendix F to select an appropriate beam.

12.44 A cantilever beam 10 ft long is to be designed to support an end service load of 30 k. This service load includes a small allowance for the beam weight. For a steel with a yield stress of 36 ksi, select an appropriate beam using each of the following methods:
(a) Elastic design (Appendix E) with a factor of safety of 2 with respect to first yield.
(b) Plastic design (Appendix F) using a load factor of 2.

Are the answers to (a) and (b) the same? The cantilever is a determinate structure. For a highly redundant structure, is the answer to (a) and (b) likely to be the same?

12.45 A two-span continuous beam shown in Figure H12.45 is loaded with a uniform service load of 6 k/ft, which contains an allowance for the beam dead load. The beam is to be fabricated of a material with a yield stress of 36 ksi. Use a load factor of 1.8 and select a beam from Appendix F.

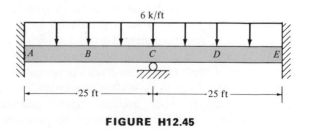

FIGURE H12.45

Chapter 13 Analysis and Design for Impact and Fatigue Loadings

*13.1 Introduction

Statically or gradually applied loadings are considered exclusively in all other chapters of this text. In this chapter, two loadings of vital interest in dynamic analysis and design are considered: impact and fatigue loadings. For example, highway and railway bridges are subjected to impact loading as vehicles cross them and allowances for such loadings are made in their design. High-speed machines contain many moving parts that are subjected to loading which varies cyclically with time. Repeated cyclical loading can lead to materials failure at stress levels substantially lower than stresses associated with static loading, provided that the number of cycles is relatively large and the range of stresses from minimum to maximum is of sizable magnitude. Such failures begin at "stress raisers" such as internal or surface flaws in the material or abrupt geometric changes. A small (often microscopic) crack forms and propagates until a section of a given component is too weak to resist further cycles of loading. The resulting failure is termed a *fatigue failure* and the associated cyclically repeated loadings are termed *fatigue loadings*.

A unified theory based upon the law of conservation of energy is presented for analysis of components subjected to axial, torsional, and bending impact loadings. These impact loadings are usually delivered by a raised weight which is dropped through a vertical distance before contacting the component to be analyzed. Potential energy of the raised weight is transformed into internal strain energy stored in the stressed body if losses are neglected. In practice, a portion of this potential energy is lost to noise and heat during the impact. Maximum stresses in the component being analyzed for impact loading occur instantaneously and are not as severe as stresses resulting from a loading that endures for a long period of time. Of course, impact stresses of very high levels or those of lower level which are repeated a number of times can lead to failure of the materials of which these components are fabricated. Failure of critical machine or structural components frequently have led to overall machine or structural failures.

Recent research on behavior of materials subjected to fatigue has centered on studies of crack propagation and use of the electron microscope to study the nature of these failures as cracks propagate through the stressed member. Empirical studies continue to be widely employed for engineering design. These studies have centered around costly and time-consuming test programs for each material which result in S–N curves, where S refers to a completely reversed cycle of stress and N refers to the

610

number of cycles associated with fatigue failure at a given stress level. Incomplete reversal of stresses, which is the usual case in practice, may be analyzed as the sum of a static stress component and a completely reversed cycle of stress. Three methods of analyzing incomplete reversal of stresses advanced by Gerber, Goodman, and Soderberg are presented and extended to design by the use of appropriate static and fatigue factors of safety. Axial, torsional, and bending loadings applied cyclically are considered in design examples.

The reader should be aware that this chapter provides an introduction to impact and fatigue loadings, but the literature on these topics, particularly fatigue, is voluminous and further study would be required before one could confidently design structures and machines for these dynamic effects.

13.2 Stress Concentrations

Stresses computed from the equations developed in previous chapters of this text are often referred to as *nominal stresses*. These equations do not account for the fact that stresses may reach much larger local values as a result of the following:

1. Loads or reactions delivered to a component over a relatively small area.
2. Internal or surface flaws in the material of which a component is fabricated.
3. Geometric changes in a component such as those discussed in the examples of this section.

To account for the occurrence of these increases in stress due to static loadings, the stress concentration factor k is defined as the ratio of the maximum stress associated with a well-defined geometry to the nominal stress computed from equations such as 3.1, 4.19, and 5.10a. In modified form to account for stress concentrations, these equations become

$$\sigma_{max} = k \frac{F_n}{A_{net}} \tag{13.1}$$

$$\tau_{max} = k \frac{T\rho}{J_{min}} \tag{13.2}$$

$$\sigma_{max} = k \frac{M_u v}{I_{net}} \tag{13.3}$$

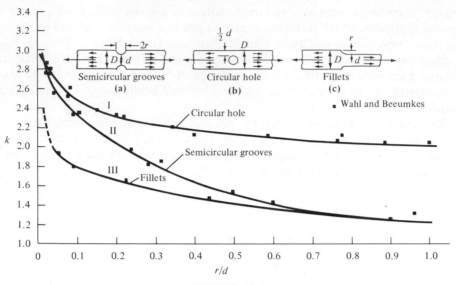

FIGURE 13.1

After M. M. Frocht, "Factors of Stress Concentration Photoelastically Determined," *Trans. ASME*, Vol. 57, 1935; by permission of the American Society of Mechanical Engineers.

where A_{net}, J_{min}, and I_{net} are, respectively, the net area, minimum polar moment of inertia, and net rectangular moment of inertia for a given cross section, and k is the stress concentration factor.

Stress concentration factors k are usually determined experimentally, although some analytical work has been done, and modern computer approaches with finite elements is usually most economical. Figure 13.1 shows stress concentration factors for axially loaded members which were determined by the experimental method referred to as *photoelasticity*. Figure 13.2 shows stress concentration factors for shearing stresses in a stepped shaft of two different diameters determined by electrical analog experiments. Figure 13.3 shows stress concentration factors for a notched flexural member in pure bending determined by photoelasticity. Stress concentration data provided in these three figures are only a small sampling of a plethora of similar information available in the literature of this subject.

Examples 13.1, 13.2, and 13.3 illustrate the use of stress concentration factors for axial, torsional, and bending loadings. Stresses such as those computed in these examples are not significant for ductile materials loaded statically, since such materials are capable of undergoing very large plastic strains prior to fracture while their stress levels remain practically constant if strain hardening is neglected. Use of stress concentration factors for static loading of components fabricated of ductile materials will usually be a very conservative approach. Methods developed in Chapter 12 are more suitable for analysis and design of components fabricated of ductile metals

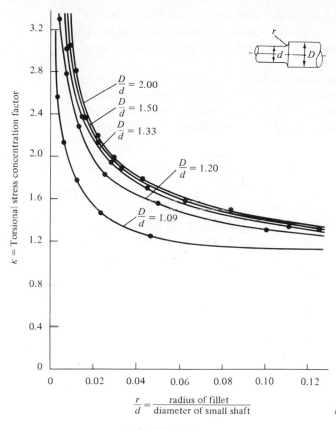

FIGURE 13.2

After L. S. Jacobsen, "Torsional Stress Concentrations in Shafts of Circular Cross-Sections and Variable Diameter," *Trans. ASME*, Vol. 47, 1925; by permission of the American Society of Mechanical Engineers.

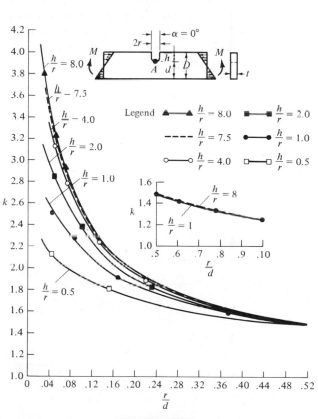

FIGURE 13.3

After M. M. Leven and M. M. Frocht, "Stress Concentration Factors for a Single Notch in a Flat Bar in Pure and Central Bending," *Trans. ASME*, Vol. 74, 1952; by permission of the American Society of Mechanical Engineers.

subjected to static loadings. Stresses such as those computed in these examples are significant for brittle materials loaded statically, since such materials fracture at relatively low strains and each stress increase is accompanied by a corresponding increase in strain until fracture occurs.

Stress concentrations are always significant for fatigue loadings, as discussed in Section 13.4. The examples and Homework Problems of Section 13.4 have static and fatigue stress concentration factors (k_s and k_f) stated as required. It should be noted that $k_f < k_s$ and that k_f would usually be determined from fatigue loading experiments.

Example 13.1

An axially loaded plate is shown in Figure 13.4. It is 3.00 in. wide and has a thickness of 0.50 in. A 1.00-in. hole is drilled through the plate at its center. Use the stress concentration factor data of Figure 13.1 to find the maximum tensile stresses along lines at the edges of the hole. These lines extend throughout the thickness of the plate and are denoted by B and C in Figure 13.4.

Solution. The stress concentration factor k is a function of the r/d ratio, where r is the radius of the hole and d is the net width of the plate. In this case $d = D - 2r$.

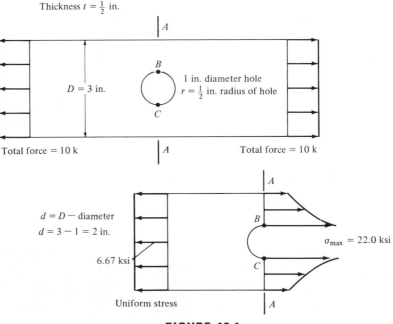

FIGURE 13.4

$$d = 3.00 - 2(0.50) = 2.00 \text{ in.}$$

$$\frac{r}{d} = \frac{0.50}{2.00} = 0.25$$

Read the ordinate to the curve for a circular hole of Figure 13.1 for $r/d = 0.25$:

$$k = 2.28$$

$$\sigma_{\max} = k \frac{F_n}{A_{\text{net}}} \qquad\qquad\qquad \textbf{(13.1)}$$

$$= 2.28 \frac{10}{2(0.50)} = 22.8 \text{ ksi}$$

The variation of stress across the plate width is uniform on sections far enough from the hole such that its effect is minimized. In this case sections 3 in. on either side of the hole would have practically uniform stress distributions. It has been shown both experimentally and analytically that the effects of stress concentrations disappear, for all practical purposes, at relatively small distances (i.e., distances approximately equal to the largest cross-sectional member dimension) from the position of stress disturbance. This concept is known in the literature as *Saint-Venant's principle*. A section cut through the hole would have a nonlinear variation of stress as shown for A–A of Figure 13.4.

Example 13.2
A stepped shaft is shown in Figure 13.5. It is subjected to a constant torque of $T - 1040$ N-m. The diameter of the smaller shaft is $d = 0.04$ m and the diameter of the larger shaft is 0.08 m. At the plane where the smaller and larger shafts are joined, a fillet of $r = 0.004$ m is used around the complete circumference of the smaller shaft. Use the stress concentration factor data of Figure 13.2 to find the shearing stress around the circumference of the smaller shaft.

Solution. The stress concentration factor k is a function of two ratios (D/d and r/d) as shown in Figure 13.2. The diametral ratio $D/d = 0.08/0.04 = 2$. The ratio

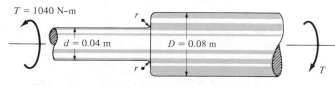

$T = 1040$ N-m $\quad r$

$d = 0.04$ m $\qquad D = 0.08$ m

$r \quad T$

Fillet radius $r = 0.004$ m

FIGURE 13.5

of the fillet radius to the diameter of the smaller shaft: $r/d = 0.004/0.04 = 0.1$. Read the ordinate to the $D/d = 2$ curve of Figure 13.2 for $r/d = 0.1$, $k = 1.45$:

$$\tau_{max} = k\,\frac{T\rho}{J_{min}} \qquad\qquad (13.2)$$

$$= 1.45\,\frac{1040(0.02)}{(\pi/32)(0.04)^4}$$

$$= \boxed{120 \text{ MPa}}$$

Example 13.3

A notched flexural member is subjected to bending moments $M = 120$ k-in. as shown in Figure 13.6. The member has a depth $D = 10$ in. and a thickness $t = 2$ in. The notch has a total depth $h = 4$ in. with a rounded bottom of $r = 1$ in. Use the stress concentration factor data of Figure 13.3 to find the flexural stress at the base of the notch along the line extending over the full thickness of the member and denoted by A in Figure 13.6.

Solution. The stress concentration factor k is a function of two ratios, h/r and r/d, as shown in Figure 13.3.

The ratio of the notch depth h to the notch radius r is

$$\frac{h}{r} = \frac{4}{1} = 4.0$$

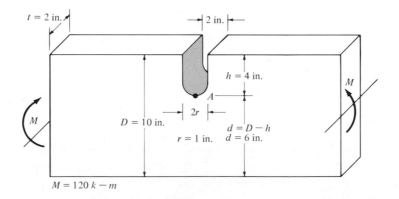

FIGURE 13.6

The ratio of the notch radius r to the net depth d is

$$\frac{r}{d} = \frac{1}{6} = 0.167$$

Read the ordinate to the $h/r = 4.0$ curve of Figure 13.3 for $r/d = 0.167$, $k = 2.06$:

$$\sigma_{max} = k \frac{M_u v}{I_{net}} \qquad (13.3)$$

where $v = d/2$ and $I_{net} = \frac{1}{12}td^3$:

$$\sigma_{max} = 2.06 \frac{120(3)}{\frac{1}{12}(2)(6)^3}$$

$$= \boxed{10.6 \text{ ksi}}$$

★13.3 Axial, Torsional, and Bending Impact Analyses

The three impact situations depicted in Figure 13.7 have much in common. First consider assumptions and limitations placed upon the impact solutions to follow.

1. Assume that all the potential energy of the elevated weight W is transformed into elastic strain energy as it falls through the height h, upon release from rest. Actually, some of this energy will be transformed to heat and sound during the impact as W strikes the body at the end of its vertical travel. Energy of each system is conserved, and assuming that the potential energy is transformed into elastic strain energy is an assumption on the safe side, since it will lead to higher stresses than an analysis that allows for energy losses to heat and sound.
2. Calculated stresses must have magnitudes less than the proportional limit stresses of the materials of which the impacted bodies are fabricated. Linear elastic behavior of the stressed bodies is assumed in each case.
3. Assume that the dead load stresses associated with the weights of the impacted bodies are negligible compared to the stresses associated with impact.
4. Neglect the work done by their own weights as the impacted bodies are deformed.

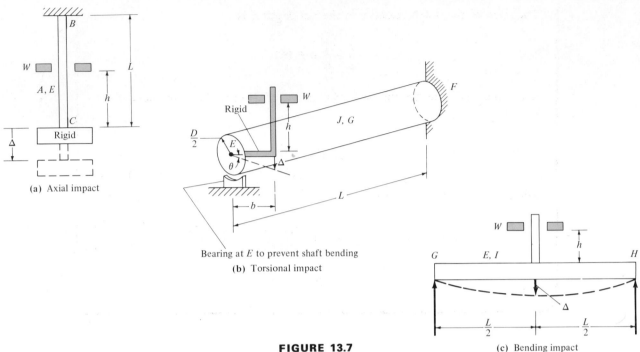

FIGURE 13.7

(a) Axial impact

(b) Torsional impact

Bearing at E to prevent shaft bending

(c) Bending impact

5. Rigid body elements of any system do not absorb energy because they are nondeformable.

6. Maximum stresses and deflections occur instantaneously in the impacted bodies, which continue to oscillate until the vibrations associated with the impact are damped out.

In Figure 13.7(a), (b), and (c), the impacted bodies are the axially loaded rod BC, the torsionally loaded shaft EF, and the transversely loaded beam GH, respectively. The shaft EF is assumed to be supported such that the development of bending stresses is prevented.

The fundamental equation for these impact analyses is a statement of the *law of conservation of energy*, which states:

> *potential energy of W = internal elastic strain energy stored in the stressed body at maximum deformation*

Mathematically, this becomes

$$W(h + \Delta) = U \tag{13.4}$$

where W = weight dropped from rest, which impacts the stressed body

h = height from which the weight is dropped

Δ = maximum deformation of the stressed body at the point where the impact takes place

U = internal elastic strain energy stored in the stressed body at maximum deformation

This quantity U depends upon the nature of the loading. Equations are developed below for axial, torsional, and transverse bending loadings.

Consider an element of volume dV taken from the rod BC of Figure 13.7(a) at the instant when W has come to rest at the end of its downward travel and the rod has been stretched its maximum amount Δ. At this instant the stress in the volume element dV has been built up from zero to a maximum value σ, which means that the average stress during the deformation of the rod is $\frac{1}{2}\sigma$. The differential strain energy dU stored in the volume element dV is given by $\frac{1}{2}\sigma\varepsilon$, where ε is the maximum strain of the rod associated with the maximum stress σ and the maximum total elongation Δ.

$$\frac{dU}{dV} = \frac{1}{2}\sigma\varepsilon \tag{13.5}$$

A graphical interpretation of Eq. 13.5 is shown in Figure 13.8, and students will find this a convenient way to understand and recall the equations. To determine the quantity U, integrate Eq. 13.5:

$$\int dU = U = \int_{vol} \tfrac{1}{2}\sigma\varepsilon \, dV \tag{13.6}$$

The integration extends throughout the volume of the rod and the integration is simplified by the fact that σ and ε are constant throughout the volume. The maximum stress in the rod is given by

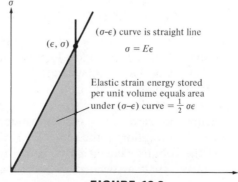

FIGURE 13.8

$$\sigma = \frac{F}{A} \tag{13.7}$$

where F is the maximum force in the rod when W has reached its lowest point during the impact and A is the cross-sectional area of the rod.

The corresponding maximum strain ε in the rod is given by

$$\varepsilon = \frac{\Delta}{L} \tag{13.8}$$

where L is the length of the rod. Substitution of Eqs. 13.7 and 13.8 into Eq. 13.6 gives the following equation for U:

$$U = \frac{1}{2}\frac{F}{A}\frac{\Delta}{L}\int_{vol} dV = \frac{1}{2}\frac{F}{A}\frac{\Delta}{L}AL = \frac{1}{2}F\Delta \tag{13.9}$$

Equation 13.9 substituted into Eq. 13.4 gives the basic equation for solving axial impact problems:

$$W(h + \Delta) = \tfrac{1}{2}F\Delta \tag{13.10}$$

The *impact factor* is defined as the ratio of the maximum force F exerted on the rod during the impact to the static force W if it were applied without impact.

$$IF = \frac{F}{W} \tag{13.11}$$

In addition to Eqs. 13.10 and 13.11, the relationship between F and Δ will be required for solving axial impact problems:

$$\Delta = \frac{FL}{AE} \tag{13.12}$$

To analyze torsional impact depicted in Figure 13.7(b), consider an element of volume dV taken from the shaft EF at the instant when W has come to rest at the end of its downward travel and the shaft has rotated through its maximum angle $\theta = \Delta/b$, provided that θ is a small angle. At this instant the shearing stress in the volume element dV has been built up from zero to a maximum value τ, which means that average shearing stress during the rotation of the rod is $\tfrac{1}{2}\tau$. The differential shearing strain energy dU stored in the volume element dV is given by $\tfrac{1}{2}\tau\gamma$, where γ is the maximum shearing strain of the shaft associated with the maximum deflection Δ measured at the point of impact.

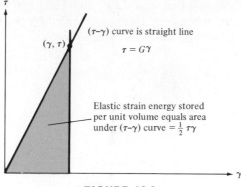

FIGURE 13.9

$$\frac{dU}{dV} = \frac{1}{2}\tau\gamma \tag{13.13}$$

A graphical interpretation of Eq. 13.13 is shown in Figure 13.9. To determine the quantity U, integrate Eq. 13.13:

$$\int dU = U = \int_{\text{vol}} \tfrac{1}{2}\tau\gamma \, dV \tag{13.14}$$

It can be shown that, as for the axially loaded rod, U is given by

$$U = \tfrac{1}{2}F\Delta \tag{13.15}$$

Details of this development are not presented here. Refer to Problem 13.1, which deals with this development.

In summary, Eqs. 13.10 and 13.11 are also valid for solving torsional impact problems when they are supplemented by

$$T = Fb \tag{13.16}$$

$$\Delta = \theta b = \frac{TLb}{JG} = \frac{Fb^2L}{JG} \tag{13.17}$$

$$\tau = \frac{T\rho}{J} = \frac{FbD}{2J} \tag{13.18}$$

Similarly, it can be shown that Eqs. 13.10 and 13.11 are also valid for solving bending impact problems when they are supplemented by the following equations, which apply to a simply supported beam impacted at its center, as shown in Figure 13.7(c):

$$M = \frac{FL}{4} \tag{13.19}$$

$$\Delta = \frac{FL^3}{48EI_u} \tag{13.20}$$

$$\sigma = \frac{M_u v}{I_u} = \frac{FLD/2}{4I_u} \tag{13.21}$$

Equations for other loading positions and other beam supports will be presented as required in the examples. Several Homework Problems require the development of the appropriate equations.

Example 13.4

Refer to Figure 13.10 and for $W = 100$ N, $h = 0.1$ m, determine the maximum stress in the rods, the impact factor, and the maximum deflection of the rods. Assume that 75 percent of the potential energy is effective in stressing the rods. Each rod has a length of 3.0 m and has a cross-sectional area of 2×10^{-4} m^2. Use $E = 200$ GPa for the rods.

Solution. Modify Eq. 13.10 to account for energy losses: $0.75W(h + \Delta) = \frac{1}{2}F\Delta$. From Eq. 13.12:

$$\Delta = \frac{FL}{AE}$$

$$= \frac{F(3.0)}{2 \times 10^{-4}(200 \times 10^9)} = 7.5 \times 10^{-8}F$$

With this latter relationship, express the energy equation in terms of F.

Substitute numerical values with an additional factor of $\frac{1}{2}$ introduced on the left side of Eq. 13.10 to account for the fact that two rods resist the impact loading:

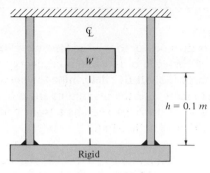

FIGURE 13.10

$$\tfrac{1}{2}(0.75)100(0.1 + 7.5 \times 10^{-8}F) = \tfrac{1}{2}F(7.5 \times 10^{-8}F)$$

$$F^2 - 75F - 10^8 = 0$$

Solve this quadratic for F:

$$F = 10{,}038 \text{ N} \qquad \text{(impact maximum force in each rod)}$$

Since the force is axially applied, the maximum stress in each rod is

$$\sigma = \frac{F}{A} = \frac{10038}{2 \times 10^{-4}} = \boxed{50.2 \text{ MPa}}$$

From Eq. 13.11, the impact factor is

$$IF = \frac{F}{W} = \frac{10038}{100} \doteq 100$$

From Eq. 13.12, the maximum deflection is

$$\Delta = 7.5 \times 10^{-8}F = 7.5 \times 10^{-8}(10038) = \boxed{7.53 \times 10^{-4} \text{ m}}$$

Note that $h = 0.1$ m is much larger than $\Delta = 7.53 \times 10^{-4}$ m, which means that the linear term $75F$ of the quadratic for F could have been neglected without serious error. An impact factor of 100 means that the stress and deflection are 100 times the values for a static application of W. However, these impact stresses and deflections occur only at an instant of time which somewhat mitigates their effect on the materials as compared to long-time loading effects. The temperature of the rods is assumed to be not low enough that brittle fracture may occur.

Example 13.5

A weight $W = 120$ lb is dropped onto the rigid member AB as shown in Figure 13.11. Assuming that 80 percent of the potential energy of W is transformed into shearing strain energy in shaft BC, determine the maximum shearing stress in the shaft, the impact factor, and the maximum angle of rotation of the shaft. Note that the bearing at B prevents the shaft from bending as a cantilever beam. Use $G = 12 \times 10^6$ psi for the shaft.

Solution. Equations 13.10, 13.11, and 13.16 to 13.18 will be employed in solving this torsional impact problem. Modify Eq. 13.10 to account for energy losses:

$$0.80W(h + \Delta) = \tfrac{1}{2}F\Delta$$

From Eq. 13.17:

$$\Delta = \frac{Fb^2L}{JG}$$

Substitute given numerical values into these equations:

$$0.80(120)(0.80 + \Delta) = \tfrac{1}{2}F\Delta$$

$$\Delta = \frac{F(10)^2(36)}{\pi/32(4.5)^4(12 \times 10^6)} = 7.452 \times 10^{-6}F$$

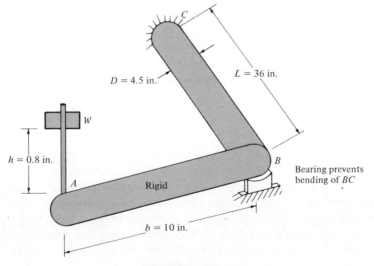

FIGURE 13.11

With this latter relationship, express the energy equation in terms of F:

$$76.8 + 715.392 \times 10^{-6}F = 3.726 \times 10^{-6}F^2$$

$$F^2 - 192F - 2.0612 \times 10^7 = 0$$

$$F = \frac{192 \pm \sqrt{(-192)^2 - 4(1)(-2.0612 \times 10^7)}}{2}$$

$$= 4637 \text{ lb} \qquad \text{(maximum force developed at impact point } A\text{)}$$

The maximum torque applied to the shaft equals Fb:

$$T = Fb = 4637(10) = 46370 \text{ lb-in.}$$

The maximum shearing stress in the shaft is given by Eq. 13.18:

$$\tau = \frac{T\rho}{J} = \frac{FbD}{2J} = \frac{46,370(4.5)}{2\pi/32(4.5)^4} = 2592 \text{ psi}$$

The impact factor is, from Eq. 13.11,

$$IF = \frac{F}{W} = \frac{4637}{120} = 38.6$$

The maximum angle of rotation is given by

$$\theta = \frac{\Delta}{b} = \frac{7.452 \times 10^{-6}F}{10}$$

$$= \frac{7.452 \times 10^{-6}(4637)}{10} = 3.46 \times 10^{-3} \text{ rad} \quad \text{or} \quad 0.198°$$

Example 13.6

In Figure 13.12 a 20-lb weight W is dropped through a distance of $h = 2$ in. before striking a 5-ft-long fixed-end beam. The steel beam has a depth of 2 in. and a width of 4 in., a modulus of elasticity of 30×10^6 psi, and a yield stress of 33,000 psi. If 90 percent of the potential energy of W is effective in stressing the beam, find the maximum stress in the beam, the impact factor, and maximum deflection of the beam. Consider bending effects to predominate when compared to shear effects. Comment on the validity of the solution.

Solution. If the beam were simply supported, Eqs. 13.19, 13.20, and 13.21

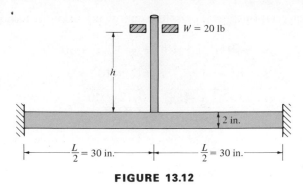

FIGURE 13.12

would be directly applicable; but the beam is fixed at both ends, and these equations must be modified as shown below.

Modify Eq. 13.10 to account for energy losses: $0.90W(h + \Delta) = \frac{1}{2}F\Delta$. For a fixed-ended beam loaded at the center:

$$\Delta = \frac{FL^3}{192EI} = \frac{F(60)^3}{192(30 \times 10^6)(1/12)(4)(2)^3}$$

$$= 1.406 \times 10^{-5}F$$

With this latter relationship, express the energy equation in terms of F:

$$0.90(20)(2 + 1.406 \times 10^{-5}F) = \frac{1}{2}F(1.406 \times 10^{-5}F)$$

$$36 + 2.531 \times 10^{-4}F = 0.703 \times 10^{-5}F^2$$

$$F^2 - 36.00F - 5.121 \times 10^6 = 0$$

$$F = \frac{36 \pm \sqrt{(-36.00)^2 - 4(1)(-5.121 \times 10^6)}}{2}$$

$$= 2281 \text{ lb} \qquad \text{(maximum force developed at impact point)}$$

Bending moments in this fixed-end beam reach their largest magnitudes at the ends and the center.

$$M = \frac{FL}{8} = \frac{2281 \times 60}{8} = 17{,}110 \text{ lb-in.}$$

The maximum stress is given by the flexure formula:

$$\sigma = \frac{M_v}{I_v}u = \frac{17{,}110(1)}{(1/12)(4)(2)^3} = 6420 \text{ psi}$$

The impact factor is, from Eq. 13.11,

$$IF = \frac{F}{W} = \frac{2281}{20} = 114$$

The maximum deflection is given by

$$\Delta = \frac{FL^3}{192EI} = \frac{2281(60)^3}{192(30 \times 10^6)(\frac{1}{12})(4)(2)^3}$$

$$= 0.0321 \text{ in.}$$

Since the maximum stress of 6420 psi is well below the yield stress of 33,000 psi, the elastic solution presented here is valid. Of course, any stress concentrations have been neglected in this solution. Furthermore, the maximum stresses and deflections occur at an instant of time rather than during some time interval. The effect on the material is mitigated somewhat due to the fact that these maximums occur instantaneously. If the temperature of the beam were very low, a brittle fracture may occur even though the material is ductile at ordinary temperatures.

Homework Problems

13.1 Evaluate the integral of Eq. 13.14 and hence verify Eq. 13.15. In this case the shearing stress and strain are functions of ρ, and the element of volume should be taken as $dV = \rho \, d\rho \, d\theta \, dz$, where z is measured along the length of the shaft and (ρ, θ) are polar coordinates in the cross section.

13.2 Solve Example 13.4 with W increased to 200 N and compare your answer with the answers for Example 13.4.

13.3 Neglect energy losses and solve Example 13.5 again. Compare your answers with the answers for Example 13.5.

13.4 Solve Example 13.6 by assuming that the impacted beam is simply supported rather than fixed-ended. Compare your answers with the answers for Example 13.6. Comment on the validity of your solution.

13.5 Refer to Example 13.6 and use the given information except for rotating the beam such that the 2-in. side is placed horizontally. Compare your answers with the answers for Example 13.6.

13.6 Refer to Figure 13.7(a), neglect energy losses, and derive general equations for maximum axial stress, impact factor, and maximum deflection in terms of the given quantities W, h, L, A, and E.

13.7 Refer to the solution for Problem 13.6 and let $h = 0$. This is known as a *suddenly applied load*. Compare results for maximum stress and deflection for suddenly applied loading with the results for static loading.

13.8 Refer to Figure 13.7(b), neglect energy losses, and derive general equations for maximum shearing stress, maximum rotation, and impact factor in terms of the given quantities W, h, L, b, J, and G.

13.9 Refer to Figure 13.7(c), neglect energy losses, and derive general equations for maximum bending stress, maximum deflection, and impact factor in terms of the given quantities W, h, L, I, and E.

13.10 A body of weight $W = 25$ lb falls a height $h = 4.2$ ft, as shown in Figure H13.10, and impacts the rigid plate at the base of an 8-ft-long steel rod. What must be the cross-sectional area of the rod to prevent yielding of the material? The yield stress for the steel is 40 ksi and $E = 30 \times 10^6$ psi. Neglect energy losses during impact.

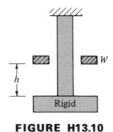

FIGURE H13.10

13.11 A body of weight $W = 112$ N falls a height $h = 1.30$ m, as shown in Figure H13.10, and impacts the rigid plate at the base of a 2.50-m-long steel rod. What must be the cross-sectional area of the rod to prevent yielding of the material? The yield stress for the steel is 276 MPa and $E = 207$ GPa.

13.12 A cantilever beam fabricated of aluminum alloy is 4 ft long and is subjected to a weight W of 80 lb which is dropped a height $h = 4$ in. before impacting the end of the beam. If the beam has a width of 3 in. and a depth of 9 in., determine the maximum bending stress due to impact. The modulus of elasticity is 10.2×10^6 psi. Neglect energy losses during impact.

13.13 A cantilever beam fabricated of aluminum alloy is 1 m long and is subjected to a weight W of

350 N, which is dropped a height $h = 0.10$ m before impacting the end of the beam. If the beam has a width of 0.08 m and a depth of 0.24 m, determine the maximum bending stress due to impact. The modulus of elasticity is 70.2 GPa. Neglect energy losses during impact.

13.14 A diving board with unyielding supports is depicted in Figure H13.14. A diver weighing 650 N jumps from a height $h = 0.05$ m onto the end of the board. Determine the maximum bending stress in the diving board and its maximum end deflection for the following set of input values: $a = 1.80$ m, $b = 1.00$ m, $c = 0.26$ m, $d = 0.05$ m, and $E = 14.0$ GPa. Compare the impact values for maximum bending stress and the deflection to corresponding static values for application of the divers weight. Neglect energy losses during impact.

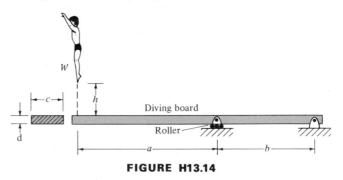

FIGURE H13.14

13.15 A diving board with unyielding supports is depicted in Figure H13.14. A diver weighing 150 lb jumps from a height $h = 2.0$ in. onto the end of the board. Determine the maximum bending stress in the diving board and its maximum end deflection for the following set of input values: $a = 72$ in., $b = 40$ in., $c = 10.5$ in., $d = 1.875$ in., and $E = 2.00 \times 10^6$ psi. Compare the impact values for the maximum bending stress and deflection to corresponding static values for application of the diver's weight. Neglect energy losses during impact.

13.16 A solid, circular shaft with a flywheel attached at one end rotates at 10 rad/sec. The shaft is suddenly stopped at the opposite end, and the kinetic energy of the flywheel is transformed into elastic strain energy in the shaft. Neglect losses of energy during the impact as well as the kinetic energy of the shaft. Determine the maximum shearing stress and maximum angle of rotation of the shaft for the following input values: flywheel weight = 500 N; idealize the flywheel as a disk of radius = 0.30 m, length of shaft = 1.20 m, diameter of shaft = 0.05 m, and $G = 80.0$ GPa.

13.17 A hollow circular shaft with a flywheel attached at one end rotates at 8.0 rad/sec. The shaft is suddenly stopped at the opposite end and the kinetic energy of the flywheel is transformed into elastic strain energy in the shaft. Neglect energy losses during the impact as well as the kinetic energy of the shaft. Determine the maximum shearing stress and the maximum angle of rotation of the shaft for the following input values: flywheel weight = 120 lb; idealize the flywheel as a disk of radius = 15 in., length of shaft = 72 in., outside diameter of shaft = 3.0 in., inside diameter of shaft = 2.5 in., and $G = 12.0 \times 10^6$ psi.

13.18 A weight $W = 20$ lb is dropped from a height $h = 4$ in onto a simply supported beam as shown in Figure H13.18. Determine the maximum bending stress in the beam and its maximum deflection at the point of impact if 90 percent of the potential energy is transformed into elastic flexural strain energy of the beam. Use the following set of input values: $a = 15$ in., $b = 25$ in., beam depth = 2.0 in., beam width = 6.0 in., and $E = 10 \times 10^6$ psi. Compare the impact values for the maximum bending stress and deflection at the point of impact to corresponding static values for gradual application of W.

13.19 A weight $W = 80$ N is dropped from a height $h = 0.10$ m onto a simply supported beam as shown in Figure H13.18. Determine the maximum bending stress in the beam and its maximum deflection at the point of impact if 85 percent of the potential energy is transformed into elastic flexural strain energy of the beam. Use the following set of input values: $a = 0.40$ m, $b = 0.60$ m, beam depth = 0.05 m, beam width = 0.15 m, and $E = 70.0$ GPa. Compare the impact values for the maximum bending stress and deflection at the point of impact to corresponding static values for gradual application of W.

13.20 Two possible axially loaded rods are shown in Figure H13.20(a) and (b) for resisting impact loadings. The rod shown in Figure H13.20(a) has a diameter $2D - 2$ in. over its top half and a diameter $D = 1$ in. over its bottom half. Stress concentrations and energy losses are to be ignored in your analysis. For $W = 10$ lb., $h = 5$ in., $L = 16$ in., and $E = 10 \times 10^6$ psi, find the maximum stress and deflection for each of the rods due to impact. Compare the results and choose one of the rods as being more effective for resisting impact loadings.

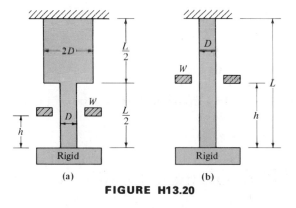

FIGURE H13.20

13.21 Two possible axially loaded rods are shown in Figure H13.20(a) and (b) for resisting impact loadings. The rod shown in Figure H13.20(b) has a uniform diameter $D = 0.04$ m, and the rod shown in Figure H13.20(a) has a diameter $2D = 0.08$ m over its top half and a diameter $D = 0.04$ m over its bottom half. Stress concentrations and energy losses are to be

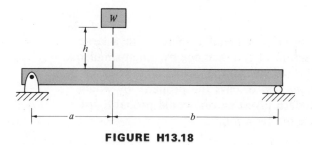

FIGURE H13.18

ignored in your analysis. For $W = 50$ N, $h = 0.12$ m, $L = 0.40$ m, and $E = 70.0$ GPa. Find the maximum stress and deflection for each of the rods due to impact. Compare results and choose one of the rods as being more effective for resisting impact loadings.

13.22 Refer to Figure H13.22 and determine:
(a) The weight W.
(b) The height h from which W is to be dropped such that the maximum stress in the beam is 20 MPa.
Given: beam width = 0.08 m, beam depth = 0.03 m, and spring constant $k = 1000$ N/m. When the weight W is statically applied to the system, the springs deform 0.02 m. Beam length $L = 5$ m. $E = 200$ GPa. Ignore beam and spring weights in your analysis.

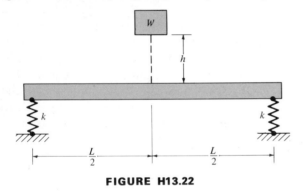

FIGURE H13.22

13.23 Refer to Figure H13.22 and determine:
(a) The weight W.
(b) The height h from which W is to be dropped such

that the maximum stress in the beam is 32,000 psi.
Given: beam width = 3 in., beam depth = 2 in., spring constant $k = 40,000$ lb/ft. When the weight W is statically applied to the system, the springs deform 0.08 in. Beam length $L = 5$ ft., $E = 30 \times 10^6$ psi. Ignore the beam and spring weights in your analysis.

13.24 An aluminum beam and a steel beam are simply supported at their ends and subjected to an impact loading by dropping a weight W from a height h onto the midpoint of each beam. The length of the aluminum beam is 0.30 m and each beam is 0.02 m in diameter. Determine the length of the steel beam required to resist the same load. The design stress of each material is 150 MPa and the moduli of elasticity of the aluminum and steel beams are 70 GPa and 200 GPa, respectively. Neglect energy lost during impact. Use $h = 0.02$ m.

13.25 An aluminum beam and a steel beam are simply supported at their ends and subjected to an impact loading by dropping a weight W from a height $h = 1.75$ in. onto the midpoint of each beam. The length of the aluminum beam is 25 in. and each beam is 1.75 in. in diameter. Determine the length of the steel beam required to resist the same load. The design stress of each material is 25,000 psi and the moduli of elasticity of the aluminum and steel beams are 10×10^6 psi and 29×10^6 psi, respectively. Neglect the energy lost during impact.

★**13.4** Fatigue Loadings—Analysis and Design

Engineers have long recognized that subjecting a metallic member to a large number of cycles of stress will produce fracture of the member by much smaller stresses than those associated with static failure of the member. Poncelet in 1839 was probably the first to introduce the term *fatigue* and discuss the property by which materials resist repeated cycles of stress. Modern investigators would probably use a term such as *progressive fracture* to replace the term *fatigue*.

Between 1852 and 1869, A. Wohler, a German engineer, designed the first repeated-load testing machines. Wohler discovered that:

1. The number of cycles of stress rather than elapsed time of testing is primary.
2. Ferrous materials stressed below a certain limiting value can withstand an infinite number of cycles of stress without fracture.

The literature of fatigue studies is voluminous. These studies have proceeded along two lines, fundamental research seeking to explain the phenomenon and empirical investigations to provide information for practical design and analysis. Recent progress in studies of crack propagation and use of the electron microscope have enhanced fundamental research efforts, but empirical studies still continue to be widely employed for engineering design. Fatigue fractures originate at points where stress concentrations occur, such as fillets, keyways, holes, and screw threads, or internal inclusions or defects in the material. Cracks begin at these points of stress concentrations and then propagate through the cross section until the remaining uncracked regions are insufficient to resist the applied forces and fracture occurs suddenly.

R. R. Moore devised the rotating beam fatigue testing machine illustrated in Figure 13.13. The rotating specimen is subjected to a constant bending moment over its full length as shown in the associated moment diagram. To minimize the effect of stress concentrations, the surface of the specimen is polished and the geometry is chosen to provide a gradual change in cross-sectional dimensions on both sides of the critical cross section located at the center. Care is also taken to remove residual stresses from the specimen. These residual stresses arise during unequal cooling of

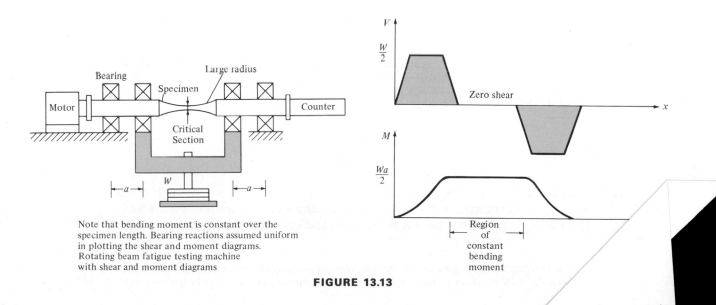

Note that bending moment is constant over the specimen length. Bearing reactions assumed uniform in plotting the shear and moment diagrams. Rotating beam fatigue testing machine with shear and moment diagrams

FIGURE 13.13

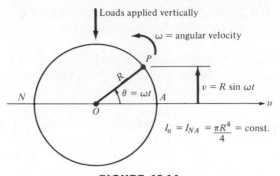

FIGURE 13.14
Cross Section of Rotating Beam Specimen.

the specimen, and annealing with gradual, carefully controlled, cooling will remove most of these stresses. In summary, important variables related to fatigue test specimens are size, shape, method of fabrication, surface finish, and previous history.

For a given test, the choice of the weight W and the set spacing a of the machine bearings determines the bending moment applied to the specimen as it rotates at a constant angular velocity ω. As shown in Figure 13.14, the loads are applied in a vertical plane, and for a given point P on the specimen circumference the distance v measured perpendicular to the neutral axis (N.A., which coincides with the u axis) varies with time. This is the one quantity that is time-variant in the flexure formula:*

$$S = \frac{M_u v}{I_u} = \frac{Wa}{2}\frac{R \sin \omega t}{\pi R^4/4}$$

$$S = \frac{2Wa}{\pi R^3} \sin \omega t \qquad\qquad \textbf{(13.22)}$$

$$S = S_r \sin \omega t \qquad\qquad \textbf{(13.23)}$$

where S_r is the amplitude given by

$$S_r = \frac{2Wa}{\pi R^3}$$

Equation 13.23 expresses the fact that the bending stress S at any point P on the specimen circumference will vary sinusoidally with time as shown in Figure 13.15. This variation between a maximum tension S_r and a maximum compression of equal

---

* S will be used as a generic symbol for stress in this section, since repeated reference is made to the $S–N$ diagram in the huge volume of information available on fatigue.

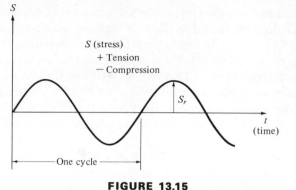

FIGURE 13.15
Completely Reversed Cycles of Stress.

magnitude S_r is known as *complete reversal* and is the most severe form of stress cycle. Provided that the amplitude S_r is not set too low, the specimen will experience a fatigue failure after a certain number of cycles N. The testing machine turns off automatically when the specimen fractures and the number of cycles to fracture is read at the counter. This test of a single specimen provides two numbers: S_r and N. A large number of such tests will provide a series of points, which are then plotted as an S–N diagram. Actually, the log S–log N plot shown in Figure 13.16 is more convenient and will be used in this text, except for Problems 13.27 and 13.32, in which S versus N and S versus log N plots are requested.

Information obtained from S–N diagrams forms the basis for design and analysis for fatigue loadings. Fatigue-testing machines are also available for axial, torsional,

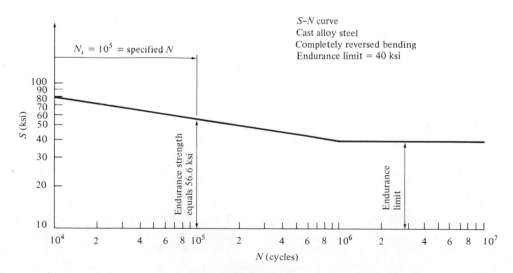

FIGURE 13.16

and combined loadings and appropriate S–N diagrams may be constructed for these other loadings. Definitions related to the S–N diagram are best understood by reference to Figure 13.16. The endurance limit S_e is the maximum stress that can be completely reversed an indefinitely large number of times without producing fracture. For the material of Figure 13.16, S_e equals 40.0 ksi. The endurance strength S_s is the highest stress in a complete reversal fatigue test which the material can withstand without rupture for a specified number of cycles N_s. In Figure 13.16, if the specified number of cycles $N_s = 10^5$, the corresponding endurance strength S_s for this material equals 56.6 ksi.

Figure 13.17 shows an S–N curve for an iron-based superalloy which illustrates the effect of data scatter. Experimental results from fatigue tests do not lie perfectly along a line, as shown in Figure 13.16, but rather in a band, as shown in Figure 13.17. The band within which the experimental points lie is indicated in Figure 13.17 by the dashed lines above and below the solid line. Stress levels corresponding to a specified number of cycles can be stated with reasonable accuracy. For example, in Figure 13.17 consider 2×10^5 cycles as specified and observe that the ordinate at this point intersects the scatter band at points C and D. The stress levels at D and C are in the ratio 8 : 7. This is to be contrasted with fatigue life predictions, which are estimates at best. For example, in Figure 13.17 consider a stress level of 150 ksi and observe that a horizontal line at this level intersects the scatter band at points A and B. The abscissas of B and A are in the ratio 4.11 : 1. For a stress level of 150 ksi, the fatigue life varies from 9.0×10^4 cycles to 3.7×10^5 cycles. These effects of data scatter are usually covered by the factor of safety for fatigue, which is usually considerably larger than the factor of safety for static loadings.

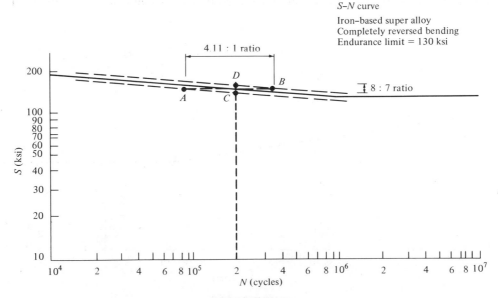

FIGURE 13.17

The S–N curves for steel, unlike those for most nonferrous metals, level off after a large number of cycles. As shown in Figures 13.16 and 13.17, this leveling off occurs at about 10^6 cycles, but there are many materials for which a continual decline in stress levels will occur with increasing number of cycles. In this case the designers are faced with the task of estimating the number of stress cycles associated with the useful life of the component to be designed and then basing their calculations upon a corresponding endurance strength. Estimating the number of stress cycles is further complicated when the component experiences high-frequency vibrations because in a very short time the number of cycles can increase by very large amounts. For example, fatigue failure of such components has been the cause of jet and helicopter aircraft crashes.

If the stress varies sinusoidally with time between a maximum tensile value S_{max} and a minimum tensile value S_{min} as shown in Figure 13.18(a), it is possible to

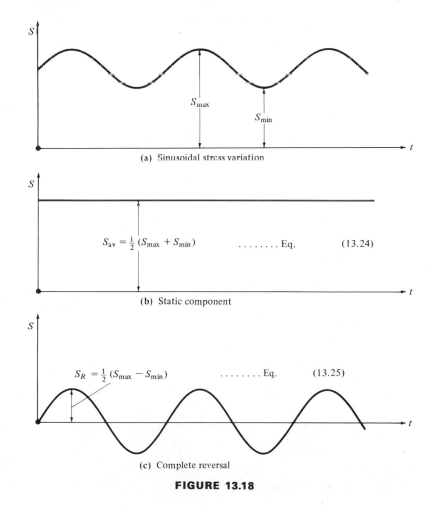

(a) Sinusoidal stress variation

$$S_{av} = \tfrac{1}{2}(S_{max} + S_{min}) \quad \ldots\ldots\ldots \text{Eq.} \quad (13.24)$$

(b) Static component

$$S_R = \tfrac{1}{2}(S_{max} - S_{min}) \quad \ldots\ldots\ldots \text{Eq.} \quad (13.25)$$

(c) Complete reversal

FIGURE 13.18

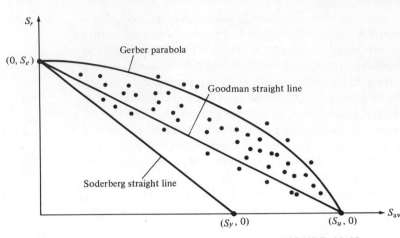

S_e based upon infinite or 5×10^8 cycle life. For shorter cycle life, point $(0, S_e)$, is replaced by a value of S taken from the appropriate S–N curve.

Equations:

Gerber parabola: $\quad S_r = S_e \left[1 - \left(\dfrac{S_{av}}{S_u} \right)^2 \right]$ (13.26)

Goodman straight line: $\dfrac{S_{av}}{S_u} + \dfrac{S_r}{S_e} = 1$ (13.27)

Soderberg straight line: $\dfrac{S_{av}}{S_y} + \dfrac{S_r}{S_e} = 1$ (13.28)

FIGURE 13.19

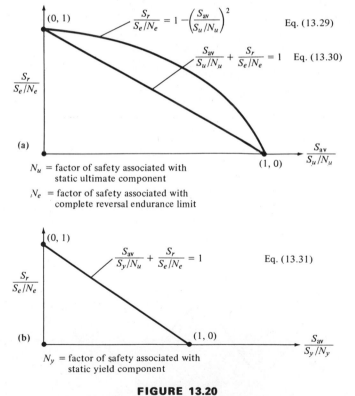

$\dfrac{S_r}{S_e/N_e} = 1 - \left(\dfrac{S_{av}}{S_u/N_u} \right)^2$ Eq. (13.29)

$\dfrac{S_{av}}{S_u/N_u} + \dfrac{S_r}{S_e/N_e} = 1$ Eq. (13.30)

N_u = factor of safety associated with static ultimate component

N_e = factor of safety associated with complete reversal endurance limit

$\dfrac{S_{av}}{S_y/N_u} + \dfrac{S_r}{S_e/N_e} = 1$ Eq. (13.31)

N_y = factor of safety associated with static yield component

FIGURE 13.20
Dimensionless Equations with Safety Factors.

decompose this stress variation into a static component, as shown in Figure 13.18(b), and a completely reversed sinusoidal variation, shown in Figure 13.18(c). This decomposition is desirable, since a very large portion of experimental fatigue data has been obtained for completely reversed cycling. Three relationships between the static component and the complete reversal component are shown in Figure 13.19. The Gerber parabola, the Goodman straight line, and the Soderberg straight line are plotted together with experimental points in this figure. The Goodman straight line has been widely used in American engineering practice, and the Soderberg straight line has been used to a lesser extent. The Gerber parabola has not been widely used in America because it is not conservative with respect to a great deal of the data. The Soderberg straight line provides a very conservative approach, since it is based upon a yield stress S_y rather than upon an ultimate stress S_u for the static component. Factors of safety are not included in Figure 13.19, but Figure 13.20 provides nondimensional forms of the three methods that incorporate the appropriate factors of safety. Further discussion of these methods follows in the examples.

A summary of the equations shown in Figures 13.18 to 13.20 follows for ready reference in solving fatigue loading problems. The reader will more readily understand these equations if he or she associates them with the figures.

SINUSOIDAL STRESS VARIATION WITH TIME (Figure 13.18)
Static component:

$$S_{av} = \tfrac{1}{2}(S_{max} + S_{min}) \qquad\qquad (13.24)$$

Completely reversed component:

$$S_r = \tfrac{1}{2}(S_{max} - S_{min}) \qquad\qquad (13.25)$$

GERBER, GOODMAN, AND SODERBERG EQUATIONS WITHOUT SAFETY FACTORS (Figure 13.19)
Gerber parabola:

$$S_r = S_e\left[1 - \left(\frac{S_{av}}{S_u}\right)^2\right] \qquad\qquad (13.26)$$

Goodman straight line:

$$\frac{S_{av}}{S_u} + \frac{S_r}{S_e} = 1 \qquad\qquad (13.27)$$

Soderberg straight line:

$$\frac{S_{av}}{S_y} + \frac{S_r}{S_e} = 1 \qquad (13.28)$$

GERBER, GOODMAN, AND SODERBERG EQUATIONS WITH SAFETY FACTORS (Figure 13.20)

Gerber parabola:

$$\frac{S_r}{S_e/N_e} = 1 - \left[\frac{S_{av}}{S_u/N_u}\right]^2 \qquad (13.29)$$

Goodman straight line:

$$\frac{S_{av}}{S_u/N_u} + \frac{S_r}{S_e/N_e} = 1 \qquad (13.30)$$

Soderberg straight line:

$$\frac{S_{av}}{S_y/N_y} + \frac{S_r}{S_e/N_e} = 1 \qquad (13.31)$$

Example 13.7

A cantilever machine component is to be fabricated of an iron-based superalloy (AISI Grade 635) for which the fatigue endurance limit is 54 ksi. This component is subjected to a completely reversed loading of 2.0 k maximum magnitude at the free end. A circular cross section is to be chosen for this 25-in.-long beam using a factor of safety $N_e = 3$. The design details will provide generous fillets and rounds, which means that stress concentrations need not be considered in this case. Determine the required diameter, D.

Solution. Since the loading is completely reversed, the allowable stress is obtained by dividing the endurance limit by the factor of safety.

$$S_r = \frac{S_e}{N_e} = \frac{54}{3} = 18 \text{ ksi}$$

The flexure formula applies if one neglects shear effects compared to bending:

$$S_r = \frac{M_u v}{I_u}$$

Substituting numerical values and D for diameter:

$$18 = \frac{2(25)(D/2)}{(\pi/32)D^4}$$

$$\boxed{D = 2.42 \text{ in.} \qquad \text{use 2.50-in.-diameter beam}}$$

Example 13.8

The beam shown in Figure 13.21(a) is to be fabricated of a nickel alloy (Monel 400—hot rolled bar) for which the fatigue endurance limit in bending is 289 MPa (10^8 cycles, 70°F), the yield strength is 300 MPa, and the tensile strength is 570 MPa. The fillet radius r at the change of cross section at the center of the beam has been chosen such that the stress concentration factors are as follows:

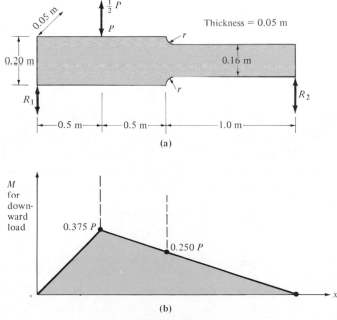

(a)

(b)

FIGURE 13.21

$$k_s = 1.5 \quad \text{(static at the net section)}$$

$$k_f = 1.3 \quad \text{(fatigue at the net section)}$$

The loading is applied at the quarter point and ranges from a downward load P to an upward load $P/2$. Use factors of safety with respect to yielding of $N_y = 2$, with respect to ultimate of $N_u = 3$ and with respect to the endurance limit of $N_e = 4$. The beam has a uniform thickness of 0.05 m. (a) Determine the maximum static downward load using first yielding as a criterion; (b) determine the load range (i.e., P and $P/2$), using (i) the Gerber parabola, (ii) the Goodman straight line, and (iii) the Soderberg straight line; and (c) compare the solutions of part (b) on a percentage basis using the value from part (a) as a basis for the comparisons.

Solution

(a) This material is ductile, and hence a static analysis based upon use of a stress concentration factor at the center section will be very conservative. At the center section, where $M = 0.25P$ for a downward load P as shown in Figure 13.21(b), the flexure formula incorporating static stress concentration factor k_s and a factor of safety N_y on the yield stress becomes

$$\frac{S}{N_y} = k_s \frac{M_u v}{I_u}$$

Substituting for the net section at the center, we obtain

$$\frac{300 \times 10^6}{2} = \frac{1.5(0.25P)(0.08)}{\frac{1}{12}(0.05)(0.16)^3}$$

$$P = 85.3 \text{ kN}$$

The section at the load application point must also be investigated. Note that stress concentrations at loading and reaction points are being neglected in this example. Such concentrations would be mitigated by distributing the loads and reactions over some reasonable area. Substituting again into the flexure formula for the gross section at the loading point and using $k_s = 1$ yields

$$\frac{300 \times 10^6}{2} = \frac{1.0(0.375P)0.10}{\frac{1}{12}(0.05)(0.20)^3}$$

$$P = 133.3 \text{ kN}$$

Since $P = 85.3 \text{ kN} < 133.3 \text{ kN}$, choose

$$\boxed{P = 85.3 \text{ kN}}$$ Maximum static downward load using first yielding as a criterion

(b) Investigate the flexural stresses at the center section using equations from Figure 13.18 and the flexure formula

$$S = k_f \frac{M_u v}{I_u}$$

Maximum stress: $\quad S_{max} = 1.3 \dfrac{0.25P(0.08)}{\frac{1}{12}(0.05)(0.16)^3}$

$$= 1520P$$

Minimum stress: $\quad S_{min} = -760P$

The static component of stress, from Eq. 13.24,

$$S_{av} = \tfrac{1}{2}(S_{max} + S_{min})$$

$$= \tfrac{1}{2}(1520P - 760P)$$

$$= 380P$$

The completely reversed component, from Eq. 13.25,

$$S_r = \tfrac{1}{2}(S_{max} - S_{min})$$

$$= \tfrac{1}{2}(1520P + 760P)$$

$$= 1140P$$

Refer to Figure 13.20 to obtain the equations in nondimensional form:
(i) Gerber parabola (Eq. 13.29)

$$\frac{S_r}{S_e/N_e} = 1 - \left(\frac{S_{av}}{S_u/N_u}\right)^2$$

Substituting in terms of P and from the given information, we obtain

$$\frac{1140P}{(289 \times 10^6/4)} = 1 - \left[\frac{380P}{(570 \times 10^6/3)}\right]^2$$

Solving this quadratic for P gives

$$\boxed{P = 62.4 \text{ kN} \qquad \text{Gerber}}$$

(ii) Goodman straight line (Eq. 13.30)

$$\frac{S_{av}}{S_u/N_u} + \frac{S_r}{S_e/N_e} = 1$$

Substituting in terms of P and from the given information, we find that

$$\frac{380P}{(570 \times 10^6/3)} + \frac{1140P}{(289 \times 10^6/4)} = 1$$

Solving this linear equation for P gives

$$\boxed{P = 56.2 \text{ kN} \qquad \text{Goodman}}$$

(iii) Soderberg straight line (Eq. 13.31)

$$\frac{S_{av}}{S_y/N_y} + \frac{S_r}{S_e/N_e} = 1$$

Substituting in terms of P and from the given information, we obtain

$$\frac{380P}{(300 \times 10^6/2)} + \frac{1140P}{(289 \times 10^6/4)} = 1$$

Solving this linear equation for P gives

$$\boxed{P = 54.6 \text{ kN} \qquad \text{Soderberg}}$$

(c) If we assume the static value of $P = 85.3$ kN as a basis for comparison, the solutions may be compared as follows:

	P (kN)	Percentage Comparison
Static	85.3	100.0
Gerber	62.4	73.2
Goodman	56.2	65.9
Soderberg	54.6	64.0

Example 13.9
A circular cylindrical shaft is to be fabricated of a material for which the S–N data are given in Figure H13.32. The maximum torque is 4000 ft-lb and the minimum torque is 1000 ft-lb. Determine the required diameter of the shaft. A

small keyway in the shaft has stress concentration factors $k_f = 1.3$, $k_s = 1.4$. The keyway is small enough to allow use of gross cross-sectional area properties. The torsional ultimate strength of this material is 90 ksi. Use the Goodman straight line and factors of safety $N_e = 4$ and $N_u = 3$.

Solution. The symbol S refers to shearing stress in this problem, since a shaft is to be designed.

$$S = k_f \frac{T\rho}{J}$$

Maximum shearing stress: $\quad S_{max} = 1.3 \dfrac{4000(12)(D/2)}{\dfrac{\pi D^4}{32}}$

$$= \frac{3.178 \times 10^5}{D^3}$$

Minimum shearing stress: $\quad S_{min} - \dfrac{7.945 \times 10^4}{D^3}$

From Eq. 13.24,

$$S_{av} = \tfrac{1}{2}(S_{max} + S_{min})$$

$$= \frac{1}{2} \frac{(3.178 \times 10^5 + 0.7945 \times 10^5)}{D^3}$$

$$= \frac{1.986 \times 10^5}{D^3}$$

From Eq. 13.25,

$$S_r = \tfrac{1}{2}(S_{max} - S_{min})$$

$$= \frac{1}{2} \frac{(3.178 \times 10^5 - 0.7945 \times 10^5)}{D^3}$$

$$= \frac{1.192 \times 10^5}{D^3}$$

From Eq. 13.30 and Figure 13.20 for the Goodman straight line equation:

$$\frac{S_{av}}{S_u/N_u} + \frac{S_r}{S_e/N_e} = 1$$

Substituting from above yields

$$\frac{1.986 \times 10^5}{D^3/(90,000/3)} + \frac{1.192 \times 10^5}{D^3/(24,000/4)} = 1$$

Solving for the diameter, we obtain

$$D = 2.98 \text{ in.}$$

Consider static application of the maximum torque:

$$S = k_s \frac{Tr}{J}$$

Substitute given values:

$$\frac{90,000}{3} = 1.4 \frac{4000(12)(D/2)}{\pi D^4/32}$$

$$D = 2.25 \text{ in.}$$

$$2.98 > 2.25 \quad \text{choose} \quad \boxed{D = 3.00 \text{ in.}}$$

Example 13.10

A rod of circular cross section is subjected to alternating tensile forces varying from a minimum of 160 kN to a maximum of 450 kN. It is to be fabricated of a material with an ultimate tensile strength of 700 MPa and an endurance limit for complete reversal of axial forces of 560 MPa. Determine the diameter of the rod using safety factors of $N_u = 3$ and $N_e = 3.5$ and stress concentration factors of $k_s = 1.6$ and $k_f = 1.5$. Use the gross area in stress calculations and the Goodman straight line as a basis for design. Check the diameter required for static loading.

Solution. The symbol S refers to axial normal stress in this problem:

$$S = k_f \frac{P}{A}$$

Maximum stress: $S_{max} = 1.5 \dfrac{450}{\pi D^2/4} = \dfrac{859.4}{D^2}$

Minimum stress: $S_{min} = \dfrac{1.5(160)}{\pi D^2/4} = \dfrac{305.6}{D^2}$

From Eq. 13.24,

$$S_{av} = \tfrac{1}{2}(S_{max} + S_{min})$$

$$= \frac{1}{2}\left(\frac{859.4}{D^2} + \frac{305.6}{D^2}\right) = \frac{582.5}{D^2}$$

From Eq. 13.25,

$$S_r = \tfrac{1}{2}(S_{max} - S_{min})$$

$$= \frac{1}{2}\left(\frac{859.4}{D^2} - \frac{305.6}{D^2}\right) = \frac{276.9}{D^2}$$

From Eq. 13.30 and Figure 13.20(a) for the Goodman straight line equation:

$$\frac{S_{av}}{S_u/N_u} + \frac{S_r}{S_e/N_e} = 1$$

$$\frac{582.5 \times 10^3/D^2}{700 \times 10^6/3} + \frac{276.9 \times 10^3/D^2}{560 \times 10^6/3.5} = 1; \qquad D = 0.065 \text{ m}$$

Consider static application of the maximum axial force:

$$S = k_s \frac{P}{A}$$

Substitute given values:

$$\frac{700 \times 10^6}{3} = \frac{1.6(450 \times 10^3)}{\pi D^2/4}$$

$$D = 0.0627 \text{ m}$$

$$0.0650 > 0.0627 \qquad \text{choose} \quad \boxed{D = 0.065 \text{ m}}$$

Homework Problems

13.26 Refer to Figure 13.16 and note that the sloping straight line on the log-log plot passes through the following points: $(10^4, 80)$ and $(10^6, 40)$. Show that the equation of this line may be written $\log S = -0.15051 \log N + 2.50513$. Then verify the endurance strength for a specified number of cycles $N_s = 10^5$. What is the endurance strength corresponding to 3×10^5 cycles?

13.27 Refer to Figure 13.16 and read stress ordinates to the two straight lines for $N = 1 \times 10^4$, $2 \times 10^4, \ldots, 1 \times 10^7$. Record these values in columns headed N and S. Then prepare the following plots:
(a) S versus N.
(b) S versus $\log N$ (on semilog paper).
Observe these plots together with Figure 13.16 and note why the $\log S$ versus $\log N$ plot is to be preferred. Plotting these data should increase your awareness of the great expenditures of time and money required for assembling fatigue test information on a wide variety of materials.

13.28
(a) A series of complete reversal bending tests were run on a steel alloy specimen and a plot of the data on log-log paper appeared as two straight lines, similar to Figure 13.16. The sloping straight line on the left passes through the following points: $(10^5, 60)$ and $(10^6, 30)$. The line on the right has a constant ordinate of 30. Stresses are given in ksi. Construct a log-log plot of these data. What is the endurance strength of this material for 3×10^5 cycles and for 8×10^5 cycles? What is the endurance limit for this material?
(b) Refer to the data from the log-log plot described in part (a). Write the equation of the sloping straight line in the form $\log S = C \log N + D$,

where C and D are constants to be determined from the data. From the equation find the endurance strength S_s corresponding to $N = 4 \times 10^5$ cycles. Check the result with the plot.

13.29 Rework Example 13.7 if the material is changed to the cast alloy steel for which the S–N data are provided in Figure 13.16.

13.30 Use the data given in Example 13.8 for the nickel alloy (Monel 400—hot rolled bar) and prepare scaled plots similar to those shown in Figure 13.19 for the Gerber parabola, the Goodman straight line, and the Soderberg straight line. Clearly note on your plots the points corresponding to the solutions to Example 13.8.

13.31 Refer to Problem 13.30 and prepare scaled plots as described there. Extend the plots to cover the case of negative static components, S_{av}.

13.32 A log-log plot of S–N data for ferrous alloy A1 is shown in Figure H13.32. These data were obtained from completely reversed torsional tests. Read stress ordinates to the two straight lines for $N = 1 \times 10^4$, $2 \times 10^4, \ldots, 1 \times 10^7$. Record these values in columns headed N and S. Then prepare the following plots:
(a) S versus N.

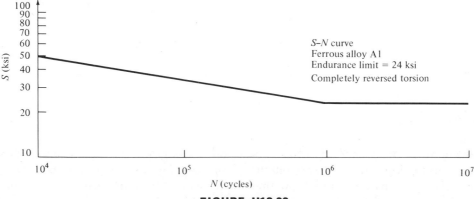

S–N curve
Ferrous alloy A1
Endurance limit = 24 ksi
Completely reversed torsion

FIGURE H13.32

(b) S versus $\log N$ (on semilog paper).

Observe these plots together with Figure H13.32 and note why the $\log S$ versus $\log N$ plot is to be preferred. Plotting these data should increase your awareness of the great expenditures of time and money required for assembling fatigue test information on a wide variety of materials.

13.33 Refer to the log-log plot of S–N data shown in Figure H13.32. Write the equation of the sloping straight line in the form $\log S = C \log N + D$, where C and D are constants to be determined from the data. From the equation find the endurance strength S_s corresponding to $N = 5 \times 10^5$ cycles. Check the result with the plot.

13.34 An axially loaded member is subjected to a tensile force P which varies from a maximum value of 400 k to a minimum value of 60 k. The member has a rectangular cross section d by $2d$. Determine the cross-sectional dimensions using an endurance limit of 40 ksi for completely reversed axial loading and a static ultimate strength of 90 ksi. Use factors of safety of $N_e = 4$ and $N_u = 3$ and stress concentration factors of $k_f = 1.20$ and $k_s = 1.35$. Your solution is to be based upon the gross cross-sectional area. Use the Goodman straight line and construct a scaled plot of this line together with the point on the line associated with your solution to this problem.

13.35 Solve Problem 13.34 using the Soderberg straight line for yield stress of 65 ksi and $N_y = 2$. Construct a scaled plot of the Soderberg straight line, together with the point on the line associated with your solution to the problem.

13.36 Solve Problem 13.34 for compressive forces and plot the Goodman straight line in the appropriate quadrant. Assume the member is short enough that buckling is not a problem.

13.37 Solve Problem 13.35 for compressive forces and plot the Soderberg straight line in the appropriate quadrant. Assume the member is short enough that buckling is not a problem.

13.38 Solve Example 13.9 using the Gerber parabola. Compare your answer to the diameter of 3.00 in., which is the solution to Example 13.9.

13.39 Refer to Example 13.9 and solve again for a maximum torque of 6000 ft-lb and a minimum torque of 2500 ft-lb.

13.40 Refer to Example 13.9 and solve again for a hollow circular cylindrical shaft with an outside diameter equal to 1.4 times the inside diameter. Assume that the same stress concentration factors apply in this case, even though design details would vary at the keyway in the case of a hollow shaft.

13.41 Refer to Example 13.10. Determine the rod diameter if the alternating tensile forces vary from a minimum of 90 kN to a maximum of 230 kN.

13.42 Refer to Example 13.10 and solve again using the following new information: The cross section is to be square with $k_s = 1.80$ and $k_f = 1.68$.

13.43 A steel rod is threaded at the ends and is subjected to axial tensile loads varying from a maximum of 320 kN to a minimum of 90 kN. The material of which the rod is to be fabricated has an ultimate strength of 560 MPa and an endurance limit for complete reversal of tensile loading of 300 MPa. Find:
(a) The cross-sectional area of the rod using the Goodman straight line with $N_e = 3$ and $N_u = 2$. Neglect stress concentrations at the threads.
(b) The cross-sectional area of the rod assuming the maximum force is applied statically.

13.44 A simply supported beam is subjected to a load at its center which varies from zero to 10,000 lb downward. It is 10 ft long and has a rectangular cross section with a depth 3 times its width. Use the Goodman straight line to determine the cross-sectional dimensions of the beam. Other given information: $S_e = 20$ ksi, $S_u = 60$ ksi, $N_e = 3$, and $N_u = 2$. Design details are such that the stress concentrations need not be considered.

13.45 Refer to Problem 13.44 and redesign the beam for a center load that varies from 10,000 lb to 20,000 lb downward.

13.46 A rod of circular cross section is subjected to alternating tensile forces varying from a minimum of 200 kN to a maximum of 500 kN. It is to be fabricated of a material with an ultimate tensile strength of 900 MPa and an endurance limit of 700 MPa. Determine the diameter of the rod using safety factors of $N_u = 3.5$ and $N_e = 4.0$ and stress concentration factors of $k_s = 1.80$ and $k_f = 1.65$. Use the gross area in stress calculations and the Goodman straight line as a basis for design. Check the diameter required for static loading.

13.47 Refer to Problem 13.46 and redesign the rod for axial tensile forces that vary from a minimum of 10 kN to a maximum of 310 kN.

13.48 A hollow circular cylindrical shaft is to be fabricated of a material for which the S–N data are given in Figure H13.32. The maximum torque is 2500 ft-lb and the minimum torque is 500 ft-lb. Determine the required diameters of the shaft if the outside diameter is 1.6 times the inside diameter. A hole drilled in the wall of the shaft gives rise to stress concentration factors $k_f = 1.4$ and $k_s = 1.5$. Neglect the effect of the hole on cross-sectional-area properties of the shaft (i.e., use gross properties). The torsional ultimate strength of this material is 88 ksi. Use the Goodman straight line and factors of safety $N_e = 3.8$ and $N_u = 2.9$.

13.49 Refer to Problem 13.48 and redesign the shaft for a maximum torque of 2000 ft-lb and a zero minimum torque.

13.50 A cantilever beam is subjected to a load at its free end which varies from 2 kN to 10 kN downward. It is 2.5 m long and has a rectangular cross section with a depth 3.2 times its width. Use the Goodman straight line to determine the cross-sectional dimensions of the beam. Other given information: $S_e = 150$ MPa, $S_u = 450$ MPa, $N_e = 3$, and $N_u = 2$. Design details are such that stress concentrations need not be considered.

Chapter 14 Selected Topics

★14.1 Introduction

Several topics of special practical interest have been chosen for inclusion in this chapter. The selection is by no means exhaustive, but it provides a concise treatment of an assortment of areas directly or indirectly related to the broad subject of mechanics of materials.

Some of the selected topics are sufficiently independent to warrant placing them either in an appendix or in a separate chapter. Others, because of their nature, could have been included as separate sections in earlier chapters. For example, the topics of shear centers and beams of two materials could have been included in Chapter 5. However, a first course in mechanics of materials does not always include coverage of such topics and, consequently, it was decided to separate them from the fundamental treatment of beam analysis presented in Chapter 5. Furthermore, such a separation would provide flexibility and convenience to instructors in the selection of topics for inclusion in a given course.

The range of the selected topics is sufficiently broad to satisfy a variety of needs. These include cylindrical pressure vessels (thin- and thick-walled), shear center, beams of two materials, reinforced concrete (elastic and ultimate-strength methods), singularity functions for beam deflections, curved beams, matrices, and computer methods.

14.2 Cylindrical Pressure Vessels

Numerous industrial applications require the use of cylindrical containers for either the storage or the transmission of gases and liquids. Examples include the cylindrical tanks used for the storage of gaseous oxygen under high pressure and the piping used to deliver high-pressure liquids to hydraulic machines.

Two types of cylindrical containers are generally identified as thick-walled and thin-walled cylindrical pressure vessels. The distinction between these two types of vessels is based upon the nature of the circumferential stress distribution over the thickness of the cylindrical vessel; if the variation of this stress over the thickness is such that it may be assumed approximately constant, the cylindrical vessel is referred to as *thin-walled*; if not, it is known as *thick-walled*. The distinction between thin-walled and thick-walled cylinders is discussed in much more detail after the development of the appropriate equation. These equations will be derived for the case of the thick-walled cylinder and then will be specialized to the case of the thin-walled cylinder.

Figure 14.1(a) shows a cylindrical pipe which is open-ended and subjected to internal pressure p_1 and external pressure p_2. Consider a circular segment of this pipe contained between the two parallel planes $A-A$ and $B-B$ which are at a distance dz apart. An end view of this circular segment is shown in Figure 14.1(b). Note that because of geometric and loading symmetry, the surfaces defined by planes $A-A$ and $B-B$ are free of shearing stresses. However, the existence of the pressures p_1 and p_2 give rise to circumferential (hoop) stresses σ_t and radial stresses σ_r at any point in the walls of the cylinder. These stresses are shown in Figure 14.1(b) acting on the surfaces of a differential element located at a radial distance r from the center of the cylindrical pipe. This differential element is bounded by two concentric cylindrical surfaces separated by the distance dr and by two longitudinal planes subtending the angle $d\theta$ between them. Obviously, no shearing stresses exist on the planes of σ_t and σ_r, and therefore they are principal stresses. Note that both σ_t and σ_r vary with the radius r and that while the radial stress on the cylindrical surface located by the radius r is σ_r, it is $\sigma_r + d\sigma_r$ on the cylindrical surface located by the radius $r + dr$.

A summation of forces acting on the differential element in the radial direction yields the following equilibrium equation.

$$(\sigma_r + d\sigma_r)(r + dr)\, d\theta\, dz - \sigma_r r\, d\theta\, dz - 2\sigma_t\, dr\, dz \sin \frac{d\theta}{2} = 0 \qquad \textbf{(a)}$$

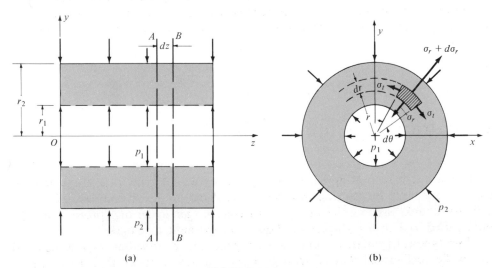

(a)   (b)

FIGURE 14.1

Considering that for small angles the function sin $(d\theta/2)$ is approximately equal to $d\theta/2$, and eliminating quantities of higher order, we obtain the relation

$$\sigma_r \, dr + r \, d\sigma_r - \sigma_t \, dr = 0 \tag{b}$$

Division by dr results in one relation between the stresses σ_t and σ_r. Thus

$$\sigma_t = \sigma_r + r\left(\frac{d\sigma_r}{dr}\right) \tag{c}$$

A second relation between σ_t and σ_r is obtained by considering the deformation of the cylinder in the longitudinal direction. A more rigorous analysis shows that axial (longitudinal) deformations are uniform for open-ended cylinders and may be assumed uniform for close-ended cylinders at locations sufficiently removed from the two ends. Therefore, the assumption may be made that the longitudinal strain ε_z is a constant. Using the third of Eq. 2.28, we obtain the following relation:

$$\varepsilon_z = -\left(\frac{\mu}{E}\right)(\sigma_t + \sigma_r) \tag{d}$$

Since μ and E are constants and since ε_z may be assumed constant, the second relation between σ_t and σ_r may be expressed as follows

$$\sigma_t = -\sigma_r - 2K \tag{e}$$

where the constant $2K$ represents the quantity $(E\varepsilon_z)/\mu$ and its exact value will be determined from the boundary conditions, which will be stated later.

Eliminating σ_t in Eqs. (c) and (e), we obtain

$$2\sigma_r + r\left(\frac{d\sigma_r}{dr}\right) = -2K \tag{f}$$

Multiply both sides of Eq. (f) by r to obtain

$$2r\sigma_r + r^2\left(\frac{d\sigma_r}{dr}\right) = -2rK \tag{g}$$

The left-hand side of Eq. (g) is the derivative with respect to r of the quantity $r^2\sigma_r$. Therefore,

$$\frac{d}{dr}(r^2\sigma_r) = -2rK \tag{h}$$

and

$$\sigma_r = -K + \frac{C}{r^2} \tag{i}$$

where C is a constant of integration to be determined from the boundary conditions. The circumferential stress σ_t may now be obtained from Eq. (e). Thus

$$\sigma_t = -K - \frac{C}{r^2} \tag{j}$$

The two constants K and C are determined from the two boundary conditions in the problem. These boundary conditions may be stated mathematically as follows:

$$\left. \begin{array}{ll} \sigma_r = -p_1 & \text{for } r = r_1 \\ \sigma_r = -p_2 & \text{for } r = r_2 \end{array} \right\} \tag{k}$$

Substitution of these conditions into Eq. (i) leads to two simultaneous equations, which may be solved for the constants K and C. Thus

$$C = (p_2 - p_1)\left(\frac{r_1^2 r_2^2}{r_2^2 - r_1^2} \right)$$

and $\tag{l}$

$$K = \frac{p_2 r_2^2 - p_1 r_1^2}{r_2^2 - r_1^2}$$

When the values of C and K from Eq. (l) are substituted into Eqs. (i) and (j), they lead to the equations for σ_r and σ_t at any point in the cylindrical pressure vessel. Thus

$$\sigma_r = \frac{p_1 r_1^2 - p_2 r_2^2}{r_2^2 - r_1^2} + \frac{r_1^2 r_2^2 / r^2}{r_2^2 - r_1^2}(p_2 - p_1) \tag{14.1}$$

and

$$\sigma_t = \frac{p_1 r_1^2 - p_2 r_2^2}{r_2^2 - r_1^2} - \frac{r_1^2 r_2^2 / r^2}{r_2^2 - r_1^2}(p_2 - p_1) \tag{14.2}$$

Note that the absolute maximum value of σ_r is the larger of the two pressures p_1 and p_2 and that the absolute maximum value of σ_t occurs on the inner surface of the cylinder, where r assumes its minimum value.

The development of Eqs. 14.1 and 14.2 has assumed an open-ended cylinder, in

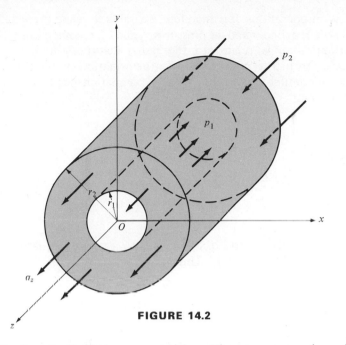

FIGURE 14.2

which case the longitudinal stress σ_z vanishes. There are a number of situations, however, in which the walls of the cylinder are subjected not only to σ_r and σ_t, but also to σ_z. This situation occurs in cylindrical containers with closed ends subjected to internal and/or external pressures.

A basic assumption made in the development of the longitudinal stress σ_z is to consider it uniformly distributed on any transverse cross section of the cylinder. This assumption is justified at least for transverse cross sections sufficiently removed from the two ends of the cylinder.

Consider one part of a closed-ended cylindrical container subjected to both p_1 and p_2 and lying to one side of a transverse cross section as shown in Figure 14.2. The value of σ_z may be obtained by summing forces in the z direction and setting them equal to zero. Thus

$$\pi(r_2^2 - r_1^2)\sigma_z + \pi r_2^2 p_2 - \pi r_1^2 p_1 = 0$$

from which

$$\sigma_z = \frac{p_1 r_1^2 - p_2 r_2^2}{r_2^2 - r_1^2} \tag{14.3}$$

The deformations of an open-ended cylinder (both radial and circumferential) due to the pressures p_1 and p_2 become of major concern in certain industrial applications, such as shrink-fit operations. It is, therefore, desirable to relate these deformations to the geometry of the cylinder and the pressures p_1 and p_2. Geometrically, the circumferential deformation δ_t is related to the radial deformation δ_r by the equation $\delta_t = 2\pi\delta_r$. Also, from the relation between deformation and strain, $\delta_t = 2\pi r\varepsilon_t$, in which ε_t is the circumferential strain. These two equations lead to a relation between δ_r and ε_t as follows:

$$\delta_r = r\varepsilon_t \tag{14.4a}$$

From Eq. 2.28,

$$\varepsilon_t = \frac{1}{E}(\sigma_t - \mu\sigma_r) \tag{14.4b}$$

Substitution of Eq. 14.4b into Eq. 14.4a leads to a general equation for the radial deformation in terms of σ_t and σ_r. Thus

$$\delta_r = \frac{r}{E}(\sigma_t - \mu\sigma_r) \tag{14.4c}$$

Substituting the values of σ_r and σ_t from Eqs. 14.1 and 14.2, respectively, we obtain the following equation for the radial deformation δ_r:

$$\delta_r = \frac{(1-\mu)r}{E}\frac{p_1 r_1^2 - p_2 r_2^2}{r_2^2 - r_1^2} - \frac{(1+\mu)}{E}\frac{r_1^2 r_2^2/r}{r_2^2 - r_1^2}(p_2 - p_1) \tag{14.5}$$

From the relation $\delta_t = 2\pi\delta_r$ and Eq. 14.5, we obtain an equation for the circumferential deformation δ_t:

$$\delta_t = \frac{2\pi(1-\mu)r}{E}\frac{p_1 r_1^2 - p_2 r_2^2}{r_2^2 - r_1^2} - \frac{2\pi(1+\mu)}{E}\frac{r_1^2 r_2^2/r}{r_2^2 - r_1^2}(p_2 - p_1) \tag{14.6}$$

Several special cases of practical interest will now be considered.

Internal Pressure Alone

There are many applications in which the cylindrical container is subjected only to the action of internal pressure p_1 (i.e., $p_2 = 0$). In such a case the radial and circumferential stresses may be obtained from Eqs. 14.1 and 14.2, respectively, by setting $p_2 = 0$. Thus

$$\sigma_r = \frac{p_1 r_1^2}{r_2^2 - r_1^2}\left(1 - \frac{r_2^2}{r^2}\right) \tag{14.7}$$

and

$$\sigma_t = \frac{p_1 r_1^2}{r_2^2 - r_1^2}\left(1 + \frac{r_2^2}{r^2}\right) \tag{14.8}$$

Since r_2^2/r^2 is always larger than unity, Eq. 14.7 always yields a negative value for σ_r, which indicates that σ_r is always a compressive stress with a maximum absolute value on the inner surface of the cylinder where $r = r_1$. The circumferential stress σ_t, however, is always positive or tensile and, as in the case of σ_r, it assumes its maximum value at the inner surface where $r = r_1$.

If the cylindrical container is open-ended, as, for example, in the case of a gun barrel, the walls of the cylinder are not subjected to longitudinal stress and $\sigma_z = 0$. If, on the other hand, the cylindrical container is closed-ended (i.e., it has caps or covers at its two ends), as, for example, in the case of an oxygen tank, the walls of the cylinder are subjected to a longitudinal stress σ_z which may be determined by specializing Eq. 14.3 to the case when $p_2 = 0$. Therefore,

$$\sigma_z = -\frac{p_1 r_1^2}{r_2^2 - r_1^2} \tag{14.9}$$

The radial and circumferential deformations experienced by the cylinder are given as special cases of Eqs. 14.5 and 14.6, respectively, by setting $p_2 = 0$. Thus

$$\delta_r = \frac{p_1 r_1^2}{E(r_2^2 - r_1^2)}\left[(1 - \mu)r + (1 + \mu)\frac{r_2^2}{r}\right] \tag{14.10}$$

and

$$\delta_t = \frac{2\pi p_1 r_1^2}{E(r_2^2 - r_1^2)}\left[(1 - \mu)r + (1 + \mu)\frac{r_2^2}{r}\right] \tag{14.11}$$

External Pressure Alone

The stresses induced in a cylindrical container subjected only to external pressure p_2 (i.e., $p_1 = 0$) may be quickly obtained as special cases of the general expressions derived earlier. The radial and circumferential stresses are obtained from Eqs. 14.1 and 14.2, respectively, by setting $p_1 = 0$. Therefore,

$$\sigma_r = -\frac{p_2 r_2^2}{r_2^2 - r_1^2}\left(1 - \frac{r_1^2}{r^2}\right) \tag{14.12}$$

and

$$\sigma_t = -\frac{p_2 r_2^2}{r_2^2 - r_1^2}\left(1 + \frac{r_1^2}{r^2}\right) \tag{14.13}$$

Note that both σ_r and σ_t are compressive stresses.

If the cylindrical container is open-ended, the longitudinal stress $\sigma_z = 0$. If, however, the container is closed-ended, the value of σ_z is given by a special case of Eq. 14.3. Thus, by setting $p_1 = 0$, we obtain

$$\sigma_z = \frac{p_2 r_2^2}{r_2^2 - r_1^2} \tag{14.14}$$

which is also a compressive stress.

The radial and circumferential deformations developed under the action of external pressure only are obtained by setting $p_1 = 0$ in Eqs. 14.5 and 14.6, respectively; thus

$$\delta_r = -\frac{p_2 r_2^2}{E(r_2^2 - r_1^2)}\left[(1 - \mu)r + (1 + \mu)\frac{r_1^2}{r}\right] \tag{14.15}$$

and

$$\delta_t = -\frac{2\pi p_2 r_2^2}{E(r_2^2 - r_1^2)}\left[(1 - \mu)r + (1 + \mu)\frac{r_1^2}{r}\right] \tag{14.16}$$

Example 14.1
A cylindrical pipe has a 10-in. inner diameter and a 20-in. outer diameter. It is subjected only to internal pressure $p_1 = 5000$ psi. Show the radial and circumferential stress variation across the wall thickness of the pipe.

Solution. Since the pipe is subjected only to internal pressure, the radial and circumferential stresses are given by Eqs. 14.7 and 14.8. Therefore,

$$\sigma_r = \frac{5000(25)}{100 - 25}\left(1 - \frac{100}{r^2}\right)$$

$$= \boxed{1666.7\left(1 - \frac{100}{r^2}\right)}$$

and

$$\sigma_t = \frac{5000(25)}{100 - 25}\left(1 + \frac{100}{r^2}\right)$$

$$= \boxed{1666.7\left(1 + \frac{100}{r^2}\right)}$$

A plot of these two equations is shown in Figure 14.3. Note that while the radial stress σ_r is compressive everywhere, the circumferential stress σ_t is tensile throughout. Note also that the inner fibers of the pipe are very heavily stressed when compared to the outer fibers and that the variations of σ_r and σ_t from the inner to the outer fibers are nonlinear.

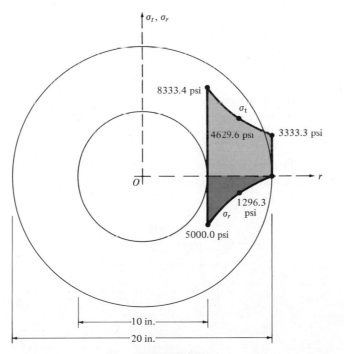

FIGURE 14.3

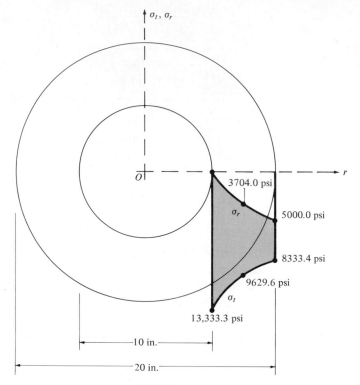

FIGURE 14.4

Example 14.2

The cylindrical pipe described in Example 14.1 is subjected only to external pressure $p_2 = 5000$ psi. Show the radial and circumferential stress variations across the wall thickness of the pipe.

Solution. Since the pipe is subjected only to external pressure, the radial and circumferential stresses are given by Eqs. 14.12 and 14.13, respectively. Therefore,

$$\sigma_r = -\frac{5000(100)}{100 - 25}\left(1 - \frac{25}{r^2}\right)$$

$$= \boxed{= 6666.7\left(1 - \frac{25}{r^2}\right)}$$

and

$$\sigma_t = -\frac{5000(100)}{100 - 25}\left(1 + \frac{25}{r^2}\right)$$

$$= \boxed{-6666.7\left(1 + \frac{25}{r^2}\right)}$$

These relations are plotted in Figure 14.4. Note that in this case both σ_r and σ_t are compressive stresses. The radial stress σ_r assumes a maximum value equal to p_2 on the outside surface, and the maximum circumferential stress occurs on the inner surface of the pipe.

Example 14.3

Frequently, in order to increase their load-carrying capacity, composite cylinders are made by shrinking an outer cylinder or jacket onto an inner cylinder. Initially, the inside radius of the jacket is a little smaller than the outside radius of the inner cylinder, and the jacket is expanded by heating, for easy placement on the inner cylinder. If the difference between the inside radius of the jacket and the outside radius of the inner cylinder is Δr, develop an expression for the interface pressure p_i that is created between the jacket and the inner cylinder after the shrink-fitting assembly is completed. Also, sketch the circumferential stress distributions across the wall thicknesses for both the jacket and the inner cylinder produced by p_i.

Solution. Consider the assembly shown in Figure 14.5. Let r_1 and r_2 be the inside and outside radii, respectively, for the inner cylinder. Let r_2 and r_3 be the

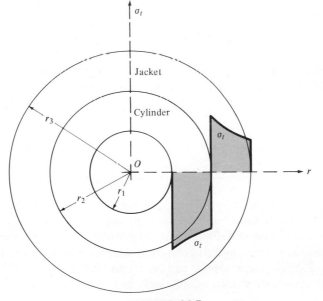

FIGURE 14.5

inside and outside radii, respectively, for the jacket. The symbol r_2 may be used to represent the outside radius of the inner cylinder and the inside radius of the jacket because their difference Δr is very small. Assume both jacket and the inner cylinder to be made of the same material.

When the shrink-fitting process is completed, the interface pressure p_i expands the inside radius of the jacket by the amount δ_{rj} and shrinks the outside radius of the inner cylinder by the amount δ_{rc}. The jacket is subjected only to the internal pressure p_i and δ_{rj} is obtained from Eq. 14.10. Thus

$$\delta_{rj} = \frac{p_i r_2^2}{E(r_3^2 - r_2^2)} \left[(1 - \mu)r_2 + (1 + \mu)\frac{r_3^2}{r_2} \right]$$

The inner cylinder is subjected only to the external pressure p_i and δ_{rc} is given by Eq. 14.15. Therefore,

$$\delta_{rc} = -\frac{p_i r_2^2}{E(r_2^2 - r_1^2)} \left[(1 - \mu)r_2 + (1 + \mu)\frac{r_1^2}{r_2} \right]$$

The interface pressure p_i is found from the condition that the sum of the absolute magnitudes of δ_{rj} and δ_{rc} must be equal to the radial difference Δr. Therefore,

$$\delta_{rj} + \delta_{rc} = \Delta r$$

which, after simplifications, leads to

$$\frac{2p_i r_2^3}{E} \frac{r_3^2 - r_1^2}{(r_3^2 - r_2^2)(r_2^2 - r_1^2)} = \Delta r$$

Therefore, the interference pressure p_i may be expressed in terms of the radial difference Δr, the geometry of the cylinders, and the modulus of elasticity E as follows:

$$\boxed{p_i = \frac{(\Delta r)E}{2r_2^3} \frac{(r_3^2 - r_2^2)(r_2^2 - r_1^2)}{r_3^2 - r_1^2}}$$

The jacket is subjected to the internal pressure p_i only, and therefore the circumferential stress distribution across its wall thickness is comparable to that found for the pipe in Example 14.1. The inner cylinder is subjected to the external pressure p_i only, and therefore the circumferential stress distribution across its wall thickness is similar to that found for the pipe in Example 14.2. These distributions are shown qualitatively in Figure 14.5.

Internal Pressure on Thin-Walled Cylinders

The case of a cylindrical container subjected only to the internal pressure p_1 was discussed earlier and Eqs. 14.7, 14.8, and 14.9 were developed for σ_r, σ_t, and σ_z, respectively. These equations will now be specialized to the case of cylindrical containers with relatively thin walls.

Equation 14.8, which gives the circumferential stress σ_t at any point within the walls of a cylinder, may be specialized to obtain the circumferential stress σ_{t1} on the inner surface, where $r = r_1$, and σ_{t2} on the outer surface, where $r = r_2$. Thus

$$\sigma_{t1} = \frac{p_1 r_1^2}{r_2^2 - r_1^2}\left(1 + \frac{r_2^2}{r_1^2}\right) \tag{14.17a}$$

and

$$\sigma_{t2} = \frac{2p_1 r_1^2}{r_2^2 - r_1^2} \tag{14.17b}$$

Dividing Eq. 14.17a by Eq. 14.17b, we obtain the dimensionless ratio σ_{t1}/σ_{t2} as follows:

$$\frac{\sigma_{t1}}{\sigma_{t2}} = \frac{1}{2}\left[1 + \left(\frac{r_2}{r_1}\right)^2\right] \tag{14.18}$$

A graph of σ_{t1}/σ_{t2} as a function of r_2/r_1 is shown in Figure 14.6. It is evident that as the ratio r_2/r_1 (i.e., as the wall thickness of the cylinder) becomes larger, the ratio

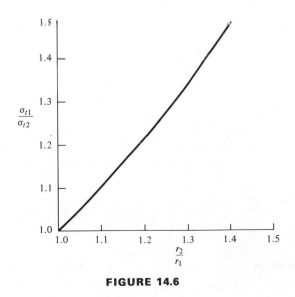

FIGURE 14.6

σ_{t1}/σ_{t2} increases. A range of values of the ratio r_2/r_1 may, however, be chosen for which σ_{t1}/σ_{t2} is, for all practical purposes, sufficiently close to unity. For such cases the circumferential stress distribution across the wall thickness may be assumed uniform without introducing appreciable error and the cylinder is said to be thin-walled.

The range of values of r_2/r_1 for which a cylinder may be assumed thin-walled (i.e., for which $\sigma_{t1}/\sigma_{t2} \doteq 1$) depends upon specific applications. If, for example, a 5 percent maximum variation in the circumferential stress is permissible across the wall thickness (i.e., $\sigma_{t1}/\sigma_{t2} = 1.05$), the ratio r_2/r_1 may be no more than about 1.05. This means that the wall thickness of the cylinder (i.e., $r_2 - r_1$) cannot exceed a value of $0.05r_1$, or 5 percent of the inner radius. If, on the other hand, a 10 percent maximum variation in the circumferential stress is allowable, then r_2/r_1 is about 1.09, and the wall thickness may be as large as 9 percent of the inner radius of the cylinder.

If the stress σ_{t1} is assumed to be equal to the stress σ_{t2}, Eq. 14.17a yields the value of the uniform circumferential stress in the wall of a thin-walled cylinder. This equation may be expressed in simpler form as follows:

$$\sigma_t = \sigma_{t1} = \frac{p_1 r_1^2}{r_2^2 - r_1^2}\left(1 + \frac{r_2^2}{r_1^2}\right) \tag{14.17a}$$

from which

$$\sigma_t = p_1 \frac{r_2^2 + r_1^2}{(r_2 - r_1)(r_2 + r_1)}$$

Since the wall thickness is relatively small, r_2 may be assumed equal to r_1 and, consequently, the quantities $r_2^2 + r_1^2$ and $r_2 + r_1$ may be written, respectively, as $2r_1^2$ and $2r_1$. Also, the quantity $r_2 - r_1$ represents the wall thickness t of the cylinder. Therefore,

$$\sigma_t = p_1 \frac{2r_1^2}{t(2r_1)} = \frac{p_1 r_1}{t} \tag{14.19}$$

In general, the radial stress σ_r, because of its relative magnitude, is not as significant a stress quantity as the circumferential stress σ_t in the case of thin-walled cylinders. However, its value may be obtained from Eq. 14.7. The longitudinal stress, σ_z, when the thin-walled cylinder has closed ends, may be obtained by specializing Eq. 14.9. Thus

$$\sigma_z = \frac{p_1 r_1^2}{r_2^2 - r_1^2} \tag{14.9}$$

This equation leads to

$$\sigma_z = \frac{p_1 r_1^2}{(r_2 - r_1)(r_2 + r_1)}$$

$$= \frac{p_1 r_1^2}{t(2r_1)}$$

from which

$$\sigma_z = \frac{p_1 r_1}{2t} \qquad\qquad (14.20)$$

Note that the longitudinal stress σ_z is exactly one-half that of the circumferential stress σ_t and that both of these stresses are always tensile.

Example 14.4

A thin-walled cylinder with closed ends for which $r_1 = 0.50$ m and $r_2 = 0.54$ m is subjected to internal pressure $p_1 - 2$ MPa. Determine (a) the absolute maximum shearing stress on the inner surface of the cylinder, (b) the absolute maximum shearing stress on the outer surface of the cylinder, and (c) the normal and shearing stresses in the wall of the cylinder on a plane inclined to the longitudinal axis of the cylinder through a 30° angle.

Solution

(a) Consider a three-dimensional stress element on the inner surface of the cylinder such that two of its planes are parallel to the cylinder longitudinal axis and two perpendicular to this axis. The two planes parallel to the longitudinal axis are subjected to the circumferential stress σ_t given by Eq. 14.19. Thus

$$\sigma_t = \frac{p_1 r_1}{t} = \frac{2 \times 0.50}{0.02} = 50 \text{ MPa}$$

This stress is tensile and, as mentioned earlier, it is a principal stress. The two planes perpendicular to the longitudinal axis of the cylinder are subjected to the longitudinal stress σ_z given by Eq. 14.20. Therefore,

$$\sigma_z = \frac{p_1 r_1}{2t} = 25 \text{ MPa}$$

This stress is also a tensile principal stress. The third set of two parallel planes, one of which is the inner cylindrical surface, is subjected to a radial stress which is equal in magnitude to the applied pressure $p_1 = 2$ MPa, as may be verified from Eq. 14.7. This last stress is obviously a compressive principal stress. Thus the stress element on the inner surface of the cylinder is subjected to the following three principal stresses, which are consistent with the condition that algebraically, $\sigma_1 > \sigma_2 > \sigma_3$:

$$\sigma_1 = \sigma_t = 50 \text{ MPa}$$

$$\sigma_2 = \sigma_z = 25 \text{ MPa}$$

$$\sigma_3 = \sigma_r = -2 \text{ MPa}$$

Therefore, the absolute maximum shearing stress is given by Eq. 2.13b. Thus

$$\left| \tau_{\max} \right| = \frac{\sigma_1 - \sigma_3}{2} = \frac{50 + 2}{2}$$

$$= \boxed{26 \text{ MPa}}$$

This stress acts on a plane that bisects the angle between σ_1 and σ_3.

(b) Consider a three-dimensional stress element on the outer surface of the cylinder whose planes are defined in the same manner as those in part (a). The circumferential and longitudinal stresses are identical to those found in part (a). However, the radial stress on the last set of two parallel planes, one of which is the outer cylindrical surface, is zero, since it is a free surface. This conclusion may be also verified from Eq. 14.7. Thus the three principal stresses on the outer surface of the cylinder are

$$\sigma_1 = \sigma_t = 50 \text{ MPa}$$

$$\sigma_2 = \sigma_z = 25 \text{ MPa}$$

$$\sigma_3 = \sigma_r = 0$$

Hence the absolute maximum shearing stress is given by Eq. 2.13b:

$$\left| \tau_{\max} \right| = \frac{\sigma_1 - \sigma_3}{2} = \frac{50 + 0}{2}$$

$$= \boxed{25 \text{ MPa}}$$

This stress acts on a plane that bisects the 90° angle between σ_1 and σ_3.

(c) The Mohr's circle for the two-dimensional stress condition defined by σ_1 and σ_2 is constructed as shown in Figure 14.7(a). As shown in the two-dimensional stress element of Figure 14.7(b), the plane of interest *B–B* is inclined to the plane of σ_1 through a 30° angle shown clockwise in the sketch, although a counterclockwise angle would serve the same purpose. This plane is

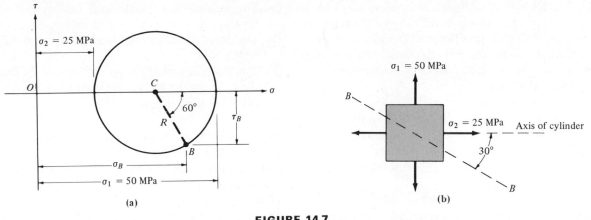

FIGURE 14.7

represented on the Mohr's circle by point B, whose coordinates give the values of the normal and shearing stresses on this plane. Thus

$$\sigma_B = OC + R \cos 60 = \frac{50 + 25}{2} + \left(\frac{50 - 25}{2}\right) \cos 60$$

$$= 37.50 + 6.25$$

$$= \boxed{43.75 \text{ MPa}}$$

$$\tau_B = -R \sin 60$$

$$= \boxed{-10.83 \text{ MPa}}$$

Homework Problems

14.1 A thick-walled cylinder with closed ends is subjected only to internal pressure $p_1 = 8000$ psi. If the inner radius $r_1 = 5$ in. and the outer radius $r_2 = 10$ in., determine the maximum values of σ_t, σ_r, and σ_z at a section sufficiently far away from the ends of the cylinder.

14.2 A thick-walled cylinder with closed ends for which $r_1 = 0.25$ m and $r_2 = 0.50$ m is submerged in seawater to a depth of 20 m. Assume the density of seawater to be 1100 kg/m³ and determine the maximum values of σ_t, σ_r, and σ_z at a section sufficiently removed from the two ends of the cylinder. The pressure inside the tank is atmospheric.

14.3 A thick-walled cylinder with $r_1 = 0.20$ m is to be subjected only to the internal pressure $p_1 = 50$ MPa. If the maximum tensile stress in the cylinder is not to exceed 100 MPa, determine the necessary wall thickness.

14.4 A thick-walled cylindrical tank with $r_2 = 3$ ft has closed ends and is subjected only to the external pressure $p_2 = 20{,}000$ psi. If the maximum compressive stress, at distances far removed from the two ends, is not to exceed 50,000 psi, determine the needed wall thickness.

14.5 The internal pressure in the thick-walled cylinder described in Problem 14.2 is increased from atmospheric to 2 MPa. Determine the maximum values of σ_t, σ_r, and σ_z.

14.6 A composite thick-walled cylinder is made by shrinking a jacket onto a main cylinder as described in Example 14.3. The following data are provided: $E = 30 \times 10^6$, $r_1 = 20$ in., $r_2 = 30$ in., $r_3 = 40$ in., and $\Delta r = 0.05$ in. Determine the absolute maximum values of σ_t and σ_r produced by the shrink-fitting operation.

14.7 The composite thick-walled cylinder described in Problem 14.6 is subjected to the internal pressure $p_1 = 30{,}000$ psi. Assume the external pressure $p_2 = 0$ and determine the absolute maximum value of σ_t.

14.8 A gas storage thin-walled tank with $r_1 = 2.50$ m and $r_2 = 2.60$ m is subjected to the internal pressure $p_1 = 3$ MPa. Determine on a section far removed from the two ends:

(a) The maximum normal stress in the wall of the tank.

(b) The absolute maximum shearing stress in the wall of the tank.

(c) The normal and shearing stresses in the wall of the cylinder on a plane inclined to the axis of the cylinder through a 45° angle.

14.9 A thin-walled cylinder with closed ends is constructed by butt-welding $\frac{3}{4}$-in. plates such that the weld makes an angle of 30° with the axis of the cylinder. If $r_1 = 50$ in., $p_1 = 300$ psi, and the normal and shearing stresses in the weld are not to exceed 15,000 psi and 10,000 psi, respectively, determine the minimum wall thickness.

14.10 A water storage cylindrical thin-walled stand tank is 20 m high and 2 m inside diameter and has a wall thickness of 0.04 m. The density of water is 1000 kg/m^3. Determine the absolute maximum normal and shearing stresses at the bottom of the tank when it is full of water.

14.11 Consider a thin-walled spherical shell subjected to the internal pressure p_1. If the inner radius is r_1 and the wall thickness is t, determine expressions for the absolute maximum normal and shearing stresses in the shell.

*14.3 Shear Center for Thin-Walled, Open Cross Sections

In discussing the flexure of beams in Chapter 5, the assumption was made that whether the beam was symmetrically or unsymmetrically loaded, the bending loads produced no twisting action. For the usual type of cross section, this assumption is satisfied if the plane of the loads passes through the centroid of the section. For open cross sections such as channels and angles which consist of thin rectangles, application of the bending loads through the centroid of the section produces twisting of these sections. This twisting may be avoided by shifting the bending loads from the centroid of the section to a point in the plane of the section known as the *shear center*. Locating the shear center for a given cross section consisting of relatively thin rectangles requires the determination of the shearing stresses produced in the thin rectangles by the flexural loads. These shearing stresses produce a twisting moment that has to be balanced, to avoid twisting of the section, by properly locating the plane of the bending loads through the shear center. Locating the shear center in the plane of the section requires finding two coordinates with respect to the two principal centroidal axes of the section. These concepts are illustrated by the following three examples. Examples 14.5 and 14.6 deal with open cross sections consisting of thin rectangles and possessing one axis of symmetry. As was shown in Section 5.2, an axis of symmetry coincides with one of the two principal centroidal axes of inertia. In such a case, as will be shown in Example 14.5, only one coordinate need be computed, because the shear center lies on the axis of symmetry for the section. Example 14.7 deals with an open cross section consisting of thin rectangles but possessing no axis of symmetry. In such a case, the principal centroidal axes need to be located before proceeding to the determination of the shear center, which requires two separate computations to locate its two coordinates.

Example 14.5

A cantilever beam has a channel cross section as shown in Figure 14.8(a). This cross section is symmetric about the centroidal u axis. Thus the centroidal u axis is one of the two principal axes of inertia and the centroidal v axis the second. Assume the flange and web thicknesses t to be very small in comparison to the dimensions w and h. (a) Consider a concentrated load P at the free end of the cantilever acting in a direction perpendicular to the centroidal u axis and in the negative v direction. Determine the shearing stresses and the shearing forces in the two flanges and in the web of the section. What should be the location of the applied load P so that these shearing forces produce no twisting of the section? (b) Repeat part (a) if the load P is applied in a direction perpendicular to the centroidal v axis and in the positive u direction.

Solution

(a) Because of the way the load P is applied, the centroidal u axis is, itself, the neutral axis for bending of the cantilever beam. Consider a small segment of the upper flange at distance x from the load P and having a differential length dx and a finite width u. This small segment is isolated and its free-body diagram constructed as shown magnified in Figure 14.8(b). The force F_R represents the resultant of the flexural stresses acting over the flange segment of dimensions tu at a distance x from the load P, where the moment is M_u. Therefore,

$$F_R = \int \sigma_R \, dA = \int \frac{M_u v}{I_u} \, dA$$

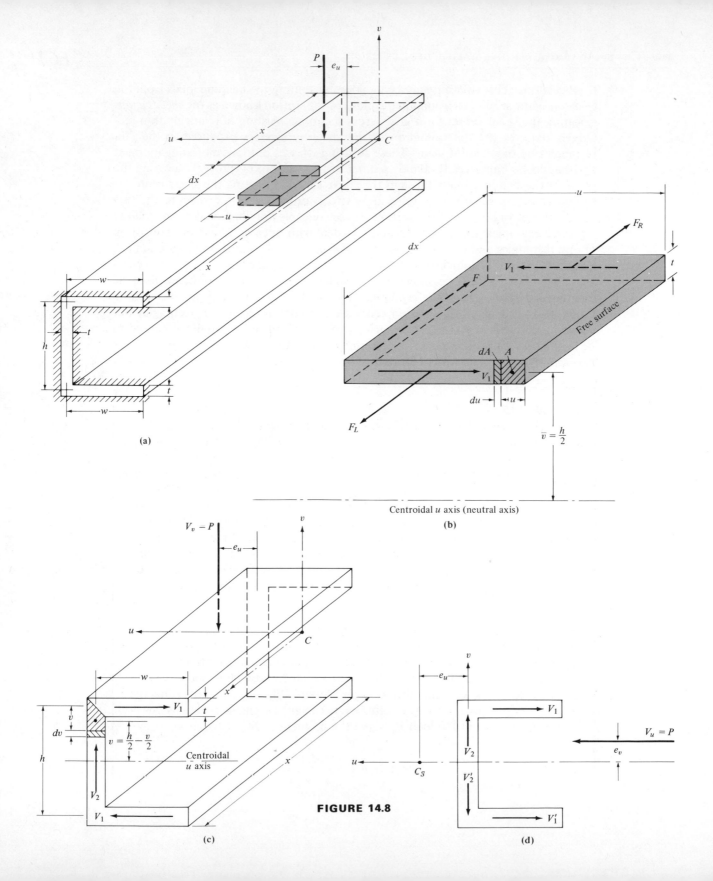

FIGURE 14.8

The force F_L represents the resultant of the flexural stresses acting over the flange segment of dimensions tu at a distance $x + dx$ from the load P, where the moment is $M_u + dM_u$. Thus

$$F_L = \int \sigma_L \, dA = \int \frac{M_u + dM_u}{I_u} v \, dA$$

The force F represents the resultant of the shearing stresses τ_{ux} on the inner longitudinal surface of the block shown in Figure 14.8(b), which must exist in order to balance forces in the x direction. Hence

$$F_L - F_R - F = 0$$

and

$$F = F_L - F_R = \frac{dM_u}{I_u} \int v \, dA$$

The shearing stress τ_{ux} is assumed constant over the small thickness t, and consequently $F = \tau_{ux} t \, dx$. Therefore,

$$\tau_{ux} = \frac{dM_u/dx}{I_u t} \int v \, dA = \frac{V_v}{I_u t} \int v \, dA \qquad \textbf{(a)}$$

where dM_u/dx was replaced by the shear force V_v (see Eq. 1.6) at the position in the beam under consideration.

The existence of the shear stress τ_{ux} on longitudinal flange planes requires that a shear stress τ_{xu} of equal magnitude exist on transverse flange planes. Thus

$$\tau_{xu} = \tau_{ux} = \frac{V_v}{I_u t} \int v \, dA = \frac{V_v Q}{I_u t} \qquad \textbf{(b)}$$

where the quantity Q is used to represent $\int v \, dA$. Note that Eq. (b) is identical with Eqs. 5.12a and 5.12b for horizontal and vertical shear stresses due to transverse loads in beams. As discussed in Section 5.4, the quantity $Q = \int v \, dA$ represents the first moment of an area with respect to the neutral axis for bending and may be conveniently written in the form $Q = A\bar{v}$. In this form A is the area between the free edge and the location on the upper flange where τ_{xu} occurs, and $\bar{v}$ is the moment arm between the centroid of the area A and the neutral axis. These quantities are shown in Figure 14.8(b). Therefore, using Eq. (b), we can write the shearing stress τ_{xu} in the upper flange in the form

$$\tau_{xu} = \frac{V_v}{I_u t} A\bar{v} \qquad \textbf{(c)}$$

As shown in Figure 14.8(b), $A = tu$ and $\bar{v} = h/2$, and consequently the shearing stress τ_{xu} in the upper flange of the section is given by the equation

$$\tau_{xu} = \frac{V_v h}{2I_u} u \qquad \text{(d)}$$

Thus the shearing stress τ_{xu} in the upper flange of the channel increases linearly from zero at the free surface to its maximum value $V_v hw/2I_u$ at the junction between the upper flange and the web of the channel where $u = w$.

The shear force V_1 on the transverse flange plane [see Figure 14.8(b)] is the resultant of the shear stresses τ_{xu} which are assumed constant over the flange thickness t. Hence, by Eq. (c),

$$V_1 = \int \tau_{xu}\, dA = \frac{V_v}{I_u t} \int \bar{v} A\, dA \qquad \text{(e)}$$

By substituting the values for $\bar{v}$, A, and dA as shown in Figure 14.8(b), one obtains the magnitude of the shear force V_1 in the upper flange. Hence

$$V_1 = \frac{V_v}{I_u t} \int_0^w \frac{h}{2} (tu)(t\, du)$$

from which

$$\boxed{V_1 = \frac{V_v thw^2}{4I_u}} \qquad \text{(f)}$$

A free-body diagram of a segment of the cantilever beam between the free end and the location defined by the coordinate x is shown in Figure 14.8(c). Note that the shear force V_1 in the upper flange is pointed in the negative u direction. Note also that equilibrium of forces in the u direction requires that both transverse flange planes located at a distance dx from each other in Figure 14.8(b) be subjected to the same shear force V_1.

By considering a small segment of the lower flange, as was done for the upper flange, we conclude that the magnitudes of the shear stress and the shear force in this lower flange are given, respectively, by Eqs. (d) and (f). As shown in Figure 14.8(c), the shear force V_1 in the lower flange has to point in the positive u direction. This conclusion is reached either from examination of the shear stress τ_{xu} in the lower flange as was done for the upper flange, or simply by considering equilibrium of forces in the u direction in the free-body diagram of Figure 14.8(c).

By assuming that the shear V_v at any location along the beam is carried entirely by the web of the channel, we may quickly conclude on the basis of the equilibrium of forces in the v direction that the shear force V_2 in the web, shown in Figure 14.8(c), must be equal to the applied force P. Thus $V_2 = V_v = P$ and must be directed in the positive v direction in order to resist the action of the force P. The assumption that the flanges do not contribute in resisting the vertical shear is valid for all practical purposes.

The fact that V_2 is equal to the shear force V_v, which in turn is equal to the applied force P, can also be established by the use of Eq. (e). Thus referring to Figure 14.8(c), we see that

$$V_2 = \frac{V_v}{I_u t} \int \left[(\bar{v}A)_{\text{flange}} + (\bar{v}A)_{\text{web}} \right] dA$$

where the quantity $(\bar{v}A)_{\text{flange}}$ represents Q for the entire flange and is equal to $twh/2$, and the quantity $(\bar{v}A)_{\text{web}}$ represents Q for that part of the web between the flange and the location defined by the coordinate v at which the shear stress is considered. Thus, as discussed earlier, the entire quantity $[(\bar{v}A)_{\text{flange}} + (\bar{v}A)_{\text{web}}]$ represents the first moment of the area between the free surface of the flange and the location in the web defined by the coordinate at which the shear stress is considered. Therefore, using the information in Figure 14.8(c),

$$V_2 = \frac{V_v}{I_u t} \int_0^h \left[\frac{twh}{2} + \frac{(vt)(h-v)}{2} \right] t \, dv$$

$$= \frac{V_v}{I_u} \left(\frac{twh^2}{2} + \frac{th^3}{12} \right)$$

By neglecting terms containing the quantity t^3 in comparison to the other terms, the moment of inertia of the section with respect to the u principal centroidal axis becomes

$$I_u = \frac{twh^2}{2} + \frac{th^3}{12}$$

Thus, as before, the shear force in the web is shown to be equal to V_v, the entire shear force at the section, which, in turn, is equal to the applied force P.

Equilibrium of the free-body diagram shown in Figure 14.8(c) requires that the torque produced by the forces V_1 about the x axis be balanced by an equal and opposite torque about this axis produced by $V_2 = V_v$ and $P = V_v$. Hence

$$V_v e_u - V_1 h = 0$$

from which

$$e_u = \frac{V_1 h}{V_v}$$

and by substituting for V_1 from Eq. (f), we obtain

$$e_u = \frac{th^2w^2}{4I_u} \qquad \textbf{(g)}$$

which could be expressed in terms of the dimensions of the channel section if the value of I_u is substituted. Thus, to avoid twisting action when the load P is applied normal to the u axis, it should be placed at a distance e_u from the centerline of the web, as shown in Figure 14.8(c).

(b) If the force P is applied in a direction perpendicular to the centroidal v axis, and in the positive u direction, shear forces would be induced in both flanges and in the web of the channel section, as shown in Figure 14.8(d). By inspection and considering the fact that the channel is symmetric about the centroidal u axis, it can be concluded that $V_1 = V'_1$ and $V_2 = V'_2$ only when the load P is coincident with the centroidal u axis. Therefore, to avoid twisting of the channel section when the applied load is perpendicular to the v axis, it should be placed such that

$$e_v = 0 \qquad \textbf{(h)}$$

Thus the two coordinates of the shear center of the channel section are established as given by Eqs. (g) and (h), and its location is depicted by point C_s in Figure 14.8(d). Note that the shear center of a given cross section is a function only of its geometry and does not depend upon the loads that are applied.

Example 14.6

Locate the shear center C_s for the cross section shown in Figure 14.9(a). The thickness $t = 0.2$ in. is the same for all rectangles in the cross section.

Solution. The centroid C of the cross section is located by the usual methods. However, because of the symmetric nature of the cross section, only one coordinate for the centroid C need be determined. Also, the axis of symmetry is one of the two principal axes of inertia. Establish the u and v principal centroidal axes as shown in Figure 14.9(a).

As was shown in Example 14.5, the shear center for this symmetric cross section lies along the axis of symmetry (i.e., along the u principal centroidal axis). Thus only the u coordinate for the shear center need be determined. This may be accomplished by assuming an applied load acting parallel to the v principal centroidal axis and positioning it such that it produces no twisting action in the cross section. Assume for purposes of this solution that the applied load acts in the negative v direction. However, it should be pointed out

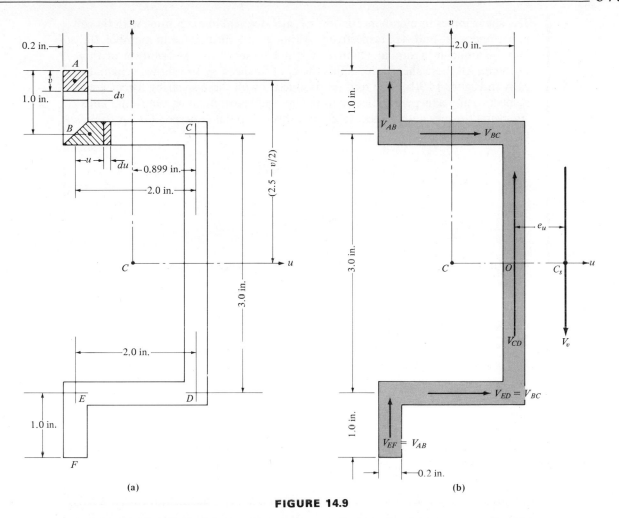

(a)   (b)

FIGURE 14.9

that the same answer is reached even if the load were assumed to act in the positive v direction.

The shear force induced in any rectangle in the section of Figure 14.9(a) may be found by using Eq. (e) of Example 14.5 and the information contained in Figure 14.9(a). Thus

$$V_{AB} = \frac{V_v}{I_u t} \int (\bar{v} A)\, dA$$

$$= \frac{V_v}{0.2 I_u} \int_0^{1.0} 0.2 v \left(2.5 - \frac{v}{2}\right)(0.2\, dv) = 0.22 \frac{V_v}{I_u}$$

$$V_{BC} = \frac{V_v}{0.2 I_u} \int_0^{2.0} [0.2(1.0)(2.0) + 0.2 u(1.5)](0.2\, du) = 1.4 \frac{V_v}{I_u}$$

The shear forces in members DE and EF are identical in magnitude with those in members BC and AB, respectively. Although the shear force in member DC may be found in a similar manner, it is not required for the solution of this problem. All these shear forces, plus the applied shear V_v, are shown schematically in Figure 14.9(b). The torques produced by all these shearing forces with respect to any axis perpendicular to the uv plane must be in equilibrium. Thus summing torques with respect to an axis through point O perpendicular to the uv plane, we obtain

$$e_u V_v + 2V_{AB}(2.0) - 2V_{BC}(1.5) = 0$$

Therefore,

$$e_u = \frac{2V_{BC}(1.5) - 2V_{AB}(2.0)}{V_v}$$

Substituting the values obtained earlier for V_{BC} and V_{AB} in terms of V_v, we obtain

$$e_u = \frac{3(1.4V_v/I_u) - 4(0.22V_v/I_u)}{V_v} = \frac{3.32}{I_u}$$

If the value of $I_u = 3.89$ in.[4] is substituted, the value of e_u is found to be

$$\boxed{e_u = 0.85 \text{ in.}}$$

Thus the shear center C_s for this symmetric cross section lies along the u principal centroidal axis at a distance $e_u = 0.85$ in. to the right of the centerline of vertical rectangle CD, as shown in Figure 14.9(b).

Example 14.7

Using the methods developed in Examples 14.5 and 14.6, outline a procedure for locating the shear center for an open cross section that possesses no axes of symmetry, such as the one shown in Figure 14.10(a).

Solution. The first step in the solution of this type of problem is to locate the centroid of the section and establish the principal centroidal axes as shown in Figure 14.10(a). An applied shear load V_v is then assumed acting parallel to the v principal centroidal axis and, using the procedures discussed in Examples 14.5 and 14.6, the shear forces in the various rectangles of the cross section are determined using Eq. (e) of Example 14.5. These shear forces as well as the applied force V_v are shown schematically in Figure 14.10(a). Examination of the forces in Figure 14.10(a) shows that a judicious choice of the axis about which

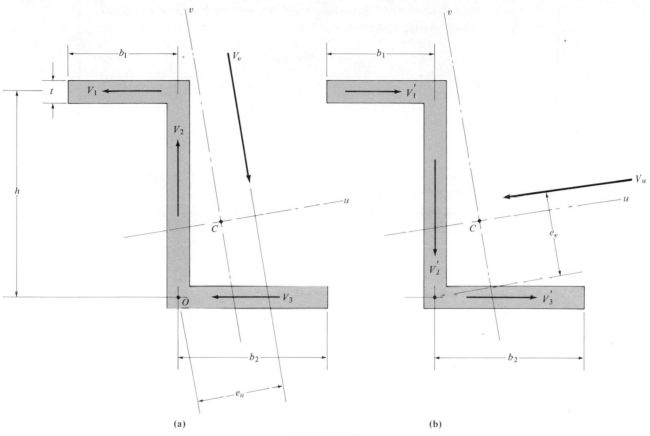

(a) (b)

FIGURE 14.10

the torques are summed in determining e_u would eliminate the need for finding all the shear forces. For example, summing torques about an axis through point O would eliminate the need for determining V_2 and V_3, since they pass through this axis and their torques would be zero. Thus

$$e_u V_v - V_1 h = 0$$

from which

$$\boxed{e_u = \frac{V_1 h}{V_v}} \qquad \textbf{(a)}$$

Since V_1 is expressible in terms of V_v, Eq. (a) yields the value of e_u in terms of the cross-sectional dimensions.

An applied shear force V_u is now assumed acting parallel to the u principal centroidal axis and the shear force V_1' shown in Figure 14.10(b) is determined

676

using Eq. (e) of Example 14.5. Summing torques about an axis through point O in Figure 14.10(b) yields the relation

$$e_v V_u - V'_1 h = 0$$

from which

$$\boxed{e_v = \frac{V'_1 h}{V_u}} \qquad \textbf{(b)}$$

Since V'_1 may be expressed in terms of V_u, the quantity e_v may be given entirely in terms of the cross-sectional dimensions.

Homework Problems

14.12 Locate the shear center for:
(a) An angle section.
(b) A tee section with equal flanges.

14.13 Use Eq. (e) of Example 14.5 to determine the shear force V_{CD} in rectangle CD of Figure 14.9. Check your answer by applying the conditions of equilibrium.

14.14
(a) Assume that a load is applied normal to the axis of symmetry of the section shown in Figure H14.14 and determine the shear forces in rectangles AB, BC, and CD.
(b) Locate the shear center for this cross section.

14.15 A hollow equilateral triangular tube is slit longitudinally at the midpoint of one of the three sides, as shown in Figure H14.15. Assume the width

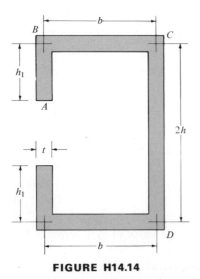

FIGURE H14.14

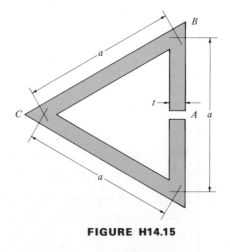

FIGURE H14.15

of the cut to be insignificantly small. This tube is to be used as a cantilever to carry a load normal to the axis of symmetry of the cross section.

(a) Determine the shear forces induced in rectangles *AB* and *BC*.

(b) Locate the shear center for this cross section.

14.16 If in Problem 14.15, the applied force *P* acts along rectangle *CB* instead of being normal to the axis of symmetry, determine the amount of torque to which this cross section is subjected.

14.17 The cross section shown in Figure H14.17 is that for a cantilever beam subjected to a load $P = 5$ k at the free end and acting perpendicular to the axis of symmetry of the cross section. The load is placed so that it coincides with the centerline of the web. How much torque is induced in the cross section?

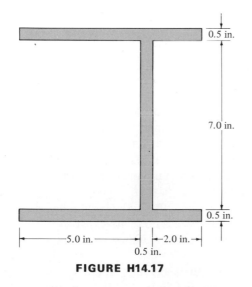

FIGURE H14.17

14.18–14.20 Compute the location of the shear center for the following symmetric cross sections. Assume the load to act perpendicular to the axis of symmetry.

14.18 Cross section shown in Figure H14.18.
14.19 Cross section shown in Figure H14.19.
14.20 Cross section shown in Figure H14.20.

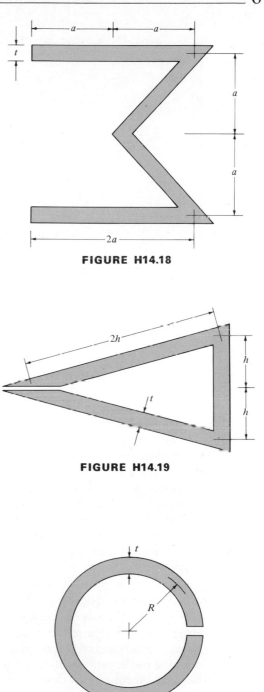

FIGURE H14.18

FIGURE H14.19

FIGURE H14.20

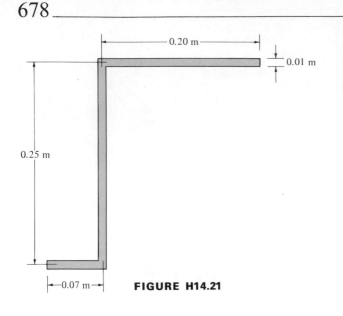

FIGURE H14.21

14.21–14.22 Compute the location of the shear center for the following unsymmetric cross sections.
14.21 Cross section shown in Figure H14.21.
14.22 Cross section shown in Figure H14.22.

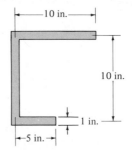

FIGURE H14.22

14.4 Beams of Two Materials

Beams are often constructed of two different materials in order to achieve certain desirable effects. Such is the case, for example, with reinforced concrete beams, which will be dealt with in some detail in Sections 14.5 and 14.6. In this section the general theory underlying the method of solution for beams of two materials will be developed and discussed. This theory is based upon the same fundamental assumptions used in Chapter 5 for deriving the flexure equation, except that, in general, the moduli of two different materials have to be assumed unequal. Also, the two materials are assumed to be securely fastened together so that there is no relative displacement between them, and hence the two-material beam can be assumed to act as a single unit.

Consider a segment of a beam subjected to symmetric bending about the u principal centroidal axis by the moments M_u, as shown in Figure 14.11(a). In this case the neutral axis for bending coincides with the u principal centroidal axis. The cross-sectional area of the beam is shown in Figure 14.11(b) and indicates that the beam is composed of two materials, A and B. One of the basic assumptions made in deriving the flexure equation is that plane sections before bending remain plane after bending. According to this assumption, a plane section such as a–a before bending remains plane but rotates into the new position a'–a' as shown in Figure 14.1(a). Thus for any distance v from the neutral axis (i.e., the u axis) the deformations δ, and hence the strains ε, of both materials are identical. Therefore,

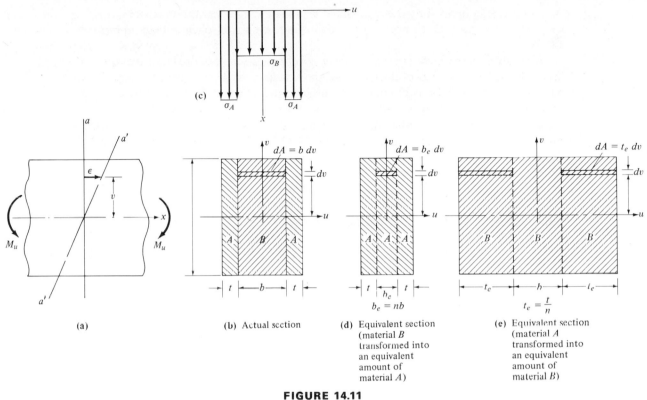

FIGURE 14.11

$$c_A = c_B \qquad \text{(14.21a)}$$

By introducing the assumption that both materials obey Hooke's law, we conclude that

$$\frac{\sigma_A}{E_A} = \frac{\sigma_B}{E_B} \qquad \text{(14.21b)}$$

where E_A, σ_A and E_B, σ_B represent the modulus of elasticity and the stress for materials A and B, respectively. Equation 14.21(b) can be written as follows:

$$\sigma_B = \frac{E_B}{E_A}\sigma_A = n\sigma_A \qquad \text{(14.22)}$$

in which $n = E_B/E_A$ is a shrinking or magnifying factor, depending upon whether $E_B < E_A$ or $E_B > E_A$. In either case there is an abrupt change in the magnitude of the bending stress at the junction between material A and material B. For purposes of discussion, assume that $E_B < E_A$, and therefore n is less than unity and by Eq. 14.22, for any position defined by the coordinate v, the stress σ_B is less than the stress σ_A. A stress distribution with respect to the u coordinate is shown schematically in Figure 14.11(c).

In analyzing beams of two materials, it is very convenient to deal with an equivalent cross section instead of dealing with the actual cross section of the beam. This equivalent cross section is obtained by transforming either of the two materials into an equivalent amount of the second material. Consider the differential quantity of normal force dF_B acting over the differential area $dA = b\,dv$ in material B of the original cross section as shown in Figure 14.11(b). This element of force may be written

$$dF_B = \sigma_B b\,dv \tag{14.23a}$$

Substituting for σ_B its value from Eq. 14.22, we obtain

$$dF_B = \sigma_A(nb)\,dv = \sigma_A b_e\,dv \tag{14.23b}$$

Equation 14.23b shows that that portion of the actual cross section that is made of material B may be transformed into an equivalent amount of material A by changing the actual width b of material B into an equivalent width b_e of material A. It is evident from Eq. 14.23b that the relation between the actual and the equivalent widths is given by

$$b_e = nb \tag{14.24}$$

Thus in obtaining the equivalent cross section shown in Figure 14.11(d), material B was transformed into an equivalent amount of material A by Eq. 14.24 to obtain a

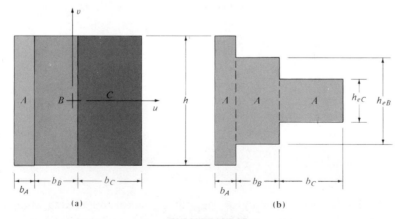

FIGURE 14.12
(a) Actual Section. (b) Equivalent Section (Materials B and C Transformed into Equivalent Amounts of Material A).

new cross section of the same material. This equivalent cross section is capable of developing the same resistance to bending as the actual cross section. Obviously, another equivalent cross section may be used by transforming material A to an equivalent amount of material B to obtain a new cross section of the same material. Such an equivalent cross section is shown in Figure 14.11(e).

Beams of more than two materials may be analyzed in basically the same fashion. For example, the cross section shown in Figure 14.12(a) consists of the three materials A, B, and C. If it is assumed that $E_A > E_B > E_C$, and bending is to be about the v axis, the equivalent cross section would be that shown in Figure 14.12(b).

Once an equivalent cross section of a homogeneous material is obtained, it may be analyzed for stresses and deflections by the methods developed in previous chapters. Some of these ideas and concepts are illustrated in Example 14.8.

Example 14.8

A composite cantilever beam 10 ft long is to support a uniform load of 100 lb/ft perpendicular to the u axis along its entire length. The cross-sectional area of the beam is shown in Figure 14.13(a). Let $E_{oak} = 2 \times 10^6$ psi and $E_{al} = 10 \times 10^6$ psi and determine the maximum flexural stresses in the oak and in the aluminum. Also determine the deflection at the end of the cantilever beam.

Solution. Transform the oak into an equivalent amount of aluminum by using Eq. 14.24, in which $n = E_{oak}/E_{al} = \frac{1}{5}$. Thus

$$b_e = nb = \tfrac{1}{5}(2) = 0.4 \text{ in.}$$

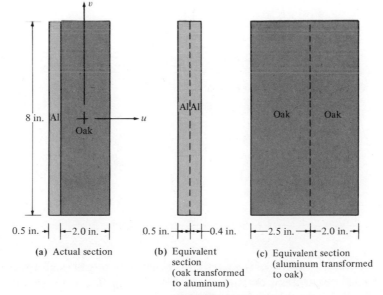

0.5 in. →| |←—2.0 in.—→| 0.5 in. →|—|—|←—0.4 in. |←—2.5 in.—→|←—2.0 in.—→|

(a) Actual section   **(b)** Equivalent section (oak transformed to aluminum)   **(c)** Equivalent section (aluminum transformed to oak)

FIGURE 14.13

The equivalent cross section is shown in Figure 14.13(b). The maximum flexural stresses for both the oak and the aluminum occur where the bending moment assumes its maximum value at the fixed end of the beam. Thus

$$(M_u)_{\max} = \frac{wL^2}{2} = 5000 \text{ lb-ft} = 60,000 \text{ lb-in.}$$

Using the flexure equation, we can obtain the flexural stress in the aluminum as follows:

$$(\sigma_{al})_{\max} = \frac{M_u}{I_u} v = \frac{60,000 \times 4}{\frac{1}{12}(0.9)(8)^3}$$

$$= \boxed{6250 \text{ psi}}$$

The maximum flexural stress in the oak is now obtained from Eq. 14.22. Thus

$$(\sigma_{oak})_{\max} = n(\sigma_{al})_{\max} = \frac{1}{5}(6250)$$

$$= \boxed{1250 \text{ psi}}$$

The same answers may be obtained by transforming the aluminum into an equivalent amount of oak. Thus

$$t_e = \frac{t}{n} = \left(\frac{1}{2}\right)\left(\frac{1}{5}\right) = 2.5 \text{ in.}$$

The equivalent section thus obtained is shown in Figure 14.13(c). Hence

$$(\sigma_{oak})_{\max} = \frac{60,000 \times 4}{(\frac{1}{12})(4.5)(8)^3}$$

and

$$= \boxed{1250 \text{ psi}}$$

$$(\sigma_{al})_{\max} = \frac{(\sigma_{oak})_{\max}}{n} = \frac{1250}{\frac{1}{5}}$$

$$= \boxed{6250 \text{ psi}}$$

The deflection at the end of the cantilever beam may be obtained by using either of the two equivalent cross sections. Thus, using the equivalent cross section shown in Figure 14.13(b), we obtain

$$v = \frac{wL^4}{8E_{al}(I_u)_{al}} = \frac{(100/12)(10 \times 12)^4}{8 \times 10 \times 10^6(0.9/12)(8)^3}$$

$$= \boxed{0.563 \text{ in.}}$$

The same answer can be reached by using the equivalent section shown in Figure 14.13(c). Thus

$$v = \frac{wL^4}{8E_{oak}(I_u)_{oak}} = \frac{(100/12)(10 \times 12)^4}{8 \times 2 \times 10^6(4.5/12)(8)^3}$$

$$= \boxed{0.563 \text{ in.}}$$

Example 14.9

A beam of two materials is constructed such that the two materials are placed on top of each other instead of side by side as was assumed in the development of Eqs. 14.22 and 14.24. Derive the applicable equations for this type of arrangement.

Solution. A segment of a beam subjected to symmetric bending about the u principal centroidal axis by the moments M_u is shown in Figure 14.14(a). The same assumptions that led to Eq. 14.22 lead to the following relations:

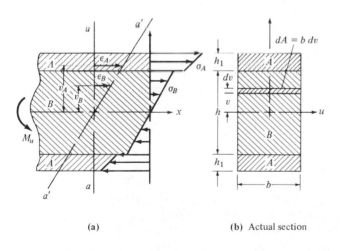

(a)

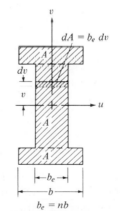

(b) Actual section

(c) Equivalent section (material B transformed into an equivalent amount of material A)

$b_e = nb$

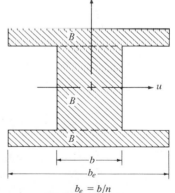

(d) Equivalent section (material A transformed into an equivalent amount of material B)

$b_e = b/n$

FIGURE 14.14

$$\varepsilon_B = \left(\frac{v_B}{v_A}\right)\varepsilon_A \tag{a}$$

from which, by using Hooke's law, one may obtain

$$\sigma_B = \left(\frac{v_B}{v_A}\right)\left(\frac{E_B}{E_A}\right)\sigma_A = \left(\frac{v_B}{v_A}\right)n\sigma_A \tag{b}$$

where $n = E_B/E_A$ is, again, a shrinking or magnifying factor, depending upon whether $E_B < E_A$ or $E_B > E_A$. In either case there is an abrupt change in the magnitude of the flexure stress at the junction between material A and material B, where $v_A = v_B$. For purposes of discussion, assume that $E_B < E_A$ so that n is less than unity. It follows from Eq. (b) that at the junction (i.e., at $v_A = v_B$), σ_B is less than σ_A. A stress distribution with respect to the v coordinate is shown schematically in Figure 14.14(a).

The development of the equivalent cross section proceeds in essentially the same manner as was followed in deriving Eq. 14.24. Thus consider the actual beam cross section shown in Figure 14.14(b). Assume that material B of this actual section is to be transformed into an equivalent amount of material A. The differential element of normal force dF_B acting over a differential area $dA = b\, dv$ becomes

$$dF_B = \sigma_B b\, dv \tag{c}$$

Substituting for σ_B its value in terms of σ_A from Eq. (b), we obtain

$$dF_B = \left(\frac{\sigma_A}{v_A}\right)v_B(nb)\, dv = \left(\frac{\sigma_A}{v_A}\right)v_B b_e\, dv \tag{d}$$

where the quantity $(\sigma_A/v_A)v_B$ represents the stress at a distance v_B from the neutral axis in a homogeneous material A. Therefore, Eq. (d) shows that the portion of the actual cross section made of material B may be transformed into an equivalent amount of material A by changing the actual width b of material B into an equivalent width b_e of material A. This equivalent width is obtained from the relation

$$b_e = nb \tag{e}$$

which is identical with Eq. 14.24. The equivalent cross section is capable of developing the same bending resistance as the actual cross section. Also, either material may be transformed into an equivalent amount of the second material. Transforming material B into an equivalent amount of material A yields the equivalent section of homogeneous material A shown in Figure 14.14(c). Transforming material A into an equivalent amount of material B leads to the equivalent section of homogeneous material B shown in Figure 14.14(d).

Homework Problems

14.23 A simply supported beam 5 m long carries a uniform load of intensity 2000 N/m and acting along the v principal centroidal axis. The cross section for the beam is shown in Figure H14.23. Assume that $E_{wood} = 10 \times 10^9$ N/m^2 and $E_{steel} = 200 \times 10^9$ N/m^2 and determine the maximum flexural stresses induced in the steel and in the wood by transforming the wood into steel. Determine also the deflection of the beam at midspan.

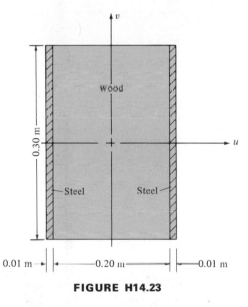

FIGURE H14.23

14.24 Repeat Problem 14.23 by transforming the steel into wood.

14.25 Repeat Problem 14.23 if the load of 2000 N/m acts along the u instead of the v principal centroidal axis.

14.26 A cantilever beam 15 ft long has the cross section shown in Figure H14.26. A concentrated load $P = 5000$ lb is placed at the free end acting in a plane that contains the v principal axis of inertia. Let $E_{al} = 10 \times 10^6$ psi and $E_{br} = 15 \times 10^6$ psi and deter-

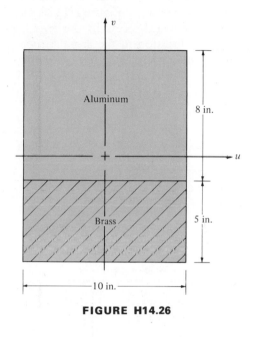

FIGURE H14.26

mine the maximum flexural stresses in the aluminum and in the brass. Solve the problem by transforming the brass into aluminum. Determine also the deflection at the end of the cantilever.

14.27 Repeat Problem 14.26 by transforming the aluminum into brass.

14.28 A rectangular wood beam 10 in. deep and 5 in. wide is reinforced by securely fastening aluminum plates 5 in. wide and $\frac{3}{4}$ in. deep at both top and bottom. If the beam is simply supported over a span of 20 ft, determine the safe value of a concentrated load placed at midspan. Let $E_W = 1.5 \times 10^6$ psi, $E_A = 10 \times 10^6$ psi and the allowable stresses for wood and aluminum are, respectively, 1000 and 10,000 psi.

14.29 Repeat Problem 14.28 with the midspan load acting parallel to the 5-in. dimension. All other conditions remain unchanged.

14.30 A simply supported beam 5 m long is acted upon by a concentrated force $P = 10$ kN acting at midspan and making a 30° angle with the u principal centroidal axis as shown in Figure H14.30. Determine the maximum flexural stress in the two materials, specifying their location in the cross section. Determine also the maximum deflection in the beam. Let $E_S = 200 \times 10^9$ N/m² and $E_{D.F.} = 12 \times 10^9$ N/m².

14.31 A cantilever beam 12 ft long is to carry a uniform load of intensity $w = 500$ lb/ft along its entire length, acting in a plane that contains the v principal centroidal axis. The cross-sectional area of the beam is shown in Figure H14.31. Let $E_S = 30 \times 10^6$ psi, $E_A = 10 \times 10^6$ psi, and $E_O = 2 \times 10^6$ psi, and determine the maximum deflection of the cantilever and the maximum flexural stresses in the three materials.

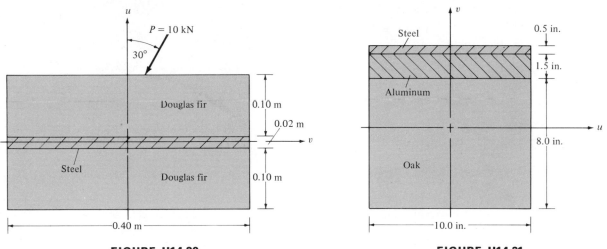

FIGURE H14.30   **FIGURE H14.31**

⋆**14.5** Reinforced Concrete Beams—Elastic Method

The flexural stress equation $\sigma_x = M_u v / I_u$ applies to beams fabricated of one material, but reinforced concrete beams comprise an example of a broad class of composite materials. In order to apply the equation, the steel rods used to reinforce concrete beams on the tension side are replaced by fictional concrete, which will resist tension. Since the tensile strength of concrete is small compared to its compressive strength, the tensile resistance of concrete is ignored in the computations. Refer to Figure 14.15 to understand how this replacement based upon the following two concepts is to be accomplished.

1. The force resisted by the fictional concrete fiber must equal the force resisted by the steel rod or fiber:

$$F_c = F_s \tag{14.25}$$

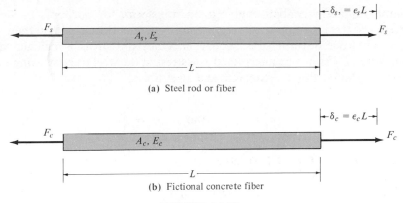

(a) Steel rod or fiber

(b) Fictional concrete fiber

FIGURE 14.15

2. The deformation of the fictional concrete fiber must equal the deformation of the steel rod or fiber. Steel rods embedded in concrete are bonded to the concrete by adhesion forces, friction forces, and forces between the concrete and lugs formed on the steel rod surfaces. Strains in the steel rods and adjacent concrete surrounding the steel rods are equal as long as the bond is not broken.

$$\delta_c = \delta_s \tag{14.26}$$

In general, $\delta = FL/AE$. Applying this equation for concrete and steel as expressed by Eq. 14.26 gives

$$\frac{F_c L_c}{A_c E_c} = \frac{F_s L_s}{A_s E_s} \tag{14.27}$$

Solving for A_c yields

$$A_c = \frac{F_c}{F_s} \frac{L_c}{L_s} \frac{E_s}{E_c} A_s \tag{14.28}$$

By Eq. 14.25, $F_c = F_s$ and $L_c = L_s = L$, which reduces Eq. 14.28 to

$$A_c = \frac{E_s}{E_c} A_s \tag{14.29}$$

The ratio of the modulus of elasticity of steel to the modulus of elasticity of concrete is referred to as the *modular ratio* and is denoted by the symbol n. Equation 14.29 may be written

$$A_c = nA_s \tag{14.30}$$

The simple equation enables engineers to transform, for purposes of analysis, a reinforced concrete beam to an equivalent concrete beam, which then may be analyzed with the flexural stress equation. Once the stress is calculated in the fictional concrete it must be interpreted as a stress in the steel rods by using Eq. 14.27 modified as follows. For the concrete and steel fibers,

$$\sigma_c = \frac{F_c}{A_c} \quad \text{and} \quad \sigma_s = \frac{F_s}{A_s}$$

Substitute these into Eq. 14.27 to give

$$\frac{\sigma_c L_c}{E_c} = \frac{\sigma_s L_s}{E_s} \tag{14.31}$$

Note that $L_c = L_s = L$ and solve for σ_s:

$$\sigma_s = \frac{E_s}{E_c} \sigma_c \tag{14.32}$$

or

$$\sigma_s = n\sigma_c \tag{14.33}$$

To find the stress in the steel rods, the stress in the fictional concrete must be multiplied by the modular ratio, n. Equations 14.30 and 14.33, together with the flexural stress Equation 5.10, form the basis for the elastic analysis of reinforced concrete beams.

Example 14.10
Refer to Figure 14.16* and determine the stresses in the concrete and steel of this reinforced concrete beam associated with an applied moment of 100 k-ft.

Solution. Use Eq. 14.30 to transform the steel area to equivalent concrete as depicted in Figure 14.16.

$$nA_s = 10(3.20) = 32.0 \text{ in.}^2$$

Note that the transformed cross section consists of the concrete in compression above the neutral axis and the equivalent concrete having an area of 32 in.², which replaces the steel rods. Concrete below the neutral axis, which would be

---

* The symbol f appearing in Figure 14.16 is traditionally used for stress in reinforced concrete. This symbol will be used occasionally in Sections 14.5 and 14.6.

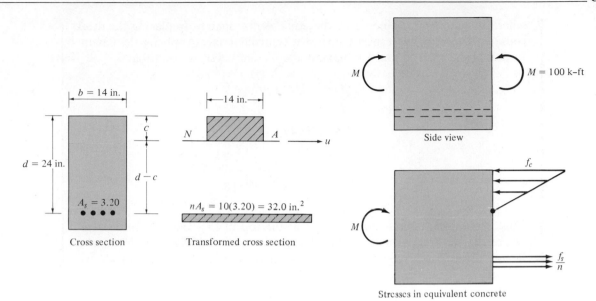

FIGURE 14.16

in tension, is ignored in the analysis, since the tensile strength of concrete is only 10 to 20 percent of the compressive strength of concrete. In fact, tensile cracks develop in reinforced concrete beams at relatively low load levels, and the engineer is justified in ignoring concrete on the tension side of flexural members.

The next step is to calculate c, which locates the neutral axis with respect to the top of the beam.

$$\sum_{i=1}^{2} A_i \bar{v}_i = 0$$

$$A_1 \bar{v}_1 - A_2 \bar{v}_2 = 0$$

$$14c \frac{c}{2} - 32(24 - c) = 0$$

$$7c^2 + 32c - 768 = 0$$

This quadratic equation in c has roots $c = 8.44$ or -13.0. The negative root has no physical significance; hence $c = 8.44$ in.

The moment of inertia of this transformed area with respect to the neutral axis is given by the following:

$$I_u = I_{NA} = \tfrac{1}{3}(14)c^3 + 32(24 - c)^2$$

$$= I_{NA} = \tfrac{1}{3}(14)(8.44)^3 + 32(24 - 8.44)^2 = 10{,}550 \text{ in.}^4$$

where the moment of inertia of the equivalent concrete (replacing the steel rods) is ignored with respect to its own centroidal axis. Applying the flexure formula (Eq. 5.10) to find the maximum concrete stress, we find that

$$\sigma_x = \frac{M_u v}{I_u}$$

$$\sigma_c = \frac{100(12)(8.44)}{10{,}550}$$

$$= \boxed{0.96 \text{ ksi} \quad \text{or} \quad 960 \text{ psi}}$$

This maximum concrete stress occurs at the top of the beam at the section, where the bending moment reaches the given value of 100 k-ft. Applying the flexure formula together with Eq. 14.33 to find the stress in the steel rods, we obtain

$$\sigma_s = n\frac{M_u(d - c)}{I_{NA}}$$

$$= 10\frac{100(12)(24 - 8.44)}{10{,}550}$$

$$= \boxed{17.7 \text{ ksi} \quad \text{or} \quad 17{,}700 \text{ psi}}$$

Homework Problems

14.32 Refer to the reinforced concrete beam depicted in Figure 14.16, change d from 24 in. to 22 in. and b from 14 in. to 15 in., and determine the stresses in the steel and concrete. Use a modular ratio $n = 10$ and $M_u = 50$ k-ft.

14.33 A reinforced concrete beam has a rectangular cross section with $b = 0.36$ m, $d = 0.60$ m, and $A_s = 20.6 \times 10^{-4}$ m^2. If a moment of 140 kN-m is applied to the beam, find the steel and concrete stresses. Use $n = 8$.

14.34 Refer to the reinforced concrete beam depicted in Figure 14.16. Determine the allowable

bending moment that may be applied to this beam if the allowable stress in the concrete is 2000 psi and the allowable steel stress is 30,000 psi. Note that these allowable stresses do not occur simultaneously unless the beam has been designed for what are termed *balanced conditions*.

14.35 Refer to Problem 14.33. Determine the allowable bending moment that may be applied to this beam if the allowable stress in the concrete is 14 MPa and the allowable steel stress is 200 MPa. Note that these allowable stresses do not occur simultaneously unless the beam has been designed for what are termed *balanced conditions*.

14.36 A reinforced concrete beam has a rectangular cross section with $b = 18$ in., $d = 30$ in., and $A_s = 5.00$ in.^{2}. If a moment of 180 k-ft is applied to the beam, find the steel and concrete stresses. Use $n = 10$.

14.37 The cross section described in Problem 14.36 is to be used for a 10-ft-long cantilever beam. Note that the steel would have to be placed near the top of the cantilever beam in order to resist the tensile forces developed there. If the beam is subjected to a uniform load of intensity w and the allowable stresses are 2000 psi and 25,000 psi for concrete and steel, respectively, find w. This value of w would include the weight of the beam. Deduct this weight per foot to determine w', which is the allowable uniform load that this beam would carry. Use a unit weight of 150 lb/ft^3 for reinforced concrete. Note that the given allowable stresses do not occur simultaneously unless the beam has been designed for what are termed *balanced conditions*. Use $n = 10$.

14.38 A reinforced concrete beam has a rectangular cross section with $b = 0.46$ m, $d = 0.75$ m, and $A_s = 32.0 \times 10^{-4}$ m^2. If a moment of 250 kN-m is applied to the beam, find the steel and concrete stresses. Use $n = 10$.

14.39 The cross section described in Problem 14.38 is to be used for a simply supported beam that is 6 m long. If the beam is subjected to a uniform load of intensity w and the allowable stresses are 15 MPa and 200 MPa for concrete and steel, respectively, find w. This value of w would include the weight of the beam. Deduct this weight per foot to determine w', which is the allowable uniform load that the beam would carry. Use a unit weight of 2.36×10^4 N/m^3 for reinforced concrete. Which material, steel or concrete, is stressed critically? Use $n = 10$.

*14.6 Reinforced Concrete Beams—Ultimate-Strength Method

Ultimate-strength methods are now widely employed for analysis and design of reinforced concrete members. Working or service loads are multiplied by load factors to obtain ultimate loads associated with failure of a given structural component. To design a beam at the ultimate stage, these ultimate loads are used to determine the corresponding ultimate moment diagram. From this moment diagram critical ordinates are selected for which the reinforced concrete beam cross section is then proportioned. As for all structures, the external forces, which are natural or man-made, are to be resisted by a system designed by engineers.

In this section it will be assumed that the rectangular beam cross section together with the steel area has been determined. In other words, the focus will be on analysis rather than design.

At ultimate loading, the maximum concrete strain will be assumed equal to 0.003. Even though variations occur for concrete strengths from 2000 to 6000 psi, this value of 0.003 is reasonably conservative. On the tension side, yielding of the steel rods will be associated with ultimate loading, and strain hardening will be ignored. Balanced design for ultimate strength occurs when the concrete reaches a maximum strain of 0.003 simultaneously with yielding of the steel rods on the tension side. In practice, such a balanced design is uneconomical, because excessive amounts of steel are required, and steel together with cement are the expensive materials of reinforced concrete construction. If the amount of steel is reduced, together with an increase of beam depth, failure will occur first on the tension side by yielding of the steel, which leads to larger and larger deflections of the beam. Such failure is termed *ductile* and often provides adequate warning. Failure of concrete on the compression side provides little warning and such a failure is termed *brittle*. Ductile failures are preferred in modern design practices. Even if failure is initiated on the tension side, a modest increase in moment capacity will occur until the concrete stresses are distributed on the compression side as shown in Figure 14.17(a). This actual distribution is replaced

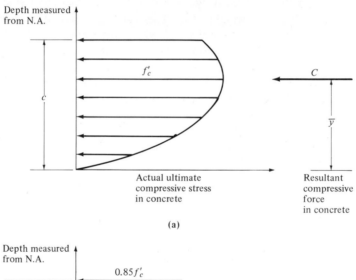

(a)

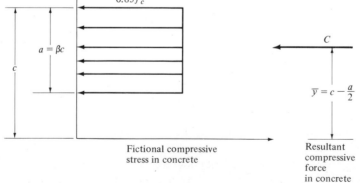

(b)

FIGURE 14.17

by the fictional constant one shown in Figure 14.17(b). These two very different distributions have approximately the same resultant force applied at the same distance above the neutral axis. The uniform distribution is referred to as a *rectangular stress block* and it has the same resultant because the intensity of stress $0.85f'_c$ and the depth a have been chosen for this purpose. Obviously, the rectangular stress block is simpler to handle mathematically. The depth of the stress block related to the neutral axis location varies with the concrete ultimate strength f'_c as follows:

1. For $f'_c \leq 4000$ psi, $a = 0.85c$ or $\beta = 0.85$.
2. For $f'_c > 4000$ psi, reduce the β value linearly at the rate of 0.05 per 1000 psi excess over 4000 psi. For example, $\beta = 0.80$ for $f'_c = 5000$ psi.

On the tension side, the resultant ultimate force is obtained by multiplying the area of the steel rods by the yield stress of the steel. Commonly employed reinforcement steels have yield stresses of 40, 50, and 60 ksi. Since reinforcing bar diameters are usually small compared to beam depths, the stress variation over the bar diameters is ignored and the steel stress is assumed to be constant over the bar areas.

Example 14.11
A rectangular beam shown in Figure 14.18 has a width $b = 12$ in., depth to the centroid of the steel layer $d = 20$ in., steel area $A_s = 1.80$ in.^{2}, concrete ultimate strength $f'_c = 4000$ psi, and steel yield stress $f_y = 50$ ksi. Calculate the ultimate moment capacity.

Solution. Determine the total compressive and tensile forces:

$$C = 0.85f'_c ab = 0.85(4000)a(12) = 40{,}800a \text{ lb}$$

$$T = f_y A_s = 50{,}000(1.80) = 90{,}000 \text{ lb}$$

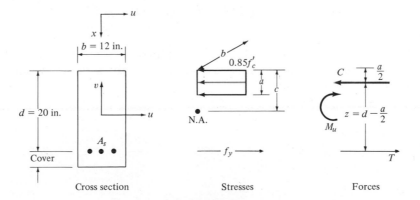

Cross section   Stresses   Forces

FIGURE 14.18

Since the beam is subjected only to a bending moment,* the algebraic sum of the horizontal forces acting on the beam equals zero.

$$\rightarrow \ \sum F_x = 0$$

$$-C + T = 0$$

$$C = T$$

$$40{,}800a = 90{,}000$$

$$a = 2.21 \text{ in.}$$

This determines the depth of the compressive stress block. At ultimate loading, concrete on the tension side will be cracked and is ignored in this analysis.

The neutral axis is located by calculating c from $a = \beta c$, where $\beta = 0.85$ for $f'_c = 4000$ psi:

$$2.21 = 0.85c$$

$$c = 2.60 \text{ in.}$$

Compressive force C and tensile force T form a couple that resists M_u, the applied moment associated with external forces.

$$Cz = Tz = M_u$$

where

$$z = d - \frac{a}{2} = 20 - \frac{2.21}{2} = 18.9 \text{ in.}$$

$$M_u = 90{,}000(18.9)$$

$$= \boxed{1{,}701{,}000 \text{ lb-in.} \quad \text{or} \quad 141.8 \text{ k-ft}}$$

In practice, this ideal value of M_u would be reduced 10 percent, to account for imperfect fabrication and assumptions made in the analysis. In this section only ideal values will be considered.

---

* Shear reinforcement, referred to as *stirrups*, will not be considered here.

Homework Problems

14.40 Determine the ultimate moment capacity of a rectangular cross-sectioned beam if $b = 14$ in., $d = 22$ in., $A_s = 3.00$ in.^{2}, $f'_c = 4000$ psi, and $f_y = 50$ ksi. Locate the neutral axis.

14.41 Refer to Example 14.11 and determine the total ultimate uniform load that may be applied to a simply supported beam of 20 ft length with this cross section. Subtract the uniform loading due to the weight of the beam and thus determine the ultimate live load intensity. Use a unit weight of 150 lb/ft^3 for reinforced concrete.

14.42 Find the depth d of a reinforced concrete beam of rectangular cross section to resist an ultimate moment of 200 k-ft. Given: $b = 13.5$ in., $A_s = 3.00$ in.^{2}, $f'_c = 5000$ psi, and $f_y = 60$ ksi. Locate the neutral axis.

14.43 Determine the ultimate moment capacity of a rectangular cross-sectioned beam if $b = 17$ in., $d = 30$ in., $A_s = 4.68$ in.^{2}, $f'_c = 3000$ psi, and $f_y = 50$ ksi. Locate the neutral axis.

14.44 A cantilever beam, which requires reinforcement at the top, is subjected to an ultimate uniform load intensity of 2.80 k/ft, which includes an allowance for the beam weight. If the beam is 10 ft long, find the required beam depth d. Other given quantities: $b = 12$ in., $A_s = 1.80$ in.^{2}, $f'_c = 4000$ psi, and $f_y = 50$ ksi.

14.45 Determine the depth d of a reinforced concrete beam of rectangular cross section to resist an ultimate moment of 300 k-ft. Given: $b = 16.0$ in., $A_s = 4.00$ in.^{2}, $f'_c = 6000$ psi, and $f_y = 50$ ksi. Locate the neutral axis.

*14.7 Beam Deflections Using Singularity Functions

The method of two successive integrations is a relatively simple technique for determining beam deflections when the loading on the beam is continuous and expressible by a single continuous function. This method, however, becomes extremely cumbersome for cases where the beam is subjected to concentrated forces and moments or to distributed loads that change in intensity abruptly. This type of problem can be handled much more conveniently and also much more elegantly by the use of special mathematical tools known as *singularity functions*, specifically designed to handle discontinuous functions. Only those properties and operational procedures of singularity functions needed for beam deflections will be given in this section.

A singularity function $f(x)$ may be written in the following form, using angle brackets to distinguish it from ordinary functions:

$$f(x) = \langle x - a \rangle^n \tag{14.34}$$

where x is any distance along the beam, a is a specific value of x, and n is any positive or negative integer, including zero. For cases where $n \geq 0$, the singularity function expressed in Eq. 14.34 obeys the following operational rules:

1. When $x < a$ the quantity $\langle x - a \rangle^n$ vanishes.
2. When $x \geq a$, the quantity $\langle x - a \rangle^n$ becomes the ordinary function $(x - a)^n$ and obeys the same rules as any other ordinary function.

3. $\displaystyle \int \langle x - a \rangle^n \, dx = \frac{\langle x - a \rangle^{n+1}}{n+1} + C.$

The use of singularity functions in solving beam deflection problems will be illustrated in Examples 14.12 and 14.13.

Example 14.12
The cantilever beam shown in Figure 14.19 is subjected to the concentrated load P, the concentrated moment Q, and the uniformly distributed load w. Determine the deflection at the free end of the cantilever using singularity functions and the method of two successive integrations.

Solution. Without the use of singularity functions, one would write a moment equation for each of the three segments of the cantilever beam defined by the ranges $0 \leq x \leq L/3$, $L/3 \leq x \leq 2L/3$, and $2L/3 \leq x \leq L$. The method of two successive integrations would then be applied to each of the three beam segments, yielding three slope and three deflection equations. These six equations would contain six constants of integration. Consequently, six boundary and matching conditions would be required to obtain six simultaneous equations, the solution for which would be tedious. Obviously, the tediousness of the solution would increase for beams with more concentrated loads and/or moments and segmented distributed loads. With the use of singularity functions, most of these objections are eliminated, since only one moment equation

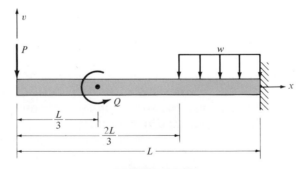

FIGURE 14.19

is needed for the entire beam. By applying the method of two successive integrations, this single moment equation yields one slope and one deflection equation, containing two constants of integration. These two constants of integration require only two boundary conditions, which lead to the solution of only two simultaneous equations. Thus, using the coordinate system shown in Figure 14.19,

$$M_u = -Px - Q\left\langle x - \frac{L}{3} \right\rangle^0 - \left(\frac{w}{2} \right)\left\langle x - \frac{2L}{3} \right\rangle^2 \qquad \text{(a)}$$

Note that when $x < L/3$, the quantities $\langle x - L/3 \rangle^0$ and $\langle x - 2L/3 \rangle^2$ vanish according to the first operational rule of singularity functions, and all that remains in the moment equation is the term $-Px$. Thus $M_u = -Px$ for $0 \le x \le L/3$. When $L/3 \le x \le 2L/3$, the quantity $\langle x - L/3 \rangle^0$ becomes $(x - L/3)^0$ according to the second rule of singularity functions, and therefore it is equal to unity. The quantity $\langle x - 2L/3 \rangle^2$, however, is zero according to the first rule. Hence $M_u = -Px - Q$ for $L/3 \le x \le 2L/3$. Similarly, one may show that $M_u = -Px - Q - (w/2)(x - 2L/3)^2$ in the interval $2L/3 \le x \le L$. Therefore, Eq. (a) defines the bending moment M_u for the entire beam and the differential equation of the elastic curve may be written in the form

$$EI_u v'' = M_u = -Px - Q\left\langle x - \frac{L}{3} \right\rangle^0 - \left(\frac{w}{2} \right)\left\langle x - 2\frac{L}{3} \right\rangle^2 \qquad \text{(b)}$$

Using the third operational rule, we can integrate Eq. (b) to obtain the slope equation. Thus

$$EI_u v' = -\frac{Px^2}{2} - Q\left\langle x - \frac{L}{3} \right\rangle^1 - \left(\frac{w}{6} \right)\left\langle x - 2\frac{L}{3} \right\rangle^3 + C_1$$

One boundary condition in this problem is the fact that $v' = 0$ when $x = L$. Applying the second operational rule for singularity functions, we find that this boundary condition leads to

$$0 = -\frac{PL^2}{2} - Q\left(\frac{2L}{3} \right)^1 - \frac{w}{6}\left(\frac{L}{3} \right)^3 + C_1$$

from which

$$C_1 = \frac{PL^2}{2} + \frac{2QL}{3} + \frac{wL^3}{162} \qquad \text{(c)}$$

Therefore, the slope equation may now be written as follows:

$$EI_u v' = -\frac{Px^2}{2} - Q\left\langle x - \frac{L}{3} \right\rangle^1 - \frac{w}{6}\left\langle x - \frac{2L}{3} \right\rangle^3 + \frac{PL^2}{2} + \frac{2QL}{3} + \frac{wL^3}{162} \tag{d}$$

Integrating Eq. (d), we obtain the deflection equation. Thus

$$EI_u v = -\frac{Px^3}{6} - \frac{Q}{2}\left\langle x - \frac{L}{3} \right\rangle^2 - \frac{w}{24}\left\langle x - \frac{2L}{3} \right\rangle^4 + \frac{PL^2}{2}x$$

$$+ \frac{2QL}{3}x + \frac{wL^3}{162}x + C_2$$

The second boundary condition requires that $v = 0$ when $x = L$. Hence

$$0 = -\frac{PL^3}{6} - \left(\frac{Q}{2}\right)\left(\frac{2L}{3}\right)^2 - \left(\frac{w}{24}\right)\left(\frac{L}{3}\right)^4 + \frac{PL^3}{2} + \frac{2QL^2}{3} + \frac{wL^4}{162} + C_2$$

from which

$$C_2 = -\frac{PL^3}{3} - \frac{4QL^2}{9} - \frac{11wL^4}{1944} \tag{e}$$

Substituting Eq. (e) into the deflection equation yields

$$EI_u v = -\frac{Px^3}{6} - \frac{Q}{2}\left\langle x - \frac{L}{3} \right\rangle^2 - \frac{w}{24}\left\langle x - \frac{2L}{3} \right\rangle^4$$

$$+ \left(\frac{PL^2}{2} + \frac{2QL}{3} + \frac{wL^3}{162}\right)x - \frac{PL^3}{3} - \frac{4QL^2}{9} - \frac{11wL^4}{1944} \tag{f}$$

The deflection at the free end of the cantilever is obtained from Eq. (f) by setting $x = 0$ and by applying the first operational rule, which necessitates that both $\langle -L/3 \rangle^2$ and $\langle -2L/3 \rangle^4$ become zero. Therefore,

$$\boxed{v_{x=0} = -\frac{1}{EI_u}\left(\frac{PL^3}{3} + \frac{4QL^2}{9} + \frac{11wL^4}{1944}\right)}$$

Example 14.13
The simply supported beam shown in Figure 14.20(a) is subjected to a uniformly distributed load of intensity w over a portion of the span. Using singularity functions and the method of two successive integrations, determine the equation of the elastic curve and the deflection of the beam at midspan.

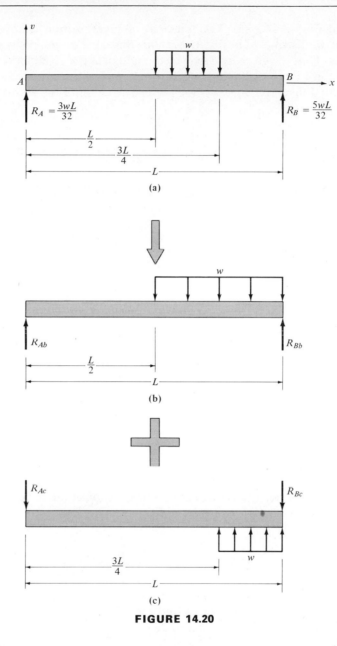

FIGURE 14.20

Solution. The support reactions are found in the usual manner to be

$$R_A = \frac{3wL}{32} \quad \text{and} \quad R_B = \frac{5wL}{32}$$

as shown in Figure 14.20(a). The principal of superposition is now used to break down the distributed load into segments that can be handled by singularity functions. Thus the given load may be thought of as the algebraic sum of the loads shown in Figure 14.20(b) and (c). Using the given coordinate system, we can write the differential equation for the elastic curve of the entire beam as follows:

$$EI_u v'' = M_u = \frac{3wL}{32} x - \left(\frac{w}{2}\right)\left\langle x - \frac{L}{2}\right\rangle^2 + \left(\frac{w}{2}\right)\left\langle x - \frac{3L}{4}\right\rangle^2 \qquad \textbf{(a)}$$

Two successive integrations lead to the slope and deflection equations as follows:

$$EI_u v' = \frac{3wL}{64} x^2 - \left(\frac{w}{6}\right)\left\langle x - \frac{L}{2}\right\rangle^3 + \left(\frac{w}{6}\right)\left\langle x - \frac{3L}{4}\right\rangle^3 + C_1 \qquad \textbf{(b)}$$

$$EI_u v = \frac{3wL}{192} x^3 - \left(\frac{w}{24}\right)\left\langle x - \frac{L}{2}\right\rangle^4 + \left(\frac{w}{24}\right)\left\langle x - \frac{3L}{4}\right\rangle^4 + C_1 x + C_2 \qquad \textbf{(c)}$$

The two boundary conditions needed for the determination of C_1 and C_2 are $v = 0$ at $x = 0$ and $v = 0$ at $x = L$. Therefore, substituting the first condition into Eq. (c) and utilizing the first operational rule for singularity functions, we find that $C_2 = 0$. By substituting the second condition into Eq. (c) and making use of the second operational rule for singularity functions, the value of C_1 is determined. Thus

$$C_1 = - \frac{81wL^3}{6144}$$

Hence the equation for the elastic curve of the beam may now be determined from Eq. (c) by substituting the values of C_1 and C_2. Therefore,

$$EI_u v = \frac{3wL}{192} x^3 - \left(\frac{w}{24}\right)\left\langle x - \frac{L}{2}\right\rangle^4 + \left(\frac{w}{24}\right)\left\langle x - \frac{3L}{4}\right\rangle^4 - \frac{81wL^3}{6144} x$$

The deflection of the simply supported beam at midspan is found by setting $x = L/2$. Since both terms $\langle x - L/2\rangle^4$ and $\langle x - 3L/4\rangle^4$ vanish at this value of x, the deflection at midspan is found to be

$$v = - \frac{57wL^4}{12,288 EI_u}$$

Homework Problems

Use singularity functions in the solution of all of the following problems.

14.46 Determine the deflection at the free end of the cantilever beam shown in Figure H14.46 in terms of P, L, E, and I_u.

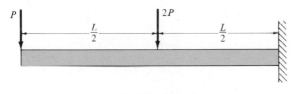

FIGURE H14.46

14.47 Refer to the simply supported beam shown in Figure H14.47 and let $P = 5000$ lb, $L = 20$ ft, $E = 30 \times 10^6$ psi, and $I_u = 400$ in.^{4}. Determine the deflection at midspan.

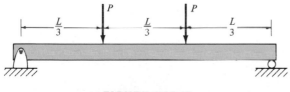

FIGURE H14.47

14.48 A cantilever beam is loaded as shown in Figure H14.48. If $w = 100$ N/m, $P = 5000$ N, $L = 5$ m, $E = 70 \times 10^9$ N/m^2, and $I_u = 30 \times 10^{-6}$ m^4, determine the deflection at the free end of the cantilever beam.

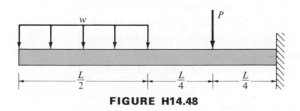

FIGURE H14.48

14.49 The cantilever beam shown in Figure H14.49 is subjected to a concentrated couple Q at the left end and to the uniformly varying load whose maximum intensity is w. Determine the maximum deflection of the cantilever in terms Q, w, L, E, and I_u.

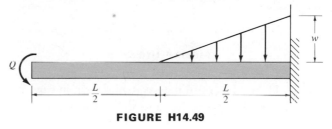

FIGURE H14.49

14.50 Compute the deflection of the overhanging beam shown in Figure H14.50:
(a) At the midpoint between supports A and B.
(b) At end C.
Express your answers in terms P, w, L, E, and I_u.

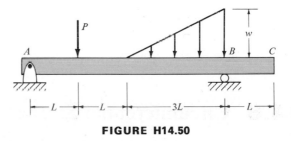

FIGURE H14.50

14.51 Compute, in terms of w, P, L, E and I_u, the deflection under the load P for the beam shown in Figure H14.51.

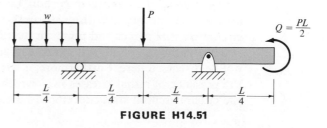

FIGURE H14.51

14.52 Compute the deflection at midspan for the simply supported beam shown in Figure H14.52. Let $L = 18$ ft, $E = 30 \times 10^6$ psi and $I_u = 500$ in.^{4}.

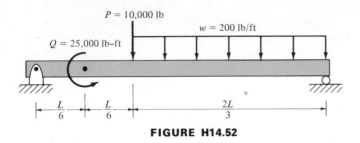

FIGURE H14.52

14.53 Using the coordinate system shown in Figure H14.53, develop the slope and deflection equations

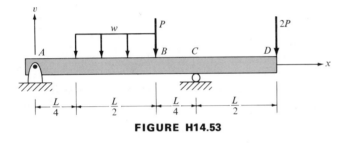

FIGURE H14.53

for the entire beam. Also compute the slopes and deflections at points B and D.

14.54 Compute the deflections at A and B for the cantilever beam loaded as shown in Figure H14.54. The following values are given: $E = 200 \times 10^9$ N/m^2 and $I_u = 5 \times 10^{-4}$ m^4.

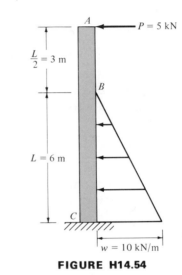

FIGURE H14.54

★**14.8** Circumferential Stresses in Curved Beams

When an initially curved beam is subjected to bending action, two normal stresses are induced at every point in the member. One of these stresses acts on radial planes and points in the circumferential direction. These stresses, referred to as *circumferential stresses* are, in general, the most significant stresses in a curved beam and are the only stresses discussed in this section. The second normal stress acts on circumferential planes and points in the radial direction and are referred to as *radial stresses*. Although of interest in many situations, radial stresses in curved beams will not be discussed here. Furthermore, it has been shown in more refined solutions that the presence of radial stresses in curved beams does not seriously affect the value of the circumferential stress developed here.

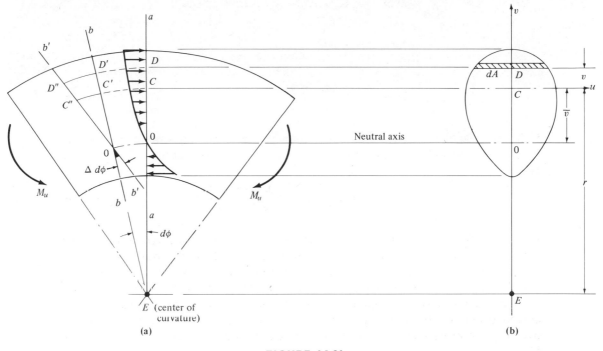

FIGURE 14.21

Consider the curved beam shown in Figure 14.21(a). This curved beam has the cross-sectional area shown in Figure 14.21(b) in which the v principal centroidal axis is an axis of symmetry. The initial radius of curvature for the curved beam, measured from the centroidal axis of the section, is denoted by the symbol r. The curved beam is subjected to the pure bending moments M_u, which are assumed to act in the plane of symmetry. Bending moments that tend to decrease the radius of curvature are, by convention, assumed positive. Thus those in Figure 14.21(a) are positive moments.

Except for the fact that the beam is not initially straight, the fundamental assumptions used in deriving the equation for the circumferential stress are the same as those used in deriving the flexure equation in Section 5.3. Thus consider two adjacent cross sections (radial planes) a–a and b–b in Figure 14.21(a) separated by the angle $d\phi$. Using the assumption that plane sections before bending remain plane after bending, plane section b–b remains plane but rotates about the axis of zero stress represented by point O, into the position b'–b', rotating through the small angle $\Delta d\phi$. The circumferential deformation C'–C'' at the centroidal axis may be expressed in terms of ε_c, the circumferential strain at this axis by the relation $C'C'' = (\varepsilon_c)CC' = (\varepsilon_c)r\,d\phi$. The circumferential strain ε at any point D a distance v from the centroidal axis may be written as follows:

$$\varepsilon = \frac{D'D''}{DD'} = \frac{C'C'' + v\Delta d\phi}{DD'} = \frac{\varepsilon_c r \, d\phi + v\Delta d\phi}{(r + v)\, d\phi}$$

$$= \frac{\varepsilon_c r + v(\Delta d\phi/d\phi)}{r + v}$$

$$= \varepsilon_c + \frac{v}{r + v}\left(\frac{\Delta d\phi}{d\phi} - \varepsilon_c\right) \qquad \textbf{(14.35a)}$$

If the material obeys Hooke's law, the circumferential stress at any point D a distance v from the centroidal axis may be expressed in terms of the quantities of Eq. 14.35a. Hence

$$\sigma = E\left[\varepsilon_c + \frac{v}{r + v}\left(\frac{\Delta d\phi}{d\phi} - \varepsilon_c\right)\right] \qquad \textbf{(14.35b)}$$

Equation 14.35b indicates that the circumferential stress in curved beams is not directly proportional to the distance v measured from the centroidal axis, as is the case with straight beams. Its exact variation with the coordinate v will be determined after finding the two unknown quantities ε_c and $\Delta d\phi/d\phi$. These two unknown quantities are determined by applying the conditions of equilibrium. Thus, since the curved beam is subjected to pure bending, the circumferential stresses distributed over a given cross section lead to circumferential forces that are self-equilibrating. Hence

$$\int \sigma \, dA = 0$$

Substituting from Eq. 14.35b and rearranging terms, we obtain

$$\varepsilon_c A = -\left(\frac{\Delta d\phi}{d\phi} - \varepsilon_c\right) \int \frac{v}{r + v} \, dA \qquad \textbf{(14.35c)}$$

Let

$$K = -\frac{1}{A} \int \frac{v}{r + v} \, dA$$

where K is a dimensionless quantity representing a property of the cross-sectional area of the curved beam. Substituting into Eq. 14.35c and simplifying, we obtain

$$\varepsilon_c = \left(\frac{\Delta d\phi}{d\phi} - \varepsilon_c\right) K \qquad \textbf{(14.35d)}$$

The second condition of equilibrium is expressed by the fact that the resisting moment is equal to the applied moment. Therefore,

$$\int \sigma v \, dA = M_u \qquad \textbf{(14.35e)}$$

Substituting for σ its value from Eq. 14.35b yields

$$M_u = E\left[\varepsilon_c \int v \, dA + \left(\frac{\Delta d\phi}{d\phi} - \varepsilon_c\right) \int \frac{v^2}{r+v} \, dA\right] \tag{14.35f}$$

Since v is measured from the centroidal u axis, it follows that $\int v \, dA = 0$. Also, the quantity $\int v^2/(r+v) \, dA$ may be expressed in terms of the property K as follows:

$$\int \frac{v^2}{r+v} \, dA = \int v \, dA - r \int \frac{v}{r+v} \, dA = KAr \tag{14.35g}$$

Thus Eq. 14.35f may now be written as

$$M_u = \left(\frac{\Delta d\phi}{d\phi} - \varepsilon_c\right) EKAr \tag{14.35h}$$

From a simultaneous solution of Eqs. 14.35d and 14.35h, we obtain the values of the unknown quantities $\Delta d\phi/d\phi$ and ε_c as follows:

$$\frac{\Delta d\phi}{d\phi} = \frac{M_u}{EAr}\left(1 + \frac{1}{K}\right) \tag{14.35i}$$

and

$$\varepsilon_c = \frac{M_u}{EAr} \tag{14.35j}$$

Substitution of Eqs. 14.35i and 14.35j into Eq. 14.35b yields the value of the circumferential stress in terms of known quantities. Hence

$$\sigma = \frac{M_u}{Ar}\left[1 + \frac{v}{K(r+v)}\right] \tag{14.36}$$

A sketch of the stress distribution given by Eq. 14.36 is shown in Figure 14.21(a). Note that point O, representing the axis of zero stress, is located at a distance $\bar{v}$ from the centroidal axis, which may be determined, for pure bending, from Eq. 14.36 by setting $\sigma = 0$. Thus

$$\bar{v} = -\frac{Kr}{K+1} \tag{14.37}$$

where the negative sign signifies that the location of point O is toward the center of curvature of the curved beam from the centroidal axis of the cross section.

Determination of the factor K will be illustrated in Example 14.14 for a rectangular cross section and numerical values of K for use in the solution of Homework

Problems are provided within the problem statements. The use of Eq. 14.36 to determine the circumferential stress at various locations in a curved beam will be illustrated in Example 14.15.

Example 14.14

A curved beam has a rectangular cross-sectional area with the dimensions shown in Figure 14.22. The radius of curvature for the beam measured from the centroid of the section is r. Determine the value of K by the use of its basic definition.

Solution. A differential element of area dA needs to be defined so that the integral in the equation for K can be evaluated. Such an element of area is shown in Figure 14.22. Thus

$$K = -\frac{1}{A} \int \frac{v}{r+v}\, dA = -\frac{1}{2ba} \int_{-a}^{+a} \frac{v}{r+v}\, b\, dv$$

$$= -\frac{1}{2a} \int_{-a}^{+a} \frac{v}{r+v}\, dv$$

$$= -\frac{1}{2a} \left[\int_{-a}^{+a} dv - \int_{-a}^{+a} \frac{v}{r+v}\, dv \right]$$

$$= -\frac{1}{2a} [v - r \ln (r+v)]_{-a}^{+a}$$

$$= \boxed{-1 + \frac{r}{2a} \ln \left(\frac{r+a}{r-a} \right)}$$

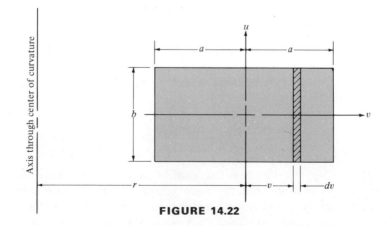

FIGURE 14.22

Example 14.15

A curved beam is fixed at one end and subjected to the load $P = 8000$ lb as shown in Figure 14.23(a). The cross-sectional area of the beam is a 2×4 in. rectangle. Determine the value of the distance d, measured from the centroid of the section, so that the maximum tensile and compressive stresses along section n–n do not exceed 5000 and 15,000 psi, respectively. Determine also the location of the point of zero stress at section n–n for the value found for distance d.

Solution. The value of K for the rectangular cross section of this curved beam can be obtained by using the equation derived in solution of Example 14.14. Thus

$$K = -1 + \frac{6}{4} \ln \left(\frac{6 + 2}{6 - 2} \right)$$

$$= -1 + 1.5 \ln 2 = 0.0397$$

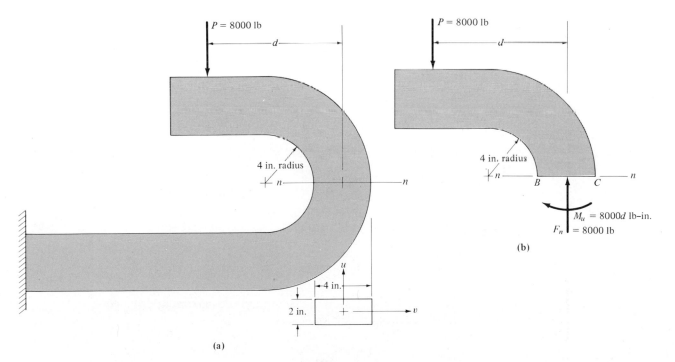

(a)

(b)

FIGURE 14.23

Note that the value of K is, in general, a very small quantity, and care must be exercised in obtaining an accurate value.

Construct the free-body diagram of that portion of the curved beam above section $n–n$ as shown in Figure 14.23(b). Equilibrium requires that at section $n–n$, this portion of the curved beam be subjected to a compressive axial force $F_n = 8000$ lb and a bending moment $M_u = 8000d$ lb-in., as shown in Figure 14.23(b). Thus the curved beam at this particular cross section is subjected not only to a positive bending moment resulting in a normal circumferential stress given by Eq. 14.36, but also to a compressive axial force F_n leading to a uniform compressive normal stress given by F_n/A. Hence, for any location at this section of the curved beam, the normal stress is given by the algebraic sum of these two stresses. Thus

$$\sigma = \frac{F_n}{A} + \frac{M_u}{Ar}\left[1 + \frac{v}{K(r + v)}\right]$$

Consider point B at section $n–n$ on the inner surface of the curved beam. This point is subjected to compressive stresses due to both F_n and M_u. Hence

$$\sigma_B = -15{,}000 = -\frac{8000}{8} + \frac{8000d}{8(6)}\left[1 + \frac{-2}{(0.0397)(6 - 2)}\right]$$

Solution of this relation for d yields

$$d = 7.25 \text{ in.}$$

Consider now point C at section $n–n$ on the outer surface of the curved beam. At this point the bending moment causes tension, while the axial force produces compression, resulting in a net tensile stress.

$$\sigma_C = 5000 = -\frac{8000}{8} + \frac{8000d}{8(6)}\left[1 + \frac{2}{(0.0397)(6 + 2)}\right]$$

This equation leads to a value for d as follows:

$$d = 4.93 \text{ in.}$$

Therefore, the tensile stress is the governing factor and the maximum distance of the load P from the centroid of section $n–n$ is

$$\boxed{d = 4.93 \text{ in.}}$$

The point of zero stress is located by finding the position in section n–n (i.e., the value of v) at which the algebraic sum of the axial and bending stresses vanishes.

$$-\frac{8000}{8} + \frac{8000(4.93)}{8(6)}\left[1 + \frac{v}{(0.0397)(6 + v)}\right] = 0$$

Solution of this relation for v yields

$$\boxed{v = 0.052 \text{ in.}}$$

measured from the centroidal axis away from the center of curvature of the curved beam.

Homework Problems

14.55 The curved beam shown in Figure H14.55(a) has a rectangular cross section as shown in Figure H14.55(b). Determine the maximum tensile and maximum compressive circumferential stresses in the member. Determine the value of K by using the equation derived in Example 14.14.

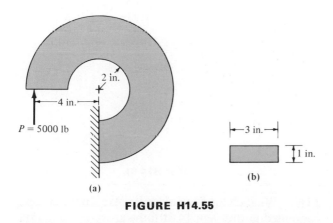

FIGURE H14.55

14.56 At the location considered in Problem 14.55, construct a complete circumferential stress distribution computing the circumferential stress at $\frac{1}{2}$-in.

intervals and accurately locating the point of zero stress.

14.57 The curved member shown in Figure H14.57 has a solid circular cross section 0.10 m in diameter. If the maximum tensile and compressive stresses in the member are not to exceed 150 MPa and 200 MPa, respectively, determine the value of the load P which may be carried safely by the member. Use $K = 0.1459$.

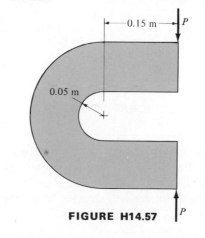

FIGURE H14.57

14.58 Repeat Problem 14.57 if the member has a trapezoidal cross-sectional area as shown in Figure H14.58. Use $K = 0.1588$.

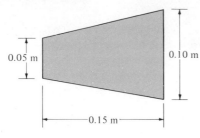

FIGURE H14.58

14.59 The curved member shown in Figure H14.59(a) has a triangular cross section as shown in Figure H14.59(b). Compute the maximum tensile and compressive stresses at:
(a) An infinitesinal distance above section a–a.
(b) At section b–b. Use $K = 0.0583$.

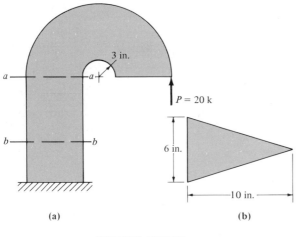

(a)

(b)

FIGURE H14.59

14.60 The clamp shown schematically in Figure H14.60(a) has the T section shown in Figure H14.60(b). The material is such that the allowable tensile and compressive stresses are, respectively,

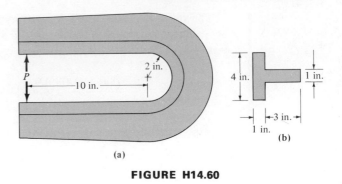

(a)

(b)

FIGURE H14.60

5000 psi and 15,000 psi. Determine the maximum capacity of the clamp. Use $K = 0.1102$.

14.61 The hook shown in Figure H14.61(a) has the cross section shown in Figure H14.61(b). The hook has a rated capacity of 20 kN. Compute the maximum tensile and compressive stresses in the hook when subjected to its rated capacity load. Use $K = 0.0886$.

(a)

FIGURE H14.61

14.62 What changes would take place in the stresses computed in Problem 14.61 if the cross section of the hook were the trapezoid shown in Figure H14.61(c)? Use $K = 0.1306$.

*14.9 Matrices—Flexibility Method

The method of two successive integrations will be applied to formulate the equations for the beam of Figure 14.24(a), and these equations will then be written in matrix form and interpreted in terms of the flexibility method, which is one of the two basic modern methods of structural analysis. The second method, known as the *stiffness method*, is discussed in Section 14.10. Force unknowns for this statically indeterminate beam are V_A, M_A, V_B, and M_B. If horizontal forces are ignored, only two equations of equilibrium are available to solve for four unknowns, and the beam is statically indeterminate to the second degree. The two equilibrium equations will be supplemented by compatibility equations arising from the boundary conditions at both ends of the beam in order to write the four equations required.

Consider the free-body diagram shown in Figure 14.24(a) and write the equations of equilibrium:

$\uparrow \quad \sum F_v = 0$

$$V_A + V_B = wL \qquad (14.38)$$

$\circlearrowleft \quad \sum M_B = 0$

$$M_A + M_B + V_A L = \frac{wL^2}{2} \qquad (14.39)$$

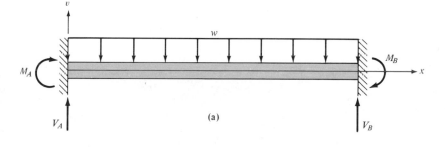

(a)

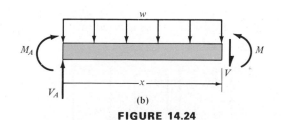

(b)

FIGURE 14.24

The free-body diagram of Figure 14.24(b) is utilized for writing the basic differential equation:

$$EIv'' = M_A + V_A x - \frac{wx^2}{2} \tag{14.40}$$

Integrate Eq. 14.40 term by term and introduce the constant of integration C_1:

$$EIv' = M_A x + V_A \frac{x^2}{2} - \frac{wx^3}{6} + C_1 \tag{14.41}$$

Integrate Eq. 14.41 term by term and introduce the constant of integration C_2:

$$EIv = M_A \frac{x^2}{2} + V_A \frac{x^3}{6} - \frac{wx^4}{24} + C_1 x + C_2 \tag{14.42}$$

Four boundary conditions may be stated for this fixed-ended beam as follows:

$$x = 0 \qquad v = 0 \quad \text{(zero deflection at } A) \tag{14.43}$$
$$x = 0 \qquad v' = 0 \quad \text{(zero slope at } A) \tag{14.44}$$
$$x = L \qquad v = 0 \quad \text{(zero deflection at } B) \tag{14.45}$$
$$x = L \qquad v' = 0 \quad \text{(zero slope at } B) \tag{14.46}$$

The conditions given by Eqs. 14.43 and 14.44 substituted into Eqs. 14.41 and 14.42 reveal that both constants of integration C_1 and C_2 are equal to zero. The conditions given by Eqs. 14.45 and 14.46 substituted into Eqs. 14.41 and 14.42 give the following equations:

$$M_A L + V_A \frac{L^2}{2} - \frac{wL^3}{6} = 0 \tag{14.47}$$

$$M_A \frac{L^2}{2} + V_A \frac{L^3}{6} - \frac{wL^4}{24} = 0 \tag{14.48}$$

The bending moment M_A and the shear V_A were chosen as the redundants when the free-body diagram of Figure 14.24(b) was utilized in writing Eqs. 14.47 and 14.48, and they will be written in matrix form:

$$\begin{bmatrix} L & \dfrac{L^2}{2} \\ \dfrac{L^2}{2} & \dfrac{L^3}{6} \end{bmatrix} \begin{bmatrix} M_A \\ V_A \end{bmatrix} = \begin{bmatrix} \dfrac{wL^3}{6} \\ \dfrac{wL^4}{24} \end{bmatrix} \tag{14.49}$$

Arbitrarily multiply Eq. 14.49 through by $1/EI$ to put it into a readily recognizable matrix form:

$$\begin{bmatrix} \dfrac{L}{EI} & \dfrac{L^2}{2EI} \\[2mm] \dfrac{L^2}{2EI} & \dfrac{L^3}{6EI} \end{bmatrix} \begin{bmatrix} M_A \\[2mm] V_A \end{bmatrix} = \begin{bmatrix} \dfrac{wL^3}{6EI} \\[2mm] \dfrac{wL^4}{24EI} \end{bmatrix} \qquad \textbf{(14.50)}$$

Equation 14.50 may be written more compactly using matrix notation:

$$fF = W \qquad \textbf{(14.51)}$$

where f represents the 2×2 flexibility matrix, F is the 2×1 column vector of unknown forces and bending moments, and W is a 2×1 column vector which depends upon the applied loading, geometry, and the material property E. Solution of Eq. 14.51 is obtained by premultiplying both sides by the inverse of the flexibility matrix:

$$f^{-1}fF = f^{-1}W \qquad \textbf{(14.52)}$$

But $f^{-1}f = I$ and $IF = F$, where I is the identity matrix, which gives

$$F = f^{-1}W \qquad \textbf{(14.53)}$$

Details of the matrix inversion are given in Problem 14.65. Matrix multiplication of f^{-1} by W gives the solution vector:

$$F = \begin{bmatrix} \dfrac{-wL^2}{12} \\[2mm] \dfrac{wL}{2} \end{bmatrix} \qquad \textbf{(14.54)}$$

which means that $M_A = -wL^2/12$ and $V_A = wL/2$. Substitution of Eq. 14.54 into Eq. 14.50 reveals that Eq. 14.54 provides the correct solution. The reader is urged to verify the solution by performing this matrix multiplication of Eq. 14.50.

In general, Eq. 14.51 for a system containing N unknown redundant forces consists of an $N \times N$ flexibility matrix f, an $N \times 1$ column vector F of the N unknown redundant forces, and an $N \times 1$ column vector W which depends upon the applied loadings. This set of N simultaneous linear algebraic equations for large-order systems would not be formulated as shown above, but the matrix approach is widely used for such systems and solutions are obtained with the aid of digital computers. Completion of the solution follows by substitution of values of M_A and V_A from Eq. 14.54 into the equilibrium equations 14.38 and 14.39. This yields: $M_B = wL^2/12$ and $V_B = wL/2$.

Homework Problems

14.63 Draw the shear and moment diagrams for the fixed-ended beam shown in Figure 14.24(a). Use the results of Eq. 14.54.

14.64 Refer to Figure 14.24(a) and replace the uniform load w by a loading that varies linearly from zero intensity at the left end to a downward intensity q at the right end. Derive the flexibility matrix for this beam using the method of two successive integrations. Consider the redundant forces to be the shear and moment at the left end of the beam.

14.65 Refer to Eqs. 14.47 and 14.48 and write these two equations in matrix form. Note that M_A and V_A are the only unknowns in these equations. Invert the flexibility matrix using the following information on a 2×2 matrix. Given

$$\begin{bmatrix} a_{11} & a_{12} \\ a_{21} & a_{22} \end{bmatrix}$$

The inverse is given by

$$\frac{1}{a_{11}a_{22} - a_{12}a_{21}} \begin{bmatrix} a_{22} & -a_{12} \\ -a_{21} & a_{11} \end{bmatrix}$$

Solve for the matrix F (i.e., M_A and V_A) by applying Eq. 14.53 to this 2×2 system. Determine M_B and V_B using the equations of statics.

14.66 Refer to Figure 14.24(a) and replace the uniform load w by a loading that varies linearly from zero intensity at the left end to a downward intensity

q at the right end. Use the approach of Problem 14.65 and solve for M_A and V_A by dealing with a 2×2 system. Use the method of two successive integrations to formulate the equations. Determine M_B and V_B using the equations of statics.

14.67 Refer to Figure 14.24(a) and remove M_B at the right end (i.e., consider the fixed end at the right to be replaced by a roller). Derive the flexibility matrix for this beam using the method of two successive integrations.

14.68 A two-span continuous beam is loaded as shown in Figure H14.68. Choose an origin at the left end A and determine the 2×2 flexibility matrix using the method of two successive integrations. Note that once R_A and R_B are determined by this procedure, supplemented with the equilibrium equation for moments with respect to C, the reaction R_C follows from the vertical equilibrium-of-forces equation. Use left free-body diagrams in your analysis.

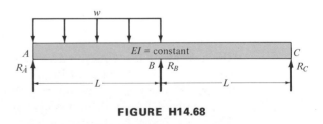

FIGURE H14.68

14.69 Refer to Problem 14.68, double the length of the right span BC, and solve again.

14.70 A known couple C_B is applied to the center of the left span of the continuous beam depicted in Figure H14.70. Choose an origin at the left end A and determine the 2×2 flexibility matrix using the method of two successive integrations. Note that once R_A and R_C are determined by this procedure, supplemented with the equilibrium equation for moments with respect to D, the reaction R_D follows from the vertical equilibrium-of-forces equation. Use left free-body diagrams for segments AB, BC, and CD of this two-span continuous beam.

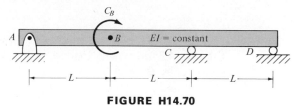

FIGURE H14.70

*14.10 Matrices—Stiffness Method

The other modern method of structural analysis, termed the *stiffness method*, will be introduced by starting with the classical slope-deflection equation, which expresses bending moments in terms of joint rotations and fixed-ended beam moments. Derivation of this equation may be found in structural analysis texts. It is derived with the aid of the principle of superposition, but the details will not be given here. Reference to Figure 14.25 will clarify the meaning of symbols used in the basic slope deflection equation, Eq. 14.55 (p. 716).

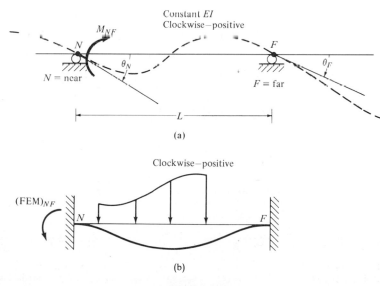

FIGURE 14.25

$$M_{NF} = \frac{2EI}{L}(2\theta_N + \theta_F) + (FEM)_{NF} \qquad (14.55)$$

Any given span of a continuous beam with all supports at the same level as shown in Figure 14.25(a), when loaded, will rotate at its near and far ends through angles denoted by θ_N and θ_F, respectively. These angles are referred to as *kinematic unknowns* and are the unknown quantities first determined when applying the stiffness method of analysis.

The symbols E, I, and L denote the modulus of elasticity, the area moment of inertia of the beam cross section, and the span length, respectively. The fixed-end beam moment at the near end is denoted by $(FEM)_{NF}$, as depicted in Figure 14.25(b). Note that the span is assumed fixed at both near (N) and far (F) ends when this quantity is computed for the equation. First, Eq. 14.55 is written for both ends of each span of the continuous beam. Next, joint equilibrium equations are written which are solved for the kinematic unknowns. Once the kinematic unknowns (i.e., the theta values) are determined, they are substituted back into the basic slope deflection equations to determine the bending moments. Completion of the solution for shear and moment diagrams follows from consideration of appropriate free-body diagrams.

The two-span continuous beam of Figure 14.26(a) will be used to illustrate the stiffness method in detail. There are two kinematic unknowns, θ_A and θ_B, and each is assumed to be positive or clockwise, as shown in Figure 14.26(a). Equation 14.55 will be written four times for the beam as follows, using the fixed-end beam moments shown in Figure 14.26(b):

$$M_{AB} = \frac{2EI}{L}(2\theta_A + \theta_B) - \frac{PL}{8} \qquad (14.56)$$

$$M_{BA} = \frac{2EI}{L}(2\theta_B + \theta_A) + \frac{PL}{8} \qquad (14.57)$$

$$M_{BC} = \frac{2EI}{L/2}(2\theta_B + 0) + 0 \qquad (14.58)$$

$$M_{CB} = \frac{2EI}{L/2}(0 + \theta_B) + 0 \qquad (14.59)$$

Joint equilibrium equations for joints A and B are given by

$$M_{AB} = 0 \qquad (14.60)$$

$$M_{BA} + M_{BC} = 0 \qquad (14.61)$$

Substitute Eq. 14.56 into Eq. 14.60. Also, substitute Eqs. 14.57 and 14.58 into Eq. 14.61:

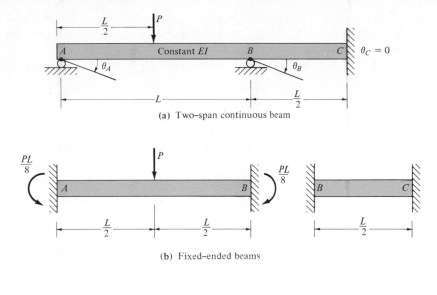

(a) Two-span continuous beam

(b) Fixed-ended beams

(c) Deflected shape

FIGURE 14.26

$$\frac{2EI}{L}(2\theta_A + \theta_B) - \frac{PL}{8} = 0 \qquad (14.62)$$

$$\frac{2EI}{L}(2\theta_R + \theta_A) + \frac{PL}{8} + \frac{4EI}{L}(2\theta_R) = 0 \qquad (14.63)$$

Rearrange Eqs. 14.62 and 14.63 and express them in matrix form:

$$\begin{bmatrix} \dfrac{4EI}{L} & \dfrac{2EI}{L} \\[2mm] \dfrac{2EI}{L} & \dfrac{12EI}{L} \end{bmatrix} \begin{bmatrix} \theta_A \\[2mm] \theta_B \end{bmatrix} = \begin{bmatrix} \dfrac{PL}{8} \\[2mm] -\dfrac{PL}{8} \end{bmatrix} \qquad (14.64)$$

Equation 14.64 may be written more compactly using matrix notation:

$$SD = P \qquad (14.65)$$

where S represents the 2×2 symmetric stiffness matrix, D is a 2×1 column vector of unknown displacements, and P is a 2×1 column vector which depends upon the applied loading and the span length L.

Solution of Eq. 14.65 is obtained by premultiplying both sides by the inverse of the stiffness matrix:

$$S^{-1}SD = S^{-1}P \tag{14.66}$$

But $S^{-1}S = I$ and $ID = D$, where I is the identity matrix, which gives

$$D = S^{-1}P \tag{14.67}$$

Details of the inversion of a 2×2 matrix are given in Problem 14.72. Matrix multiplication of S^{-1} by P gives the solution vector:

$$D = \begin{bmatrix} \dfrac{7PL^2}{176EI} \\[2ex] \dfrac{-3PL^2}{176EI} \end{bmatrix} \tag{14.68}$$

which means that

$$\theta_A = \frac{7}{176} \frac{PL^2}{EI} \qquad (+ \text{ means clockwise})$$

and

$$\theta_B = \frac{-3}{176} \frac{PL^2}{EI} \qquad (- \text{ means counterclockwise})$$

The deflected shape of the beam is shown to a greatly exaggerated scale in Figure 14.26(c).

Substitution of these values of θ_A and θ_B into Eqs. 14.56, 14.57, 14.58, and 14.59 will give the bending moments in the beam:

$$M_{AB} = \frac{2EI}{L} \left[2\left(\frac{7}{176} \frac{PL^2}{EI} \right) + \left(\frac{-3PL^2}{176EI} \right) \right] - \frac{PL}{8} = 0 \tag{14.69}$$

which checks the boundary condition at the pinned end A.

$$M_{BA} = \frac{2EI}{L} \left[2\left(\frac{-3}{176} \frac{PL^2}{EI} \right) + \frac{7PL^2}{176EI} \right] + \frac{PL}{8}$$

$$M_{BA} = \frac{3}{22} PL \qquad (+ \text{ means clockwise on the member}) \qquad \textbf{(14.70)}$$

$$M_{BC} = \frac{4EI}{L}\left[2\left(\frac{-3PL^2}{176EI}\right)\right]$$

$$M_{BC} = -\frac{3}{22} PL \qquad (- \text{ means counterclockwise on the member}) \quad \textbf{(14.71)}$$

$$M_{CB} = \frac{4EI}{L}\left[\left(\frac{-3}{176}\frac{PL^2}{EI}\right)\right]$$

$$M_{CB} = -\tfrac{3}{44}PL \qquad (- \text{ means counterclockwise on the member}) \qquad \textbf{(14.72)}$$

Construction of the shear and moment diagrams follows from analysis of the appropriate free-body diagrams. Their construction is required in Problem 14.71.

In general, Eq. 14.65 for a system containing N unknown displacements consists of an $N \times N$ stiffness matrix S, an $N \times 1$ column vector D of the N unknown displacements, and an $N \times 1$ column vector P, depending upon the applied loadings and geometry of the system. This set of N simultaneous linear algebraic equations for large-order systems would not usually be formulated as shown above, but the matrix approach is widely used for such systems and solutions are obtained with the aid of digital computers.

Homework Problems

14.71 Construct the shear and moment diagrams for the beam shown in Figure 14.26(a). Use the bending moments given in Eqs. 14.69 to 14.72.

14.72 The 2×2 matrix:

$$\begin{bmatrix} a_{11} & a_{12} \\ a_{21} & a_{22} \end{bmatrix}$$

has an inverse given by

$$\frac{1}{a_{11}a_{22} - a_{12}a_{21}}\begin{bmatrix} a_{22} & -a_{12} \\ -a_{21} & a_{11} \end{bmatrix}$$

Utilize this result to obtain the inverse of S given in Eq. 14.64, perform the matrix multiplication of Eq. 14.67, and verify the displacements of Eq. 14.68.

14.73 Refer to Figure 14.26(a), move the load P from the left span AB to the center of the right span BC, and solve this problem using the stiffness method. Construct the shear and moment diagrams for this continuous beam.

14.74 Refer to Figure H14.74 and determine the rotation at joint B of this continuous beam, then find the bending moments at both ends of spans AB and BC. Use the stiffness method.

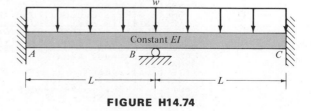

FIGURE H14.74

14.75 Refer to Problem 14.74, remove the uniform load w on the right span BC, and solve using the stiffness method.

14.76 Refer to Problem 14.74, remove the uniform load w on the left span AB, and solve using the stiffness method.

14.77 Refer to Figure H14.77 and determine the rotations at joints B and C of this continuous beam, then find the bending moments at both ends of spans AB, BC, and CD. Use the stiffness method.

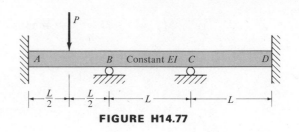

FIGURE H14.77

14.78 Refer to Problem 14.77, replace the concentrated force P by a uniform load of intensity w over the left span AB, and solve using the stiffness method.

*14.11 Computer Methods with Sample Programs

Fortran IV source programs for two typical problems will illustrate the versatility and great utility of digital computers for solving mechanics of materials problems. In addition to complete source programs, typical output is provided and briefly discussed.

A source program for strain rosettes is shown in Figure 14.27. Section 2.9 provides a discussion of strain rosettes, and this program accepts input strain measurements and material properties for four types of rosettes described in the initial comment statements. Typical output for rectangular and equiangular rosettes is shown in Figure 14.28. Principal strains and stresses, together with the input measured strains and material properties, comprise the output.

```
C     TYPE = KIND OF STRAIN ROSETTE
C     IF TYPE = -1.0, EQUIANGULAR, 60 DEGREE DELTA, 3 - GAGE
C     IF TYPE = 0.0, RECTANGULAR, 3 - GAGE 45 DEGREE, STAR
C     IF TYPE = 1.0, 4 GAGE 45 DEGREE, FAN
C     IF TYPE = 2.0, T - DELTA, 4 GAGE
C     GAGE 1 IS THE GAGE ON THE LEFT, THEN READ THE GAGES TO THE RIGHT.
C     E = MODULUS OF ELASTICITY (PSI),   U = POISSON'S RATIO
C     BB = MANUFACTURER'S AUXILIARY SENSITIVITY COEFFICIENT
C     N = NUMBER OF GAGES AND STRAINS RECORDED
C     NROSE = NUMBER OF THE ROSETTE FOR WHICH OUTPUT IS DESIRED
C     NMROSE = TOTAL NUMBER OF ROSETTES FOR WHICH OUTPUT IS DESIRED
C     ARRAY IS READ AS THE NAME OF THE STRAIN ROSETTE
C     SHEAR STRESS CALCULATIONS ARE BASED ON 3-DIMENSIONAL EFFECTS
C     IF DESIRE TO HAVE OUTPUT GROUPED ACCORDING TO WAY DATA READ IN,
C     PUT BLANK CARD AFTER EACH DATA SET EXCEPT THE LAST
      DIMENSION STRESS(5),ARRAY(20),X(5),$ALPA(5)
      DOUBLE PRECISION STRAIN(5),A,B,TOP,BOT,STRAMX,STRAMN,SHSTRA,E,DSQRT
70    READ (1,3,END=80) ARRAY
3     FORMAT(20A4)
      WRITE (3,4) ARRAY
4     FORMAT('1',///,T15,20A4,/)
      READ(1,1) TYPE,E,U,BB,NMROSE
1     FORMAT(4F10.2,I5)
      WRITE(3,110)U,E
```

FIGURE 14.27 (continued on pp. 721 and 722)

```
  110 FORMAT(//,T10,'POISSON,S RATIO = ',F6.3,/T10,'MODULUS OF ELASTICIT
     1Y = ',F11.1,2X,'PSI',/T10,'GAGE NO = GAGES ON ROSETTE ARE NUMBERED
     2 IN CLOCKWISE DIRECTION',/T10,'ALPHA IS THE ANGLE FROM GAGE 1 TO T
     3HE LINE OF PRINCIPAL STRESS MEASURED CLOCKWISE FROM GAGE 1',/T10,'
     4THETA = ANGLE FROM RESPECTIVE GAGE TO LINE OF PRINCIPAL STRESS')
        IF(NMROSE.GT.1)WRITE(3,111)
  111 FORMAT(T10,'ROSE NO = NUMBER OF THE ROSETTE FOR WHICH OUTPUT HAS B
     1EEN CALCULATED')
        IF(NMROSE.EQ.1)GO TO 112
   82 WRITE(3,100)
  100 FORMAT(//T4,'ROSE NO',3X,'GAGE NO',3X,'THETA',5X,'STRN',5X,'STRS',
     15X,'ALPHA',5X,'MAX STRN',3X,'MIN STRN',3X,'MAX SHR STRN',3X,'MAX S
     2TRS',3X,'MIN STRS',3X,'MAX SHR STRS',/T24,'(DEG)',4X,'(IN/IN)',3X,
     3'(PSI)',4X,'(DEG)',6X,'(IN/IN)',4X,'(IN/IN)',7X,'(RAD)',8X,'(PSI)'
     4,6X,'(PSI)',7X,'(PSI)',//)
  112 JJJ=1
        IF(TYPE.LE.0.0) N=3
        IF(TYPE.GT.0.0) N=4
   71 READ(1,2)NROSE,(STRAIN(I),I=1,N)
    2 FORMAT(I5,7F10.8)
        IF(NROSE.EQ.0)WRITE(3,99)
   99 FORMAT(/)
        IF(NROSE.EQ.0)GO TO 71
        IF(TYPE.LT.0.0) XK=3.0/(2.0*BB-1.0)
        IF(TYPE.GE.0.0.AND.BB.NE.0.0) XK=1.0/BB
        IF(BB.EQ.0.0) XK=0.0
        IF(TYPE.NE.-1.0) GO TO 10
        A=(STRAIN(2)+STRAIN(3))/3.0
        B=DSQRT((STRAIN(1)-A)**2+((STRAIN(2)-STRAIN(3))**2)/3.0)
        TOP=SQRT(3.0)*(STRAIN(2)-STRAIN(3))
        BOT=3.0*(STRAIN(1)-A)
        Y1=60.0
        Y2=120.0
   10 IF(TYPE.NE.0.0) GO TO 20
        A=(STRAIN(1)+STRAIN(3))/2.0
        B=DSQRT((STRAIN(1)-A)**2+(STRAIN(2)-A)**2)

        TOP=STRAIN(2)-A
        BOT=STRAIN(1)-A
        Y1=45.0
        Y2=90.0
   20 IF(TYPE.NE.1.0) GO TO 30
        A=(STRAIN(1)+STRAIN(2)+STRAIN(3)+STRAIN(4))/4.0
        B=DSQRT((STRAIN(1)-STRAIN(3))**2+(STRAIN(2)-STRAIN(4))**2)/2.0
        TOP=STRAIN(2)-STRAIN(4)
        BOT=STRAIN(1)-STRAIN(3)
        Y1=45.0
        Y2=90.0
        Y3=135.0
   30 IF(TYPE.NE.2.0) GO TO 40
        A=(STRAIN(3)+STRAIN(4))/2.0
        B=DSQRT(((STRAIN(1)-STRAIN(2))**2)/3.0+((STRAIN(3)-STRAIN(4))**2)
     1(4.0)
        TOP=(STRAIN(2)-STRAIN(3))*SQRT(3.0)
        BOT=2.0*STRAIN(1)-STRAIN(2)-STRAIN(3)
        Y1=60.0
        Y2=120.0
        Y3=30.0
   40 B=B*(1.0+XK)/(1.0-XK)
        IF(BOT.EQ.0.0) ANGLE1=3.14159/2.0
        IF(BOT.EQ.0.0) GO TO 90
        Q=TOP/BOT
        Q=ABS(Q)
        ANGLE1=ATAN(Q)
   90 STRAMX=A+B
        STRAMN=A-B
        SHSTRA=2.0*B
        AP=A*E/(1.0-U)
        BP=B*E/(1.0+U)
        STRSMX=AP+BP
        STRSMN=AP-BP
C
C*****  NEXT 3 CARDS DETERMINE MAX SHSTSS BASED ON 3-DIMENSIONAL ELEMENT
        IF(STRSMX.GT.0.0.AND.STRSMN.GE.0.0)SHSTSS=STRSMX/2.0
        IF(STRSMX.GT.0.0.AND.STRSMN.LT.0.0)SHSTSS=(STRSMX-STRSMN)/2.0
        IF(STRSMX.LE.0.0.AND.STRSMN.LT.0.0)SHSTSS=-STRSMN/2.0
C
        X(1)=COS(ANGLE1)
        IF(TOP)11,12,12
   12 IF(BOT)13,14,14
   14 GO TO 55
   13 ANGLE1=3.14159-ANGLE1
        X(1)=-X(1)
        GO TO 55
   11 IF(BOT)15,16,16
```

FIGURE 14.27 (continued)

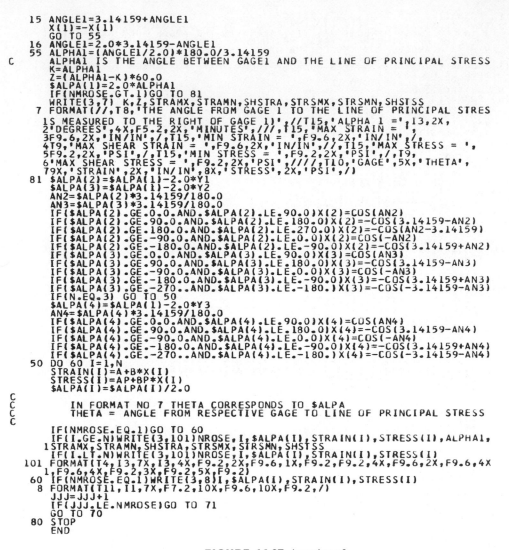

```
   15 ANGLE1=3.14159+ANGLE1
      X(1)=-X(1)
      GO TO 55
   16 ANGLE1=2.0*3.14159-ANGLE1
   55 ALPHA1=(ANGLE1/2.0)*180.0/3.14159
C     ALPHA1 IS THE ANGLE BETWEEN GAGE1 AND THE LINE OF PRINCIPAL STRESS
      K=ALPHA1
      Z=(ALPHA1-K)*60.0
      $ALPA(1)=2.0*ALPHA1
      IF(NMROSE.GT.1)GO TO 81
      WRITE(3,7) K,Z,STRAMX,STRAMN,SHSTRA,STRSMX,STRSMN,SHSTSS
    7 FORMAT(///,T8,'THE ANGLE FROM GAGE 1 TO THE LINE OF PRINCIPAL STRES
     1S MEASURED TO THE RIGHT OF GAGE 1)',//T15,'ALPHA 1 =',I3,2X,
     2'DEGREES',4X,F5.2,2X,'MINUTES',///,T15,'MAX STRAIN = ',
     3F9.6,2X,'IN/IN',/,T15,'MIN STRAIN = ',F9.6,2X,'IN/IN',/,
     4T9,'MAX SHEAR STRAIN = ',F9.6,2X,'IN/IN',/,T15,'MAX STRESS = ',
     5F9.2,2X,'PSI',/,T15,'MIN STRESS = ',F9.2,2X,'PSI',/,T9,
     6'MAX SHEAR STRESS = ',F9.2,2X,'PSI',////,T10,'GAGE',5X,'THETA',
     79X,'STRAIN',2X,'IN/IN',8X,'STRESS',2X,'PSI',/)
   81 $ALPA(2)=$ALPA(1)-2.0*Y1
      $ALPA(3)=$ALPA(1)-2.0*Y2
      AN2=$ALPA(2)*3.14159/180.0
      AN3=$ALPA(3)*3.14159/180.0
      IF($ALPA(2).GE.0.0.AND.$ALPA(2).LE.90.0)X(2)=COS(AN2)
      IF($ALPA(2).GE.90.0.AND.$ALPA(2).LE.180.0)X(2)=-COS(3.14159-AN2)
      IF($ALPA(2).GE.180.0.AND.$ALPA(2).LE.270.0)X(2)=-COS(AN2-3.14159)
      IF($ALPA(2).GE.-90.0.AND.$ALPA(2).LE.0.0)X(2)=COS(-AN2)
      IF($ALPA(2).GE.-180.0.AND.$ALPA(2).LE.-90.0)X(2)=-COS(3.14159+AN2)
      IF($ALPA(3).GE.0.0.AND.$ALPA(3).LE.90.0)X(3)=COS(AN3)
      IF($ALPA(3).GE.90.0.AND.$ALPA(3).LE.180.0)X(3)=-COS(3.14159-AN3)
      IF($ALPA(3).GE.-90.0.AND.$ALPA(3).LE.0.0)X(3)=COS(-AN3)
      IF($ALPA(3).GE.-180.0.AND.$ALPA(3).LE.-90.0)X(3)=-COS(3.14159+AN3)
      IF($ALPA(3).GE.-270..AND.$ALPA(3).LE.-180.)X(3)=-COS(-3.14159-AN3)
      IF(N.EQ.3) GO TO 50
      $ALPA(4)=$ALPA(1)-2.0*Y3
      AN4=$ALPA(4)*3.14159/180.0
      IF($ALPA(4).GE.0.0.AND.$ALPA(4).LE.90.0)X(4)=COS(AN4)
      IF($ALPA(4).GE.90.0.AND.$ALPA(4).LE.180.0)X(4)=-COS(3.14159-AN4)
      IF($ALPA(4).GE.-90.0.AND.$ALPA(4).LE.0.0)X(4)=COS(-AN4)
      IF($ALPA(4).GE.-180.0.AND.$ALPA(4).LE.-90.0)X(4)=-COS(3.14159+AN4)
      IF($ALPA(4).GE.-270..AND.$ALPA(4).LE.-180.)X(4)=-COS(-3.14159-AN4)
   50 DO 60 I=1,N
      STRAIN(I)=A+B*X(I)
      STRESS(I)=AP+BP*X(I)
      $ALPA(I)=$ALPA(I)/2.0
C
C     IN FORMAT NO 7 THETA CORRESPONDS TO $ALPA
C     THETA = ANGLE FROM RESPECTIVE GAGE TO LINE OF PRINCIPAL STRESS
C
      IF(NMROSE.EQ.1)GO TO 60
      IF(I.GE.N)WRITE(3,101)NROSE,I,$ALPA(I),STRAIN(I),STRESS(I),ALPHA1,
     1STRAMX,STRAMN,SHSTRA,STRSMX,STRSMN,SHSTSS
      IF(I.LT.N)WRITE(3,101)NROSE,I,$ALPA(I),STRAIN(I),STRESS(I)
  101 FORMAT(T4,I3,7X,I3,4X,F9.2,2X,F9.6,1X,F9.2,F9.2,4X,F9.6,2X,F9.6,4X
     1,F9.6,4X,F9.2,3X,F9.2,5X,F9.2)
   60 IF(NMROSE.EQ.1)WRITE(3,8)I,$ALPA(I),STRAIN(I),STRESS(I)
    8 FORMAT(T11,I1,7X,F7.2,10X,F9.6,10X,F9.2,/)
      JJJ=JJJ+1
      IF(JJJ.LE.NMROSE)GO TO 71
      GO TO 70
   80 STOP
      END
```

FIGURE 14.27 *(continued)*

A source program for flexural stresses and deflections of unsymmetric beams is shown in Figure 14.29. Sections 5.5 and 6.10 provide the fundamental theory upon which this Fortran IV program is based. This program is restricted to cantilever and simply supported beams that are loaded with either a uniform load over their full lengths or a concentrated load applied at the end of the cantilever or at the center of the simply supported beam. Modifications for other support and loading conditions would be relatively simple.

Typical output from this program for a cantilever beam loaded with a concentrated load at its free end is shown in Figure 14.30. Figures 14.31 and 14.32 were

RECTANGULAR ROSETTE

POISSON'S RATIO = 0.270
MODULUS OF ELASTICITY = 30000000.0 PSI

GAGE NO = GAGES ON ROSETTE ARE NUMBERED IN CLOCKWISE DIRECTION

ALPHA IS THE ANGLE FROM GAGE 1 TO THE LINE OF PRINCIPAL STRESS MEASURED CLOCKWISE FROM GAGE 1

THETA = ANGLE FROM RESPECTIVE GAGE TO LINE OF PRINCIPAL STRESS

ROSE NO = NUMBER OF THE ROSETTE FOR WHICH OUTPUT HAS BEEN CALCULATED

ROSE NO	GAGE NO	THETA (DEG)	STRN (IN/IN)	STRS (PSI)	ALPHA (DEG)	MAX STRN (IN/IN)	MIN STRN (IN/IN)
1	1	116.57	−0.000400	−5954.06			
1	2	71.57	−0.000600	−10678.45			
1	3	26.57	0.000800	22392.44	116.57	0.001200	−0.000800
2	1	168.69	0.001800	53004.00			
2	2	123.69	0.000100	12846.29			
2	3	78.69	−0.000600	−3688.87	168.69	0.001900	−0.000700
3	1	91.68	−0.000600	−9804.77			
3	2	46.68	0.000200	9092.91			
3	3	1.68	0.001100	30352.71	91.68	0.001101	−0.000601

ROSE NO	GAGE NO	THETA (DEG)	MAX SHR STRN (RAD)	MAX STRS (PSI)	MIN STRS (PSI)	MAX SHR STRS (PSI)
1	1	116.57				
1	2	71.57				
1	3	26.57	0.002000	31841.23	−15402.88	23622.
2	1	168.69				
2	2	123.69				
2	3	78.69	0.002600	55366.20	−6051.14	30708.
3	1	91.68				
3	2	46.68				
3	3	1.68	0.001703	30387.42	9839.48	20113.

EQUIANGULAR ROSETTE

POISSON,S RATIO = 0.270
MODULUS OF ELASTICITY = 30000000.0 PSI
GAGE NO = GAGES ON ROSETTE ARE NUMBERED IN CLOCKWISE DIRECTION
ALPHA IS THE ANGLE FROM GAGE 1 TO THE LINE OF PRINCIPAL STRESS MEASURED CLOCKWISE FROM GAGE 1
THETA = ANGLE FROM RESPECTIVE GAGE TO LINE OF PRINCIPAL STRESS

THE ANGLE FROM GAGE 1 TO THE LINE OF PRINCIPAL STRESS MEASURED TO THE RIGHT OF GAGE 1)

ALPHA 1 =116   DEGREES   28.71   MINUTES

MAX STRAIN = 0.006436 IN/IN
MIN STRAIN = −0.003170 IN/IN
MAX SHEAR STRAIN = 0.009606 IN/IN

MAX STRESS = 180577.75 PSI
MIN STRESS = −46331.25 PSI
MAX SHEAR STRESS = 113454.50 PSI

GAGE	THETA	STRAIN IN/IN	STRESS PSI
1	116.48	−0.001260	−1223.25
2	56.48	−0.000240	22871.64
3	−3.52	0.006400	179721.63

FIGURE 14.28

723

```
C       STRESS AND DEFLECTION ANALYSES
C       THETA EQUALS PHI IN DEGREES
        INTEGER CCANT,UCANT,CSIMP,USIMP
        REAL IX,IY,IXY,INT,IN,MT,MN,MX,MY,M,L
        DIMENSION X(12),Y(12),T(12),SIG(12)
        DIMENSION ARRAY(20)
        PRINT 800
  800   FORMAT ('1',T10,'BEAM STRESSES AND DEFLECTIONS')
        READ (1,850) ARRAY
  850   FORMAT(20A4)
        WRITE (3,851) ARRAY
  851   FORMAT('0',20A4)
C       ZERO THE COORDINATE ARRAYS.
        DO 92 I=1,12
        X(I)=0.0
        Y(I)=0.0
        T(I)=0.0
   92   CONTINUE
C       NPOINT= NUMBER OF CRITICAL POINTS SELECTED. MAXIMUM OF 12 POINTS.
        READ (1,200) NPOINT
  200   FORMAT(I5)
   16   READ (1,100,END=80) (X(I),Y(I),I=1,NPOINT)
  100   FORMAT (6F10.0)
        PRINT 852
  852   FORMAT ('0',T3,'CROSS SECTION POINT COORDINATES IN X-Y ORDER')
        PRINT 853
  853   FORMAT (' ',T3,'POINT 1,2,3......N TH POINT')
        WRITE(3,600)(X(I),Y(I),I=1,NPOINT)
  600   FORMAT('0',6F10.2)
        READ(1,400)P,W,L,E
  400   FORMAT(3F10.0,E14.7)
        READ(1,105)IX,IY,IXY
  105   FORMAT(3F15.0)
        READ(1,300) CCANT,UCANT,CSIMP,USIMP
  300   FORMAT(4I5)
        IF(CCANT.EQ.1) M=P*L
        IF(UCANT.EQ.1) M=W*L**2/2
        IF(CSIMP.EQ.1) M=0.25*P*L
        IF(USIMP.EQ.1) M=W*L**2/8
        WRITE(3,700) IX,IY,IXY,M
  700   FORMAT ('0',T3,'IX=',F8.0,T18,'IY=',F8.0,T33,'IXY=',F8.0,T48,'M=',
       $F8.0)
        WRITE (3,950) P,W,L,E
  950   FORMAT (' ',T4,'P=',F7.0,T19,'W=',F7.0,T35,'L=',F6.0,T48,'E= ',
       $E14.7)
        PI=3.1415927
        THETA=-180.
   85   IF (THETA.GE.180.) GO TO 45
        IF (ABS(ABS(THETA)-90.).LE.0.1) GO TO 95
        PHI=THETA*PI/180.
        MX=M*COS(PHI)
        MY=M*SIN(PHI)
        DO 99 I=1,NPOINT
   99   SIG(I)=(MX*(X(I)*IXY-Y(I)*IY)+MY*(Y(I)*IXY-X(I)*IX))/(IXY**2
       $-IX*IY)
        Y1=IXY-IX*TAN(PHI)
        Y2=IXY*TAN(PHI)-IY
        ALPHA=ATAN(Y1/Y2)
        DALPHA=2*ALPHA
        INT=0.5*(IY-IX)*SIN(DALPHA)+IXY*COS(DALPHA)
        IN=0.5*(IX+IY)+0.5*(IX-IY)*COS(DALPHA)+IXY*SIN(DALPHA)
        PT=P*COS(PHI-ALPHA)
        WT=W*COS(PHI-ALPHA)
        IF(CCANT.EQ.1) DEFL=PT*L**3/(3*E*IN)
        IF(UCANT.EQ.1) DEFL=WT*L**4/(8*E*IN)
        IF(CSIMP.EQ.1) DEFL=PT*L**3/(48*E*IN)
        IF(USIMP.EQ.1) DEFL=WT*L**4/(76.8*E*IN)
        MT=M *SIN(ALPHA-PHI)
        MN=M *COS(ALPHA-PHI)
        DO 20 I=1,NPOINT
   20   T(I)=Y(I)*COS(ALPHA)+X(I)*SIN(ALPHA)
        TMAX=AMAX1(T(1),T(2),T(3),T(4),T(5),T(6),T(7),T(8),T(9),T(10),
       $T(11),T(12))
        TMIN=AMIN1(T(1),T(2),T(3),T(4),T(5),T(6),T(7),T(8),T(9),T(10),
       $T(11),T(12))
        DO 30 I=1,NPOINT
        IF(ABS(T(I)-TMAX).LE.0.01) IP1=I
        IF(ABS(T(I)-TMIN).LE.0.01) IP2=I
   30   CONTINUE
        TMAX=SIGN(TMAX,SIG(IP1))
        TMIN=SIGN(TMIN,SIG(IP2))
        CN=SQRT(MT**2+MN**2)
        CD=SQRT(IN**2+INT**2)
        C=CN/CD
        STRESA=C*TMAX
        STRESB=C*TMIN
        ALPHA=ALPHA*180./PI
        PRINT 854
  854   FORMAT ('0',T3,'ALPHA',T13,'PHI',T21,'STRESS AT POINT',T38,'STRESS
       3 AT POINT',T55,'DEFLECTION')
        WRITE (3,855) ALPHA,THETA,STRESA,IP1,STRESB,IP2,DEFL
  855   FORMAT(' ',F7.1,F9.1,F12.1,I5,F12.1,I5,F13.3)
   95   THETA=THETA+2
        GO TO 85
   45   GO TO 16
   80   STOP
        END                         FIGURE 14.29
```

```
BEAM STRESSES AND DEFLECTIONS

CANTILEVER BEAM - CONCENTRATED LOAD AT END

CROSS SECTION POINT COORDINATES IN X-Y ORDER
POINT 1,2,3......N TH POINT

    11.00      18.50     -11.00      18.50     -11.00     -18.50

    11.00     -18.50

IX=  38958.    IY=  16793.    IXY=       0.    M=  50000.
P=     200.    W=      0.      L=   1000.    E=  0.2000000E 05

ALPHA     PHI     STRESS AT POINT   STRESS AT POINT   DEFLECTION
  0.0   -180.0       -23.7    2        23.7    4        -5.348

ALPHA     PHI     STRESS AT POINT   STRESS AT POINT   DEFLECTION
  4.6   -178.0       -24.9    1        24.9    3        -5.362

ALPHA     PHI     STRESS AT POINT   STRESS AT POINT   DEFLECTION
  9.2   -176.0       -26.0    1        26.0    3        -5.404

ALPHA     PHI     STRESS AT POINT   STRESS AT POINT   DEFLECTION
 13.7   -174.0       -27.0    1        27.0    3        -5.474

ALPHA     PHI     STRESS AT POINT   STRESS AT POINT   DEFLECTION
 18.1   -172.0       -28.1    1        28.1    3        -5.570

ALPHA     PHI     STRESS AT POINT   STRESS AT POINT   DEFLECTION
 22.2   -170.0       -29.1    1        29.1    3        -5.690

ALPHA     PHI     STRESS AT POINT   STRESS AT POINT   DEFLECTION
 26.2   -168.0       -30.0    1        30.0    3        -5.832

ALPHA     PHI     STRESS AT POINT   STRESS AT POINT   DEFLECTION
 30.0   -166.0       -31.0    1        31.0    3        -5.994

ALPHA     PHI     STRESS AT POINT   STRESS AT POINT   DEFLECTION
 33.6   -164.0       -31.9    1        31.9    3        -6.174

ALPHA     PHI     STRESS AT POINT   STRESS AT POINT   DEFLECTION
 37.0   -162.0       -32.7    1        32.7    3        -6.369

ALPHA     PHI     STRESS AT POINT   STRESS AT POINT   DEFLECTION
 40.2   -160.0       -33.5    1        33.5    3        -6.577

ALPHA     PHI     STRESS AT POINT   STRESS AT POINT   DEFLECTION
 43.1   -158.0       -34.3    1        34.3    3        -6.796

ALPHA     PHI     STRESS AT POINT   STRESS AT POINT   DEFLECTION
 45.9   -156.0       -35.0    1        35.0    3        -7.023

ALPHA     PHI     STRESS AT POINT   STRESS AT POINT   DEFLECTION
 48.5   -154.0       -35.7    1        35.7    3        -7.258

ALPHA     PHI     STRESS AT POINT   STRESS AT POINT   DEFLECTION
 51.0   -152.0       -36.3    1        36.3    3        -7.498

ALPHA     PHI     STRESS AT POINT   STRESS AT POINT   DEFLECTION
 53.3   -150.0       -36.9    1        36.9    3        -7.741

ALPHA     PHI     STRESS AT POINT   STRESS AT POINT   DEFLECTION
 55.4   -148.0       -37.5    1        37.5    3        -7.987
```

FIGURE 14.30

prepared from output from this unsymmetric beam program. The beam with the I cross section is uniformly loaded over a simply supported span of 30 ft. Critical flexural stresses in this beam are plotted versus the loading direction angle ϕ in Figure 14.31. Magnitudes of this beam's deflections are plotted versus the loading direction angle ϕ in Figure 14.32. Similar plots may be readily prepared for other beams of interest by running this program with appropriate input.

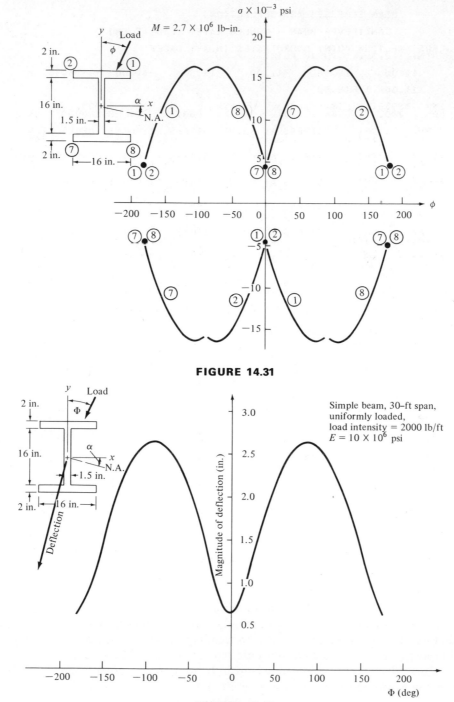

FIGURE 14.31

FIGURE 14.32

Appendix A Selected References

ARGES, K. P., et al. *Engineering Materials*. Pitman Publishing Corporation, Belmont, Calif., 1958.

ARGES, K. P., et al. *Mechanics of Materials*. McGraw-Hill Book Company, New York, 1963.

AU, T. *Elementary Structural Mechanics*. Prentice-Hall, Inc., Englewood Cliffs, N.J., 1963.

BEEDLE, L. S. *Plastic Design of Steel Frames*. John Wiley & Sons, Inc., New York, 1958.

BEER, F. P., et al. *Vector Mechanics for Engineers: Statics and Dynamics*. McGraw-Hill Book Company, New York, 1977.

BORESI, A. P., et al. *Advanced Mechanics of Materials*. John Wiley & Sons, Inc., New York, 1978.

BYARS, EDWARD F., et al. *Engineering Mechanics of Deformable Bodies*, 2nd ed. International Textbook Company, Scranton, Pa., 1969.

CHAJES, ALEXANDER. *Principles of Structural Stability Theory*. Prentice-Hall, Inc., Englewood Cliffs, N.J., 1974.

CHOU, PEI CHI, et al. *Elasticity*. Van Nostrand Reinhold Company, New York, 1967.

CRANDALL, THOMAS J., et al. *An Introduction to the Mechanics of Solids*, 2nd ed. (with SI units). McGraw-Hill Book Company, New York, 1978.

DEN HARTOG, J. P. *Advanced Strength of Materials*. McGraw-Hill Book Company, New York, 1952.

DEN HARTOG, J. P. *Strength of Materials*. Dover Publications, Inc., New York, 1961.

DRUCKER, DANIEL C. *Introduction to Mechanics of Deformable Solids*. McGraw-Hill Book Company, New York, 1967.

EISENSTADT, MELVIN M. *Introduction to Mechanical Properties of Materials*. Macmillan Publishing Co., Inc., New York, 1971.

FERGUSON, P. M. *Reinforced Concrete Fundamentals*. John Wiley & Sons, Inc., New York, 1973.

FORD, HUGH. *Advanced Mechanics of Materials*. John Wiley & Sons, Inc., New York, 1963.

FREUDENTHAL, ALFRED M. *Introduction to Mechanics of Solids*. John Wiley & Sons, Inc., New York, 1966.

FROCHT, M. M. *Photo-Elasticity*. John Wiley & Sons, Inc., New York, 1941.

GINSBERG, J. H., et al. *Combined Statics–Dynamics*. John Wiley & Sons, Inc., New York, 1977.

GOL'DENBLAT, I. I. *Some Principles of the Mechanics of Deformable Media*. P. Noordhoff, Groningen, Holland, 1962.

HIBBELER, R. C. *Engineering Mechanics: Statics*, 2nd ed. Macmillan Publishing Co., Inc., New York, 1978.

HIGDON, A., et al. *Mechanics of Materials*. John Wiley & Sons, Inc., New York, 1976.

HOADLEY, A. *Essentials of Structural Design*. John Wiley & Sons, Inc., New York, 1964.

ILYUSHIN, A. A., et al. *Strength of Materials*. Pergamon Press, Inc., New York, 1967.

JAMES, M. L., et al. *Applied Numerical Methods for Digital Computations—With FORTRAN and CSMP*. Dun-Donnelley Publishing Corporation, New York, 1977.

LEITHOLD, L. *The Calculus with Analytic Geometry*. Harper & Row, Inc., New York, 1968.

LEVINSON, IRVING J. *Mechanics of Materials*. Prentice-Hall, Inc., Englewood Cliffs, N.J., 1970.

LIN, T. H. *Theory of Inelastic Structures*. John Wiley & Sons, Inc., New York, 1968.

LUBAHN, J. D., et al. *Plasticity and Creep of Metals*. John Wiley & Sons, Inc., New York, 1961.

McCORMAC, J. C. *Structural Analysis*. Intext Educational Publishers, New York, 1975.

McCUEN, R. H. *Fortran Programming for Civil Engineers*. Prentice-Hall, Inc., Englewood Cliffs, N.J., 1975.

MERIAM, J. T. *Statics—SI Version*. John Wiley & Sons, Inc., New York, 1975.

MILLER, F. E., et al. *Mechanics of Materials*. International Textbook Company, Scranton, Pa., 1962.

POLAKOWSKI, N. H., et al. *Strength and Structure of Engineering Materials*. Prentice-Hall, Inc., Englewood Cliffs, N.J., 1966.

POPOV, EGOR P. *Introduction to Mechanics of Solids*. Prentice-Hall, Inc., Englewood Cliffs, N.J., 1968.

RIMROTT, F. P. J., et al., eds. *Mechanics of the Solid State.* University of Toronto Press, Toronto, 1968.

ROARK, RAYMOND J. *Formulas for Stress and Strain,* 4th ed., McGraw-Hill Book Company, New York, 1965.

SEELY, FRED B., et al. *Resistance of Materials.* John Wiley & Sons, Inc., New York, 1956.

SHAMES, IRVING H. *Mechanics of Deformable Solids.* Prentice-Hall, Inc., Englewood Cliffs, N.J., 1964.

SHANLEY, F. R. *Strength of Materials.* McGraw-Hill Book Company, New York, 1957.

SHIGLEY, J. E. *Mechanical Engineering Design.* McGraw-Hill Book Company, New York, 1972.

SLOANE, ALVIN. *Mechanics of Materials.* Macmillan Publishing Co., Inc., New York, 1952.

SMITH, J. O., et al. *Inelastic Behavior of Load-Carrying Members.* John Wiley & Sons, Inc., New York, 1965.

SNYDER, R. D., et al. *Engineering Mechanics* (*Statics and Strength of Materials*). McGraw-Hill Book Company, 1973.

TIMOSHENKO, STEPHEN P. *History of Strength of Materials.* McGraw-Hill Book Company, New York, 1953.

TIMOSHENKO, STEPHEN P., et al. *Theory of Elastic Stability.* McGraw-Hill Book Company, New York, 1936.

TIMOSHENKO, STEPHEN P., et al. *Mechanics of Materials.* Van Nostrand Reinhold Company, New York, 1972.

TIMOSHENKO, STEPHEN P., et al. *Theory of Elasticity,* 3rd ed., McGraw-Hill Book Company, New York, 1970.

VOLTERRA, ENRICO, et al. *Advanced Strength of Materials.* Prentice-Hall, Inc., Englewood Cliffs, N.J., 1971.

WHITE, R. N., et al. *Structural Engineering.* John Wiley & Sons, Inc., New York, 1972.

WONG, C. T. *Applied Elasticity.* McGraw-Hill Book Company, New York, 1953.

American Institute of Steel Construction, Manual of Steel Construction, AISC, 1973.

ASME Journal of Applied Mechanics, 1934, 1935, 1936, and 1951.

Machine Design, Reference Issues, 1975, 1976, and 1977.

Appendix # B Reactions at Connections and Supports

Connection or Support	Force Unknowns
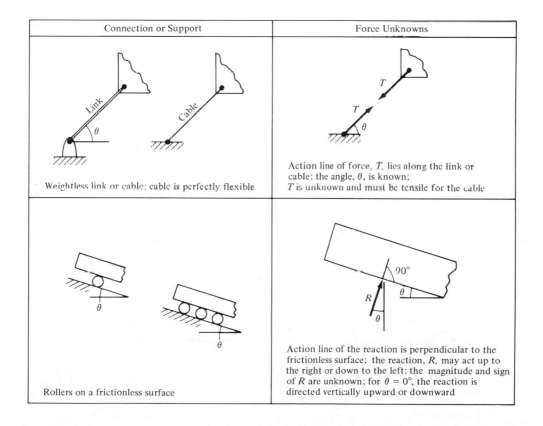	

Weightless link or cable; cable is perfectly flexible

Action line of force, T, lies along the link or cable; the angle, θ, is known; T is unknown and must be tensile for the cable

Rollers on a frictionless surface

Action line of the reaction is perpendicular to the frictionless surface; the reaction, R, may act up to the right or down to the left; the magnitude and sign of R are unknown; for $\theta = 0°$, the reaction is directed vertically upward or downward

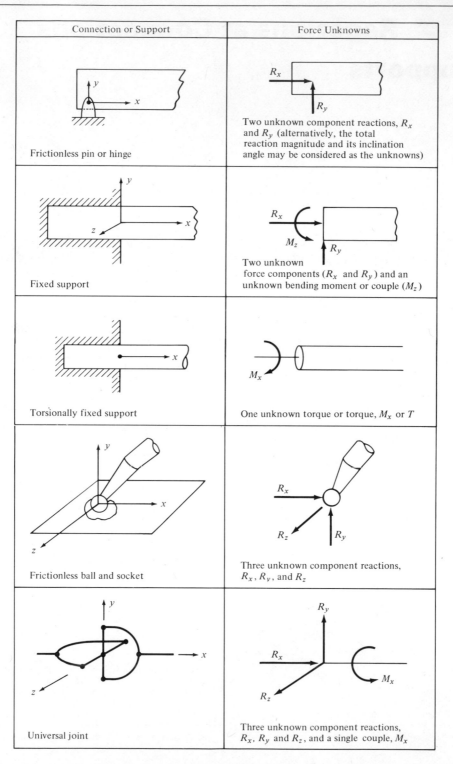

Connection or Support	Force Unknowns
Frictionless pin or hinge	Two unknown component reactions, R_x and R_y (alternatively, the total reaction magnitude and its inclination angle may be considered as the unknowns)
Fixed support	Two unknown force components (R_x and R_y) and an unknown bending moment or couple (M_z)
Torsionally fixed support	One unknown torque or torque, M_x or T
Frictionless ball and socket	Three unknown component reactions, R_x, R_y, and R_z
Universal joint	Three unknown component reactions, R_x, R_y and R_z, and a single couple, M_x

C Typical Properties of Selected Materials*

Typical mechanical properties of carbon steels*

	1015/1020/1022 AISI Type			1035/1040			1045/1050			1095		
	Hot Rolled	Cold Drawn	Annealed (1,600F)	Hot Rolled	Cold Drawn	Quenched and Tempered†	Hot Rolled	Cold Drawn	Quenched and Tempered†	Hot Rolled	Cold Drawn and Annealed (1,450F)	Quenched and Tempered†
Yield strength (10^3 psi)	40	51	43	42	71	63-96	49	84	68-117	66	76	80-152
Tensile strength (10^3 psi)	65	61	60	76	85	96-130	90	100	105-137	130 min.	99	130-216
Impact strength, Izod (ft-lb)	85	—	80	36	—	36-72	23	—	16-53	3	2	5-6
Creep strength, 0.0001%/hr. (10^3 psi)· 800F	—	12.3	—	—	—	—	—	—	—	—	—	—
900F	—	5.7	—	—	—	—	—	—	—	—	—	—
1,000F	—	2.6	—	—	—	—	—	—	—	—	—	—
Elongation in 2 in. (%)	25	15	38	18	12	17-24	15	10	25-15	9	13	10-84
Modulus of elasticity (10^6 psi)	29-30	29-30	29-30	29-30	29-30	29-30	29-30	29-30	29-30	29-30	29-30	29-30

*Approximate properties based on 1-in. cross section. Larger sections have lower mechanical properties. These steels generally are not selected for fatigue applications, so data is not available. Properties vary with quench medium and temper time and temperature. Values shown represent the range of properties available.

Typical mechanical properties of alloy steels—carburizing grades*

	E3310 AISI Type		4320		4620		8620		E9310	
	Annealed	Carburized†	Annealed	Carburized†	Annealed	Carburized†	Annealed	Carburized†	Annealed	Carburized†
Yield strength (10^3 psi)	56	110-112	61	105-107	54	66-67	55	80-83	63	122-123
Tensile strength (10^3 psi)	104	145-146	94	148-152	74	98	77	124-126	119	157-159
Impact strength, Charpy (ft-lb)	65	68-74	—	—	—	84	—		65	68-74
Elongation in 2 in. (%)	21.3	16-18	29	17-18	31	27	31	19-21	17	15-16
Modulus of elasticity (10^6 psi)					29-30					

Typical mechanical properties of cast alloy steels*

	A352-68a ASTM Classification	A219-6	A148-65	A148-1	A148-65	A148-65	A148-65	A148-65
	LCI	WC4	80-50	90-60 Grade	105-85	120-95	150-125	175-145
	Normalized & tempered	Normalized & tempered	Normalized & tempered	Normalized & tempered	Quenched & tempered Bhn 217	Quenched & tempered Bhn 262	Quenched & tempered Bhn 311	Quenched & tempered Bhn 352
Yield strength (10^3 psi)	35	40	50	60	85	95	125	145
Tensile strength (10^3 psi)	65	70	80	90	106	120	150	175
Impact strength, Charpy (ft-lb)	60	55	48	40	58	45	30	24
Fatigue endurance limit, polished specimen (10^3 psi)	20	23	25	31	34	37	44	48
Elongation in 2 in. (%)	24	20	22	20	17	14	9	6

*These steels are seldom specified for high-temperature applications, so creep data are not readily available.

Typical mechanical properties of rolled carbon steels

	1015/1020/1022 AISI Type			1035/1040			1045/1050			1095		
	Hot Rolled	Cold Drawn	Annealed (1,600F)	Hot Rolled	Cold Drawn	Quenched and Tempered	Hot Rolled	Cold Drawn	Quenched and Tempered	Hot Rolled	Cold Drawn and Annealed (1,450F)	Quenched and Tempered
Yield strength (10^3 psi)	40	51	43	42	71	63-96	49	84	68-117	66	76	80-152
Tensile strength (10^3 psi)	65	61	60	76	85	96-130	90	100	105-137	130 min.	99	130-216
Creep strength, 0.0001%/hr. (10^3 psi) 800F	—	12.3	—	—	—	—	—	—	—	—	—	—
900F	—	5.7	—	—	—	—	—	—	—	—	—	—
1,000F	—	2.6	—	—	—	—	—	—	—	—	—	—
Elongation in 2 in. (%)	25	15	38	18	12	17-24	15	10	25-15	9	13	10-84
Modulus of elasticity (10^6 psi)						29-30						

Approximate properties based on 1-in. cross section. Larger sections have lower mechanical properties. These steels generally are not selected for fatigue applications, so data is not available. Properties vary with quench medium and temper time and temperature. Values shown represent the range of properties available.

* From the 1975, 1976, and 1977 reference issues with the permission of the magazine *Machine Design*.

Typical mechanical properties of HSLA steels

	Semikilled or Killed	Semikilled or Killed—Improved Corrosion Resistance	Semikilled or Killed	Inclusion Control—Improved Formability, Killed	Semikilled or Killed	Semikilled or Killed	Inclusion Control—Improved Formability, Killed
Specifications	SAE J410C ASTM A607	ASTM A606 Type 2 & 4 (weathering)	ASTM A607 SAE J410C	ASTM 715 (Sheet) ASTM A656 (Plate)	ASTM A607 SAE J410C	ASTM A607 SAE J410C	ASTM 715 (Sheet) ASTM A656 (Plate)
Yield strength (10^3 psi)	45	45-50	50	50	55	60	60
Tensile strength (10^3 psi)	60*	65-70	65*	60	70*	75*	70
Cold bending†	1t	1t	1t	0t	1½t	2t	0t
Elongation in 2 in. (%)	25	22	22	24§	20	18	22§

Adding 0.20% (minimum) Cu to HSLA steels provides approximately two times the atmospheric corrosion resistance of carbon steel.
*Reduce by 5,000 psi when lower carbon composition is required.
†ASTM A370 longitudinal test specimen up to 0.229 in. inclusive. For transverse bending: use values ½ to 2t higher; or 0 to ½t higher for improved-formability grades.
§Subtract 2% for thicknesses 0.097 in. or less.

Typical mechanical properties of iron–based superalloys

	Martensitic Low Alloy		Martensitic Secondary Hardening		Martensitic Chromium		Semiaustenitic	
AISI Grade	601	604	610	613	616	619	633	635
Typical Designation	17-22A	Chromalloy	H-11	M-50	422	Lapelloy	AM350	Stainless W
Yield strength (10^3 psi) 70F	100	95-108	100-240	335	125-175	85-220	60-175	215-290
1,000F	54	85	140	250	126	85	108	37-50
Tensile strength (10^3 psi) 70F	120	125-138	135-310	410	150-240	125-250	160-205	220-225
1,000F	77	110	180	309	170	95	160	75-80
Impact strength, Charpy (ft-lb) 70F	—	—	10-32	—	10-38 (T700-1400)	—	14	4-106
Fatigue endurance limit (10^3 psi)	—	—	130 (at 260 t.s.)	—	90-110 (T980-1100)	—	70-100	54-96
Creep strength, 0.0001 %/hr at 1,000F (10^3 psi)	—	—	—	—	—	—	—	—
Elongation in 2 in. (%) 70F	30	7	3-17	2	16-19	10-20	12-38	1-5
1,000F	29	—	11	6	16	15	9	47-58
Modulus of elasticity (10^6 psi)	30.8	31.7	30.5	29.5	29.0	30.0	29.4	30.2
Rupture strength, 100hr, 1,000F (10^6 psi)	49	75	95-115	—	58	65	103	32

Typical mechanical properties of wrought stainless steels

	Austenitic								Ferritic			Martensitic (annealed)			
AISI Grades	202	301	302	304/ 304L	309/ 309S	310/ 310S	316/ 316L	347/ 348	405	430	446	403/ 410	416/ 416Se	440A	440C
Yield strength (10^3 psi)	42	40	35	35	40	40	30	40	40	45	50	40	40	60	65
Tensile strength (10^3 psi) 70F	105	105	90	85	90	95	80	95	65	65	80	75	75	105	110
−320F	200	275	220	220	—	152	185	200	—	90	—	158	—	—	—
Impact strength, Izod (ft-lb) 70F	115	100	110	110	110	110	110	110	20-35	35	5-15	85	20-64	—	—
−300F	42-120	110	110	110	—	85	—	95	—	2	—	5	—	—	—
Fatigue endurance limit (10^3 psi)	—	36	34	35	—	—	39	39	40	40	47	40	40	40	40
Creep strength, 0.0001 %/hr at 1,000F (10^3 psi)	—	19	20	20	16.5	33	25	32	8.4	8.5	6.4	11	11	—	—
Elongation in 2 in. (%)	55	55	55	55	45	45	55	45	25	30	25	30	25	20	14
Modulus of elasticity (10^6 psi)	29	28	28	28	29	30	28	28	29	29	29	29	29	29	29

*Charpy V-notch. †For 10^8 cycles, condition H900, room temperature. §At 600F. **Condition H900, in tension.
‡For 15 × 10^6 cycles, condition RH 950, room temperature. ††Condition RH 950, in tension. §§For 10^8 cycles, aged.

Typical mechanical properties of cast iron

	Ductile				White	Gray					
	0-55-06	60-40-18	100-70-03	120-90-02		20	25	30	40	50	60
Yield strength (10^3 psi)	60-75	45-60	75-90	90-125	——	✷	✷	✷	✷	✷	✷
Tensile strength (10^3 psi)	90-110	60-80	100-120	120-150	20-50	20-25†	25-30†	30-35†	40-48†	50-57†	60-66†
Impact strength, Charph unnotched (ft-lb)	15-65	60-115	35-50	25-40	3.5-10.0	—	—	—	—	—	—
Fatigue endurance limit (10^3 psi)	40	30	44	49	——	10	12.5	14.5	19	22	24
Creep strength, 0.0001 %/hr at 800 F (10^3 psi)	19	14	—	——	——	——	——	——	18	30	38
Elongation in 2 in. (%)	3-10	10-25	6-10	2-7	——	≈1%	≈1%	≈1%	≈0.8%	≈0.5%	≈0.5%
Modulus of elasticity (10^6 psi)	22-25	22-25	22-25	22-25	——	12	13	15	17	19	20

✷Yield strength for gray iron is considered to be strength required to produce 0.1% permanent strain, usually about 65-80% of tensile strength.

Typical properties of wrought aluminum alloys

	AA Designation										
	1100		2024		3003	5052		6061		7075	
	Temper										
	−0	−H18	−0	−T4	−0	−0	−H38	−0	−T6	−0	−T6
Yield strength (10^3 psi)	5	22	11	47	6	13	37	8	40	15	73
Tensile strength (10^3 psi)	13	24	27	68	16	28	42	18	45	33	83
Fatigue endurance limit (10^3 psi)	5	9	13	20	7	16	20	9	14	---	23
Elongation in 2 in. (%)	45	15	22	19	40	30	8	30	17	16	11
Modulus of elasticity (10^6 psi)	10.0	10.0	10.6	10.6	10.0	10.2	10.2	10.0	10.0	10.4	10.4
Melting temperature (F)	1,190-1,215	1,190-1,215	935-1,180	935-1,180	1,190-1,210	1,125-1,200	1,125-1,200	1,080-1,205	1,080-1,205	890-1,175	890-1,175
Coefficient of thermal expansion (in./in.-° F× 10^{-6})	13.1	13.1	12.9	12.9	12.9	13.2	13.2	13.1	13.1	13.1	13.1
Thermal conductivity (Btu-in./hr-ft²-° F)	1540	1510	1340	840	1340	960	960	1250	1160	---	900
Electrical resistivity (Microhm-cm)	2.9	3.0	3.4	5.7	3.4	4.9	4.9	3.7	4.0	---	5.2
Density (lb/in.³)	0.098	0.098	0.100	0.100	0.099	0.097	0.097	0.098	0.098	0.101	0.101

Aluminum alloys are rarely used at elevated temperature so creep data are unavailable.

Cast aluminum—which alloy for which process?

| Alloy | Alloying elements | Casting process | | |
		Sand	Permanent mold	Die casting
319	Si-Cu	□	□	
333	Si-Cu		□	
355	Si-Mg-Cu	□		
C355	Si-Mg-Cu		□	
356	Si-Mg	□		
A356	Si-Mg		□	
360				□
380	Si-Mg-Cu-Fe			□
A413	Si-Mg-Cu-Fe			□
443	Si-Cu-Fe	□	□	
C443	Si-Cu-Fe			□
518	Mg-Fe			□
520	Mg	□		
712	Zn-Mg-Cr-Ti	□		
850	Sn-Ni-Cu-Si	□		

Typical properties of cast aluminum

	AA Designation				
	355.0	356.0	360.0	380.0	B443.0
			Temper		
	T6	T6	F (as die cast)	F (as die cast)	F (as cast)
Yield strength (10^3 psi)	20	20	25	24	8
Tensile strength (10^3 psi)	32	30	47	48	19
Impact strength, Charpy unnotched (ft-lb)	—	—	—	3.5	—
Fatigue endurance limit (10^3 psi)	10.0	13	19	21	—
Elongation in 2 in. (%)	2	3	3	3	3-5
Modulus of elasticity (10^6 psi)	—	—	—	10.3	—
Melting temperature (F)	1,150-1,015	1,135-1,035	1,035-1,105	1,000-1,100	1,070-1,175
Coefficient of thermal expansion (In./In.-°F × 10^{-6})	12.4	11.9	—	12.5	12.2
Thermal conductivity ()Btu-in./hr-ft²-°F)	87	92	—	58	84
Electrical resistivity (Microhm-cm)	6.20	6.80	4.80	4.00	6.40
Density (lb/in.³)	0.098	0.097	0.095	0.098	0.097

Typical properties of magnesium

	Die castings	Extruded shapes	Forgings	Sheet and plate
	AZ91B-F	AZ31B-F ZK60A-T5	AZ31B-F HM21A-T5 AZ80A-T5 ZK60A-T6	AZ31B-H24 HK31A-H24 HM21A-T8
Melting temperature (F)	875- 1,105	970- 1,175	970- 1,202	1,050- 1,202
Coefficient of thermal expansion (in./in.-°F)	14.5	14.3-14.5	14.3-14.5	14.3-14.5
Thermal conductivity (Btu-ft/hr-ft²-°F)	31	44-70	—	44-79
Electrical resistivity (Microhm-cm)	12.9	5.7-9.2	5.0-9.2	5.0-9.2
Density (lb/in.³)	0.0653	0.0642- 0.0660	0.0655- 0.0660	0.0642- 0.0648
Yield strength (10^3 psi)	23	28-44	22-39	21-32
Tensile strength (10^3 psi)	34	38-53	34-50	33-42
Impact strength, Charpy (ft-lb)	2.0	3.2	3.2	—
Fatigue endurance limit (10^3 psi)	14	16-23	15-18	16-24
Creep strength, 0.1% in 100 hr (10^3 psi)	3-4 (250F)	up to 5 (250F)	5 (500F)	up to 8 (500F)
Elongation in 2 in. (%)	3	11-15	6-11	9-21
Modulus of elasticity (10^6 psi)	6.5	6.5	6.5	6.5

Typical mechanical properties of titanium and titanium alloys

	Commercially Pure Alpha Ti			Alpha Alloy Ti-0.2Pd	Alpha-Beta Alloy		Beta Alloy	
	Ti-35A	Ti-50A	Ti-65A		Ti-6Al-4V	Ti-6Al-4V	Ti-3Al-13V-11Cr	Ti-3Al-13V-11Cr
	Wrought	Wrought	Wrought	Wrought	Annealed	Heat treated *	Heat treated†	Heat treated§
Yield strength, 0.2% offset (10^3 psi)	25	40	55	40	120	150	175	130
Tensile strength (10^3 psi)	35	50	65	50	130	160	185	135
Impact strength, Charpy (ft-lb)	11-40	11-40	11-40	—	10-20	—	5-15	—
Fatigue endurance limit (10^3 psi)	63	63	63	—	85	—	—	—
Elongation in 2 in. (%)	24	20	18	20	10	7	6	16
Modulus of elasticity, min (10^6 psi)	15.0	15.0	15.0	14.9	16.5	16.5	16.0	14.7

*Heated 1,700 F, water quenched. †Heated 1,400 F, air-cooled and aged. §Heated 1,400 F, water quenched.

Typical properties of wrought copper alloys

	Tellurium-copper	Leaded beryllium-copper	Med. leaded brass (65Cu-34Zn)	Free-cutting brass	Leaded phos.-bronze (88Cu-4Zn)	Aluminum-silicon-bronze (91Cu-7Al-2Si)	Silicon-bronze (97Cu-3Si)	Manganese-bronze
CDA Designation	145	173	340	360	544	642	655	675
Melting temperature (F)	1,967	1,800	1,700	1,650	1,830	1,840	1,880	1,650
Coefficient of thermal expansion (in./in.-°F × 10⁻⁷)	95	99	113	114	96	100	100	118
Thermal conductivity (Btu-in./hr-ft²-°F)	205	62-75	67	67	50	26	21	61
Electrical resistivity (Microhm-cm)	1.86	7.68	6.63	6.63	9.07	18.6	24.6	7.18
Density (lb/in.³)	0.323	0.298	0.306	0.307	0.321	0.278	0.308	0.302
Yield strength (10^3 psi)	10-44	25-178	19-42	18-45	na-57	55-68	22-60	30-60
Tensile strength (10^3 psi)	32-48	68-195	50-55	49-58	na-68	90-102	58-108	65-84
Fatigue endurance limit (10^3 psi)	—	44*	—	17	—	50	29†	—
Elongation in 2 in. (%)	50-20	48-5	60-40	53-25	na-20	28-22	60-13	33-19
Modulus of elasticity (10^6 psi)	17	18.5	15	14	15	16	15	15

*Strip. †Wire. Fatigue data, obtained by rotating beam method, vary markedly for copper alloys. Structure, temper, and surface condition affect test results. Data here represent typical values selected from the wide range found in literature.

Appendix **D** Properties of Plane Areas

Shape and Dimensions	Area	Centroid Location	Area moment of Inertia	Radius of Gyration
Rectangular area 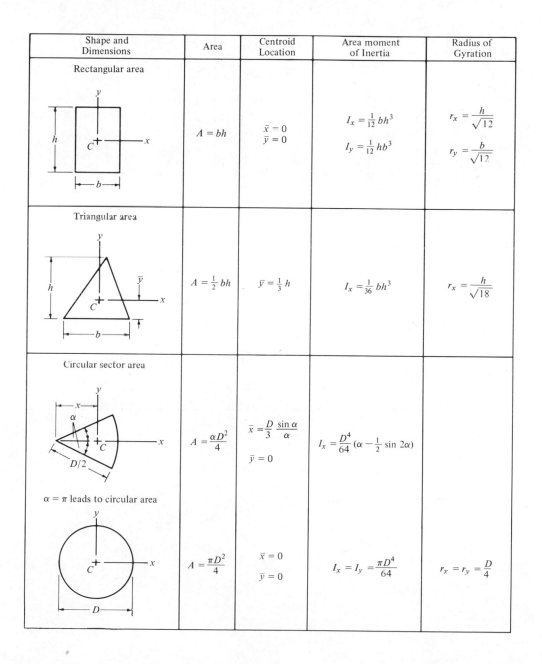	$A = bh$	$\bar{x} = 0$ $\bar{y} = 0$	$I_x = \frac{1}{12} bh^3$ $I_y = \frac{1}{12} hb^3$	$r_x = \dfrac{h}{\sqrt{12}}$ $r_y = \dfrac{b}{\sqrt{12}}$
Triangular area	$A = \frac{1}{2} bh$	$\bar{y} = \frac{1}{3} h$	$I_x = \frac{1}{36} bh^3$	$r_x = \dfrac{h}{\sqrt{18}}$
Circular sector area $\alpha = \pi$ leads to circular area	$A = \dfrac{\alpha D^2}{4}$	$\bar{x} = \dfrac{D}{3} \dfrac{\sin \alpha}{\alpha}$ $\bar{y} = 0$	$I_x = \dfrac{D^4}{64} \left(\alpha - \dfrac{1}{2} \sin 2\alpha\right)$	
	$A = \dfrac{\pi D^2}{4}$	$\bar{x} = 0$ $\bar{y} = 0$	$I_x = I_y = \dfrac{\pi D^4}{64}$	$r_x = r_y = \dfrac{D}{4}$

Shape and Dimensions	Area	Centroid Location	Area Moment of Inertia	Radius of Gyration
Elliptical area $A = \pi ab$	$A = \pi ab$	$\bar{x} = 0$ $\bar{y} = 0$	$I_x = \dfrac{\pi ab^3}{4}$ $I_y = \dfrac{\pi ba^3}{4}$	$r_x = \dfrac{b}{2}$ $r_y = \dfrac{a}{2}$
nth–degree parabolic quadrant Vertex	$A = \dfrac{nab}{n+1}$	$\bar{x} = \dfrac{(n+1)a}{2(n+2)}$ $\bar{y} = \dfrac{(n+1)b}{2n+1}$		
nth–degree parabolic spandrel Vertex	$A = \dfrac{ab}{n+1}$	$\bar{x} = \dfrac{n+1}{n+2}\,a$ $\bar{y} = \dfrac{n+1}{2(2n+1)}\,b$		

E Design Properties for Selected Structural Sections*

Appendix

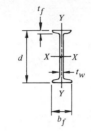

S Shapes
Properties for Designing

| | | | Flange | | Web | Elastic Properties | | | | | |
| | Area, A | Depth, D | Width, b_f | Web Thickness, t_f | Thickness, t_w | Axis X–X | | | Axis Y–Y | | |
Designation	(in.²)	(in.)	(in.)	(in.)	(in.)	I (in.⁴)	Z (in.³)	r (in.)	I (in.⁴)	Z (in.³)	r (in.)
S24 × 120	35.3	24.00	8.048	1.102	0.798	3030	252	9.26	84.2	20.9	1.54
× 105.9	31.1	24.00	7.875	1.102	0.625	2830	236	9.53	78.2	19.8	1.58
S24 × 100	29.4	24.00	7.247	0.871	0.747	2390	199	9.01	47.8	13.2	1.27
× 90	26.5	24.00	7.124	0.871	0.624	2250	187	9.22	44.9	12.6	1.30
× 79.9	23.5	24.00	7.001	0.871	0.501	2110	175	9.47	42.3	12.1	1.34
S20 × 95	27.9	20.00	7.200	0.916	0.800	1610	161	7.60	49.7	13.8	1.33
× 85	25.0	20.00	7.053	0.916	0.653	1520	152	7.79	46.2	13.1	1.36
S20 × 75	22.1	20.00	6.391	0.789	0.641	1280	128	7.60	29.6	9.28	1.16
× 65.4	19.2	20.00	6.250	0.789	0.500	1180	118	7.84	27.4	8.77	1.19
S18 × 70	20.6	18.00	6.251	0.691	0.711	926	103	6.71	24.1	7.72	1.08
× 54.7	16.1	18.00	6.001	0.691	0.461	804	89.4	7.07	20.8	6.94	1.14
S15 × 50	14.7	15.00	5.640	0.622	0.550	486	64.8	5.75	15.7	5.57	1.03
× 42.9	12.6	15.00	5.501	0.622	0.411	447	59.6	5.95	14.4	5.23	1.07
S12 × 50	14.7	12.00	5.477	0.659	0.687	305	50.8	4.55	15.7	5.74	1.03
× 40.8	12.0	12.00	5.252	0.659	0.462	272	45.4	4.77	13.6	5.16	1.06
S12 × 35	10.3	12.00	5.078	0.544	0.428	229	38.2	4.72	9.87	3.89	0.980
× 31.8	9.35	12.00	5.000	0.544	0.350	218	36.4	4.83	9.36	3.74	1.00
S10 × 35	10.3	10.00	4.944	0.491	0.594	147	29.4	3.78	8.36	3.38	0.901
× 25.4	7.46	10.00	4.661	0.491	0.311	124	24.7	4.07	6.79	2.91	0.954
S8 × 23	6.77	8.00	4.171	0.425	0.441	64.9	16.2	3.10	4.31	2.07	0.798
× 18.4	5.41	8.00	4.001	0.425	0.271	57.6	14.4	3.26	3.73	1.86	0.831
S7 × 20	5.88	7.00	3.860	0.392	0.450	42.4	12.1	2.69	3.17	1.64	0.734
× 15.3	4.50	7.00	3.662	0.392	0.252	36.7	10.5	2.86	2.64	1.44	0.766
S6 × 17.25	5.07	6.00	3.565	0.359	0.465	26.3	8.77	2.28	2.31	1.30	0.675
× 12.5	3.67	6.00	3.332	0.359	0.232	22.1	7.37	2.45	1.82	1.09	0.705
S5 × 14.75	4.34	5.00	3.284	0.326	0.494	15.2	6.09	1.87	1.67	1.01	0.620
× 10	2.94	5.00	3.004	0.326	0.214	12.3	4.92	2.05	1.22	0.809	0.643
S4 × 9.5	2.79	4.00	2.796	0.293	0.326	6.79	3.39	1.56	0.903	0.646	0.569
× 7.7	2.26	4.00	2.663	0.293	0.193	6.08	3.04	1.64	0.764	0.574	0.581
S3 × 7.5	2.21	3.00	2.509	0.260	0.349	2.93	1.95	1.15	0.586	0.468	0.516
× 5.7	1.67	3.00	2.330	0.260	0.170	2.52	1.68	1.23	0.455	0.390	0.522

Reprinted with permission of the American Institute of Steel Construction.

Pipe
Dimensions and Properties

Dimension					Properties			
Nominal Diameter (in.)	Outside Diameter (in.)	Inside Diameter (in.)	Wall Thickness (in.)	Weight per Foot-Lb, Plain Ends	A (in.^{2})	I (in.^{4})	Z (in.^{3})	r (in.)
Standard Weight								
$\frac{1}{2}$	0.840	.622	0.109	0.85	0.250	0.017	0.041	0.261
$\frac{3}{4}$	1.050	.824	0.113	1.13	0.333	0.037	0.071	0.334
1	1.315	1.049	0.133	1.68	0.494	0.087	0.133	0.421
$1\frac{1}{4}$	1.660	1.380	0.140	2.27	0.669	0.195	0.235	0.540
$1\frac{1}{2}$	1.900	1.610	0.145	2.72	0.799	0.310	0.326	0.623
2	2.375	2.067	0.154	3.65	1.07	0.666	0.561	0.787
$2\frac{1}{2}$	2.875	2.469	0.203	5.79	1.70	1.53	1.06	0.947
3	3.500	3.068	0.216	7.58	2.23	3.02	1.72	1.16
$3\frac{1}{2}$	4.000	3.548	0.226	9.11	2.68	4.79	2.39	1.34
4	4.500	4.026	0.237	10.79	3.17	7.23	3.21	1.51
5	5.563	5.047	0.258	14.62	4.30	15.2	5.45	1.88
6	6.625	6.065	0.280	18.97	5.58	28.1	8.50	2.25
8	8.625	7.981	0.322	28.55	8.40	72.5	16.8	2.94
10	10.750	10.020	0.365	40.48	11.9	161	29.9	3.67
12	12.750	12.000	0.375	49.56	14.6	279	43.8	4.38
Extra Strong								
$\frac{1}{2}$	0.840	0.546	0.147	1.09	0.320	0.020	0.048	0.250
$\frac{3}{4}$	1.050	0.742	0.154	1.47	0.433	0.045	0.085	0.321
1	1.315	0.957	0.179	2.17	0.639	0.106	0.161	0.407
$1\frac{1}{4}$	1.660	1.278	0.191	3.00	0.881	0.242	0.291	0.524
$1\frac{1}{2}$	1.900	1.500	0.200	3.63	1.07	0.391	0.412	0.605
2	2.375	1.939	0.218	5.02	1.48	0.868	0.731	0.766
$2\frac{1}{2}$	2.875	2.323	0.276	7.66	2.25	1.92	1.34	0.924
3	3.500	2.900	0.300	10.25	3.02	3.89	2.23	1.14
$3\frac{1}{2}$	4.000	3.364	0.318	12.50	3.68	6.28	3.14	1.31
4	4.500	3.826	0.337	14.98	4.41	9.61	4.27	1.48
5	5.563	4.813	0.375	20.78	6.11	20.7	7.43	1.84
6	6.625	5.761	0.432	28.57	8.40	40.5	12.2	2.19
8	8.625	7.625	0.500	43.39	12.8	106	24.5	2.88
10	10.750	9.750	0.500	54.74	16.1	212	39.4	3.63
12	12.750	11.750	0.500	65.42	19.2	362	56.7	4.33
Double-Extra Strong								
2	2.375	1.503	0.436	9.03	2.66	1.31	1.10	0.703
$2\frac{1}{2}$	2.875	1.771	0.552	13.69	4.03	2.87	2.00	0.844
3	3.500	2.300	0.600	18.58	5.47	5.99	3.42	1.05
4	4.500	3.152	0.674	27.54	8.10	15.3	6.79	1.37
5	5.563	4.063	0.750	38.55	11.3	33.6	12.1	1.72
6	6.625	4.897	0.864	53.16	15.6	66.3	20.0	2.06
8	8.625	6.875	0.875	72.42	21.3	162	37.6	2.76

Reprinted with permission of the American Institute of Steel Construction.

Note: The listed sections are available in conformance with ASTM Specification A53 Grade B or A501. Other sections are made to these specifications. Consult with pipe manufacturers or distributors for availability.

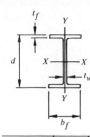

W Shapes
Properties for Designing

			Flange		Web Thickness,	Elastic Properties					
						Axis X-X			Axis Y-Y		
	Area, A	Depth, d	Width, b_f	Thickness t_f	t_w	I	Z	r	I	Z	r
Designation	(in.²)	(in.)	(in.)	(in.)	(in.)	(in.⁴)	(in.³)	(in.)	(in.⁴)	(in.³)	(in.)
W36 × 300	88.3	36.72	16.655	1.680	0.945	20300	1110	15.2	1300	156	3.83
× 280	82.4	36.50	16.595	1.570	0.885	18900	1030	15.1	1200	144	3.81
W36 × 194	57.2	36.48	12.117	1.260	0.770	12100	665	14.6	375	61.9	2.56
× 182	53.6	36.32	12.072	1.180	0.725	11300	622	14.5	347	57.5	2.55
W33 × 240	70.6	33.50	15.865	1.400	0.830	13600	813	13.9	933	118	3.64
× 220	64.8	33.25	15.810	1.275	0.775	12300	742	13.8	841	106	3.60
W33 × 152	44.8	33.50	11.565	1.055	0.635	8160	487	13.5	273	47.2	2.47
× 141	41.6	33.31	11.535	0.960	0.605	7460	448	13.4	246	42.7	2.43
W30 × 210	61.9	30.38	15.105	1.315	0.775	9890	651	12.6	757	100	3.50
× 190	56.0	30.12	15.040	1.185	0.710	8850	587	12.6	673	89.5	3.47
W30 × 132	38.9	30.30	10.551	1.000	0.615	5760	380	12.2	196	37.2	2.25
× 124	36.5	30.16	10.521	0.930	0.585	5360	355	12.1	181	34.4	2.23
W27 × 177	52.2	27.31	14.090	1.190	0.725	6740	494	11.4	556	78.9	3.26
× 160	47.1	27.08	14.023	1.073	0.658	6030	446	11.3	493	70.6	3.24
W27 × 114	33.6	27.28	10.070	0.932	0.570	4090	300	11.0	159	31.6	2.18
× 102	30.0	27.07	10.018	0.827	0.518	3610	267	11.0	139	27.7	2.15
W24 × 160	47.1	24.72	14.091	1.135	0.656	5120	414	10.4	530	75.2	3.35
× 145	42.7	24.49	14.043	1.020	0.608	4570	373	10.3	471	67.1	3.32
W24 × 120	35.4	24.31	12.088	0.930	0.556	3650	300	10.2	274	45.4	2.78
× 110	32.5	24.16	12.042	0.855	0.510	3330	276	10.1	249	41.4	2.77
W24 × 94	27.7	24.29	9.061	0.872	0.516	2690	221	9.86	108	23.9	1.98
× 84	24.7	24.09	9.015	0.772	0.470	2370	197	9.79	94.5	21.0	1.95
W24 × 61	18.0	23.72	7.023	0.591	0.419	1540	130	9.25	34.3	9.76	1.38
× 55	16.2	23.55	7.000	0.503	0.396	1340	114	9.10	28.9	8.25	1.34
W21 × 142	41.8	21.46	13.132	1.095	0.659	3410	317	9.03	414	63.0	3.15
× 127	37.4	21.24	13.061	0.985	0.588	3020	284	8.99	366	56.1	3.13
W21 × 96	28.3	21.14	9.038	0.935	0.575	2100	198	8.61	115	25.5	2.02
× 82	24.2	20.86	8.962	0.795	0.499	1760	169	8.53	95.6	21.3	1.99
W21 × 73	21.5	21.24	8.295	0.740	0.455	1600	151	8.64	70.6	17.0	1.81
× 68	20.0	21.13	8.270	0.685	0.430	1480	140	8.60	64.7	15.7	1.80

Reprinted with permission of the American Institute of Steel Construction.

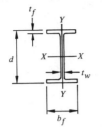

W Shapes
Properties for Designing

Designation	Area, A (in.^{2})	Depth, d (in.)	Flange Width, b_f (in.)	Flange Thickness, t_f (in.)	Web Thickness, t_w (in.)	Axis X-X I (in.^{4})	Axis X-X Z (in.^{3})	Axis X-X r (in.)	Axis Y-Y I (in.^{4})	Axis Y-Y Z (in.^{3})	Axis Y-Y r (in.)
W21 × 49	14.4	20.82	6.520	0.532	0.368	971	93.3	8.21	24.7	7.57	1.31
× 44	13.0	20.66	6.500	0.451	0.348	843	81.6	8.07	20.7	6.38	1.27
W18 × 114	33.5	18.48	11.833	0.991	0.595	2040	220	7.79	274	46.3	2.86
× 105	30.9	18.32	11.792	0.911	0.554	1850	202	7.75	249	42.3	2.84
W18 × 85	25.0	18.32	8.838	0.911	0.526	1440	157	7.57	105	23.8	2.05
× 77	22.7	18.16	8.787	0.831	0.475	1290	142	7.54	94.1	21.4	2.04
W18 × 60	17.7	18.25	7.558	0.695	0.416	986	108	7.47	50.1	13.3	1.68
× 55	16.2	18.12	7.532	0.630	0.390	891	98.4	7.42	45.0	11.9	1.67
W18 × 40	11.8	17.90	6.018	0.524	0.316	612	68.4	7.21	19.1	6.34	1.27
× 35	10.3	17.71	6.000	0.429	0.298	513	57.9	7.05	15.5	5.16	1.23
W16 × 96	28.2	16.32	11.533	0.875	0.535	1360	166	6.93	224	38.8	2.82
× 88	25.9	16.16	11.502	0.795	0.504	1220	151	6.87	202	35.1	2.79
W16 × 78	23.0	16.32	8.586	0.875	0.529	1050	128	6.75	92.5	21.6	2.01
× 71	20.9	16.16	8.543	0.795	0.486	941	116	6.71	82.8	19.4	1.99
W16 × 50	14.7	16.25	7.073	0.628	0.380	657	80.8	6.68	37.1	10.5	1.59
× 45	13.3	16.12	7.039	0.563	0.346	584	72.5	6.64	32.8	9.32	1.57
W16 × 31	9.13	15.84	5.525	0.442	0.275	374	47.2	6.40	12.5	4.51	1.17
× 26	7.67	15.65	5.500	0.345	0.250	300	38.3	6.25	9.59	3.49	1.12
W14 × 730	215	22.44	17.889	4.910	3.069	14400	1280	8.18	4720	527	4.69
× 665	196	21.67	17.646	4.522	2.826	12500	1150	7.99	4170	472	4.62
W14 × 426	125	18.69	16.695	3.033	1.875	6610	707	7.26	2360	283	4.34
× 398	117	18.31	16.590	2.843	1.770	6010	657	7.17	2170	262	4.31
W14 × 320	94.1	16.81	16.710	2.093	1.890	4140	493	6.63	1640	196	4.17
W14 × 237	69.7	16.12	15.910	1.748	1.090	3080	382	6.65	1170	148	4.11
× 228	67.1	16.00	15.865	1.688	1.045	2940	368	6.62	1120	142	4.10
W14 × 136	40.0	14.75	14.740	1.063	0.660	1590	216	6.31	568	77.0	3.77
× 127	37.3	14.62	14.690	0.998	0.610	1480	202	6.29	528	71.8	3.76
× 119	35.0	14.50	14.650	0.938	0.570	1370	189	6.26	492	67.1	3.75
W14 × 84	24.7	14.18	12.023	0.778	0.451	928	131	6.13	225	37.5	3.02
× 78	22.9	14.06	12.000	0.718	0.428	851	121	6.09	207	34.5	3.00

Reprinted with permission of the American Institute of Steel Construction.

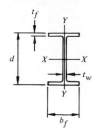

W Shapes
Properties for Designing

Designation	Area, A (in.²)	Depth, d (.in.)	Flange Width, b_f (in.)	Flange Thickness, t_f (in.)	Web Thickness, t_w (in.)	Axis X-X I (in.⁴)	Axis X-X Z (in.³)	Axis X-X r (in.)	Axis Y-Y I (in.⁴)	Axis Y-Y Z (in.³)	Axis Y-Y r (in.)
W14 × 74	21.8	14.19	10.072	0.783	0.450	797	112	6.05	133	26.5	2.48
× 68	20.0	14.06	10.040	0.718	0.418	724	103	6.02	121	24.1	2.46
W14 × 53	15.6	13.94	8.062	0.658	0.370	542	77.8	5.90	57.5	14.3	1.92
× 48	14.1	13.81	8.031	0.593	0.339	485	70.2	5.86	51.3	12.8	1.91
W14 × 38	11.2	14.12	6.776	0.513	0.313	386	54.7	5.88	26.6	7.86	1.54
× 34	10.0	14.00	6.750	0.453	0.287	340	48.6	5.83	23.3	6.89	1.52
W14 × 26	7.67	13.89	5.025	0.418	0.255	244	35.1	5.64	8.86	3.53	1.08
× 22	6.49	13.72	5.000	0.335	0.230	198	28.9	5.53	7.00	2.80	1.04
W12 × 190	55.9	14.38	12.670	1.736	1.060	1890	263	5.82	590	93.1	3.25
× 161	47.4	13.88	12.515	1.486	0.905	1540	222	5.70	486	77.7	3.20
W12 × 58	17.1	12.19	10.014	0.641	0.359	476	78.1	5.28	107	21.4	2.51
× 53	15.6	12.06	10.000	0.576	0.345	426	70.7	5.23	96.1	19.2	2.48
W12 × 50	14.7	12.19	8.077	0.641	0.371	395	64.7	5.18	56.4	14.0	1.96
× 45	13.2	12.06	8.042	0.576	0.336	351	58.2	5.15	50.0	12.4	1.94
W12 × 36	10.6	12.24	6.565	0.540	0.305	281	46.0	5.15	25.5	7.77	1.55
× 31	9.13	12.09	6.525	0.465	0.265	239	39.5	5.12	21.6	6.61	1.54
W12 × 22	6.47	12.31	4.030	0.424	0.260	156	25.3	4.91	4.64	2.31	0.847
× 19	5.59	12.16	4.007	0.349	0.237	130	21.3	4.82	3.76	1.88	0.820
W10 × 112	32.9	11.38	10.415	1.248	0.755	719	126	4.67	235	45.2	2.67
× 100	29.4	11.12	10.345	1.118	0.685	625	112	4.61	207	39.9	2.65
W10 × 45	13.2	10.12	8.022	0.618	0.350	249	49.1	4.33	53.2	13.2	2.00
× 39	11.5	9.94	7.990	0.528	0.318	210	42.2	4.27	44.9	11.2	1.98
W10 × 29	8.54	10.22	5.799	0.500	0.289	158	30.8	4.30	16.3	5.61	1.38
× 25	7.36	10.08	5.762	0.430	0.252	133	26.5	4.26	13.7	4.76	1.37
W10 × 19	5.61	10.25	4.020	0.394	0.250	96.3	18.8	4.14	4.28	2.13	0.874
× 17	4.99	10.12	4.010	0.329	0.240	81.9	16.2	4.05	3.55	1.77	0.844
W8 × 67	19.7	9.00	8.287	0.933	0.575	272	60.4	3.71	88.6	21.4	2.12
× 58	17.1	8.75	8.222	0.808	0.510	227	52.0	3.65	74.9	18.2	2.10
W8 × 28	8.23	8.06	6.540	0.463	0.285	97.8	24.3	3.45	21.6	6.61	1.62
× 24	7.06	7.93	6.500	0.398	0.245	82.5	20.8	3.42	18.2	5.61	1.61

Reprinted with permission of the American Institute of Steel Construction.

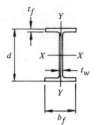

M Shapes
Properties for Designing

Designation	Area, A (in.^{2})	Depth, d (in.)	Flange Width b_f (in.)	Flange Thickness, t_f (in.)	Web Thickness, t_w (in.)	Axis X-X I (in.^{4})	Axis X-X Z (in.^{3})	Axis X-X r (in.)	Axis Y-Y I (in.^{4})	Axis Y-Y Z (in.^{2})	Axis Y-Y r (in.)
M14 × 17.2	5.05	14.00	4.000	0.272	0.210	147	21.1	5.40	2.65	1.33	0.725
M12 × 11.8	3.47	12.00	3.065	0.225	0.177	71.9	12.0	4.55	0.980	0.639	0.532
M10 × 29.1	8.56	9.88	5.937	0.389	0.427	131	26.6	3.92	11.2	3.76	1.14
× 22.9	6.73	9.88	5.752	0.389	0.242	117	23.6	4.16	10.0	3.48	1.22
M10 × 9	2.65	10.00	2.690	0.206	0.157	38.8	7.76	3.83	0.609	0.453	0.480
M8 × 34.3	10.1	8.00	8.003	0.459	0.378	116	29.1	3.40	34.9	8.73	1.86
× 32.6	9.58	8.00	7.940	0.459	0.315	114	28.4	3.44	34.1	8.58	1.89
M8 × 22.5	6.60	8.00	5.395	0.353	0.375	68.2	17.1	3.22	7.48	2.77	1.06
× 18.5	5.44	8.00	5.250	0.353	0.230	62.0	15.5	3.38	6.82	2.60	1.12
M8 × 6.5	1.92	8.00	2.281	0.189	0.135	18.5	4.62	3.10	0.343	0.301	0.423
M7 × 5.5	1.62	7.00	2.080	0.180	0.128	12.0	3.44	2.73	0.249	0.239	0.392
M6 × 22.5	6.62	6.00	6.060	0.379	0.372	41.2	13.7	2.49	12.4	4.08	1.37
× 20	5.89	6.00	5.938	0.379	0.250	39.0	13.0	2.57	11.6	3.90	1.40
M6 × 4.4	1.29	6.00	1.844	0.171	0.114	7.20	2.40	2.36	0.165	0.179	0.358
M5 × 18.9	5.55	5.00	5.003	0.416	0.316	24.1	9.63	2.08	7.86	3.14	1.19
M4 × 13.8	4.06	4.00	4.000	0.371	0.313	10.8	5.42	1.63	3.58	1.79	0.939
× 13	3.81	4.00	3.940	0.371	0.254	10.5	5.24	1.66	3.36	1.71	0.939

Reprinted with permission of the American Institute of Steel Construction.

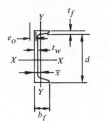

Channels
American Standard
Properties for Designing

			Flange				Axis X–X		
Designation	Area, A (in.²)	Depth, d (in.)	Width b_f (in.)	Average Thickness, t_f (in.)	Web Thickness, t_w (in.)	$\dfrac{d}{A_f}$	I (in.⁴)	Z (in.³)	r (in.)
C15 × 50	14.7	15.00	3.716	0.650	0.716	6.21	404	53.8	5.24
× 40	11.8	15.00	3.520	0.650	0.520	6.56	349	46.5	5.44
× 33.9	9.96	15.00	3.400	0.650	0.400	6.79	315	42.0	5.62
C12 × 30	8.82	12.00	3.170	0.501	0.510	7.55	162	27.0	4.29
× 25	7.35	12.00	3.047	0.501	0.387	7.85	144	24.1	4.43
× 20.7	6.09	12.00	2.942	0.501	0.282	8.13	129	21.5	4.61
C10 × 30	8.82	10.00	3.033	0.436	0.673	7.55	103	20.7	3.42
× 25	7.35	10.00	2.886	0.436	0.526	7.94	91.2	18.2	3.52
× 20	5.88	10.00	2.739	0.436	0.379	8.36	78.9	15.8	3.66
× 15.3	4.49	10.00	2.600	0.436	0.240	8.81	67.4	13.5	3.87
C9 × 20	5.88	9.00	2.648	0.413	0.448	8.22	60.9	13.5	3.22
× 15	4.41	9.00	2.485	0.413	0.285	8.76	51.0	11.3	3.40
× 13.4	3.94	9.00	2.433	0.413	0.233	8.95	47.9	10.6	3.48
C8 × 18.75	5.51	8.00	2.527	0.390	0.487	8.12	44.0	11.0	2.82
× 13.75	4.04	8.00	2.343	0.390	0.303	8.75	36.1	9.03	2.99
× 11.5	3.38	8.00	2.260	0.390	0.220	9.08	32.6	8.14	3.11
C7 × 14.75	4.33	7.00	2.299	0.366	0.419	8.31	27.2	7.78	2.51
× 12.25	3.60	7.00	2.194	0.366	0.314	8.71	24.2	6.93	2.60
× 9.8	2.87	7.00	2.090	0.366	0.210	9.14	21.3	6.08	2.72
C6 × 13	3.83	6.00	2.157	0.343	0.437	8.10	17.4	5.80	2.13
× 10.5	3.09	6.00	2.034	0.343	0.314	8.59	15.2	5.06	2.22
× 8.2	2.40	6.00	1.920	0.343	0.200	9.10	13.1	4.38	2.34
C5 × 9	2.64	5.00	1.885	0.320	0.325	8.29	8.90	3.56	1.83
× 6.7	1.97	5.00	1.750	0.320	0.190	8.93	7.49	3.00	1.95
C4 × 7.25	2.13	4.00	1.721	0.296	0.321	7.84	4.59	2.29	1.47
× 5.4	1.59	4.00	1.584	0.296	0.184	8.52	3.85	1.93	1.56
C3 × 6	1.76	3.00	1.596	0.273	0.356	6.87	2.07	1.38	1.08
× 5	1.47	3.00	1.498	0.273	0.258	7.32	1.85	1.24	1.12
× 4.1	1.21	3.00	1.410	0.273	0.170	7.78	1.66	1.10	1.17

Reprinted with permission of the American Institute of Steel Construction.

Angles
Equal Legs
Properties for Designing

Size and Thickness (in.)	k (in.)	Weight per Foot (lb)	Area (in.^{2})	Axis $X\text{-}X$ and Axis $Y\text{-}Y$				Axis $Z\text{-}Z$
				I (in.^{4})	Z (in.^{3})	r (in.)	x or y (in.)	r (in.)
L8 × 8 × $1\frac{1}{8}$	$1\frac{3}{4}$	56.9	16.7	98.0	17.5	2.42	2.41	1.56
1	$1\frac{5}{8}$	51.0	15.0	89.0	15.8	2.44	2.37	1.56
$\frac{7}{8}$	$1\frac{1}{2}$	45.0	13.2	79.6	14.0	2.45	2.32	1.57
$\frac{3}{4}$	$1\frac{3}{8}$	38.9	11.4	69.7	12.2	2.47	2.28	1.58
$\frac{5}{8}$	$1\frac{1}{4}$	32.7	9.61	59.4	10.3	2.49	2.23	1.58
$\frac{9}{16}$	$1\frac{3}{16}$	29.6	8.68	54.1	9.34	2.50	2.21	1.59
$\frac{1}{2}$	$1\frac{1}{8}$	26.4	7.75	48.6	8.36	2.50	2.19	1.59
L6 × 6 × 1	$1\frac{1}{2}$	37.4	11.0	35.5	8.57	1.80	1.86	1.17
$\frac{7}{8}$	$1\frac{3}{8}$	33.1	9.73	31.9	7.63	1.81	1.82	1.17
$\frac{3}{4}$	$1\frac{1}{4}$	28.7	8.44	28.2	6.66	1.83	1.78	1.17
$\frac{5}{8}$	$1\frac{1}{8}$	24.2	7.11	24.2	5.66	1.84	1.73	1.18
$\frac{9}{16}$	$1\frac{1}{16}$	21.9	6.43	22.1	5.14	1.85	1.71	1.18
$\frac{1}{2}$	1	19.6	5.75	19.9	4.61	1.86	1.68	1.18
$\frac{7}{16}$	$\frac{15}{16}$	17.2	5.06	17.7	4.08	1.87	1.66	1.19
$\frac{3}{8}$	$\frac{7}{8}$	14.9	4.36	15.4	3.53	1.88	1.64	1.19
$\frac{5}{16}$	$\frac{13}{16}$	12.4	3.65	13.0	2.97	1.89	1.62	1.20
L5 × 5 × $\frac{7}{8}$	$1\frac{3}{8}$	27.2	7.98	17.8	5.17	1.49	1.57	.973
$\frac{3}{4}$	$1\frac{1}{4}$	23.6	6.94	15.7	4.53	1.51	1.52	.975
$\frac{5}{8}$	$1\frac{1}{8}$	20.0	5.86	13.6	3.86	1.52	1.48	.978
$\frac{1}{2}$	1	16.2	4.75	11.3	3.16	1.54	1.43	.983
$\frac{7}{16}$	$\frac{15}{16}$	14.3	4.18	10.0	2.79	1.55	1.41	.986
$\frac{3}{8}$	$\frac{7}{8}$	12.3	3.61	8.74	2.42	1.56	1.39	.990
$\frac{5}{16}$	$\frac{13}{16}$	10.3	3.03	7.42	2.04	1.57	1.37	.994
L4 × 4 × $\frac{3}{4}$	$1\frac{1}{8}$	18.5	5.44	7.67	2.81	1.19	1.27	.778
$\frac{5}{8}$	1	15.7	4.61	6.66	2.40	1.20	1.23	.779
$\frac{1}{2}$	$\frac{7}{8}$	12.8	3.75	5.56	1.97	1.22	1.18	.782
$\frac{7}{16}$	$\frac{13}{16}$	11.3	3.31	4.97	1.75	1.23	1.16	.785
$\frac{3}{8}$	$\frac{3}{4}$	9.8	2.86	4.36	1.52	1.23	1.14	.788
$\frac{5}{16}$	$\frac{11}{16}$	8.2	2.40	3.71	1.29	1.24	1.12	.791
$\frac{1}{4}$	$\frac{5}{8}$	6.6	1.94	3.04	1.05	1.25	1.09	.795

Reprinted with permission of the American Institute of Steel Construction.

Plastic Design Selection Table
For Shapes Used as Beams or Columns

Z_x

Z_x	Shape	A	$\dfrac{d}{t_w}$	r_x	r_y	$F_y = 36$ ksi		$F_y = 50$ ksi	
						M_p	P_y	M_p	P_y
(in.³)		(in.²)		(in.)	(in.)	(k-ft)	(k)	(k-ft)	(k)
253	W24 × 94	27.7	47.1	9.86	1.98	759	* 997	1050	*1390
248	W18 × 114	33.5	31.1	7.79	2.86	744	1210	1030	1680
244	W27 × 84	24.8	57.6	10.7	2.06	732	* 893	–	–
243	W14 × 136	40.0	22.3	6.31	3.77	729	1440	1010	2000
240	S24 × 100	29.4	32.1	9.01	1.27	720	1060	1000	1470
227	W21 × 96	28.3	36.8	8.61	2.02	681	1020	946	*1420
227	W18 × 105	30.9	33.1	7.75	2.84	681	1110	946	1550
226	W14 × 127	37.3	24.0	6.29	3.76	678	1340	–	–
224	W24 × 84	24.7	51.3	9.79	1.95	672	* 889	933	*1240
222	S24 × 90	26.5	38.5	9.22	1.30	666	954	925	*1330
211	W14 × 119	35.0	25.4	6.26	3.75	633	1260	–	–
210	W12 × 133	39.1	17.7	5.59	3.16	630	1410	875	1960
206	W18 × 96	28.2	35.5	7.70	2.82	618	1020	–	
205	S24 × 79.9	23.5	47.9	9.47	1.34	615	* 846	854	*1180
201	W24 × 76	22.4	54.3	9.69	1.92	603	* 806	838	*1120
196	W14 × 111	32.7	26.6	6.23	3.73	588	1180		
194	S20 × 95	27.9	25.0	7.60	1.33	582	1000	808	1400
192	W21 × 82	24.2	41.8	8.53	1.99	576	871	800	*1210
186	W16 × 96	28.2	30.5	6.93	2.82	558	1020	775	1410
186	W12 × 120	35.3	18.5	5.51	3.13	558	1270	775	1770
179	S20 × 85	25.0	30.6	7.79	1.36	537	900	746	1250
178	W18 × 85	25.0	34.8	7.57	2.05	534	900	742	1250
176	W24 × 68	20.0	57.0	9.53	1.87	528	* 720	–	–
172	W21 × 73	21.5	46.7	8.64	1.81	516	* 774	717	*1080
169	W16 × 88	25.9	32.1	6.87	2.79	507	932	–	
164	W12 × 106	31.2	20.8	5.46	3.11	492	1120	683	1560
161	W18 × 77	22.7	38.2	7.54	2.04	483	817	671	*1140
160	W21 × 68	20.0	49.1	8.60	1.80	480	* 720	667	*1000
153	S20 × 75	22.1	31.2	7.60	1.16	459	796	638	1110
152	W24 × 61	18.0	56.6	9.25	1.38	456	* 648	633	* 900
152	W12 × 99	29.1	21.9	5.43	3.09	456	1050	633	1460
148	W10 × 112	32.9	15.1	4.67	2.67	444	1180	617	1650
146	W16 × 78	23.0	30.9	6.75	2.01	438	828	608	1150
145	W18 × 70	20.6	41.1	7.50	2.02	435	742	604	*1030
145	W14 × 84	24.7	31.4	6.13	3.02	435	889	–	–
144	W21 × 62	18.3	52.5	8.54	1.77	432	* 659	600	* 915
140	W12 × 92	27.1	23.2	5.40	3.08	420	976	–	–
138	S20 × 65.4	19.2	40.0	7.84	1.19	414	691	575	* 960

*Check shape for compliance with Formulas (2.7-1a) or (2.7-1b), Section 2.7, AISC Specification, as applicable, when subjected to combined axial force and bending moment at ultimate loading.

Reprinted with permission of the American Institute of Steel Construction.

G SI Units

Selection of Derived Units and Conversion Factors

Quantity	Symbol	English to SI
Area	m^2	1 ft^2 = 0.0929 m^2
Volume	m^3	1 ft^3 = 0.0283 m^3
Moment of inertia	m^4	1 $in.^4$ = 0.4162 × 10^{-6} m^4
Section modulus	m^3	1 $in.^3$ = 16.39 × 10^{-6} m^3
Density	kg/m^3	1 lb/ft^3 = 16.03 kg/m^3
Force	N (Newton)	1 lb = 4.448 N
Moment or torque	N · m	1 lb-ft = 1.356 N · m
Stress or pressure	N/m^2 (Pascal)	1 psi = 0.006895 MPa
Energy or work	m · N	1 ft-lb = 1.356 m · N
Coefficient of thermal expansion	$m/m/°C$	1 $in./in./°F$ = 1.8 $m/m/°C$

Fundamental Units in Mechanics

Quantity	Unit	Symbol
Mass	kilogram	kg
Length	meter	m
Time	second	s

SI Unit Prefixes

Multiplication Factor	Prefix	Symbol	Pronunciation (U.S.)	Meaning (U.S.)
1 000 000 000 = 10^9	giga	G	jig′ a (a as in about)	One billion times
1 000 000 = 10^6	mega	M	as in mega-phone	One million times
1000 = 10^3	kilo	k	as in kilo-watt	One thousand times
100 = 10^2	hecto	h	heck′ toe	One hundred times
10 = 10^1	deka	da	deck′ a (a as in about)	Ten times
0.1 = 10^{-1}	deci	d	as in decimal	One tenth of
0.01 = 10^{-2}	centi	c	as in senti-ment	One hundredth of
0.001 = 10^{-3}	milli	m	as in mili-tary	One thousandth of
0.000 001 = 10^{-6}	micro	μ	as in micro-phone	One millionth of
0.000 000 001 = 10^{-9}	nano	n	nan′ oh (an as in ant)	One billionth of

Answers to Even-Numbered Problems

(*In some cases, partial answers are given.*)

Chapter 1

1.2   Maximum $F = 25$ lb at $y = 8$ ft

1.4   Maximum $F = 390$ lb at $y = 17$ ft

1.6   Maximum $F = 40$ kN, x from 6 to 8 m

1.8 $F = 102y$ (origin at bottom for y)

1.10 $F = 15y$ (origin at bottom for y)

1.12   Maximum $T = 10$ k-in., between 0 and 10 in. and 25 and 40 in.

1.14   Maximum $T = 7.5$ k-ft between 5 and 7 ft

1.16   Maximum $T = 650$ k-in. at $x = 90$ in.

1.18   Maximum $T = 127.3$ N-m at $x = 4$ m

1.20   Maximum $T = 4$ k-ft between 0 and 4 ft

1.22   Maximum $T = 30$ k-in. between 0 and 20 in.

1.24   4 ft from left end: $V = 8.75$ k, $M = 35$ k-ft; 9 ft from left end: $V = 3.75$ k, $M = 58.75$ k-ft; 15 ft from left end: $V = -6.25$ k, $M = 56.25$ k-ft

1.26 $V = 15$ k, $M = 67.5$ k-ft

1.28   Sec. A-A: $V = 3$ k, $M = 29$ k-ft; Sec. B-B: $V = -3$ k, $M = 29$ k-ft

1.30 $V = -1.8$ k, $M = -42.8$ k-ft

1.32 $V = 5.57$ k, $M = 47.8$ k-ft

1.34   Sec. A-A: $V = -16$ lb, $M = -10.7$ lb-ft

1.36   Maximum $V = 5460$ N at left end; maximum $M = 12420$ N-m, 4.55 m from left end

1.38 x from 0 to 6: $V = 24$, $M = 24x$; x from 6 to 10: $V = -36$, $M = 360 - 36x$

1.40   Maximum $V = 45$ k, 20 ft from left end; $M = -425$ k-ft, 20 ft from left end

1.42   Maximum $V = 800$ N, 2 m from left end; maximum $M = 1400$ N-m, 4 m from left end

1.44   1 m to left of B: $V = 0$, $M = 2000$ N-m; 3 m to right of B: $V = 0$, $M = 2000$ N-m

1.46   Maximum $V = 2P$, 4-ft segment at left end; maximum $M = 14.7P$, 8 ft from left end

1.48 $V = (-w_B/(2L))x^2$; $M = (w_B/(6L))x^3 - (w_B/2)L^2$

1.50 $V = -1.25Q$ $(0 < x < L)$; $V = -0.25Q$ $(L < x < 4L)$; maximum $M = 0.25QL$, just to the right of $x = 3L$

1.52 $V = -5P$ $(0 < x < 2L)$; $M = -5Px$ $(0 \leq x \leq 2L)$; $V = 0$ $(2L < x < 3L)$; $M = -10PL$ $(2L \leq x \leq 3L)$

1.54 $V = 2.29$ wb $(4b < x < 8b)$; maximum $M = 1.16$ wb^2, $x = 8b$

1.56 $F = 2$ k; maximum $T = 15$ k-in $(0 < x < 16$ in.$)$;

1.58 $F = 600$ lb, $V = 200$ lb $(0 < x < 16$ ft$)$; maximum $M = 3200$ lb-ft at $x = 16$ ft

1.60   Maximum values: $F = 4$ k, $V = 9.8$ k; maximum $M = 39.2$ k-ft

1.62 $F = 20$ k, $V = -2$ k; maximum $M = 20$ k-ft

1.64   Maximum values: $F = 9$ k, $V = 4$ k, $M = 52.3$ k-ft

1.66   Maximum values: $F = 6$ k, $V = 12.5$ k, $M = 50$ k-ft

1.68   Replace 4 k left by 5.6 k left; maximum values: $F = 5.6$ k, $V = 5.86$ k, $M = 46.9$ k-ft

Chapter 2

2.2 $\sigma = -50$ MN/m^2; $\tau = 30$ MN/m^2

2.4 $\sigma = -59.7$ MPa; $\tau = 28.1$ MPa

2.6 $\sigma = -43.0$ MPa; $\tau = 0.7$ MPa

2.8 $\sigma_1 = \sigma_2 = 150$ MN/m^2; $\sigma_3 = 0$; $|\tau_{max}| = 75$ MN/m^2

2.10 $\sigma_1 = 146.9$ MPa; $\sigma_2 = 23.2$ MPa; $\sigma_3 = 0$; $\tau_1 = -\tau_2 = 61.9$ MPa; $|\tau_{max}| = 73.5$ MPa

2.12 $\sigma_1 = 166.5$ MN/m^2; $\sigma_2 = 0$; $\sigma_3 = -86.5$ MN/m^2; $\tau_1 = -\tau_2 = 126.5$ MN/m^2; $|\tau_{max}| = 126.5$ MN/m^2

2.14 $\sigma_1 = 40$ MN/m^2; $\sigma_2 = 0$; $\sigma_3 = -160$ MN/m^2

2.16 $\gamma_{nt} = -160 \times 10^{-6}$

2.18 $\gamma_{nt} = 0$

2.20 $\gamma_{nt} = -400 \times 10^{-6}$

2.22 $\varepsilon_n = -150 \times 10^{-6}$; $\varepsilon_t = 350 \times 10^{-6}$; $\gamma_{nt} = -600 \times 10^{-6}$

2.24 $\varepsilon_x = 537 \times 10^{-6}$; $\varepsilon_y = -337 \times 10^{-6}$; $\gamma_{xy} = -1112 \times 10^{-6}$

2.26 $\varepsilon_x = 750 \times 10^{-6}$; $\varepsilon_y = -50 \times 10^{-6}$; $\gamma_{xy} = 300 \times 10^{-6}$

2.28 $\alpha = 42.0°\circlearrowright$ or $7.0° \circlearrowleft$

2.30 $\varepsilon_1 = 780 \times 10^{-6}$, $\varepsilon_2 = 0$; $\varepsilon_3 = -80 \times 10^{-6}$; $\gamma_1 = -\gamma_2 = 860 \times 10^{-6}$; $|\gamma_{max}| = 860 \times 10^{-6}$

2.32 $\varepsilon_1 = 0$, $\varepsilon_2 = -120 \times 10^{-6}$; $\varepsilon_3 = -480 \times 10^{-6}$; $\gamma_1 = -\gamma_2 = 361 \times 10^{-6}$; $|\gamma_{max}| = 480 \times 10^{-6}$

2.34 $\varepsilon_1 = 500 \times 10^{-6}$; $\varepsilon_2 = 0$; $\varepsilon_3 = -800 \times 10^{-6}$; $\gamma_1 = -\gamma_2 = 1300 \times 10^{-6}$; $|_{max}| = 1300 \times 10^{-6}$

2.36(a) $\gamma_{xy} = 1200 \times 10^{-6}$ **(b)** $\varepsilon_1 = 400 \times 10^{-6}$; $\varepsilon_2 = 0$; $\varepsilon_3 = -1600 \times 10^{-6}$ **(c)** $|\gamma_{max}| = 2000 \times 10^{-6}$

2.38 $\varepsilon_1 = 800 \times 10^{-6}$; $\varepsilon_3 = -400 \times 10^{-6}$; $\gamma_1 = -\gamma_2 = 1200 \times 10^{-6}$

2.40 $\varepsilon_1 = 812 \times 10^{-6}$; $\varepsilon_3 = -12 \times 10^{-6}$; $\gamma_1 = -\gamma_2 = 825 \times 10^{-6}$

2.42 $\varepsilon_1 = 224 \times 10^{-6}$; $\varepsilon_3 = -357 \times 10^{-6}$; $\gamma_1 = -\gamma_2 = 581 \times 10^{-6}$

2.44 $\varepsilon_x = 2\varepsilon_a - 2\varepsilon_b + \varepsilon_c$; $\varepsilon_y = \varepsilon_c$; $\gamma_{xy} = \left(\dfrac{2}{\sqrt{3}}\right)(\varepsilon_a - 3\varepsilon_b + 2\varepsilon_c)$

2.46(a) $E = 187.5 \text{ GN/m}^2$ **(b)** $\mu = 0.313$ **(c)** $|\tau_{max}| = 75 \text{ MPa}$
(d) $|\gamma_{max}| = 1050 \times 10^{-6}$

2.48(a) $\sigma_1 = 24.28 \text{ MPa}$; $\sigma_2 = 0$; $\sigma_3 - -34.58 \text{ MPa}$
(b) $|\tau_{max}| = 29.43 \text{ MPa}$

2.50(a) $\sigma_1 = 6.90 \text{ MPa}$; $\sigma_2 = 0$; $\sigma_3 = -37.76 \text{ MPa}$
(b) $|\tau_{max}| = 22.33 \text{ MPa}$

2.52(a) $\sigma_1 = 22,530 \text{ psi}$; $\sigma_3 = 0$; $\sigma_2 = 10,810 \text{ psi}$ **(b)** $|\tau_{max}| = 11,260 \text{ psi}$

2.54(a) $\varepsilon_1 = 813 \times 10^{-6}$; $\varepsilon_2 = 500 \times 10^{-6}$; $\varepsilon_3 = -438 \times 10^{-6}$
(b) $\gamma_1 = -\gamma_2 = 313 \times 10^{-6}$ **(c)** $|\gamma_{max}| = 1251 \times 10^{-6}$

2.56(a) $\varepsilon_1 = 313 \times 10^{-6}$; $\varepsilon_2 = 176 \times 10^{-6}$; $\varepsilon_3 = -1113 \times 10^{-6}$
(b) $\gamma_1 = -\gamma_2 = 1289 \times 10^{-6}$ **(c)** $|\gamma_{max}| = 1426 \times 10^{-6}$

2.58(a) $\sigma_x = 11.4 \text{ MPa}$; $\sigma_y = -56.2 \text{ MPa}$; $\tau_{xy} = -37.6 \text{ MPa}$
(b) $\sigma_1 = 28.2 \text{ MPa}$; $\sigma_2 = 0$; $\sigma_3 = -73.0 \text{ MPa}$ **(c)** $\tau_1 = -\tau_2 = 50.6 \text{ MPa}$
(d) $|\tau_{max}| = 50.6 \text{ MPa}$

2.60 $\sigma_1 = 7846 \text{ psi}$; $\sigma_2 = 0$; $\sigma_3 = -1846 \text{ psi}$; $|\tau_{max}| = 4846 \text{ psi}$

2.62 $\sigma_1 = 9330 \text{ psi}$; $\sigma_3 = 0$; $\sigma_2 = 2670 \text{ psi}$; $|\tau_{max}| - 4665 \text{ psi}$

2.64 $\sigma_1 = 28.8 \text{ MPa}$; $\sigma_2 = 0$; $\sigma_3 = -64.2 \text{ MPa}$; $|\tau_{max}| = 46.5 \text{ MPa}$

Chapter 3

3.2(a) $\sigma_{CD} = 100 \text{ MN/m}^2$ **(b)** $|\tau_{max}| = 50 \text{ MN/m}^2$
(c) $\delta_{BC} = 5.48 \times 10^{-4} \text{ m}$ **(d)** $\delta = 7.12 \times 10^{-3} \text{ m}$

3.4 $(\sigma_1)_{max} = 655.7$ psi; $(|\tau_{max}|)_{max} = 327.8$ psi; $\delta = 5.72 \times 10^{-3}$ in.

3.6(a) $\sigma_1 = 21.24$ ksi **(b)** $|\tau_{max}| = 10.62$ ksi **(c)** $\tau_B = 84.95$ ksi
(d) $\tau_D = 73.26$ ksi

3.8(a) $\sigma_{AB} = 35.35$ MPa (T); $\sigma_{DE} = 50.00$ MPa (T); $\sigma_{CF} = 25.00$ MPa (C);
$\sigma_{BE} = 0$

3.10 $\sigma_{AB} = 94.68$ MPa (T); $\delta_{AB} = 0.0047$ m; $\sigma_{CD} = 92.28$ MPa (T);
$\delta_{CD} = 0.0040$ m

3.12(a) $\sigma_b = 11.32$ ksi (T) **(b)** $\delta = 8.49 \times 10^{-3}$ in.

3.14(a) $\mu = 0.333$ **(b)** $w_{final} = 9.99667 \times 10^{-3}$ m
(c) $\sigma = 200$ MPa (T) **(d)** $|\tau_{max}| = 100$ MPa
(e) $E = 200 \times 10^9$ N/m²

3.16(a) $E = 10.19 \times 10^6$ psi **(b)** $\mu = 0.32$ **(c)** M.R. $= 31.83$ lb-in./in.³

3.18 $\sigma_u = 101,860$ psi; $\sigma_f = 84,030$ psi

3.20(a) Percent reduction of area $= 50.4$ **(b)** Percent elongation $= 22.5$

Chapter 4

4.2 AB: 20 k-ft, 5660 psi; BC: 16 k-ft, 1910 psi

4.4 AB: 199 MPa; BC: 102 MPa; CD: 199 MPa; 0.162 rad

4.6 BC: 62.2 MPa; AC: 0.0194 rad; BC: 0.0194 rad

4.8 117 MPa; DA: 0.0 rad

4.10 AB: 9550 psi; BC: 0.0 psi; CD: 7824 psi; DA: 0.0158 rad ccw, observer looking D to A

4.12 1430 psi

4.14 $\sigma_1 = 5660$; $\sigma_2 = 0$; $\sigma_3 = -5660$ psi

4.16 $\sigma_1 = 102$ MPa; $\sigma_2 = 0$; $\sigma_3 = -102$ MPa

4.18 $\sigma_1 = 62.2$ MPa; $\sigma_2 = 0$; $\sigma_3 = -62.2$ MPa

4.20 Outside: $\sigma_1 = 117$ MPa, $\sigma_2 = 0$, $\sigma_3 = -117$ MPa; inside: $\sigma_1 = 70.2$ MPa, $\sigma_2 = 0$, $\sigma_3 = -70.2$ MPa

4.22 AB: $\sigma_1 = 9550$ psi, $\sigma_2 = 0$, $\sigma_3 = -9550$ psi, $\varepsilon_1 = 4.26 \times 10^{-4}$, $\varepsilon_2 = 0$,
$\varepsilon_3 = -4.26 \times 10^{-4}$; BC: zero torque. Stresses and strains vanish.

4.24 $\sigma_1 = 716$, $\sigma_2 = 0$, $\sigma_3 = -716$ psi; $\varepsilon_1 = 2.98 \times 10^{-5}$, $\varepsilon_2 = 0$,
$\varepsilon_3 = -2.98 \times 10^{-5}$

4.26 $\tau_{max} = 9860$ psi. AB: 0.189 rad ccw, observer looks A to B; BC: 0.237 rad cw, observer looks B to C.

4.28 $\tau_{max} = 2675$ psi. AB: 0.0241 rad ccw, observer looks A to B; BC: 0.0481 rad cw, observer looks B to C.

4.30 $\tau_{max} = \sigma_1 = 5707$ psi (AB). AB: 0.073 rad cw, observer looks A to B; BC: 0.0197 rad ccw, observer looks B to C. CD: 0.00493 rad ccw, observer looks C to D.

4.32 $\tau_{max} = \sigma_1 = 6340$ psi (CD). AB: 0.00704 rad cw, observer looks A to B; BC: 0.0321 rad ccw, observer looks B to C; CD: 0.0676 rad ccw, observer looks C to D.

4.34 $\tau_{max} = \sigma_1 = 43.2$ MPa. AB: 0.288 rad cw, observer looks A to B; BC: 0.288 rad ccw, observer looks B to C.

4.36 AB: 3.24 MPa, 0.00251 rad cw, observer looks B to A; BC: 1.88 MPa, 0.00161 rad ccw, observer looks C to B.

4.38 29.6 ksi, 0.147 rad

4.40 $\sigma_1 = 99.5$ MPa, $\sigma_2 = 0$, $\sigma_3 = -99.5$ MPa; $\varepsilon_1 = 0.000663$, $\varepsilon_2 = 0$, $\varepsilon_3 = -0.000663$

4.42 82.0 k-in , 20.0 ksi

4.44 15.0 ksi, 23.6 k-in.

4.46 6925 psi, 0.0350 rad, 693 lb/in.

4.48 2370 psi

4.50 17.1 MPa, 51,300 N/m

4.52 1050 psi, 0.00165 rad

4.54 $G = 12.0 \times 10^6$ psi, $u = 0.0664$ k-in./in.^{3}

4.56 $G = 85.8$ GPa, $u = 0.523$ MN-m/m^3

4.58 $G = 10.5 \times 10^6$ psi; $u = 0.112$ k-in./in.^{3} ; $\tau_u = 101$ ksi

Chapter 5

5.2 $I_X = 73.0$ in.^{4}; $I_Y = 23.0$ in.^{4}; $P_{XY} = 12.8$ in.^{4}

5.4 $A = 20$ in.^{2}; $I_x = I_y = 2700$ in.^{4}; $J_C = 5400$ in.^{4}; $J_0 = 7900$ in.^{4}

5.6 $I_u = 432$ in.^{4}; $I_v = 27$ in.^{4}

5.8 $I_u = 2.78 \times 10^{-6}$ m^4; $I_v = 2.08 \times 10^{-6}$ m^4

5.10 $I_u = 290.7$ in.^{4}; $I_v = 90.7$ in.^{4}

5.12 $I_{x'} = 1047.5$ in.^{4}; $I_{y'} = 657.0$ in.^{4}; $P_{x'y'} = -338.2$ in.^{4}

5.14 $r_u = 0.0433$ m; $r_v = 0.0433$ m; $r_C = 0.0612$ m

5.16 $r_u = 4.27$ in.; $r_v = 2.61$ in.; $r_C = 5.00$ in.

5.18 $r_u = 6.90 \times 10^{-2}$ m; $r_v = 3.46 \times 10^{-2}$ m; $r_C = 7.72 \times 10^{-2}$ m

5.20(a) $I_u = 23.49 \times 10^{-6}$ m^4; $I_v = 6.89 \times 10^{-6}$ m^4 **(b)** $r_u = 0.053$ m; $r_v = 0.029$ m

5.22(a) $I_u = 11.00 \times 10^{-6}$ m^4; $I_v = 8.56 \times 10^{-6}$ m^4 **(b)** $r_u = 0.035$ m; $r_v = 0.031$ m

5.24(a) $\sigma_x = 201.2 \times 10^6 v$ **(b)** $(\sigma_x)_{max} = 50.3$ MPa; tension on bottom, compression on top

5.26 $(\sigma_x)_{max} = 75$ ksi; tension on top, compression on bottom

5.28 $(\sigma_x)_{max} = 106.67$ MN/m^2; tension on top, compression on bottom

5.30 $(\sigma_x)_{max} = 178.69$ ksi; tension on top, compression on bottom

5.32 At 8.45 m from left end of beam: $(\sigma_x)_{max} = 101.14$ MPa (T) at bottom; $(\sigma_x)_{max} = 65.12$ MPa (C) at top

5.34 In 8-ft central section: $(\sigma_x)_{max} = 3.10$ ksi (T) at top; $(\sigma_x)_{max} = 2.48$ ksi (C) at bottom

5.36(a) Rotate section 32.9° $\circlearrowright$ **(b)** $(\sigma_{max})_{tension} = (\sigma_{max})_{compression} = 1319$ psi

5.38(a) $\tau_{vx} = 250$ psi **(b)** $\tau_{vx} = 333$ psi

5.40(a) $\tau_{vx} = 344$ psi **(b)** $\tau_{vx} = 363$ psi

5.42(a) $\tau_{vx} = 15.29$ MPa **(b)** $\tau_{vx} = 13.46$ MPa

5.44(a) $\tau_{vx} = 391$ psi **(b)** $\tau_{vx} = 521$ psi

5.46(a) $\tau_{vx} = 860$ psi **(b)** $\tau_{vx} = 908$ psi

5.48(a) $\tau_{vx} = 6.12$ MPa **(b)** $\tau_{vx} = 5.30$ MPa

5.50 $(\tau_{xv})_{N.A.} = \dfrac{4}{3}\left(\dfrac{V}{A}\right)$

5.52(a) $\tau_{xv} = 3.95$ MPa **(b)** $(\tau_{vx})_{max} = 4.43$ MPa
(c) $(\tau_{vx})_{max} = 6.86$ MPa

5.54(a) $\tau_{xv} = 2.240$ ksi **(b)** $(\tau_{xv})_{max} = 2.32$ ksi **(c)** $(\tau_{vx})_{max} = 2.32$ ksi

5.56(a) $\tau_{xv} = 19$ psi **(b)** $(\tau_{vx})_{max} = 21$ psi **(c)** $(\tau_{vx})_{max} = 150$ psi

5.58 $\sigma_1 = 17.18$ MPa; $\sigma_2 = 0$; $\sigma_3 = -0.02$ MPa; $|\tau_{max}| = 8.60$ MPa

5.60 $(\sigma_{max})_{tension} = (\sigma_{max})_{compression} = 14{,}030$ psi

5.62 $|\sigma_{max}|_{tension} = 167.50$ MPa; $|\sigma_{max}|_{compression} = 196.90$ MPa

5.64 $|\sigma_{max}|_{tension} = 193.70$ MPa; $|\sigma_{max}|_{compression} = 169.30$ MPa

5.66 $|\sigma_{max}|_{tension} = |\sigma_{max}|_{compression} = 1600$ psi

5.68 $|\sigma_{max}|_{tension} = 3.76$ ksi; $|\sigma_{max}|_{compression} = 4.29$ ksi

Chapter 6

6.2 3000 in.

6.4 $EI_u v_1 = (P/18)x^3 - (\frac{4}{9})PL^2 x$; $EI_u v_2 = (P/18)x^3 - (P/6)(x - 2L)^2 - (\frac{4}{9})PL^2 x$. At load: $v = -(\frac{4}{9})PL^3/(EI_u)$

6.6 Slope at center $= M_1 L/(12EI_u)$

6.8 Maximum $v = -0.06415\, M_1 L^2/(EI_u)$ at $x = L/\sqrt{3}$

6.10 Maximum $v = 0.00652 wL^4/(EI_u)$ at $x = 0.52L$

6.12 Maximum $v = -0.0866 wL^4/(EI_u)$ at $x = 0.843L$; $v_C = -wL^4/(8EI_u)$

6.14 $EI_u v = \dfrac{-16w_1 L^4}{\pi^4}\cos\dfrac{\pi x}{2L} + \dfrac{C_1 x^3}{6} + \dfrac{C_2 x^2}{2} + C_3 x + C_4$, where $C_1 = \dfrac{2w_1 L}{\pi}$,

$C_2 = \dfrac{-2w_1 L^2}{\pi^2}$, $C_3 = 0$, and $C_4 = \dfrac{16L^4}{\pi^4}w_1$

6.16 $EI_u v$ as for 6.14 with the following values for constants: $C_1 = \dfrac{4w_1 L}{\pi^2}$,

$C_2 = \dfrac{-4w_1 L}{\pi^2}$, $C_3 = \dfrac{4w_1 L^3}{3\pi^2} - \dfrac{16w_1 L^3}{\pi^4}$, $C_4 = \dfrac{16w_1 L^4}{\pi^4}$

6.18 $EI_u v = \dfrac{-w_1 x^5}{120L} + C_1 \dfrac{x^3}{6} + C_2 \dfrac{x^2}{2} + C_3 x + C_4$, where $C_1 = 0$, $C_2 = 0$,

$C_3 = \dfrac{1}{24}w_1 L^3$, and $C_4 = \dfrac{-w_1 L^4}{30}$

6.20 $EI_u v = \dfrac{-w_1 x^4}{24} - \dfrac{w_2 x^5}{120L} + \dfrac{w_1 x^5}{120L} + C_1 \dfrac{x^3}{6} + C_2 \dfrac{x^2}{2} + C_3 x + C_4$, where

$C_1 = \dfrac{w_1 L}{3} + \dfrac{w_2 L}{6}$, $C_2 = 0$, $C_3 = \dfrac{-w_1 L^3}{45} - \dfrac{7w_2 L^3}{360}$, and $C_4 = 0$

6.22 $EI_u v = \dfrac{-w_1 x^7}{840L^3} + \dfrac{C_1}{6}x^3 + C_2 \dfrac{x^2}{2} + C_3 x + C_4$, where $C_1 = 0$, $C_2 = 0$,

$C_3 = \dfrac{w_1 L^3}{120}$, and $C_4 = \dfrac{-w_1 L^4}{140}$

6.24 $EI_u v = \dfrac{-w_1 x^4}{24} + \dfrac{w_1 L^4}{\pi^4}\sin\dfrac{\pi x}{L} + \dfrac{C_1}{6}x^3 + \dfrac{C_2}{2}x^2 + C_3 x + C_4$, where

$C_1 = \dfrac{w_1 L}{2}$, $C_2 = 0$, $C_3 = \dfrac{-w_1 L^3}{24}$, and $C_4 = 0$

6.26 0.0208 m, up

6.28   0.259 in., down

6.30   0.0192 m, down; 0.010 rad, cw at left end

6.32   0.0954 m, down; 0.023 rad, cw at B

6.34   0.0653 in., down; 0.001296 rad, cw

6.36   0.435 in., down

6.38   See Table 6.1, case 2

6.40   See Table 6.1, case 4

6.42   See Table 6.1, case 6

6.44   At D: $-PL^3/(16EI_u)$; at B: $-PL^3/(96EI_u)$

6.46 $v_B = \dfrac{M_B a}{EI_u}\left(\dfrac{-b^2}{2L} + \dfrac{L}{6} + \dfrac{a^2}{6L}\right)$

6.48   Maximum deflection (at $x = 1.5a$) $= -9\, Pa^3/(8EI_u)$; deflection (at D) $= -2\, Pa^3/(EI_u)$

6.50   Maximum deflection (at $x = 1.4724a$) $= 0.740\, wa^4/(EI_u)$, where x is measured from D to E; deflection at $C = 47\, wa^4/(72EI_u)$

6.52 $-91Pa^3/(6EI_1)$

6.54 $-319\, wa^4/(16EI_1)$

6.56 $-65\, Pa^3/(18EI_1)$

6.58   Rotation at $A = PL^2/(24EI_u)$, cw

6.60   Rotation at $B = M_B(a^3 + b^3)/(3EI_u L^2)$, cw, where $L = a + b$

6.62 $11\, Pa^3/(6EI_u)$, down

6.64 $47\, wa^4/(72EI_u)$, down

6.66 $\dfrac{v_c'}{v_c} = \dfrac{4}{15}\left(\dfrac{L}{h}\right)^2$

6.68 $\dfrac{v_c'}{v_c} = \dfrac{4}{5}\left(\dfrac{L}{h}\right)^2$

6.70   0.186 in.

6.72   zero

6.74   0.0201 in.

6.76   0.753 in.

Chapter 7

7.2(a) $\sigma_1 = 22.84$ MPa; $\sigma_2 = 0$; $\sigma_3 = -2.62$ MPa; $|\tau_{max}| = 12.73$ MPa
(b) $\delta = 5.05 \times 10^{-4}$ m; $\theta = 0.0065$ rad

7.4(a) $\sigma_1 = 18.37$ ksi; $\sigma_2 = 0$; $\sigma_3 = -8.82$ ksi; $|\tau_{max}| = 13.60$ ksi
(b) $\delta = 0.019$ in.; $\theta = 0.0118$ rad

7.6 $P = 398.1$ kN

7.8 $T = 36.5$ k-in.

7.10 $\sigma_1 = 108.2$ MPa (T) at bottom of section

7.12 $P = 271.6$ kN

7.14 $\sigma_1 = 112.0$ MPa; $\sigma_2 = \sigma_3 = 0$; $|\tau_{max}| = 56.0$ MPa

7.16 $\sigma_1 = 0.21$ MPa; $\sigma_2 = 0$; $\sigma_3 = -135.0$ MPa; $|\tau_{max}| = 67.6$ MPa

7.18 $\sigma_1 = 63.2$ ksi; $\sigma_2 = \sigma_3 = 0$; $|\tau_{max}| = 31.6$ ksi

7.20 $\sigma_1 = 0.83$ ksi; $\sigma_2 = 0$; $\sigma_3 = -1.76$ ksi; $|\tau_{max}| = 1.29$ ksi

7.22 CW from lower left corner of section: $\sigma_A = 1.30$ ksi; $\sigma_B = -0.26$ ksi; $\sigma_C = -2.34$ ksi; $\sigma_D = -0.78$ ksi

7.24 $F = 44,120$ lb

7.26 $\sigma_1 = 50.34$ MPa; $\sigma_2 = 0$; $\sigma_3 = -1.08$ MPa; $|\tau_{max}| = 25.71$ MPa

7.28 $\sigma_1 = 74.63$ MPa; $\sigma_2 = 0$; $\sigma_3 = -0.73$ MPa; $|\tau_{max}| = 37.68$ MPa

7.30 $\sigma_1 = 3911$ psi; $\sigma_2 = \sigma_3 = 0$; $|\tau_{max}| = 1956$ psi

7.32 Diameter $= 0.087$ m

7.34 Diameter $= 8.53$ in.

7.36 $\sigma_1 = 283.22$ MPa; $\sigma_2 = 0$; $\sigma_3 = -12.58$ MPa; $|\tau_{max}| = 147.90$ MPa

7.38 $P = 97,370$ lb; $F = 13,910$ lb

7.40 $P = 55,410$ lb; $F = 7.915$ lb

7.42 $\sigma_1 = 1.03$ MPa; $\sigma_2 = 0$; $\sigma_3 = -52.67$ MPa; $|\tau_{max}| = 26.85$ MPa

7.44 $\sigma_1 = 4472$ psi; $\sigma_2 = 0$; $\sigma_3 = -1924$ psi; $|\tau_{max}| = 3198$ psi

7.46 $\sigma_1 = \sigma_2 = 0$; $\sigma_3 = -1364$ psi; $|\tau_{max}| = 682$ psi

7.48 Diameter $= 5.50$ in.

7.50 Diameter $= 2.67$ in.

7.52 $T = 10.77$ kN-m

7.54 Diameter $= 3.96$ in.

7.56 Diameter $= 3.64$ in.

7.58 $T = 5.39$ kN-m

Chapter 8

8.2 See Example 8.2.

8.4 Critical $W = 4\,k/L$

8.6(a) 1324 k   **(c)** 32.1 ft

8.8(a) 533 k   **(c)** 18.6 ft

8.10 $(L/r, \sigma_{CR})$: (88.86, 25.0); (100, 19.74); (150, 8.77); (200, 4.93); (250, 3.16)

8.12 12.42 in.

8.14 7.17 in.

8.16 166.3 k, 195.8 in. < 264 in.

8.18 $L_e = 2L$

8.20 See Eq. 8.31.

8.22 $P/P_e = 0.8$, $\delta = 1.770$

8.24 0.188 in.

8.26 $\dfrac{d^2u}{dx^2} + k^2(u + u_0 + e) = 0$

8.28 139 k

8.30 287 k

8.32 2.89 MN

8.34 30.7 kN

8.36 22.8 k

8.38 Hollow square: 31.1 k
Hollow circle: 28.8 k

8.40(c) $(L_e/r_v, P/A)$ ksi; (4.05, 35); (35.7, 25); (57.4, 15); (99.35, 5)

8.42(c) $(L_e/r_v, P/A)$ ksi; (0.24, 40.0); (4.32, 30.0); (43.75, 20.0); (88.86, 5.0)

8.44 52.8 k

8.46 1210 lb

8.48(a) $D_0 = 4.68$ in.   **(b)** $D_0 = 4.59$ in.

8.50 $\sigma = 10,000\,\varepsilon$ $(\sigma \leq 23)$; $\varepsilon = 7.2184 \times 10^{-3} - \frac{1}{750}\ln(109 - 3\sigma)$
$(23 \leq \sigma \leq 35)$

Chapter 9

9.2 $P = 11,140$ lb; $\delta_M = 2.86 \times 10^{-3}$ in.

9.4(a) $\theta = 34.3°$ **(b)** 0.120 m to left of right hinge
(c) $\sigma_A = 215.0$ MPa (T); $\sigma_B = 240.0$ MPa (T) **(d)** $\delta_{\text{vertical}} = 1.3 \times 10^{-4}$ m

9.6(a) 0.056 m to left of right end of magnesium block
(b) $\sigma_M = 24.80$ MPa (C); $\sigma_W = 4.96$ MPa (C) **(c)** $\delta = 9.9 \times 10^{-5}$ m

9.8 $P = 226.2$ k; $\sigma_A = 5.0$ ksi (C); $\sigma_B = 3.0$ ksi (C); $\sigma_C = 1.6$ ksi (C)

9.10 $\sigma_{AB} = 8.29$ ksi (T); $\delta_{AB} = 1.4 \times 10^{-3}$ in.; $\sigma_{BC} = 1.97$ ksi (C);
$\delta_{BC} = -1.4 \times 10^{-3}$ in.

9.12 $P = 23,410$ lb; $\sigma_A = 16,880$ psi; $\delta_A = 1.69 \times 10^{-2}$ in.; $\delta_S = 0.75 \times 10^{-2}$ in.

9.14 $\sigma_{AB} = 11.07$ ksi (T); $\sigma_{BC} = 0.67$ ksi (T)

9.16 $|T_{max}| = 13,260$ psi; $\theta_{max} = 5.5 \times 10^{-3}$ rad

9.18(a) $T_{AB} = 8.53$ kN-m; $T_{BC} = 1.47$ kN-m   **(b)** $\theta_{B/C} = 2.90 \times 10^{-2}$ rad
(c) $|\tau_{max}| = 43.43$ MPa

9.20 $|\tau_{max}| = 7,660$ psi; $\theta_C = 2.58 \times 10^{-2}$ rad

9.22 $T = 15,710$ lb-ft

9.24 Diameter $= 3.0$ in.

9.26(a) $|\tau_{max}|_X = 7.77$ MPa; $|\tau_{max}|_Y = 1.11$ MPa; $|\tau_{max}|_Z = 6.66$ MPa
(b) $\theta_C = 1.1 \times 10^{-3}$ rad

9.28 $|\tau_{max}| = 6,250$ psi

9.30 $\Delta T = 251°$ drop

9.32 $A_v = \dfrac{3wL}{8} \uparrow$; $B_v = \dfrac{5wL}{8} \uparrow$; $M_B = \dfrac{wL^2}{8} \circlearrowleft$

9.34 $A_v = \dfrac{3wL}{8} - \dfrac{3EI_u\delta}{L^3} \uparrow$; $B_v = \dfrac{5wL}{8} + \dfrac{3EI_u\delta}{L^3} \uparrow$; $M_B = \dfrac{wL^2}{8} - \dfrac{3EI_u\delta}{L^2} \circlearrowleft$

9.36(a) $M_A = 0.0833wL^2 \circlearrowleft$ **(b)** $M_A = 0.0432wL^2 \circlearrowleft$
(c) $M_A = 0.1632wL^2 \circlearrowleft$

9.38 $\sigma_M = 122$ psi (C); $C_v = 9560$ lb $\uparrow$; $M_C = 18,750$ lb-ft $\circlearrowleft$

9.40(a) $\sigma_S = 29,140$ psi   **(b)** $\delta_S = 0.140$ in.   **(c)** $C_v = 1430$ lb $\uparrow$;
$M_C = 14,290$ lb-ft $\circlearrowleft$

9.42 See answers to Problem 9.32.

9.44 See answers to Problem 9.34.

9.46 See answers to Problem 9.36.

9.48 See answers to Problem 9.40.

9.50 $F_S = \dfrac{AwL^4}{10[AL^3 + 3I_u l]}$(tension); $B_v = \dfrac{wL}{2} - \dfrac{AwL^4}{10[AL^3 + 3I_u l]} \uparrow$;

$M_B = \dfrac{wL^2}{6} - \dfrac{AwL^5}{10[AL^3 + 3I_u l]} \circlearrowright$

9.52 $A_v = B_v = \dfrac{73wL}{88} \uparrow;\ C_v = \dfrac{10wL}{44} \uparrow;\ D_v = \dfrac{5wL}{44} \uparrow$

9.54   See answers to Problem 9.32.

9.56   See answers to Problem 9.34.

9.58   See answers to Problem 9.36.

9.60   See answers to Problem 9.40.

9.62   See answers to Example 9.8.

9.64 $B_v = 30.68$ kN $\downarrow;\ A_v = 69.32$ kN $\uparrow;\ M_A = 115.90$ kN-m $\circlearrowleft;$
$|v_{\max}|_{CD} = 9.1 \times 10^{-3}$ m $\downarrow$

9.66 $A_v = 0.709wL + 0.633P \uparrow;\ B_v = 0.291wL + 0.367P \uparrow;$
$M_B = 0.078PL + 0.027wL^2 \circlearrowright$

9.68 $F = \dfrac{7wL^3bA}{48(b^2AL + lI_u)}$ (tension); $B_v = wL \uparrow;\ M_B = \dfrac{wL^2}{2} - \dfrac{7wL^3b^2A}{24(b^2AL + lI_u)} \circlearrowleft$

9.70 $A_v = \dfrac{3wL}{20} \uparrow;\ M_A = \dfrac{wL^2}{30} \circlearrowleft;\ B_v = \dfrac{7wL}{20} \uparrow;\ M_B = \dfrac{wL^2}{20} \circlearrowright$

9.72 $\sigma_M = 120$ psi; $C_v = 9600$ lb$\uparrow;\ M_C = 19{,}200$ lb-ft $\circlearrowright$

9.74   See answers to Problem 9.40.

9.76   See answers to Problem 9.64.

9.78 $A_v = \dfrac{31wL}{80} - \dfrac{3P}{2} \uparrow;\ M_A = \dfrac{13wL^2}{240} - \dfrac{PL}{2} \circlearrowleft;\ B_v = \dfrac{49wL}{80} + \dfrac{5P}{2} \uparrow$

Chapter 10

10.2 $D_0 = 0.240$ m; $D_i = 0.080$ m

10.4   Let side of the square cross section be a: **(a)** $a = 0.026$ and 0.037 m
(b) $a = 0.022$ and 0.032 m

10.6 $S12 \times 50$ for AB; $S7 \times 20$ for BC

10.8 $A = 4.69$ in.^{2}; $N_u = 2.25$

10.10   0.0960 m

10.12 $D_0 = 0.112$ m; $D_i = 0.084$ m

10.14 $D_1 = 1.77$ in.; $D_2 = 3.94$ in.; $D_3 = 2.56$ in.

10.16 $D_1 = 0.0161$ m; $D_2 = 0.0135$ m

10.18   3.84 in. $\times$ 1.92 in.

10.20   5.36 in. $\times$ 2.86 in. (outside)

10.22 $b = 0.0570$ m; $h = 0.2280$ m

10.24 $t = 0.0856$ m

10.26 $L5 \times 5 \times \frac{3}{8}$

10.28 $d = 0.0876$ m

10.30 $S20 \times 85$

10.32 $D_0 = 4.94$ in.; $D_i = 3.95$ in.

10.34 0.64 in. $\times$ 0.64 in.

10.36 0.0344 m $\times$ 0.0172 m

10.38 4.38 in. outside

10.40 5-in. standard steel pipe

10.42 0.0963-m diameter

10.44 $D_0 = 0.1986$ m; $D_i = 0.1490$ m

10.46 $d = w = 0.2054$ m

10.48 Flange width $a = 0.224$ m

Chapter 11

11.2 $7 - \frac{3}{4}$-in. bolts (shear)

11.4 $6 - \frac{5}{8}$-in. bolts (shear)

11.6 $4 - \frac{3}{4}$-in. bolts (shear)

11.8 9.41 in.

11.10 $2\frac{1}{2}$-in. standard weight pipe

11.12 $C3 \times 4.1$

11.14 $S24 \times 105.9$

11.16 $D = 0.0365$ m

11.18 $S10 \times 25.4$

11.20 4-in. standard weight pipe

11.22 $D_{CE} = D_{CD} = 0.0354$ m

11.24 $D = 0.98$ in.

11.26 $D_{BE} = D_{BD} = 0.0176$ m

11.28 $D_0 = 4.81$ in.; $D_i = 3.61$ in.

11.30 $F_1 = 23.652$ k (T); $F_2 = 6.922$ k (T)

11.32 Member length is 7.5 ft and equilibrium equations are satisfied.

11.34 Equations of equilibrium are satisfied.

11.36 Remove loadings shown in program of Fig. 11.5 and add under MEMBER LOADS: 2 FORCE Y UNIFORM -0.1. (The JOINT LOADS heading should also be removed.)

Chapter 12

12.2(a) B **(b)** 97.5 k   **(c)** (0.0356 in., 66.65 k); (0.200 in., 97.5 k)

12.4 A fails first, 153 kN, 0.00064 m at D.

12.6 A fails first, 132 kN, 0.00120 m at D.

12.8 0.000577 m^2

12.10 0.00400 m^2

12.12 0.00438 m^2

12.14 2121 k-in., 2827 k-in., 1.33

12.16 Answer given

12.18 $16/(3\pi)$

12.20(a) 0.05357 m from top   **(b)** 0.00375 m from bottom of flange **(c)** 93.97 kN-m   **(d)** 169.9 kN-m   **(e)** 1.81

12.22 1.48

12.24 2.00

12.26 Mechanism ABC, 15.0 k

12.28 Mechanism ABC is critical, 12.0 k.

12.30 Mechanism CDE is critical, 300 kN.

12.32 540 kN

12.34 Mechanism ABC is critical, 375 kN

12.36 Required $Z = 250$ in.^{3}, $W27 \times 102$

12.38 6-in. extra strong pipe

12.40 $D = 0.0362$ m

12.42 Required $M_p = 825$ k-ft, $W24 \times 76$

12.44(a) Required $Z = 220$ in.^{3}, $W24 \times 94$ **(b)** Required $M_p = 600$ k-ft, $W24 \times 76$

Chapter 13

13.2 71.1 MPa, 71.1, 0.00107 m

13.4 6470 psi, 57.5, 0.0647 in.

13.6 $\sigma = \dfrac{W + \sqrt{W^2 + 2AEWh/L}}{A}$

$IF = 1 + \sqrt{1 + 2AEh/(WL)}$

$\Delta = \dfrac{(W + \sqrt{W^2 + 2AEWh/L})L}{AE}$

13.8 $\tau = \dfrac{[W + \sqrt{W^2 + 2JGWh/(b^2L)}]bD}{2J}$

$\theta = [W + \sqrt{W^2 + 2JGWh/(b^2L)}]\dfrac{bL}{JG}$

$IF = 1 + \sqrt{1 + 2JGh/(Wb^2L)}$

13.10 0.495 in.^{2}

13.12 6.83 ksi

13.14

	Stress	Deflection
Static	10.8 MPa	0.0516 m
Impact	29.3 MPa	0.140 m

13.16 124.8 MPa, 0.075 rad

13.18

	Stress	Deflection
Static	46.9 psi	0.000586 in.
Impact	5230 psi	0.0654 in.

13.20

	Part (a)	Part (b)
Stress	11300 psi	8930 psi
Deflection	0.0113 in.	0.0143 in.

Choose rod of part (b).

13.22 40 N, 0.154 m

13.24 Roots of quadratic: 1.095 and 2.88 m

13.26 Endurance strengths: 56.57 and 47.95 ksi

13.28(a) 43.1, 32.1, and 30.0 ksi   **(b)** $C = -0.301$; $D = 3.2832$, 39.53 ksi

13.30 Gerber parabola:

$$S_r = 289\left[1 - \left(\frac{S_{av}}{570}\right)^2\right]$$

Goodman straight line in first quadrant:

$$S_r = -0.507\,S_{av} + 289$$

Soderberg straight line in first quadrant:

$$S_r = -0.963 \, S_{av} + 289$$

13.32 Values for plotting:

$N \times 10^{-4}$	$\log N$	$\log S$ (ksi)
1	4.000	50.00
5	4.699	38.69
10	5.000	34.64
50	5.699	26.80
100	6.000	24.00
300	6.477	24.00
600	6.778	24.00
1000	7.000	24.00

13.34 Goodman: $d = 3.85$ in., static: $d = 3.00$ in. Choose $d = 3.85$ in.

13.36 Goodman straight line in third quadrant. Answer: See 13.34, $d = 3.85$ in.

13.38 Gerber: $D = 2.80$ in., static: $D = 2.25$ in. Choose $D = 2.80$ in.

13.40 $D_0 = 3.30$ in.; $D_i = 2.36$ in. (Goodman)

13.42 $D = 0.0610$ m (Goodman)

13.44 Width $= 2.64$ in. (Goodman)

13.46 $D = 0.0683$ m (Goodman)

13.48 $D_0 = 2.75$ in.; $D_i = 1.72$ in. (Goodman)

13.50 Width $= 0.0539$ m (Goodman)

Chapter 14

14.2 $(\sigma_t)_{max} = 0.4065$ MPa (C); $(\sigma_r)_{max} = 0.2158$ MPa (C); $(\sigma_z)_{max} = 0.2539$ MPa (C)

14.4 $t = 19.90$ in.

14.6 $|\sigma_{t\,max}| = 29{,}170$ psi (C); $|\sigma_{r\,max}| = 8102$ psi (C)

14.8(a) $|\sigma_{max}| = \sigma_1 = 75$ MPa (T) **(b)** $|\tau_{max}| = 39$ MPa
(c) $\sigma = 56.25$ MPa (T); $\tau = -18.75$ MPa

14.10 $|\sigma_{max}| = \sigma_1 = 4.905$ MPa (T); $|\tau_{max}| = 2.551$ MPa

14.12(a) At intersection of two legs of angle section   **(b)**   at intersection of flange and web of tee section

14.14(a) $V_{AB} = \dfrac{h_1^2 t V_v}{6 I_u}(3h - 2h_1)$; $V_{BC} = \dfrac{b t V_v}{2 I_u}(2h_1 h + hb - h_1^2)$;

$$V_{CD} = \frac{htV_v}{3I_u}(6h_1h - 3h_1^2 + 6bh + 2h^2) \qquad \textbf{(b)} \quad e_u = \frac{2bt}{I_u}(2h_1h^2 + h^2b - 2h_1^3)$$

To the right of vertical member CD

14.16 $\quad T = \left(\dfrac{\sqrt{3}ta^4}{96I_u}\right)P$

14.18 $\quad e_u = \dfrac{4ta^4}{I_u}$ to the right of intersection of two inclined members

14.20 $\quad e_u = \dfrac{2\pi tR^4}{I_u}$ to the left of geometric center of section

14.22 $\quad e_u = 5.42$ in.; $e_v = 6.33$ in. above and to the right of lower left-hand corner of section

14.24 $\quad (\sigma_S)_{max} = 13.890$ MPa; $(\sigma_W)_{max} = 0.694$ MPa; $v = 0.0012$ m

14.26 $\quad (\sigma_A)_{max} = 2900$ psi; $(\sigma_B)_{max} = 3560$ psi; $v = 0.44$ in.

14.28 $\quad P = 6210$ lb

14.30 $\quad (\sigma_S)_{max} = 12.35$ MPa; $(\sigma_{D.F.})_{max} = 3.76$ MPa; $\delta_{max} = 0.0052$ m

14.32 $\quad \sigma_c = 0.53$ ksi; $\sigma_s = 9.66$ ksi

14.34 $\quad 169.5$ k-ft; steel is critical.

14.36 $\quad \sigma_c = 0.87$ ksi; $\sigma_s = 16.3$ ksi

14.38 $\quad \sigma_c = 6.29$ MPa; $\sigma_s = 118$ MPa

14.40 $\quad 3065$ k-in.

14.42 $\quad 14.9$ in.

14.44 $\quad 19.8$ in.

14.46 $\quad v = \dfrac{13PL^3}{24EI_u} \downarrow$

14.48 $\quad v = 0.012$ m $\downarrow$

14.50(a) $\quad v = \dfrac{-1}{EI_u}(1.49PL^3 + 2.01wL^4) \downarrow \qquad \textbf{(b)} \quad v_C = \dfrac{1}{EI_u}(0.80PL^3 + 1.51wL^4) \uparrow$

14.52 $\quad v = 0.091$ in. $\downarrow$

14.54 $\quad v_A = 0.0192$ m $\leftarrow$; $v_B = 0.0106$ m $\leftarrow$

14.56 $\quad (\sigma_{max})_{tension} = 36{,}920$ psi; $(\sigma_{max})_{compression} = 17{,}430$ psi; $v_0 = 0.124$ in. to left of section centroid

14.58 $\quad P = 125{,}200$ N

14.60 $\quad P = 5570$ lb

14.62 $(\sigma_{max})_{tension} = 11.27$ MPa, compared to 21.95 MPa; $(\sigma_{max})_{compression} = 5.70$ MPa, compared to 12.72 MPa

14.64 $f = \dfrac{L}{EI}\begin{bmatrix} 1 & \dfrac{L}{2} \\ \dfrac{L}{2} & \dfrac{L^2}{6} \end{bmatrix}$

14.66 $M_A = -(\frac{1}{30})qL^2; \; V_A = (\frac{3}{20})qL$

14.68 $R_A = (\frac{7}{16})wL; \; R_B = (\frac{5}{8})wL; \; R_C = (-\frac{1}{16})wL$

14.70 $R_A = (-\frac{13}{24})(C_B/L); \; R_C = (\frac{5}{8})(C_B/L)$

14.72 See Eq. 14.68.

14.74 $M_{AB} = M_{BC} = -wL^2/12$

14.76 $M_{AB} = wL^2/48; \; M_{BC} = -wL^2/24$

14.78 $M_{AB} = -19wL^2/180; \; M_{BC} = -7wL^2/180; \; M_{CD} = wL^2/90; \; M_{DC} = wL^2/180$

INDEX

FUNDAMENTAL UNITS IN MECHANICS

Quantity	Unit	Symbol
Mass	kilogram	kg
Length	meter	m
Time	second	s

SELECTION OF DERIVED UNITS AND CONVERSION FACTORS

Quantity	Symbol	English to SI
Area	m^2	$1\ ft^2 = 0.0929\ m^2$
Volume	m^3	$1\ ft^3 = 0.0283\ m^3$
Moment of inertia	m^4	$1\ in.^4 = 0.4162 \times 10^{-6}\ m^4$
Section modulus	m^3	$1\ in.^3 = 16.39 \times 10^{-6}\ m^3$
Density	kg/m^3	$1\ lb/ft^3 = 16.03\ kg/m^3$
Force	N (Newton)	$1\ lb = 4.448\ N$
Moment of torque	$N \cdot m$	$1\ lb\text{-}ft = 1.356\ N \cdot m$
Stress or pressure	N/m^2 (Pascal)	$1\ psi = 0.006895\ MPa$
Energy or work	$m \cdot N$	$1\ ft\text{-}lb = 1.356\ m \cdot N$
Coefficient of thermal expansion	$m/m/°C$	$1\ in./in./°F = 1.8\ m/m/°C$

SI UNIT PREFIXES

Multiplication Factor	Prefix	Symbol	Pronunciation (U.S.)	Meaning (U.S.)
$1\,000\,000\,000 = 10^9$	giga	G	jig' a (*a* as in *a*bout)	One billion times
$1\,000\,000 = 10^6$	mega	M	as in *mega* phone	One million times
$1000 = 10^3$	kilo	k	as in *kilo* watt	One thousand times
$100 = 10^2$	hecto	h	heck' toe	One hundred times
$10 = 10^1$	deka	da	deck' a (*a* as in *a*bout)	Ten times
$0.1 = 10^{-1}$	deci	d	as in *deci*mal	One tenth of
$0.01 = 10^{-2}$	centi	c	as in *senti*ment	One hundredth of
$0.001 = 10^{-3}$	milli	m	as in *mili*tary	One thousandth of
$0.000\,001 = 10^{-6}$	micro	μ	as in *micro*phone	One millionth of
$0.000\,000\,001 = 10^{-9}$	nano	n	nan' oh (*an* as in ant)	One billionth of